Texts in Computer Science

Series Editors

David Gries, Department of Computer Science, Cornell University, Ithaca, NY, USA

Orit Hazzan, Faculty of Education in Technology and Science, Technion—Israel Institute of Technology, Haifa, Israel

Steven S. Skiena

The Algorithm Design Manual

Third Edition

 Springer

Steven S. Skiena
Department of Computer Science
Stony Brook University
Stony Brook, NY, USA

ISSN 1868-0941   ISSN 1868-095X (electronic)
Texts in Computer Science
ISBN 978-3-030-54258-0   ISBN 978-3-030-54256-6 (eBook)
https://doi.org/10.1007/978-3-030-54256-6

This Springer imprint is published by the registered company Springer Nature Switzerland AG
The registered company address is: Gewerbestrasse 11, 6330 Cham, Switzerland

Preface

Many professional programmers are not well prepared to tackle algorithm design problems. This is a pity, because the techniques of algorithm design form one of the core practical *technologies* of computer science.

This book is intended as a manual on algorithm design, providing access to combinatorial algorithm technology for both students and computer professionals. It is divided into two parts: *Techniques* and *Resources*. The former is a general introduction to the design and analysis of computer algorithms. The Resources section is intended for browsing and reference, and comprises the catalog of algorithmic resources, implementations, and an extensive bibliography.

To the Reader

I have been gratified by the warm reception previous editions of *The Algorithm Design Manual* have received, with over 60,000 copies sold in various formats since first being published by Springer-Verlag in 1997. Translations have appeared in Chinese, Japanese, and Russian. It has been recognized as a unique guide to using algorithmic techniques to solve problems that often arise in practice.

Much has changed in the world since the second edition of *The Algorithm Design Manual* was published in 2008. The popularity of my book soared as software companies increasingly emphasized algorithmic questions during employment interviews, and many successful job candidates have trusted *The Algorithm Design Manual* to help them prepare for their interviews.

Although algorithm design is perhaps the most classical area of computer science, it continues to advance and change. Randomized algorithms and data structures have become increasingly important, particularly techniques based on hashing. Recent breakthroughs have reduced the algorithmic complexity of the best algorithms known for such fundamental problems as finding minimum spanning trees, graph isomorphism, and network flows. Indeed, if we date the origins of modern algorithm design and analysis to about 1970, then roughly 20% of modern algorithmic history has happened since the second edition of *The Algorithm Design Manual*.

The time has come for a new edition of my book, incorporating changes in the algorithmic and industrial world plus the feedback I have received from hundreds of readers. My major goals for the third edition are.

- To introduce or expand coverage of important topics like hashing, randomized algorithms, divide and conquer, approximation algorithms, and quantum computing in the first part of the book (Practical Algorithm Design).

- To update the reference material for all the catalog problems in the second part of the book (The Hitchhiker's Guide to Algorithms).

- To take advantage of advances in color printing to produce more informative and eye-catching illustrations.

Three aspects of *The Algorithm Design Manual* have been particularly beloved: (1) the hitchhiker's guide to algorithms, (2) the war stories, and (3) the electronic component of the book. These features have been preserved and strengthened in this third edition:

- *The Hitchhiker's Guide to Algorithms* – Since finding out what is known about an algorithmic problem can be a difficult task, I provide a catalog of the seventy-five most important algorithmic problems arising in practice. By browsing through this catalog, the student or practitioner can quickly identify what their problem is called, what is known about it, and how they should proceed to solve it.

  I have updated every section in response to the latest research results and applications. Particular attention has been paid to updating discussion of available software implementations for each problem, reflecting sources such as GitHub, which have emerged since the previous edition.

- *War stories* – To provide a better perspective on how algorithm problems arise in the real world, I include a collection of "war stories", tales from my experience on real problems. The moral of these stories is that algorithm design and analysis is not just theory, but an important tool to be pulled out and used as needed.

  The new edition of the book updates the best of the old war stories, plus adds new tales on randomized algorithms, divide and conquer, and dynamic programming.

- *Online component* – Full lecture notes and a problem solution Wiki is available on my website `www.algorist.com`. My algorithm lecture videos have been watched over 900,000 times on YouTube. This website has been updated in parallel with the book.

Equally important is what is not done in this book. I do not stress the mathematical analysis of algorithms, leaving most of the analysis as informal arguments. You will not find a single theorem anywhere in this book. When more details are needed, the reader should study the cited programs or references. The goal of this manual is to get you going in the right direction as quickly as possible.

To the Instructor

This book covers enough material for a standard *Introduction to Algorithms* course. It is assumed that the reader has completed the equivalent of a second programming course, typically titled *Data Structures* or *Computer Science II*.

A full set of lecture slides for teaching this course are available online at www.algorist.com. Further, I make available online video lectures using these slides to teach a full-semester algorithm course. Let me help teach your course, through the magic of the Internet!

I have made several pedagogical improvements throughout the book, including:

- *New material* – To reflect recent developments in algorithm design, I have added new chapters on randomized algorithms, divide and conquer, and approximation algorithms. I also delve deeper into topics such as hashing. But I have been careful to heed the readers who begged me to keep the book of modest length. I have (painfully) removed less important material to keep total expansion by page count under 10% over the previous edition.

- *Clearer exposition* – Reading through my text ten years later, I was thrilled to find many sections where my writing seemed ethereal, but other places that were a muddled mess. Every page in this manuscript has been edited or rewritten for greater clarity, correctness and flow.

- *More interview resources* – *The Algorithm Design Manual* remains very popular for interview prep, but this is a fast-paced world. I include more and fresher interview problems, plus coding challenges associated with interview sites like LeetCode and Hackerrank. I also include a new section with advice on how to best prepare for interviews.

- *Stop and think* – Each of my course lectures begins with a "Problem of the Day," where I illustrate my thought process as I solve a topic-specific homework problem – false starts and all. This edition had more Stop and Think sections, which perform a similar mission for the reader.

- *More and better homework problems* – The third edition of *The Algorithm Design Manual* has more and better homework exercises than the previous one. I have added over a hundred exciting new problems, pruned some less interesting problems, and clarified exercises that proved confusing or ambiguous.

- *Updated code style* – The second edition featured algorithm implementations in C, replacing or augmenting pseudocode descriptions. These have generally been well received, although certain aspects of my programming have been condemned by some as old fashioned. All programs have been revised and updated, and are structurally highlighted in color.

- *Color images* – My companion book *The Data Science Design Manual* was printed with color images, and I was excited by how much this made concepts clearer. Every single image in the *The Algorithm Design Manual* is now rendered in living color, and the process of review has improved the contents of most figures in the text.

Acknowledgments

Updating a book dedication every ten years focuses attention on the effects of time. Over the lifespan of this book, Renee became my wife and then the mother of our two children, Bonnie and Abby, who are now no longer children. My father has left this world, but Mom and my brothers Len and Rob remain a vital presence in my life. I dedicate this book to my family, new and old, here and departed.

I would like to thank several people for their concrete contributions to this new edition. Michael Alvin, Omar Amin, Emily Barker, and Jack Zheng were critical to building the website infrastructure and dealing with a variety of manuscript preparation issues. Their roles were played by Ricky Bradley, Andrew Gaun, Zhong Li, Betson Thomas, and Dario Vlah on previous editions. The world's most careful reader, Robert Piché of Tampere University, and Stony Brook students Peter Duffy, Olesia Elfimova, and Robert Matsibekker read early versions of this edition, and saved both you and me the trouble of dealing with many errata. Thanks also to my Springer-Verlag editors, Wayne Wheeler and Simon Rees.

Several exercises were originated by colleagues or inspired by other texts. Reconstructing the original sources years later can be challenging, but credits for each problem (to the best of my recollection) appear on the website.

Much of what I know about algorithms I learned along with my graduate students. Several of them (Yaw-Ling Lin, Sundaram Gopalakrishnan, Ting Chen, Francine Evans, Harald Rau, Ricky Bradley, and Dimitris Margaritis) are the real heroes of the war stories related within. My Stony Brook friends and algorithm colleagues Estie Arkin, Michael Bender, Jing Chen, Rezaul Chowdhury, Jie Gao, Joe Mitchell, and Rob Patro have always been a pleasure to work with.

Caveat

It is traditional for the author to magnanimously accept the blame for whatever deficiencies remain. I don't. Any errors, deficiencies, or problems in this book are somebody else's fault, but I would appreciate knowing about them so as to determine who is to blame.

Steven S. Skiena
Department of Computer Science
Stony Brook University
Stony Brook, NY 11794-2424
http://www.cs.stonybrook.edu/~skiena
August 2020

Contents

II The Hitchhiker's Guide to Algorithms   435

Part I

Practical Algorithm Design

Chapter 1

Introduction to Algorithm Design

What is an algorithm? An algorithm is a procedure to accomplish a specific task. An algorithm is the idea behind any reasonable computer program.

To be interesting, an algorithm must solve a general, well-specified *problem*. An algorithmic problem is specified by describing the complete set of *instances* it must work on and of its output after running on one of these instances. This distinction, between a problem and an instance of a problem, is fundamental. For example, the algorithmic *problem* known as *sorting* is defined as follows:

Problem: Sorting

Input: A sequence of n keys $a_1, \ldots, a_n$.

Output: The permutation (reordering) of the input sequence such that $a'_1 \leq a'_2 \leq \cdots \leq a'_{n-1} \leq a'_n$.

An *instance* of sorting might be an array of names, like {*Mike, Bob, Sally, Jill, Jan*}, or a list of numbers like {*154, 245, 568, 324, 654, 324*}. Determining that you are dealing with a general problem instead of an instance is your first step towards solving it.

An *algorithm* is a procedure that takes any of the possible input instances and transforms it to the desired output. There are many different algorithms that can solve the problem of sorting. For example, *insertion sort* is a method that starts with a single element (thus trivially forming a sorted list) and then incrementally inserts the remaining elements so that the list remains sorted. An animation of the logical flow of this algorithm on a particular instance (the letters in the word "INSERTIONSORT") is given in Figure 1.1.

This algorithm, implemented in C, is described below:

© The Editor(s) (if applicable) and The Author(s), under exclusive license to
Springer Nature Switzerland AG 2020
S. S. Skiena, *The Algorithm Design Manual*, Texts in Computer Science,
https://doi.org/10.1007/978-3-030-54256-6_1

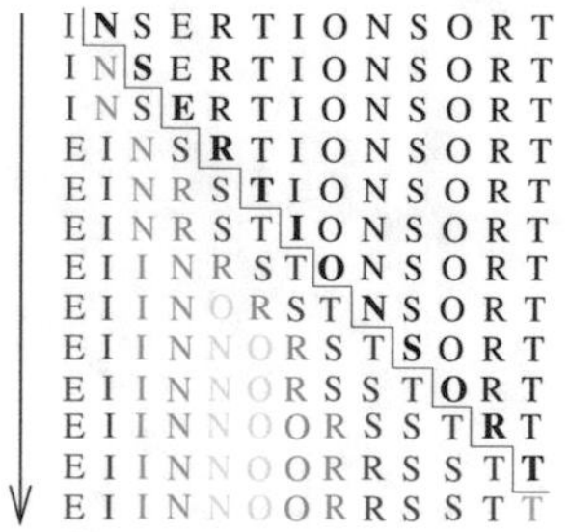

```
I N S E R T I O N S O R T
I N S E R T I O N S O R T
I N S E R T I O N S O R T
E I N S R T I O N S O R T
E I N R S T I O N S O R T
E I N R S T I O N S O R T
E I I N R S T O N S O R T
E I I N O R S T N S O R T
E I I N N O R S T S O R T
E I I N N O R S S T O R T
E I I N N O O R S S T R T
E I I N N O O R R S S T T
E I I N N O O R R S S T T
```

Figure 1.1: Animation of insertion sort in action (time flows downward).

```
void insertion_sort(item_type s[], int n) {
    int i, j;     /* counters */

    for (i = 1; i < n; i++) {
        j = i;
        while ((j > 0) && (s[j] < s[j - 1])) {
            swap(&s[j], &s[j - 1]);
            j = j-1;
        }
    }
}
```

Note the generality of this algorithm. It works just as well on names as it does on numbers. Or anything else, given the appropriate comparison operation (<) to test which of two keys should appear first in sorted order. It can be readily verified that this algorithm correctly orders every possible input instance according to our definition of the sorting problem.

There are three desirable properties for a good algorithm. We seek algorithms that are *correct* and *efficient*, while being *easy to implement*. These goals may not be simultaneously achievable. In industrial settings, any program that seems to give good enough answers without slowing the application down is often acceptable, regardless of whether a better algorithm exists. The issue of finding the best possible answer or achieving maximum efficiency usually arises in industry only after serious performance or legal troubles.

This chapter will focus on algorithm correctness, with our discussion of efficiency concerns deferred to Chapter 2. It is seldom obvious whether a given algorithm correctly solves a given problem. Correct algorithms usually come with a proof of correctness, which is an explanation of *why* we know that the algorithm must take every instance of the problem to the desired result. But before we go further, it is important to demonstrate why *it's obvious* never suffices as a proof of correctness, and is usually flat-out wrong.

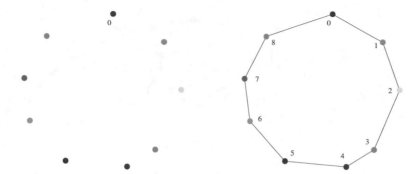

Figure 1.2: A good instance for the nearest-neighbor heuristic. The rainbow coloring (red to violet) reflects the order of incorporation.

1.1 Robot Tour Optimization

Let's consider a problem that arises often in manufacturing, transportation, and testing applications. Suppose we are given a robot arm equipped with a tool, say a soldering iron. When manufacturing circuit boards, all the chips and other components must be fastened onto the substrate. More specifically, each chip has a set of contact points (or wires) that need be soldered to the board. To program the robot arm for this job, we must first construct an ordering of the contact points so that the robot visits (and solders) the first contact point, then the second point, third, and so forth until the job is done. The robot arm then proceeds back to the first contact point to prepare for the next board, thus turning the tool-path into a closed tour, or cycle.

Robots are expensive devices, so we want the tour that minimizes the time it takes to assemble the circuit board. A reasonable assumption is that the robot arm moves with fixed speed, so the time to travel between two points is proportional to their distance. In short, we must solve the following algorithm problem:

Problem: Robot Tour Optimization
Input: A set S of n points in the plane.
Output: What is the shortest cycle tour that visits each point in the set S?

You are given the job of programming the robot arm. Stop right now and think up an algorithm to solve this problem. I'll be happy to wait for you...

Several algorithms might come to mind to solve this problem. Perhaps the most popular idea is the *nearest-neighbor* heuristic. Starting from some point p_0, we walk first to its nearest neighbor p_1. From p_1, we walk to its nearest unvisited neighbor, thus excluding only p_0 as a candidate. We now repeat this process until we run out of unvisited points, after which we return to p_0 to close off the tour. Written in pseudo-code, the nearest-neighbor heuristic looks like

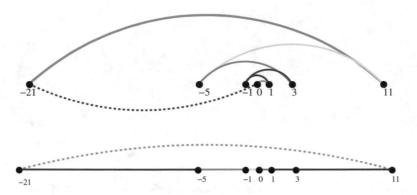

Figure 1.3: A bad instance for the nearest-neighbor heuristic, with the optimal solution. Colors are sequenced as ordered in the rainbow.

this:

> NearestNeighbor(P)
>   Pick and visit an initial point p_0 from P
> $p = p_0$
> $i = 0$
>   While there are still unvisited points
> $i = i + 1$
>   Select p_i to be the closest unvisited point to p_{i-1}
>   Visit p_i
>   Return to p_0 from p_{n-1}

This algorithm has a lot to recommend it. It is simple to understand and implement. It makes sense to visit nearby points before we visit faraway points to reduce the total travel time. The algorithm works perfectly on the example in Figure 1.2. The nearest-neighbor rule is reasonably efficient, for it looks at each pair of points (p_i, p_j) at most twice: once when adding p_i to the tour, the other when adding p_j. Against all these positives there is only one problem. This algorithm is completely wrong.

Wrong? How can it be wrong? The algorithm always finds a tour, but it doesn't necessarily find the shortest possible tour. It doesn't necessarily even come close. Consider the set of points in Figure 1.3, all of which lie along a line. The numbers describe the distance that each point lies to the left or right of the point labeled "0". When we start from the point "0" and repeatedly walk to the nearest unvisited neighbor, we might keep jumping left–right–left–right over "0" as the algorithm offers no advice on how to break ties. A much better (indeed optimal) tour for these points starts from the left-most point and visits each point as we walk right before returning at the left-most point.

Try now to imagine your boss's delight as she watches a demo of your robot arm hopscotching left–right–left–right during the assembly of such a simple board.

"But wait," you might be saying. "The problem was in starting at point "0." Instead, why don't we start the nearest-neighbor rule using the left-most point as the initial point p_0? By doing this, we will find the optimal solution on this instance."

That is 100% true, at least until we rotate our example by 90 degrees. Now all points are equally left-most. If the point "0" were moved just slightly to the left, it would be picked as the starting point. Now the robot arm will hopscotch up–down–up–down instead of left–right–left–right, but the travel time will be just as bad as before. No matter what you do to pick the first point, the nearest-neighbor rule is doomed to work incorrectly on certain point sets.

Maybe what we need is a different approach. Always walking to the closest point is too restrictive, since that seems to trap us into making moves we didn't want. A different idea might repeatedly connect the closest pair of endpoints whose connection will not create a problem, such as premature termination of the cycle. Each vertex begins as its own single vertex chain. After merging everything together, we will end up with a single chain containing all the points in it. Connecting the final two endpoints gives us a cycle. At any step during the execution of this *closest-pair heuristic*, we will have a set of single vertices and the end of vertex-disjoint chains available to merge. In pseudocode:

ClosestPair(P)
    Let n be the number of points in set P.
    For $i = 1$ to $n - 1$ do
        $d = \infty$
        For each pair of endpoints (s, t) from distinct vertex chains
            if $dist(s, t) \leq d$ then $s_m = s$, $t_m = t$, and $d = dist(s, t)$
        Connect (s_m, t_m) by an edge
    Connect the two endpoints by an edge

This closest-pair rule does the right thing in the example in Figure 1.3. It starts by connecting "0" to its two immediate neighbors, the points 1 and -1. Subsequently, the next closest pair will alternate left–right, growing the central path by one link at a time. The closest-pair heuristic is somewhat more complicated and less efficient than the previous one, but at least it gives the right answer in this example.

But not on all examples. Consider what this algorithm does on the point set in Figure 1.4(l). It consists of two rows of equally spaced points, with the rows slightly closer together (distance $1 - \epsilon$) than the neighboring points are spaced within each row (distance $1 + \epsilon$). Thus, the closest pairs of points stretch across the gap, not around the boundary. After we pair off these points, the closest remaining pairs will connect these pairs alternately around the boundary. The total path length of the closest-pair tour is $3(1-\epsilon)+2(1+\epsilon)+\sqrt{(1 - \epsilon)^2 + (2 + 2\epsilon)^2}$. Compared to the tour shown in Figure 1.4(r), we travel over 20% farther than necessary when $\epsilon \to 0$. Examples exist where the penalty is considerably worse than this.

Thus, this second algorithm is also wrong. Which one of these algorithms

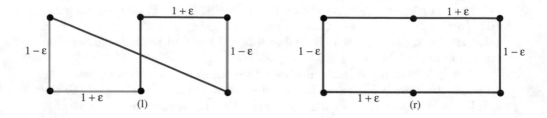

Figure 1.4: A bad instance for the closest-pair heuristic, with the optimal solution.

performs better? You can't tell just by looking at them. Clearly, both heuristics can end up with very bad tours on innocent-looking input.

At this point, you might wonder what a correct algorithm for our problem looks like. Well, we could try enumerating *all* possible orderings of the set of points, and then select the one that minimizes the total length:

> OptimalTSP(P)
> $\quad d = \infty$
> $\quad$ For each of the $n!$ permutations P_i of point set P
> $\quad\quad$ If $(cost(P_i) \le d)$ then $d = cost(P_i)$ and $P_{min} = P_i$
> $\quad$ Return P_{min}

Since all possible orderings are considered, we are guaranteed to end up with the shortest possible tour. This algorithm is correct, since we pick the best of all the possibilities. But it is also extremely slow. Even a powerful computer couldn't hope to enumerate all the 20! = 2,432,902,008,176,640,000 orderings of 20 points within a day. For real circuit boards, where $n \approx 1,000$, forget about it. All of the world's computers working full time wouldn't come close to finishing the problem before the end of the universe, at which point it presumably becomes moot.

The quest for an efficient algorithm to solve this problem, called the *traveling salesman problem* (TSP), will take us through much of this book. If you need to know how the story ends, check out the catalog entry for TSP in Section 19.4 (page 594).

> *Take-Home Lesson:*   There is a fundamental difference between *algorithms*, procedures that always produce a correct result, and *heuristics*, which may usually do a good job but provide no guarantee of correctness.

1.2   Selecting the Right Jobs

Now consider the following scheduling problem. Imagine you are a highly in demand actor, who has been presented with offers to star in n different movie projects under development. Each offer comes specified with the first and last

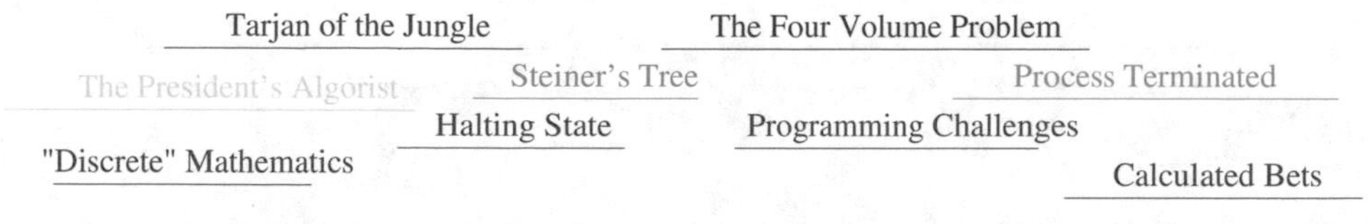

Tarjan of the Jungle		The Four Volume Problem	
The President's Algorist	Steiner's Tree		Process Terminated
	Halting State	Programming Challenges	
"Discrete" Mathematics			Calculated Bets

Figure 1.5: An instance of the non-overlapping movie scheduling problem. The four red titles define an optimal solution.

day of filming. Whenever you accept a job, you must commit to being available throughout this entire period. Thus, you cannot accept two jobs whose intervals overlap.

For an artist such as yourself, the criterion for job acceptance is clear: you want to make as much money as possible. Because each film pays the same fee, this implies you seek the largest possible set of jobs (intervals) such that no two of them conflict with each other.

For example, consider the available projects in Figure 1.5. You can star in at most four films, namely *"Discrete" Mathematics*, *Programming Challenges*, *Calculated Bets*, and one of either *Halting State* or *Steiner's Tree*.

You (or your agent) must solve the following algorithmic scheduling problem:

Problem: Movie Scheduling Problem
Input: A set I of n intervals on the line.
Output: What is the largest subset of mutually non-overlapping intervals that can be selected from I?

Now you (the algorist) are given the job of developing a scheduling algorithm for this task. Stop right now and try to find one. Again, I'll be happy to wait...

There are several ideas that may come to mind. One is based on the notion that it is best to work whenever work is available. This implies that you should start with the job with the earliest start date – after all, there is no other job you can work on then, at least during the beginning of this period:

> EarliestJobFirst(I)
>> Accept the earliest starting job j from I that does not overlap any previously accepted job, and repeat until no more such jobs remain.

This idea makes sense, at least until we realize that accepting the earliest job might block us from taking many other jobs if that first job is long. Check out Figure 1.6(l), where the epic *War and Peace* is both the first job available and long enough to kill off all other prospects.

This bad example naturally suggests another idea. The problem with *War and Peace* is that it is too long. Perhaps we should instead start by taking the shortest job, and keep seeking the shortest available job at every turn. Maximizing the number of jobs we do in a given period is clearly connected to the notion of banging them out as quickly as possible. This yields the heuristic:

Figure 1.6: Bad instances for the (l) earliest job first and (r) shortest job first heuristics. The optimal solutions are in red.

> ShortestJobFirst(I)
>   While ($I \neq \emptyset$) do
>   Accept the shortest possible job j from I.
>   Delete j, and any interval that intersects j, from I.

Again this idea makes sense, at least until we realize that accepting the shortest job might block us from taking two other jobs, as shown in Figure 1.6(r). While the maximum potential loss here seems smaller than with the previous heuristic, it can still limit us to half the optimal payoff.

At this point, an algorithm where we try all possibilities may start to look good. As with the TSP problem, we can be certain exhaustive search is correct. If we ignore the details of testing whether a set of intervals are in fact disjoint, it looks something like this:

> ExhaustiveScheduling(I)
> $j = 0$
> $S_{max} = \emptyset$
>   For each of the 2^n subsets S_i of intervals I
>   If (S_i is mutually non-overlapping) and ($size(S_i) > j$)
>   then $j = size(S_i)$ and $S_{max} = S_i$.
>   Return S_{max}

But how slow is it? The key limitation is enumerating the 2^n subsets of n things. The good news is that this is *much* better than enumerating all $n!$ orders of n things, as proposed for the robot tour optimization problem. There are only about one million subsets when $n = 20$, which can be enumerated within seconds on a decent computer. However, when fed $n = 100$ movies, we get 2^{100} subsets, which is much much greater than the 20! that made our robot cry "uncle" in the previous problem.

The difference between our scheduling and robotics problems is that there *is* an algorithm that solves movie scheduling both correctly and efficiently. Think about the first job to terminate—that is, the interval x whose right endpoint is left-most among all intervals. This role is played by *"Discrete" Mathematics* in Figure 1.5. Other jobs may well have started before x, but all of these must at least partially overlap each other. Thus, we can select at most one from the group. The first of these jobs to terminate is x, so any of the overlapping jobs potentially block out other opportunities to the right of it. Clearly we can never lose by picking x. This suggests the following correct, efficient algorithm:

    OptimalScheduling(I)
       While ($I \neq \emptyset$) do
            Accept the job j from I with the earliest completion date.
            Delete j, and any interval which intersects j, from I.

Ensuring the optimal answer over all possible inputs is a difficult but often achievable goal. Seeking counterexamples that break pretender algorithms is an important part of the algorithm design process. Efficient algorithms are often lurking out there; this book will develop your skills to help you find them.

Take-Home Lesson: Reasonable-looking algorithms can easily be incorrect. Algorithm correctness is a property that must be carefully demonstrated.

1.3 Reasoning about Correctness

Hopefully, the previous examples have opened your eyes to the subtleties of algorithm correctness. We need tools to distinguish correct algorithms from incorrect ones, the primary one of which is called a *proof.*

A proper mathematical proof consists of several parts. First, there is a clear, precise statement of what you are trying to prove. Second, there is a set of assumptions of things that are taken to be true, and hence can be used as part of the proof. Third, there is a chain of reasoning that takes you from these assumptions to the statement you are trying to prove. Finally, there is a little square (∎) or *QED* at the bottom to denote that you have finished, representing the Latin phrase for "thus it is demonstrated."

This book is not going to emphasize formal proofs of correctness, because they are very difficult to do right and quite misleading when you do them wrong. A proof is indeed a *demonstration.* Proofs are useful only when they are honest, crisp arguments that explain why an algorithm satisfies a non-trivial correctness property. Correct algorithms require careful exposition, and efforts to show both correctness and *not incorrectness.*

1.3.1 Problems and Properties

Before we start thinking about algorithms, we need a careful description of the problem that needs to be solved. Problem specifications have two parts: (1) the set of allowed input instances, and (2) the required properties of the algorithm's output. It is impossible to prove the correctness of an algorithm for a fuzzily-stated problem. Put another way, ask the wrong question and you will get the wrong answer.

Some problem specifications allow too broad a class of input instances. Suppose we had allowed film projects in our movie scheduling problem to have gaps in production (e.g. filming in September and November but a hiatus in October). Then the schedule associated with any particular film would consist of a given *set* of intervals. Our star would be free to take on two interleaving but not overlapping projects (such as the above-mentioned film nested with one filming in August and October). The earliest completion algorithm would not work for such a generalized scheduling problem. Indeed, *no* efficient algorithm exists for this generalized problem, as we will see in Section 11.3.2.

Take-Home Lesson: An important and honorable technique in algorithm design is to narrow the set of allowable instances until there *is* a correct and efficient algorithm. For example, we can restrict a graph problem from general graphs down to trees, or a geometric problem from two dimensions down to one.

There are two common traps when specifying the output requirements of a problem. The first is asking an ill-defined question. Asking for the *best* route between two places on a map is a silly question, unless you define what *best* means. Do you mean the shortest route in total distance, or the fastest route, or the one minimizing the number of turns? All of these are liable to be different things.

The second trap involves creating compound goals. The three route-planning criteria mentioned above are all well-defined goals that lead to correct, efficient optimization algorithms. But you must pick a single criterion. A goal like *Find the shortest route from a to b that doesn't use more than twice as many turns as necessary* is perfectly well defined, but complicated to reason about and solve.

I encourage you to check out the problem statements for each of the seventy-five catalog problems in Part II of this book. Finding the right formulation for your problem is an important part of solving it. And studying the definition of all these classic algorithm problems will help you recognize when someone else has thought about similar problems before you.

1.3.2   Expressing Algorithms

Reasoning about an algorithm is impossible without a careful description of the sequence of steps that are to be performed. The three most common forms of algorithmic notation are (1) English, (2) pseudocode, or (3) a real programming language. Pseudocode is perhaps the most mysterious of the bunch, but it is best defined as a programming language that never complains about syntax errors.

All three methods are useful because there is a natural tradeoff between greater ease of expression and precision. English is the most natural but least precise programming language, while Java and C/C++ are precise but difficult to write and understand. Pseudocode is generally useful because it represents a happy medium.

The choice of which notation is best depends upon which method you are most comfortable with. I usually prefer to describe the *ideas* of an algorithm in English (with pictures!), moving to a more formal, programming-language-like pseudocode or even real code to clarify sufficiently tricky details.

A common mistake my students make is to use pseudocode to dress up an ill-defined idea so that it looks more formal. Clarity should be the goal. For example, the ExhaustiveScheduling algorithm on page 10 would have better been written in English as:

ExhaustiveScheduling(I)
>Test all 2^n subsets of intervals from I, and return the largest subset consisting of mutually non-overlapping intervals.

Take-Home Lesson: The heart of any algorithm is an *idea*. If your idea is not clearly revealed when you express an algorithm, then you are using too low-level a notation to describe it.

1.3.3 Demonstrating Incorrectness

The best way to prove that an algorithm is *incorrect* is to produce an instance on which it yields an incorrect answer. Such instances are called *counterexamples*. No rational person will ever defend the correctness of an algorithm after a counter-example has been identified. Very simple instances can instantly defeat reasonable-looking heuristics with a quick *touché*. Good counterexamples have two important properties:

- *Verifiability* – To demonstrate that a particular instance is a counterexample to a particular algorithm, you must be able to (1) calculate what answer your algorithm will give in this instance, and (2) display a better answer so as to prove that the algorithm didn't find it.

- *Simplicity* – Good counter-examples have all unnecessary details stripped away. They make clear exactly *why* the proposed algorithm fails. Simplicity is important because you must be able to hold the given instance in your head in order to reason about it. Once a counterexample has been found, it is worth simplifying it down to its essence. For example, the counterexample of Figure 1.6(l) could have been made simpler and better by reducing the number of overlapped segments from five to two.

Hunting for counterexamples is a skill worth developing. It bears some similarity to the task of developing test sets for computer programs, but relies more on inspiration than exhaustion. Here are some techniques to aid your quest:

- *Think small* – Note that the robot tour counter-examples I presented boiled down to six points or less, and the scheduling counter-examples to only three intervals. This is indicative of the fact that when algorithms fail, there is usually a very simple example on which they fail. Amateur algorists tend to draw a big messy instance and then stare at it helplessly. The pros look carefully at several small examples, because they are easier to verify and reason about.

- *Think exhaustively* – There are usually only a small number of possible instances for the first non-trivial value of n. For example, there are only three distinct ways two intervals on the line can occur: as disjoint intervals, as overlapping intervals, and as properly nesting intervals, one within the other. All cases of three intervals (including counter-examples to both of the movie heuristics) can be systematically constructed by adding a third segment in each possible way to these three instances.

- *Hunt for the weakness* – If a proposed algorithm is of the form "always take the biggest" (better known as the *greedy algorithm*), think about why that might prove to be the wrong thing to do. In particular, . . .

- *Go for a tie* – A devious way to break a greedy heuristic is to provide instances where everything is the same size. Suddenly the heuristic has nothing to base its decision on, and perhaps has the freedom to return something suboptimal as the answer.

- *Seek extremes* – Many counter-examples are mixtures of huge and tiny, left and right, few and many, near and far. It is usually easier to verify or reason about extreme examples than more muddled ones. Consider two tightly bunched clouds of points separated by a much larger distance d. The optimal TSP tour will be essentially $2d$ regardless of the number of points, because what happens within each cloud doesn't really matter.

> *Take-Home Lesson:* Searching for counterexamples is the best way to disprove the correctness of a heuristic.

Stop and Think: Greedy Movie Stars?

Problem: Recall the movie star scheduling problem, where we seek to find the largest possible set of non-overlapping intervals in a given set S. A natural greedy heuristic selects the interval i, which overlaps the smallest number of other intervals in S, removes them, and repeats until no intervals remain.

Give a counter-example to this proposed algorithm.

Solution: Consider the counter-example in Figure 1.7. The largest possible independent set consists of the four intervals in red, but the interval of lowest degree (shown in pink) overlaps two of these, After we grab it, we are doomed to finding a solution of only three intervals.

But how would you go about constructing such an example? My thought process started with an odd-length chain of intervals, each of which overlaps one interval to the left and one to the right. Picking an even-length chain would mess up the optimal solution (*hunt for the weakness*). All intervals overlap two others, except for the left and right-most intervals (*go for the tie*). To make these terminal intervals unattractive, we can pile other intervals on top of them

Figure 1.7: Counter-example to the greedy heuristic for movie star scheduling. Picking the pink interval, which intersects the fewest others, blocks us from the optimal solution (the four red intervals).

(*seek extremes*). The length of our chain (7) is the shortest that permits this construction to work. ∎

1.4 Induction and Recursion

Failure to find a counterexample to a given algorithm does not mean "it is obvious" that the algorithm is correct. A proof or demonstration of correctness is needed. Often mathematical induction is the method of choice.

When I first learned about mathematical induction it seemed like complete magic. You proved a formula like $\sum_{i=1}^{n} i = n(n+1)/2$ for some basis case like $n = 1$ or 2, then *assumed* it was true all the way to $n - 1$ before proving it was in fact true for general n using the assumption. That was a proof? Ridiculous!

When I first learned the programming technique of recursion it also seemed like complete magic. The program tested whether the input argument was some basis case like 1 or 2. If not, you solved the bigger case by breaking it into pieces and *calling the subprogram itself* to solve these pieces. That was a program? Ridiculous!

The reason both seemed like magic is because recursion *is* mathematical induction in action. In both, we have general and boundary conditions, with the general condition breaking the problem into smaller and smaller pieces. The *initial* or boundary condition terminates the recursion. Once you understand either recursion or induction, you should be able to see why the other one also works.

I've heard it said that a computer scientist is a mathematician who only knows how to prove things by induction. This is partially true because computer scientists are lousy at proving things, but primarily because so many of the algorithms we study are either recursive or incremental.

Consider the correctness of *insertion sort*, which we introduced at the beginning of this chapter. The *reason* it is correct can be shown inductively:

Figure 1.8: Large-scale changes in the optimal solution (red boxes) after inserting a single interval (dashed) into the instance.

- The basis case consists of a single element, and by definition a one-element array is completely sorted.

- We assume that the first $n - 1$ elements of array A are completely sorted after $n - 1$ iterations of insertion sort.

- To insert one last element x to A, we find where it goes, namely the unique spot between the biggest element less than or equal to x and the smallest element greater than x. This is done by moving all the greater elements back by one position, creating room for x in the desired location. ∎

One must be suspicious of inductive proofs, however, because very subtle reasoning errors can creep in. The first are *boundary errors*. For example, our insertion sort correctness proof above boldly stated that there was a unique place to insert x between two elements, when our basis case was a single-element array. Greater care is needed to properly deal with the special cases of inserting the minimum or maximum elements.

The second and more common class of inductive proof errors concerns cavalier extension claims. Adding one extra item to a given problem instance might cause the entire optimal solution to change. This was the case in our scheduling problem (see Figure 1.8). The optimal schedule after inserting a new segment may contain none of the segments of any particular optimal solution prior to insertion. Boldly ignoring such difficulties can lead to very convincing inductive proofs of incorrect algorithms.

Take-Home Lesson: Mathematical induction is usually the right way to verify the correctness of a recursive or incremental insertion algorithm.

Stop and Think: Incremental Correctness

Problem: Prove the correctness of the following recursive algorithm for incrementing natural numbers, that is, $y \rightarrow y + 1$:

```
Increment(y)
     if (y = 0) then return(1) else
          if (y mod 2) = 1 then
               return(2 · Increment(⌊y/2⌋))
          else return(y + 1)
```

Solution: The correctness of this algorithm is certainly *not* obvious to me. But as it is recursive and I am a computer scientist, my natural instinct is to try to prove it by induction. The basis case of $y = 0$ is obviously correctly handled. Clearly the value 1 is returned, and $0 + 1 = 1$.

Now assume the function works correctly for the general case of $y = n - 1$. Given this, we must demonstrate the truth for the case of $y = n$. The cases corresponding to even numbers are obvious, because $y + 1$ is explicitly returned when $(y \bmod 2) = 0$.

For odd numbers, the answer depends on what Increment$(\lfloor y/2 \rfloor)$ returns. Here we want to use our inductive assumption, but it isn't quite right. We have assumed that Increment worked correctly for $y = n - 1$, but not for a value that is about half of it. We can fix this problem by strengthening our assumption to declare that the general case holds for all $y \le n - 1$. This costs us nothing in principle, but is necessary to establish the correctness of the algorithm.

Now, the case of odd y (i.e. $y = 2m + 1$ for some integer m) can be dealt with as:

$$
\begin{aligned}
2 \cdot \text{Increment}(\lfloor (2m+1)/2 \rfloor) &= 2 \cdot \text{Increment}(\lfloor m + 1/2 \rfloor) \\
&= 2 \cdot \text{Increment}(m) \\
&= 2(m + 1) \\
&= 2m + 2 = y + 1
\end{aligned}
$$

and the general case is resolved. ∎

1.5 Modeling the Problem

Modeling is the art of formulating your application in terms of precisely described, well-understood problems. Proper modeling is the key to applying algorithmic design techniques to real-world problems. Indeed, proper modeling can eliminate the need to design or even implement algorithms, by relating your application to what has been done before. Proper modeling is the key to effectively using the "Hitchhiker's Guide" in Part II of this book.

Real-world applications involve real-world objects. You might be working on a system to route traffic in a network, to find the best way to schedule classrooms in a university, or to search for patterns in a corporate database. Most algorithms, however, are designed to work on rigorously defined *abstract* structures such as permutations, graphs, and sets. To exploit the algorithms literature, you must learn to describe your problem abstractly, in terms of procedures on such fundamental structures.

1.5.1 Combinatorial Objects

Odds are very good that others have probably stumbled upon any algorithmic problem you care about, perhaps in substantially different contexts. But to find

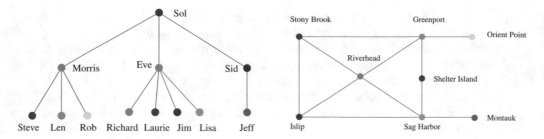

Figure 1.9: Modeling real-world structures with trees and graphs.

out what is known about your particular "widget optimization problem," you can't hope to find it in a book under *widget*. You must first formulate widget optimization in terms of computing properties of common structures such as those described below:

- *Permutations* are arrangements, or orderings, of items. For example, $\{1, 4, 3, 2\}$ and $\{4, 3, 2, 1\}$ are two distinct permutations of the same set of four integers. We have already seen permutations in the robot optimization problem, and in sorting. Permutations are likely the object in question whenever your problem seeks an "arrangement," "tour," "ordering," or "sequence."

- *Subsets* represent selections from a set of items. For example, $\{1, 3, 4\}$ and $\{2\}$ are two distinct subsets of the first four integers. Order does not matter in subsets the way it does with permutations, so the subsets $\{1, 3, 4\}$ and $\{4, 3, 1\}$ would be considered identical. Subsets arose as candidate solutions in the movie scheduling problem. They are likely the object in question whenever your problem seeks a "cluster," "collection," "committee," "group," "packaging," or "selection."

- *Trees* represent hierarchical relationships between items. Figure 1.9(a) shows part of the family tree of the Skiena clan. Trees are likely the object in question whenever your problem seeks a "hierarchy," "dominance relationship," "ancestor/descendant relationship," or "taxonomy."

- *Graphs* represent relationships between arbitrary pairs of objects. Figure 1.9(b) models a network of roads as a graph, where the vertices are cities and the edges are roads connecting pairs of cities. Graphs are likely the object in question whenever you seek a "network," "circuit," "web," or "relationship."

- *Points* define locations in some geometric space. For example, the locations of McDonald's restaurants can be described by points on a map/plane. Points are likely the object in question whenever your problems work on "sites," "positions," "data records," or "locations."

- *Polygons* define regions in some geometric spaces. For example, the borders of a country can be described by a polygon on a map/plane. Polygons and polyhedra are likely the object in question whenever you are working on "shapes," "regions," "configurations," or "boundaries."

- *Strings* represent sequences of characters, or patterns. For example, the names of students in a class can be represented by strings. Strings are likely the object in question whenever you are dealing with "text," "characters," "patterns," or "labels."

These fundamental structures all have associated algorithm problems, which are presented in the catalog of Part II. Familiarity with these problems is important, because they provide the language we use to model applications. To become fluent in this vocabulary, browse through the catalog and study the *input* and *output* pictures for each problem. Understanding these problems, even at a cartoon/definition level, will enable you to know where to look later when the problem arises in your application.

Examples of successful application modeling will be presented in the war stories spaced throughout this book. However, some words of caution are in order. The act of modeling reduces your application to one of a small number of existing problems and structures. Such a process is inherently constraining, and certain details might not fit easily into the given target problem. Also, certain problems can be modeled in several different ways, some much better than others.

Modeling is only the first step in designing an algorithm for a problem. Be alert for how the details of your applications differ from a candidate model, but don't be too quick to say that your problem is unique and special. Temporarily ignoring details that don't fit can free the mind to ask whether they really were fundamental in the first place.

Take-Home Lesson: Modeling your application in terms of well-defined structures and algorithms is the most important single step towards a solution.

1.5.2 Recursive Objects

Learning to think recursively is learning to look for big things that are made from smaller things of *exactly the same type as the big thing*. If you think of houses as sets of rooms, then adding or deleting a room still leaves a house behind.

Recursive structures occur everywhere in the algorithmic world. Indeed, each of the abstract structures described above can be thought about recursively. You just have to see how you can break them down, as shown in Figure 1.10:

- *Permutations* – Delete the first element of a permutation of n things $\{1, \ldots, n\}$ and you get a permutation of the remaining $n-1$ things. This may require renumbering to keep the object a permutation of consecutive integers. For example, removing the first element of $\{4, 1, 5, 2, 3\}$ and

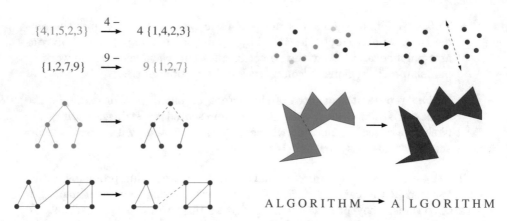

Figure 1.10:   Recursive decompositions of combinatorial objects. Permutations, subsets, trees, and graphs (left column). Point sets, polygons, and strings (right column). Note that the elements of a permutation of $\{1, \ldots, n\}$ get renumbered after element deletion in order to remain a permutation of $\{1, \ldots, n-1\}$.

renumbering gives $\{1, 4, 2, 3\}$, a permutation of $\{1, 2, 3, 4\}$. Permutations are recursive objects.

- *Subsets* – Every subset of $\{1, \ldots, n\}$ contains a subset of $\{1, \ldots, n-1\}$ obtained by deleting element n, if it is present. Subsets are recursive objects.

- *Trees* – Delete the root of a tree and what do you get? A collection of smaller trees. Delete any leaf of a tree and what do you get? A slightly smaller tree. Trees are recursive objects.

- *Graphs* – Delete any vertex from a graph, and you get a smaller graph. Now divide the vertices of a graph into two groups, left and right. Cut through all edges that span from left to right, and what do you get? Two smaller graphs, and a bunch of broken edges. Graphs are recursive objects.

- *Points* – Take a cloud of points, and separate them into two groups by drawing a line. Now you have two smaller clouds of points. Point sets are recursive objects.

- *Polygons* – Inserting any internal chord between two non-adjacent vertices of a simple polygon cuts it into two smaller polygons. Polygons are recursive objects.

- *Strings* – Delete the first character from a string, and what do you get? A shorter string. Strings are recursive objects.[1]

---

[1]An alert reader of the previous edition observed that salads are recursive objects, in ways that hamburgers are not. Take a bite (or an ingredient) out of a salad, and what is left is a smaller salad. Take a bite out of a hamburger, and what is left is something disgusting.

Recursive descriptions of objects require both decomposition rules and *basis cases*, namely the specification of the smallest and simplest objects where the decomposition stops. These basis cases are usually easily defined. Permutations and subsets of zero things presumably look like {}. The smallest interesting tree or graph consists of a single vertex, while the smallest interesting point cloud consists of a single point. Polygons are a little trickier; the smallest genuine simple polygon is a triangle. Finally, the empty string has zero characters in it. The decision of whether the basis case contains zero or one element is more a question of taste and convenience than any fundamental principle.

Recursive decompositions will define many of the algorithms we will see in this book. Keep your eyes open for them.

1.6 Proof by Contradiction

Although some computer scientists only know how to prove things by induction, this isn't true of everyone. The best sometimes use *contradiction*.

The basic scheme of a contradiction argument is as follows:

- Assume that the hypothesis (the statement you want to prove) *is false*.

- Develop some logical consequences of this assumption.

- Show that one consequence is demonstrably false, thereby showing that the assumption is incorrect and the hypothesis is true.

The classic contradiction argument is Euclid's proof that there are an infinite number of prime numbers: integers n like $2, 3, 5, 7, 11, \ldots$ that have no non-trivial factors, only 1 and n itself. The negation of the claim would be that there are only a finite number of primes, say m, which can be listed as $p_1, \ldots, p_m$. So let's assume this is the case and work with it.

Prime numbers have particular properties with respect to division. Suppose we construct the integer formed as the product of "all" of the listed primes:

$$N = \prod_{i=1}^{m} p_i$$

This integer N has the property that it is divisible by each and every one of the known primes, because of how it was built.

But consider the integer $N+1$. It can't be divisible by $p_1 = 2$, because N is. The same is true for $p_2 = 3$ and every other listed prime. Since $N + 1$ doesn't have any non-trivial factor, this means it must be prime. But you asserted that there are exactly m prime numbers, none of which are $N + 1$. This assertion is false, so there cannot be a bounded number of primes. *Touché!*

For a contradiction argument to be convincing, the final consequence must be clearly, ridiculously false. Muddy outcomes are not convincing. It is also important that this contradiction be a logical consequence of the assumption. We will see contradiction arguments for minimum spanning tree algorithms in Section 8.1.

1.7   About the War Stories

The best way to learn how careful algorithm design can have a huge impact on performance is to look at real-world case studies. By carefully studying other people's experiences, we learn how they might apply to our work.

Scattered throughout this text are several of my own algorithmic war stories, presenting our successful (and occasionally unsuccessful) algorithm design efforts on real applications. I hope that you will be able to internalize these experiences so that they will serve as models for your own attacks on problems.

Every one of the war stories is true. Of course, the stories improve somewhat in the retelling, and the dialogue has been punched up to make them more interesting to read. However, I have tried to honestly trace the process of going from a raw problem to a solution, so you can watch how this process unfolded.

The *Oxford English Dictionary* defines an *algorist* as "one skillful in reckonings or figuring." In these stories, I have tried to capture some of the mindset of the algorist in action as they attack a problem.

The war stories often involve at least one problem from the problem catalog in Part II. I reference the appropriate section of the catalog when such a problem occurs. This emphasizes the benefits of modeling your application in terms of standard algorithm problems. By using the catalog, you will be able to pull out what is known about any given problem whenever it is needed.

1.8   War Story: Psychic Modeling

The call came for me out of the blue as I sat in my office.

"Professor Skiena, I hope you can help me. I'm the President of Lotto Systems Group Inc., and we need an algorithm for a problem arising in our latest product."

"Sure," I replied. After all, the dean of my engineering school is always encouraging our faculty to interact more with industry.

"At Lotto Systems Group, we market a program designed to improve our customers' psychic ability to predict winning lottery numbers.[2] In a standard lottery, each ticket consists of six numbers selected from, say, 1 to 44. However, after proper training, our clients can visualize (say) fifteen numbers out of the 44 and be certain that at least four of them will be on the winning ticket. Are you with me so far?"

"Probably not," I replied. But then I recalled how my dean encourages us to interact with industry.

"Our problem is this. After the psychic has narrowed the choices down to fifteen numbers and is certain that at least four of them will be on the winning ticket, we must find the most efficient way to exploit this information. Suppose a cash prize is awarded whenever you pick at least three of the correct numbers on your ticket. We need an algorithm to construct the smallest set of tickets that we must buy in order to guarantee that we win at least one prize."

---

[2] Yes, this is a true story.

Tickets			Winning Pairs			
1 2 3			1 2	2 3	3 4	
1 4 5			1 3	2 4	3 5	
2 4 5			1 4	2 5	4 5	
3 4 5			1 5			

Figure 1.11: Covering all pairs of $\{1,2,3,4,5\}$ with tickets $\{1,2,3\}$, $\{1,4,5\}$, $\{2,4,5\}$, $\{3,4,5\}$. Pair color reflects the covering ticket.

"Assuming the psychic is correct?"

"Yes, assuming the psychic is correct. We need a program that prints out a list of all the tickets that the psychic should buy in order to minimize their investment. Can you help us?"

Maybe they did have psychic ability, for they had come to the right place. Identifying the best subset of tickets to buy was very much a combinatorial algorithm problem. It was going to be some type of covering problem, where each ticket bought would "cover" some of the possible 4-element subsets of the psychic's set. Finding the absolute smallest set of tickets to cover everything was a special instance of the NP-complete problem *set cover* (discussed in Section 21.1 (page 678)), and presumably computationally intractable.

It was indeed a special instance of set cover, completely specified by only four numbers: the size n of the candidate set S (typically $n \approx 15$), the number of slots k for numbers on each ticket (typically $k \approx 6$), the number of psychically-promised correct numbers j from S (say $j = 4$), and finally, the number of matching numbers l necessary to win a prize (say $l = 3$). Figure 1.11 illustrates a covering of a smaller instance, where $n = 5$, $k = 3$, and $l = 2$, and no psychic contribution (meaning $j = 5$).

"Although it will be hard to find the *exact* minimum set of tickets to buy, with heuristics I should be able to get you pretty close to the cheapest covering ticket set," I told him. "Will that be good enough?"

"So long as it generates better ticket sets than my competitor's program, that will be fine. His system doesn't always guarantee a win. I really appreciate your help on this, Professor Skiena."

"One last thing. If your program can train people to pick lottery winners, why don't you use it to win the lottery yourself?"

"I look forward to talking to you again real soon, Professor Skiena. Thanks for the help."

I hung up the phone and got back to thinking. It seemed like the perfect project to give to a bright undergraduate. After modeling it in terms of sets and subsets, the basic components of a solution seemed fairly straightforward:

- We needed the ability to generate all subsets of k numbers from the candidate set S. Algorithms for generating and ranking/unranking subsets of sets are presented in Section 17.5 (page 521).

- We needed the right formulation of what it meant to have a covering set of purchased tickets. The obvious criteria would be to pick a small set of tickets such that we have purchased at least one ticket containing each of the $\binom{n}{l}$ l-subsets of S that might pay off with the prize.

- We needed to keep track of which prize combinations we have thus far covered. We seek tickets to cover as many thus-far-uncovered prize combinations as possible. The currently covered combinations are a subset of all possible combinations. Data structures for subsets are discussed in Section 15.5 (page 456). The best candidate seemed to be a bit vector, which would answer in constant time "is this combination already covered?"

- We needed a search mechanism to decide which ticket to buy next. For small enough set sizes, we could do an exhaustive search over all possible subsets of tickets and pick the smallest one. For larger problems, a randomized search process like simulated annealing (see Section 12.6.3 (page 406)) would select tickets-to-buy to cover as many uncovered combinations as possible. By repeating this randomized procedure several times and picking the best solution, we would be likely to come up with a good set of tickets.

The bright undergraduate, Fayyaz Younas, rose to the challenge. Based on this framework, he implemented a brute-force search algorithm and found optimal solutions for problems with $n \leq 5$ in a reasonable time. He implemented a random search procedure to solve larger problems, tweaking it for a while before settling on the best variant. Finally, the day arrived when we could call Lotto Systems Group and announce that we had solved the problem.

"Our program found an optimal solution for $n = 15$, $k = 6$, $j = 4$, $l = 3$ meant buying 28 tickets."

"Twenty-eight tickets!" complained the president. "You must have a bug. Look, these five tickets will suffice to cover everything *twice* over: $\{2, 4, 8, 10, 13, 14\}$, $\{4, 5, 7, 8, 12, 15\}$, $\{1, 2, 3, 6, 11, 13\}$, $\{3, 5, 6, 9, 10, 15\}$, $\{1, 7, 9, 11, 12, 14\}$." We fiddled with this example for a while before admitting that he was right.

We hadn't modeled the problem correctly! In fact, we didn't need to explicitly cover all possible winning combinations. Figure 1.12 illustrates the principle by giving a two-ticket solution to our previous four-ticket example. Although the pairs $\{2, 4\}$, $\{2, 5\}$, $\{3, 4\}$, or $\{3, 5\}$ do not explicitly appear in one of our two tickets, these pairs plus any possible third ticket number must create a pair in either $\{1, 2, 3\}$ or $\{1, 4, 5\}$. We were trying to cover too many combinations, and the penny-pinching psychics were unwilling to pay for such extravagance.

Fortunately, this story has a happy ending. The general outline of our search-based solution still holds for the real problem. All we must fix is which subsets we get credit for covering with a given set of tickets. After this modification, we obtained the kind of results they were hoping for. Lotto Systems Group gratefully accepted our program to incorporate into their product, and we hope they hit the jackpot with it.

Tickets	Winning Pairs		
1 2 3	1 2	2 3	3 4
1 4 5	1 3	2 4	3 5
	1 4	2 5	4 5
	1 5		

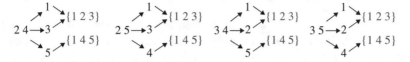

Figure 1.12: Guaranteeing a winning pair from $\{1, 2, 3, 4, 5\}$ using only tickets $\{1, 2, 3\}$ and $\{1, 4, 5\}$. The bottom figure shows how all missing pairs imply a covered pair on expansion.

The moral of this story is to make sure that you model your problem correctly before trying to solve it. In our case, we came up with a reasonable model, but didn't work hard enough to validate it before we started to program. Our misinterpretation would have become obvious had we worked out a small example by hand and bounced it off our sponsor before beginning work. Our success in recovering from this error is a tribute to the basic correctness of our initial formulation, and our use of well-defined abstractions for such tasks as (1) ranking/unranking k-subsets, (2) the set data structure, and (3) combinatorial search.

1.9   Estimation

When you don't know the right answer, the best thing to do is guess. Principled guessing is called *estimation*. The ability to make back-of-the-envelope estimates of diverse quantities such as the running time of a program is a valuable skill in algorithm design, as it is in any technical enterprise.

Estimation problems are best solved through some kind of logical reasoning process, typically a mix of principled calculations and analogies. *Principled calculations* give the answer as a function of quantities that either you already know, can look up on Google, or feel confident enough to guess. *Analogies* reference your past experiences, recalling those that seem similar to some aspect of the problem at hand.

I once asked my class to estimate the number of pennies in a hefty glass jar, and got answers ranging from 250 to 15,000. Both answers will seem pretty silly if you make the right analogies:

- A penny roll holds 50 coins in a tube roughly the length and width of your biggest finger. So five such rolls can easily be held in your hand, with no need for a hefty jar.

- 15,000 pennies means $150 in value. I have never managed to accumulate that much value in coins even in a hefty pile of change—and here I am only using pennies!

But the class average estimate proved very close to the right answer, which turned out to be 1879. There are at least three principled ways I can think of to estimate the number of coins in the jar:

- *Volume* – The jar was a cylinder with about 5 inches diameter, and the coins probably reached a level equal to about ten pennies tall stacked end to end. Figure a penny is ten times longer than it is thick. The bottom layer of the jar was a circle of radius about five pennies. So

$$(10 \times 10) \times (\pi \times 2.5^2) \approx 1962.5$$

- *Weight* – Lugging the jar felt like carrying around a bowling ball. Multiplying the number of US pennies in a pound (181 when I looked it up) times a 10 lb. ball gave a frighteningly accurate estimate of 1810.

- *Analogy* – The coins had a height of about 8 inches in the jar, or twice that of a penny roll. Figure I could stack about two layers of ten rolls per layer in the jar, or a total estimate of 1,000 coins.

A best practice in estimation is to try to solve the problem in different ways and see if the answers generally agree in magnitude. All of these are within a factor of two of each other, giving me confidence that my answer is about right.

Try some of the estimation exercises at the end of this chapter, and see how many different ways you can approach them. If you do things right, the ratio between your high and low estimates should be somewhere within a factor of two to ten, depending upon the nature of the problem. A sound reasoning process matters a lot more here than the actual numbers you get.

Chapter Notes

Every algorithm book reflects the design philosophy of its author. For students seeking alternative presentations and viewpoints, I particularly recommend the books of Cormen, et al. [CLRS09], Kleinberg and Tardos [KT06], Manber [Man89], and Roughgarden [Rou17].

Formal proofs of algorithm correctness are important, and deserve a fuller discussion than this chapter is able to provide. See Gries [Gri89] for a thorough introduction to the techniques of program verification.

The movie scheduling problem represents a very special case of the general *independent set* problem, which will be discussed in Section 19.2 (page 589). The restriction limits the allowable input instances to *interval* graphs, where the vertices of the graph G can be represented by intervals on the line and (i, j) is an edge of G iff the intervals overlap. Golumbic [Gol04] provides a full treatment of this interesting and important class of graphs.

Jon Bentley's *Programming Pearls* columns are probably the best known collection of algorithmic "war stories," collected in two books [Ben90, Ben99]. Brooks's *The Mythical Man Month* [Bro95] is another wonderful collection of war stories that, although focused more on software engineering than algorithm design, remain a source of considerable wisdom. Every programmer should read these books for pleasure as well as insight.

Our solution to the lotto ticket set covering problem is presented in more detail in Younas and Skiena [YS96].

1.10   Exercises

Finding Counterexamples

1-1. *[3]* Show that $a + b$ can be less than $\min(a, b)$.

1-2. *[3]* Show that $a \times b$ can be less than $\min(a, b)$.

1-3. *[5]* Design/draw a road network with two points a and b such that the fastest route between a and b is not the shortest route.

1-4. *[5]* Design/draw a road network with two points a and b such that the shortest route between a and b is not the route with the fewest turns.

1-5. *[4]* The *knapsack problem* is as follows: given a set of integers $S = \{s_1, s_2, \ldots, s_n\}$, and a target number T, find a subset of S that adds up exactly to T. For example, there exists a subset within $S = \{1, 2, 5, 9, 10\}$ that adds up to $T = 22$ but not $T = 23$.

Find counterexamples to each of the following algorithms for the knapsack problem. That is, give an S and T where the algorithm does not find a solution that leaves the knapsack completely full, even though a full-knapsack solution exists.

(a) Put the elements of S in the knapsack in left to right order if they fit, that is, the first-fit algorithm.

(b) Put the elements of S in the knapsack from smallest to largest, that is, the best-fit algorithm.

(c) Put the elements of S in the knapsack from largest to smallest.

1-6. *[5]* The *set cover problem* is as follows: given a set S of subsets $S_1, \ldots, S_m$ of the universal set $U = \{1, \ldots, n\}$, find the smallest subset of subsets $T \subseteq S$ such that $\cup_{t_i \in T} t_i = U$. For example, consider the subsets $S_1 = \{1, 3, 5\}$, $S_2 = \{2, 4\}$, $S_3 = \{1, 4\}$, and $S_4 = \{2, 5\}$. The set cover of $\{1, \ldots, 5\}$ would then be S_1 and S_2.

Find a counterexample for the following algorithm: Select the largest subset for the cover, and then delete all its elements from the universal set. Repeat by adding the subset containing the largest number of uncovered elements until all are covered.

1-7. *[5]* The *maximum clique* problem in a graph $G = (V, E)$ asks for the largest subset C of vertices V such that there is an edge in E between every pair of vertices in C. Find a counterexample for the following algorithm: Sort the vertices of G from highest to lowest degree. Considering the vertices in order of degree, for each vertex add it to the clique if it is a neighbor of all vertices currently in the clique. Repeat until all vertices have been considered.

Proofs of Correctness

1-8. *[3]* Prove the correctness of the following recursive algorithm to multiply two natural numbers, for all integer constants $c \geq 2$.

$$\text{Multiply}(y, z)$$
$$\text{if } z = 0 \text{ then return}(0) \text{ else}$$
$$\text{return}(\text{Multiply}(cy, \lfloor z/c \rfloor) + y \cdot (z \bmod c))$$

1-9. *[3]* Prove the correctness of the following algorithm for evaluating a polynomial $a_n x^n + a_{n-1} x^{n-1} + \cdots + a_1 x + a_0$.

$$\text{Horner}(a, x)$$
$$p = a_n$$
$$\text{for } i \text{ from } n - 1 \text{ to } 0$$
$$p = p \cdot x + a_i$$
$$\text{return } p$$

1-10. *[3]* Prove the correctness of the following sorting algorithm.

$$\text{Bubblesort } (A)$$
$$\text{for } i \text{ from } n \text{ to } 1$$
$$\text{for } j \text{ from } 1 \text{ to } i - 1$$
$$\text{if } (A[j] > A[j + 1])$$
$$\text{swap the values of } A[j] \text{ and } A[j + 1]$$

1-11. *[5]* The *greatest common divisor* of positive integers x and y is the largest integer d such that d divides x and d divides y. Euclid's algorithm to compute $\gcd(x, y)$ where $x > y$ reduces the task to a smaller problem:

$$\gcd(x, y) = \gcd(y, x \bmod y)$$

Prove that Euclid's algorithm is correct.

Induction

1-12. *[3]* Prove that $\sum_{i=1}^{n} i = n(n + 1)/2$ for $n \geq 0$, by induction.

1-13. *[3]* Prove that $\sum_{i=1}^{n} i^2 = n(n + 1)(2n + 1)/6$ for $n \geq 0$, by induction.

1-14. *[3]* Prove that $\sum_{i=1}^{n} i^3 = n^2(n + 1)^2/4$ for $n \geq 0$, by induction.

1-15. *[3]* Prove that

$$\sum_{i=1}^{n} i(i + 1)(i + 2) = n(n + 1)(n + 2)(n + 3)/4$$

1-16. *[5]* Prove by induction on $n \geq 1$ that for every $a \neq 1$,

$$\sum_{i=0}^{n} a^i = \frac{a^{n+1} - 1}{a - 1}$$

1-17. *[3]* Prove by induction that for $n \geq 1$,

$$\sum_{i=1}^{n} \frac{1}{i(i+1)} = \frac{n}{n+1}$$

1-18. *[3]* Prove by induction that $n^3 + 2n$ is divisible by 3 for all $n \geq 0$.

1-19. *[3]* Prove by induction that a tree with n vertices has exactly $n-1$ edges.

1-20. *[3]* Prove by induction that the sum of the cubes of the first n positive integers is equal to the square of the sum of these integers, that is,

$$\sum_{i=1}^{n} i^3 = (\sum_{i=1}^{n} i)^2$$

Estimation

1-21. *[3]* Do all the books you own total at least one million pages? How many total pages are stored in your school library?

1-22. *[3]* How many words are there in this textbook?

1-23. *[3]* How many hours are one million seconds? How many days? Answer these questions by doing all arithmetic in your head.

1-24. *[3]* Estimate how many cities and towns there are in the United States.

1-25. *[3]* Estimate how many cubic miles of water flow out of the mouth of the Mississippi River each day. Do not look up any supplemental facts. Describe all assumptions you made in arriving at your answer.

1-26. *[3]* How many Starbucks or McDonald's locations are there in your country?

1-27. *[3]* How long would it take to empty a bathtub with a drinking straw?

1-28. *[3]* Is disk drive access time normally measured in milliseconds (thousandths of a second) or microseconds (millionths of a second)? Does your RAM memory access a word in more or less than a microsecond? How many instructions can your CPU execute in one year if the machine is left running all the time?

1-29. *[4]* A sorting algorithm takes 1 second to sort 1,000 items on your machine. How long will it take to sort 10,000 items...

   (a) if you believe that the algorithm takes time proportional to n^2, and

   (b) if you believe that the algorithm takes time roughly proportional to $n \log n$?

Implementation Projects

1-30. *[5]* Implement the two TSP heuristics of Section 1.1 (page 5). Which of them gives better solutions in practice? Can you devise a heuristic that works better than both of them?

1-31. *[5]* Describe how to test whether a given set of tickets establishes sufficient coverage in the Lotto problem of Section 1.8 (page 22). Write a program to find good ticket sets.

Interview Problems

1-32. *[5]* Write a function to perform integer division without using either the / or *
operators. Find a fast way to do it.

1-33. *[5]* There are twenty-five horses. At most, five horses can race together at a
time. You must determine the fastest, second fastest, and third fastest horses.
Find the minimum number of races in which this can be done.

1-34. *[3]* How many piano tuners are there in the entire world?

1-35. *[3]* How many gas stations are there in the United States?

1-36. *[3]* How much does the ice in a hockey rink weigh?

1-37. *[3]* How many miles of road are there in the United States?

1-38. *[3]* On average, how many times would you have to flip open the Manhattan
phone book at random in order to find a specific name?

LeetCode

1-1. `https://leetcode.com/problems/daily-temperatures/`

1-2. `https://leetcode.com/problems/rotate-list/`

1-3. `https://leetcode.com/problems/wiggle-sort-ii/`

HackerRank

1-1. `https://www.hackerrank.com/challenges/array-left-rotation/`

1-2. `https://www.hackerrank.com/challenges/kangaroo/`

1-3. `https://www.hackerrank.com/challenges/hackerland-radio-transmitters/`

Programming Challenges

These programming challenge problems with robot judging are available at
`https://onlinejudge.org`:

1-1. "The $3n + 1$ Problem"—Chapter 1, problem 100.

1-2. "The Trip"—Chapter 1, problem 10137.

1-3. "Australian Voting"—Chapter 1, problem 10142.

Chapter 2

Algorithm Analysis

Algorithms are the most important and durable part of computer science because they can be studied in a language- and machine-independent way. This means we need techniques that let us compare the efficiency of algorithms without implementing them. Our two most important tools are (1) the RAM model of computation and (2) the asymptotic analysis of computational complexity.

Assessing algorithmic performance makes use of the "Big Oh" notation that proves essential to compare algorithms, and design better ones. This method of keeping score will be the most mathematically demanding part of this book. But once you understand the intuition behind this formalism it becomes a lot easier to deal with.

2.1 The RAM Model of Computation

Machine-independent algorithm design depends upon a hypothetical computer called the *Random Access Machine*, or RAM. Under this model of computation, we are confronted with a computer where:

- Each *simple* operation (+, *, −, =, if, call) takes exactly one time step.

- Loops and subroutines are *not* considered simple operations. Instead, they are the composition of many single-step operations. It makes no sense for *sort* to be a single-step operation, since sorting 1,000,000 items will certainly take much longer than sorting ten items. The time it takes to run through a loop or execute a subprogram depends upon the number of loop iterations or the specific nature of the subprogram.

- Each memory access takes exactly one time step. Furthermore, we have as much memory as we need. The RAM model takes no notice of whether an item is in cache or on the disk.

Under the RAM model, we measure run time by counting the number of steps an algorithm takes on a given problem instance. If we assume that our

© The Editor(s) (if applicable) and The Author(s), under exclusive license to
Springer Nature Switzerland AG 2020
S. S. Skiena, *The Algorithm Design Manual*, Texts in Computer Science,
https://doi.org/10.1007/978-3-030-54256-6_2

RAM executes a given number of steps per second, this operation count converts naturally to the actual running time.

The RAM is a simple model of how computers perform. Perhaps it sounds too simple. After all, multiplying two numbers takes more time than adding two numbers on most processors, which violates the first assumption of the model. Fancy compiler loop unrolling and hyperthreading may well violate the second assumption. And certainly memory-access times differ greatly depending on where your data sits in the storage hierarchy. This makes us zero for three on the truth of our basic assumptions.

And yet, despite these objections, the RAM proves an *excellent* model for understanding how an algorithm will perform on a real computer. It strikes a fine balance by capturing the essential behavior of computers while being simple to work with. We use the RAM model because it is useful in practice.

Every model in science has a size range over which it is useful. Take, for example, the model that the Earth is flat. You might argue that this is a bad model, since it is quite well established that the Earth is round. But, when laying the foundation of a house, the flat Earth model is sufficiently accurate that it can be reliably used. It is so much easier to manipulate a flat-Earth model that it is inconceivable that you would try to think spherically when you don't have to.[1]

The same situation is true with the RAM model of computation. We make an abstraction that is generally very useful. It is difficult to design an algorithm where the RAM model gives you substantially misleading results. The robustness of this model enables us to analyze algorithms in a machine-independent way.

Take-Home Lesson: Algorithms can be understood and studied in a language- and machine-independent manner.

2.1.1 Best-Case, Worst-Case, and Average-Case Complexity

Using the RAM model of computation, we can count how many steps our algorithm takes on any given input instance by executing it. However, to understand how good or bad an algorithm is in general, we must know how it works over *all* possible instances.

To understand the notions of the best, worst, and average-case complexity, think about running an algorithm over all possible instances of data that can be fed to it. For the problem of sorting, the set of possible input instances includes every possible arrangement of n keys, for all possible values of n. We can represent each input instance as a point on a graph (shown in Figure 2.1) where the x-axis represents the size of the input problem (for sorting, the number of items to sort), and the y-axis denotes the number of steps taken by the algorithm in this instance.

---

[1]The Earth is not completely spherical either, but a spherical Earth provides a useful model for such things as longitude and latitude.

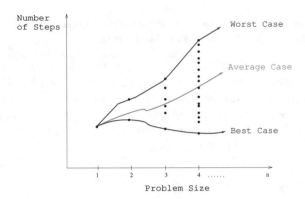

Figure 2.1: Best-case, worst-case, and average-case complexity.

These points naturally align themselves into columns, because only integers represent possible input sizes (e.g., it makes no sense to sort 10.57 items). We can define three interesting functions over the plot of these points:

- The *worst-case complexity* of the algorithm is the function defined by the maximum number of steps taken in any instance of size n. This represents the curve passing through the highest point in each column.

- The *best-case complexity* of the algorithm is the function defined by the minimum number of steps taken in any instance of size n. This represents the curve passing through the lowest point of each column.

- The *average-case complexity* or *expected time* of the algorithm, which is the function defined by the average number of steps over all instances of size n.

The worst-case complexity generally proves to be most useful of these three measures in practice. Many people find this counterintuitive. To illustrate why, try to project what will happen if you bring $\$n$ into a casino to gamble. The best case, that you walk out owning the place, is so unlikely that you should not even think about it. The worst case, that you lose all $\$n$, is easy to calculate and distressingly likely to happen.

The average case, that the typical bettor loses 87.32% of the money that he or she brings to the casino, is both difficult to establish and its meaning subject to debate. What exactly does *average* mean? Stupid people lose more than smart people, so are you smarter or stupider than the average person, and by how much? Card counters at blackjack do better on average than customers who accept three or more free drinks. We avoid all these complexities and obtain a very useful result by considering the worst case.

That said, average-case analysis for expected running time will prove very important with respect to *randomized algorithms*, which use random numbers to make decisions within the algorithm. If you make n independent $1 red-black bets on roulette in the casino, your expected loss is indeed well defined at

$(2n/38)$, because American roulette wheels have eighteen red, eighteen black, and two green slots 0 and 00 where every bet loses.

Take-Home Lesson: Each of these time complexities defines a numerical function for any given algorithm, representing running time as a function of input size. These functions are just as well defined as any other numerical function, be it $y = x^2 - 2x + 1$ or the price of Alphabet stock as a function of time. But time complexities are such complicated functions that we must simplify them for analysis using the "Big Oh" notation.

2.2   The Big Oh Notation

The best-case, worst-case, and average-case time complexities for any given algorithm are numerical functions over the size of possible problem instances. However, it is very difficult to work precisely with these functions, because they tend to:

- *Have too many bumps* – An algorithm such as binary search typically runs a bit faster for arrays of size exactly $n = 2^k - 1$ (where k is an integer), because the array partitions work out nicely. This detail is not particularly important, but it warns us that the *exact* time complexity function for any algorithm is liable to be very complicated, with lots of little up and down bumps as shown in Figure 2.2.

- *Require too much detail to specify precisely* – Counting the exact number of RAM instructions executed in the worst case requires the algorithm be specified to the detail of a complete computer program. Furthermore, the precise answer depends upon uninteresting coding details (e.g. did the code use a case statement or nested ifs?). Performing a precise worst-case analysis like

$$T(n) = 12754n^2 + 4353n + 834 \lg_2 n + 13546$$

  would clearly be very difficult work, but provides us little extra information than the observation that "the time grows quadratically with n."

It proves to be much easier to talk in terms of simple upper and lower bounds of time-complexity functions using the Big Oh notation. The Big Oh simplifies our analysis by ignoring levels of detail that do not impact our comparison of algorithms.

The Big Oh notation ignores the difference between multiplicative constants. The functions $f(n) = 2n$ and $g(n) = n$ are identical in Big Oh analysis. This makes sense given our application. Suppose a given algorithm in (say) C language ran twice as fast as one with the same algorithm written in Java. This multiplicative factor of two can tell us nothing about the algorithm itself, because both programs implement exactly the same algorithm. We should ignore such constant factors when comparing two algorithms.

The formal definitions associated with the Big Oh notation are as follows:

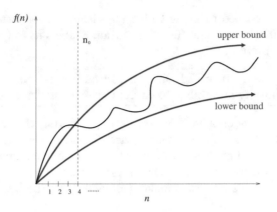

Figure 2.2: Upper and lower bounds valid for $n > n_0$ smooth out the behavior of complex functions.

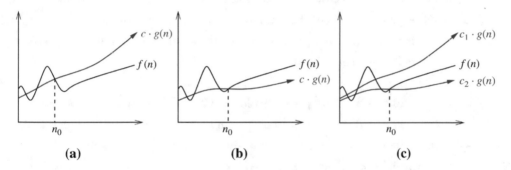

Figure 2.3: Illustrating notations: (a) $f(n) = O(g(n))$, (b) $f(n) = \Omega(g(n))$, and (c) $f(n) = \Theta(g(n))$.

- $f(n) = O(g(n))$ means $c \cdot g(n)$ is an *upper bound* on $f(n)$. Thus, there exists some constant c such that $f(n) \leq c \cdot g(n)$ for every large enough n (that is, for all $n \geq n_0$, for some constant n_0).

- $f(n) = \Omega(g(n))$ means $c \cdot g(n)$ is a *lower bound* on $f(n)$. Thus, there exists some constant c such that $f(n) \geq c \cdot g(n)$ for all $n \geq n_0$.

- $f(n) = \Theta(g(n))$ means $c_1 \cdot g(n)$ is an upper bound on $f(n)$ and $c_2 \cdot g(n)$ is a lower bound on $f(n)$, for all $n \geq n_0$. Thus, there exist constants c_1 and c_2 such that $f(n) \leq c_1 \cdot g(n)$ and $f(n) \geq c_2 \cdot g(n)$ for all $n \geq n_0$. This means that $g(n)$ provides a nice, tight bound on $f(n)$.

Got it? These definitions are illustrated in Figure 2.3. Each of these definitions assumes there is a constant n_0 beyond which they are satisfied. We are not concerned about small values of n, anything to the left of n_0. After all, we don't really care whether one sorting algorithm sorts six items faster than another, but we do need to know which algorithm proves faster when sorting 10,000 or 1,000,000 items. The Big Oh notation enables us to ignore details and focus on the big picture.

> *Take-Home Lesson:*   The Big Oh notation and worst-case analysis are tools that greatly simplify our ability to compare the efficiency of algorithms.

Make sure you understand this notation by working through the following examples. Certain constants (c and n_0) are chosen in the explanations below because they work and make a point, but other pairs of constants will do exactly the same job. You are free to choose any constants that maintain the same inequality (ideally constants that make it obvious that the inequality holds):

$f(n) = 3n^2 - 100n + 6 = O(n^2)$, because for $c = 3$, $3n^2 > f(n)$;

$f(n) = 3n^2 - 100n + 6 = O(n^3)$, because for $c = 1$, $n^3 > f(n)$ when $n > 3$;

$f(n) = 3n^2 - 100n + 6 \neq O(n)$, because for any $c > 0$, $cn < f(n)$ when $n > (c + 100)/3$,

$\qquad$ since $n > (c + 100)/3 \Rightarrow 3n > c + 100 \Rightarrow 3n^2 > cn + 100n > cn + 100n - 6$

$\qquad \Rightarrow 3n^2 - 100n + 6 = f(n) > cn$;

$f(n) = 3n^2 - 100n + 6 = \Omega(n^2)$, because for $c = 2$, $2n^2 < f(n)$ when $n > 100$;

$f(n) = 3n^2 - 100n + 6 \neq \Omega(n^3)$, because for any $c > 0$, $f(n) < c \cdot n^3$ when $n > 3/c + 3$;

$f(n) = 3n^2 - 100n + 6 = \Omega(n)$, because for any $c > 0$, $f(n) < 3n^2 + 6n^2 = 9n^2$,

$\qquad$ which is $< cn^3$ when $n > \max(9/c, 1)$;

$f(n) = 3n^2 - 100n + 6 = \Theta(n^2)$, because both O and Ω apply;

$f(n) = 3n^2 - 100n + 6 \neq \Theta(n^3)$, because only O applies;

$f(n) = 3n^2 - 100n + 6 \neq \Theta(n)$, because only Ω applies.

The Big Oh notation provides for a rough notion of equality when comparing functions. It is somewhat jarring to see an expression like $n^2 = O(n^3)$, but its meaning can always be resolved by going back to the definitions in terms of upper and lower bounds. It is perhaps most instructive to read the "=" here as meaning *one of the functions that are*. Clearly, n^2 is one of the functions that are $O(n^3)$.

Stop and Think: Back to the Definition

Problem: Is $2^{n+1} = \Theta(2^n)$?

Solution: Designing novel algorithms requires cleverness and inspiration. However, applying the Big Oh notation is best done by swallowing any creative instincts you may have. All Big Oh problems can be correctly solved by going back to the definition and working with that.

- Is $2^{n+1} = O(2^n)$? Well, $f(n) = O(g(n))$ iff there exists a constant c such that for all sufficiently large n, $f(n) \leq c \cdot g(n)$. Is there? Yes, because $2^{n+1} = 2 \cdot 2^n$, and clearly $2 \cdot 2^n \leq c \cdot 2^n$ for any $c \geq 2$.

- Is $2^{n+1} = \Omega(2^n)$? Go back to the definition. $f(n) = \Omega(g(n))$ iff there exists a constant $c > 0$ such that for all sufficiently large n $f(n) \geq c \cdot g(n)$. This would be satisfied for any $0 < c \leq 2$. Together the Big Oh and Ω bounds imply $2^{n+1} = \Theta(2^n)$.

∎

Stop and Think: Hip to the Squares?

Problem: Is $(x + y)^2 = O(x^2 + y^2)$?

Solution: Working with the Big Oh means going back to the definition at the slightest sign of confusion. By definition, this expression is valid iff we can find some c such that $(x + y)^2 \leq c(x^2 + y^2)$ for all sufficiently large x and y.

My first move would be to expand the left side of the equation, that is, $(x + y)^2 = x^2 + 2xy + y^2$. If the middle $2xy$ term wasn't there, the inequality would clearly hold for any $c > 1$. But it is there, so we need to relate $2xy$ to $x^2 + y^2$. What if $x \leq y$? Then $2xy \leq 2y^2 \leq 2(x^2 + y^2)$. What if $x \geq y$? Then $2xy \leq 2x^2 \leq 2(x^2 + y^2)$. Either way, we now can bound $2xy$ by two times the right-side function $x^2 + y^2$. This means that $(x + y)^2 \leq 3(x^2 + y^2)$, and so the result holds. ∎

2.3 Growth Rates and Dominance Relations

With the Big Oh notation, we cavalierly discard the multiplicative constants. Thus, the functions $f(n) = 0.001n^2$ and $g(n) = 1000n^2$ are treated identically, even though $g(n)$ is a million times larger than $f(n)$ for all values of n.

The reason why we are content with such coarse Big Oh analysis is provided by Figure 2.4, which shows the growth rate of several common time analysis functions. In particular, it shows how long algorithms that use $f(n)$ operations take to run on a fast computer, where each operation costs one nanosecond (10^{-9} seconds). The following conclusions can be drawn from this table:

- All such algorithms take roughly the same time for $n - 10$.

- Any algorithm with $n!$ running time becomes useless for $n \geq 20$.

n	$\lg n$	n	$n \lg n$	n^2	2^n	$n!$
10	0.003 μs	0.01 μs	0.033 μs	0.1 μs	1 μs	3.63 ms
20	0.004 μs	0.02 μs	0.086 μs	0.4 μs	1 ms	77.1 years
30	0.005 μs	0.03 μs	0.147 μs	0.9 μs	1 sec	8.4×10^{15} yrs
40	0.005 μs	0.04 μs	0.213 μs	1.6 μs	18.3 min	
50	0.006 μs	0.05 μs	0.282 μs	2.5 μs	13 days	
100	0.007 μs	0.1 μs	0.644 μs	10 μs	4×10^{13} yrs	
1,000	0.010 μs	1.00 μs	9.966 μs	1 ms		
10,000	0.013 μs	10 μs	130 μs	100 ms		
100,000	0.017 μs	0.10 ms	1.67 ms	10 sec		
1,000,000	0.020 μs	1 ms	19.93 ms	16.7 min		
10,000,000	0.023 μs	0.01 sec	0.23 sec	1.16 days		
100,000,000	0.027 μs	0.10 sec	2.66 sec	115.7 days		
1,000,000,000	0.030 μs	1 sec	29.90 sec	31.7 years		

Figure 2.4:   Running times of common functions measured in nanoseconds. The function $\lg n$ denotes the base-2 logarithm of n.

- Algorithms whose running time is 2^n have a greater operating range, but become impractical for $n > 40$.

- Quadratic-time algorithms, whose running time is n^2, remain usable up to about $n = 10,000$, but quickly deteriorate with larger inputs. They are likely to be hopeless for $n > 1,000,000$.

- Linear-time and $n \lg n$ algorithms remain practical on inputs of one billion items.

- An $O(\lg n)$ algorithm hardly sweats for any imaginable value of n.

The bottom line is that even ignoring constant factors, we get an excellent idea of whether a given algorithm is appropriate for a problem of a given size.

2.3.1   Dominance Relations

The Big Oh notation groups functions into a set of classes, such that all the functions within a particular class are essentially equivalent. Functions $f(n) = 0.34n$ and $g(n) = 234{,}234n$ belong in the same class, namely those that are order $\Theta(n)$. Furthermore, when two functions f and g belong to different classes, they are *different* with respect to our notation, meaning either $f(n) = O(g(n))$ or $g(n) = O(f(n))$, but not both.

We say that a faster growing function *dominates* a slower growing one, just as a faster growing company eventually comes to dominate the laggard. When f and g belong to different classes (i.e. $f(n) \neq \Theta(g(n))$), we say g *dominates* f when $f(n) = O(g(n))$. This is sometimes written $g \gg f$.

The good news is that only a few different function classes tend to occur in the course of basic algorithm analysis. These suffice to cover almost all the algorithms we will discuss in this text, and are listed in order of increasing dominance:

- *Constant functions, $f(n) = 1$:* Such functions might measure the cost of adding two numbers, printing out "The Star Spangled Banner," or the growth realized by functions such as $f(n) = \min(n, 100)$. In the big picture, there is no dependence on the parameter n.

- *Logarithmic functions, $f(n) = \log n$:* Logarithmic time complexity shows up in algorithms such as binary search. Such functions grow quite slowly as n gets big, but faster than the constant function (which is standing still, after all). Logarithms will be discussed in more detail in Section 2.7 (page 48).

- *Linear functions, $f(n) = n$:* Such functions measure the cost of looking at each item once (or twice, or ten times) in an n-element array, say to identify the biggest item, the smallest item, or compute the average value.

- *Superlinear functions, $f(n) = n \lg n$:* This important class of functions arises in such algorithms as quicksort and mergesort. They grow just a little faster than linear (recall Figure 2.4), but enough so to rise to a higher dominance class.

- *Quadratic functions, $f(n) = n^2$:* Such functions measure the cost of looking at most or all *pairs* of items in an n-element universe. These arise in algorithms such as insertion sort and selection sort.

- *Cubic functions, $f(n) = n^3$:* Such functions enumerate all *triples* of items in an n-element universe. These also arise in certain dynamic programming algorithms, to be developed in Chapter 10.

- *Exponential functions, $f(n) = c^n$ for a given constant $c > 1$:* Functions like 2^n arise when enumerating all subsets of n items. As we have seen, exponential algorithms become useless fast, but not as fast as. . .

- *Factorial functions, $f(n) = n!$:* Functions like $n!$ arise when generating all permutations or orderings of n items.

The intricacies of dominance relations will be further discussed in Section 2.10.2 (page 58). However, all you really need to understand is that:

$$n! \gg 2^n \gg n^3 \gg n^2 \gg n \log n \gg n \gg \log n \gg 1$$

> *Take-Home Lesson:* Although esoteric functions arise in advanced algorithm analysis, a small set of time complexities suffice for most algorithms we will see in this book.

2.4 Working with the Big Oh

You learned how to do simplifications of algebraic expressions back in high school. Working with the Big Oh requires dusting off these tools. *Most* of what you learned there still holds in working with the Big Oh, but not everything.

2.4.1   Adding Functions

The sum of two functions is governed by the dominant one, namely:

$$f(n) + g(n) \rightarrow \Theta(\max(f(n), g(n)))$$

This is very useful in simplifying expressions: for example it gives us that $n^3 + n^2 + n + 1 = \Theta(n^3)$. Everything else is small potatoes besides the dominant term.

The intuition is as follows. At least half the bulk of $f(n) + g(n)$ must come from the larger value. The dominant function will, by definition, provide the larger value as $n \rightarrow \infty$. Thus, dropping the smaller function from consideration reduces the value by at most a factor of $1/2$, which is just a multiplicative constant. For example, if $f(n) = O(n^2)$ and $g(n) = O(n^2)$, then $f(n) + g(n) = O(n^2)$ as well.

2.4.2   Multiplying Functions

Multiplication is like repeated addition. Consider multiplication by any constant $c > 0$, be it 1.02 or 1,000,000. Multiplying a function by a constant cannot affect its asymptotic behavior, because we can multiply the bounding constants in the Big Oh analysis to account for it. Thus,

$$O(c \cdot f(n)) \rightarrow O(f(n))$$
$$\Omega(c \cdot f(n)) \rightarrow \Omega(f(n))$$
$$\Theta(c \cdot f(n)) \rightarrow \Theta(f(n))$$

Of course, c must be strictly positive (i.e. $c > 0$) to avoid any funny business, since we can wipe out even the fastest growing function by multiplying it by zero.

On the other hand, when two functions in a product are increasing, both are important. An $O(n! \log n)$ function dominates $n!$ by just as much as $\log n$ dominates 1. In general,

$$O(f(n)) \cdot O(g(n)) \rightarrow O(f(n) \cdot g(n))$$
$$\Omega(f(n)) \cdot \Omega(g(n)) \rightarrow \Omega(f(n) \cdot g(n))$$
$$\Theta(f(n)) \cdot \Theta(g(n)) \rightarrow \Theta(f(n) \cdot g(n))$$

Stop and Think: Transitive Experience

Problem: Show that Big Oh relationships are transitive. That is, if $f(n) = O(g(n))$ and $g(n) = O(h(n))$, then $f(n) = O(h(n))$.

Solution: We always go back to the definition when working with the Big Oh. What we need to show here is that $f(n) \leq c_3 \cdot h(n)$ for $n > n_3$ given that

$f(n) \leq c_1 \cdot g(n)$ and $g(n) \leq c_2 \cdot h(n)$, for $n > n_1$ and $n > n_2$, respectively. Cascading these inequalities, we get that

$$f(n) \leq c_1 \cdot g(n) \leq c_1 c_2 \cdot h(n)$$

for $n > n_3 = \max(n_1, n_2)$. ∎

2.5 Reasoning about Efficiency

Coarse reasoning about an algorithm's running time is usually easy, given a precise description of the algorithm. In this section, I will work through several examples, perhaps in greater detail than necessary.

2.5.1 Selection Sort

Here we'll analyze the selection sort algorithm, which repeatedly identifies the smallest remaining unsorted element and puts it at the end of the sorted portion of the array. An animation of selection sort in action appears in Figure 2.5, and the code is shown below:

```
void selection_sort(item_type s[], int n) {
    int i, j;    /* counters */
    int min;     /* index of minimum */

    for (i = 0; i < n; i++) {
        min = i;
        for (j = i + 1; j < n; j++) {
            if (s[j] < s[min]) {
                min = j;
            }
        }
        swap(&s[i], &s[min]);
    }
}
```

The outer `for` loop goes around n times. The nested inner loop goes around $n - (i + 1)$ times, where i is the index of the outer loop. The exact number of times the *if* statement is executed is given by:

$$T(n) = \sum_{i=0}^{n-1} \sum_{j=i+1}^{n-1} 1 = \sum_{i=0}^{n-1} n - i - 1$$

What this sum is doing is adding up the non-negative integers in decreasing order starting from $n - 1$, that is,

$$T(n) = (n - 1) + (n - 2) + (n - 3) + \ldots + 2 + 1$$

```
S E L E C T I O N S O R T
C E L E S T I O N S O R T
C E L   E S T I O N S O R T
C E E   L S T I O N S O R T
C E E I S T L O N S O R T
C E E I   T S O N S O R T
C E E I   N S O T S O R T
C E E I L   N O S T S O R T
C E E I L   N O O T S S R T
C E E I L   N O O R S S T T
C E E I L   N O O R S S T T
C E E I L   N O O R S S T T
C E E I L   N O O R S S T T
C E E I L   N O O R S S T T
```

Figure 2.5: Animation of selection sort in action.

How can we reason about such a formula? We must solve the summation formula using the techniques of Section 2.6 (page 46) to get an exact value. But, with the Big Oh, we are only interested in the *order* of the expression. One way to think about it is that we are adding up $n-1$ terms, whose average value is about $n/2$. This yields $T(n) \approx (n-1)n/2 = O(n^2)$.

Proving the Theta

Another way to think about this algorithm's running time is in terms of upper and lower bounds. We have n terms at most, each of which is at most $n-1$. Thus, $T(n) \leq n(n-1) = O(n^2)$. The Big Oh is an upper bound.

The Big Ω is a lower bound. Looking at the sum again, we have $n/2$ terms each of which is bigger than $n/2$, followed by $n/2$ terms each greater than zero.. Thus, $T(n) \geq (n/2) \cdot (n/2) + (n/2) \cdot 0 = \Omega(n^2)$. Together with the Big Oh result, this tells us that the running time is $\Theta(n^2)$, meaning that selection sort is quadratic.

Generally speaking, turning a Big Oh worst-case analysis into a Big Θ involves identifying a bad input instance that forces the algorithm to perform as poorly as possible. But selection sort is distinctive among sorting algorithms in that it takes exactly the same time on all $n!$ possible input instances. Since $T(n) = n(n-1)/2$ for all $n \geq 0$, $T(n) = \Theta(n^2)$.

2.5.2   Insertion Sort

A basic rule of thumb in Big Oh analysis is that worst-case running time follows from multiplying the largest number of times each nested loop can iterate. Consider the insertion sort algorithm presented on page 3, whose inner loops are repeated here:

```
for (i = 1; i < n; i++) {
    j = i;
    while ((j > 0) && (s[j] < s[j - 1])) {
        swap(&s[j], &s[j - 1]);
        j = j-1;
    }
}
```

How often does the inner *while* loop iterate? This is tricky because there are two different stopping conditions: one to prevent us from running off the bounds of the array $(j > 0)$ and the other to mark when the element finds its proper place in sorted order $(s[j] < s[j-1])$. Since worst-case analysis seeks an upper bound on the running time, we ignore the early termination and assume that this loop *always* goes around i times. In fact, we can simplify further and assume it *always* goes around n times since $i < n$. Since the outer loop goes around n times, insertion sort must be a quadratic-time algorithm, that is, $O(n^2)$.

This crude "round it up" analysis always does the job, in that the Big Oh running time bound you get will always be correct. Occasionally, it might be too pessimistic, meaning the actual worst-case time might be of a lower order than implied by such analysis. Still, I strongly encourage this kind of reasoning as a basis for simple algorithm analysis.

Proving the Theta

The worst case for insertion sort occurs when each newly inserted element must slide all the way to the front of the sorted region. This happens if the input is given in reverse sorted order. Each of the last $n/2$ elements of the input must slide over at least $n/2$ elements to find the correct position, taking at least $(n/2)^2 = \Omega(n^2)$ time.

2.5.3 String Pattern Matching

Pattern matching is the most fundamental algorithmic operation on text strings. This algorithm implements the find command available in any web browser or text editor:

Problem: Substring Pattern Matching
Input: A text string t and a pattern string p.
Output: Does t contain the pattern p as a substring, and if so, where?

Perhaps you are interested in finding where "Skiena" appears in a given news article (well, *I* would be interested in such a thing). This is an instance of string pattern matching with t as the news article and $p =$ "Skiena".

There is a fairly straightforward algorithm for string pattern matching that considers the possibility that p may start at each possible position in t and then tests if this is so.

```
a  b  b  a
   a  b  b  a
      a  b  b  a
         a  b  b  a
            a  b  b  a
               a  b  b  a
```
———————————————————————————
```
a  b  a  a  b  a  b  b  a  b  a
```

Figure 2.6: Searching for the substring *abba* in the text *abaababbaba*. Blue characters represent pattern-text matches, with red characters mismatches. The search stops as soon as a match is found.

```
int findmatch(char *p, char *t) {
    int i, j;          /* counters */
    int plen, tlen;    /* string lengths */

    plen = strlen(p);
    tlen = strlen(t);

    for (i = 0; i <= (tlen-plen); i = i + 1) {
        j = 0;
        while ((j < plen) && (t[i + j] == p[j])) {
            j = j + 1;
        }
        if (j == plen) {
            return(i);  /* location of the first match */
        }
    }
    return(-1);         /* there is no match */
}
```

What is the worst-case running time of these two nested loops? The inner *while* loop goes around at most m times, and potentially far less when the pattern match fails. This, plus two other statements, lies within the outer *for* loop. The outer loop goes around at most $n - m$ times, since no complete alignment is possible once we get too far to the right of the text. The time complexity of nested loops multiplies, so this gives a worst-case running time of $O((n - m)(m + 2))$.

We did not count the time it takes to compute the length of the strings using the function *strlen*. Since the implementation of *strlen* is not given, we must guess how long it should take. If it explicitly counts the number of characters until it hits the end of the string, this will take time linear in the length of the string. Thus, the total worst-case running time is $O(n + m + (n - m)(m + 2))$.

Let's use our knowledge of the Big Oh to simplify things. Since $m + 2 = \Theta(m)$, the "+2" isn't interesting, so we are left with $O(n + m + (n - m)m)$.

Multiplying this out yields $O(n + m + nm - m^2)$, which still seems kind of ugly.

However, in any interesting problem we know that $n \geq m$, since p can't be a substring of t when the pattern is longer than the text itself. One consequence of this is that $n + m \leq 2n = \Theta(n)$. Thus, our worst-case running time simplifies further to $O(n + nm - m^2)$.

Two more observations and we are done. First, note that $n \leq nm$, since $m \geq 1$ in any interesting pattern. Thus, $n + nm = \Theta(nm)$, and we can drop the additive n, simplifying our analysis to $O(nm - m^2)$.

Finally, observe that the $-m^2$ term is negative, and thus only serves to lower the value within. Since the Big Oh gives an upper bound, we can drop any negative term without invalidating the upper bound. The inequality $n \geq m$ implies that $mn \geq m^2$, so the negative term is not big enough to cancel the term that is left. Thus, we can express the worst-case running time of this algorithm simply as $O(nm)$.

After you get enough experience, you will be able to do such an algorithm analysis in your head without even writing the algorithm down. After all, algorithm design for a given task involves mentally rifling through different possibilities and selecting the best approach. This kind of fluency comes with practice, but if you are confused about why a given algorithm runs in $O(f(n))$ time, start by writing the algorithm out carefully and then employ the kind of reasoning we used in this section.

Proving the Theta

The analysis above gives a quadratic-time upper bound on the running time of this simple pattern matching algorithm. To prove the theta, we must show an example where it actually does take $\Omega(mn)$ time.

Consider what happens when the text $t = $ "$aaaa \ldots aaaa$" is a string of n a's, and the pattern $p = $ "$aaaa \ldots aaab$" is a string of $m - 1$ a's followed by a b. Wherever the pattern is positioned on the text, the `while` loop will successfully match the first $m - 1$ characters before failing on the last one. There are $n - m + 1$ possible positions where p can sit on t without overhanging the end, so the running time is:

$$(n - m + 1)(m) = mn - m^2 + m = \Omega(mn)$$

Thus, this string matching algorithm runs in worst-case $\Theta(nm)$ time. Faster algorithms do exist: indeed we will see an expected linear-time algorithm for this problem in Section 6.7.

2.5.4   Matrix Multiplication

Nested summations often arise in the analysis of algorithms with nested loops. Consider the problem of matrix multiplication:

Problem: Matrix Multiplication
Input: Two matrices, A (of dimension $x \times y$) and B (dimension $y \times z$).
Output: An $x \times z$ matrix C where $C[i][j]$ is the dot product of the ith row of A and the jth column of B.

Matrix multiplication is a fundamental operation in linear algebra, presented with an entry in the catalog in Section 16.3 (page 472). That said, the elementary algorithm for matrix multiplication is implemented as three nested loops:

```
for (i = 1; i <= a->rows; i++) {
    for (j = 1; j <= b->columns; j++) {
        c->m[i][j] = 0;
        for (k = 1; k <= b->rows; k++) {
            c->m[i][j] += a->m[i][k] * b->m[k][j];
        }
    }
}
```

How can we analyze the time complexity of this algorithm? Three nested loops should smell $O(n^3)$ to you by this point, but let's be precise. The number of multiplications $M(x, y, z)$ is given by the following summation:

$$M(x, y, z) = \sum_{i=1}^{x} \sum_{j=1}^{y} \sum_{k=1}^{z} 1$$

Sums get evaluated from the right inward. The sum of z ones is z, so

$$M(x, y, z) = \sum_{i=1}^{x} \sum_{j=1}^{y} z$$

The sum of y z's is just as simple, yz, so

$$M(x, y, z) = \sum_{i=1}^{x} yz$$

Finally, the sum of x yz's is xyz.

Thus, the running of this matrix multiplication algorithm is $O(xyz)$. If we consider the common case where all three dimensions are the same, this becomes $O(n^3)$. The same analysis holds for an $\Omega(n^3)$ lower bound, because the matrix dimensions govern the number of iterations of the **for** loops. Simple matrix multiplication is a cubic algorithm that runs in $\Theta(n^3)$ time. Faster algorithms exist: see Section 16.3.

2.6   Summations

Mathematical summation formulae are important to us for two reasons. First, they often arise in algorithm analysis. Second, proving the correctness of such formulae is a classic application of mathematical induction. Several exercises on inductive proofs of summations appear as exercises at the end of this chapter. To make them more accessible, I review the basics of summations here.

Summation formulae are concise expressions describing the addition of an arbitrarily large set of numbers, in particular the formula

$$\sum_{i=1}^{n} f(i) = f(1) + f(2) + \ldots + f(n)$$

Simple closed forms exist for summations of many algebraic functions. For example, since the sum of n ones is n,

$$\sum_{i=1}^{n} 1 = n$$

When n is even, the sum of the first $n = 2k$ integers can be seen by pairing up the ith and $(n - i + 1)$th integers:

$$\sum_{i=1}^{n} i = \sum_{i=1}^{k} (i + (2k - i + 1)) = k(2k + 1) = n(n + 1)/2$$

The same result holds for odd n with slightly more careful analysis.

Recognizing two basic classes of summation formulae will get you a long way in algorithm analysis:

- *Sum of a power of integers* – We encountered the sum of the first n positive integers $S(n) = \sum_{i=1}^{n} i = n(n + 1)/2$ in the analysis of selection sort. From the big picture perspective, the important thing is that the sum is quadratic, not that the constant is $1/2$. In general,

$$S(n, p) = \sum_{i=1}^{n} i^p = \Theta(n^{p+1})$$

  for $p \geq 0$. Thus, the sum of squares is cubic, and the sum of cubes is quartic (if you use such a word).

  For $p < -1$, this sum $S(n, p)$ always converges to a constant as $n \to \infty$, while for $p \geq 0$ it diverges. The interesting case between these is the Harmonic numbers, $H(n) = \sum_{i=1}^{n} 1/i = \Theta(\log n)$.

- *Sum of a geometric progression* – In geometric progressions, the index of the loop affects the exponent, that is,

$$G(n, a) = \sum_{i=0}^{n} a^i = (a^{n+1} - 1)/(a - 1)$$

  How we interpret this sum depends upon the *base* of the progression, in this case a. When $|a| < 1$, $G(n, a)$ converges to a constant as $n \to \infty$.

This series convergence proves to be the great "free lunch" of algorithm analysis. It means that the sum of a linear number of things can be constant, not linear. For example, $1 + 1/2 + 1/4 + 1/8 + \ldots \le 2$ no matter how many terms we add up.

When $a > 1$, the sum grows rapidly with each new term, as in $1 + 2 + 4 + 8 + 16 + 32 = 63$. Indeed, $G(n, a) = \Theta(a^{n+1})$ for $a > 1$.

Stop and Think: Factorial Formulae

Problem: Prove that $\sum_{i=1}^{n} i \times i! = (n+1)! - 1$ by induction.

Solution: The inductive paradigm is straightforward. First verify the basis case. The case $n = 0$ gives an empty sum, which by definition evaluates to 0. Alternately we can do $n = 1$:

$$\sum_{i=1}^{1} i \times i! = 1 \text{ and } (1+1)! - 1 = 2 - 1 = 1$$

Now assume the statement is true up to n. To prove the general case of $n + 1$, observe that separating out the largest term

$$\sum_{i=1}^{n+1} i \times i! = (n+1) \times (n+1)! + \sum_{i=1}^{n} i \times i!$$

reveals the left side of our inductive assumption. Substituting the right side gives us

$$\sum_{i=1}^{n+1} i \times i! = (n+1) \times (n+1)! + (n+1)! - 1$$

$$= (n+1)! \times ((n+1) + 1) - 1$$
$$= (n+2)! - 1$$

This general trick of separating out the largest term from the summation to reveal an instance of the inductive assumption lies at the heart of all such proofs. ∎

2.7   Logarithms and Their Applications

Logarithm is an anagram of algorithm, but that's not why we need to know what logarithms are. You've seen the button on your calculator, but may have forgotten why it is there. A *logarithm* is simply an inverse exponential function. Saying $b^x = y$ is equivalent to saying that $x = \log_b y$. Further, this equivalence is the same as saying $b^{\log_b y} = y$.

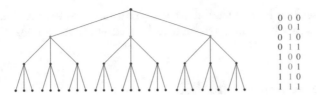

Figure 2.7: A height h tree with d children per node has d^h leaves. Here $h = 3$ and $d = 3$ (left). The number of bit patterns grows exponentially with pattern length (right). These would be described by the root-to-leaf paths of a *binary* tree of height $h = 3$.

Exponential functions grow at a distressingly fast rate, as anyone who has ever tried to pay off a credit card balance understands. Thus, inverse exponential functions (logarithms) grow refreshingly slowly. Logarithms arise in any process where things are repeatedly halved. We'll now look at several examples.

2.7.1 Logarithms and Binary Search

Binary search is a good example of an $O(\log n)$ algorithm. To locate a particular person p in a telephone book[2] containing n names, you start by comparing p against the middle, or $(n/2)$nd name, say *Monroe, Marilyn*. Regardless of whether p belongs before this middle name (*Dean, James*) or after it (*Presley, Elvis*), after just one comparison you can discard one half of all the names in the book. The number of steps the algorithm takes equals the number of times we can halve n until only one name is left. By definition, this is exactly $\log_2 n$. Thus, twenty comparisons suffice to find any name in the million-name Manhattan phone book!

Binary search is one of the most powerful ideas in algorithm design. This power becomes apparent if we imagine trying to find a name in an unsorted telephone book.

2.7.2 Logarithms and Trees

A binary tree of height 1 can have up to 2 leaf nodes, while a tree of height 2 can have up to 4 leaves. What is the height h of a rooted binary tree with n leaf nodes? Note that the number of leaves doubles every time we increase the height by 1. To account for n leaves, $n = 2^h$, which implies that $h = \log_2 n$.

What if we generalize to trees that have d children, where $d = 2$ for the case of binary trees? A tree of height 1 can have up to d leaf nodes, while one of height 2 can have up to d^2 leaves. The number of possible leaves multiplies by d every time we increase the height by 1, so to account for n leaves, $n = d^h$, which implies that $h = \log_d n$, as shown in Figure 2.7.

---

[2]If necessary, ask your grandmother what telephone books were.

The punch line here is that short trees can have very many leaves, which is the main reason why binary trees prove fundamental to the design of fast data structures.

2.7.3 Logarithms and Bits

There are two bit patterns of length 1 (0 and 1), four of length 2 (00, 01, 10, and 11), and eight of length 3 (see Figure 2.7 (right)). How many bits w do we need to represent any one of n different possibilities, be it one of n items or the integers from 0 to $n - 1$?

The key observation is that there must be at least n different bit patterns of length w. Since the number of different bit patterns doubles as you add each bit, we need at least w bits where $2^w = n$. In other words, we need $w = \log_2 n$ bits.

2.7.4 Logarithms and Multiplication

Logarithms were particularly important in the days before pocket calculators. They provided the easiest way to multiply big numbers by hand, either implicitly using a slide rule or explicitly by using a book of logarithms.

Logarithms are still useful for multiplication, particularly for exponentiation. Recall that $\log_a(xy) = \log_a(x) + \log_a(y)$; that is, the log of a product is the sum of the logs. A direct consequence of this is

$$\log_a n^b = b \cdot \log_a n$$

How can we compute a^b for any a and b using the $\exp(x)$ and $\ln(x)$ functions on your calculator, where $\exp(x) = e^x$ and $\ln(x) = \log_e(x)$? We know

$$a^b = \exp(\ln(a^b)) = \exp(b \ln(a))$$

so the problem is reduced to one multiplication plus one call to each of these functions.

2.7.5 Fast Exponentiation

Suppose that we need to *exactly* compute the value of a^n for some reasonably large n. Such problems occur in primality testing for cryptography, as discussed in Section 16.8 (page 490). Issues of numerical precision prevent us from applying the formula above.

The simplest algorithm performs $n - 1$ multiplications, by computing $a \times a \times \ldots \times a$. However, we can do better by observing that $n = \lfloor n/2 \rfloor + \lceil n/2 \rceil$. If n is even, then $a^n = (a^{n/2})^2$. If n is odd, then $a^n = a(a^{\lfloor n/2 \rfloor})^2$. In either case, we have halved the size of our exponent at the cost of, at most, two multiplications, so $O(\lg n)$ multiplications suffice to compute the final value.

```
function power(a, n)
    if (n = 0) return(1)
    x = power(a, ⌊n/2⌋)
    if (n is even) then return(x²)
        else return(a × x²)
```

This simple algorithm illustrates an important principle of *divide and conquer*. It always pays to divide a job as evenly as possible. When n is not a power of two, the problem cannot always be divided perfectly evenly, but a difference of one element between the two sides as shown here cannot cause any serious imbalance.

2.7.6 Logarithms and Summations

The *Harmonic numbers* arise as a special case of a sum of a power of integers, namely $H(n) = S(n, -1)$. They are the sum of the progression of simple reciprocals, namely,

$$H(n) = \sum_{i=1}^{n} 1/i = \Theta(\log n)$$

The Harmonic numbers prove important, because they usually explain "where the log comes from" when one magically pops out from algebraic manipulation. For example, the key to analyzing the average-case complexity of quicksort is the summation $n \sum_{i=1}^{n} 1/i$. Employing the Harmonic number's Θ bound immediately reduces this to $\Theta(n \log n)$.

2.7.7 Logarithms and Criminal Justice

Figure 2.8 will be our final example of logarithms in action. This table appears in the Federal Sentencing Guidelines, used by courts throughout the United States. These guidelines are an attempt to standardize criminal sentences, so that a felon convicted of a crime before one judge receives the same sentence that they would before a different judge. To accomplish this, the judges have prepared an intricate point function to score the depravity of each crime and map it to time-to-serve.

Figure 2.8 gives the actual point function for fraud—a table mapping dollars stolen to points. Notice that the punishment increases by one level each time the amount of money stolen roughly doubles. That means that the level of punishment (which maps roughly linearly to the amount of time served) grows logarithmically with the amount of money stolen.

Think for a moment about the consequences of this. Many a corrupt CEO certainly has. It means that your total sentence grows *extremely* slowly with the amount of money you steal. Embezzling an additional $100,000 gets you 3 additional punishment levels if you've already stolen $10,000, adds only 1 level if you've stolen $50,000, and has no effect if you've stolen a million. The corresponding benefit of stealing really large amounts of money becomes even

Loss (apply the greatest)	Increase in level
(A) $2,000 or less	no increase
(B) More than $2,000	add 1
(C) More than $5,000	add 2
(D) More than $10,000	add 3
(E) More than $20,000	add 4
(F) More than $40,000	add 5
(G) More than $70,000	add 6
(H) More than $120,000	add 7
(I) More than $200,000	add 8
(J) More than $350,000	add 9
(K) More than $500,000	add 10
(L) More than $800,000	add 11
(M) More than $1,500,000	add 12
(N) More than $2,500,000	add 13
(O) More than $5,000,000	add 14
(P) More than $10,000,000	add 15
(Q) More than $20,000,000	add 16
(R) More than $40,000,000	add 17
(S) More than $80,000,000	add 18

Figure 2.8:   The Federal Sentencing Guidelines for fraud

greater. The moral of logarithmic growth is clear: *If you're gonna do the crime, make it worth the time!*[3]

> *Take-Home Lesson:*   Logarithms arise whenever things are repeatedly halved or doubled.

2.8   Properties of Logarithms

As we have seen, stating $b^x = y$ is equivalent to saying that $x = \log_b y$. The b term is known as the *base* of the logarithm. Three bases are of particular importance for mathematical and historical reasons:

- *Base $b = 2$*: The *binary logarithm*, usually denoted $\lg x$, is a base 2 logarithm. We have seen how this base arises whenever repeated halving (i.e., binary search) or doubling (i.e., nodes in trees) occurs. Most algorithmic applications of logarithms imply binary logarithms.

- *Base $b = e$*: The *natural logarithm*, usually denoted $\ln x$, is a base $e = 2.71828\ldots$ logarithm. The inverse of $\ln x$ is the exponential function $\exp(x) = e^x$ on your calculator. Thus, composing these functions gives us the identity function,

$$\exp(\ln x) = x \text{ and } \ln(\exp x) = x$$

---

[3]Life imitates art. After publishing this example in the previous edition, I was approached by the U.S. Sentencing Commission seeking insights to improve these guidelines.

- *Base $b = 10$*: Less common today is the base-10 or *common logarithm*, usually denoted as $\log x$. This base was employed in slide rules and logarithm books in the days before pocket calculators.

We have already seen one important property of logarithms, namely that

$$\log_a(xy) = \log_a(x) + \log_a(y)$$

The other important fact to remember is that it is easy to convert a logarithm from one base to another. This is a consequence of the following formula:

$$\log_a b = \frac{\log_c b}{\log_c a}$$

Thus, changing the base of $\log b$ from base-a to base-c simply involves multiplying by $\log_c a$. It is easy to convert a common log function to a natural log function, and vice versa.

Two implications of these properties of logarithms are important to appreciate from an algorithmic perspective:

- *The base of the logarithm has no real impact on the growth rate*: Compare the following three values: $\log_2(1,000,000) = 19.9316$, $\log_3(1,000,000) = 12.5754$, and $\log_{100}(1,000,000) = 3$. A big change in the base of the logarithm produces little difference in the value of the log. Changing the base of the log from a to c involves multiplying by $\log_c a$. This conversion factor is absorbed in the Big Oh notation whenever a and c are constants. Thus, we are usually justified in ignoring the base of the logarithm when analyzing algorithms.

- *Logarithms cut any function down to size*: The growth rate of the logarithm of any polynomial function is $O(\lg n)$. This follows because

$$\log_a n^b = b \cdot \log_a n$$

The effectiveness of binary search on a wide range of problems is a consequence of this observation. Note that performing a binary search on a sorted array of n^2 things requires only twice as many comparisons as a binary search on n things.

Logarithms efficiently cut any function down to size. It is hard to do arithmetic on factorials except after taking logarithms, since

$$n! = \prod_{i=1}^{n} i \rightarrow \log n! = \sum_{i=1}^{n} \log i = \Theta(n \log n)$$

provides another way logarithms pop up in algorithm analysis.

Stop and Think: Importance of an Even Split

Problem: How many queries does binary search take on the million-name Manhattan phone book if each split were 1/3-to-2/3 instead of 1/2-to-1/2?

Solution: When performing binary searches in a telephone book, how important is it that each query split the book exactly in half? Not very much. For the Manhattan telephone book, we now use $\log_{3/2}(1,000,000) \approx 35$ queries in the worst case, not a significant change from $\log_2(1,000,000) \approx 20$. Changing the base of the log does not affect the asymptotic complexity. The effectiveness of binary search comes from its logarithmic running time, not the base of the log.

∎

2.9   War Story: Mystery of the Pyramids

That look in his eyes should have warned me off even before he started talking.

"We want to use a parallel supercomputer for a numerical calculation up to 1,000,000,000, but we need a faster algorithm to do it."

I'd seen that distant look before. Eyes dulled from too much exposure to the raw horsepower of supercomputers—machines so fast that brute force seemed to eliminate the need for clever algorithms; at least until the problems got hard enough.

"I am working with a Nobel prize winner to use a computer on a famous problem in number theory. Are you familiar with Waring's problem?"

I knew some number theory. "Sure. Waring's problem asks whether every integer can be expressed at least one way as the sum of at most four integer squares. For example, $78 = 8^2 + 3^2 + 2^2 + 1^2 = 7^2 + 5^2 + 2^2$. I remember proving that four squares suffice to represent any integer in my undergraduate number theory class. Yes, it's a famous problem but one that got solved 200 years ago."

"No, we are interested in a different version of Waring's problem. A *pyramidal number* is a number of the form $(m^3 - m)/6$, for $m \geq 2$. Thus, the first several pyramidal numbers are 1, 4, 10, 20, 35, 56, 84, 120, and 165. The conjecture since 1928 is that every integer can be represented by the sum of at most five such pyramidal numbers. We want to use a supercomputer to prove this conjecture on all numbers from 1 to 1,000,000,000."

"Doing a billion of anything will take a substantial amount of time," I warned. "The time you spend to compute the minimum representation of each number will be critical, since you are going to do it one billion times. Have you thought about what kind of an algorithm you are going to use?"

"We have already written our program and run it on a parallel supercomputer. It works very fast on smaller numbers. Still, it takes much too much time as soon as we get to 100,000 or so."

"Terrific," I thought. Our supercomputer junkie had discovered asymptotic growth. No doubt his algorithm ran in something like quadratic time, and went into vapor lock as soon as n got large.

"We need a faster program in order to get to a billion. Can you help us? Of course, we can run it on our parallel supercomputer when you are ready."

I am a sucker for this kind of challenge, finding fast algorithms to speed up programs. I agreed to think about it and got down to work.

I started by looking at the program that the other guy had written. He had built an array of all the $\Theta(n^{1/3})$ pyramidal numbers from 1 to n inclusive.[4] To test each number k in this range, he did a brute force test to establish whether it was the sum of two pyramidal numbers. If not, the program tested whether it was the sum of three of them, then four, and finally five, until it first got an answer. About 45% of the integers are expressible as the sum of three pyramidal numbers. Most of the remaining 55% require the sum of four, and usually each of these can be represented in many different ways. Only 241 integers are known to require the sum of five pyramidal numbers, the largest being 343,867. For about half of the n numbers, this algorithm presumably went through all of the three-tests and at least some of the four-tests before terminating. Thus, the total time for this algorithm would be at least $O(n \times (n^{1/3})^3) = O(n^2)$ time, where $n = 1,000,000,000$. No wonder his program cried "Uncle."

Anything that was going to do significantly better on a problem this large had to avoid explicitly testing all triples. For each value of k, we were seeking the smallest set of pyramidal numbers that add up to exactly to k. This problem is called the *knapsack problem*, and is discussed in Section 16.10 (page 497). In our case, the weights are the pyramidal numbers no greater than n, with an additional constraint that the knapsack holds exactly k items.

A standard approach to solving knapsack precomputes the sum of smaller subsets of the items for use in computing larger subsets. If we have a table of all sums of two numbers and want to know whether k is expressible as the sum of three numbers, we can ask whether k is expressible as the sum of a single number plus a number in this two-table.

Therefore, I needed a table of all integers less than n that can be expressed as the sum of two of the 1,816 non-trivial pyramidal numbers less than 1,000,000,000. There can be at most $1,816^2 = 3,297,856$ of them. Actually, there are only about half this many, after we eliminate duplicates and any sum bigger than our target. Building a sorted array storing these numbers would be no big deal. Let's call this sorted data structure of all pair-sums the *two*-table.

To find the minimum decomposition for a given k, I would first check whether it was one of the 1,816 pyramidal numbers. If not, I would then check whether k was in the sorted table of the sums of two pyramidal numbers. To see whether k was expressible as the sum of three such numbers, all I had to do was check whether $k - p[i]$ was in the *two*-table for $1 \leq i \leq 1,816$. This could be done quickly using binary search. To see whether k was expressible as the sum of four pyramidal numbers, I had to check whether $k - two[i]$ was in the two-table

---

[4]Why $n^{1/3}$? Recall that pyramidal numbers are of the form $(m^3 - m)/6$. The largest m such that the resulting number is at most n is roughly $\sqrt[3]{6n}$, so there are $\Theta(n^{1/3})$ such numbers.

for any $1 \leq i \leq |two|$. However, since almost every k was expressible in many ways as the sum of four pyramidal numbers, this test would terminate quickly, and the total time taken would be dominated by the cost of the threes. Testing whether k was the sum of three pyramidal numbers would take $O(n^{1/3} \lg n)$. Running this on each of the n integers gives an $O(n^{4/3} \lg n)$ algorithm for the complete job. Comparing this to his $O(n^2)$ algorithm for $n = 1,000,000,000$ suggested that my algorithm was a cool *30,000* times faster than his original!

My first attempt to code this solved up to $n = 1,000,000$ on my ancient Sparc ELC in about 20 minutes. From here, I experimented with different data structures to represent the sets of numbers and different algorithms to search these tables. I tried using hash tables and bit vectors instead of sorted arrays, and experimented with variants of binary search such as interpolation search (see Section 17.2 (page 510)). My reward for this work was solving up to $n = 1,000,000$ in under three minutes, a factor of six improvement over my original program.

With the real thinking done, I worked to tweak a little more performance out of the program. I avoided doing a sum-of-four computation on any k when $k - 1$ was the sum-of-three, since 1 is a pyramidal number, saving about 10% of the total run time using this trick alone. Finally, I got out my profiler and tried some low-level tricks to squeeze a little more performance out of the code. For example, I saved another 10% by replacing a single procedure call with inline code.

At this point, I turned the code over to the supercomputer guy. What he did with it is a depressing tale, which is reported in Section 5.8 (page 161).

In writing up this story, I went back to rerun this program, which is now older than my current graduate students. Even though single-threaded, it ran in 1.113 seconds. Turning on the compiler optimizer reduced this to a mere 0.334 seconds: this is why you need to remember to turn your optimizer on when you are trying to make your program run fast. This code has gotten hundreds of times faster by doing nothing, except waiting for 25 years of hardware improvements. Indeed a server in our lab can now run up to a billion in under three hours (174 minutes and 28.4 seconds) using only a single thread. Even more amazingly, I can run this code to completion in 9 hours, 37 minutes, and 34.8 seconds on the same crummy Apple MacBook laptop that I am writing this book on, despite its keys falling off as I type.

The primary lesson of this war story is to show the enormous potential for algorithmic speedups, as opposed to the fairly limited speedup obtainable via more expensive hardware. I sped his program up by about 30,000 times. His million-dollar computer (at that time) had 16 processors, each reportedly five times faster on integer computations than the $3,000 machine on my desk. That gave him a maximum potential speedup of less than 100 times. Clearly, the algorithmic improvement was the big winner here, as it is certain to be in any sufficiently large computation.

2.10   Advanced Analysis (*)

Ideally, each of us would be fluent in working with the mathematical techniques of asymptotic analysis. And ideally, each of us would be rich and good looking as well.

In this section I will survey the major techniques and functions employed in advanced algorithm analysis. I consider this optional material—it will not be used elsewhere in the textbook section of this book. That said, it will make some of the complexity functions reported in the Hitchhiker's Guide a little less mysterious.

2.10.1   Esoteric Functions

The bread-and-butter classes of complexity functions were presented in Section 2.3.1 (page 38). More esoteric functions also make appearances in advanced algorithm analysis. Although we will not see them much in this book, it is still good business to know what they mean and where they come from.

- *Inverse Ackermann's function $f(n) = \alpha(n)$*: This function arises in the detailed analysis of several algorithms, most notably the Union-Find data structure discussed in Section 8.1.3 (page 250). It is sufficient to think of this as geek talk for the slowest growing complexity function. Unlike the constant function $f(n) = 1$, $\alpha(n)$ eventually gets to infinity as $n \to \infty$, but it certainly takes its time about it. The value of $\alpha(n)$ is smaller than 5 for any value of n that can be written in this physical universe.

- $f(n) = \log \log n$: The "log log" function is just that—the logarithm of the logarithm of n. One natural example of how it might arise is doing a binary search on a sorted array of only $\lg n$ items.

- $f(n) = \log n / \log \log n$: This function grows a little slower than $\log n$, because it is divided by an even slower growing function. To see where this arises, consider an n-leaf rooted tree of degree d. For binary trees, that is, when $d = 2$, the height h is given

$$n = 2^h \to h = \lg n$$

by taking the logarithm of both sides of the equation. Now consider the height of such a tree when the degree $d = \log n$. Then

$$n = (\log n)^h \to h = \log n / \log \log n$$

- $f(n) = \log^2 n$: This is the product of two log functions, $(\log n) \times (\log n)$. It might arise if we wanted to count the bits looked at when doing a binary search on n items, each of which was an integer from 1 to (say) n^2. Each such integer requires a $\lg(n^2) = 2 \lg n$ bit representation, and we look at $\lg n$ of them, for a total of $2 \lg^2 n$ bits.

The "log squared" function typically arises in the design of intricate nested data structures, where each node in (say) a binary tree represents another data structure, perhaps ordered on a different key.

- $f(n) = \sqrt{n}$: The square root is not very esoteric, but represents the class of "sublinear polynomials" since $\sqrt{n} = n^{1/2}$. Such functions arise in building d-dimensional grids that contain n points. A $\sqrt{n} \times \sqrt{n}$ square has area n, and an $n^{1/3} \times n^{1/3} \times n^{1/3}$ cube has volume n. In general, a d-dimensional hypercube of length $n^{1/d}$ on each side has volume n.

- $f(n) = n^{(1+\epsilon)}$: Epsilon (ϵ) is the mathematical symbol to denote a constant that can be made arbitrarily small but never quite goes away.

  It arises in the following way. Suppose I design an algorithm that runs in $2^c \cdot n^{(1+1/c)}$ time, and I get to pick whichever c I want. For $c = 2$, this is $4n^{3/2}$ or $O(n^{3/2})$. For $c = 3$, this is $8n^{4/3}$ or $O(n^{4/3})$, which is better. Indeed, the exponent keeps getting better the larger I make c.

  The problem is that I cannot make c arbitrarily large before the 2^c term begins to dominate. Instead, we report this algorithm as running in $O(n^{1+\epsilon})$, and leave the best value of ϵ to the beholder.

2.10.2   Limits and Dominance Relations

The dominance relation between functions is a consequence of the theory of limits, which you may recall from taking calculus. We say that $f(n)$ *dominates* $g(n)$ if $\lim_{n\to\infty} g(n)/f(n) = 0$.

Let's see this definition in action. Suppose $f(n) = 2n^2$ and $g(n) = n^2$. Clearly $f(n) > g(n)$ for all n, but it does not dominate since

$$\lim_{n\to\infty} \frac{g(n)}{f(n)} = \lim_{n\to\infty} \frac{n^2}{2n^2} = \lim_{n\to\infty} \frac{1}{2} \neq 0$$

This is to be expected because both functions are in the class $\Theta(n^2)$. What about $f(n) = n^3$ and $g(n) = n^2$? Since

$$\lim_{n\to\infty} \frac{g(n)}{f(n)} = \lim_{n\to\infty} \frac{n^2}{n^3} = \lim_{n\to\infty} \frac{1}{n} = 0$$

the higher-degree polynomial dominates. This is true for any two polynomials, that is, n^a dominates n^b if $a > b$ since

$$\lim_{n\to\infty} \frac{n^b}{n^a} = \lim_{n\to\infty} n^{b-a} \to 0$$

Thus, $n^{1.2}$ dominates $n^{1.1999999}$.

Now consider two exponential functions, say $f(n) = 3^n$ and $g(n) = 2^n$. Since

$$\lim_{n\to\infty} \frac{g(n)}{f(n)} = \frac{2^n}{3^n} = \lim_{n\to\infty} \left(\frac{2}{3}\right)^n = 0$$

the exponential with the higher base dominates.

Our ability to prove dominance relations from scratch depends upon our ability to prove limits. Let's look at one important pair of functions. Any polynomial (say $f(n) = n^\epsilon$) dominates logarithmic functions (say $g(n) = \lg n$). Since $n = 2^{\lg n}$,

$$f(n) = (2^{\lg n})^\epsilon = 2^{\epsilon \lg n}$$

Now consider

$$\lim_{n \to \infty} \frac{g(n)}{f(n)} = \lg n / 2^{\epsilon \lg n}$$

In fact, this does go to 0 as $n \to \infty$.

Take-Home Lesson: By interleaving the functions here with those of Section 2.3.1 (page 38), we see where everything fits into the dominance pecking order:

$$n! \gg c^n \gg n^3 \gg n^2 \gg n^{1+\epsilon} \gg n \log n \gg n \gg \sqrt{n} \gg$$
$$\log^2 n \gg \log n \gg \log n / \log \log n \gg \log \log n \gg \alpha(n) \gg 1$$

Chapter Notes

Most other algorithm texts devote considerably more efforts to the formal analysis of algorithms than I have here, and so I refer the theoretically inclined reader elsewhere for more depth. Algorithm texts more heavily stressing analysis include Cormen et al. [CLRS09] and Kleinberg and Tardos [KT06].

The book *Concrete Mathematics* by Graham, Knuth, and Patashnik [GKP89] offers an interesting and thorough presentation of mathematics for the analysis of algorithms. Niven and Zuckerman [NZM91] is my favorite introduction to number theory, including Waring's problem, discussed in the war story.

The notion of dominance also gives rise to the "Little Oh" notation. We say that $f(n) = o(g(n))$ iff $g(n)$ dominates $f(n)$. Among other things, the Little Oh proves useful for asking questions. Asking for an $o(n^2)$ algorithm means you want one that is better than quadratic in the worst case—and means you would be willing to settle for $O(n^{1.999} \log^2 n)$.

2.11 Exercises

Program Analysis

2-1. *[3]* What value is returned by the following function? Express your answer as a function of n. Give the worst-case running time using the Big Oh notation.

$$\text{Mystery}(n)$$
$$r = 0$$
$$\text{for } i = 1 \text{ to } n - 1 \text{ do}$$
$$\quad \text{for } j = i + 1 \text{ to } n \text{ do}$$
$$\quad\quad \text{for } k = 1 \text{ to } j \text{ do}$$

$$r = r + 1$$
$$\text{return}(r)$$

2-2. *[3]* What value is returned by the following function? Express your answer as a function of n. Give the worst-case running time using Big Oh notation.

> Pesky(n)
>> $r = 0$
>> for $i = 1$ to n do
>>> for $j = 1$ to i do
>>>> for $k = j$ to $i + j$ do
>>>>> $r = r + 1$
>> return(r)

2-3. *[5]* What value is returned by the following function? Express your answer as a function of n. Give the worst-case running time using Big Oh notation.

> Pestiferous(n)
>> $r = 0$
>> for $i = 1$ to n do
>>> for $j = 1$ to i do
>>>> for $k = j$ to $i + j$ do
>>>>> for $l = 1$ to $i + j - k$ do
>>>>>> $r = r + 1$
>> return(r)

2-4. *[8]* What value is returned by the following function? Express your answer as a function of n. Give the worst-case running time using Big Oh notation.

> Conundrum(n)
>> $r = 0$
>> for $i = 1$ to n do
>>> for $j = i + 1$ to n do
>>>> for $k = i + j - 1$ to n do
>>>>> $r = r + 1$
>> return(r)

2-5. *[5]* Consider the following algorithm: (the print operation prints a single asterisk; the operation $x = 2x$ doubles the value of the variable x).

> for $k = 1$ to n:
>> $x = k$
>> while $(x < n)$:
>>> print '*'
>>> $x = 2x$

Let $f(n)$ be the time complexity of this algorithm (or equivalently the number of times * is printed). Provide correct bounds for $O(f(n))$ and $\Omega(f(n))$, ideally converging on $\Theta(f(n))$.

2-6. *[5]* Suppose the following algorithm is used to evaluate the polynomial

$$p(x) = a_n x^n + a_{n-1} x^{n-1} + \ldots + a_1 x + a_0$$

$$p = a_0;$$
$$xpower = 1;$$
for $i = 1$ to n do
$$\cdot \quad xpower = x \cdot xpower;$$
$$p = p + a_i * xpower$$

(a) How many multiplications are done in the worst case? How many additions?

(b) How many multiplications are done on the average?

(c) Can you improve this algorithm?

2-7. *[3]* Prove that the following algorithm for computing the maximum value in an array $A[1..n]$ is correct.

$$\max(A)$$
$$m = A[1]$$
for $i = 2$ to n do
$$\text{if } A[i] > m \text{ then } m = A[i]$$
$$\text{return } (m)$$

Big Oh

2-8. *[3]* (a) Is $2^{n+1} = O(2^n)$?
(b) Is $2^{2n} = O(2^n)$?

2-9. *[3]* For each of the following pairs of functions, $f(n)$ is in $O(g(n))$, $\Omega(g(n))$, or $\Theta(g(n))$. Determine which relationships are correct and briefly explain why.

(a) $f(n) = \log n^2$; $g(n) = \log n + 5$

(b) $f(n) = \sqrt{n}$; $g(n) = \log n^2$

(c) $f(n) = \log^2 n$; $g(n) = \log n$

(d) $f(n) = n$; $g(n) = \log^2 n$

(e) $f(n) = n \log n + n$; $g(n) = \log n$

(f) $f(n) = 10$; $g(n) = \log 10$

(g) $f(n) = 2^n$; $g(n) = 10n^2$

(h) $f(n) = 2^n$; $g(n) = 3^n$

2-10. *[3]* For each of the following pairs of functions $f(n)$ and $g(n)$, determine whether $f(n) = O(g(n))$, $g(n) = O(f(n))$, or both.

(a) $f(n) = (n^2 - n)/2$, $g(n) = 6n$

(b) $f(n) = n + 2\sqrt{n}$, $g(n) = n^2$

(c) $f(n) = n \log n$, $g(n) = n\sqrt{n}/2$

(d) $f(n) = n + \log n$, $g(n) = \sqrt{n}$

(e) $f(n) = 2(\log n)^2$, $g(n) = \log n + 1$

(f) $f(n) = 4n \log n + n$, $g(n) = (n^2 - n)/2$

2-11. *[5]* For each of the following functions, which of the following asymptotic bounds hold for $f(n)$: $O(g(n))$, $\Omega(g(n))$, or $\Theta(g(n))$?

    (a) $f(n) = 3n^2$, $g(n) = n^2$

    (b) $f(n) = 2n^4 - 3n^2 + 7$, $g(n) = n^5$

    (c) $f(n) = \log n$, $g(n) = \log n + \frac{1}{n}$

    (d) $f(n) = 2^{k \log n}$, $g(n) = n^k$

    (e) $f(n) = 2^n$, $g(n) = 2^{2n}$

2-12. *[3]* Prove that $n^3 - 3n^2 - n + 1 = \Theta(n^3)$.

2-13. *[3]* Prove that $n^2 = O(2^n)$.

2-14. *[3]* Prove or disprove: $\Theta(n^2) = \Theta(n^2 + 1)$.

2-15. *[3]* Suppose you have algorithms with the five running times listed below. (Assume these are the exact running times.) How much slower do each of these algorithms get when you (a) double the input size, or (b) increase the input size by one?

    (a) n^2 (b) n^3 (c) $100n^2$ (d) $n \log n$ (e) 2^n

2-16. *[3]* Suppose you have algorithms with the six running times listed below. (Assume these are the exact number of operations performed as a function of the input size n.) Suppose you have a computer that can perform 10^{10} operations per second. For each algorithm, what is the largest input size n that you can complete within an hour? (a) n^2 (b) n^3 (c) $100n^2$ (d) $n \log n$ (e) 2^n (f) 2^{2^n}

2-17. *[3]* For each of the following pairs of functions $f(n)$ and $g(n)$, give an appropriate positive constant c such that $f(n) \le c \cdot g(n)$ for all $n > 1$.

    (a) $f(n) = n^2 + n + 1$, $g(n) = 2n^3$

    (b) $f(n) = n\sqrt{n} + n^2$, $g(n) = n^2$

    (c) $f(n) = n^2 - n + 1$, $g(n) = n^2/2$

2-18. *[3]* Prove that if $f_1(n) = O(g_1(n))$ and $f_2(n) = O(g_2(n))$, then $f_1(n) + f_2(n) = O(g_1(n) + g_2(n))$.

2-19. *[3]* Prove that if $f_1(n) = \Omega(g_1(n))$ and $f_2(n) = \Omega(g_2(n))$, then $f_1(n) + f_2(n) = \Omega(g_1(n) + g_2(n))$.

2-20. *[3]* Prove that if $f_1(n) = O(g_1(n))$ and $f_2(n) = O(g_2(n))$, then $f_1(n) \cdot f_2(n) = O(g_1(n) \cdot g_2(n))$.

2-21. *[5]* Prove that for all $k \ge 0$ and all sets of real constants $\{a_k, a_{k-1}, \ldots, a_1, a_0\}$,

$$a_k n^k + a_{k-1} n^{k-1} + \ldots + a_1 n + a_0 = O(n^k)$$

2-22. *[5]* Show that for any real constants a and b, $b > 0$

$$(n + a)^b = \Theta(n^b)$$

2-23. *[5]* List the functions below from the lowest to the highest order. If any two or more are of the same order, indicate which.

$$n \qquad 2^n \qquad n \lg n \qquad \ln n$$
$$n - n^3 + 7n^5 \qquad \lg n \qquad \sqrt{n} \qquad e^n$$
$$n^2 + \lg n \qquad n^2 \qquad 2^{n-1} \qquad \lg \lg n$$
$$n^3 \qquad (\lg n)^2 \qquad n! \qquad n^{1+\varepsilon} \text{ where } 0 < \varepsilon < 1$$

2-24. *[8]* List the functions below from the lowest to the highest order. If any two or more are of the same order, indicate which.

$$n^\pi \qquad \pi^n \qquad \binom{n}{5} \qquad \sqrt{2^{\sqrt{n}}}$$
$$\binom{n}{n-4} \qquad 2^{\log^4 n} \qquad n^{5(\log n)^2} \qquad n^4 \binom{n}{n-4}$$

2-25. *[8]* List the functions below from the lowest to the highest order. If any two or more are of the same order, indicate which.

$$\sum_{i=1}^{n} i^i \qquad n^n \qquad (\log n)^{\log n} \qquad 2^{(\log n^2)}$$
$$n! \qquad 2^{\log^4 n} \qquad n^{(\log n)^2} \qquad \binom{n}{n-4}$$

2-26. *[5]* List the functions below from the lowest to the highest order. If any two or more are of the same order, indicate which.

$$\sqrt{n} \qquad n \qquad 2^n$$
$$n \log n \qquad n - n^3 + 7n^5 \qquad n^2 + \log n$$
$$n^2 \qquad n^3 \qquad \log n$$
$$n^{\frac{1}{3}} + \log n \qquad (\log n)^2 \qquad n!$$
$$\ln n \qquad \frac{n}{\log n} \qquad \log \log n$$
$$(1/3)^n \qquad (3/2)^n \qquad 6$$

2-27. *[5]* Find two functions $f(n)$ and $g(n)$ that satisfy the following relationship. If no such f and g exist, write "None."

(a) $f(n) = o(g(n))$ and $f(n) \neq \Theta(g(n))$

(b) $f(n) = \Theta(g(n))$ and $f(n) = o(g(n))$

(c) $f(n) = \Theta(g(n))$ and $f(n) \neq O(g(n))$

(d) $f(n) = \Omega(g(n))$ and $f(n) \neq O(g(n))$

2-28. *[5]* True or False?

(a) $2n^2 + 1 = O(n^2)$

(b) $\sqrt{n} = O(\log n)$

(c) $\log n = O(\sqrt{n})$

(d) $n^2(1 + \sqrt{n}) = O(n^2 \log n)$

(e) $3n^2 + \sqrt{n} = O(n^2)$

(f) $\sqrt{n} \log n = O(n)$

(g) $\log n = O(n^{-1/2})$

2-29. *[5]* For each of the following pairs of functions $f(n)$ and $g(n)$, state whether $f(n) = O(g(n))$, $f(n) = \Omega(g(n))$, $f(n) = \Theta(g(n))$, or none of the above.

(a) $f(n) = n^2 + 3n + 4$, $g(n) = 6n + 7$

(b) $f(n) = n\sqrt{n}$, $g(n) = n^2 - n$

(c) $f(n) = 2^n - n^2$, $g(n) = n^4 + n^2$

2-30. *[3]* For each of these questions, answer *yes* or *no* and briefly explain your answer.

(a) If an algorithm takes $O(n^2)$ worst-case time, is it possible that it takes $O(n)$ on some inputs?

(b) If an algorithm takes $O(n^2)$ worst-case time, is it possible that it takes $O(n)$ on all inputs?

(c) If an algorithm takes $\Theta(n^2)$ worst-case time, is it possible that it takes $O(n)$ on some inputs?

(d) If an algorithm takes $\Theta(n^2)$ worst-case time, is it possible that it takes $O(n)$ on all inputs?

(e) Is the function $f(n) = \Theta(n^2)$, where $f(n) = 100n^2$ for even n and $f(n) = 20n^2 - n\log_2 n$ for odd n?

2-31. *[3]* For each of the following, answer *yes*, *no*, or *can't tell*. Explain your reasoning.

(a) Is $3^n = O(2^n)$?

(b) Is $\log 3^n = O(\log 2^n)$?

(c) Is $3^n = \Omega(2^n)$?

(d) Is $\log 3^n = \Omega(\log 2^n)$?

2-32. *[5]* For each of the following expressions $f(n)$ find a simple $g(n)$ such that $f(n) = \Theta(g(n))$.

(a) $f(n) = \sum_{i=1}^n \frac{1}{i}$.

(b) $f(n) = \sum_{i=1}^n \lceil \frac{1}{i} \rceil$.

(c) $f(n) = \sum_{i=1}^n \log i$.

(d) $f(n) = \log(n!)$.

2-33. *[5]* Place the following functions into increasing order: $f_1(n) = n^2 \log_2 n$, $f_2(n) = n(\log_2 n)^2$, $f_3(n) = \sum_{i=0}^n 2^i$ and, $f_4(n) = \log_2(\sum_{i=0}^n 2^i)$.

2-34. *[5]* Which of the following are true?

(a) $\sum_{i=1}^n 3^i = \Theta(3^{n-1})$.

(b) $\sum_{i=1}^n 3^i = \Theta(3^n)$.

(c) $\sum_{i=1}^n 3^i = \Theta(3^{n+1})$.

2-35. *[5]* For each of the following functions f find a simple function g such that $f(n) = \Theta(g(n))$.

(a) $f_1(n) = (1000)2^n + 4^n$.

(b) $f_2(n) = n + n\log n + \sqrt{n}$.

(c) $f_3(n) = \log(n^{20}) + (\log n)^{10}$.

(d) $f_4(n) = (0.99)^n + n^{100}$.

2-36. *[5]* For each pair of expressions (A, B) below, indicate whether A is O, o, Ω, ω, or Θ of B. Note that zero, one, or more of these relations may hold for a given pair; list all correct ones.

	A	B
(a)	n^{100}	2^n
(b)	$(\lg n)^{12}$	$\sqrt{n}$
(c)	$\sqrt{n}$	$n^{\cos(\pi n/8)}$
(d)	10^n	100^n
(e)	$n^{\lg n}$	$(\lg n)^n$
(f)	$\lg(n!)$	$n \lg n$

Summations

2-37. *[5]* Find an expression for the sum of the ith row of the following triangle, and prove its correctness. Each entry is the sum of the three entries directly above it. All non-existing entries are considered 0.

```
                    1
               1    1    1
          1    2    3    2    1
     1    3    6    7    6    3    1
1    4   10   16   19   16   10    4    1
```

2-38. *[3]* Assume that Christmas has n days. Exactly how many presents did my "true love" send to me? (Do some research if you do not understand this question.)

2-39. *[5]* An unsorted array of size n contains distinct integers between 1 and $n+1$, with one element missing. Give an $O(n)$ algorithm to find the missing integer, without using any extra space.

2-40. *[5]* Consider the following code fragment:

```
for i=1 to n do
    for j=i to 2*i do
        output ''foobar''
```

Let $T(n)$ denote the number of times 'foobar' is printed as a function of n.

    a. Express $T(n)$ as a summation (actually two nested summations).

    b. Simplify the summation. Show your work.

2-41. *[5]* Consider the following code fragment:

```
for i=1 to n/2 do
    for j=i to n-i do
        for k=1 to j do
            output ''foobar''
```

Assume n is even. Let $T(n)$ denote the number of times "foobar" is printed as a function of n.

    (a) Express $T(n)$ as three nested summations.

    (b) Simplify the summation. Show your work.

2-42. *[6]* When you first learned to multiply numbers, you were told that $x \times y$ means adding x a total of y times, so $5 \times 4 = 5 + 5 + 5 + 5 = 20$. What is the time complexity of multiplying two n-digit numbers in base b (people work in base 10, of course, while computers work in base 2) using the repeated addition method, as a function of n and b. Assume that single-digit by single digit addition or multiplication takes $O(1)$ time. (Hint: how big can y be as a function of n and b?)

2-43. *[6]* In grade school, you learned to multiply long numbers on a digit-by-digit basis, so that $127 \times 211 = 127 \times 1 + 127 \times 10 + 127 \times 200 = 26,797$. Analyze the time complexity of multiplying two n-digit numbers with this method as a function of n (assume constant base size). Assume that single-digit by single-digit addition or multiplication takes $O(1)$ time.

Logarithms

2-44. *[5]* Prove the following identities on logarithms:

    (a) $\log_a(xy) = \log_a x + \log_a y$

    (b) $\log_a x^y = y \log_a x$

    (c) $\log_a x = \frac{\log_b x}{\log_b a}$

    (d) $x^{\log_b y} = y^{\log_b x}$

2-45. *[3]* Show that $\lceil \lg(n+1) \rceil = \lfloor \lg n \rfloor + 1$

2-46. *[3]* Prove that that the binary representation of $n \geq 1$ has $\lfloor \lg_2 n \rfloor + 1$ bits.

2-47. *[5]* In one of my research papers I give a comparison-based sorting algorithm that runs in $O(n \log(\sqrt{n}))$. Given the existence of an $\Omega(n \log n)$ lower bound for sorting, how can this be possible?

Interview Problems

2-48. *[5]* You are given a set S of n numbers. You must pick a subset S' of k numbers from S such that the probability of each element of S occurring in S' is equal (i.e., each is selected with probability k/n). You may make only one pass over the numbers. What if n is unknown?

2-49. *[5]* We have 1,000 data items to store on 1,000 nodes. Each node can store copies of exactly three different items. Propose a replication scheme to minimize data loss as nodes fail. What is the expected number of data entries that get lost when three random nodes fail?

2-50. *[5]* Consider the following algorithm to find the minimum element in an array of numbers $A[0, \ldots, n]$. One extra variable tmp is allocated to hold the current minimum value. Start from A[0]; tmp is compared against $A[1]$, $A[2]$, $\ldots$, $A[N]$ in order. When $A[i] < tmp$, $tmp = A[i]$. What is the expected number of times that the assignment operation $tmp = A[i]$ is performed?

2-51. *[5]* You are given ten bags of gold coins. Nine bags contain coins that each weigh 10 grams. One bag contains all false coins that weigh 1 gram less. You must identify this bag in just one weighing. You have a digital balance that reports the weight of what is placed on it.

2-52. *[5]* You have eight balls all of the same size. Seven of them weigh the same, and one of them weighs slightly more. How can you find the ball that is heavier by using a balance and only two weighings?

2-53. *[5]* Suppose we start with n companies that eventually merge into one big company. How many different ways are there for them to merge?

2-54. *[7]* Six pirates must divide $300 among themselves. The division is to proceed as follows. The senior pirate proposes a way to divide the money. Then the pirates vote. If the senior pirate gets at least half the votes he wins, and that division remains. If he doesn't, he is killed and then the next senior-most pirate gets a chance to propose the division. Now tell what will happen and why (i.e. how many pirates survive and how the division is done)? All the pirates are intelligent and the first priority is to stay alive and the next priority is to get as much money as possible.

2-55. *[7]* Reconsider the pirate problem above, where we start with only one indivisible dollar. Who gets the dollar, and how many are killed?

LeetCode

2-1. `https://leetcode.com/problems/remove-k-digits/`

2-2. `https://leetcode.com/problems/counting-bits/`

2-3. `https://leetcode.com/problems/4sum/`

HackerRank

2-1. `https://www.hackerrank.com/challenges/pangrams/`

2-2. `https://www.hackerrank.com/challenges/the-power-sum/`

2-3. `https://www.hackerrank.com/challenges/magic-square-forming/`

Programming Challenges

These programming challenge problems with robot judging are available at `https://onlinejudge.org`:

2-1. "Primary Arithmetic"—Chapter 5, problem 10035.

2-2. "A Multiplication Game"—Chapter 5, problem 847.

2-3. "Light, More Light"—Chapter 7, problem 10110.

Chapter 3

Data Structures

Putting the right data structure into a slow program can work the same wonders as transplanting fresh parts into a sick patient. Important classes of *abstract data types* such as containers, dictionaries, and priority queues have many functionally equivalent *data structures* that implement them. Changing the data structure does not affect the correctness of the program, since we presumably replace a correct implementation with a different correct implementation. However, the new implementation may realize different trade-offs in the time to execute various operations, so the total performance can improve dramatically. Like a patient in need of a transplant, only one part might need to be replaced in order to fix the problem.

But it is better to be born with a good heart than have to wait for a replacement. The maximum benefit from proper data structures results from designing your program around them in the first place. We assume that the reader has had some previous exposure to elementary data structures and pointer manipulation. Still, data structure courses (CS II) focus more on data abstraction and object orientation than the nitty-gritty of how structures should be represented in memory. This material will be reviewed here to make sure you have it down.

As with most subjects, in data structures it is more important to really understand the basic material than to have exposure to more advanced concepts. This chapter will focus on each of the three fundamental abstract data types (containers, dictionaries, and priority queues) and show how they can be implemented with arrays and lists. Detailed discussion of the trade-offs between more sophisticated implementations is deferred to the relevant catalog entry for each of these data types.

3.1 Contiguous vs. Linked Data Structures

Data structures can be neatly classified as either *contiguous* or *linked*, depending upon whether they are based on arrays or pointers. *Contiguously allocated*

S. S. Skiena, *The Algorithm Design Manual*, Texts in Computer Science,
https://doi.org/10.1007/978-3-030-54256-6_3

structures are composed of single slabs of memory, and include arrays, matrices, heaps, and hash tables. *Linked data structures* are composed of distinct chunks of memory bound together by *pointers*, and include lists, trees, and graph adjacency lists.

In this section, I review the relative advantages of contiguous and linked data structures. These trade-offs are more subtle than they appear at first glance, so I encourage readers to stick with me here even if you may be familiar with both types of structures.

3.1.1   Arrays

The *array* is the fundamental contiguously allocated data structure. Arrays are structures of fixed-size data records such that each element can be efficiently located by its *index* or (equivalently) address.

A good analogy likens an array to a street full of houses, where each array element is equivalent to a house, and the index is equivalent to the house number. Assuming all the houses are of equal size and numbered sequentially from 1 to n, we can compute the exact position of each house immediately from its address.[1]

Advantages of contiguously allocated arrays include:

- *Constant-time access given the index* – Because the index of each element maps directly to a particular memory address, we can access arbitrary data items instantly provided we know the index.

- *Space efficiency* – Arrays consist purely of data, so no space is wasted with links or other formatting information. Further, end-of-record information is not needed because arrays are built from fixed-size records.

- *Memory locality* – Many programming tasks require iterating through all the elements of a data structure. Arrays are good for this because they exhibit excellent memory locality. Physical continuity between successive data accesses helps exploit the high-speed *cache memory* on modern computer architectures.

The downside of arrays is that we cannot adjust their size in the middle of a program's execution. Our program will fail as soon as we try to add the $(n+1)$st customer, if we only allocated room for n records. We can compensate by allocating extremely large arrays, but this can waste space, again restricting what our programs can do.

Actually, we *can* efficiently enlarge arrays as we need them, through the miracle of *dynamic arrays*. Suppose we start with an array of size 1, and double its size from m to $2m$ whenever we run out of space. This doubling process allocates a new contiguous array of size $2m$, copies the contents of the old array

---

[1] Houses in Japanese cities are traditionally numbered in the order they were built, not by their physical location. This makes it extremely difficult to locate a Japanese address without a detailed map.

to the lower half of the new one, and then returns the space used by the old array to the storage allocation system.

The apparent waste in this procedure involves recopying the old contents on each expansion. How much work do we really do? It will take $\log_2 n$ (also known as $\lg n$) doublings until the array gets to have n positions, plus one final doubling on the last insertion when $n = 2^j$ for some j. There are recopying operations after the first, second, fourth, eighth, ..., nth insertions. The number of copy operations at the ith doubling will be 2^{i-1}, so the total number of movements M will be:

$$M = n + \sum_{i=1}^{\lg n} 2^{i-1} = 1 + 2 + 4 + \ldots + \frac{n}{2} + n = \sum_{i=i}^{\lg n} \frac{n}{2^i} \leq n \sum_{i=0}^{\infty} \frac{1}{2^i} = 2n$$

Thus, each of the n elements move only two times on average, and the total work of managing the dynamic array is the same $O(n)$ as it would have been if a single array of sufficient size had been allocated in advance!

The primary thing lost in using dynamic arrays is the guarantee that each insertion takes constant time *in the worst case*. Note that all accesses and *most* insertions will be fast, except for those relatively few insertions that trigger array doubling. What we get instead is a promise that the nth element insertion will be completed quickly enough that the *total* effort expended so far will still be $O(n)$. Such *amortized* guarantees arise frequently in the analysis of data structures.

3.1.2 Pointers and Linked Structures

Pointers are the connections that hold the pieces of linked structures together. Pointers represent the address of a location in memory. A variable storing a pointer to a given data item can provide more freedom than storing a copy of the item itself. A cell-phone number can be thought of as a pointer to its owner as they move about the planet.

Pointer syntax and power differ significantly across programming languages, so we begin with a quick review of pointers in C language. A pointer `p` is assumed to give the address in memory where a particular chunk of data is located.[2] Pointers in C have types declared at compile time, denoting the data type of the items they can point to. We use `*p` to denote the item that is pointed to by pointer `p`, and `&x` to denote the address of (i.e. pointer to) a particular variable `x`. A special `NULL` pointer value is used to denote structure-terminating or unassigned pointers.

All linked data structures share certain properties, as revealed by the following type declaration for linked lists:

---

[2]C permits direct manipulation of memory addresses in ways that may horrify Java programmers, but I will avoid doing any such tricks.

Figure 3.1: Linked list example showing data and pointer fields.

```
typedef struct list {
    item_type item;              /* data item */
    struct list *next;           /* point to successor */
} list;
```

In particular:

- Each node in our data structure (here `list`) contains one or more data fields (here `item`) that retain the data that we need to store.

- Each node contains a pointer field to at least one other node (here `next`). This means that much of the space used in linked data structures is devoted to pointers, not data.

- Finally, we need a pointer to the head of the structure, so we know where to access it.

The list here is the simplest linked structure. The three basic operations supported by lists are searching, insertion, and deletion. In *doubly linked lists*, each node points both to its predecessor and its successor element. This simplifies certain operations at a cost of an extra pointer field per node.

Searching a List

Searching for item x in a linked list can be done iteratively or recursively. I opt for recursively in the implementation below. If x is in the list, it is either the first element or located in the rest of the list. Eventually, the problem is reduced to searching in an empty list, which clearly cannot contain x.

```
list *search_list(list *l, item_type x) {
    if (l == NULL) {
        return(NULL);
    }

    if (l->item == x) {
        return(l);
    } else {
        return(search_list(l->next, x));
    }
}
```

Insertion into a List

Insertion into a singly linked list is a nice exercise in pointer manipulation, as shown below. Since we have no need to maintain the list in any particular order, we might as well insert each new item in the most convenient place. Insertion at the beginning of the list avoids any need to traverse the list, but does require us to update the pointer (denoted l) to the head of the data structure.

```
void insert_list(list **l, item_type x) {
    list *p;     /* temporary pointer */

    p = malloc(sizeof(list));
    p->item = x;
    p->next = *l;
    *l = p;
}
```

Two C-isms to note. First, the `malloc` function allocates a chunk of memory of sufficient size for a new node to contain x. Second, the funny double star in **l denotes that l is a *pointer to a pointer* to a list node. Thus, the last line, *l=p; copies p to the place pointed to by l, which is the external variable maintaining access to the head of the list.

Deletion From a List

Deletion from a linked list is somewhat more complicated. First, we must find a pointer to the *predecessor* of the item to be deleted. We do this recursively:

```
list *item_ahead(list *l, list *x) {
    if ((l == NULL) || (l->next == NULL)) {
        return(NULL);
    }

    if ((l->next) == x) {
        return(l);
    } else {
        return(item_ahead(l->next, x));
    }
}
```

The predecessor is needed because it points to the doomed node, so its `next` pointer must be changed. The actual deletion operation is simple, once ruling out the case that the to-be-deleted element does not exist. Special care must be taken to reset the pointer to the head of the list (l) when the first element is deleted:

```
void delete_list(list **l, list **x) {
    list *p;                /* item pointer */
    list *pred;             /* predecessor pointer */

    p = *l;
    pred = item_ahead(*l, *x);

    if (pred == NULL) { /* splice out of list */
        *l = p->next;
    } else {
        pred->next = (*x)->next;
    }
    free(*x);               /* free memory used by node */
}
```

C language requires explicit deallocation of memory, so we must **free** the deleted node after we are finished with it in order to return the memory to the system. This leaves the incoming pointer as a *dangling reference* to a location that no longer exists, so care must be taken not to use this pointer again. Such problems can generally be avoided in Java because of its stronger memory management model.

3.1.3   Comparison

The advantages of linked structures over static arrays include:

- Overflow on linked structures never occurs unless the memory is actually full.

- Insertion and deletion are *simpler* than for static arrays.

- With large records, moving pointers is easier and faster than moving the items themselves.

Conversely, the relative advantages of arrays include:

- Space efficiency: linked structures require extra memory for storing pointer fields.

- Efficient random access to items in arrays.

- Better memory locality and cache performance than random pointer jumping.

Take-Home Lesson: Dynamic memory allocation provides us with flexibility on how and where we use our limited storage resources.

One final thought about these fundamental data structures is that both arrays and linked lists can be thought of as recursive objects:

- *Lists* – Chopping the first element off a linked list leaves a smaller linked list. This same argument works for strings, since removing characters from a string leaves a string. Lists are recursive objects.

- *Arrays* – Splitting the first k elements off of an n element array gives two smaller arrays, of size k and $n - k$, respectively. Arrays are recursive objects.

This insight leads to simpler list processing, and efficient divide-and-conquer algorithms such as quicksort and binary search.

3.2 Containers: Stacks and Queues

I use the term *container* to denote an abstract data type that permits storage and retrieval of data items *independent of content*. By contrast, dictionaries are abstract data types that retrieve based on key values or content, and will be discussed in Section 3.3 (page 76).

Containers are distinguished by the particular retrieval order they support. In the two most important types of containers, this retrieval order depends on the insertion order:

- *Stacks* support retrieval by last-in, first-out (LIFO) order. Stacks are simple to implement and very efficient. For this reason, stacks are probably the right container to use when retrieval order doesn't matter at all, such as when processing batch jobs. The *put* and *get* operations for stacks are usually called *push* and *pop*:

  - *Push(x,s)*: Insert item x at the top of stack s.
  - *Pop(s)*: Return (and remove) the top item of stack s.

  LIFO order arises in many real-world contexts. People crammed into a subway car exit in LIFO order. Food inserted into my refrigerator usually exits the same way, despite the incentive of expiration dates. Algorithmically, LIFO tends to happen in the course of executing recursive algorithms.

- *Queues* support retrieval in first-in, first-out (FIFO) order. This is surely the fairest way to control waiting times for services. Jobs processed in FIFO order minimize the *maximum* time spent waiting. Note that the *average* waiting time will be the same regardless of whether FIFO or LIFO is used. Many computing applications involve data items with infinite patience, which renders the question of maximum waiting time moot.

  Queues are somewhat trickier to implement than stacks and thus are most appropriate for applications (like certain simulations) where the order is important. The *put* and *get* operations for queues are usually called *enqueue* and *dequeue*.

- *Enqueue(x,q)*: Insert item x at the back of queue q.

- *Dequeue(q)*: Return (and remove) the front item from queue q.

We will see queues later as the fundamental data structure controlling breadth-first search (BFS) in graphs.

Stacks and queues can be effectively implemented using either arrays or linked lists. The key issue is whether an upper bound on the size of the container is known in advance, thus permitting the use of a statically allocated array.

3.3   Dictionaries

The *dictionary* data type permits access to data items by content. You stick an item into a dictionary so you can find it when you need it. The primary operations dictionaries support are:

- *Search(D,k)* – Given a search key k, return a pointer to the element in dictionary D whose key value is k, if one exists.

- *Insert(D,x)* – Given a data item x, add it to the dictionary D.

- *Delete(D,x)* – Given a pointer x to a given data item in the dictionary D, remove it from D.

Certain dictionary data structures also efficiently support other useful operations:

- *Max(D)* or *Min(D)* – Retrieve the item with the largest (or smallest) key from D. This enables the dictionary to serve as a priority queue, as will be discussed in Section 3.5 (page 87).

- *Predecessor(D,x)* or *Successor(D,x)* – Retrieve the item from D whose key is immediately before (or after) item x in *sorted* order. These enable us to iterate through the elements of the data structure in sorted order.

Many common data processing tasks can be handled using these dictionary operations. For example, suppose we want to remove all duplicate names from a mailing list, and print the results in sorted order. Initialize an empty dictionary D, whose search key will be the record name. Now read through the mailing list, and for each record *search* to see if the name is already in D. If not, *insert* it into D. After reading through the mailing list, we print the names in the dictionary. By starting from the first item *Min(D)* and repeatedly calling *Successor* until we obtain *Max(D)*, we traverse all elements in sorted order.

By defining such problems in terms of abstract dictionary operations, we can ignore the details of the data structure's representation and focus on the task at hand.

In the rest of this section, we will carefully investigate simple dictionary implementations based on arrays and linked lists. More powerful dictionary implementations such as binary search trees (see Section 3.4 (page 81)) and hash tables (see Section 3.7 (page 93)) are also attractive options in practice. A complete discussion of different dictionary data structures is presented in the catalog in Section 15.1 (page 440). I encourage the reader to browse through the data structures section of the catalog to better learn what your options are.

Stop and Think: Comparing Dictionary Implementations (I)

Problem: What are the asymptotic worst-case running times for all seven fundamental dictionary operations (search, insert, delete, successor, predecessor, minimum, and maximum) when the data structure is implemented as:

- An unsorted array.

- A sorted array.

Solution: This problem (and the one following it) reveals some of the inherent trade-offs of data structure design. A given data representation may permit efficient implementation of certain operations at the cost that other operations are expensive.

In addition to the array in question, we will assume access to a few extra variables such as n, the number of elements currently in the array. Note that we must *maintain* the value of these variables in the operations where they change (e.g., insert and delete), and charge these operations the cost of this maintenance.

The basic dictionary operations can be implemented with the following costs on unsorted and sorted arrays. The starred element indicates cleverness.

Dictionary operation	Unsorted array	Sorted array
Search(A, k)	$O(n)$	$O(\log n)$
Insert(A, x)	$O(1)$	$O(n)$
Delete(A, x)	$O(1)^*$	$O(n)$
Successor(A, x)	$O(n)$	$O(1)$
Predecessor(A, x)	$O(n)$	$O(1)$
Minimum(A)	$O(n)$	$O(1)$
Maximum(A)	$O(n)$	$O(1)$

We must understand the implementation of each operation to see why. First, let's discuss the operations when maintaining an *unsorted* array A.

- *Search* is implemented by testing the search key k against (potentially) each element of an unsorted array. Thus, search takes linear time in the worst case, which is when key k is not found in A.

- *Insertion* is implemented by incrementing n and then copying item x to the nth cell in the array, $A[n]$. The bulk of the array is untouched, so this operation takes constant time.

- *Deletion* is somewhat trickier, hence the asterisk in the table above. The definition states that we are given a pointer x to the element to delete, so we need not spend any time searching for the element. But removing the xth element from the array A leaves a hole that must be filled. We could fill the hole by moving each of the elements from $A[x+1]$ to $A[n]$ down one position, but this requires $\Theta(n)$ time when the first element is deleted. The following idea is better: just overwrite $A[x]$ with $A[n]$, and decrement n. This only takes constant time.

- The definitions of the traversal operations, *Predecessor* and *Successor*, refer to the item appearing before/after x *in sorted order*. Thus, the answer is not simply $A[x-1]$ (or $A[x+1]$), because in an unsorted array an element's physical predecessor (successor) is not necessarily its logical predecessor (successor). Instead, the predecessor of $A[x]$ is the biggest element smaller than $A[x]$. Similarly, the successor of $A[x]$ is the smallest element larger than $A[x]$. Both require a sweep through all n elements of A to determine the winner.

- *Minimum* and *Maximum* are similarly defined with respect to sorted order, and so require linear-cost sweeps to identify in an unsorted array. It is tempting to set aside extra variables containing the current minimum and maximum values, so we can report them in $O(1)$ time. But this is incompatible with constant-time deletion, as deleting the minimum valued item mandates a linear-time search to find the new minimum.

Implementing a dictionary using a *sorted* array completely reverses our notions of what is easy and what is hard. Searches can now be done in $O(\log n)$ time, using binary search, because we know the median element sits in $A[n/2]$. Since the upper and lower portions of the array are also sorted, the search can continue recursively on the appropriate portion. The number of halvings of n until we get to a single element is $\lceil \lg n \rceil$.

The sorted order also benefits us with respect to the other dictionary retrieval operations. The minimum and maximum elements sit in $A[1]$ and $A[n]$, while the predecessor and successor to $A[x]$ are $A[x-1]$ and $A[x+1]$, respectively.

Insertion and deletion become more expensive, however, because making room for a new item or filling a hole may require moving many items arbitrarily. Thus, both become linear-time operations. ∎

Take-Home Lesson: Data structure design must balance all the different operations it supports. The fastest data structure to support *both* operations A and B may well not be the fastest structure to support just operation A or B.

Stop and Think: Comparing Dictionary Implementations (II)

Problem: What are the asymptotic worst-case running times for each of the seven fundamental dictionary operations when the data structure is implemented as

- A singly linked unsorted list.

- A doubly linked unsorted list.

- A singly linked sorted list.

- A doubly linked sorted list.

Solution: Two different issues must be considered in evaluating these implementations: singly vs. doubly linked lists and sorted vs. unsorted order. Operations with subtle implementations are denoted with an asterisk:

| | Singly linked | | Doubly linked | |
Dictionary operation	unsorted	sorted	unsorted	sorted
Search(L, k)	$O(n)$	$O(n)$	$O(n)$	$O(n)$
Insert(L, x)	$O(1)$	$O(n)$	$O(1)$	$O(n)$
Delete(L, x)	$O(n)^*$	$O(n)^*$	$O(1)$	$O(1)$
Successor(L, x)	$O(n)$	$O(1)$	$O(n)$	$O(1)$
Predecessor(L, x)	$O(n)$	$O(n)^*$	$O(n)$	$O(1)$
Minimum(L)	$O(1)^*$	$O(1)$	$O(n)$	$O(1)$
Maximum(L)	$O(1)^*$	$O(1)^*$	$O(n)$	$O(1)$

As with unsorted arrays, search operations are destined to be slow while maintenance operations are fast.

- *Insertion/Deletion* – The complication here is deletion from a singly linked list. The definition of the *Delete* operation states we are given a pointer x to the item to be deleted. But what we *really* need is a pointer to the element pointing to x in the list, because that is the node that needs to be changed. We can do nothing without this list predecessor, and so must spend linear time searching for it on a singly linked list. Doubly linked lists avoid this problem, since we can immediately retrieve the list predecessor of x.[3]

---

[3] Actually, there *is* a way to delete an element from a singly linked list in constant time, as shown in Figure 3.2. Overwrite the node that x points to with the contents of what x.next points to, then deallocate the node that x.next originally pointed to. Special care must be taken if x is the first node in the list, or the last node (by employing a permanent sentinel element that is always the last node in the list). But this would prevent us from having constant-time minimum/maximum operations, because we no longer have time to find new extreme elements after deletion.

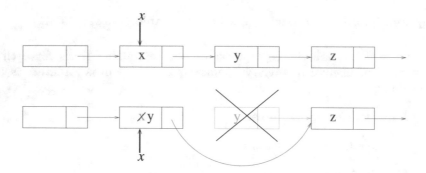

Figure 3.2: By overwriting the contents of a node-to-delete, and deleting its original successor, we can delete a node without access to its list predecessor.

Deletion is faster for sorted doubly linked lists than sorted arrays, because splicing out the deleted element from the list is more efficient than filling the hole by moving array elements. The predecessor pointer problem again complicates deletion from singly linked sorted lists.

- *Search* – Sorting provides less benefit for linked lists than it did for arrays. Binary search is no longer possible, because we can't access the median element without traversing all the elements before it. What sorted lists *do* provide is quick termination of unsuccessful searches, for if we have not found *Abbott* by the time we hit *Costello* we can deduce that he doesn't exist in the list. Still, searching takes linear time in the worst case.

- *Traversal operations* – The predecessor pointer problem again complicates implementing *Predecessor* with singly linked lists. The logical successor is equivalent to the node successor for both types of sorted lists, and hence can be implemented in constant time.

- *Minimum/Maximum* – The minimum element sits at the head of a sorted list, and so is easily retrieved. The maximum element is at the tail of the list, which normally requires $\Theta(n)$ time to reach in either singly or doubly linked lists.

  However, we can maintain a separate pointer to the list tail, provided we pay the maintenance costs for this pointer on every insertion and deletion. The tail pointer can be updated in constant time on doubly linked lists: on insertion check whether `last->next` still equals `NULL`, and on deletion set `last` to point to the list predecessor of `last` if the last element is deleted.

  We have no efficient way to find this predecessor for singly linked lists. So why can we implement *Maximum* in $O(1)$? The trick is to charge the cost to each deletion, which *already* took linear time. Adding an extra linear sweep to update the pointer does not harm the asymptotic complexity of *Delete*, while gaining us *Maximum* (and similarly *Minimum*) in constant time as a reward for clear thinking. ∎

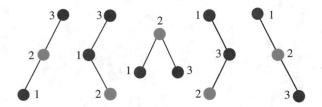

Figure 3.3: The five distinct binary search trees on three nodes. All nodes in the left (resp. right) subtree of node x have keys $< x$ (resp. $> x$).

3.4 Binary Search Trees

We have seen data structures that allow fast search or flexible update, but not fast search *and* flexible update. Unsorted, doubly linked lists supported insertion and deletion in $O(1)$ time but search took linear time in the worst case. Sorted arrays support binary search and logarithmic query times, but at the cost of linear-time update.

Binary search requires that we have fast access to *two elements*—specifically the median elements above and below the given node. To combine these ideas, we need a "linked list" with two pointers per node. This is the basic idea behind binary search trees.

A *rooted binary tree* is recursively defined as either being (1) empty, or (2) consisting of a node called the *root*, together with two rooted binary trees called the left and right subtrees, respectively. The order among "sibling" nodes matters in rooted trees, that is, left is different from right. Figure 3.3 gives the shapes of the five distinct binary trees that can be formed on three nodes.

A binary *search* tree labels each node in a binary tree with a single key such that for any node labeled x, all nodes in the left subtree of x have keys $< x$ while all nodes in the right subtree of x have keys $> x$.[4] This search tree labeling scheme is very special. For any binary tree on n nodes, and any set of n keys, there is *exactly* one labeling that makes it a binary search tree. Allowable labelings for three-node binary search trees are given in Figure 3.3.

3.4.1 Implementing Binary Search Trees

Binary tree nodes have *left* and *right* pointer fields, an (optional) *parent* pointer, and a data field. These relationships are shown in Figure 3.4; a type declaration for the tree structure is given below:

---

[4] Allowing duplicate keys in a binary search tree (or any other dictionary structure) is bad karma, often leading to very subtle errors. To better support repeated items, we can add a third pointer to each node, explicitly maintaining a list of all items with the given key.

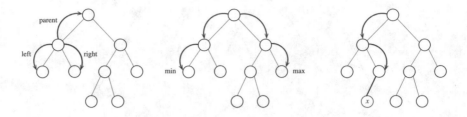

Figure 3.4: Relationships in a binary search tree. Parent and sibling pointers (left). Finding the minimum and maximum elements in a binary search tree (center). Inserting a new node in the correct position (right).

```
typedef struct tree {
    item_type item;              /* data item */
    struct tree *parent;         /* pointer to parent */
    struct tree *left;           /* pointer to left child */
    struct tree *right;          /* pointer to right child */
} tree;
```

The basic operations supported by binary trees are searching, traversal, insertion, and deletion.

Searching in a Tree

The binary search tree labeling uniquely identifies where each key is located. Start at the root. Unless it contains the query key x, proceed either left or right depending upon whether x occurs before or after the root key. This algorithm works because both the left and right subtrees of a binary search tree are themselves binary search trees. This recursive structure yields the recursive search algorithm below:

```
tree *search_tree(tree *l, item_type x) {
    if (l == NULL) {
        return(NULL);
    }

    if (l->item == x) {
        return(l);
    }

    if (x < l->item) {
        return(search_tree(l->left, x));
    } else {
        return(search_tree(l->right, x));
    }
}
```

This search algorithm runs in $O(h)$ time, where h denotes the height of the tree.

Finding Minimum and Maximum Elements in a Tree

By definition, the smallest key must reside in the left subtree of the root, since all keys in the left subtree have values less than that of the root. Therefore, as shown in Figure 3.4 (center), the minimum element must be the left-most descendant of the root. Similarly, the maximum element must be the right-most descendant of the root.

```
tree *find_minimum(tree *t) {
    tree *min;     /* pointer to minimum */

    if (t == NULL) {
        return(NULL);
    }

    min = t;
    while (min->left != NULL) {
        min = min->left;
    }
    return(min);
}
```

Traversal in a Tree

Visiting all the nodes in a rooted binary tree proves to be an important component of many algorithms. It is a special case of traversing all the nodes and edges in a graph, which will be the foundation of Chapter 7.

A prime application of tree traversal is listing the labels of the tree nodes. Binary search trees make it easy to report the labels in *sorted* order. By definition, all the keys smaller than the root must lie in the left subtree of the root, and all keys bigger than the root in the right subtree. Thus, visiting the nodes recursively, in accord with such a policy, produces an *in-order* traversal of the search tree:

```
void traverse_tree(tree *l) {
    if (l != NULL) {
        traverse_tree(l->left);
        process_item(l->item);
        traverse_tree(l->right);
    }
}
```

Each item is processed only once during the course of traversal, so it runs in $O(n)$ time, where n denotes the number of nodes in the tree.

Different traversal orders come from changing the position of process_item relative to the traversals of the left and right subtrees. Processing the item first yields a *pre-order* traversal, while processing it last gives a *post-order* traversal. These make relatively little sense with search trees, but prove useful when the rooted tree represents arithmetic or logical expressions.

Insertion in a Tree

There is exactly one place to insert an item x into a binary search tree T so we can be certain where to find it again. We must replace the NULL pointer found in T after an unsuccessful query for the key of x.

This implementation uses recursion to combine the search and node insertion stages of key insertion. The three arguments to insert_tree are (1) a pointer l to the pointer linking the search subtree to the rest of the tree, (2) the key x to be inserted, and (3) a parent pointer to the parent node containing l. The node is allocated and linked in after hitting the NULL pointer. Note that we pass the *pointer* to the appropriate left/right pointer in the node during the search, so the assignment *l = p; links the new node into the tree:

```
void insert_tree(tree **l, item_type x, tree *parent) {
    tree *p;      /* temporary pointer */

    if (*l == NULL) {
        p = malloc(sizeof(tree));
        p->item = x;
        p->left = p->right = NULL;
        p->parent = parent;
        *l = p;
        return;
    }

    if (x < (*l)->item) {
        insert_tree(&((*l)->left), x, *l);
    } else {
        insert_tree(&((*l)->right), x, *l);
    }
}
```

Allocating the node and linking it into the tree is a constant-time operation, after the search has been performed in $O(h)$ time. Here h denotes the height of the search tree.

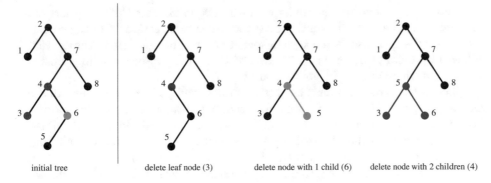

initial tree   delete leaf node (3)   delete node with 1 child (6)   delete node with 2 children (4)

Figure 3.5: Deleting tree nodes with 0, 1, and 2 children. Colors show the nodes affected as a result of the deletion.

Deletion from a Tree

Deletion is somewhat trickier than insertion, because removing a node means appropriately linking its two descendant subtrees back into the tree somewhere else. There are three cases, illustrated in Figure 3.5. Study this figure. Leaf nodes have no children, and so may be deleted simply by clearing the pointer to the given node.

The case of a doomed node having one child is also straightforward. We can link this child to the deleted node's parent without violating the in-order labeling property of the tree.

But what of a node with two children? Our solution is to relabel this node with the key of its immediate successor in sorted order. This successor must be the smallest value in the right subtree, specifically the left-most descendant in the right subtree p. Moving this descendant to the point of deletion results in a properly labeled binary search tree, and reduces our deletion problem to physically removing a node with at most one child—a case that has been resolved above. The full implementation has been omitted here because it looks a little ghastly, but the code follows logically from the description above.

The worst-case complexity analysis is as follows. Every deletion requires the cost of at most two search operations, each taking $O(h)$ time where h is the height of the tree, plus a constant amount of pointer manipulation.

3.4.2   How Good are Binary Search Trees?

When implemented using binary search trees, all three dictionary operations take $O(h)$ time, where h is the height of the tree. The smallest height we can hope for occurs when the tree is perfectly balanced, meaning that $h = \lceil \log n \rceil$. This is very good, but the tree must be perfectly balanced.

Our insertion algorithm puts each new item at a leaf node where it should have been found. This makes the shape (and more importantly height) of the tree determined by the order in which we insert the keys.

Unfortunately, bad things can happen when building trees through insertion. The data structure has no control over the order of insertion. Consider what happens if the user inserts the keys in sorted order. The operations insert(a), followed by insert(b), insert(c), insert(d), ... will produce a skinny, linear-height tree where only right pointers are used.

Thus, binary trees can have heights ranging from $\lg n$ to n. But how tall are they on average? The average case analysis of algorithms can be tricky because we must carefully specify what we mean by *average*. The question is well defined if we assume each of the $n!$ possible insertion orderings to be equally likely, and average over those. If this assumption is valid then we are in luck, because with high probability the resulting tree will have $\Theta(\log n)$ height. This will be shown in Section 4.6 (page 130).

This argument is an important example of the power of *randomization*. We can often develop simple algorithms that offer good performance with high probability. We will see that a similar idea underlies the fastest known sorting algorithm, quicksort.

3.4.3   Balanced Search Trees

Random search trees are *usually* good. But if we get unlucky with our order of insertion, we can end up with a linear-height tree in the worst case. This worst case is outside of our direct control, since we must build the tree in response to the requests given by our potentially nasty user.

What would be better is an insertion/deletion procedure that *adjusts* the tree a little after each insertion, keeping it close enough to be balanced that the maximum height is logarithmic. Sophisticated *balanced* binary search tree data structures have been developed that guarantee the height of the tree always to be $O(\log n)$. Therefore, all dictionary operations (insert, delete, query) take $O(\log n)$ time each. Implementations of balanced tree data structures such as red–black trees and splay trees are discussed in Section 15.1 (page 440).

From an algorithm design viewpoint, it is important to know that these trees exist and that they can be used as black boxes to provide an efficient dictionary implementation. When figuring the costs of dictionary operations for algorithm analysis, we can assume the worst-case complexities of balanced binary trees to be a fair measure.

> *Take-Home Lesson:*   Picking the wrong data structure for the job can be disastrous in terms of performance. Identifying the very best data structure is usually not as critical, because there can be several choices that perform in a similar manner.

Stop and Think: Exploiting Balanced Search Trees

Problem: You are given the task of reading n numbers and then printing them out in sorted order. Suppose you have access to a balanced dictionary data structure, which supports the operations search, insert, delete, minimum, maximum, successor, and predecessor each in $O(\log n)$ time. How can you sort in $O(n \log n)$ time using only:

1. insert and in-order traversal?

2. minimum, successor, and insert?

3. minimum, insert, and delete?

Solution: Every algorithm for sorting items using a binary search tree *has* to start by building the actual tree. This involves initializing the tree (basically setting the pointer t to NULL), and then reading/inserting each of the n items into t. This costs $O(n \log n)$, since each insertion takes at most $O(\log n)$ time. Curiously, just building the data structure is a rate-limiting step for each of our sorting algorithms!

The first problem allows us to do insertion and in-order traversal. We can build a search tree by inserting all n elements, then do a traversal to access the items in sorted order.

The second problem allows us to use the minimum and successor operations after constructing the tree. We can start from the minimum element, and then repeatedly find the successor to traverse the elements in sorted order.

The third problem does not give us successor, but does allow us delete. We can repeatedly find and delete the minimum element to once again traverse all the elements in sorted order.

In summary, the solutions to the three problems are:

```
Sort1()                 Sort2()                 Sort3()
    initialize-tree(t)      initialize-tree(t)      initialize-tree(t)
    While (not EOF)         While (not EOF)         While (not EOF)
        read(x);               read(x);                read(x);
        insert(x,t)            insert(x,t);            insert(x,t);
    Traverse(t)            y = Minimum(t)          y = Minimum(t)
                           While (y ≠ NULL) do     While (y ≠ NULL) do
                               print(y→item)           print(y→item)
                               y = Successor(y,t)      Delete(y,t)
                                                       y = Minimum(t)
```

Each of these algorithms does a linear number of logarithmic-time operations, and hence runs in $O(n \log n)$ time. The key to exploiting balanced binary search trees is using them as black boxes. ∎

3.5   Priority Queues

Many algorithms need to process items in a specific order. For example, suppose you must schedule jobs according to their importance relative to other jobs. Such scheduling requires sorting the jobs by importance, and then processing them in this sorted order.

The *priority queue* is an abstract data type that provides more flexibility than simple sorting, because it allows new elements to enter a system at arbitrary intervals. It can be much more cost-effective to insert a new job into a priority queue than to re-sort everything on each such arrival.

The basic priority queue supports three primary operations:

- *Insert(Q,x)*– Given item x, insert it into the priority queue Q.

- *Find-Minimum(Q)* or *Find-Maximum(Q)*– Return a pointer to the item whose key value is smallest (or largest) among all keys in priority queue Q.

- *Delete-Minimum(Q)* or *Delete-Maximum(Q)*– Remove the item whose key value is minimum (or maximum) from priority queue Q.

Naturally occurring processes are often informally modeled by priority queues. Single people maintain a priority queue of potential dating candidates, mentally if not explicitly. One's impression on meeting a new person maps directly to an attractiveness or desirability score, which serves as the *key* field for inserting this new entry into the "little black book" priority queue data structure. Dating is the process of extracting the most desirable person from the data structure (*Find-Maximum*), spending an evening to evaluate them better, and then reinserting them into the priority queue with a possibly revised score.

Take-Home Lesson: Building algorithms around data structures such as dictionaries and priority queues leads to both clean structure and good performance.

Stop and Think: Basic Priority Queue Implementations

Problem: What is the worst-case time complexity of the three basic priority queue operations (insert, find-minimum, and delete-minimum) when the basic data structure is as follows:

- An unsorted array.

- A sorted array.

- A balanced binary search tree.

Solution: There is surprising subtlety when implementing these operations, even using a data structure as simple as an unsorted array. The unsorted array dictionary (discussed on page 77) implements insertion and deletion in constant time, and search and minimum in linear time. A linear-time implementation of delete-minimum can be composed from *find-minimum*, followed by *delete*.

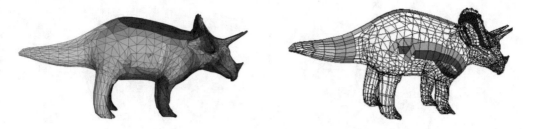

Figure 3.6: A triangulated model of a dinosaur (l), with several triangle strips peeled off the model (r).

For sorted arrays, we can implement insert and delete in linear time, and minimum in constant time. However, priority queue deletions involve only the minimum element. By storing the sorted array in reverse order (largest value on top), the minimum element will always be the last one in the array. Deleting the tail element requires no movement of any items, just decrementing the number of remaining items n, and so delete-minimum can be implemented in constant time.

All this is fine, yet the table below claims we can implement *find-minimum* in constant time for each data structure:

	Unsorted array	Sorted array	Balanced tree
Insert(Q, x)	$O(1)$	$O(n)$	$O(\log n)$
Find-Minimum(Q)	$O(1)$	$O(1)$	$O(1)$
Delete-Minimum(Q)	$O(n)$	$O(1)$	$O(\log n)$

The trick is using an extra variable to store a pointer/index to the minimum entry in each of these structures, so we can simply return this value whenever we are asked to find-minimum. Updating this pointer on each insertion is easy—we update it iff the newly inserted value is less than the current minimum. But what happens on a delete-minimum? We can delete that minimum element we point to, and then do a search to restore this canned value. The operation to identify the new minimum takes linear time on an unsorted array and logarithmic time on a tree, and hence can be folded into the cost of each deletion. ∎

Priority queues are very useful data structures. Indeed, they will be the hero of two of our war stories. A particularly nice priority queue implementation (the heap) will be discussed in the context of sorting in Section 4.3 (page 115). Further, a complete set of priority queue implementations is presented in Section 15.2 (page 445) of the catalog.

3.6 War Story: Stripping Triangulations

Geometric models used in computer graphics are commonly represented by a triangulated surface, as shown in Figure 3.6(l). High-performance rendering

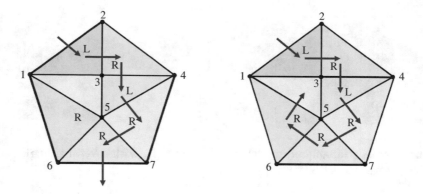

Figure 3.7: Extracting a triangle strip from a triangular mesh: A strip with partial coverage using alternating left and right turns (left), and a strip with complete coverage by exploiting the flexibility of arbitrary turns (right).

engines have special hardware for rendering and shading triangles. This rendering hardware is so fast that the computational bottleneck becomes the cost of feeding the triangulation structure into the hardware engine.

Although each triangle can be described by specifying its three endpoints, an alternative representation proves more efficient. Instead of specifying each triangle in isolation, suppose that we partition the triangles into *strips* of adjacent triangles and walk along the strip. Since each triangle shares two vertices in common with its neighbors, we save the cost of retransmitting the two extra vertices and any associated information. To make the description of the triangles unambiguous, the *OpenGL* triangular-mesh renderer assumes that all turns alternate left and right. The strip in Figure 3.7 (left) completely describes five triangles of the mesh with the vertex sequence *[1,2,3,4,5,7,6]* and the implied left/right order. The strip on the right describes all seven triangles with specified turns: *[1,2,3,l-4,r-5,l-7,r-6,r-1,r-3]*.

The task was to find a small number of strips that together cover all the triangle in a mesh, without overlap. This can be thought of as a graph problem. The graph of interest has a vertex for every *triangle* of the mesh, and an edge between every pair of vertices representing adjacent triangles. This *dual graph* representation captures all the information about the triangulation (see Section 18.12 (page 581)) needed to partition it into strips.

Once we had the dual graph, the project could begin in earnest. We sought to partition the dual graph's vertices into as few paths or strips as possible. Partitioning it into one path implied that we had discovered a Hamiltonian path, which by definition visits each vertex exactly once. Since finding a Hamiltonian path is NP-complete (see Section 19.5 (page 598)), we knew not to look for an optimal algorithm, but concentrate instead on heuristics.

The simplest approach for strip cover would start from an arbitrary triangle and then do a left–right walk until the walk ends, either by hitting the boundary of the object or a previously visited triangle. This heuristic had the advantage

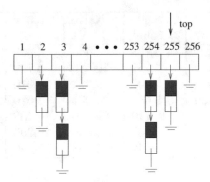

Figure 3.8: A bounded-height priority queue for triangle strips.

that it would be fast and simple, although there is no reason to expect that it must find the smallest possible set of left–right strips for a given triangulation.

The *greedy* heuristic should result in a relatively small number of strips, however. Greedy heuristics always try to grab the best possible thing first. In the case of the triangulation, the natural greedy heuristic would be to identify the starting triangle that yields the longest left–right strip, and peel that one off first.

Being greedy does not guarantee you the best possible solution overall, since the first strip you peel off might break apart a lot of potential strips we would have wanted to use later. Still, being greedy is a good rule of thumb if you want to get rich. Removing the longest strip leaves the fewest number of triangles remaining for later strips, so greedy should outperform the naive heuristic of pick anything.

But how much time does it take to find the largest strip to peel off next? Let k be the length of the walk possible from an average vertex. Using the simplest possible implementation, we could walk from each of the n vertices to find the largest remaining strip to report in $O(kn)$ time. Repeating this for each of the roughly n/k strips we extract yields an $O(n^2)$-time implementation, which would be hopelessly slow on even a small model of 20,000 triangles.

How could we speed this up? It seems wasteful to re-walk from each triangle after deleting a single strip. We could maintain the lengths of all the possible future strips in a data structure. However, whenever we peel off a strip, we must update the lengths of all affected strips. These strips will be shortened because they walked through a triangle that now no longer exists. There are two aspects of such a data structure:

- *Priority queue* – Since we were repeatedly identifying the longest remaining strip, we needed a priority queue to store the strips ordered according to length. The next strip to peel always sits at the top of the queue. Our priority queue had to permit reducing the priority of arbitrary elements of the queue whenever we updated the strip lengths to reflect what triangles were peeled away. Because all of the strip lengths were bounded by a fairly

Model name	Triangle count	Naive cost	Greedy cost	Greedy time
Diver	3,798	8,460	4,650	6.4 sec
Heads	4,157	10,588	4,749	9.9 sec
Framework	5,602	9,274	7,210	9.7 sec
Bart Simpson	9,654	24,934	11,676	20.5 sec
Enterprise	12,710	29,016	13,738	26.2 sec
Torus	20,000	40,000	20,200	272.7 sec
Jaw	75,842	104,203	95,020	136.2 sec

Figure 3.9: A comparison of the naive and greedy heuristics for several triangular meshes. Cost is the number of strips. Running time generally scales with triangle count, except for the highly symmetric torus with very long strips.

small integer (hardware constraints prevent any strip from having more than 256 vertices), we used a bounded-height priority queue (an array of buckets, shown in Figure 3.8 and described in Section 15.2 (page 445)). An ordinary heap would also have worked just fine.

To update the queue entry associated with each triangle, we needed to quickly find where it was. This meant that we also needed a ...

- *Dictionary* – For each triangle in the mesh, we had to find where it was in the queue. This meant storing a pointer to each triangle in a dictionary. By integrating this dictionary with the priority queue, we built a data structure capable of a wide range of operations.

Although there were various other complications, such as quickly recalculating the length of the strips affected by the peeling, the key idea needed to obtain better performance was to use the priority queue. Run time improved by several orders of magnitude after employing this data structure.

How much better did the greedy heuristic do than the naive heuristic? Study the table in Figure 3.9. In all cases, the greedy heuristic led to a set of strips that cost less, as measured by the total number of vertex occurrences in the strips. The savings ranged from about 10% to 50%, which is quite remarkable since the greatest possible improvement (going from three vertices per triangle down to one) yields a savings of only 66.6%.

After implementing the greedy heuristic with our priority queue data structure, the program ran in $O(n \cdot k)$ time, where n is the number of triangles and k is the length of the average strip. Thus, the torus, which consisted of a small number of very long strips, took longer than the jaw, even though the latter contained over three times as many triangles.

There are several lessons to be gleaned from this story. First, when working with a large enough data set, only linear or near-linear algorithms are likely to be fast enough. Second, choosing the right data structure is often the key to getting the time complexity down. Finally, using the greedy heuristic can significantly improve performance over the naive approach. How much this improvement will be can only be determined by experimentation.

3.7 Hashing

Hash tables are a *very* practical way to maintain a dictionary. They exploit the fact that looking an item up in an array takes constant time once you have its index. A *hash function* is a mathematical function that maps keys to integers. We will use the value of our hash function as an index into an array, and store our item at that position.

The first step of the hash function is usually to map each key (here the string S) to a big integer. Let α be the size of the alphabet on which S is written. Let char(c) be a function that maps each symbol of the alphabet to a unique integer from 0 to $\alpha - 1$. The function

$$H(S) = \alpha^{|S|} + \sum_{i=0}^{|S|-1} \alpha^{|S|-(i+1)} \times char(s_i)$$

maps each string to a unique (but large) integer by treating the characters of the string as "digits" in a base-α number system.

This creates unique identifier numbers, but they are so large they will quickly exceed the number of desired slots in our hash table (denoted by m). We must reduce this number to an integer between 0 and $m-1$, by taking the remainder $H'(S) = H(S) \bmod m$. This works on the same principle as a roulette wheel. The ball travels a long distance, around and around the circumference-m wheel $\lfloor H(S)/m \rfloor$ times before settling down to a random bin. If the table size is selected with enough finesse (ideally m is a large prime not too close to $2^i - 1$), the resulting hash values should be fairly uniformly distributed.

3.7.1 Collision Resolution

No matter how good our hash function is, we had better be prepared for collisions, because two distinct keys will at least occasionally hash to the same value. There are two different approaches for maintaining a hash table:

- *Chaining* represents a hash table as an array of m linked lists ("buckets"), as shown in Figure 3.10. The ith list will contain all the items that hash to the value of i. Search, insertion, and deletion thus reduce to the corresponding problem in linked lists. If the n keys are distributed uniformly in a table, each list will contain roughly n/m elements, making them a constant size when $m \approx n$.

  Chaining is very natural, but devotes a considerable amount of memory to pointers. This is space that could be used to make the table larger, which reduces the likelihood of collisions. In fact, the highest-performing hash tables generally rely on an alternative method called open addressing.

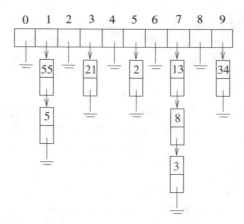

Figure 3.10: Collision resolution by chaining, after hashing the first eight Fibonacci numbers in increasing order, with hash function $H(x) = (2x + 1)$ mod 10. Insertions occur at the head of each list in this figure.

0	1	2	3	4	5	6	7	8	9
34	5	55	21		2		3	8	13

Figure 3.11: Collision resolution by open addressing and sequential probing, after inserting the first eight Fibonacci numbers in increasing order with $H(x) = (2x + 1)$ mod 10. The red elements have been bumped to the first open slot after the desired location.

- *Open addressing* maintains the hash table as a simple array of elements (not buckets). Each cell is initialized to null, as shown in Figure 3.11. On each insertion, we check to see whether the desired cell is empty; if so, we insert the item there. But if the cell is already occupied, we must find some other place to put the item. The simplest possibility (called *sequential probing*) inserts the item into the next open cell in the table. Provided the table is not too full, the contiguous runs of non-empty cells should be fairly short, hence this location *should* be only a few cells away from its intended position.

  Searching for a given key now involves going to the appropriate hash value and checking to see if the item there is the one we want. If so, return it. Otherwise we must keep checking through the length of the run. Deletion in an open addressing scheme can get ugly, since removing one element might break a chain of insertions, making some elements inaccessible. We have no alternative but to reinsert all the items in the run that follows the new hole.

Chaining and open addressing both cost $O(m)$ to initialize an m-element hash table to null elements prior to the first insertion.

When using chaining with doubly linked lists to resolve collisions in an m-element hash table, the dictionary operations for n items can be implemented in the following expected and worst case times:

	Hash table (expected)	Hash table (worst case)
Search(L, k)	$O(n/m)$	$O(n)$
Insert(L, x)	$O(1)$	$O(1)$
Delete(L, x)	$O(1)$	$O(1)$
Successor(L, x)	$O(n+m)$	$O(n+m)$
Predecessor(L, x)	$O(n+m)$	$O(n+m)$
Minimum(L)	$O(n+m)$	$O(n+m)$
Maximum(L)	$O(n+m)$	$O(n+m)$

Traversing all the elements in the table takes $O(n+m)$ time for chaining, since we have to scan all m buckets looking for elements, even if the actual number of inserted items is small. This reduces to $O(m)$ time for open addressing, since n must be at most m.

Pragmatically, a hash table often is the best data structure to maintain a dictionary. The applications of hashing go far beyond dictionaries, however, as we will see below.

3.7.2 Duplicate Detection via Hashing

The key idea of hashing is to represent a large object (be it a key, a string, or a substring) by a single number. We get a representation of the large object by a value that can be manipulated in constant time, such that it is relatively unlikely that two different large objects map to the same value.

Hashing has a variety of clever applications beyond just speeding up search. I once heard Udi Manber—at one point responsible for all search products at Google—talk about the algorithms employed in industry. The three most important algorithms, he explained, were "hashing, hashing, and hashing."

Consider the following problems with nice hashing solutions:

- *Is a given document unique within a large corpus?* – Google crawls yet another webpage. How can it tell whether this is new content never seen before, or just a duplicate page that exists elsewhere on the web?

  Explicitly comparing the new document D against all n previous documents is hopelessly inefficient for a large corpus. However, we can hash D to an integer, and compare $H(D)$ to the hash codes of the rest of the corpus. Only if there is a collision *might* D be a possible duplicate. Since we expect few spurious collisions, we can explicitly compare the few documents sharing a particular hash code with little total effort.

- *Is part of this document plagiarized?* – A lazy student copies a portion of a web document into their term paper. "The web is a big place," he smirks. "How will anyone ever find the page I stole this from?"

This is a more difficult problem than the previous application. Adding, deleting, or changing even one character from a document will completely change its hash code. The hash codes produced in the previous application thus cannot help for this more general problem.

However, we *could* build a hash table of all overlapping windows (substrings) of length w in all the documents in the corpus. Whenever there is a match of hash codes, there is likely a common substring of length w between the two documents, which can then be further investigated. We should choose w to be long enough so such a co-occurrence is very unlikely to happen by chance.

The biggest downside of this scheme is that the size of the hash table becomes as large as the document corpus itself. Retaining a small but well-chosen subset of these hash codes is exactly the goal of min-wise hashing, discussed in Section 6.6.

- *How can I convince you that a file isn't changed?* – In a closed-bid auction, each party submits their bid in secret before the announced deadline. If you knew what the other parties were bidding, you could arrange to bid $1 more than the highest opponent and walk off with the prize as cheaply as possible. The "right" auction strategy would thus be to hack into the computer containing the bids just prior to the deadline, read the bids, and then magically emerge as the winner.

  How can this be prevented? What if everyone submits a hash code of their actual bid prior to the deadline, and then submits the full bid only after the deadline? The auctioneer will pick the largest full bid, but checks to make sure the hash code matches what was submitted prior to the deadline. Such *cryptographic hashing* methods provide a way to ensure that the file you give me today is the same as the original, because any change to the file will change the hash code.

Although the worst-case bounds on anything involving hashing are dismal, with a proper hash function we can confidently expect good behavior. Hashing is a fundamental idea in randomized algorithms, yielding linear expected-time algorithms for problems that are $\Theta(n \log n)$ or $\Theta(n^2)$ in the worst case.

3.7.3   Other Hashing Tricks

Hash functions provide useful tools for many things beyond powering hash tables. The fundamental idea is of many-to-one mappings, where *many* is controlled so it is very unlikely to be *too many*.

3.7.4   Canonicalization

Consider a word game that gives you a set of letters S, and asks you to find all dictionary words that can be made by reordering them. For example, I can make three words from the four letters in $S = (a, e, k, l)$, namely *kale*, *lake*, and *leak*.

Think how you might write a program to find the matching words for S, given a dictionary D of n words. Perhaps the most straightforward approach is to test each word $d \in D$ against the characters of S. This takes time linear in n for each S, and the test is somewhat tricky to program.

What if we instead hash every word in D to a string, by sorting the word's letters. Now *kale* goes to *aekl*, as do *lake* and *leak*. By building a hash table with the sorted strings as keys, all words with the same letter distribution get hashed to the same bucket. Once you have built this hash table, you can use it for different query sets S. The time for each query will be proportional to the number of matching words in D, which is a lot smaller than n.

Which set of k letters can be used to make the most dictionary words? This seems like a much harder problem, because there are α^k possible letter sets, where α is the size of the alphabet. But observe that the answer is simply the hash code with the largest number of collisions. Sweeping over a sorted array of hash codes (or walking through each bucket in a chained hash table) makes this fast and easy.

This is a good example of the power of *canonicalization*, reducing complicated objects to a standard (i.e. "canonical") form. String transformations like reducing letters to lower case or stemming (removing word suffixes like -ed, -s, or -ing) result in increased matches, because multiple strings collide on the same code. *Soundex* is a canonicalization scheme for names, so spelling variants of "Skiena" like "Skina,", "Skinnia," and "Schiena" all get hashed to the same Soundex code, *S25*. Soundex is described in more detail in Section 21.4.

For hash tables, collisions are very bad. But for pattern matching problems like these, collisions are exactly what we want.

3.7.5 Compaction

Suppose that you wanted to sort all n books in the library, not by their titles but by the contents of the actual text. Bulwer-Lytton's [BL30] "It was a dark and stormy night..." would appear before this book's "What is an algorithm?..." Assuming the average book is $m \approx 100{,}000$ words long, doing this sort seems an expensive and clumsy job since each comparison involves two books.

But suppose we instead represent each book by the first (say) 100 characters, and sort these strings. There will be collisions involving duplicates of the same prefix, involving multiple editions or perhaps plagiarism, but these will be quite rare. After sorting the prefixes, we can then resolve the collisions by comparing the full texts. The world's fastest sorting programs use this idea, as discussed in Section 17.1.

This is an example of hashing for compaction, also called *fingerprinting*, where we representing large objects by small hash codes. It is easier to work with small objects than large ones, and the hash code generally preserves the identity of each item. The hash function here is trivial (just take the prefix) but it is designed to accomplish a specific goal—not to maintain a hash table. More sophisticated hash functions can make the probability of collisions between even slightly different objects vanishingly low.

3.8 Specialized Data Structures

The basic data structures described thus far all represent an unstructured set of items so as to facilitate retrieval operations. These data structures are well known to most programmers. Not as well known are data structures for representing more specialized kinds of objects, such as points in space, strings, and graphs.

The design principles of these data structures are the same as for basic objects. There exists a set of basic operations we need to perform repeatedly. We seek a data structure that allows these operations to be performed very efficiently. These specialized data structures are important for efficient graph and geometric algorithms, so one should be aware of their existence:

- *String data structures* – Character strings are typically represented by arrays of characters, perhaps with a special character to mark the end of the string. Suffix trees/arrays are special data structures that preprocess strings to make pattern matching operations faster. See Section 15.3 (page 448) for details.

- *Geometric data structures* – Geometric data typically consists of collections of data points and regions. Regions in the plane can be described by polygons, where the boundary of the polygon is a closed chain of line segments. A polygon P can be represented using an array of points $(v_1, \ldots, v_n, v_1)$, such that (v_i, v_{i+1}) is a segment of the boundary of P. Spatial data structures such as kd-trees organize points and regions by geometric location to support fast search operations. See Section 15.6 (page 460).

- *Graph data structures* – Graphs are typically represented using either adjacency matrices or adjacency lists. The choice of representation can have a substantial impact on the design of the resulting graph algorithms, and will be discussed in Chapter 7 and in the catalog in Section 15.4.

- *Set data structures* – Subsets of items are typically represented using a dictionary to support fast membership queries. Alternatively, *bit vectors* are Boolean arrays such that the ith bit is 1 if i is in the subset. Data structures for manipulating sets is presented in the catalog in Section 15.5.

3.9 War Story: String 'em Up

The human genome encodes all the information necessary to build a person. Sequencing the genome has had an enormous impact on medicine and molecular

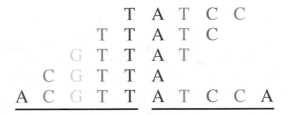

Figure 3.12: The concatenation of two fragments can be in S only if all sub-fragments are.

biology. Algorists like me have become interested in bioinformatics, for several reasons:

- DNA sequences can be accurately represented as strings of characters on the four-letter alphabet (A,C,T,G). The needs of biologist have sparked new interest in old algorithmic problems such as string matching (see Section 21.3 (page 685)) as well as creating new problems such as shortest common superstring (see Section 21.9 (page 709)).

- DNA sequences are very *long* strings. The human genome is approximately three billion base pairs (or characters) long. Such large problem size means that asymptotic (Big Oh) complexity analysis is fully justified on biological problems.

- Enough money is being invested in genomics for computer scientists to want to claim their piece of the action.

One of my interests in computational biology revolved around a proposed technique for DNA sequencing called sequencing by hybridization (SBH). This procedure attaches a set of probes to an array, forming a *sequencing chip*. Each of these probes determines whether or not the probe string occurs as a substring of the DNA target. The target DNA can now be sequenced based on the constraints of which strings are (and are not) substrings of the target.

We sought to identify all the strings of length $2k$ that are possible substrings of an unknown string S, given the set of all length-k substrings of S. For example, suppose we know that AC, CA, and CC are the only length-2 substrings of S. It is possible that $ACCA$ is a substring of S, since the center substring is one of our possibilities. However, $CAAC$ *cannot* be a substring of S, since AA is not a substring of S. We needed to find a fast algorithm to construct all the consistent length-$2k$ strings, since S could be very long.

The simplest algorithm to build the $2k$ strings would be to concatenate all $O(n^2)$ pairs of k-strings together, and then test to make sure that all $(k-1)$ length-k substrings spanning the boundary of the concatenation were in fact substrings, as shown in Figure 3.12. For example, the nine possible concatenations of AC, CA, and CC are $ACAC$, $ACCA$, $ACCC$, $CAAC$, $CACA$, $CACC$,

$CCAC$, $CCCA$, and $CCCC$. Only $CAAC$ can be eliminated, because of the absence of AA as a substring of S.

We needed a fast way of testing whether the $k - 1$ substrings straddling the concatenation were members of our dictionary of permissible k-strings. The time it takes to do this depends upon which dictionary data structure we use. A binary search tree could find the correct string within $O(\log n)$ comparisons, where each comparison involved testing which of two length-k strings appeared first in alphabetical order. The total time using such a binary search tree would be $O(k \log n)$.

That seemed pretty good. So my graduate student, Dimitris Margaritis, used a binary search tree data structure for our implementation. It worked great up until the moment we ran it.

"I've tried the fastest computer we have, but our program is too slow," Dimitris complained. "It takes forever on string lengths of only 2,000 characters. We will never get up to $n = 50,000$."

We profiled our program and discovered that almost all the time was spent searching in this data structure. This was no surprise, since we did this $k - 1$ times for each of the $O(n^2)$ possible concatenations. We needed a faster dictionary data structure, since search was the innermost operation in such a deep loop.

"How about using a hash table?" I suggested. "It should take $O(k)$ time to hash a k-character string and look it up in our table. That should knock off a factor of $O(\log n)$."

Dimitris went back and implemented a hash table implementation for our dictionary. Again, it worked great, up until the moment we ran it.

"Our program is still too slow," Dimitris complained. "Sure, it is now about ten times faster on strings of length 2,000. So now we can get up to about 4,000 characters. Big deal. We will never get up to 50,000."

"We should have expected this," I mused. "After all, $\lg_2(2,000) \approx 11$. We need a faster data structure to search in our dictionary of strings."

"But what can be faster than a hash table?" Dimitris countered. "To look up a k-character string, you must read all k characters. Our hash table already does $O(k)$ searching."

"Sure, it takes k comparisons to test the first substring. But maybe we can do better on the second test. Remember where our dictionary queries are coming from. When we concatenate $ABCD$ with $EFGH$, we are first testing whether $BCDE$ is in the dictionary, then $CDEF$. These strings differ from each other by only one character. We should be able to exploit this so each subsequent test takes constant time to perform..."

"We can't do that with a hash table," Dimitris observed. "The second key is not going to be anywhere near the first in the table. A binary search tree won't help, either. Since the keys $ABCD$ and $BCDE$ differ according to the first character, the two strings will be in different parts of the tree."

"But we can use a suffix tree to do this," I countered. "A suffix tree is a trie containing all the suffixes of a given set of strings. For example, the suffixes of $ACAC$ are $\{ACAC, CAC, AC, C\}$. Coupled with suffixes of string $CACT$,

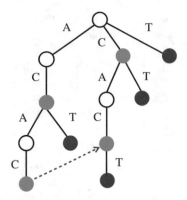

Figure 3.13: Suffix tree on $ACAC$ and $CACT$, with the pointer to the suffix of $ACAC$. Green nodes correspond to suffixes of $ACAC$, with blue nodes to suffixes of $CACT$.

we get the suffix tree of Figure 3.13. By following a pointer from $ACAC$ to its longest proper suffix CAC, we get to the right place to test whether $CACT$ is in our set of strings. One character comparison is all we need to do from there."

Suffix trees are amazing data structures, discussed in considerably more detail in Section 15.3 (page 448). Dimitris did some reading about them, then built a nice suffix tree implementation for our dictionary. Once again, it worked great up until the moment we ran it.

"Now our program is faster, but it runs out of memory," Dimitris complained. "The suffix tree builds a path of length k for each suffix of length k, so all told there can be $\Theta(n^2)$ nodes in the tree. It crashes when we go beyond 2,000 characters. We will never get up to strings with 50,000 characters."

I wasn't ready to give up yet. "There is a way around the space problem, by using compressed suffix trees," I recalled. "Instead of explicitly representing long paths of character nodes, we can refer back to the original string." Compressed suffix trees always take linear space, as described in Section 15.3 (page 448).

Dimitris went back one last time and implemented the compressed suffix tree data structure. *Now* it worked great! As shown in Figure 3.14, we ran our simulation for strings of length $n = 65,536$ without incident. Our results showed that interactive SBH could be a very efficient sequencing technique. Based on these simulations, we were able to arouse interest in our technique from biologists. Making the actual wet laboratory experiments feasible provided another computational challenge, which is reported in Section 12.8 (page 414).

The take-home lessons for programmers should be apparent. We isolated a single operation (dictionary string search) that was being performed repeatedly and optimized the data structure to support it. When an improved dictionary structure still did not suffice, we looked deeper into the kind of queries we were performing, so that we could identify an even better data structure. Finally, we didn't give up until we had achieved the level of performance we needed. In algorithms, as in life, persistence usually pays off.

String length	Binary tree	Hash table	Suffix tree	Compressed tree
8	0.0	0.0	0.0	0.0
16	0.0	0.0	0.0	0.0
32	0.1	0.0	0.0	0.0
64	0.3	0.4	0.3	0.0
128	2.4	1.1	0.5	0.0
256	17.1	9.4	3.8	0.2
512	31.6	67.0	6.9	1.3
1,024	1,828.9	96.6	31.5	2.7
2,048	11,441.7	941.7	553.6	39.0
4,096	> 2 days	5,246.7	out of	45.4
8,192		> 2 days	memory	642.0
16,384				1,614.0
32,768				13,657.8
65,536				39,776.9

Figure 3.14: Run times (in seconds) for the SBH simulation using various data structures, as a function of the string length n.

Chapter Notes

Optimizing hash table performance is surprisingly complicated for such a conceptually simple data structure. The importance of short runs in open addressing has led to more sophisticated schemes than sequential probing for optimal hash table performance. For more details, see Knuth [Knu98].

My thinking on hashing was profoundly influenced by a talk by Mihai Pătraşcu, a brilliant theoretician who sadly died before he turned 30. More detailed treatments on hashing and randomized algorithms include Motwani and Raghavan [MR95] and Mitzenmacher and Upfal [MU17].

Our triangle strip optimizing program, *stripe*, is described in Evans et al. [ESV96]. Hashing techniques for plagiarism detection are discussed in Schlieimer et al. [SWA03].

Surveys of algorithmic issues in DNA sequencing by hybridization include Chetverin and Kramer [CK94] and Pevzner and Lipshutz [PL94]. Our work on interactive SBH reported in the war story is reported in Margaritis and Skiena [MS95a].

3.10 Exercises

Stacks, Queues, and Lists

3-1. *[3]* A common problem for compilers and text editors is determining whether the parentheses in a string are balanced and properly nested. For example, the string ((())()() contains properly nested pairs of parentheses, while the strings)()(and () do not. Give an algorithm that returns true if a string contains properly nested and balanced parentheses, and false if otherwise. For full credit, identify the position of the first offending parenthesis if the string is not properly nested and balanced.

3-2. *[5]* Give an algorithm that takes a string S consisting of opening and closing parentheses, say)()(())()())()))(, and finds the length of the longest balanced parentheses in S, which is 12 in the example above. (Hint: The solution is not necessarily a contiguous run of parenthesis from S.)

3-3. *[3]* Give an algorithm to reverse the direction of a given singly linked list. In other words, after the reversal all pointers should now point backwards. Your algorithm should take linear time.

3-4. *[5]* Design a stack S that supports $S.push(x)$, $S.pop()$, and $S.findmin()$, which returns the minimum element of S. All operations should run in constant time.

3-5. *[5]* We have seen how dynamic arrays enable arrays to grow while still achieving constant-time amortized performance. This problem concerns extending dynamic arrays to let them both grow and shrink on demand.

   (a) Consider an underflow strategy that cuts the array size in half whenever the array falls below half full. Give an example sequence of insertions and deletions where this strategy gives a bad amortized cost.

   (b) Then, give a better underflow strategy than that suggested above, one that achieves constant amortized cost per deletion.

3-6. *[3]* Suppose you seek to maintain the contents of a refrigerator so as to minimize food spoilage. What data structure should you use, and how should you use it?

3-7. *[5]* Work out the details of supporting constant-time deletion from a singly linked list as per the footnote from page 79, ideally to an actual implementation. Support the other operations as efficiently as possible.

Elementary Data Structures

3-8. *[5]* Tic-tac-toe is a game played on an $n \times n$ board (typically $n = 3$) where two players take consecutive turns placing "O" and "X" marks onto the board cells. The game is won if n consecutive "O" or "X" marks are placed in a row, column, or diagonal. Create a data structure with $O(n)$ space that accepts a sequence of moves, and reports in constant time whether the last move won the game.

3-9. *[3]* Write a function which, given a sequence of digits 2–9 and a dictionary of n words, reports all words described by this sequence when typed in on a standard telephone keypad. For the sequence 269 you should return *any*, *box*, *boy*, and *cow*, among other words.

3-10. *[3]* Two strings X and Y are anagrams if the letters of X can be rearranged
to form Y. For example, *silent/listen*, and *incest/insect* are anagrams. Give an
efficient algorithm to determine whether strings X and Y are anagrams.

Trees and Other Dictionary Structures

3-11. *[3]* Design a dictionary data structure in which search, insertion, and deletion
can all be processed in $O(1)$ time in the worst case. You may assume the set
elements are integers drawn from a finite set $1, 2, .., n$, and initialization can take
$O(n)$ time.

3-12. *[3]* The maximum depth of a binary tree is the number of nodes on the path
from the root down to the most distant leaf node. Give an $O(n)$ algorithm to
find the maximum depth of a binary tree with n nodes.

3-13. *[5]* Two elements of a binary search tree have been swapped by mistake. Give
an $O(n)$ algorithm to identify these two elements so they can be swapped back.

3-14. *[5]* Given two binary search trees, merge them into a doubly linked list in sorted
order.

3-15. *[5]* Describe an $O(n)$-time algorithm that takes an n-node binary search tree
and constructs an equivalent height-balanced binary search tree. In a height-
balanced binary search tree, the difference between the height of the left and
right subtrees of every node is never more than 1.

3-16. *[3]* Find the storage efficiency ratio (the ratio of data space over total space)
for each of the following binary tree implementations on n nodes:

  (a) All nodes store data, two child pointers, and a parent pointer. The data
      field requires 4 bytes and each pointer requires 4 bytes.

  (b) Only leaf nodes store data; internal nodes store two child pointers. The
      data field requires four bytes and each pointer requires two bytes.

3-17. *[5]* Give an $O(n)$ algorithm that determines whether a given n-node binary tree
is height-balanced (see Problem 3-15).

3-18. *[5]* Describe how to modify any balanced tree data structure such that search,
insert, delete, minimum, and maximum still take $O(\log n)$ time each, but successor and predecessor now take $O(1)$ time each. Which operations have to be
modified to support this?

3-19. *[5]* Suppose you have access to a balanced dictionary data structure that supports each of the operations search, insert, delete, minimum, maximum, successor, and predecessor in $O(\log n)$ time. Explain how to modify the insert
and delete operations so they still take $O(\log n)$ but now minimum and maximum take $O(1)$ time. (Hint: think in terms of using the abstract dictionary
operations, instead of mucking about with pointers and the like.)

3-20. *[5]* Design a data structure to support the following operations:

  • *insert(x, T)* – Insert item x into the set T.

  • *delete(k, T)* – Delete the kth smallest element from T.

  • *member(x, T)* – Return true iff $x \in T$.

All operations must take $O(\log n)$ time on an n-element set.

3-21. *[8]* A *concatenate* operation takes two sets S_1 and S_2, where every key in S_1 is smaller than any key in S_2, and merges them. Give an algorithm to concatenate two binary search trees into one binary search tree. The worst-case running time should be $O(h)$, where h is the maximal height of the two trees.

Applications of Tree Structures

3-22. *[5]* Design a data structure that supports the following two operations:

- *insert(x)* – Insert item x from the data stream to the data structure.
- *median()* – Return the median of all elements so far.

All operations must take $O(\log n)$ time on an n-element set.

3-23. *[5]* Assume we are given a standard dictionary (balanced binary search tree) defined on a set of n strings, each of length at most l. We seek to print out all strings beginning with a particular prefix p. Show how to do this in $O(ml \log n)$ time, where m is the number of strings.

3-24. *[5]* An array A is called k-unique if it does not contain a pair of duplicate elements within k positions of each other, that is, there is no i and j such that $A[i] = A[j]$ and $|j - i| \le k$. Design a worst-case $O(n \log k)$ algorithm to test if A is k-unique.

3-25. *[5]* In the *bin-packing problem*, we are given n objects, each weighing at most 1 kilogram. Our goal is to find the smallest number of bins that will hold the n objects, with each bin holding 1 kilogram at most.

- The *best-fit heuristic* for bin packing is as follows. Consider the objects in the order in which they are given. For each object, place it into the partially filled bin with the *smallest* amount of extra room after the object is inserted. If no such bin exists, start a new bin. Design an algorithm that implements the best-fit heuristic (taking as input the n weights $w_1, w_2, ..., w_n$ and outputting the number of bins used) in $O(n \log n)$ time.

- Repeat the above using the *worst-fit heuristic*, where we put the next object into the partially filled bin with the *largest* amount of extra room after the object is inserted.

3-26. *[5]* Suppose that we are given a sequence of n values $x_1, x_2, ..., x_n$ and seek to quickly answer repeated queries of the form: given i and j, find the smallest value in $x_i, \ldots, x_j$.

(a) Design a data structure that uses $O(n^2)$ space and answers queries in $O(1)$ time.

(b) Design a data structure that uses $O(n)$ space and answers queries in $O(\log n)$ time. For partial credit, your data structure can use $O(n \log n)$ space and have $O(\log n)$ query time.

3-27. *[5]* Suppose you are given an input set S of n integers, and a black box that if given any sequence of integers and an integer k instantly and correctly answers whether there is a subset of the input sequence whose sum is exactly k. Show how to use the black box $O(n)$ times to find a subset of S that adds up to k.

3-28. *[5]* Let $A[1..n]$ be an array of real numbers. Design an algorithm to perform any sequence of the following operations:

- *Add(i,y)* – Add the value y to the ith number.
- *Partial-sum(i)* – Return the sum of the first i numbers, that is, $\sum_{j=1}^{i} A[j]$.

There are no insertions or deletions; the only change is to the values of the numbers. Each operation should take $O(\log n)$ steps. You may use one additional array of size n as a work space.

3-29. *[8]* Extend the data structure of the previous problem to support insertions and deletions. Each element now has both a *key* and a *value*. An element is accessed by its key, but the addition operation is applied to the values. The *Partial_sum* operation is different.

- *Add(k,y)* – Add the value y to the item with key k.
- *Insert(k,y)* – Insert a new item with key k and value y.
- *Delete(k)* – Delete the item with key k.
- *Partial-sum(k)* – Return the sum of all the elements currently in the set whose key is less than k, that is, $\sum_{i<k} x_i$.

The worst-case running time should still be $O(n \log n)$ for any sequence of $O(n)$ operations.

3-30. *[8]* You are consulting for a hotel that has n one-bed rooms. When a guest checks in, they ask for a room whose number is in the range $[l, h]$. Propose a data structure that supports the following data operations in the allotted time:

(a) *Initialize(n)*:   Initialize the data structure for empty rooms numbered $1, 2, \ldots, n$, in polynomial time.

(b) *Count(l, h)*:   Return the number of available rooms in $[l, h]$, in $O(\log n)$ time.

(c) *Checkin(l, h)*:   In $O(\log n)$ time, return the first empty room in $[l, h]$ and mark it occupied, or return NIL if all the rooms in $[l, h]$ are occupied.

(d) *Checkout(x)*:   Mark room x as unoccupied, in $O(\log n)$ time.

3-31. *[8]* Design a data structure that allows one to search, insert, and delete an integer X in $O(1)$ time (i.e., constant time, independent of the total number of integers stored). Assume that $1 \leq X \leq n$ and that there are $m + n$ units of space available, where m is the maximum number of integers that can be in the table at any one time. (Hint: use two arrays $A[1..n]$ and $B[1..m]$.) You are not allowed to initialize either A or B, as that would take $O(m)$ or $O(n)$ operations. This means the arrays are full of random garbage to begin with, so you must be very careful.

Implementation Projects

3-32. *[5]* Implement versions of several different dictionary data structures, such as linked lists, binary trees, balanced binary search trees, and hash tables. Conduct experiments to assess the relative performance of these data structures in a simple application that reads a large text file and reports exactly one instance of each word that appears within it. This application can be efficiently implemented by maintaining a dictionary of all distinct words that have appeared thus far in the text and inserting/reporting each new word that appears in the stream. Write a brief report with your conclusions.

3-33. *[5]* A Caesar shift (see Section 21.6 (page 697)) is a very simple class of ciphers for secret messages. Unfortunately, they can be broken using statistical properties of English. Develop a program capable of decrypting Caesar shifts of sufficiently long texts.

Interview Problems

3-34. *[3]* What method would you use to look up a word in a dictionary?

3-35. *[3]* Imagine you have a closet full of shirts. What can you do to organize your shirts for easy retrieval?

3-36. *[4]* Write a function to find the middle node of a singly linked list.

3-37. *[4]* Write a function to determine whether two binary trees are identical. Identical trees have the same key value at each position and the same structure.

3-38. *[4]* Write a program to convert a binary search tree into a linked list.

3-39. *[4]* Implement an algorithm to reverse a linked list. Now do it without recursion.

3-40. *[5]* What is the best data structure for maintaining URLs that have been visited by a web crawler? Give an algorithm to test whether a given URL has already been visited, optimizing both space and time.

3-41. *[4]* You are given a search string and a magazine. You seek to generate all the characters in the search string by cutting them out from the magazine. Give an algorithm to efficiently determine whether the magazine contains all the letters in the search string.

3-42. *[4]* Reverse the words in a sentence—that is, "My name is Chris" becomes "Chris is name My." Optimize for time and space.

3-43. *[5]* Determine whether a linked list contains a loop as quickly as possible without using any extra storage. Also, identify the location of the loop.

3-44. *[5]* You have an unordered array X of n integers. Find the array M containing n elements where M_i is the product of all integers in X except for X_i. You may not use division. You can use extra memory. (Hint: there are solutions faster than $O(n^2)$.)

3-45. *[6]* Give an algorithm for finding an ordered word pair (e.g. "New York") occurring with the greatest frequency in a given webpage. Which data structures would you use? Optimize both time and space.

LeetCode

3-1. `https://leetcode.com/problems/validate-binary-search-tree/`

3-2. `https://leetcode.com/problems/count-of-smaller-numbers-after-self/`

3-3. `https://leetcode.com/problems/construct-binary-tree-from-preorder-and-inorder-traversal/`

HackerRank

3-1. `https://www.hackerrank.com/challenges/is-binary-search-tree/`

3-2. `https://www.hackerrank.com/challenges/queue-using-two-stacks/`

3-3. `https://www.hackerrank.com/challenges/detect-whether-a-linked-list-contains-a-cycle/problem`

Programming Challenges

These programming challenge problems with robot judging are available at `https://onlinejudge.org`:

3-1. "Jolly Jumpers"—Chapter 2, problem 10038.

3-2. "Crypt Kicker"—Chapter 2, problem 843.

3-3. "Where's Waldorf?"—Chapter 3, problem 10010.

3-4. "Crypt Kicker II"—Chapter 3, problem 850.

Chapter 4

Sorting

Typical computer science students study the basic sorting algorithms at least three times before they graduate: first in introductory programming, then in data structures, and finally in their algorithms course. Why is sorting worth so much attention? There are several reasons:

- Sorting is the basic building block that many other algorithms are built around. By understanding sorting, we obtain an amazing amount of power to solve other problems.

- Most of the interesting ideas used in the design of algorithms appear in the context of sorting, such as divide-and-conquer, data structures, and randomized algorithms.

- Sorting is the most thoroughly studied problem in computer science. Literally dozens of different algorithms are known, most of which possess some particular advantage over all other algorithms in certain situations.

This chapter will discuss sorting, stressing how sorting can be applied to solving other problems. In this sense, sorting behaves more like a data structure than a problem in its own right. I then give detailed presentations of several fundamental algorithms: heapsort, mergesort, quicksort, and distribution sort as examples of important algorithm design paradigms. Sorting is also represented by Section 17.1 (page 506) in the problem catalog.

4.1 Applications of Sorting

I will review several sorting algorithms and their complexities over the course of this chapter. But the punch line is this: clever sorting algorithms exist that run in $O(n \log n)$. This is a *big* improvement over naive $O(n^2)$ sorting algorithms, for large values of n. Consider the number of steps done by two different sorting algorithms for reasonable values of n:

S. S. Skiena, *The Algorithm Design Manual*, Texts in Computer Science,
https://doi.org/10.1007/978-3-030-54256-6_4

n	$n^2/4$	$n \lg n$
10	25	33
100	2,500	664
1,000	250,000	9,965
10,000	25,000,000	132,877
100,000	2,500,000,000	1,660,960

You might survive using a quadratic-time algorithm even if $n = 10,000$, but the slow algorithm clearly gets ridiculous once $n \geq 100,000$.

Many important problems can be reduced to sorting, so we can use our clever $O(n \log n)$ algorithms to do work that might otherwise seem to require a quadratic algorithm. An important algorithm design technique is to use sorting as a basic building block, because many other problems become easy once a set of items is sorted.

Consider the following applications:

- *Searching* – Binary search tests whether an item is in a dictionary in $O(\log n)$ time, provided the keys are all sorted. Search preprocessing is perhaps the single most important application of sorting.

- *Closest pair* – Given a set of n numbers, how do you find the pair of numbers that have the smallest difference between them? Once the numbers are sorted, the closest pair of numbers must lie next to each other somewhere in sorted order. Thus, a linear-time scan through the sorted list completes the job, for a total of $O(n \log n)$ time including the sorting.

- *Element uniqueness* – Are there any duplicates in a given set of n items? This is a special case of the closest-pair problem, where now we ask if there is a pair separated by a gap of zero. An efficient algorithm sorts the numbers and then does a linear scan checking all adjacent pairs.

- *Finding the mode* – Given a set of n items, which element occurs the largest number of times in the set? If the items are sorted, we can sweep from left to right and count the number of occurrences of each element, since all identical items will be lumped together after sorting.

  To find out how often an arbitrary element k occurs, look up k using binary search in a sorted array of keys. By walking to the left of this point until the first element is not k and then doing the same to the right, we can find this count in $O(\log n + c)$ time, where c is the number of occurrences of k. Even better, the number of instances of k can be found in $O(\log n)$ time by using binary search to look for the positions of both $k - \epsilon$ and $k + \epsilon$ (where ϵ is suitably small) and then taking the difference of these positions.

- *Selection* – What is the kth largest item in an array? If the keys are placed in sorted order, the kth largest item can be found in constant time because it must sit in the kth position of the array. In particular,

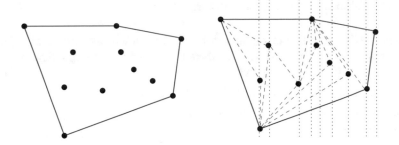

Figure 4.1: The convex hull of a set of points (left), constructed by left-to-right insertion (right).

the median element (see Section 17.3 (page 514)) appears in the $(n/2)$nd position in sorted order.

- *Convex hulls* – What is the polygon of smallest perimeter that contains a given set of n points in two dimensions? The convex hull is like a rubber band stretched around the points in the plane and then released. It shrinks to just enclose the points, as shown in Figure 4.1(l). The convex hull gives a nice representation of the shape of the point set and is an important building block for more sophisticated geometric algorithms, as discussed in the catalog in Section 20.2 (page 626).

  But how can we use sorting to construct the convex hull? Once you have the points sorted by x-coordinate, the points can be inserted from left to right into the hull. Since the right-most point is always on the boundary, we know that it must appear in the hull. Adding this new right-most point may cause others to be deleted, but we can quickly identify these points because they lie inside the polygon formed by adding the new point. See the example in Figure 4.1(r). These points will be neighbors of the previous point we inserted, so they will be easy to find and delete. The total time is linear after the sorting has been done.

While a few of these problems (namely median and selection) can be solved in linear time using more sophisticated algorithms, sorting provides quick and easy solutions to all of these problems. It is a rare application where the running time of sorting proves to be the bottleneck, especially a bottleneck that could have otherwise been removed using more clever algorithmics. Never be afraid to spend time sorting, provided you use an efficient sorting routine.

Take-Home Lesson: Sorting lies at the heart of many algorithms. Sorting the data is one of the first things any algorithm designer should try in the quest for efficiency.

Stop and Think: Finding the Intersection

Problem: Give an efficient algorithm to determine whether two sets (of size m and n, respectively) are disjoint. Analyze the worst-case complexity in terms of m and n, considering the case where $m \ll n$.

Solution: At least three algorithms come to mind, all of which are variants of sorting and searching:

- *First sort the big set* – The big set can be sorted in $O(n \log n)$ time. We can now do a binary search with each of the m elements in the second, looking to see if it exists in the big set. The total time will be $O((n + m) \log n)$.

- *First sort the small set* – The small set can be sorted in $O(m \log m)$ time. We can now do a binary search with each of the n elements in the big set, looking to see if it exists in the small one. The total time will be $O((n + m) \log m)$.

- *Sort both sets* – Observe that once the two sets are sorted, we no longer have to do binary search to detect a common element. We can compare the smallest elements of the two sorted sets, and discard the smaller one if they are not identical. By repeating this idea recursively on the now smaller sets, we can test for duplication in linear time after sorting. The total cost is $O(n \log n + m \log m + n + m)$.

So, which of these is the fastest method? Clearly small-set sorting trumps big-set sorting, since $\log m < \log n$ when $m < n$. Similarly, $(n + m) \log m$ must be asymptotically smaller than $n \log n$, since $n + m < 2n$ when $m < n$. Thus, sorting the small set is the best of these options. Note that this is linear when m is constant in size.

Note that *expected* linear time can be achieved by hashing. Build a hash table containing the elements of both sets, and then explicitly check whether collisions in the same bucket are in fact identical elements. In practice, this may be the best solution. ∎

Stop and Think: Making a Hash of the Problem?

Problem: Fast sorting is a wonderful thing. But which of these tasks can be done as fast or faster (in expected time) using hashing instead of sorting?

- *Searching?*

- *Closest pair?*

- *Element uniqueness?*

- *Finding the mode?*

- *Finding the median?*

- *Convex hull?*

Solution: Hashing can solve some these problems efficiently, but is inappropriate for others. Let's consider them one by one:

- *Searching* – Hash tables are a great answer here, enabling you to search for items in constant expected time, as opposed to $O(\log n)$ with binary search.

- *Closest pair* – Hash tables as so far defined cannot help at all. Normal hash functions scatter keys around the table, so a pair of similar numerical values are unlikely to end up in the same bucket for comparison. Bucketing values by numerical ranges will ensure that the closest pair lie within the same bucket, or at worst neighboring buckets. But we cannot also force only a small number of items to lie in this bucket, as will be discussed with respect to bucketsort in Section 4.7.

- *Element uniqueness* – Hashing is even faster than sorting for this problem. Build a hash table using chaining, and then compare each of the (expected constant) pairs of items within a bucket. If no bucket contains a duplicate pair, then all the elements must be unique. The table construction and sweeping can be completed in linear expected time.

- *Finding the mode* – Hashing leads to a linear expected-time algorithm here. Each bucket should contain a small number of *distinct* elements, but may have many duplicates. We start from the first element in a bucket and count/delete all copies of it, repeating this sweep the expected constant number of passes until the bucket is empty.

- *Finding the median* – Hashing does not help us, I am afraid. The median might be in any bucket of our table, and we have no way to judge how many items lie before or after it in sorted order.

- *Convex hull* – Sure, we can build a hash table on points just as well as any other data type. But it isn't clear what good that does us for this problem: certainly it can't help us order the points by x-coordinate.

∎

4.2 Pragmatics of Sorting

We have seen many algorithmic applications of sorting, and we will see several efficient sorting algorithms. One issue stands between them: in what order do we want our items sorted? The answer to this basic question is application specific. Consider the following issues:

- *Increasing or decreasing order?* – A set of keys S are sorted in *ascending* order when $S_i \leq S_{i+1}$ for all $1 \leq i < n$. They are in *descending order* when $S_i \geq S_{i+1}$ for all $1 \leq i < n$. Different applications call for different orders.

- *Sorting just the key or an entire record?* – Sorting a data set requires maintaining the integrity of complex data records. A mailing list of names, addresses, and phone numbers may be sorted by names as the key field, but it had better retain the linkage between names and addresses. Thus, we need to specify which is the key field in any complex record, and understand the full extent of each record.

- *What should we do with equal keys?* – Elements with equal key values all bunch together in any total order, but sometimes the relative order among these keys matters. Suppose an encyclopedia contains articles on both *Michael Jordan (the basketball player)* and *Michael Jordan (the actor)*.[1] Which entry should appear first? You may need to resort to secondary keys, such as article size, to resolve ties in a meaningful way.

  Sometimes it is required to leave the items in the same relative order as in the original permutation. Sorting algorithms that automatically enforce this requirement are called *stable*. Few fast algorithms are naturally stable. Stability can be achieved for any sorting algorithm by adding the initial position as a secondary key.

  Of course we could make no decision about equal key order and let the ties fall where they may. But beware, certain efficient sort algorithms (such as quicksort) can run into quadratic performance trouble unless explicitly engineered to deal with large numbers of ties.

- *What about non-numerical data?* – Alphabetizing defines the sorting of text strings. Libraries have very complete and complicated rules concerning the relative *collating sequence* of characters and punctuation. Is *Skiena* the same key as *skiena*? Is *Brown-Williams* before or after *Brown America*, and before or after *Brown, John*?

The right way to specify such details to your sorting algorithm is with an application-specific pairwise-element *comparison function*. Such a comparison function takes pointers to record items a and b and returns "<" if $a < b$, ">" if $a > b$, or "=" if $a = b$.

By abstracting the pairwise ordering decision to such a comparison function, we can implement sorting algorithms independently of such criteria. We simply pass the comparison function in as an argument to the sort procedure. Any reasonable programming language has a built-in sort routine as a library function. You are usually better off using this than writing your own routine. For example, the standard library for C contains the `qsort` function for sorting:

---

[1]Not to mention *Michael Jordan (the statistician)*.

```
#include <stdlib.h>
```

```
void qsort(void *base, size_t nel, size_t width,
           int (*compare) (const void *, const void *));
```

The key to using **qsort** is realizing what its arguments do. It sorts the first **nel** elements of an array (pointed to by **base**), where each element is **width**-bytes long. We can thus sort arrays of 1-byte characters, 4-byte integers, or 100-byte records, all by changing the value of **width**.

The desired output order is determined by the **compare** function. It takes as arguments pointers to two **width**-byte elements, and returns a negative number if the first belongs before the second in sorted order, a positive number if the second belongs before the first, or zero if they are the same. Here is a comparison function to sort integers in increasing order:

```
int intcompare(int *i, int *j)
{
    if (*i > *j) return (1);
    if (*i < *j) return (-1);
    return (0);
}
```

This comparison function can be used to sort an array **a**, of which the first **n** elements are occupied, as follows:

```
qsort(a, n, sizeof(int), intcompare);
```

The name **qsort** suggests that quicksort is the algorithm implemented in this library function, although this is usually irrelevant to the user.

4.3 Heapsort: Fast Sorting via Data Structures

Sorting is a natural laboratory for studying algorithm design paradigms, since many useful techniques lead to interesting sorting algorithms. The next several sections will introduce algorithmic design techniques motivated by particular sorting algorithms.

The alert reader may ask why I review all the standard sorting algorithms after saying that you are usually better off *not* implementing them, and using library functions instead. The answer is that the underlying design techniques are very important for other algorithmic problems you are likely to encounter.

We start with data structure design, because one of the most dramatic algorithmic improvements via appropriate data structures occurs in sorting. Selection sort is a simple-to-code algorithm that repeatedly extracts the smallest remaining element from the unsorted part of the set:

SelectionSort(A)
    For $i = 1$ to n do
        $Sort[i]$ = Find-Minimum from A
        Delete-Minimum from A
    Return($Sort$)

A C language implementation of selection sort appeared back in Section 2.5.1 (page 41). There we partitioned the input array into sorted and unsorted regions. To find the smallest item, we performed a linear sweep through the unsorted portion of the array. The smallest item is then swapped with the ith item in the array before moving on to the next iteration. Selection sort performs n iterations, where the average iteration takes $n/2$ steps, for a total of $O(n^2)$ time.

But what if we improve the data structure? It takes $O(1)$ time to remove a particular item from an unsorted array after it has been located, but $O(n)$ time to find the smallest item. These are exactly the operations supported by priority queues. So what happens if we replace the data structure with a better priority queue implementation, either a heap or a balanced binary tree? The operations within the loop now take $O(\log n)$ time each, instead of $O(n)$. Using such a priority queue implementation speeds up selection sort from $O(n^2)$ to $O(n \log n)$.

The name typically given to this algorithm, *heapsort*, obscures the fact that the algorithm is nothing but an implementation of selection sort using the right data structure.

4.3.1   Heaps

Heaps are a simple and elegant data structure for efficiently supporting the priority queue operations *insert* and *extract-min*. Heaps work by maintaining a partial order on the set of elements that is weaker than the sorted order (so it can be efficient to maintain) yet stronger than random order (so the minimum element can be quickly identified).

Power in any hierarchically structured organization is reflected by a tree, where each node in the tree represents a person, and edge (x, y) implies that x directly supervises (or dominates) y. The person at the root sits at the "top of the heap."

In this spirit, a *heap-labeled tree* is defined to be a binary tree such that the key of each node *dominates* the keys of its children. In a *min-heap*, a node dominates its children by having a smaller key than they do, while in a *max-heap* parent nodes dominate by being bigger. Figure 4.2(l) presents a min-heap ordered tree of noteworthy years in American history.

The most natural implementation of this binary tree would store each key in a node with pointers to its two children. But as with binary search trees, the memory used by the pointers can easily outweigh the size of keys, which is the data we are really interested in. The *heap* is a slick data structure that enables us to represent binary trees without using any pointers. We store data as an

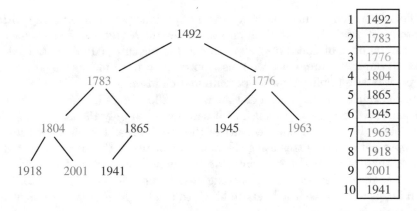

Figure 4.2: A heap-labeled tree of important years from American history (left), with the corresponding implicit heap representation (right).

array of keys, and use the position of the keys to *implicitly* play the role of the pointers.

We store the root of the tree in the first position of the array, and its left and right children in the second and third positions, respectively. In general, we store the 2^{l-1} keys of the lth level of a complete binary tree from left to right in positions 2^{l-1} to $2^l - 1$, as shown in Figure 4.2(right). We assume that the array starts with index 1 to simplify matters.

```
typedef struct {
    item_type q[PQ_SIZE+1];     /* body of queue */
    int n;                       /* number of queue elements */
} priority_queue;
```

What is especially nice about this representation is that the positions of the parent and children of the key at position k are readily determined. The *left* child of k sits in position $2k$ and the right child in $2k + 1$, while the parent of k holds court in position $\lfloor k/2 \rfloor$. Thus, we can move around the tree without any pointers.

```
int pq_parent(int n) {
    if (n == 1) {
        return(-1);
    }
    return((int) n/2);          /* implicitly take floor(n/2) */
}


int pq_young_child(int n) {
    return(2 * n);
}
```

This approach means that we can store any binary tree in an array without pointers. What is the catch? Suppose our height h tree was sparse, meaning that the number of nodes $n \ll 2^h - 1$. All missing internal nodes still take up space in our structure, since we must represent a full binary tree to maintain the positional mapping between parents and children.

Space efficiency thus demands that we not allow holes in our tree—meaning that each level be packed as much as it can be. Then only the last level may be incomplete. By packing the elements of the last level as far left as possible, we can represent an n-key tree using the first n elements of the array. If we did not enforce these structural constraints, we might need an array of size $2^n - 1$ to store n elements: consider a right-going twig using positions 1, 3, 7, 15, 31.... With heaps all but the last level are filled, so the height h of an n element heap is logarithmic because:

$$\sum_{i=0}^{h-1} 2^i = 2^h - 1 \geq n$$

implying that $h = \lceil \lg(n+1) \rceil$.

This implicit representation of binary trees saves memory, but is less flexible than using pointers. We cannot store arbitrary tree topologies without wasting large amounts of space. We cannot move subtrees around by just changing a single pointer, only by explicitly moving all the elements in the subtree. This loss of flexibility explains why we cannot use this idea to represent binary search trees, but it works just fine for heaps.

Stop and Think: Who's Where in the Heap?

Problem: How can we efficiently search for a particular key k in a heap?

Solution: We can't. Binary search does not work because a heap is not a binary search tree. We know almost nothing about the relative order of the $n/2$ leaf elements in a heap—certainly nothing that lets us avoid doing linear search through them. ∎

4.3.2   Constructing Heaps

Heaps can be constructed incrementally, by inserting each new element into the left-most open spot in the array, namely the $(n+1)$st position of a previously n-element heap. This ensures the desired balanced shape of the heap-labeled tree, but does not maintain the dominance ordering of the keys. The new key might be less than its parent in a min-heap, or greater than its parent in a max-heap.

The solution is to swap any such dissatisfied element with its parent. The old parent is now happy, because it is properly dominated. The other child of the old parent is still happy, because it is now dominated by an element even more extreme than before. The new element is now happier, but may still dominate its new parent. So we recur at a higher level, *bubbling up* the new key to its proper position in the hierarchy. Since we replace the root of a subtree by a larger one at each step, we preserve the heap order elsewhere.

```
void pq_insert(priority_queue *q, item_type x) {
    if (q->n >= PQ_SIZE) {
        printf("Warning: priority queue overflow! \n");
    } else {
        q->n = (q->n) + 1;
        q->q[q->n] = x;
        bubble_up(q, q->n);
    }
}


void bubble_up(priority_queue *q, int p) {
    if (pq_parent(p) == -1) {
        return;     /* at root of heap, no parent */
    }

    if (q->q[pq_parent(p)] > q->q[p]) {
        pq_swap(q, p, pq_parent(p));
        bubble_up(q, pq_parent(p));
    }
}
```

This swap process takes constant time at each level. Since the height of an n-element heap is $\lfloor \lg n \rfloor$, each insertion takes at most $O(\log n)$ time. A heap of n elements can thus be constructed in $O(n \log n)$ time through n such insertions:

```
void pq_init(priority_queue *q) {
    q->n = 0;
}


void make_heap(priority_queue *q, item_type s[], int n) {
    int i;          /* counter */

    pq_init(q);
    for (i = 0; i < n; i++) {
        pq_insert(q, s[i]);
    }
]
```

4.3.3 Extracting the Minimum

The remaining priority queue operations are identifying and deleting the dominant element. Identification is easy, since the top of the heap sits in the first position of the array.

Removing the top element leaves a hole in the array. This can be filled by moving the element from the *right-most* leaf (sitting in the *n*th position of the array) into the first position.

The shape of the tree has been restored but (as after insertion) the labeling of the root may no longer satisfy the heap property. Indeed, this new root may be dominated by both of its children. The root of this min-heap should be the smallest of three elements, namely the current root and its two children. If the current root is dominant, the heap order has been restored. If not, the dominant child should be swapped with the root and the problem pushed down to the next level.

This dissatisfied element *bubbles down* the heap until it dominates all its children, perhaps by becoming a leaf node and ceasing to have any. This percolate-down operation is also called *heapify*, because it merges two heaps (the subtrees below the original root) with a new key.

```
item_type extract_min(priority_queue *q) {
    int min = -1;       /* minimum value */

    if (q->n <= 0) {
        printf("Warning: empty priority queue.\n");
    } else {
        min = q->q[1];

        q->q[1] = q->q[q->n];
        q->n = q->n - 1;
        bubble_down(q, 1);
    }
    return(min);
}
```

```
void bubble_down(priority_queue *q, int p) {
    int c;            /* child index */
    int i;            /* counter */
    int min_index;    /* index of lightest child */

    c = pq_young_child(p);
    min_index = p;

    for (i = 0; i <= 1; i++) {
```

```
        if ((c + i) <= q->n) {
            if (q->q[min_index] > q->q[c + i]) {
                min_index = c + i;
            }
        }
    }

    if (min_index != p) {
        pq_swap(q, p, min_index);
        bubble_down(q, min_index);
    }
}
```

We will reach a leaf after $\lfloor \lg n \rfloor$ **bubble_down** steps, each constant time. Thus, root deletion is completed in $O(\log n)$ time.

Repeatedly exchanging the maximum element with the last element and calling heapify yields an $O(n \log n)$ sorting algorithm, named *heapsort*.

```
void heapsort_(item_type s[], int n) {
    int i;                   /* counters */
    priority_queue q;        /* heap for heapsort */

    make_heap(&q, s, n);

    for (i = 0; i < n; i++) {
        s[i] = extract_min(&q);
    }
}
```

Heapsort is a great sorting algorithm. It is simple to program; indeed, the complete implementation has been presented above. It runs in worst-case $O(n \log n)$ time, which is the best that can be expected from any sorting algorithm. It is an *in-place* sort, meaning it uses no extra memory over the array containing the elements to be sorted. Admittedly, as implemented here, my heapsort is *not* in-place because it creates the priority queue in q, not s. But each newly extracted element fits perfectly in the slot freed up by the shrinking heap, leaving behind a sorted array. Although other algorithms prove slightly faster in practice, you won't go wrong using heapsort for sorting data that sits in the computer's main memory.

Priority queues are very useful data structures. Recall they were the hero of the war story described in Section 3.6 (page 89). A complete set of priority queue implementations is presented in the catalog, in Section 15.2 (page 445).

4.3.4   Faster Heap Construction (*)

As we have seen, a heap can be constructed on n elements by incremental insertion in $O(n \log n)$ time. Surprisingly, heaps can be constructed even faster, by using our bubble_down procedure and some clever analysis.

Suppose we pack the n keys destined for our heap into the first n elements of our priority-queue array. The shape of our heap will be right, but the dominance order will be all messed up. How can we restore it?

Consider the array in reverse order, starting from the last (nth) position. It represents a leaf of the tree and so dominates its non-existent children. The same is the case for the last $n/2$ positions in the array, because all are leaves. If we continue to walk backwards through the array we will eventually encounter an internal node with children. This element may not dominate its children, but its children represent well-formed (if small) heaps.

This is exactly the situation the bubble_down procedure was designed to handle, restoring the heap order of an arbitrary root element sitting on top of two sub-heaps. Thus, we can create a heap by performing $n/2$ non-trivial calls to the bubble_down procedure:

```
void make_heap_fast(priority_queue *q, item_type s[], int n) {
    int i;                  /* counter */

    q->n = n;
    for (i = 0; i < n; i++) {
        q->q[i + 1] = s[i];
    }

    for (i = q->n/2; i >= 1; i--) {
        bubble_down(q, i);
    }
}
```

Multiplying the number of calls to bubble_down (n) times an upper bound on the cost of each operation ($O(\log n)$) gives us a running time analysis of $O(n \log n)$. This would make it no faster than the incremental insertion algorithm described above.

But note that it is indeed an *upper bound*, because only the last insertion will actually take $\lfloor \lg n \rfloor$ steps. Recall that bubble_down takes time proportional to the height of the heaps it is merging. Most of these heaps are extremely small. In a full binary tree on n nodes, there are $n/2$ nodes that are leaves (i.e. height 0), $n/4$ nodes that are height 1, $n/8$ nodes that are height 2, and so on. In general, there are at most $\lceil n/2^{h+1} \rceil$ nodes of height h, so the cost of building a heap is:

$$\sum_{h=0}^{\lfloor \lg n \rfloor} \lceil n/2^{h+1} \rceil h \leq n \sum_{h=0}^{\lfloor \lg n \rfloor} h/2^h \leq 2n$$

Since this sum is not quite a geometric series, we can't apply the usual identity to get the sum, but rest assured that the puny contribution of the numerator (h) is crushed by the denominator (2^h). The series quickly converges to linear.

Does it matter that we can construct heaps in linear time instead of $O(n \log n)$? Not really. The construction time did not dominate the complexity of heapsort, so improving the construction time does not improve its worst-case performance. Still, it is an impressive display of the power of careful analysis, and the free lunch that geometric series convergence can sometimes provide.

Stop and Think: Where in the Heap?

Problem: Given an array-based heap on n elements and a real number x, efficiently determine whether the kth smallest element in the heap is greater than or equal to x. Your algorithm should be $O(k)$ in the worst case, independent of the size of the heap. (Hint: you do not have to find the kth smallest element; you need only to determine its relationship to x.)

Solution: There are at least two different ideas that lead to correct but inefficient algorithms for this problem:

- Call extract-min k times, and test whether all of these are less than x. This explicitly sorts the first k elements and so gives us more information than the desired answer, but it takes $O(k \log n)$ time to do so.

- The kth smallest element cannot be deeper than the kth level of the heap, since the path from it to the root must go through elements of decreasing value. We can thus look at all the elements on the first k levels of the heap, and count how many of them are less than x, stopping when we either find k of them or run out of elements. This is correct, but takes $O(\min(n, 2^k))$ time, since the top k elements have $2^k - 1$ elements.

An $O(k)$ solution can look at only k elements smaller than x, plus at most $O(k)$ elements greater than x. Consider the following recursive procedure, called at the root with $i = 1$ and *count* $= k$:

```
int heap_compare(priority_queue *q, int i, int count, int x) {
    if ((count <= 0) || (i > q->n)) {
        return(count);
    }

    if (q->q[i] < x) {
        count = heap_compare(q, pq_young_child(i), count-1, x);
        count = heap_compare(q, pq_young_child(i)+1, count, x);
    }

    return(count);
}
```

If the root of the min-heap is $\geq x$, then no elements in the heap can be less than x, as by definition the root must be the smallest element. This procedure searches the children of all nodes of weight smaller than x until either (a) we have found k of them, when it returns 0, or (b) they are exhausted, when it returns a value greater than zero. Thus, it will find enough small elements if they exist.

But how long does it take? The only nodes whose children we look at are those $< x$, and there are at most k of these in total. Each have visited at most two children, so we visit at most $2k + 1$ nodes, for a total time of $O(k)$. ∎

4.3.5   Sorting by Incremental Insertion

Now consider a different approach to sorting via efficient data structures. Select the next element from the unsorted set, and put it into it's proper position in the sorted set:

```
for (i = 1; i < n; i++) {
    j = i;
    while ((j > 0) && (s[j] < s[j - 1])) {
        swap(&s[j], &s[j - 1]);
        j = j-1;
    }
}
```

Although insertion sort takes $O(n^2)$ in the worst case, it performs considerably better if the data is almost sorted, since few iterations of the inner loop suffice to sift it into the proper position.

Insertion sort is perhaps the simplest example of the *incremental insertion* technique, where we build up a complicated structure on n items by first building it on $n - 1$ items and then making the necessary changes to add the last item.

Note that faster sorting algorithms based on incremental insertion follow from more efficient data structures. Insertion into a balanced search tree takes

$O(\log n)$ per operation, or a total of $O(n \log n)$ to construct the tree. An in-order traversal reads through the elements in sorted order to complete the job in linear time.

4.4 War Story: Give me a Ticket on an Airplane

I came into this particular job seeking justice. I'd been retained by an air travel company to help design an algorithm to find the cheapest available airfare from city x to city y. Like most of you, I suspect, I'd been baffled at the crazy price fluctuations of ticket prices under modern "yield management." The price of flights seems to soar far more efficiently than the planes themselves. The problem, it seemed to me, was that airlines never wanted to show the true cheapest price. If I did my job right, I could make damned sure they would show it to me next time.

"Look," I said at the start of the first meeting. "This can't be so hard. Construct a graph with vertices corresponding to airports, and add an edge between each airport pair (u, v) that shows a direct flight from u to v. Set the weight of this edge equal to the cost of the cheapest available ticket from u to v. Now the cheapest fair from x to y is given by the shortest x–y path in this graph. This path/fare can be found using Dijkstra's shortest path algorithm. Problem solved!" I announced, waving my hand with a flourish.

The assembled cast of the meeting nodded thoughtfully, then burst out laughing. It was I who needed to learn something about the overwhelming complexity of air travel pricing. There are literally millions of different fares available at any time, with prices changing several times daily. Restrictions on the availability of a particular fare in a particular context are enforced by a vast set of pricing rules. These rules are an industry-wide kludge—a complicated structure with little in the way of consistent logical principles. My favorite rule exceptions applied only to the country of Malawi. With a population of only 18 million and per-capita income of $1,234 (180th in the world), they prove to be an unexpected powerhouse shaping world aviation price policy. Accurately pricing any air itinerary requires at least implicit checks to ensure the trip doesn't take us through Malawi.

The real problem is that there can easily be 100 different fares for the first flight leg, say from Los Angeles (LAX) to Chicago (ORD), and a similar number for each subsequent leg, say from Chicago to New York (JFK). The cheapest possible LAX–ORD fare (maybe an AARP children's price) might not be combinable with the cheapest ORD–JFK fare (perhaps a pre-Ramadan special that can only be used with subsequent connections to Mecca).

After being properly chastised for oversimplifying the problem, I got down to work. I started by reducing the problem to the simplest interesting case. "So, you need to find the cheapest two-hop fare that passes your rule tests. Is there a way to decide in advance which pairs will pass without explicitly testing them?"

"No, there is no way to tell," they assured me. "We can only consult a

		X+Y
		$150 (1,1)
		$160 (2,1)
X	**Y**	$175 (1,2)
——	——	$180 (3,1)
$100	$50	$185 (2,2)
		$205 (3,2)
$110	$75	$225 (1,3)
		$235 (2,3)
$130	$125	$255 (3,3)

Figure 4.3: Sorting the pairwise sums of lists X and Y.

black box routine to decide whether a particular price is available for the given itinerary/travelers."

"So our goal is to call this black box on the fewest number of combinations. This means evaluating all possible fare combinations in order from cheapest to most expensive, and stopping as soon as we encounter the first legal combination."

"Right."

"Why not construct the $m \times n$ possible price pairs, sort them in terms of cost, and evaluate them in sorted order? Clearly this can be done in $O(nm \log(nm))$ time."[2]

"That is basically what we do now, but it is quite expensive to construct the full set of $m \times n$ pairs, since the first one might be all we need."

I caught a whiff of an interesting problem. "So what you really want is an efficient data structure to repeatedly return the *next* most expensive pair without constructing all the pairs in advance."

This was indeed an interesting problem. Finding the largest element in a set under insertion and deletion is *exactly* what priority queues are good for. The catch here is that we could not seed the priority queue with all values in advance. We had to insert new pairs into the queue after each evaluation.

I constructed some examples, like the one in Figure 4.3. We could represent each fare by the list indexes of its two components. The cheapest single fare will certainly be constructed by adding up the cheapest component from both lists, described $(1, 1)$. The second cheapest fare would be made from the head of one list and the second element of another, and hence would be either $(1, 2)$ or $(2, 1)$. Then it gets more complicated. The third cheapest could either be the unused pair above or $(1, 3)$ or $(3, 1)$. Indeed it would have been $(3, 1)$ in the example above if the third fare of X had been $120.

"Tell me," I asked. "Do we have time to sort the two respective lists of fares

---

[2]The question of whether all such sums can be sorted faster than nm arbitrary integers is a notorious open problem in algorithm theory. See [Fre76, Lam92] for more on $X + Y$ sorting, as the problem is known.

in increasing order?"

"Don't have to," the leader replied. "They come out in sorted order from the database."

That was good news! It meant there was a natural order to the pair values. We never need to evaluate the pairs $(i + 1, j)$ or $(i, j + 1)$ before (i, j), because they clearly defined more expensive fares.

"Got it!," I said. "We will keep track of index pairs in a priority queue, with the sum of the fare costs as the key for the pair. Initially we put only the pair $(1, 1)$ in the queue. If it proves not to be feasible, we put in its two successors—namely $(1, 2)$ and $(2, 1)$. In general, we enqueue pairs $(i + 1, j)$ and $(i, j + 1)$ after evaluating/rejecting pair (i, j). We will get through all the pairs in the right order if we do so."

The gang caught on quickly. "Sure. But what about duplicates? We will construct pair (x, y) two different ways, both when expanding $(x - 1, y)$ and $(x, y - 1)$."

"You are right. We need an extra data structure to guard against duplicates. The simplest might be a hash table to tell us whether a given pair exists in the priority queue before we insert a duplicate. In fact, we will never have more than n active pairs in our data structure, since there can only be one pair for each distinct value of the first coordinate."

And so it went. Our approach naturally generalizes to itineraries with more than two legs, with complexity that grows with the number of legs. The best-first evaluation inherent in our priority queue enabled the system to stop as soon as it found the provably cheapest fare. This proved to be fast enough to provide interactive response to the user. That said, I never noticed airline tickets getting cheaper as a result.

4.5 Mergesort: Sorting by Divide and Conquer

Recursive algorithms reduce large problems into smaller ones. A recursive approach to sorting involves partitioning the elements into two groups, sorting each of the smaller problems recursively, and then interleaving the two sorted lists to totally order the elements. This algorithm is called *mergesort*, recognizing the importance of the interleaving operation:

> Mergesort($A[1, \ldots, n]$)
> Merge(MergeSort($A[1, \ldots, \lfloor n/2 \rfloor]$), MergeSort($A[\lfloor n/2 \rfloor + 1, \ldots, n]$))

The basis case of the recursion occurs when the subarray to be sorted consists of at most one element, so no rearrangement is necessary. A trace of the execution of mergesort is given in Figure 4.4. Picture the action as it happens during an in-order traversal of the tree, with each merge occurring after the two child calls return sorted subarrays.

The efficiency of mergesort depends upon how efficiently we can combine the two sorted halves into a single sorted list. We could concatenate them into one

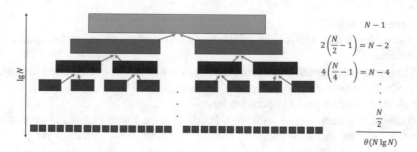

Figure 4.4: The recursion tree for mergesort. The tree has height $\lceil \log_2 n \rceil$, and the cost of the merging operations on each level are $\Theta(n)$, yielding an $\Theta(n \log n)$ time algorithm.

list and call heapsort or some other sorting algorithm to do it, but that would just destroy all the work spent sorting our component lists.

Instead we can *merge* the two lists together. Observe that the smallest overall item in two lists sorted in increasing order (as above) must sit at the top of one of the two lists. This smallest element can be removed, leaving two sorted lists behind—one slightly shorter than before. The second smallest item overall must now be atop one of these lists. Repeating this operation until both lists are empty will merge two sorted lists (with a total of n elements between them) into one, using at most $n - 1$ comparisons or $O(n)$ total work.

What is the total running time of mergesort? It helps to think about how much work is done at each level of the execution tree, as shown in Figure 4.4. If we assume for simplicity that n is a power of two, the kth level consists of all the 2^k calls to `mergesort` processing subranges of $n/2^k$ elements.

The work done on the zeroth level (the top) involves merging one pair of sorted lists, each of size $n/2$, for a total of at most $n - 1$ comparisons. The work done on the first level (one down) involves merging two pairs of sorted lists, each of size $n/4$, for a total of at most $n - 2$ comparisons. In general, the work done on the kth level involves merging 2^k pairs of sorted lists, each of size $n/2^{k+1}$, for a total of at most $n - 2^k$ comparisons. *Linear work is done merging all the elements on each level.* Each of the n elements appears in exactly one subproblem on each level. The most expensive case (in terms of comparisons) is actually the top level.

The number of elements in a subproblem gets halved at each level. The number of times we can halve n until we get to 1 is $\lceil \lg n \rceil$. Because the recursion goes $\lg n$ levels deep, and a linear amount of work is done per level, mergesort takes $O(n \log n)$ time in the worst case.

Mergesort is a great algorithm for sorting linked lists, because it does not rely on random access to elements like heapsort and quicksort. Its primary disadvantage is the need for an auxiliary buffer when sorting arrays. It is easy to merge two sorted linked lists without using any extra space, just by rearranging the pointers. However, to merge two sorted arrays (or portions of an array), we need to use a third array to store the result of the merge to avoid stepping on the component arrays. Consider merging $\{4, 5, 6\}$ with $\{1, 2, 3\}$, packed from left to right in a single array. Without the buffer, we would overwrite the elements of the left half during merging and lose them.

Mergesort is a classic divide-and-conquer algorithm. We are ahead of the game whenever we can break one large problem into two smaller problems, because the smaller problems are easier to solve. The trick is taking advantage of the two partial solutions to construct a solution of the full problem, as we did with the merge operation. Divide and conquer is an important algorithm design paradigm, and will be the subject of Chapter 5.

Implementation

The divide-and-conquer `mergesort` routine follows naturally from the pseudocode:

```
void merge_sort(item_type s[], int low, int high) {
    int middle;      /* index of middle element */

    if (low < high) {
        middle = (low + high) / 2;
        merge_sort(s, low, middle);
        merge_sort(s, middle + 1, high);

        merge(s, low, middle, high);
    }
}
```

More challenging turns out to be the details of how the merging is done. The problem is that we must put our merged array somewhere. To avoid losing an element by overwriting it in the course of the merge, we first copy each subarray to a separate queue and merge these elements back into the array. In particular:

```
void merge(item_type s[], int low, int middle, int high) {
    int i;                    /* counter */
    queue buffer1, buffer2;   /* buffers to hold elements for merging */

    init_queue(&buffer1);
    init_queue(&buffer2);

    for (i = low; i <= middle; i++) enqueue(&buffer1, s[i]);
    for (i = middle + 1; i <= high; i++) enqueue(&buffer2, s[i]);

    i = low;
    while (!(empty_queue(&buffer1) || empty_queue(&buffer2))) {
        if (headq(&buffer1) <= headq(&buffer2)) {
            s[i++] = dequeue(&buffer1);
```

```
    } else {
        s[i++] = dequeue(&buffer2);
    }
}

while (!empty_queue(&buffer1)) {
    s[i++] = dequeue(&buffer1);
}

while (!empty_queue(&buffer2)) {
    s[i++] = dequeue(&buffer2);
}
}
```

4.6   Quicksort: Sorting by Randomization

Suppose we select an arbitrary item p from the n items we seek to sort. *Quicksort* (shown in action in Figure 4.5) separates the $n-1$ other items into two piles: a low pile containing all the elements that are $< p$, and a high pile containing all the elements that are $\geq p$. Low and high denote the array positions into which we place the respective piles, leaving a single slot between them for p.

Such partitioning buys us two things. First, the pivot element p ends up in the exact array position it will occupy in the final sorted order. Second, after partitioning no element flips to the other side in the final sorted order. *Thus, we can now sort the elements to the left and the right of the pivot independently!* This gives us a recursive sorting algorithm, since we can use the partitioning approach to sort each subproblem. The algorithm must be correct, because each element ultimately ends up in the proper position:

```
void quicksort(item_type s[], int l, int h) {
    int p;      /* index of partition */

    if (l < h) {
        p = partition(s, l, h);
        quicksort(s, l, p - 1);
        quicksort(s, p + 1, h);
    }
}
```

We can partition the array in one linear scan for a particular pivot element by maintaining three sections of the array: less than the pivot (to the left of `firsthigh`), greater than or equal to the pivot (between `firsthigh` and `i`), and unexplored (to the right of `i`), as implemented below:

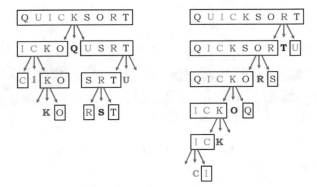

Figure 4.5: Animations of quicksort in action: first selecting the first element in each subarray as pivot (on left), and then selecting the last element as pivot (on right).

```
int partition(item_type s[], int l, int h) {
    int i;          /* counter */
    int p;          /* pivot element index */
    int firsthigh;  /* divider position for pivot element */

    p = h;          /* select last element as pivot */
    firsthigh = l;
    for (i = l; i < h; i++) {
        if (s[i] < s[p]) {
            swap(&s[i], &s[firsthigh]);
            firsthigh++;
        }
    }
    swap(&s[p], &s[firsthigh]);

    return(firsthigh);
}
```

Since the partitioning step consists of at most n swaps, it takes time linear in the number of keys. But how long does the entire quicksort take? As with mergesort, quicksort builds a recursion tree of nested subranges of the n-element array. As with mergesort, quicksort spends linear time processing (now **partitioning** instead of **merge**ing) the elements in each subarray on each level. As with mergesort, quicksort runs in $O(n \cdot h)$ time, where h is the height of the recursion tree.

The difficulty is that the height of the tree depends upon where the pivot element ends up in each partition. If we get very lucky and *happen* to repeatedly pick the median element as our pivot, the subproblems are always half the size of those at the previous level. The height represents the number of times we

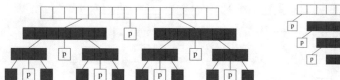

Figure 4.6: The best-case (left) and worst-case (right) recursion trees for quicksort. The left partition is in blue and the right partition in red.

can halve n until we get down to 1, meaning $h = \lceil \lg n \rceil$. This happy situation is shown in Figure 4.6(left), and corresponds to the best case of quicksort.

Now suppose we consistently get unlucky, and our pivot element *always* splits the array as unequally as possible. This implies that the pivot element is always the biggest or smallest element in the sub-array. After this pivot settles into its position, we will be left with one subproblem of size $n - 1$. After doing linear work we have reduced the size of our problem by just one measly element, as shown in Figure 4.6(right). It takes a tree of height $n - 1$ to chop our array down to one element per level, for a worst case time of $\Theta(n^2)$.

Thus, the worst case for quicksort is worse than heapsort or mergesort. To justify its name, quicksort had better be good in the average case. Understanding why requires some intuition about random sampling.

4.6.1   Intuition: The Expected Case for Quicksort

The expected performance of quicksort depends upon the height of the partition tree constructed by pivot elements at each step. Mergesort splits the keys into two equal halves, sorts both of them recursively, and then merges the halves in linear time—and hence runs in $O(n \log n)$ time. Thus, whenever our pivot element is near the center of the sorted array (meaning the pivot is close to the median element), we get the same good split realizing the same running time as mergesort.

I will give an intuitive explanation of why quicksort runs in $O(n \log n)$ time in the average case. How likely is it that a randomly selected pivot is a good one? The best possible selection for the pivot would be the median key, because exactly half of elements would end up left, and half the elements right, of the pivot. However, we have only a probability of $1/n$ that a randomly selected pivot is the median, which is quite small.

Suppose we say a key is a *good enough* pivot if it lies in the center half of the sorted space of keys—those ranked from $n/4$ to $3n/4$ in the space of all keys to be sorted. Such *good enough* pivot elements are quite plentiful, since half the elements lie closer to the middle than to one of the two ends (see Figure 4.7). Thus, on each selection we will pick a *good enough* pivot with probability of $1/2$. We will make good progress towards sorting whenever we pick a good enough

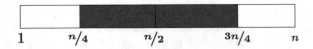

Figure 4.7: Half the time, a randomly chosen pivot is close to the median element.

pivot.

The worst possible *good enough* pivot leaves the bigger of the two partitions with $3n/4$ items. This happens also to be the expected size of the larger partition left after picking a random pivot p, at the median between the worst possible pivot ($p = 1$ or $p = n$ leaving a partition of size $n - 1$) and the best possible pivot ($p = n/2$ leaving two partitions of size $n/2$. So what is the height h_g of a quicksort partition tree constructed repeatedly from the expected pivot value? The deepest path through this tree passes through partitions of size $n, (3/4)n, (3/4)^2 n, \ldots$, down to 1. How many times can we multiply n by $3/4$ until it gets down to 1?

$$(3/4)^{h_g} n = 1 \implies n = (4/3)^{h_g}$$

so $h_g = \log_{4/3} n$.

On average, random quicksort partition trees (and by analogy, binary search trees under random insertion) are very good. More careful analysis shows the average height after n insertions is approximately $2 \ln n$. Since $2 \ln n \approx 1.386 \lg n$, this is only 39% taller than a perfectly balanced binary tree. Since quicksort does $O(n)$ work partitioning on each level, the average time is $O(n \log n)$. If we are *extremely* unlucky, and our randomly selected elements are always among the largest or smallest element in the array, quicksort turns into selection sort and runs in $O(n^2)$. But the odds against this are vanishingly small.

4.6.2   Randomized Algorithms

There is an important subtlety about the expected case $O(n \log n)$ running time for quicksort. Our quicksort implementation above selected the last element in each sub-array as the pivot. Suppose this program were given a sorted array as input. Then at each step it would pick the worst possible pivot, and run in quadratic time.

For any deterministic method of pivot selection, there exists a worst-case input instance which will doom us to quadratic time. The analysis presented above made no claim stronger than:

> "Quicksort runs in $\Theta(n \log n)$ time, with high probability, *if* you give it randomly ordered data to sort."

But now suppose we add an initial step to our algorithm where we randomly permute the order of the n elements before we try to sort them. Such a permutation can be constructed in $O(n)$ time (see Section 16.7 for details).

This might seem like wasted effort, but it provides the guarantee that we can expect $\Theta(n \log n)$ running time *whatever* the initial input was. The worst case performance still can happen, but it now depends only upon how unlucky we are. There is no longer a well-defined "worst-case" input. We now can claim that:

> "Randomized quicksort runs in $\Theta(n \log n)$ time on *any* input, with high probability."

Alternately, we could get the same guarantee by selecting a random element to be the pivot at each step.

Randomization is a powerful tool to improve algorithms with bad worst-case but good average-case complexity. It can be used to make algorithms more robust to boundary cases and more efficient on highly structured input instances that confound heuristic decisions (such as sorted input to quicksort). It often lends itself to simple algorithms that provide expected-time performance guarantees, which are otherwise obtainable only using complicated deterministic algorithms. Randomized algorithms will be the topic of Chapter 6.

Proper analysis of randomized algorithms requires some knowledge of probability theory, and will be deferred to Chapter 6. However, some of the basic approaches to designing efficient randomized algorithms are readily explainable:

- *Random sampling* – Want to get an idea of the median value of n things, but don't have either the time or space to look at them all? Select a small random sample of the input and find the median of those. The result should be representative for the full set.

  This is the idea behind opinion polling, where we sample a small number of people as a proxy for the full population. Biases creep in unless you take a truly *random* sample, as opposed to the first x people you happen to see. To avoid bias, actual polling agencies typically dial random phone numbers and hope someone answers.

- *Randomized hashing* – We have claimed that hashing can be used to implement dictionary search in $O(1)$ "expected time." However, for any hash function there is a given worst-case set of keys that all get hashed to the same bucket. But now suppose we randomly select our hash function from a large family of good ones as the first step of our algorithm. We get the same type of improved guarantee that we did with randomized quicksort.

- *Randomized search* – Randomization can also be used to drive search techniques such as simulated annealing, as will be discussed in detail in Section 12.6.3 (page 406).

Stop and Think: Nuts and Bolts

Problem: The *nuts and bolts* problem is defined as follows. You are given a collection of n bolts of different widths, and n corresponding nuts. You can test whether a given nut and bolt fit together, from which you learn whether the nut is too large, too small, or an exact match for the bolt. The differences in size between pairs of nuts or bolts are too small to see by eye, so you cannot compare the sizes of two nuts or two bolts directly. You are asked to match each bolt to each nut as efficiently as possible.

Give an $O(n^2)$ algorithm to solve the nuts and bolts problem. Then give a randomized $O(n \log n)$ expected-time algorithm for the same problem.

Solution: The brute force algorithm for matching nuts and bolts starts with the first bolt and compares it to each nut until a match is found. In the worst case, this will require n comparisons. Repeating this for each successive bolt on all remaining nuts yields an algorithm with a quadratic number of comparisons.

But what if we pick a random bolt and try it? On average, we would expect to get about halfway through the set of nuts before we found the match, so this randomized algorithm would do half the work on average as the worst case. That counts as some kind of improvement, although not an asymptotic one.

Randomized quicksort achieves the desired expected-case running time, so a natural idea is to emulate it on the nuts and bolts problem. The fundamental step in quicksort is partitioning elements around a pivot. Can we partition nuts and bolts around a randomly selected bolt b?

Certainly we can partition the nuts into those of size less than b and greater than b. But decomposing the problem into two halves requires partitioning the bolts as well, and we cannot compare bolt against bolt. But once we find the matching nut to b, we can use it to partition the bolts accordingly. In $2n - 2$ comparisons, we partition the nuts and bolts, and the remaining analysis follows directly from randomized quicksort.

What is interesting about this problem is that no simple deterministic algorithm for nut and bolt sorting is known. It illustrates how randomization makes the bad case go away, leaving behind a simple and beautiful algorithm. ∎

4.6.3 Is Quicksort Really Quick?

There is a clear, asymptotic difference between a $\Theta(n \log n)$ algorithm and one that runs in $\Theta(n^2)$. Only the most obstinate reader would doubt my claim that mergesort, heapsort, and quicksort will all outperform insertion sort or selection sort on large enough instances.

But how can we compare two $\Theta(n \log n)$ algorithms to decide which is faster? How can we prove that quicksort is really quick? Unfortunately, the RAM model and Big Oh analysis provide too coarse a set of tools to make that type of distinction. When faced with algorithms of the same asymptotic complexity, implementation details and system quirks such as cache performance and memory size often prove to be the decisive factor.

What we can say is that experiments show that when quicksort is implemented well, it is typically two to three times faster than mergesort or heapsort. The primary reason is that the operations in the innermost loop are simpler.

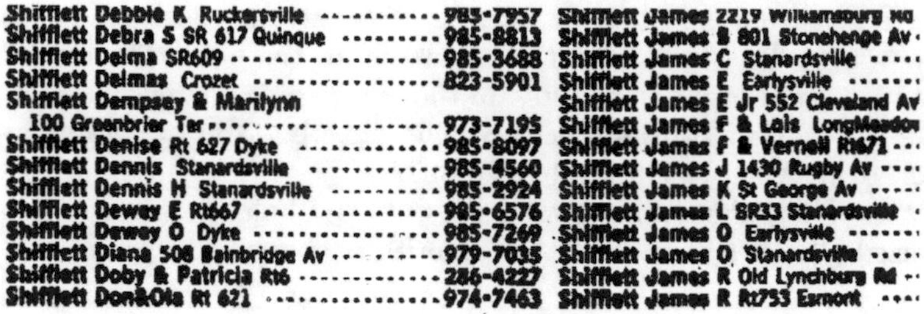

Figure 4.8: A small subset of Charlottesville Shiffletts.

But I can't argue if you don't believe me when I say quicksort is faster. It is a question whose solution lies outside the analytical tools we are using. The best way to tell is to implement both algorithms and experiment.

4.7   Distribution Sort: Sorting via Bucketing

To sort names for a class roster or the telephone book, we could first partition them according to the first letter of the last name. This will create twenty-six different piles, or buckets, of names. Observe that any name in the *J* pile must occur after all names in the *I* pile, and before any name in the *K* pile. Therefore, we can proceed to sort each pile individually and just concatenate the sorted piles together at the end.

Assuming the names are distributed evenly among the buckets, the resulting twenty-six sorting problems should each be substantially smaller than the original problem. By now further partitioning each pile based on the *second* letter of each name, we can generate smaller and smaller piles. The set of names will be completely sorted as soon as every bucket contains only a single name. Such an algorithm is commonly called *bucketsort* or *distribution sort*.

Bucketing is a very effective idea whenever we are confident that the distribution of data will be roughly uniform. It is the idea that underlies hash tables, *kd*-trees, and a variety of other practical data structures. The downside of such techniques is that the performance can be terrible when the data distribution is not what we expected. Although data structures such as balanced binary trees offer guaranteed worst-case behavior for any input distribution, no such promise exists for heuristic data structures on unexpected input distributions.

Non-uniform distributions do occur in real life. Consider Americans with the uncommon last name of Shifflett. When last I looked, the Manhattan telephone directory (with over one million names) contained exactly five Shiffletts. So how many Shiffletts should there be in a small city of 50,000 people? Figure 4.8 shows a small portion of the *two and a half pages* of Shiffletts in the Charlottesville, Virginia telephone book. The Shifflett clan is a fixture of the region, but it would play havoc with any distribution sort program, as refining buckets from *S* to *Sh* to *Shi* to *Shif* to ... to *Shifflett* results in no significant partitioning.

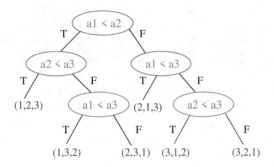

Figure 4.9: Interpreting insertion sort of input array a as a decision tree. Each leaf represents a given input permutation, while the root–to–leaf path describes the sequence of comparisons the algorithm does to sort it.

Take-Home Lesson: Sorting can be used to illustrate many algorithm design paradigms. Data structure techniques, divide and conquer, randomization, and incremental construction all lead to efficient sorting algorithms.

4.7.1 Lower Bounds for Sorting

One last issue on the complexity of sorting. We have seen several sorting algorithms that run in worst-case $O(n \log n)$ time, but none that is linear. To sort n items certainly requires looking at all of them, so any sorting algorithm must be $\Omega(n)$ in the worst case. Might sorting be possible in linear time?

The answer is no, presuming that your algorithm is based on comparing pairs of elements. An $\Omega(n \log n)$ lower bound can be shown by observing that any sorting algorithm must behave differently during execution on each of the $n!$ possible permutations of n keys. If an algorithm did *exactly* the same thing with two different input permutations, there is no way that both of them could correctly come out sorted. The outcome of each pairwise comparison governs the run-time behavior of any comparison-based sorting algorithm. We can think of the set of all possible executions for such an algorithm as a tree with $n!$ leaves, each of which correspond to one input permutation, and each root–to–leaf path describes the comparisons performed to sort the given input. The minimum height tree corresponds to the fastest possible algorithm, and it happens that $\lg(n!) = \Theta(n \log n)$.

Figure 4.9 presents the decision tree for insertion sort on three elements. To interpret it, simulate what insertion sort does on the input $a = (3, 1, 2)$. Because $a_1 \geq a_2$, these elements must be swapped to produce a sorted order. Insertion sort then compares the end of the sorted array (the original input a_1) against a_3. If $a_1 \geq a_3$, the final test of a_3 against the head of the sorted part (original input a_2) decides whether to put a_2 first or second in sorted order.

This lower bound is important for several reasons. First, the idea can be extended to give lower bounds for many applications of sorting, including element uniqueness, finding the mode, and constructing convex hulls. Sorting has one of the few non-trivial lower bounds among algorithmic problems. We will present a different approach to arguing that fast algorithms are unlikely to exist in Chapter 11.

Note that hashing-based algorithms do not perform such element comparisons, putting them outside the scope of this lower bound. But hashing-based algorithms can get unlucky, and with worst-case luck the running time of any randomized algorithm for one of these problems will be $\Omega(n \log n)$.

4.8   War Story: Skiena for the Defense

I lead a quiet, reasonably honest life. One reward for this is that I don't often find myself on the business end of surprise calls from lawyers. Thus, I was astonished to get a call from a lawyer who not only wanted to talk with me, but wanted to talk to me about sorting algorithms.

It turned out that her firm was working on a case involving high-performance programs for sorting, and needed an expert witness who could explain technical issues to the jury. They knew I knew something about algorithms, but before taking me on they demanded to see my teaching evaluations to prove that I could explain things to people.[3] It promised to be an opportunity to learn about how fast sorting programs *really* worked. I figured I would learn which in-place sorting algorithm was fastest in practice. Was it heapsort or quicksort? What subtle, secret algorithmics made the difference to minimize the number of comparisons in practice?

The answer was quite humbling. *Nobody cared about in-place sorting.* The name of the game was sorting *huge* files, much bigger than can fit in main memory. All the important action was in getting the data on and off a disk. Cute algorithms for doing internal (in-memory) sorting were not the bottleneck, because the real problem lies in sorting gigabytes at a time.

Recall that disks have relatively long seek times, reflecting how long it takes the desired part of the disk to rotate under the read/write head. Once the head is in the right place, the data moves relatively quickly, and it costs about the same to read a large data block as it does to read a single byte. Thus, the goal is minimizing the number of blocks read/written, and coordinating these operations so the sorting algorithm is never waiting to get the data it needs.

The disk-intensive nature of sorting is best revealed by the annual *Minutesort* competition. The goal is to sort as much data in one minute as possible. As of this writing, the current champion is Tencent Sort, which managed to sort 55 terabytes of data in under a minute on a little old 512-node cluster, each with 20 cores and 512 GB RAM. You can check out the current records at http://sortbenchmark.org/.

---

[3] One of my more cynical faculty colleagues said this was the first time anyone, anywhere, had ever actually looked at university teaching evaluations.

That said, which algorithm is best for external sorting? It basically turns out to be a multiway mergesort, employing a lot of engineering and special tricks. You build a heap with members of the top block from each of k sorted lists. By repeatedly plucking the top element off this heap, you build a sorted list merging these k lists. Because this heap is sitting in main memory, these operations are fast. When you have a large enough sorted run, you write it to disk, and free up memory for more data. When you get close to emptying out the elements from the top block of one of the k sorted lists you are merging, load the next block.

It proves very hard to benchmark sorting programs/algorithms at this level and decide which *really* is fastest. Is it fair to compare a commercial program designed to handle general files with a stripped-down code optimized for integers? The *Minutesort* competition employs randomly generated 100-byte records. This is a different world than sorting names: Shiffletts are not randomly distributed. For example, one widely employed trick is to strip off a relatively short prefix of the key and initially sort only on that, just to avoid lugging all those extra bytes around.

What lessons did I learn from this? The most important, by far, is to do everything you can to avoid being involved in a lawsuit either as a plaintiff or defendant.[4] Courts are not instruments for resolving disputes quickly. Legal battles have a lot in common with military battles: they escalate very quickly, become very expensive in time, money, and soul, and usually end only when both sides are exhausted and compromise. Wise are the parties who can work out their problems without going to court. Properly absorbing this lesson now could save you thousands of times the cost of this book.

On technical matters, it is important to worry about external memory performance whenever you combine very large datasets with low-complexity algorithms (say linear or $\Theta(n \log n)$). Constant factors of even 5 or 10 can make a big difference here between what is feasible and what is hopeless. Of course, quadratic-time algorithms are doomed to fail on large datasets, regardless of data access times.

Chapter Notes

Several interesting sorting algorithms have not been discussed in this section including *shellsort*, a substantially more efficient version of insertion sort, and *radix sort*, an efficient algorithm for sorting strings. You can learn more about these and every other sorting algorithm by browsing through Knuth [Knu98]. This includes external sorting, the subject of this chapter's legal war story.

As implemented here, mergesort copies the merged elements into an auxiliary buffer, to avoid overwriting the original elements to be sorted. Through clever but complicated buffer manipulation, mergesort can be implemented in an array without using too much extra storage. Kronrod's algorithm for in-place merging is presented in [Knu98].

---

[4]However, it is actually quite interesting serving as an expert witness.

Randomized algorithms are discussed in greater detail in the books by Motwani and Raghavan [MR95] and Mitzenmacher and Upfal [MU17]. The problem of nut and bolt sorting was introduced by Rawlins [Raw92]. A complicated but deterministic $\Theta(n \log n)$ algorithm is due to Komlos, Ma, and Szemeredi [KMS98].

4.9   Exercises

Applications of Sorting: Numbers

4-1. *[3]* The Grinch is given the job of partitioning $2n$ players into two teams of n players each. Each player has a numerical rating that measures how good he or she is at the game. The Grinch seeks to divide the players as *unfairly* as possible, so as to create the biggest possible talent imbalance between the teams. Show how the Grinch can do the job in $O(n \log n)$ time.

4-2. *[3]* For each of the following problems, give an algorithm that finds the desired numbers within the given amount of time. To keep your answers brief, feel free to use algorithms from the book as subroutines. For the example, $S = \{6, 13, 19, 3, 8\}$, $19 - 3$ maximizes the difference, while $8 - 6$ minimizes the difference.

(a) Let S be an *unsorted* array of n integers. Give an algorithm that finds the pair $x, y \in S$ that *maximizes* $|x - y|$. Your algorithm must run in $O(n)$ worst-case time.

(b) Let S be a *sorted* array of n integers. Give an algorithm that finds the pair $x, y \in S$ that *maximizes* $|x - y|$. Your algorithm must run in $O(1)$ worst-case time.

(c) Let S be an *unsorted* array of n integers. Give an algorithm that finds the pair $x, y \in S$ that *minimizes* $|x - y|$, for $x \neq y$. Your algorithm must run in $O(n \log n)$ worst-case time.

(d) Let S be a *sorted* array of n integers. Give an algorithm that finds the pair $x, y \in S$ that *minimizes* $|x - y|$, for $x \neq y$. Your algorithm must run in $O(n)$ worst-case time.

4-3. *[3]* Take a list of $2n$ real numbers as input. Design an $O(n \log n)$ algorithm that partitions the numbers into n pairs, with the property that the partition minimizes the maximum sum of a pair. For example, say we are given the numbers (1,3,5,9). The possible partitions are ((1,3),(5,9)), ((1,5),(3,9)), and ((1,9),(3,5)). The pair sums for these partitions are (4,14), (6,12), and (10,8). Thus, the third partition has 10 as its maximum sum, which is the minimum over the three partitions.

4-4. *[3]* Assume that we are given n pairs of items as input, where the first item is a number and the second item is one of three colors (red, blue, or yellow). Further assume that the items are sorted by number. Give an $O(n)$ algorithm to sort the items by color (all reds before all blues before all yellows) such that the numbers for identical colors stay sorted.

For example: (1,blue), (3,red), (4,blue), (6,yellow), (9,red) should become (3,red), (9,red), (1,blue), (4,blue), (6,yellow).

4-5. *[3]* The *mode* of a bag of numbers is the number that occurs most frequently in the set. The set $\{4, 6, 2, 4, 3, 1\}$ has a mode of 4. Give an efficient and correct algorithm to compute the mode of a bag of n numbers.

4-6. *[3]* Given two sets S_1 and S_2 (each of size n), and a number x, describe an $O(n \log n)$ algorithm for finding whether there exists a pair of elements, one from S_1 and one from S_2, that add up to x. (For partial credit, give a $\Theta(n^2)$ algorithm for this problem.)

4-7. *[5]* Give an efficient algorithm to take the array of citation counts (each count is a non-negative integer) of a researcher's papers, and compute the researcher's h-index. By definition, a scientist has index h if h of his or her n papers have been cited at least h times, while the other $n - h$ papers each have no more than h citations.

4-8. *[3]* Outline a reasonable method of solving each of the following problems. Give the order of the worst-case complexity of your methods.

   (a) You are given a pile of thousands of telephone bills and thousands of checks sent in to pay the bills. Find out who did not pay.

   (b) You are given a printed list containing the title, author, call number, and publisher of all the books in a school library and another list of thirty publishers. Find out how many of the books in the library were published by each company.

   (c) You are given all the book checkout cards used in the campus library during the past year, each of which contains the name of the person who took out the book. Determine how many distinct people checked out at least one book.

4-9. *[5]* Given a set S of n integers and an integer T, give an $O(n^{k-1} \log n)$ algorithm to test whether k of the integers in S add up to T.

4-10. *[3]* We are given a set of S containing n real numbers and a real number x, and seek efficient algorithms to determine whether two elements of S exist whose sum is exactly x.

   (a) Assume that S is unsorted. Give an $O(n \log n)$ algorithm for the problem.

   (b) Assume that S is sorted. Give an $O(n)$ algorithm for the problem.

4-11. *[8]* Design an $O(n)$ algorithm that, given a list of n elements, finds all the elements that appear more than $n/2$ times in the list. *Then*, design an $O(n)$ algorithm that, given a list of n elements, finds all the elements that appear more than $n/4$ times.

Applications of Sorting: Intervals and Sets

4-12. *[3]* Give an efficient algorithm to compute the union of sets A and B, where $n = \max(|A|, |B|)$. The output should be an array of distinct elements that form the union of the sets.

   (a) Assume that A and B are unsorted arrays. Give an $O(n \log n)$ algorithm for the problem.

   (b) Assume that A and B are sorted arrays. Give an $O(n)$ algorithm for the problem.

4-13. *[5]* A camera at the door tracks the entry time a_i and exit time b_i (assume $b_i > a_i$) for each of n persons p_i attending a party. Give an $O(n \log n)$ algorithm that analyzes this data to determine the time when the most people were simultaneously present at the party. You may assume that all entry and exit times are distinct (no ties).

4-14. *[5]* Given a list I of n intervals, specified as (x_i, y_i) pairs, return a list where the overlapping intervals are merged. For $I = \{(1,3), (2,6), (8,10), (7,18)\}$ the output should be $\{(1,6), (7,18)\}$. Your algorithm should run in worst-case $O(n \log n)$ time complexity.

4-15. *[5]* You are given a set S of n intervals on a line, with the ith interval described by its left and right endpoints (l_i, r_i). Give an $O(n \log n)$ algorithm to identify a point p on the line that is in the largest number of intervals.

As an example, for $S = \{(10, 40), (20, 60), (50, 90), (15, 70)\}$ no point exists in all four intervals, but $p = 50$ is an example of a point in three intervals. You can assume an endpoint counts as being in its interval.

4-16. *[5]* You are given a set S of n segments on the line, where segment S_i ranges from l_i to r_i. Give an efficient algorithm to select the fewest number of segments whose union completely covers the interval from 0 to m.

Heaps

4-17. *[3]* Devise an algorithm for finding the k smallest elements of an unsorted set of n integers in $O(n + k \log n)$.

4-18. *[5]* Give an $O(n \log k)$-time algorithm that merges k sorted lists with a total of n elements into one sorted list. (Hint: use a heap to speed up the obvious $O(kn)$-time algorithm).

4-19. *[5]* You wish to store a set of n numbers in either a max-heap or a sorted array. For each application below, state which data structure is better, or if it does not matter. Explain your answers.

(a) Find the maximum element quickly.

(b) Delete an element quickly.

(c) Form the structure quickly.

(d) Find the minimum element quickly.

4-20. *[5]* (a) Give an efficient algorithm to find the second-largest key among n keys. You can do better than $2n - 3$ comparisons.

(b) Then, give an efficient algorithm to find the third-largest key among n keys. How many key comparisons does your algorithm do in the worst case? Must your algorithm determine which key is largest and second-largest in the process?

Quicksort

4-21. *[3]* Use the partitioning idea of quicksort to give an algorithm that finds the *median* element of an array of n integers in expected $O(n)$ time. (Hint: must you look at both sides of the partition?)

4-22. *[3]* The *median* of a set of n values is the $\lceil n/2 \rceil$th smallest value.

(a) Suppose quicksort always pivoted on the median of the current sub-array. How many comparisons would quicksort make then in the worst case?

(b) Suppose quicksort always pivoted on the $\lceil n/3 \rceil$th smallest value of the current sub-array. How many comparisons would be made then in the worst case?

4-23. *[5]* Suppose an array A consists of n elements, each of which is *red, white,* or *blue.* We seek to sort the elements so that all the *reds* come before all the *whites,* which come before all the *blues.* The only operations permitted on the keys are:

- *Examine(A,i)* – report the color of the ith element of A.
- *Swap(A,i,j)* – swap the ith element of A with the jth element.

Find a correct and efficient algorithm for red–white–blue sorting. There is a linear-time solution.

4-24. *[3]* Give an efficient algorithm to rearrange an array of n keys so that all the negative keys precede all the non-negative keys. Your algorithm must be in-place, meaning you cannot allocate another array to temporarily hold the items. How fast is your algorithm?

4-25. *[5]* Consider a given pair of different elements in an input array to be sorted, say z_i and z_j. What is the most number of times z_i and z_j might be compared with each other during an execution of quicksort?

4-26. *[5]* Define the recursion depth of quicksort as the maximum number of successive recursive calls it makes before hitting the base case. What are the minimum and maximum possible recursion depths for randomized quicksort?

4-27. *[8]* Suppose you are given a permutation p of the integers 1 to n, and seek to sort them to be in increasing order $[1, \ldots, n]$. The only operation at your disposal is *reverse(p,i,j)*, which reverses the elements of a subsequence $p_i, \ldots, p_j$ in the permutation. For the permutation $[1, 4, 3, 2, 5]$ one reversal (of the second through fourth elements) suffices to sort.

- Show that it is possible to sort any permutation using $O(n)$ reversals.
- Now suppose that the cost of *reverse(p,i,j)* is equal to its length, the number of elements in the range, $|j - i| + 1$. Design an algorithm that sorts p in $O(n \log^2 n)$ cost. Analyze the running time and cost of your algorithm and prove correctness.

Mergesort

4-28. *[5]* Consider the following modification to merge sort: divide the input array into thirds (rather than halves), recursively sort each third, and finally combine the results using a three-way merge subroutine. What is the worst-case running time of this modified merge sort?

4-29. *[5]* Suppose you are given k sorted arrays, each with n elements, and you want to combine them into a single sorted array of kn elements. One approach is to use the merge subroutine repeatedly, merging the first two arrays, then merging the result with the third array, then with the fourth array, and so on until you merge in the kth and final input array. What is the running time?

4-30. *[5]* Consider again the problem of merging k sorted length-n arrays into a single sorted length-kn array. Consider the algorithm that first divides the k arrays into $k/2$ pairs of arrays, and uses the merge subroutine to combine each pair, resulting in $k/2$ sorted length-$2n$ arrays. The algorithm repeats this step until there is only one length-kn sorted array. What is the running time as a function of n and k?

Other Sorting Algorithms

4-31. *[5]* Stable sorting algorithms leave equal-key items in the same relative order as in the original permutation. Explain what must be done to ensure that mergesort is a stable sorting algorithm.

4-32. *[5]* Wiggle sort: Given an unsorted array A, reorder it such that $A[0] < A[1] > A[2] < A[3] \ldots$. For example, one possible answer for input $[3, 1, 4, 2, 6, 5]$ is $[1, 3, 2, 5, 4, 6]$. Can you do it in $O(n)$ time using only $O(1)$ space?

4-33. *[3]* Show that n positive integers in the range 1 to k can be sorted in $O(n \log k)$ time. The interesting case is when $k \ll n$.

4-34. *[5]* Consider a sequence S of n integers with many duplications, such that the number of distinct integers in S is $O(\log n)$. Give an $O(n \log \log n)$ worst-case time algorithm to sort such sequences.

4-35. *[5]* Let $A[1..n]$ be an array such that the first $n - \sqrt{n}$ elements are already sorted (though we know nothing about the remaining elements). Give an algorithm that sorts A in substantially better than $n \log n$ steps.

4-36. *[5]* Assume that the array $A[1..n]$ only has numbers from $\{1, \ldots, n^2\}$ but that at most $\log \log n$ of these numbers ever appear. Devise an algorithm that sorts A in substantially less than $O(n \log n)$.

4-37. *[5]* Consider the problem of sorting a sequence of n 0's and 1's using comparisons. For each comparison of two values x and y, the algorithm learns which of $x < y$, $x = y$, or $x > y$ holds.

   (a) Give an algorithm to sort in $n - 1$ comparisons in the worst case. Show that your algorithm is optimal.

   (b) Give an algorithm to sort in $2n/3$ comparisons in the average case (assuming each of the n inputs is 0 or 1 with equal probability). Show that your algorithm is optimal.

4-38. *[6]* Let P be a simple, but not necessarily convex, n-sided polygon and q an arbitrary point not necessarily in P. Design an efficient algorithm to find a line segment originating from q that intersects the maximum number of edges of P. In other words, if standing at point q, in what direction should you aim a gun so the bullet will go through the largest number of walls. A bullet through a vertex of P gets credit for only one wall. An $O(n \log n)$ algorithm is possible.

Lower Bounds

4-39. *[5]* In one of my research papers [Ski88], I discovered a comparison-based sorting algorithm that runs in $O(n \log(\sqrt{n}))$. Given the existence of an $\Omega(n \log n)$ lower bound for sorting, how can this be possible?

4-40. *[5]* Mr. B. C. Dull claims to have developed a new data structure for priority queues that supports the operations *Insert*, *Maximum*, and *Extract-Max*—all in $O(1)$ worst-case time. Prove that he is mistaken. (Hint: the argument does not involve a lot of gory details—just think about what this would imply about the $\Omega(n \log n)$ lower bound for sorting.)

Searching

4-41. *[3]* A company database consists of 10,000 sorted names, 40% of whom are known as good customers and who together account for 60% of the accesses to the database. There are two data structure options to consider for representing the database:

- Put all the names in a single array and use binary search.

- Put the good customers in one array and the rest of them in a second array. Only if we do not find the query name on a binary search of the first array do we do a binary search of the second array.

Demonstrate which option gives better expected performance. Does this change if linear search on an unsorted array is used instead of binary search for both options?

4-42. *[5]* A *Ramanujan number* can be written two different ways as the sum of two cubes—meaning there exist distinct positive integers a, b, c, and d such that $a^3 + b^3 = c^3 + d^3$. For example, 1729 is a Ramanujan number because $1729 = 1^3 + 12^3 = 9^3 + 10^3$.

(a) Give an efficient algorithm to test whether a given single integer n is a Ramanujan number, with an analysis of the algorithm's complexity.

(b) Now give an efficient algorithm to generate *all* the Ramanujan numbers between 1 and n, with an analysis of its complexity.

Implementation Challenges

4-43. *[5]* Consider an $n \times n$ array A containing integer elements (positive, negative, and zero). Assume that the elements in each row of A are in strictly increasing order, and the elements of each column of A are in strictly decreasing order. (Hence there cannot be two zeros in the same row or the same column.) Describe an efficient algorithm that counts the number of occurrences of the element 0 in A. Analyze its running time.

4-44. *[6]* Implement versions of several different sorting algorithms, such as selection sort, insertion sort, heapsort, mergesort, and quicksort. Conduct experiments to assess the relative performance of these algorithms in a simple application that reads a large text file and reports exactly one instance of each word that appears within it. This application can be efficiently implemented by sorting all the words that occur in the text and then passing through the sorted sequence to identify one instance of each distinct word. Write a brief report with your conclusions.

4-45. *[5]* Implement an external sort, which uses intermediate files to sort files bigger than main memory. Mergesort is a good algorithm to base such an implementation on. Test your program both on files with small records and on files with large records.

4-46. *[8]* Design and implement a parallel sorting algorithm that distributes data across several processors. An appropriate variation of mergesort is a likely candidate. Measure the speedup of this algorithm as the number of processors increases. Then compare the execution time to that of a purely sequential mergesort implementation. What are your experiences?

Interview Problems

4-47. *[3]* If you are given a million integers to sort, what algorithm would you use to sort them? How much time and memory would that consume?

4-48. *[3]* Describe advantages and disadvantages of the most popular sorting algorithms.

4-49. *[3]* Implement an algorithm that takes an input array and returns only the unique elements in it.

4-50. *[5]* You have a computer with only 4 GB of main memory. How do you use it to sort a large file of 500 GB that is on disk?

4-51. *[5]* Design a stack that supports push, pop, and retrieving the minimum element in constant time.

4-52. *[5]* Given a search string of three words, find the smallest snippet of the document that contains all three of the search words—that is, the snippet with the smallest number of words in it. You are given the index positions where these words occur in the document, such as *word1: (1, 4, 5)*, *word2: (3, 9, 10)*, and *word3: (2, 6, 15)*. Each of the lists are in sorted order, as above.

4-53. *[6]* You are given twelve coins. One of them is heavier or lighter than the rest. Identify this coin in just three weighings with a balance scale.

LeetCode

4-1. `https://leetcode.com/problems/sort-list/`

4-2. `https://leetcode.com/problems/queue-reconstruction-by-height/`

4-3. `https://leetcode.com/problems/merge-k-sorted-lists/`

4-4. `https://leetcode.com/problems/find-k-pairs-with-smallest-sums/`

HackerRank

4-1. `https://www.hackerrank.com/challenges/quicksort3/`

4-2. `https://www.hackerrank.com/challenges/mark-and-toys/`

4-3. `https://www.hackerrank.com/challenges/organizing-containers-of-balls/`

Programming Challenges

These programming challenge problems with robot judging are available at `https://onlinejudge.org`:

4-1. "Vito's Family"—Chapter 4, problem 10041.

4-2. "Stacks of Flapjacks"—Chapter 4, problem 120.

4-3. "Bridge"—Chapter 4, problem 10037.

4-4. "ShoeMaker's Problem"—Chapter 4, problem 10026.

4-5. "ShellSort"—Chapter 4, problem 10152.

Chapter 5

Divide and Conquer

One of the most powerful techniques for solving problems is to break them down into smaller, more easily solved pieces. Smaller problems are less overwhelming, and they permit us to focus on details that are lost when we are studying the whole thing. A recursive algorithm starts to become apparent whenever we can break the problem into smaller instances of the same type of problem. Multicore processors now sit in almost every computer, but effective parallel processing requires decomposing jobs into at least as many tasks as the number of processors.

Two important algorithm design paradigms are based on breaking problems down into smaller problems. In Chapter 10, we will see dynamic programming, which typically removes one element from the problem, solves the smaller problem, and then adds back the element to the solution of this smaller problem in the proper way. *Divide and conquer* instead splits the problem into (say) halves, solves each half, then stitches the pieces back together to form a full solution.

Thus, to use divide and conquer as an algorithm design technique, we must divide the problem into two smaller subproblems, solve each of them recursively, and then meld the two partial solutions into one solution to the full problem. Whenever the merging takes less time than solving the two subproblems, we get an efficient algorithm. Mergesort, discussed in Section 4.5 (page 127), is the classic example of a divide-and-conquer algorithm. It takes only linear time to merge two sorted lists of $n/2$ elements, each of which was obtained in $O(n \lg n)$ time.

Divide and conquer is a design technique with many important algorithms to its credit, including mergesort, the fast Fourier transform, and Strassen's matrix multiplication algorithm. Beyond binary search and its many variants, however, I find it to be a difficult design technique to apply in practice. Our ability to analyze divide and conquer algorithms rests on our proficiency in solving the recurrence relations governing the cost of such recursive algorithms, so we will introduce techniques for solving recurrences here.

S. S. Skiena, *The Algorithm Design Manual*, Texts in Computer Science, https://doi.org/10.1007/978-3-030-54256-6_5

5.1   Binary Search and Related Algorithms

The mother of all divide-and-conquer algorithms is *binary search*, which is a fast algorithm for searching in a sorted array of keys S. To search for key q, we compare q to the middle key $S[n/2]$. If q appears before $S[n/2]$, it must reside in the left half of S; if not, it must reside in the right half of S. By repeating this process recursively on the correct half, we locate the key in a total of $\lceil \lg n \rceil$ comparisons—a big win over the $n/2$ comparisons expected using sequential search:

```c
int binary_search(item_type s[], item_type key, int low, int high) {
    int middle;      /* index of middle element */

    if (low > high) {
        return (-1);     /* key not found */
    }

    middle = (low + high) / 2;

    if (s[middle] == key) {
        return(middle);
    }

    if (s[middle] > key) {
        return(binary_search(s, key, low, middle - 1));
    } else {
        return(binary_search(s, key, middle + 1, high));
    }
}
```

This much you probably know. What is important is to understand is just how fast binary search is. *Twenty questions* is a popular children's game where one player selects a word and the other repeatedly asks true/false questions until they guess it. If the word remains unidentified after twenty questions, the first party wins. But the second player has a winning strategy, based on binary search. Take a printed dictionary, open it in the middle, select a word (say "move"), and ask whether the unknown word is before "move" in alphabetical order. Standard dictionaries contain between 50,000 to 200,000 words, so we can be certain that the process will terminate within twenty questions.

5.1.1   Counting Occurrences

Several interesting algorithms are variants of binary search. Suppose that we want to count the number of times a given key k (say "Skiena") occurs in a given sorted array. Because sorting groups all the copies of k into a contiguous block, the problem reduces to finding that block and then measuring its size.

The binary search routine presented above enables us to find the index of an element x of the correct block in $O(\lg n)$ time. A natural way to identify the boundaries of the block is to sequentially test elements to the left of x until we find one that differs from the search key, and then repeat this search to the right of x. The difference between the indices of these boundaries (plus one) gives the number of occurrences of k.

This algorithm runs in $O(\lg n + s)$, where s is the number of occurrences of the key. But this can be as bad as $\Theta(n)$ if the entire array consists of identical keys. A faster algorithm results by modifying binary search to find the *boundary* of the block containing k, instead of k itself. Suppose we delete the equality test

```
        if (s[middle] == key) return(middle);
```

from the implementation above and return the index `high` instead of -1 on each unsuccessful search. *All* searches will now be unsuccessful, since there is no equality test. The search will proceed to the right half whenever the key is compared to an identical array element, eventually terminating at the right boundary. Repeating the search after reversing the direction of the binary comparison will lead us to the left boundary. Each search takes $O(\lg n)$ time, so we can count the occurrences in logarithmic time regardless of the size of the block.

By modifying our binary search routine to return `(low+high)/2` instead of `-1` on an unsuccessful search, we obtain the location between two array elements where the key k should have been. This variant suggests another way to solve our length of run problem. We search for the positions of keys $k - \epsilon$ and $k + \epsilon$, where ϵ is a tiny enough constant that both searches are guaranteed to fail with no intervening keys. Again, doing two binary searches takes $O(\log n)$ time.

5.1.2 One-Sided Binary Search

Now suppose we have an array A consisting of a run of 0's, followed by an unbounded run of 1's, and would like to identify the exact point of transition between them. Binary search on the array would find the transition point in $\lceil \lg n \rceil$ tests, if we had a bound n on the number of elements in the array.

But in the absence of such a bound, we can test repeatedly at larger intervals ($A[1]$, $A[2]$, $A[4]$, $A[8]$, $A[16]$, ...) until we find a non-zero value. Now we have a window containing the target and can proceed with binary search. This *one-sided binary search* finds the transition point p using at most $2\lceil \lg p \rceil$ comparisons, regardless of how large the array actually is. One-sided binary search is useful whenever we are looking for a key that lies close to our current position.

5.1.3 Square and Other Roots

The square root of n is the positive number r such that $r^2 = n$. Square root computations are performed inside every pocket calculator, but it is instructive to develop an efficient algorithm to compute them.

First, observe that the square root of $n \geq 1$ must be at least 1 and at most n. Let $l = 1$ and $r = n$. Consider the midpoint of this interval, $m = (l + r)/2$. How does m^2 compare to n? If $n \geq m^2$, then the square root must be greater than m, so the algorithm repeats with $l = m$. If $n < m^2$, then the square root must be less than m, so the algorithm repeats with $r = m$. Either way, we have halved the interval using only one comparison. Therefore, after $\lceil \lg n \rceil$ rounds we will have identified the square root to within $\pm 1/2$.

This *bisection method*, as it is called in numerical analysis, can also be applied to the more general problem of finding the roots of an equation. We say that x is a *root* of the function f if $f(x) = 0$. Suppose that we start with values l and r such that $f(l) > 0$ and $f(r) < 0$. If f is a continuous function, there must exist a root between l and r. Depending upon the sign of $f(m)$, where $m = (l + r)/2$, we can cut the window containing the root in half with each test, and stop as soon as our estimate becomes sufficiently accurate.

Root-finding algorithms converging faster than binary search are known for both of these problems. Instead of always testing the midpoint of the interval, these algorithms interpolate to find a test point closer to the actual root. Still, binary search is simple, robust, and works as well as possible without additional information on the nature of the function to be computed.

Take-Home Lesson: Binary search and its variants are the quintessential divide-and-conquer algorithms.

5.2   War Story: Finding the Bug in the Bug

Yutong stood up to announce the results of weeks of hard work. "Dead," he announced defiantly. Everybody in the room groaned.

I was part of a team developing a new approach to create vaccines: *Synthetic Attenuated Virus Engineering* or SAVE. Because of how the genetic code works, there are typically about 3^n different possible genes that code for any given protein of length n. To a first approximation, all of these are the same, since they describe exactly the same protein. But each of the 3^n synonymous genes use the biological machinery in somewhat different ways, translating at somewhat different speeds.

By substituting a virus gene with a less dangerous replacement, we hoped to create a vaccine: a weaker version of the disease-causing agent that otherwise did the same thing. Your body could fight off the weak virus without you getting sick, along the way training your immune system to fight off tougher villains. But we needed weak viruses, not dead ones: you can't learn anything fighting off something that is already dead.

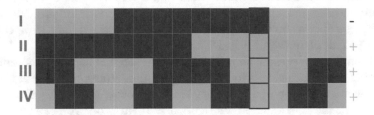

Figure 5.1: Designs of four synthetic genes to locate a specific sequence signal. The green regions are drawn from a viable sequence, while the red regions are drawn from a lethally defective sequence. Genes II, III, and IV were viable while gene I was defective, an outcome that can only be explained by a lethal signal in the region located fifth from the right.

"Dead means that there must be one place in this 1,200-base region where the virus evolved a signal, a second meaning of the sequence it needs for survival," said our senior virologist. By changing the sequence at this point, we killed the virus. "We have to find this signal to bring it back to life."

"But there are 1,200 places to look! How can we find it?" Yutong asked.

I thought about this a bit. We had to debug a bug. This sounded similar to the problem of debugging a program. I recall many a lonesome night spent trying to figure out exactly which line number was causing my program to crash. I often stooped to commenting out chunks of the code, and then running it again to test if it still crashed. It was usually easy to figure out the problem after I got the commented-out region down to a small enough chunk. The best way to search for this region was...

"Binary search!" I announced. Suppose we replace the first half of the coding sequence of the viable gene (shown in green) with the coding sequence of the dead, critical signal-deficient strain (shown in red), as in design II in Figure 5.1. If this hybrid gene is viable, it means the critical signal must occur in the right half of the gene, whereas a dead virus implies the problem must occur in the left half. Through a binary search process the signal can be located to one of n regions in $\lceil \log_2 n \rceil$ sequential rounds of experiment.

"We can narrow the area containing the signal in a length-n gene to $n/16$ by doing only four experiments," I informed them. The senior virologist got excited. But Yutong turned pale.

"Four more rounds of experiments!" he complained. "It took me a full month to synthesize, clone, and try to grow the virus the last time. Now you want me to do it again, wait to learn which half the signal is in, and then repeat three more times? Forget about it!"

Yutong realized that the power of binary search came from *interaction*: the query that we make in round r depends upon the answers to the queries in rounds 1 through $r - 1$. Binary search is an inherently sequential algorithm. When each individual comparison is a slow and laborious process, suddenly $\lg n$

comparisons doesn't look so good. But I had a very cute trick up my sleeve.

"Four successive rounds of this will be too much work for you, Yutong. But might you be able to do four different designs at the same time if we could give them to you all at once?" I asked.

"If I am doing the same things to four different sequences at the same time, it is no big deal," he said. "Not much harder than doing just one of them."

That settled, I proposed that they simultaneously synthesize the four virus designs labeled I, II, III, and IV in Figure 5.1. It turns out you *can* parallelize binary search, provided your queries can be arbitrary subsets instead of connected halves. Observe that each of the columns defined by these four designs consists of a distinct pattern of red and green. The pattern of living/dead among the four synthetic designs thus uniquely defines the position of the critical signal in one experimental round. In this example, virus I happened to be dead while the other three lived, pinpointing the location of the lethal signal to the fifth region from the right.

Yutong rose to the occasion, and after a month of toil (but not months) discovered a new signal in poliovirus [SLW+12]. He found the bug in the bug through the idea of divide and conquer, which works best when splitting the problem in half at each point. Note that all four of our designs consist of half red and half green, arranged so all sixteen regions have a distinct pattern of colors. With interactive binary search, the last test selects between just two remaining regions. By expanding each test to have half the sequence, we eliminated the need for sequential tests, making the entire process much faster.

5.3   Recurrence Relations

Many divide-and-conquer algorithms have time complexities that are naturally modeled by recurrence relations. The ability to solve such recurrences is important to understanding when divide-and-conquer algorithms perform well, and provides an important tool for analysis in general. The reader who balks at the very idea of analysis is free to skip this section, but important insights come from an understanding of the behavior of recurrence relations.

What is a recurrence relation? It is an equation in which a function is defined in terms of itself. The Fibonacci numbers are described by the recurrence relation

$$F_n = F_{n-1} + F_{n-2}$$

together with the initial values $F_0 = 0$ and $F_1 = 1$, as will be discussed in Section 10.1.1. Many other familiar functions are easily expressed as recurrences. Any polynomial can be represented by a recurrence, such as the linear function:

$$a_n = a_{n-1} + 1, a_1 = 1 \longrightarrow a_n = n$$

Any exponential can be represented by a recurrence:

$$a_n = 2a_{n-1}, a_1 = 1 \longrightarrow a_n = 2^{n-1}$$

Finally, lots of weird functions that cannot be easily described using conventional notation can be represented naturally by a recurrence, for example:

$$a_n = na_{n-1}, a_1 = 1 \quad \longrightarrow \quad a_n = n!$$

This shows that recurrence relations are a very versatile way to represent functions.

The self-reference property of recurrence relations is shared with recursive programs or algorithms, as the shared roots of both terms reflect. Essentially, recurrence relations provide a way to analyze recursive structures, such as algorithms.

5.3.1 Divide-and-Conquer Recurrences

A typical divide-and-conquer algorithm breaks a given problem into a smaller pieces, each of which is of size n/b. It then spends $f(n)$ time to combine these subproblem solutions into a complete result. Let $T(n)$ denote the worst-case time this algorithm takes to solve a problem of size n. Then $T(n)$ is given by the following recurrence relation:

$$T(n) = a \cdot T(n/b) + f(n)$$

Consider the following examples, based on algorithms we have previously seen:

- *Mergesort* – The running time of mergesort is governed by the recurrence $T(n) = 2T(n/2) + O(n)$, since the algorithm divides the data into equal-sized halves and then spends linear time merging the halves after they are sorted. In fact, this recurrence evaluates to $T(n) = O(n \lg n)$, just as we got by our previous analysis.

- *Binary search* – The running time of binary search is governed by the recurrence $T(n) = T(n/2) + O(1)$, since at each step we spend constant time to reduce the problem to an instance half its size. This recurrence evaluates to $T(n) = O(\lg n)$, just as we got by our previous analysis.

- *Fast heap construction* – The `bubble_down` method of heap construction (described in Section 4.3.4) builds an n-element heap by constructing two $n/2$ element heaps and then merging them with the root in logarithmic time. The running time is thus governed by the recurrence relation $T(n) = 2T(n/2) + O(\lg n)$. This recurrence evaluates to $T(n) = O(n)$, just as we got by our previous analysis.

Solving a recurrence means finding a nice closed form describing or bounding the result. We can use the *master theorem*, discussed in Section 5.4, to solve the recurrence relations typically arising from divide-and-conquer algorithms.

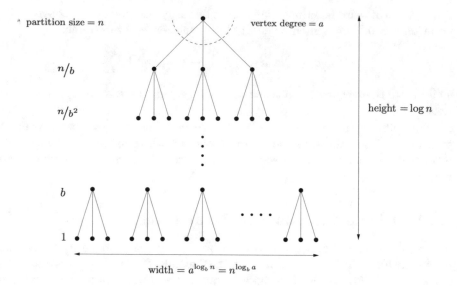

Figure 5.2:   The recursion tree resulting from decomposing each problem of size n into a problems of size n/b

5.4   Solving Divide-and-Conquer Recurrences

Divide-and-conquer recurrences of the form $T(n) = aT(n/b) + f(n)$ are generally easy to solve, because the solutions typically fall into one of three distinct cases:

1. If $f(n) = O(n^{\log_b a - \epsilon})$ for some constant $\epsilon > 0$, then $T(n) = \Theta(n^{\log_b a})$.

2. If $f(n) = \Theta(n^{\log_b a})$, then $T(n) = \Theta(n^{\log_b a} \lg n)$.

3. If $f(n) = \Omega(n^{\log_b a + \epsilon})$ for some constant $\epsilon > 0$, and if $af(n/b) \leq cf(n)$ for some $c < 1$, then $T(n) = \Theta(f(n))$.

Although this looks somewhat frightening, it really isn't difficult to apply. The issue is identifying which case of this so-called *master theorem* holds for your given recurrence. Case 1 holds for heap construction and matrix multiplication, while Case 2 holds for mergesort. Case 3 generally arises with clumsier algorithms, where the cost of combining the subproblems dominates everything.

The master theorem can be thought of as a black-box piece of machinery, invoked as needed and left with its mystery intact. However, after a little study it becomes apparent why the master theorem works.

Figure 5.2 shows the recursion tree associated with a typical $T(n) = aT(n/b) + f(n)$ divide-and-conquer algorithm. Each problem of size n is decomposed into a problems of size n/b. Each subproblem of size k takes $O(f(k))$ time to deal with internally, between partitioning and merging. The total time for the algorithm is the sum of these internal evaluation costs, plus the overhead of building the recursion tree. The height of this tree is $h = \log_b n$ and the number of leaf nodes is $a^h = a^{\log_b n}$, which happens to simplify to $n^{\log_b a}$ with some algebraic manipulation.

The three cases of the master theorem correspond to three different costs, each of which might be dominant as a function of a, b, and $f(n)$:

- *Case 1: Too many leaves* – If the number of leaf nodes outweighs the overall internal evaluation cost, the total running time is $O(n^{\log_b a})$.

- *Case 2: Equal work per level* – As we move down the tree, each problem gets smaller but there are more of them to solve. If the sums of the internal evaluation costs at each level are equal, the total running time is the cost per level $(n^{\log_b a})$ times the number of levels $(\log_b n)$, for a total running time of $O(n^{\log_b a} \lg n)$.

- *Case 3: Too expensive a root* – If the internal evaluation cost grows very rapidly with n, then the cost of the root evaluation may dominate everything. Then the total running time is $O(f(n))$.

Once you accept the master theorem, you can easily analyze any divide-and-conquer algorithm, given only the recurrence associated with it. We use this approach on several algorithms below.

5.5   Fast Multiplication

You know at least two ways to multiply integers A and B to get $A \times B$. You first learned that $A \times B$ meant adding up B copies of A, which gives an $O(n \cdot 10^n)$ time algorithm to multiply two n-digit base-10 numbers. Then you learned to multiply long numbers on a digit-by-digit basis, like

$$9256 \times 5367 = 9256 \times 7 + 9256 \times 60 + 9256 \times 300 + 9256 \times 5000 = 13{,}787{,}823$$

Recall that those zeros we pad the digit-terms by are not *really* computed as products. We implement their effect by shifting the product digits to the correct place. Assuming we perform each real digit-by-digit product in constant time, by looking it up in a times table, this algorithm multiplies two n-digit numbers in $O(n^2)$ time.

In this section I will present an even faster algorithm for multiplying large numbers. It is a classic divide-and-conquer algorithm. Suppose each number has $n = 2m$ digits. Observe that we can split each number into two pieces each of m digits, such that the product of the full numbers can easily be constructed from the products of the pieces, as follows. Let $w = 10^{m+1}$, and represent $A = a_0 + a_1 w$ and $B = b_0 + b_1 w$, where a_i and b_i are the pieces of each respective number. Then:

$$A \times B = (a_0 + a_1 w) \times (b_0 + b_1 w) = a_0 b_0 + a_0 b_1 w + a_1 b_0 w + a_1 b_1 w^2$$

This procedure reduces the problem of multiplying two n-digit numbers to four products of $(n/2)$-digit numbers. Recall that multiplication by w doesn't count: it is simply padding the product with zeros. We also have to add together these four products once computed, which is $O(n)$ work.

Let $T(n)$ denote the amount of time it takes to multiply two n-digit numbers. Assuming we use the same algorithm recursively on each of the smaller products, the running time of this algorithm is given by the recurrence:

$$T(n) = 4T(n/2) + O(n)$$

Using the master theorem (case 1), we see that this algorithm runs in $O(n^2)$ time, exactly the same as the digit-by-digit method. We divided, but we did not conquer.

Karatsuba's algorithm is an alternative recurrence for multiplication, which yields a better running time. Suppose we compute the following *three* products:

$$q_0 = a_0 b_0$$
$$q_1 = (a_0 + a_1)(b_0 + b_1)$$
$$q_2 = a_1 b_1$$

Note that

$$A \times B = (a_0 + a_1 w) \times (b_0 + b_1 w) = a_0 b_0 + a_0 b_1 w + a_1 b_0 w + a_1 b_1 w^2$$
$$= q_0 + (q_1 - q_0 - q_2)w + q_2 w^2$$

so now we have computed the full product with just three half-length multiplications and a constant number of additions. Again, the w terms don't count as multiplications: recall that they are really just zero shifts. The time complexity of this algorithm is therefore governed by the recurrence

$$T(n) = 3T(n/2) + O(n)$$

Since $n = O(n^{\log_2 3})$, this is solved by the first case of the master theorem, and $T(n) = \Theta(n^{\log_2 3}) = \Theta(n^{1.585})$. This is a substantial improvement over the quadratic algorithm for large numbers, and indeed beats the standard multiplication algorithm soundly for numbers of 500 digits or so.

This approach of defining a recurrence that uses fewer multiplications but more additions also lurks behind fast algorithms for matrix multiplication. The nested-loops algorithm for matrix multiplication discussed in Section 2.5.4 takes $O(n^3)$ for two $n \times n$ matrices, because we compute the dot product of n terms for each of the n^2 elements in the product matrix. However, Strassen [Str69] discovered a divide-and-conquer algorithm that manipulates the products of seven $n/2 \times n/2$ matrix products to yield the product of two $n \times n$ matrices. This yields a time-complexity recurrence

$$T(n) = 7 \cdot T(n/2) + O(n^2)$$

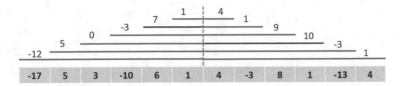

Figure 5.3: The largest subrange sum is either entirely to the left of center or entirely to the right, or (like here) the sum of the largest center-bordering ranges on left and right.

Because $\log_2 7 \approx 2.81$, $O(n^{\log_2 7})$ dominates $O(n^2)$, so Case 1 of the master theorem applies and $T(n) = \Theta(n^{2.81})$.

This algorithm has been repeatedly "improved" by increasingly complicated recurrences, and the current best is $O(n^{2.3727})$. See Section 16.3 (page 472) for more detail.

5.6 Largest Subrange and Closest Pair

Suppose you are tasked with writing the advertising copy for a hedge fund whose monthly performance this year was

$$[-17, 5, 3, -10, 6, 1, 4, -3, 8, 1, -13, 4]$$

You lost money for the year, but from May through October you had your greatest gains over any period, a net total of 17 units of gains. This gives you something to brag about.

The largest subrange problem takes an array A of n numbers, and asks for the index pair i and j that maximizes $S = \sum_{k=i}^{j} A[k]$. Summing the entire array does not necessarily maximize S because of negative numbers. Explicitly testing each possible interval start–end pair requires $\Omega(n^2)$ time. Here I present a divide-and-conquer algorithm that runs in $O(n \log n)$ time.

Suppose we divide the array A into left and right halves. Where can the largest subrange be? It is either in the left half or the right half, or includes the middle. A recursive program to find the largest subrange between $A[l]$ and $A[r]$ can easily call itself to work on the left and right subproblems. How can we find the largest subrange spanning the middle, that is, spanning positions m and $m + 1$?

The key is to observe that the largest subrange centered spanning the middle will be the union of the largest subrange on the left ending on m, and the largest subrange on the right starting from $m+1$, as illustrated in Figure 5.3. The value V_l of the largest such subrange on the left can be found in linear time with a sweep:

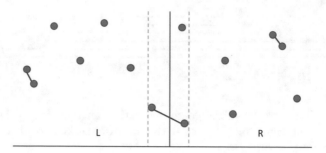

Figure 5.4:   The closest pair of points in two dimensions either lie to the left of center, to the right, or in a thin strip straddling the center.

LeftMidMaxRange(A, l, m)
    $S = M = 0$
    for $i = m$ downto l
        $S = S + A[i]$
        if $(S > M)$ then $M = S$
    return S

The corresponding value on the right can be found analogously. Dividing n into two halves, doing linear work, and recurring takes time $T(n)$, where

$$T(n) = 2 \cdot T(n/2) + \Theta(n)$$

Case 2 of the master theorem yields $T(n) = \Theta(n \log n)$.

This general approach of "find the best on each side, and then check what is straddling the middle" can be applied to other problems as well. Consider the problem of finding the smallest distance between pairs among a set of n points.

In one dimension, this problem is easy: we saw in Section 4.1 (page 109) that after sorting the points, the closest pair must be neighbors. A linear-time sweep from left to right after sorting thus yields an $\Theta(n \log n)$ algorithm. But we can replace this sweep by a cute divide-and-conquer algorithm. The closest pair is defined by the left half of the points, the right half, or the pair in the middle, so the following algorithm must find it:

ClosestPair(A, l, r)
    $mid = \lfloor (l + r)/2 \rfloor$
    l_{min} = ClosestPair(A, l, mid)
    r_{min} = ClosestPair($A, mid + 1, r$)
    return min($l_{min}, r_{min}, A[m + 1] - A[m]$)

Because this does constant work per call, its running time is given by the recurrence:

$$T(n) = 2 \cdot T(n/2) + O(1)$$

Case 1 of the master theorem tells us that $T(n) = \Theta(n)$.

This is still linear time and so might not seem very impressive, but let's generalize the idea to points in two dimensions. After we sort the n (x, y) points according to their x-coordinates, the same property must be true: the closest pair is either two points on the left half or two points on the right, or it straddles left and right. As shown in Figure 5.4, these straddling points had better be close to the dividing line (distance $d < \min(l_{min}, r_{min})$) and also have very similar y-coordinates. With clever bookkeeping, the closest straddling pair can be found in linear time, yielding a running time of

$$T(n) = 2 \cdot T(n/2) + \Theta(n) = \Theta(n \log n)$$

as defined by Case 2 of the master theorem.

5.7 Parallel Algorithms

Two heads are better than one, and more generally, n heads better than $n - 1$. Parallel processing has become increasingly prevalent, with the advent of multi-core processors and cluster computing.

5.7.1 Data Parallelism

Divide and conquer is the algorithm paradigm most suited to parallel computation. Typically, we seek to partition our problem of size n into p equal-sized parts, and simultaneously feed one to each processor. This reduces the time to completion (or *makespan*) from $T(n)$ to $T(n/p)$, plus the cost of combining the results together. If $T(n)$ is linear, this gives us a maximum possible speedup of p. If $T(n) = \Theta(n^2)$ it may look like we can do even better, but this is generally an illusion. Suppose we want to sweep through all pairs of n items. Sure we can partition the items into p independent chunks, but $n^2 - p(n/p)^2$ of the n^2 possible pairs will not ever have both elements on the same processor.

Multiple processors are typically best deployed to exploit *data parallelism*, running a single algorithm on different and independent data sets. For example, computer animation systems must render thirty frames per second for realistic animation. Assigning each frame to a distinct processor, or dividing each image into regions assigned to different processors, might be the best way to get the job done in time. Such tasks are often called *embarrassingly parallel*.

Generally speaking, such data parallel approaches are not algorithmically interesting, but they are simple and effective. There is a more advanced world of parallel algorithms where different processors synchronize their efforts so they can together solve a single problem quicker than one can. These algorithms are out of the scope of what we will cover in this book, but be aware of the challenges involved in the design and implementation of sophisticated parallel algorithms.

5.7.2   Pitfalls of Parallelism

There are several potential pitfalls and complexities associated with parallel algorithms:

- *There is often a small upper bound on the potential win* – Suppose that you have access to a machine with 24 cores that can be devoted exclusively to your job. These can potentially be used to speed up the fastest sequential program by up to a factor of 24. Sweet! But even greater performance gains can often result from developing more efficient sequential algorithms. Your time spent parallelizing a code might well be better spent enhancing the sequential version. Performance-tuning tools such as profilers are better developed for sequential machines/programs than for parallel models.

- *Speedup means nothing* – Suppose my parallel program runs 24 times faster on a 24-core machine than it does on a single processor. That's great, isn't it? If you get linear speedup and can increase the number of processors without bound, you will eventually beat any sequential algorithm. But the one-processor parallel version of your code is likely to be a crummy sequential algorithm, so measuring speedup typically provides an unfair test of the benefits of parallelism. And it is hard to buy machines with an unlimited number of cores.

  The classic example of this phenomenon occurs in the minimax game-tree search algorithm used in computer chess programs. A brute-force tree search is embarrassingly easy to parallelize: just put each subtree on a different processor. However, a lot of work gets wasted because the same positions get considered on different machines. Moving from a brute-force search to the more clever alpha–beta pruning algorithm can easily save 99.99% of the work, thus dwarfing any benefits of a parallel brute-force search. Alpha–beta can be parallelized, but not easily, and the speedups grow surprisingly slowly as a function of the number of processors you have.

- *Parallel algorithms are tough to debug* – Unless your problem can be decomposed into several independent jobs, the different processors must communicate with each other to end up with the correct final result. Unfortunately, the non-deterministic nature of this communication makes parallel programs notoriously difficult to debug, because you will get different results each time you run the code. Data parallel programs typically have no communication except copying the results at the end, which makes things much simpler.

I recommend considering parallel processing only after attempts at solving a problem sequentially prove too slow. Even then, I would restrict attention to data parallel algorithms where no communication is needed between the processors, except to collect the final results. Such large-grain, naive parallelism can be simple enough to be both implementable and debuggable, because it really reduces to producing a good sequential implementation. There can be pitfalls even in this approach, however, as shown by the following war story.

5.8   War Story: Going Nowhere Fast

In Section 2.9 (page 54), I related our efforts to build a fast program to test Waring's conjecture for pyramidal numbers. At that point, my code was fast enough that it could complete the job in a few weeks running in the background of a desktop workstation. This option did not appeal to my supercomputing colleague, however.

"Why don't we do it in parallel?" he suggested. "After all, you have an outer loop doing the same calculation on each integer from 1 to 1,000,000,000. I can split this range of numbers into different intervals and run each range on a different processor. Divide and conquer. Watch, it will be easy."

He set to work trying to do our computations on an Intel IPSC-860 hypercube using 32 nodes with 16 megabytes of memory per node—very big iron for the time. However, instead of getting answers, over the next few weeks I was treated to a regular stream of e-mail about system reliability:

- "Our code is running fine, except one processor died last night. I will rerun."

- "This time the machine was rebooted by accident, so our long-standing job was killed."

- "We have another problem. The policy on using our machine is that nobody can command the entire machine for more than 13 hours, under any condition."

Still, eventually, he rose to the challenge. Waiting until the machine was stable, he locked out 16 processors (half the computer), divided the integers from 1 to 1,000,000,000 into 16 equal-sized intervals, and ran each interval on its own processor. He spent the next day fending off angry users who couldn't get their work done because of our rogue job. The instant the first processor completed analyzing the numbers from 1 to 62,500,000, he announced to all the people yelling at him that the rest of the processors would soon follow.

But they didn't. He failed to realize that the time to test each integer increased as the numbers got larger. After all, it would take longer to test whether 1,000,000,000 could be expressed as the sum of three pyramidal numbers than it would for 100. Thus, at longer and longer intervals, each new processor would announce its completion. Because of the architecture of the hypercube, he couldn't return any of the processors until our entire job was completed. Eventually, half the machine and most of its users were held hostage by one, final interval.

What conclusions can be drawn from this? If you are going to parallelize a problem, be sure to balance the load carefully among the processors. Proper load balancing, using either back-of-the-envelope calculations or the partition algorithm we will develop in Section 10.7 (page 333), would have significantly reduced the length of time we needed the machine, and his exposure to the wrath of his colleagues.

5.9   Convolution (*)

The *convolution* of two arrays (or vectors) A and B is a new vector C such that

$$C[k] = \sum_{j=0}^{m-1} A[j] \cdot B[k-j]$$

If we assume that A and B are of length m and n respectively, and indexed starting from 0, the natural range on C is from $C[0]$ to $C[n+m-2]$. The values of all out-of-range elements of A and B are interpreted as zero, so they do not contribute to any product.

An example of convolution that you are familiar with is *polynomial multiplication*. Recall the problem of multiplying two polynomials, for example:

$$(3x^2 + 2x + 6) \times (4x^2 + 3x + 2) = (3 \cdot 4)x^4 + (3 \cdot 3 + 2 \cdot 4)x^3$$
$$+ (3 \cdot 2 + 2 \cdot 3 + 6 \cdot 4)x^2 + (2 \cdot 2 + 6 \cdot 3)x^1 + (6 \cdot 2)x^0$$

Let $A[i]$ and $B[i]$ denote the coefficients of x^i in each of the polynomials. Then multiplication is a convolution, because the coefficient of the x^k term in the product polynomial is given by the convolution $C[k]$ above. This coefficient is the sum of the products of all terms which have exponent pairs adding to k: for example, $x^5 = x^4 \cdot x^1 = x^3 \cdot x^2$.

The obvious way to implement convolution is by computing the m term dot product $C[k]$ for each $0 \le k \le n + m - 2$. This is two nested loops, running in $\Theta(nm)$ time. The inner loop does not always involve m iterations because of boundary conditions. Simpler loop bounds could have been employed if A and B were flanked by ranges of zeros.

```
for (i = 0; i < n+m-1; i++) {
    for (j = max(0,i-(n-1)); j <= min(m-1,i); j++) {
        c[i] = c[i] + a[j] * b[i-j];
    }
}
```

Convolution multiplies every possible pair of elements from A and B, and hence it seems like we *should* require quadratic time to get these $n + m - 1$ numbers. But in a miracle akin to sorting, there exists a clever divide-and-conquer algorithm that runs in $O(n \log n)$ time, assuming that $n \ge m$. And just like sorting, there are a large number of applications that take advantage of this enormous speedup for large sequences.

```
B              0 1 2 3 4                BR              4 3 2 1 0
                  ✳                                     | | | | |
A  0  1  2  3  4  5  6  7  8  9         A  0  1  2  3  4  5  6  7  8  9
```

Figure 5.5: Convolution of strings becomes equivalent to string matching when the pattern is reversed.

5.9.1 Applications of Convolution

Going from $O(n^2)$ to $O(n \log n)$ is as big a win for convolution as it was for sorting. Taking advantage of it requires recognizing when you are doing a convolution operation. Convolutions often arise when you are trying all possible ways of doing things that add up to k, for a large range of values of k, or when sliding a mask or pattern A over a sequence B and calculating at each position.

Important examples of convolution operations include:

- *Integer multiplication*: We can interpret integers as polynomials in any base b. For example, $632 = 6 \cdot b^2 + 3 \cdot b^1 + 2 \cdot b^0$, where $b = 10$. Polynomial multiplication behaves like integer multiplication without carrying.

  There are two different ways we can use fast polynomial multiplication to deal with integers. First, we can explicitly perform the carrying operation on the product polynomial, adding $\lfloor C[i]/b \rfloor$ to $C[i+1]$, and then replacing $C[i]$ with $C[i]$ (mod b). Alternatively, we could compute the product polynomial and then evaluate it at b to get the integer product $A \times B$.

  With fast convolution, either way gives us an even faster multiplication algorithm than Karatsuba, running in $O(n \log n)$ time on a RAM model of computation.

- *Cross-correlation*: For two time series A and B, the cross-correlation function measures the similarity as a function of the shift or displacement of one relative to the other. Perhaps people buy a product on average k days after seeing an advertisement for it. Then there should be high correlation between sales and advertising expenditures lagged by k days. This cross-correlation function $C[k]$ can be computed:

$$C[k] = \sum_j A[j]B[j+k]$$

  Note that the dot product here is computed over backward shifts of B instead of forward shifts, as in the original definition of a convolution. But we can still use fast convolution to compute this: simply input the reversed sequence B^R instead of B.

- *Moving average filters*: Often we are tasked with smoothing time series data by averaging over a window. Perhaps we want $C[i-1] = 0.25B[i-1] + 0.5B[i] + 0.25B[i+1]$ over all positions i. This is just another convolution, where A is the vector of weights within the window around $B[i]$.

- *String matching:*   Recall the problem of substring pattern matching, first discussed in Section 2.5.3. We are given a text string S and a pattern string P, and seek to identify all locations in P where P may be found. For $S = abaababa$ and $P = aba$, we can find P in S starting at positions 0, 3, and 5.

The $O(mn)$ algorithm described in Section 2.5.3 works by sliding the length-m pattern over each of the n possible starting points in the text. This sliding window approach is suggestive of being a convolution with the reversed pattern P^R, as shown in Figure 5.5. Can we solve string matching in $O(n \log n)$ by using fast convolution?

The answer is yes! Suppose our strings have an alphabet of size α. We can represent each character by a binary vector of length α having exactly one non-zero bit. Say $a = 10$ and $b = 01$ for the alphabet $\{a, b\}$. Then we can encode the strings S and P above as

$$S = 1001101001100110$$
$$P = 100110$$

The dot product over a window will be m on an even-numbered position of s iff p starts at that position in the text. So fast convolution can identify all locations of p in s in $O(n \log n)$ time.

Take-Home Lesson: Learn to recognize possible convolutions. A magical $\Theta(n \log n)$ algorithm instead of $O(n^2)$ is your reward for seeing this.

5.9.2   Fast Polynomial Multiplication (**)

The applications above should whet our interest in efficient ways to compute convolutions. The fast convolution algorithm uses divide and conquer, but a detailed proof of correctness relies on fairly sophisticated properties of complex numbers and linear algebra that are beyond the scope of what I want to do here. *Feel free to skip ahead!* But I will provide enough of an overview for you to understand the divide and conquer part.

We present convolution through a fast algorithm for multiplying polynomials. It is based on a series of observations:

- *Polynomials can be represented either as equations or sets of points:* You know that every pair of points defines a line. More generally, any degree-n polynomial $P(x)$ is completely defined by $n + 1$ points on the polynomial. For example, the points $(-1, -2)$, $(0, -1)$, and $(1, 2)$ define (and are defined by) the quadratic equation $y = x^2 + 2x - 1$.

- *We can find $n + 1$ such points on $P(x)$ by evaluation, but it looks expensive:* Generating a point on a given polynomial is easy—simply pick an arbitrary value x and plug it into $P(x)$. The time it takes for one such x will be linear in the degree of $P(x)$, which means n for the problems we are interested in. But doing this $n+1$ times for different values of x would take $O(n^2)$ time, which is more than we can afford if we want fast multiplication.

- *Multiplying polynomials A and B in a points representation is easy, if they have both been evaluated on the same values of x:* Suppose we want to compute the product of $(3x^2 + 2x + 6)(4x^2 + 3x + 2)$. The result will be a degree-4 polynomial, so we need five points to define it. We can evaluate both factors on the same x values:

$$A(x) = 3x^2 + 2x + 6 \longrightarrow (-2, 14), (-1, 7), (0, 6), (1, 11), (2, 22)$$

$$B(x) = 4x^2 + 3x + 2 \longrightarrow (-2, 12), (-1, 3), (0, 2), (1, 9), (2, 24)$$

Since $C(x) = A(x)B(x)$, we can now construct points on $C(x)$ by multiplying the corresponding y-values:

$$C(x) \longleftarrow (-2, 168), (-1, 21), (0, 12), (1, 99), (2, 528)$$

Thus, multiplying points in this representation takes only linear time.

- *We can evaluate a degree-n polynomial A(x) as two degree-(n/2) polynomials in x^2:* We can partition the terms of A into those of even and odd degree, for example:

$$12x^4 + 17x^3 + 36x^2 + 22x + 12 = (12x^4 + 36x^2 + 12) + x(17x^2 + 22)$$

By replacing x^2 by x', the right side gives us two smaller, lower degree polynomials as promised.

- *This suggests an efficient divide-and-conquer algorithm:* We need to evaluate n points of a degree-d polynomial. We need $n \geq 2d + 1$ points, since we will be using them to compute the product of two polynomials. We can decompose the problem into doing this evaluation on two polynomials of half the degree, plus a linear amount of work stitching the results together. This defines the recurrence $T(n) = 2T(n/2) + O(n)$, which evaluates to $O(n \log n)$.

- *Making this work correctly requires picking the right x values to evaluate on:* The trick with the squares makes it desirable for our sample points to come in pairs of the form $\pm x$, since their evaluation requires half as much work because they are identical when squared.

However, this property does not hold recursively, unless the x values are carefully chosen complex numbers. The *nth roots of unity* are the set of solutions to the equation $x^n = 1$. In reals, we only get $x \in \{-1, 1\}$, but there are n solutions with complex numbers. The kth of these n roots is given by

$$w_k = \cos(2k\pi/n) + i\sin(2k\pi/n)$$

To appreciate the magic properties of these numbers, look at what happens when we raise them to powers:

$$w = \left\{ 1, \frac{1+i}{\sqrt{2}}, i, -\frac{1-i}{\sqrt{2}}, -1, -\frac{1+i}{\sqrt{2}}, -i, \frac{1-i}{\sqrt{2}} \right\}$$
$$w^2 = \{1, i, -1, -i, 1, i, -1, -i\}$$
$$w^4 = \{1, -1, 1, -1, 1, -1, 1, -1\}$$
$$w^8 = \{1, 1, 1, 1, 1, 1, 1, 1\}$$

Observe that these terms come in positive/negative pairs, and the number of distinct terms gets halved with each squaring. These are the properties we need to make the divide and conquer work.

The best implementations of fast convolution generally compute the fast Fourier transform (FFT), so usually we seek to reduce our problems to FFTs to take advantage of existing libraries. FFTs are discussed in Section 16.11 (page 501).

> *Take-Home Lesson:* Fast convolution solves many important problems in $O(n \log n)$. The first step is to recognize your problem *is* a convolution.

Chapter Notes

Several other algorithms texts provide more substantive coverage of divide-and-conquer algorithms, including Kleinberg and Tardos [KT06] and Manber [Man89]. See Cormen et al. [CLRS09] for an excellent overview of the master theorem.

See Skiena [Ski12] for an accessible introduction to algorithmic design of vaccines. The bug searching sequences described in Section 5.2 is an example of a pooling design, enabling the identification of (say) one sick patient out of n using only $\lg n$ blood tests on pooled samples. The theory of these interesting designs is surveyed by Du and Hwang [DH00]. The left–right order of the subsets on these designs reflects a Gray code, in which neighboring subsets differ in exactly one element. Gray codes are discussed in Section 17.5.

Our parallel computations on pyramidal numbers were reported in Deng and Yang [DY94]. My treatment of convolutions and the FFT was based on Avrim Blum's 15-451/651 algorithm lecture notes.

5.10   Exercises

Binary Search

5-1. *[3]* Suppose you are given a sorted array A of size n that has been *circularly shifted* k positions to the right. For example, $[35, 42, 5, 15, 27, 29]$ is a sorted array that has been circularly shifted $k = 2$ positions, while $[27, 29, 35, 42, 5, 15]$ has been shifted $k = 4$ positions.

- Suppose you know what k is. Give an $O(1)$ algorithm to find the largest number in A.

- Suppose you *do not* know what k is. Give an $O(\lg n)$ algorithm to find the largest number in A. For partial credit, you may give an $O(n)$ algorithm.

5-2. *[3]* A sorted array of size n contains distinct integers between 1 and $n+1$, with one element missing. Give an $O(\log n)$ algorithm to find the missing integer, without using any extra space.

5-3. *[3]* Consider the numerical Twenty Questions game. In this game, the first player thinks of a number in the range 1 to n. The second player has to figure out this number by asking the fewest number of true/false questions. Assume that nobody cheats.

  (a) What is an optimal strategy if n in known?

  (b) What is a good strategy if n is not known?

5-4. *[5]* You are given a *unimodal* array of n distinct elements, meaning that its entries are in increasing order up until its maximum element, after which its elements are in decreasing order. Give an algorithm to compute the maximum element of a unimodal array that runs in $O(\log n)$ time.

5-5. *[5]* Suppose that you are given a sorted sequence of *distinct* integers $[a_1, a_2, \ldots, a_n]$. Give an $O(\lg n)$ algorithm to determine whether there exists an index i such that $a_i = i$. For example, in $[-10, -3, 3, 5, 7]$, $a_3 = 3$. In $[2, 3, 4, 5, 6, 7]$, there is no such i.

5-6. *[5]* Suppose that you are given a sorted sequence of *distinct* integers $a = [a_1, a_2, \ldots, a_n]$, drawn from 1 to m where $n < m$. Give an $O(\lg n)$ algorithm to find an integer $\leq m$ that is not present in a. For full credit, find the smallest such integer x such that $1 \leq x \leq m$.

5-7. *[5]* Let M be an $n \times m$ integer matrix in which the entries of each row are sorted in increasing order (from left to right) and the entries in each column are in increasing order (from top to bottom). Give an efficient algorithm to find the position of an integer x in M, or to determine that x is not there. How many comparisons of x with matrix entries does your algorithm use in worst case?

Divide and Conquer Algorithms

5-8. *[5]* Given two sorted arrays A and B of size n and m respectively, find the median of the $n + m$ elements. The overall run time complexity should be $O(\log(m + n))$.

5-9. *[8]* The largest subrange problem, discussed in Section 5.6, takes an array A of n numbers, and asks for the index pair i and j that maximizes $S = \sum_{k=i}^{j} A[k]$. Give an $O(n)$ algorithm for largest subrange.

5-10. *[8]* We are given n wooden sticks, each of integer length, where the ith piece has length $L[i]$. We seek to cut them so that we end up with k pieces of exactly the same length, in addition to other fragments. Furthermore, we want these k pieces to be as large as possible.

  (a) Given four wood sticks, of lengths $L = \{10, 6, 5, 3\}$, what are the largest sized pieces you can get for $k = 4$? (Hint: the answer is *not* 3).

  (b) Give a *correct* and efficient algorithm that, for a given L and k, returns the maximum possible length of the k equal pieces cut from the initial n sticks.

5-11. *[8]* Extend the convolution-based string-matching algorithm described in the text to the case of pattern matching with wildcard characters "*", which match any character. For example, "sh*t" should match both "shot" and "shut".

Recurrence Relations

5-12. *[5]* In Section 5.3, it is asserted that any polynomial can be represented by a recurrence. Find a recurrence relation that represents the polynomial $a_n = n^2$.

5-13. *[5]* Suppose you are choosing between the following three algorithms:

- Algorithm A solves problems by dividing them into five subproblems of half the size, recursively solving each subproblem, and then combining the solutions in linear time.

- Algorithm B solves problems of size n by recursively solving two subproblems of size $n - 1$ and then combining the solutions in constant time.

- Algorithm C solves problems of size n by dividing them into nine subproblems of size $n/3$, recursively solving each subproblem, and then combining the solutions in $\Theta(n^2)$ time.

What are the running times of each of these algorithms (in big O notation), and which would you choose?

5-14. *[5]* Solve the following recurrence relations and give a Θ bound for each of them:

    (a) $T(n) = 2T(n/3) + 1$

    (b) $T(n) = 5T(n/4) + n$

    (c) $T(n) = 7T(n/7) + n$

    (d) $T(n) = 9T(n/3) + n^2$

5-15. *[3]* Use the master theorem to solve the following recurrence relations:
    (a) $T(n) = 64T(n/4) + n^4$
    (b) $T(n) = 64T(n/4) + n^3$
    (c) $T(n) = 64T(n/4) + 128$

5-16. *[3]* Give asymptotically tight upper (Big Oh) bounds for $T(n)$ in each of the following recurrences. Justify your solutions by naming the particular case of the master theorem, by iterating the recurrence, or by using the substitution method:
    (a) $T(n) = T(n - 2) + 1$.
    (b) $T(n) = 2T(n/2) + n \lg^2 n$.
    (c) $T(n) = 9T(n/4) + n^2$.

LeetCode

5-1. `https://leetcode.com/problems/median-of-two-sorted-arrays/`

5-2. `https://leetcode.com/problems/count-of-range-sum/`

5-3. `https://leetcode.com/problems/maximum-subarray/`

HackerRank

5-1. `https://www.hackerrank.com/challenges/unique-divide-and-conquer`

5-2. `https://www.hackerrank.com/challenges/kingdom-division/`

5-3. `https://www.hackerrank.com/challenges/repeat-k-sums/`

Programming Challenges

These programming challenge problems with robot judging are available at `https://onlinejudge.org`:

5-1. "Polynomial Coefficients"—Chapter 5, problem 10105.

5-2. "Counting"—Chapter 6, problem 10198.

5-3. "Closest Pair Problem"—Chapter 14, problem 10245.

Chapter 6

Hashing and Randomized Algorithms

Most of the algorithms discussed in previous chapters were designed to optimize worst-case performance: they are guaranteed to return optimal solutions for every problem instance within a specified running time.

This is great, when we can do it. But relaxing the demand for either *always* correct or *always* efficient can lead to useful algorithms that still have performance guarantees. Randomized algorithms are not merely heuristics: any bad performance is due to getting unlucky on coin flips, rather than adversarial input data.

We classify randomized algorithms into two types, depending upon whether they guarantee correctness or efficiency:

- *Las Vegas algorithms*: These randomized algorithms guarantee correctness, and are usually (but not always) efficient. Quicksort is an excellent example of a Las Vegas algorithm.

- *Monte Carlo algorithms*: These randomized algorithms are provably efficient, and usually (but not always) produce the correct answer or something close to it. Representative of this class are random sampling methods discussed in Section 12.6.1, where we return the best solution found in the course of (say) 1,000,000 random samples.

We will see several examples of both types of algorithm in this chapter.

One blessing of randomized algorithms is that they tend to be very simple to describe and implement. Eliminating the need to worry about rare or unlikely situations makes it possible to avoid complicated data structures and other contortions. These clean randomized algorithms are often intuitively appealing, and relatively easy to design.

However, randomized algorithms are frequently quite difficult to analyze rigorously. Probability theory is the mathematics we need for the analysis of randomized algorithms, and is of necessity both formal and subtle. Probabilistic

© The Editor(s) (if applicable) and The Author(s), under exclusive license to
Springer Nature Switzerland AG 2020
S. S. Skiena, *The Algorithm Design Manual*, Texts in Computer Science,
https://doi.org/10.1007/978-3-030-54256-6_6

analysis often involves algebraic manipulation of long chains of inequalities that looks frightening, and relies on tricks and experience.

This makes it difficult to provide satisfying analysis on the level of this book, which maintains a strict no-theorem/proof policy. But I will try to provide intuition where I can, so you can appreciate why these algorithms are usually correct or efficient.

We have had initial peeks at randomized algorithms through our discussions of hash tables (Section 3.7) and quicksort (Section 4.6). Review these now to give yourself the best chance of understanding what is to come.

Stop and Think: Quicksort City

Problem: Why is randomized quicksort a Las Vegas algorithm, as opposed to a Monte Carlo algorithm?

Solution: Recall that Monte Carlo algorithms are always fast and usually correct, while Las Vegas algorithms are always correct and usually fast.

Randomized quicksort always produces a sorted permutation, so we know it is always correct. Picking a very bad series of pivots might cause the running to exceed $O(n \log n)$, but we are always going to end up sorted. Thus, quicksort is a nice example of a Las Vegas-style algorithm. ∎

6.1   Probability Review

I will resist the temptation to give a thorough review of probability theory here, as part of my objective to keep this book to a manageable size. My presumptions are: (1) you have had some previous exposure to probability theory, and (2) you know where to look if you feel you need more. I will therefore limit myself to a few basic definitions and properties we will use.

6.1.1   Probability

Probability theory provides a formal framework for reasoning about the likelihood of events. Because it is a formal discipline, there is a thicket of associated definitions to instantiate exactly what we are reasoning about:

- An *experiment* is a procedure that yields one of a set of possible outcomes. As our ongoing example, consider the experiment of tossing two six-sided dice, one red and one blue, with each face bearing a distinct integer $\{1, \ldots, 6\}$.

- A *sample space* S is the set of possible outcomes of an experiment. In our dice example, there are thirty-six possible outcomes, namely

$$S = \{(1,1),(1,2),(1,3),(1,4),(1,5),(1,6),$$
$$(2,1),(2,2),(2,3),(2,4),(2,5),(2,6),$$
$$(3,1),(3,2),(3,3),(3,4),(3,5),(3,6),$$
$$(4,1),(4,2),(4,3),(4,4),(4,5),(4,6),$$
$$(5,1),(5,2),(5,3),(5,4),(5,5),(5,6),$$
$$(6,1),(6,2),(6,3),(6,4),(6,5),(6,6)\}.$$

- An *event* E is a specified subset of the outcomes of an experiment. The event that the sum of the dice equals 7 or 11 (the conditions to win at craps on the first roll) is the subset

$$E = \{(1,6),(2,5),(3,4),(4,3),(5,2),(6,1),(5,6),(6,5)\}.$$

- The *probability of an outcome s*, denoted $p(s)$, is a number with the two properties:

  - For each outcome s in sample space S, $0 \leq p(s) \leq 1$.
  - The sum of probabilities of all outcomes adds to one: $\sum_{s \in S} p(s) = 1$.

  If we assume two distinct fair dice, the probability $p(s) = (1/6) \times (1/6) = 1/36$ for all outcomes $s \in S$.

- The *probability of an event E* is the sum of the probabilities of the outcomes of the event. Thus,

$$P(E) = \sum_{s \in E} p(s)$$

  An alternative formulation is in terms of the *complement* of the event $\bar{E}$, the case when E does not occur. Then

$$P(E) = 1 - P(\bar{E})$$

  This is useful, because often it is easier to analyze $P(\bar{E})$ than $P(E)$ directly.

- A *random variable V* is a numerical function on the outcomes of a probability space. The function "sum the values of two dice" $(V((a,b)) = a+b)$ produces an integer result between 2 and 12. This implies a probability distribution of the possible values of the random variable. The probability $P(V(s) = 7) = 1/6$, while $P(V(s) = 12) = 1/36$.

- The *expected value* of a random variable V defined on a sample space S, $E(V)$, is defined

$$E(V) = \sum_{s \in S} p(s) \cdot V(s)$$

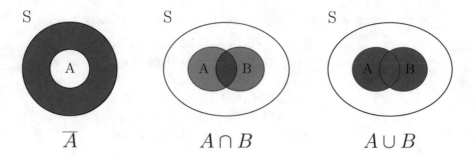

Figure 6.1: Venn diagrams illustrating set difference (left), intersection (middle), and union (right).

6.1.2   Compound Events and Independence

We will be interested in complex events computed from simpler events A and B on the same set of outcomes. Perhaps event A is that at least one of two dice be an even number, while event B denotes rolling a total of either 7 or 11. Note that there exist outcomes of A that are not outcomes of B, specifically

$$A - B = \{(1,2),(1,4),(2,1),(2,2),(2,3),(2,4),(2,6),(3,2),(3,6),(4,1),$$
$$(4,2),(4,4),(4,5),(4,6),(5,4),(6,2),(6,3),(6,4),(6,6)\}$$

This is the *set difference* operation. Observe that here $B - A = \{\}$, because every pair adding to 7 or 11 must contain one odd and one even number.

The outcomes in common between both events A and B are called the *intersection*, denoted $A \cap B$. This can be written as

$$A \cap B = A - (S - B)$$

Outcomes that appear in either A or B are called the *union*, denoted $A \cup B$. The probability of the union and intersection are related by the formula

$$P(A \cup B) = P(A) + P(B) - P(A \cap B)$$

With the complement operation $\bar{A} = S - A$, we get a rich language for combining events, shown in Figure 6.1. We can readily compute the probability of any of these sets by summing the probabilities of the outcomes in the defined sets.

The events A and B are said to be *independent* if

$$P(A \cap B) = P(A) \times P(B)$$

This means that there is no special structure of outcomes shared between events A and B. Assuming that half of the students in my class are female, and half the students in my class are above average, we would expect that a quarter of my students are both female and above average if the events are independent.

Probability theorists love independent events, because it simplifies their calculations. For example, if A_i denotes the event of getting an even number on the ith dice throw, then the probability of obtaining all evens in a throw of two dice is $P(A_1 \cap A_2) = P(A_1)P(A_2) = (1/2)(1/2) = 1/4$. Then, the probability of A, that at least one of two dice is even, is

$$P(A) = P(A_1 \cup A_2) = P(A_1) + P(A_2) - P(A_1 \cap A_2) = 1/2 + 1/2 - 1/4 = 3/4$$

That independence often doesn't hold explains much of the subtlety and difficulty of probabilistic analysis. The probability of getting n heads when tossing n independent coins is $1/2^n$. But it would be $1/2$ if the coins were perfectly correlated, since the only possibilities would be all heads or all tails. This computation would become very hard if there were complex dependencies between the outcomes of the ith and jth coins.

Randomized algorithms are typically designed around samples drawn independently at random, so that we can safely multiply probabilities to understand compound events.

6.1.3 Conditional Probability

Presuming that $P(B) > 0$, the *conditional probability* of A given B, $P(A|B)$ is defined as follows

$$P(A|B) = \frac{P(A \cap B)}{P(B)}$$

In particular, if events A and B are independent, then

$$P(A|B) = \frac{P(A \cap B)}{P(B)} = \frac{P(A)P(B)}{P(B)} = P(A)$$

and B has absolutely no impact on the likelihood of A. Conditional probability becomes interesting only when the two events have dependence on each other.

Recall the dice-rolling events from Section 6.1.2, namely:

- Event A: at least one of two dice is an even number.

- Event B: the sum of the two dice is either 7 or 11.

Observe that $P(A|B) = 1$, because *any* roll summing to an odd value must consist of one even and one odd number. Thus, $A \cap B = B$. For $P(B|A)$, note that $P(A \cap B) = P(B) = 8/36$ and $P(A) = 27/36$, so $P(B|A) = 8/27$.

Our primary tool to compute conditional probabilities will be *Bayes' theorem*, which reverses the direction of the dependencies:

$$P(B|A) = \frac{P(A|B)P(B)}{P(A)}$$

Often it proves easier to compute probabilities in one direction than another, as in this problem. By Bayes' theorem $P(B|A) = (1 \cdot 8/36)/(27/36) = 8/27$, exactly what we got before.

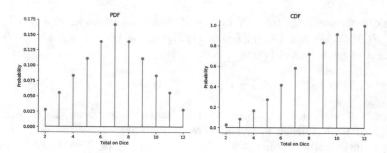

Figure 6.2: The probability density function (pdf) of the sum of two dice contains exactly the same information as the cumulative density function (cdf), but looks very different.

6.1.4   Probability Distributions

Random variables are numerical functions where the values are associated with probabilities of occurrence. In our example where $V(s)$ is the sum of two tossed dice, the function produces an integer between 2 and 12. The probability of a particular value $V(s) = X$ is the sum of the probabilities of all the outcomes whose components add up to X.

Such random variables can be represented by their *probability density function*, or pdf. This is a graph where the x-axis represents the values the random variable can take on, and the y-axis denotes the probability of each given value. Figure 6.2 (left) presents the pdf of the sum of two fair dice. Observe that the peak at $X = 7$ corresponds to the most probable dice total, with a probability of $1/6$.

6.1.5   Mean and Variance

There are two main types of summary statistics, which together tell us an enormous amount about a probability distribution or a data set:

- *Central tendency measures*, which capture the center around which the random samples or data points are distributed.

- *Variation* or *variability measures*, which describe the spread, that is, how far the random samples or data points can lie from the center.

The primary centrality measure is the *mean*. The mean of a random variable V, denoted $E(V)$ and also known as the expected value, is given by

$$E(V) = \sum_{s \in S} V(s)p(s)$$

When the elementary events are all of equal probability, the *mean* or *average*, computed as

$$\bar{X} = \frac{1}{n} \sum_{i=1}^{n} x_i$$

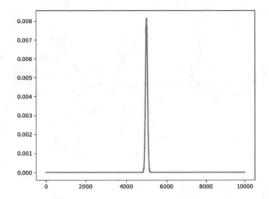

Figure 6.3: The probability distribution of getting h heads in n =10,000 tosses of a fair coin is tightly centered around the mean, $n/2$ =5,000.

The most common measure of variability is the *standard deviation* σ. The standard deviation of a random variable V is given by $\sigma = \sqrt{E((V - E(V))^2}$. For a data set, the standard deviation is computed from the sum of squared differences between the individual elements and the mean:

$$\sigma = \sqrt{\frac{\sum_{i=1}^{n}(x_i - \bar{X})^2}{n - 1}}$$

A related statistic, the *variance* $V = \sigma^2$, is the square of the standard deviation. Sometimes it is more convenient to talk about variance than standard deviation, because the term is ten characters shorter. But they measure exactly the same thing.

6.1.6 Tossing Coins

You probably have a fair degree of intuition about the distribution of the number of heads and tails when you toss a fair coin 10,000 times. You know that the expected number of heads in n tosses, each with probability $p = 1/2$ of heads, is pn, or 5,000 for this example. You likely know that the distribution for h heads out of n is a binomial distribution, where

$$P(X - h) = \frac{\binom{n}{h}}{\sum_{x=0}^{n}\binom{n}{x}} = \frac{\binom{n}{h}}{2^n}$$

and that it is a bell-shaped symmetrical distribution about the mean.

But you may not appreciate just how narrow this distribution is, as shown in Figure 6.3. Sure, anywhere from 0 to n heads *can* result from n fair coin tosses. But they won't: the number of heads we get will *almost* always be within a few standard deviations of the mean, where the standard deviation σ for the binomial distribution is given by $\sigma = \sqrt{np(1 - p)} = \Theta(\sqrt{n})$.

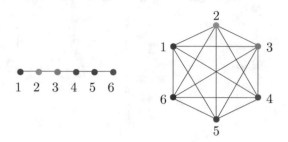

Figure 6.4: How long does a random walk take to visit all n nodes of a path (left) or a complete graph (right)?

Indeed, for any probability distribution, at least $1 - (1/k^2)$ of the mass of the distribution lies within $\pm k\sigma$ of the mean μ. Typically σ is small relative to μ for the distributions arising in the analysis of randomized algorithms and processes.

> *Take-Home Lesson:* Students often ask me "what happens" when randomized quicksort runs in $\Theta(n^2)$. The answer is that nothing happens, in exactly the same way nothing happens when you buy a lottery ticket: you almost certainly just lose. With a randomized quicksort you almost certainly just win: the probability distribution is so tight that you nearly always run in time very close to expectation.

Stop and Think: Random Walks on a Path

Problem: Random walks on graphs are important processes to understand. The expected covering time (the number of moves until we have visited all vertices) differs depending upon the topology of the graph (see Figure 6.4). What is it for a path?

Suppose we start at the left end of an m-vertex path. We repeatedly flip a fair coin, moving one step right on heads and one step left on tails (staying where we are if we were to fall off the path). How many coin flips do we expect it will take, as a function of m, until we get to the right end of the path?

Solution: To get to the right end of the path, we need $m - 1$ more heads than tails after n coin flips, assuming we don't bother to flip when on the left-most vertex where we can only move right. We expect about half of the flips to be heads, with a standard deviation of $\sigma = \Theta(\sqrt{n})$. This σ describes the spread of the difference in the number of the heads and tails we are likely to have. We must flip enough times for σ to be on the order of m, so

$$m = \Theta(\sqrt{n}) \rightarrow n = \Theta(m^2)$$

6.2 Understanding Balls and Bins

Ball and bin problems are classics of probability theory. We are given x identical balls to toss at random into y labeled bins. We are interested in the resulting distribution of balls. How many bins are expected to contain a given number of balls?

Hashing can be thought of as a ball and bin process. Suppose we hash n balls/keys into n bins/buckets. We expect an average of one ball per bin, but note that this will be true *regardless* of how good or bad the hash function is.

A good hash function should behave like a random number generator, selecting integers/bin IDs with equal probability from 1 to n. But what happens when we draw n such integers from a uniform distribution? The ideal would be for each of the n items (balls) to be assigned to a different bin, so that each bucket contains exactly one item to search. But is this really what happens?

To help develop your own intuition, I encourage you to code a little simulation and run your own experiment. I did, and got the following results, for hash table sizes from one million to one hundred million items:

	Number of Buckets with k Items		
k	$n = 10^6$	$n = 10^7$	$n = 10^8$
0	367,899	3,678,774	36,789,634
1	367,928	3,677,993	36,785,705
2	183,926	1,840,437	18,392,948
3	61,112	613,564	6,133,955
4	15,438	152,713	1,531,360
5	3130	30,517	306,819
6	499	5,133	51,238
7	56	754	7,269
8	12	107	972
9		8	89
10			10
11			1

We see that 36.78% of the buckets are empty in all three cases. That can't be a coincidence. The first bucket will be empty iff each of the n balls gets assigned to one of the other $n-1$ buckets. The probability p of missing for each particular ball is $p = (n-1)/n$, which approaches 1 as n gets large. But we must miss for *all* n balls, the probability of which is p^n. What happens when we multiply a large number of large probabilities? You actually saw the answer back when you studied limits:

$$P(|B_1| = 0) = \left(\frac{n-1}{n}\right)^n \rightarrow \frac{1}{e} = 0.367879$$

Thus, 36.78% of the buckets in a large hash table will be empty. And, as it turns out, exactly the same fraction of buckets is expected to contain one element.

If so many buckets are empty, others must be unusually full. The fullest bucket gets fuller in the table above as n increases, from 8 to 9 to 11. In fact, the expected value of the longest list is $O(\log n / \log \log n)$, which grows slowly but is not a constant. Thus, I was a little too glib when I said in Section 3.7.1

								18	
				32				16	
	24		29	31	19			15	
	21		12	25	9			11	27
30	20	14	10	23	3		26	4	22
28	13	8	5	17	1	(33)	7	2	6
1	2	3	4	5	6	7	8	9	10

Figure 6.5: Illustrating the coupon collectors problem through an experiment tossing balls at random into ten bins. It is not until the 33rd toss that all bins are non-empty.

that the worst-case access time for hashing is $O(1)$.[1]

> *Take-Home Lesson:* Precise analysis of random process requires formal probability theory, algebraic skills, and careful asymptotics. We will gloss over such issues in this chapter, but you should appreciate that they are out there.

6.2.1   The Coupon Collector's Problem

As a final hashing warmup, let's keep tossing balls into these n bins until none of them are empty, that is, until we have at least one ball in each bin. How many tosses do we expect this should take? As shown in Figure 6.5, it may require considerably more than n tosses until every bin is occupied.

We can split such a sequence of balls into n runs, where run r_i consists of the balls we toss after we have filled i buckets until the next time we hit an empty bucket. The expected number of balls to fill all n slots $E(n)$ will be the sum of the expected lengths of all runs. If you are flipping a coin with probability p of coming up heads, the expected number of flips until you get your first head is $1/p$; this is a property of the *geometric* distribution. After filling i buckets, the probability that our next toss will hit an empty bucket is $p = (n - i)/n$. Putting this together, we get the following

$$E(n) = \sum_{i=0}^{n-1} |r_i| = \sum_{i=0}^{n-1} \frac{n}{n-i} = n \sum_{i=0}^{n-1} \frac{1}{n-i} = nH_n \approx n \ln n$$

The trick here is to remember that the *harmonic number* $H_n = \sum_{i=1}^{n} 1/i$, and that $H_n \approx \ln n$.

Stop and Think: Covering Time for K_n

Problem: Suppose we start on vertex 1 of a complete n-vertex graph (see Figure 6.4). We take a random walk on this graph, at each step going to a randomly

---

[1] To be precise, the expected search time for hashing *is* $O(1)$ averaged over all n keys, but we also expect there will a few keys unlucky enough to require $\Theta(\log n / \log \log n)$ time.

selected neighbor of our current position. What is the expected number of steps until we have visited all vertices of the graph?

Solution: This is exactly the same question as the previous stop and think, but with a different graph and thus with a possibly different answer.

Indeed, the random process here of independently generating random integers from 1 to n looks essentially the same as the coupon collector's problem. This suggests that the expected length of the covering walk is $\Theta(n \log n)$.

The only hitch in this argument is that the random walk model does not permit us to stay at the same vertex for two successive steps, unless the graph has edges from a vertex to itself (self-loops). A graph without such self-loops should have a *slightly* faster covering time, since a repeat visit does not make progress in any way, but not enough to change the asymptotics. The probability of discovering one of $n - i$ untouched vertices in the next step changes from $(n-i)/n$ to $(n-i)/(n-1)$, reducing the total covering time analysis from nH_n to $(n-1)H_n$. But these are asymptotically the same. Covering the complete graph takes $\Theta(n \log n)$ steps, much faster than the covering time of the path. ∎

6.3   Why is Hashing a Randomized Algorithm?

Recall that a hash function $h(s)$ maps keys s to integers in a range from 0 to $m-1$, ideally uniformly over this interval. Because good hash functions scatter keys around this integer range in a manner similar to that of a uniform random number generator, we can analyze hashing by treating the values as drawn from tosses of an m-sided die.

But just because we can analyze hashing in terms of probabilities doesn't make it a randomized algorithm. As discussed so far, hashing is completely deterministic, involving no random numbers. Indeed, hashing *must* be deterministic, because we need $h(x)$ to produce exactly the same result whenever called with a given x, or else we can never hope to find x in a hash table.

One reason we like randomized algorithms is that they make the worst case input instance go away: bad performance should be a result of extremely bad luck, rather than some joker giving us data that makes us do bad things. But it is easy (in principle) to construct a worst case example for any hash function h. Suppose we take an arbitrary set S of nm distinct keys, and hash each $s \in S$. Because the range of this function has only m elements, there must be many collisions. Since the average number of items per bucket is $nm/m = n$, it follows from the pigeonhole principle that there must be a bucket with at least n items in it. The n items in this bucket, taken by themselves, will prove a nightmare for hash function h.

How can we make such a worst-case input go away? We are protected if we pick our *hash function* at random from a large set of possibilities, because we can only construct such a bad example by knowing the exact hash function we will be working with.

So how can we construct a family of random hash functions? Recall that typically $h(x) = f(x) \pmod{m}$, where $f(x)$ turns the key into a huge value, and taking the remainder mod m reduces it to the desired range. Our desired range is typically determined by application and memory constraints, so we would not want to select m at random. But what about first reducing with a larger integer p? Observe that in general

$$f(x) \pmod{m} \neq (f(x) \bmod p) \pmod{m}$$

For example:

$21347895537127 \pmod{17} = 8 \neq (21347895537127 \pmod{2342343}) \pmod{17} = 12$

Thus, we can select p at random to define the hash function

$$h(x) = ((f(x) \bmod p) \bmod m)$$

and things will work out just fine provided (a) $f(x)$ is large relative to p, (b) p is large relative to m, and (c) m is relatively prime to p.

This ability to select random hash functions means we can now use hashing to provide legitimate randomized guarantees, thus making the worst-case input go away. It also lets us build powerful algorithms involving multiple hash functions, such as Bloom filters, discussed in Section 6.4.

6.4   Bloom Filters

Recall the problem of detecting duplicate documents faced by search engines like Google. They seek to build an index of all the *unique* documents on the web. Identical copies of the same document often exist on many different websites, including (unfortunately) pirated copies of my book. Whenever Google crawls a new link, they need to establish whether what they found is a not previously encountered document worth adding to the index.

Perhaps the most natural solution here is to build a hash table of the documents. Should a freshly crawled document hash to an empty bucket, we know it must be new. But when there is a collision, it does not necessarily mean we have seen this document before. To be sure, we must explicitly compare the new document against all other documents in the bucket, to detect spurious collisions between a and b, where $h(a) = s$ and $h(b) = s$, but $a \neq b$. This is what was discussed back in Section 3.7.2.

But in this application, spurious collisions are not really a tragedy: they only mean that Google will fail to index a new document it has found. This can be an acceptable risk, provided the probability of it happening is low enough. Removing the need to explicitly resolve collisions has big benefits in making the table smaller. By reducing each bucket from a pointer link to a single bit (is this bucket occupied or not?), we reduce the space by a factor of 64 on typical

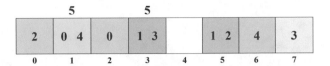

Figure 6.6: Hashing the integers 0, 1, 2, 3, and 4 into an $n = 8$ bit Bloom filter, using hash functions $h_1(x) = 2x + 1$ (blue) and $h_2(x) = 3x + 2$ (red). Searching for $x = 5$ would yield a false positive, since the two corresponding bits have been set by other elements.

machines. Some of this space can then be taken back to make the hash table larger, thus reducing the probability of collisions in the first place.

Now suppose we build such a bit-vector hash table, with a capacity of n bits. If we have distinct bits corresponding to m documents occupied, the probability that a new document will spuriously hash to one of these bits is $p = m/n$. Thus, even if the table is only 5% full, there is still a $p = 0.05$ probability that we will falsely discard a new discovery, which is much higher than is acceptable.

Much better is to employ a *Bloom filter*, which is also just a bit-vector hash table. But instead of each document corresponding to a single position in the table, a Bloom filter hashes each key k times, using k different hash functions. When we insert document s into our Bloom filter, we set all the bits corresponding to $h_1(s), h_2(s), \ldots h_k(s)$ to be 1, meaning occupied. To test whether a query document s is present in the data structure, we must test that all k of these bits equal 1. For a document to be falsely convicted of already being in the filter, it must be unlucky enough that all k of these bits were set in hashes of previous documents, as in the example of Figure 6.6.

What are the chances of this? Hashes of m documents in such a Bloom filter will occupy at most km bits, so the probability of a single collision rises to $p_1 = km/n$, which is k times greater than the single hash case. But *all* k bits must collide with those of our query document, which only happens with probability $p_k = (p_1)^k = (km/n)^k$. This is a peculiar expression, because a probability raised to the kth power quickly becomes smaller with increasing k, yet here the probability being raised simultaneously increases with k. To find the k that minimizes p_k, we could take the derivative and set it to zero.

Figure 6.7 graphs this error probability $((km/n)^k)$ as a function of load (m/n), with a separate line for each k from 1 to 5. It is clear that using a large number of hash functions (increased k) reduces false positive error substantially over a conventional hash table (the blue line, $k = 1$), at least for small loads. But observe that the error rate associated with larger k increases rapidly with load, so for any given load there is always a point where adding more hash functions becomes counter-productive.

For a 5% load, the error rate for a simple hash table of $k = 1$ will be 51.2 times larger than a Bloom filter with $k = 5$ (9.77×10^{-4}), even though they use exactly the same amount of memory. A Bloom filter is an excellent data structure for maintaining an index, provided you can live with occasionally

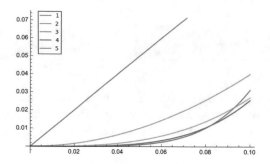

Figure 6.7: Bloom filter error probability as a function of load (m/n) for k from 1 to 5. By selecting the right k for the given load, we can dramatically reduce false positive error rate with no increase in table space.

saying yes when the answer is no.

6.5   The Birthday Paradox and Perfect Hashing

Hash tables are an excellent data structure in practice for the standard dictionary operations of insert, delete, and search. However, the $\Theta(n)$ worst-case search time for hashing is an annoyance, no matter how rare it is. Is there a way we can guarantee worst-case constant time search?

Perfect hashing offers us this possibility for *static* dictionaries. Here we are given all possible keys in one batch, and are not allowed to later insert or delete items. We can thus build the data structure once and use it repeatedly for search/membership testing. This is a fairly common use case, so why pay for the flexibility of dynamic data structures when you don't need them?

One idea for how this might work would be to try a given hash function $h(x)$ on our set of n keys S and hope it creates a hash table with no collisions, that is, $h(x) \neq h(y)$ for all pairs of distinct $x, y \in S$. It should be clear that our chances of getting lucky improve as we increase the size of our table relative to n: the more empty slots there are available for the next key, the more likely we find one.

How large a hash table m do we need before we can expect zero collisions among n keys? Suppose we start from an empty table, and repeatedly insert keys. For the $(i+1)$th insertion, the probability that we hit one of the $m - i$ still-open slots in the table is $(m - i)/m$. For a perfect hash, *all* n inserts must succeed, so

$$P(\text{no collision}) = \prod_{i=0}^{n-1} \left(\frac{m-i}{m} \right) = \frac{m!}{m^n((m-n)!)}$$

What happens when you evaluate this is famously called the *birthday paradox*. How many people do you need in a room before it is likely that at least two of them share the same birthday? Here the table size is $m = 365$. The

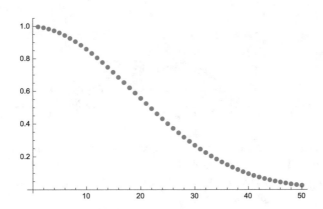

Figure 6.8: The probability of no collisions in a hash table decreases rapidly with n, the number of keys to hash. Here the hash table size is $m = 365$.

probability of no collisions drops below $1/2$ for $n = 23$ and is less than 3% when $n \geq 50$, as shown in Figure 6.8. In other words, odds are we have a collision when only 23 people are in the room. Solving this asymptotically, we begin to expect collisions when $n = \Theta(\sqrt{m})$ or equivalently when $m = \Theta(n^2)$.

But quadratic space seems like an awful large penalty to pay for constant time access to n elements. Instead, we will create a two-level hash table. First, we hash the n keys of set S into a table with n slots. We expect collisions, but unless we are very unlucky all the lists will be short enough.

Let l_i be the length of the ith list in this table. Because of collisions, many lists will be of length longer than 1. Our definition of short enough is that n items are distributed around the table such that the sum of squares of the list lengths is linear, that is,

$$N = \sum_{i=1}^{n} l_i^2 = \Theta(n)$$

Suppose that it happened that all elements were in lists of length l, meaning that we have n/l non-empty lists. The sum of squares of the list lengths is $N = (n/l)l^2 = nl$, which is linear because l is a constant. We can even get away with a fixed number of lists of length $\sqrt{n}$ and still use linear space.

In fact, it can be shown that $N \leq 4n$ with high probability. So if this isn't true on S for the first hash function we try, we can just try another. Pretty soon we will find one with short-enough list lengths that we can use.

We will use an array of length N for our second-level table, allocating l_i^2 space for the elements of the ith bucket. Note that this is big enough relative to the number of elements to avoid the birthday paradox—odds are we will have no collision in any given hash function. And if we do, simply try another hash function until all elements end up in unique places.

The complete scheme is illustrated in Figure 6.9. The contents of the ith entry in the first-level hash table include the starting and ending positions for the l_i^2 entries in the second-level table corresponding to this list. It also contains

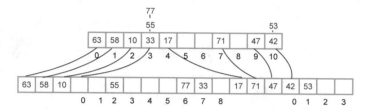

Figure 6.9:   Perfect hashing uses a two-level hash table to guarantee lookup in constant time. The first-level table encodes an index range of l_i^2 elements in the second-level table, allocated to store the l_i items in first-level bucket i.

a description (or identifier) of the hash function that we will use for this region of the second-level table.

Lookup for an item s starts by calling hash function $h_1(s)$ to compute the right spot in the first table, where we will find the appropriate start/stop position and hash function parameters. From this we can compute $start + (h_2(s) \ (\text{mod} \ (stop - start)))$, the location where item s will appear in the second table. Thus, search can always be performed in $\Theta(1)$ time, using linear storage space between the two tables.

Perfect hashing is a very useful data structure in practice, ideal for whenever you will be making large numbers of queries to a static dictionary. There is a lot of fiddling you can do with this basic scheme to minimize space demands and construction/search cost, such as working harder to find second-level hash functions with fewer holes in the table. Indeed, *minimum perfect hashing* guarantees constant time access with zero empty hash table slots, resulting in an n-element second hash table for n keys.

6.6   Minwise Hashing

Hashing lets you quickly test whether a specific word w in document D_1 also occurs in document D_2: build a hash table on the words of D_2 and then hunt for $h(w)$ in this table T. For simplicity and efficiency we can remove duplicate words from each document, so each contains only one entry for each vocabulary term used. By so looking up all the vocabulary words $w_i \in D_1$ in T, we can get a count of the intersection, and compute the *Jaccard similarity* $J(D_1, D_2)$ of the two documents, where

$$J(D_1, D_2) = \frac{|D_1 \cap D_2|}{|D_1 \cup D_2|}$$

This similarity measure ranges from 0 to 1, sort of like a probability that the two documents are similar.

But what if you want to test whether two documents are similar without looking at all the words? If we are doing this repeatedly, on a web scale, effi-

s:	The	cat	in	the	hat		The	hat	in	the	store
h(s):	17	128	56	17	**(4)**		17	**(4)**	56	17	**96**

Figure 6.10: The words associated with the minimum hash code from each of two documents are likely to be the same, if the documents are very similar.

ciency matters. Suppose we are allowed to look at only *one* word per document to make a decision. Which word should we pick?

A first thought might be to pick the most frequent word in the original document, but it is likely to be "the" and tell you very little about similarity. Picking the most representative word, perhaps according to the TD–IDF statistic, would be better. But it still makes assumptions about the distribution of words, which may be unwarranted.

The key idea is to synchronize, so we pick the same word out of the two documents while looking at the documents separately. *Minwise hashing* is a clever but simple idea to do this. We compute the hash code $h(w_i)$ of every word in D_1, and select the code with smallest value among all the hash codes. Then we do the same thing with the words in D_2, using the same hash function.

What is cool about this is that it gives a way to pick the same random word in both documents, as shown in Figure 6.10. Suppose the vocabularies of each document were identical. Then the word with minimum hash code will be the same in both D_1 and D_2, and we get a match. In contrast, suppose we had picked completely random words from each document. Then the probability of picking the same word would be only $1/v$, where v is the vocabulary size.

Now suppose that D_1 and D_2 do not have identical vocabularies. The probability that the minhash word appears in both documents depends on the number of words in common, that is, the intersection of the vocabularies, and also on the total vocabulary size of the documents. In fact, this probability is exactly the Jaccard similarity described above.

Sampling a larger number of words, say, the k smallest hash values in each document, and reporting the size of intersection over k gives us a better estimate of Jaccard similarity. But the alert reader may wonder why we bother. It takes time linear in the size of D_1 and D_2 to compute all the hash values just to find the minhash values, yet this is the same running time that it would take to compute the exact size of intersection using hashing!

The value of minhash comes in building indexes for similarity search and clustering over large corpora of documents. Suppose we have N documents, each with an average of m vocabulary words in them. We want to build an index to help us determine which of these is most similar to a new query document Q. Hashing all words in all documents gives us a table of size $O(Nm)$. Storing $k \ll m$ minwise hash values from each document will be much smaller at $O(Nk)$, but the documents at the intersection of the buckets associated with the k minwise hashes of Q are likely to contain the most similar documents—particularly if the Jaccard similarity is high.

Stop and Think: Estimating Population Size

Problem: Suppose we will receive a stream S of n numbers one by one. This stream will contain many duplicates, possibly even a single number repeated n times. How can we estimate the number of distinct values in S using only a constant amount of memory?

Solution: If space was not an issue, the natural solution would be to build a dictionary data structure on the distinct elements from the stream, each with an associated count of how often it has occurred. For the next element we see in the stream, we add one to the count if it exists in the dictionary, or insert it if it is not found. But we only have enough space to store a constant number of elements. What can we do?

Minwise hashing comes to the rescue. Suppose we hash each new element s of S as it comes in, and only save $h(s)$ if it is smaller than the previous minhash.

Why is this interesting? Suppose the range of possible hash values is between 0 and $M-1$, and we select k values in this range uniformly at a random. What is the expected minimum of these k values? If $k = 1$, the expected value will (obviously) be $M/2$. For general k, we might hand wave and say that if our k values were equally spaced in the interval, the minhash should be $M/(k+1)$.

In fact, this hand waving happens to produce the right answer. Define X as the smallest of k samples. Then

$$P(X = i) = P(X \geq i) - P(X \geq i+1) = \left(\frac{M-i}{M}\right)^k - \left(\frac{M-i-1}{M}\right)^k$$

Taking the limit of the expected value as M gets large gives the result

$$E(X) = \sum_{i=0}^{M-1} i\left(\left(\frac{M-i}{M}\right)^k - \left(\frac{M-i-1}{M}\right)^k\right) \longrightarrow \frac{M}{k+1}$$

The punch line is that M divided by the minhash value gives an excellent estimate of the number of distinct values we have seen. This method will not be fooled by repeated values in the stream, since repeated occurrences will yield precisely the same value every time we evaluate the hash function. ∎

6.7   Efficient String Matching

Strings are sequences of characters where the order of the characters matters: the string `ALGORITHM` is different than `LOGARITHM`. Text strings are fundamental to a host of computing applications, from programming language parsing/compilation, to web search engines, to biological sequence analysis.

The primary data structure for representing strings is an array of characters. This allows us constant-time access to the ith character of the string. Some

$H(s,j)$		0	1	2	5	3	6	5
$\sum C_i + j$		0	1	1	2	2	2	2
	A	A	A	B	A	B	B	A	B

Figure 6.11: The Rabin–Karp hash function $H(s, j)$ gives distinctive codes to different substrings (in blue), while a less powerful hash function that just adds the character codes yields many collisions (shown in purple). Here the pattern string (BBA) has length $m = 3$, and the character codes are $A = 0$ and $B = 1$.

auxiliary information must be maintained to mark the end of the string: either a special end-of-string character or (perhaps more usefully) a count n of the characters in the string.

The most fundamental operation on text strings is substring search, namely:

Problem: Substring Pattern Matching

Input: A text string t and a pattern string p.

Output: Does t contain the pattern p as a substring, and if so where?

The simplest algorithm to search for the presence of pattern string p in text t overlays the pattern string at every position in the text, and checks whether every pattern character matches the corresponding text character. As demonstrated in Section 2.5.3 (page 43), this runs in $O(nm)$ time, where $n = |t|$ and $m = |p|$.

This quadratic bound is worst case. More complicated, worst-case linear-time search algorithms do exist: see Section 21.3 (page 685) for a complete discussion. But here I give a linear *expected-time* algorithm for string matching, called the Rabin–Karp algorithm. It is based on hashing. Suppose we compute a given hash function on both the pattern string p and the m-character substring starting from the ith position of t. If these two strings are identical, clearly the resulting hash values must be the same. If the two strings are different, the hash values will *almost certainly* be different. These false positives should be so rare that we can easily spend the $O(m)$ time it takes to explicitly check the identity of two strings whenever the hash values agree.

This reduces string matching to $n - m + 2$ hash value computations (the $n - m + 1$ windows of t, plus one hash of p), plus what *should be* a very small number of $O(m)$ time verification steps. The catch is that it takes $O(m)$ time to compute a hash function on an m-character string, and $O(n)$ such computations seems to leave us with an $O(mn)$ algorithm again.

But let's look more closely at our previously defined (in Section 3.7) hash function, applied to the m characters starting from the jth position of string S:

$$H(S, j) = \sum_{i=0}^{m-1} \alpha^{m-(i+1)} \times char(s_{i+j})$$

What changes if we now try to compute $H(S, j + 1)$—the hash of the next window of m characters? Note that $m - 1$ characters are the same in both windows, although this differs by one in the number of times they are multiplied by α. A little algebra reveals that

$$H(S, j + 1) = \alpha(H(S, j) - \alpha^{m-1} char(s_j)) + char(s_{j+m})$$

This means that once we know the hash value from the jth position, we can find the hash value from the $(j+1)$th position for the cost of two multiplications, one addition, and one subtraction. This can be done in constant time (the value of α^{m-1} can be computed once and used for all hash value computations). This math works even if we compute $H(S, j)$ mod M, where M is a reasonably large prime number. This keeps the size of our hash values small (at most M) even when the pattern string is long.

Rabin–Karp is a good example of a randomized algorithm (if we pick M in some random way). We get no guarantee the algorithm runs in $O(n + m)$ time, because we may get unlucky and have the hash values frequently collide with spurious matches. Still, the odds are heavily in our favor—if the hash function returns values uniformly from 0 to $M - 1$, the probability of a false collision should be $1/M$. This is quite reasonable: if $M \approx n$, there should only be one false collision per string, and if $M \approx n^k$ for $k \geq 2$, the odds are great we will never see any false collisions.

6.8   Primality Testing

One of the first programming assignments students get is to test whether an integer n is a prime number, meaning that its only divisors are 1 and itself. The sequence of prime numbers starts with $2, 3, 5, 7, 11, 13, 17, \ldots$, and never ends.

That program you presumably wrote employed trial division as the algorithm: using a loop where i runs from 2 to $n - 1$, and check whether n/i is an integer. If so, then i is a factor of n, and so n must be composite. Any integer that survives this gauntlet of tests is prime. In fact, the loop only needs to run up to $\lceil \sqrt{n} \rceil$, since that is the largest possible value of the smallest non-trivial factor of n.

Still, such trial division is not cheap. If we assume that each division takes constant time this gives an $O(\sqrt{n})$ algorithm, but here n is the *value* of the integer being factored. A 1024-bit number (the size of a small RSA encryption key) encodes numbers up to $2^{1024} - 1$, with the security of RSA depending on factoring being hard. Observe that $\sqrt{2^{1024}} = 2^{512}$, which is greater than the number of atoms in the universe. So expect to spend some time waiting before you get the answer.

Randomized algorithms for primality testing (not factoring) turn out to be much faster. Fermat's little theorem states that if n is a prime number then

$$a^{n-1} = 1 (\mathrm{mod}\ n) \text{ for all } a \text{ not divisible by } n$$

For example, when $n = 17$ and $a = 3$, observe that $(3^{17-1} - 1)/17 = 2,532,160$, so $3^{17-1} = 1 (\mod 17)$. But for $n = 16$, $3^{16-1} = 11 (\mod 16)$, which proves that 16 cannot be prime.

What makes this interesting is that the mod of this big power *always* is 1 if n is prime. This is a pretty good trick, because the odds of it being 1 by chance should be very small—only $1/n$ *if* the residue was uniform in the range.

Let's say we can argue that the probability of a composite giving a residue of 1 is less than $1/2$. This suggests the following algorithm: Pick 100 random integers a_j, each between 1 and $n - 1$. Verify that none of them divide n. Then compute $(a_j)^{n-1} \pmod{n}$. If all hundred of these come out to be 1, then the probability that n is not prime must be less than $(1/2)^{100}$, which is vanishingly small. Because the number of tests (100) is fixed, the running time is always fast, which makes this a Monte Carlo type of randomized algorithm.

There is a minor issue in our probability analysis, however. It turns out that a very small fraction of integers (roughly 1 in 50 billion up to 10^{21}) are not prime, yet also satisfy the Fermat congruence for all a. Such *Carmichael numbers* like 561 and 1105 are doomed to be always be misclassified as prime. Still, this randomized algorithm proves very effective at distinguishing likely primes from composite integers.

> *Take-Home Lesson:* Monte Carlo algorithms are always fast, usually correct, and most of them are wrong in only one direction.

One issue that might concern you is the time complexity of computing $a^{n-1} \pmod{n}$. In fact, it can be done in $O(\log n)$ time. Recall that we can compute a^{2m} as $(a^m)^2$ by divide and conquer, meaning we only need a number of multiplications logarithmic in the size of the exponent. Further, we don't have to work with excessively large numbers to do it. Because of the properties of modular arithmetic,

$$(x \cdot y) \bmod n = ((x \bmod n) \cdot (y \bmod n)) \bmod n$$

so we never need multiply numbers larger than n over the course of the computation.

6.9 War Story: Giving Knuth the Middle Initial

The great Donald Knuth is the seminal figure in creating computer science as an intellectually distinct academic discipline. The first three volumes of his *Art of Computer Programming* series (now four), published between 1968 and 1973, revealed the mathematical beauty of algorithm design, and still make fun and exciting reading. Indeed, I give you my blessing to put my book aside to pick up one of his, at least for a little while.

Knuth is also a co-author of the textbook *Concrete Mathematics*, which focuses on mathematical analysis techniques for algorithms and discrete mathematics. Like his other books it contains open research questions in addition to homework problems. One problem that caught my eye concerned middle binomial coefficients, asking whether it is true that

$$\binom{2n}{n} = (-1)^n (\mathrm{mod}\ (2n+1))\ \text{iff}\ 2n+1\ \text{is prime}.$$

This is suggestive of Fermat's little theorem, discussed in Section 6.8 (page 190).

The congruence is readily shown to hold whenever $2n+1$ is prime. By basic modular arithmetic,

$$(2n)(2n-1)...(n+1) = (-1)(-2)...(-n) = (-1)^n \cdot n!\ (\mathrm{mod}\ (2n+1))$$

Since $ad = bd(\mathrm{mod}\ m)$ implies $a = b(\mathrm{mod}\ m)$ if d is relatively prime to m and $n!$ divides $(2n)!/n!$, $n!$ can be divided from both sides, giving the result.

But does this formula hold *only* when $2n+1$ is prime, as conjectured? That didn't sound right to me, for logic that is dual to the randomized primality testing algorithm. If we treat the residue mod $2n+1$ as a random integer, the probability that it would happen to be $(-1)^n$ is very small, only $1/n$. Thus, not seeing a counterexample over a small number of tests is not very impressive evidence, because chance counterexamples should be rare.

So I wrote a 16-line Mathematica program, and left it running for the weekend. When I got back, the program has stopped at $n = 2{,}953$. It turns out that $\binom{5906}{2953} \approx 7.93285 \times 10^{1775}$ is congruent to 5,906 when taken modulo 5,907. But since $5{,}907 = 3 \cdot 11 \cdot 179$, this shows that $2n+1$ is not prime and the conjecture is refuted.

It was a big thrill sending this result to Knuth himself, who said he would put my name in the next edition of his book. A notorious stickler for detail, he asked me to give him my middle initial. I proudly replied "S", and asked him when he would send me my check. Knuth famously offered $2.56 checks to anyone who found mistakes in one of his books,[2] and I wanted one as a souvenir. But he nixed it, explaining that solving an open problem did not count as fixing an error in his book. I have always regretted that I did not send him my middle initial as "T", because *then* I would have had an error for him to correct in a future printing.

6.10   Where do Random Numbers Come From?

All the clever randomized algorithms discussed in this chapter raises a question: Where do we get random numbers? What happens when you call the random number generator associated with your favorite programming language?

We are used to employing physical processes to generate randomness, such as flipping coins, tossing dice, or even monitoring radioactive decay using a Geiger counter. We trust these events to be unpredictable, and hence indicative of true randomness.

---

[2]I would go broke were I ever to make such an offer.

But this is *not* what your random number generator does. Most likely it employs what is essentially a hash function, called a *linear congruential generator*. The nth random number R_n is a simple function of the previous random number R_{n-1}:

$$R_n = (aR_{n-1} + c) \bmod m$$

where a, c, m, and R_0 are large and carefully selected constants. Essentially, we hash the previous random number (R_{n-1}) to get the next one.

The alert reader may question exactly how random such numbers really are. Indeed, they are completely predictable, because knowing R_{n-1} provides enough information to construct R_n. This predictability means that a sufficiently determined adversary *could* in principle construct a worst-case input to a randomized algorithm provided they know the current state of your random number generator.

Linear congruential generators are more accurately called *pseudo-random number generators*. The stream of numbers produced looks random, in that they have the same statistical properties as would be expected from a truly random source. This is generally good enough for randomized algorithms to work well in practice. However, there is a philosophical sense of randomness which has been lost that occasionally comes back to bite us, typically in cryptographic applications whose security guarantees rest on an assumption of true randomness.

Random number generation is a fascinating problem. Look ahead to section 16.7 in the Hitchhiker's Guide for a more detailed discussion of how random numbers should and should not be generated.

Chapter Notes

Readers interested in more formal and substantial treatments of randomized algorithms are referred to the book of Mitzenmacher and Upfal [MU17] and the older text by Motwani and Raghavan [MR95]. Minwise hashing was invented by Broder [Bro97].

6.11 Exercises

Probability

6-1. *[5]* You are given n unbiased coins, and perform the following process to generate all heads. Toss all n coins independently at random onto a table. Each round consists of picking up all the tails-up coins and tossing them onto the table again. You repeat until all coins are heads.

(a) What is the expected number of rounds performed by the process?

(b) What is the expected number of coin tosses performed by the process?

6-2. *[5]* Suppose we flip n coins each of known bias, such that p_i is the probability of the ith coin being a head. Present an efficient algorithm to determine the exact probability of getting exactly k heads given $p_1, \ldots, p_n \in [0, 1]$.

6-3. *[5]* An *inversion* of a permutation is a pair of elements that are out of order.

    (a) Show that a permutation of n items has at most $n(n-1)/2$ inversions. Which permutation(s) have exactly $n(n-1)/2$ inversions?

    (b) Let P be a permutation and P^r be the reversal of this permutation. Show that P and P^r have a total of exactly $n(n-1)/2$ inversions.

    (c) Use the previous result to argue that the expected number of inversions in a random permutation is $n(n-1)/4$.

6-4. *[8]* A *derangement* is a permutation p of $\{1, \ldots, n\}$ such that no item is in its proper position, that is, $p_i \neq i$ for all $1 \leq i \leq n$. What is the probability that a random permutation is a derangement?

Hashing

6-5. *[easy]* An all-Beatles radio station plays nothing but recordings by the Beatles, selecting the next song at random (uniformly with replacement). They get through about ten songs per hour. I listened for 90 minutes before hearing a repeated song. Estimate how many songs the Beatles recorded.

6-6. *[5]* Given strings S and T of length n and m respectively, find the shortest window in S that contains all the characters in T in expected $O(n+m)$ time.

6-7. *[8]* Design and implement an efficient data structure to maintain a *least recently used* (LRU) cache of n integer elements. A LRU cache will discard the least recently accessed element once the cache has reached its capacity, supporting the following operations:

    • *get(k)*– Return the value associated with the key k if it currently exists in the cache, otherwise return -1.

    • *put(k,v)* – Set the value associated with key k to v, or insert if k is not already present. If there are already n items in the queue, delete the least recently used item before inserting (k, v).

Both operations should be done in $O(1)$ expected time.

Randomized Algorithms

6-8. *[5]* A pair of English words (w_1, w_2) is called a *rotodrome* if one can be circularly shifted (rotated) to create the other word. For example, the words (windup, upwind) are a rotodrome pair, because we can rotate "windup" two positions to the right to get "upwind."

Give an efficient algorithm to find all rotodrome pairs among n words of length k, with a worst-case analysis. Also give a faster expected-time algorithm based on hashing.

6-9. *[5]* Given an array w of positive integers, where $w[i]$ describes the weight of index i, propose an algorithm that randomly picks an index in proportion to its weight.

6-10. *[5]* You are given a function *rand7*, which generates a uniform random integer in the range 1 to 7. Use it to produce a function *rand10*, which generates a uniform random integer in the range 1 to 10.

6-11. *[5]* Let $0 < \alpha < 1/2$ be some constant, independent of the input array length n. What is the probability that, with a randomly chosen pivot element, the partition subroutine from quicksort produces a split in which the size of both the resulting subproblems is at least α times the size of the original array?

6-12. *[8]* Show that for any given load m/n, the error probability of a Bloom filter is minimized when the number of hash functions is $k = \exp(-1)/(m/n)$.

LeetCode

6-1. `https://leetcode.com/problems/random-pick-with-blacklist/`

6-2. `https://leetcode.com/problems/implement-strstr/`

6-3. `https://leetcode.com/problems/random-point-in-non-overlapping-rectangles/`

HackerRank

6-1. `https://www.hackerrank.com/challenges/ctci-ransom-note/`

6-2. `https://www.hackerrank.com/challenges/matchstick-experiment/`

6-3. `https://www.hackerrank.com/challenges/palindromes/`

Programming Challenges

These programming challenge problems with robot judging are available at `https://onlinejudge.org`:

6-1. "Carmichael Numbers"—Chapter 10, problem 10006.

6-2. "Expressions"—Chapter 6, problem 10157.

6-3. "Complete Tree Labeling"—Chapter 6, problem 10247.

Chapter 7

Graph Traversal

Graphs are one of the unifying themes of computer science—an abstract representation that describes the organization of transportation systems, human interactions, and telecommunication networks. That so many different structures can be modeled using a single formalism is a source of great power to the educated programmer.

More precisely, a graph $G = (V, E)$ consists of a set of *vertices* V together with a set E of vertex pairs or *edges*. Graphs are important because they can be used to represent essentially *any* relationship. For example, graphs can model a network of roads, with cities as vertices and roads between cities as edges, as shown in Figure 7.1. Electrical circuits can also be modeled as graphs, with junctions as vertices and components as edges (or alternately, electrical components as vertices and direct circuit connections as edges).

The key to solving many algorithmic problems is to think of them in terms of graphs. Graph theory provides a language for talking about the properties of relationships, and it is amazing how often messy applied problems have a simple description and solution in terms of classical graph properties.

Designing truly novel graph algorithms is a difficult task, but usually unnecessary. The key to using graph algorithms effectively in applications lies in correctly modeling your problem so you can take advantage of existing algorithms. Becoming familiar with many different algorithmic graph *problems* is

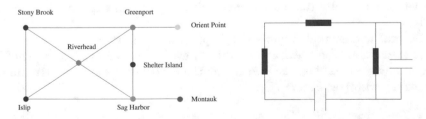

Figure 7.1: Modeling road networks and electrical circuits as graphs.

S. S. Skiena, *The Algorithm Design Manual*, Texts in Computer Science,
https://doi.org/10.1007/978-3-030-54256-6_7

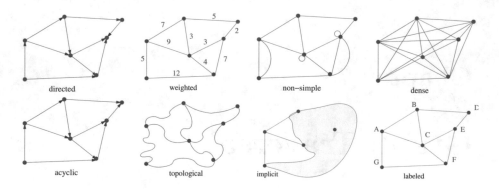

Figure 7.2: Important properties / flavors of graphs.

more important than understanding the details of particular graph algorithms, particularly since Part II of this book will point you to an implementation as soon as you know the name of your problem.

Here I present the basic data structures and traversal operations for graphs, which will enable you to cobble together solutions for elementary graph problems. Chapter 8 will present more advanced graph algorithms that find minimum spanning trees, shortest paths, and network flows, but I stress the primary importance of correctly modeling your problem. Time spent browsing through the catalog now will leave you better informed of your options when a real job arises.

7.1   Flavors of Graphs

A graph $G = (V, E)$ is defined on a set of *vertices* V, and contains a set of *edges* E of ordered or unordered pairs of vertices from V. In modeling a road network, the vertices may represent the cities or junctions, certain pairs of which are connected by roads/edges. In analyzing the source code of a computer program, the vertices may represent lines of code, with an edge connecting lines x and y if y is the next statement executed after x. In analyzing human interactions, the vertices typically represent people, with edges connecting pairs of related souls.

Several fundamental properties of graphs impact the choice of the data structures used to represent them and algorithms available to analyze them. The first step in any graph problem is determining the flavors of the graphs that you will be dealing with (see Figure 7.2):

- *Undirected vs. directed* – A graph $G = (V, E)$ is *undirected* if the presence of edge (x, y) in E implies that edge (y, x) is also in E. If not, we say that the graph is *directed*. Road networks *between* cities are typically undirected, since any large road has lanes going in both directions. Street networks *within* cities are almost always directed, because there are at least a few one-way streets lurking somewhere. Program-flow graphs are typically directed, because the execution flows from one line into the next and changes direction only at branches. Most graphs of graph-theoretic interest are undirected.

- *Weighted vs. unweighted* – Each edge (or vertex) in a *weighted* graph G is assigned a numerical value, or weight. The edges of a road network graph might be weighted with their length, drive time, or speed limit, depending upon the application. In *unweighted* graphs, there is no cost distinction between various edges and vertices.

  The difference between weighted and unweighted graphs becomes particularly apparent in finding the shortest path between two vertices. For unweighted graphs, a shortest path is one that has the fewest number of edges, and can be found using a breadth-first search (BFS) as discussed in this chapter. Shortest paths in weighted graphs requires more sophisticated algorithms, as discussed in Chapter 8.

- *Simple vs. non-simple* – Certain types of edges complicate the task of working with graphs. A *self-loop* is an edge (x, x) involving only one vertex. An edge (x, y) is a *multiedge* if it occurs more than once in the graph.

  Both of these structures require special care in implementing graph algorithms. Hence any graph that avoids them is called *simple*. I confess that all implementations in this book are designed to work **only** on simple graphs.

- *Sparse vs. dense*:   Graphs are *sparse* when only a small fraction of the possible vertex pairs actually have edges defined between them. Graphs where a large fraction of the vertex pairs define edges are called *dense*. A graph is *complete* if it contains all possible edges; for a simple undirected graph on n vertices that is $\binom{n}{2} = (n^2 - n)/2$ edges. There is no official boundary between what is called sparse and what is called dense, but dense graphs typically have $\Theta(n^2)$ edges, while sparse graphs are linear in size.

  Sparse graphs are usually sparse for application-specific reasons. Road networks must be sparse because of the complexity of road junctions. The ghastliest intersection I have ever managed to identify is the endpoint of just seven different roads. Junctions of electrical components are similarly limited to the number of wires that can meet at a point, perhaps except for power and ground.

- *Cyclic vs. acyclic* – A *cycle* is a closed path of 3 or more vertices that has no repeating vertices except the start/end point. An *acyclic* graph does not contain any cycles. *Trees* are undirected graphs that are connected and acyclic. They are the simplest interesting graphs. Trees are inherently recursive structures, because cutting any edge leaves two smaller trees.

Directed acyclic graphs are called *DAGs*. They arise naturally in scheduling problems, where a directed edge (x, y) indicates that activity x must occur before y. An operation called *topological sorting* orders the vertices of a DAG to respect these precedence constraints. Topological sorting is typically the first step of any algorithm on a DAG, as will be discussed in Section 7.10.1 (page 231).

- *Embedded vs. topological* – The edge–vertex representation $G = (V, E)$ describes the purely topological aspects of a graph. We say a graph is *embedded* if the vertices and edges are assigned geometric positions. Thus, any drawing of a graph is an *embedding*, which may or may not have algorithmic significance.

  Occasionally, the structure of a graph is completely defined by the geometry of its embedding. For example, if we are given a collection of points in the plane, and seek the minimum cost tour visiting all of them (i.e., the traveling salesman problem), the underlying topology is the *complete graph* connecting each pair of vertices. The weights are typically defined by the Euclidean distance between each pair of points.

  Grids of points are another example of topology from geometry. Many problems on an $n \times m$ rectangular grid involve walking between neighboring points, so the edges are implicitly defined from the geometry.

- *Implicit vs. explicit* – Certain graphs are not explicitly constructed and then traversed, but built as we use them. A good example is in backtrack search. The vertices of this implicit search graph are the states of the search vector, while edges link pairs of states that can be directly generated from each other. Web-scale analysis is another example, where you should try to dynamically crawl and analyze the small relevant portion of interest instead of initially downloading the entire web. The cartoon in Figure 7.2 tries to capture this distinction between the part of the graph you explicitly know from the fog that covers the rest, which dissipates as you explore it. It is often easier to work with an implicit graph than to explicitly construct and store the entire thing prior to analysis.

- *Labeled vs. unlabeled* – Each vertex is assigned a unique name or identifier in a *labeled* graph to distinguish it from all other vertices. In *unlabeled* graphs, no such distinctions have been made.

  Graphs arising in applications are often naturally and meaningfully labeled, such as city names in a transportation network. A common problem is that of *isomorphism testing*—determining whether the topological structures of two graphs are identical either respecting or ignoring any labels, as discussed in Section 19.9.

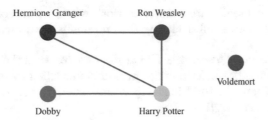

Figure 7.3: A portion of the friendship graph from *Harry Potter*.

7.1.1   The Friendship Graph

To demonstrate the importance of proper modeling, let us consider a graph where the vertices are people, and edge between two people indicates that they are friends. Such graphs are called *social networks* and are well defined on any set of people—be they the people in your neighborhood, at your school/business, or even spanning the entire world. An entire science analyzing social networks has sprung up in recent years, because many interesting aspects of people and their behavior are best understood as properties of this friendship graph.

We use this opportunity to demonstrate the graph theory terminology described above. "Talking the talk" proves to be an important part of "walking the walk":

- *If I am your friend, does that mean you are my friend?* – This question really asks whether the graph is directed. A graph is *undirected* if edge (x, y) always implies (y, x). Otherwise, the graph is said to be *directed*. The "heard-of" graph is directed, since I have heard of many famous people who have never heard of me! The "had-sex-with" graph is presumably undirected, since the critical operation always requires a partner. I'd like to think that the "friendship" graph is also an undirected graph.

- *How close a friend are you?* – In *weighted* graphs, each edge has an associated numerical attribute. We could model the strength of a friendship by associating each edge with an appropriate value, perhaps from -100 (enemies) to 100 (blood brothers). The edges of a road network graph might be weighted with their length, drive time, or speed limit, depending upon the application. A graph is said to be *unweighted* if all edges are assumed to be of equal weight.

- *Am I my own friend?* – This question addresses whether the graph is *simple*, meaning it contains no loops and no multiple edges. An edge of the form (x, x) is said to be a *self-loop*. Sometimes people are friends in several different ways. Perhaps x and y were college classmates that now work together at the same company. We can model such relationships using *multiedges*—multiple edges (x, y) perhaps distinguished by different labels.

Simple graphs really are simpler to work with in practice. Therefore, we are generally better off declaring that no one is their own friend.

- *Who has the most friends?* – The *degree* of a vertex is the number of edges adjacent to it. The most popular person is identified by finding the vertex of highest degree in the friendship graph. Remote hermits are associated with degree-zero vertices.

  Most of the graphs that one encounters in real life are sparse. The friendship graph is a good example. Even the most gregarious person on earth knows only an insignificant fraction of the world's population.

  In *dense* graphs, most vertices have high degrees, as opposed to *sparse* graphs with relatively few edges. In a *regular graph*, each vertex has exactly the same degree. A regular friendship graph is truly the ultimate in social-ism.

- *Do my friends live near me?* – Social networks are strongly influenced by geography. Many of your friends are your friends only because they happen to live near you (neighbors) or used to live near you (old college roommates).

  Thus, a full understanding of social networks requires an *embedded* graph, where each vertex is associated with the point on this world where they live. This geographic information may not be explicitly encoded, but the fact that the graph is inherently embedded on the surface of a sphere shapes our interpretation of the network.

- *Oh, you also know her?* – Social networking services such as Instagram and LinkedIn *explicitly* define friendship links between members. Such graphs consist of directed edges from person/vertex x professing his or her friendship with person/vertex y.

  That said, the actual friendship graph of the world is represented *implicitly*. Each person knows who their friends are, but cannot find out about other people's friendships except by asking them. The "six degrees of separation" theory argues that there is a short path linking every two people in the world (e.g. Skiena and the President) but offers us no help in actually finding this path. The shortest such path I know of contains three hops:

  Steven Skiena $\rightarrow$ Mark Fasciano $\rightarrow$ Michael Ashner $\rightarrow$ Donald Trump

  but there could be a shorter one (say if Trump went to college with my dentist).[1] The friendship graph is stored implicitly, so I have no way of easily checking.

---

[1] There is also a path Steven Skiena $\rightarrow$ Steve Israel $\rightarrow$ Joe Biden, so I am covered regardless of the outcome of the 2020 election.

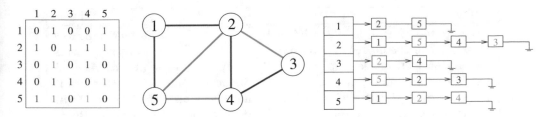

Figure 7.4: The adjacency matrix and adjacency list representation of a given graph. Colors encode specific edges.

- *Are you truly an individual, or just one of the faceless crowd?* – This question boils down to whether the friendship graph is labeled or unlabeled. Are the names or vertex IDs important for our analysis?

  Much of the study of social networks is unconcerned with labels on graphs. Often the ID number given to a vertex in the graph data structure serves as its label, either for convenience or the need for anonymity. You may assert that you are a name, not a number—but try protesting to the fellow who implements the algorithm. Someone studying how rumors or infectious diseases spread through a friendship graph might examine network properties such as connectedness, the distribution of vertex degrees, or the distribution of path lengths. These properties aren't changed by a scrambling of the vertex IDs.

Take-Home Lesson: Graphs can be used to model a wide variety of structures and relationships. Graph-theoretic terminology gives us a language to talk about them.

7.2 Data Structures for Graphs

Selecting the right graph data structure can have an enormous impact on performance. Your two basic choices are adjacency matrices and adjacency lists, illustrated in Figure 7.4. We assume the graph $G = (V, E)$ contains n vertices and m edges.

- *Adjacency matrix* – We can represent G using an $n \times n$ matrix M, where element $M[i, j] = 1$ if (i, j) is an edge of G, and 0 if it isn't. This allows fast answers to the question "is (i, j) in G?", and rapid updates for edge insertion and deletion. It may use excessive space for graphs with many vertices and relatively few edges, however.

  Consider a graph that represents the street map of Manhattan in New York City. Every junction of two streets will be a vertex of the graph. Neighboring junctions are connected by edges. How big is this graph? Manhattan is basically a grid of 15 avenues each crossing roughly 200

Comparison	Winner
Faster to test if (x, y) is in graph?	adjacency matrices
Faster to find the degree of a vertex?	adjacency lists
Less memory on sparse graphs?	adjacency lists $(m + n)$ vs. (n^2)
Less memory on dense graphs?	adjacency matrices (a small win)
Edge insertion or deletion?	adjacency matrices $O(1)$ vs. $O(d)$
Faster to traverse the graph?	adjacency lists $\Theta(m + n)$ vs. $\Theta(n^2)$
Better for most problems?	adjacency lists

Figure 7.5: Relative advantages of adjacency lists and matrices.

streets. This gives us about 3,000 vertices and 6,000 edges, since almost all vertices neighbor four other vertices and each edge is shared between two vertices. This is a modest amount of data to store, yet an adjacency matrix would have $3,000 \times 3,000 = 9,000,000$ elements, almost all of them empty!

There is some potential to save space by packing multiple bits per word or using a symmetric-matrix data structure (e.g. triangular matrix) for undirected graphs. But these methods lose the simplicity that makes adjacency matrices so appealing and, more critically, remain inherently quadratic even for sparse graphs.

- *Adjacency lists* – We can more efficiently represent sparse graphs by using linked lists to store the neighbors of each vertex. Adjacency lists require pointers, but are not frightening once you have experience with linked structures.

  Adjacency lists make it harder to verify whether a given edge (i, j) is in G, since we must search through the appropriate list to find the edge. However, it is surprisingly easy to design graph algorithms that avoid any need for such queries. Typically, we sweep through all the edges of the graph in one pass via a breadth-first or depth-first traversal, and update the implications of the current edge as we visit it. Table 7.5 summarizes the tradeoffs between adjacency lists and matrices.

Take-Home Lesson: Adjacency lists are the right data structure for most applications of graphs.

We will use adjacency lists as our primary data structure to represent graphs. We represent a graph using the following data type. For each graph, we keep a count of the number of vertices, and assign each vertex a unique identification number from 1 to `nvertices`. We represent the edges using an array of linked lists:

```
#define MAXV          100      /* maximum number of vertices */

typedef struct edgenode {
    int y;                     /* adjacency info */
    int weight;                /* edge weight, if any */
    struct edgenode *next;     /* next edge in list */
} edgenode;

typedef struct {
    edgenode *edges[MAXV+1];   /* adjacency info */
    int degree[MAXV+1];        /* outdegree of each vertex */
    int nvertices;             /* number of vertices in the graph */
    int nedges;                /* number of edges in the graph */
    int directed;              /* is the graph directed? */
} graph;
```

We represent directed edge (x, y) by an **edgenode** y in x's adjacency list. The degree field of the **graph** counts the number of meaningful entries for the given vertex. An undirected edge (x, y) appears twice in any adjacency-based graph structure, once as y in x's list, and the other as x in y's list. The Boolean flag **directed** identifies whether the given graph is to be interpreted as directed or undirected.

To demonstrate the use of this data structure, we show how to read a graph from a file. A typical graph file format consists of an initial line giving the number of vertices and edges in the graph, followed by a list of the edges, one vertex pair per line. We start by initializing the structure:

```
void initialize_graph(graph *g, bool directed) {
    int i;     /* counter */

    g->nvertices = 0;
    g->nedges = 0;
    g->directed = directed;

    for (i = 1; i <= MAXV; i++) {
        g->degree[i] = 0;
    }

    for (i = 1; i <= MAXV; i++) {
        g->edges[i] = NULL;
    }
]
```

Then we actually read the graph file, inserting each edge into this structure:

```
void read_graph(graph *g, bool directed) {
    int i;                   /* counter */
    int m;                   /* number of edges */
    int x, y;                /* vertices in edge (x,y) */

    initialize_graph(g, directed);

    scanf("%d %d", &(g->nvertices), &m);

    for (i = 1; i <= m; i++) {
        scanf("%d %d", &x, &y);
        insert_edge(g, x, y, directed);
    }
}
```

The critical routine is `insert_edge`. The new `edgenode` is inserted at the beginning of the appropriate adjacency list, since order doesn't matter. We parameterize our insertion with the `directed` Boolean flag, to identify whether we need to insert two copies of each edge or only one. Note the use of recursion to insert the copy:

```
void insert_edge(graph *g, int x, int y, bool directed) {
    edgenode *p;             /* temporary pointer */

    p = malloc(sizeof(edgenode));     /* allocate edgenode storage */

    p->weight = 0;
    p->y = y;
    p->next = g->edges[x];

    g->edges[x] = p;         /* insert at head of list */

    g->degree[x]++;

    if (!directed) {
        insert_edge(g, y, x, true);
    } else {
        g->nedges++;
    }
}
```

Printing the associated graph is just a matter of two nested loops: one through the vertices, and the second through adjacent edges:

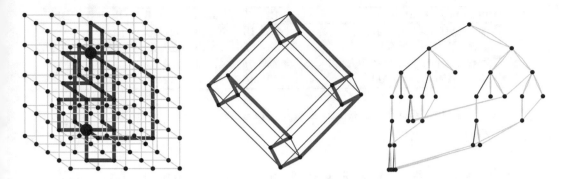

Figure 7.6: Representative *Combinatorica* graphs: edge-disjoint paths (left), Hamiltonian cycle in a hypercube (center), animated depth-first search tree traversal (right).

```
void print_graph(graph *g) {
    int i;          /* counter */
    edgenode *p;    /* temporary pointer */

    for (i = 1; i <= g->nvertices; i++) {
        printf("%d: ", i);
        p = g->edges[i];
        while (p != NULL) {
            printf(" %d", p->y);
            p = p->next;
        }
        printf("\n");
    }
}
```

It is a good idea to use a well-designed graph data type as a model for building your own, or even better as the foundation for your application. I recommend LEDA (see Section 22.1.1 (page 713)) or Boost (see Section 22.1.3 (page 714)) as the best-designed general-purpose graph data structures currently available. They may be more powerful (and hence somewhat slower/larger) than you need, but they do so many things right that you are likely to lose most of the potential do-it-yourself benefits through clumsiness.

7.3 War Story: I was a Victim of Moore's Law

I am the author of a popular library of graph algorithms called *Combinatorica* (www.combinatorica.com), which runs under the computer algebra system *Mathematica*. Efficiency is a great challenge in *Mathematica*, due to its applicative model of computation (it does not support constant-time write operations

command/machine	Sun-3	Sun-4	Sun-5	Ultra 5	Blade
PlanarQ[GridGraph[4,4]]	234.10	69.65	27.50	3.60	0.40
Length[Partitions[30]]	289.85	73.20	24.40	3.44	1.58
VertexConnectivity[GridGraph[3,3]]	239.67	47.76	14.70	2.00	0.91
RandomPartition[1000]	831.68	267.5	22.05	3.12	0.87

Table 7.1: Old *Combinatorica* benchmarks on five generations of workstations (running time in seconds).

to arrays) and the overhead of interpretation (as opposed to compilation). *Mathematica* code is typically 1,000 to 5,000 times slower than C code.

Such slowdowns can be a tremendous performance hit. Even worse, *Mathematica* is a memory hog, needing a then-outrageous 4MB of main memory to run effectively when I completed *Combinatorica* in 1990. Any computation on large structures was doomed to thrash in virtual memory. In such an environment, my graph package could only hope to work effectively on *very* small graphs.

One design decision I made as a result was to use adjacency matrices as the basic *Combinatorica* graph data structure instead of lists. This may sound peculiar. If pressed for memory, wouldn't it pay to use adjacency lists and conserve every last byte? Yes, but the answer is not so simple for very small graphs. An adjacency list representation of a weighted n-vertex, m-edge directed graph should use about $n + 2m$ words to represent; the $2m$ comes from storing the endpoint and weight components of each edge. Thus, the space advantages of adjacency lists only kick in when $n+2m$ is substantially smaller than n^2. The adjacency matrix is still manageable in size for $n \leq 100$ and, of course, about half the size of adjacency lists on dense graphs.

My more immediate concern was dealing with the overhead of using a slow interpreted language. Check out the benchmarks reported in Table 7.1. Two particularly complex but polynomial-time problems on 9 and 16 vertex graphs took several minutes to complete on my desktop machine in 1990! The quadratic-sized data structure certainly could not have had much impact on these running times, since 9×9 equals only 81. From experience, I knew the *Mathematica* programming language handled regular structures like adjacency matrices better than irregular-sized adjacency lists.

Still, *Combinatorica* proved to be a very good thing despite these performance problems. Thousands of people have used my package to do all kinds of interesting things with graphs. *Combinatorica* was never intended to be a high-performance algorithms library. Most users quickly realized that computations on large graphs were out of the question, but were eager to take advantage of *Combinatorica* as a mathematical research tool and prototyping environment. Everyone was happy.

But over the years, my users started asking why it took so long to do a modest-sized graph computation. The mystery wasn't that my program was slow, because it had always been slow. The question was why did it take them so many years to figure this out?

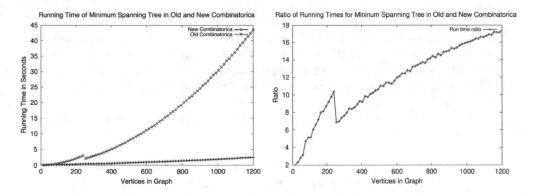

Figure 7.7: Performance comparison between old and new *Combinatorica*: absolute running times (left), and the ratio of these times (right).

The reason is that computers keep doubling in speed every two years or so. People's *expectation* of how long something should take moves in concert with these technology improvements. Partially because of *Combinatorica*'s dependence on a quadratic-size graph data structure, it didn't scale as well as it should on sparse graphs.

As the years rolled on, user demands became more insistent. *Combinatorica* needed to be updated. My collaborator, Sriram Pemmaraju, rose to the challenge. We (mostly he) completely rewrote *Combinatorica* to take advantage of faster graph data structures ten years after the initial version.

The new *Combinatorica* uses a list of edges data structure for graphs, largely motivated by increased efficiency. Edge lists are linear in the size of the graph (edges plus vertices), just like adjacency lists. This makes a huge difference on most graph-related functions—for large enough graphs. The improvement is most dramatic in "fast" graph algorithms—those that run in linear or near-linear time, such as graph traversal, topological sort, and finding connected or biconnected components. The implications of this change are felt throughout the package in running time improvements and memory savings. *Combinatorica* can now work with graphs that are fifty to a hundred times larger than what the old package could deal with.

Figure 7.7(left) plots the running time of the `MinimumSpanningTree` functions for both *Combinatorica* versions. The test graphs were sparse (grid graphs), designed to highlight the difference between the two data structures. Yes, the new version is *much* faster, but note that the difference only becomes important for graphs larger than the old *Combinatorica* was designed for. However, the relative difference in run time keeps growing with increasing n. Figure 7.7(right) plots the ratio of the running times as a function of graph size. The difference between linear size and quadratic size is asymptotic, so the consequences become ever more important as n gets larger.

What is the weird bump in running times that occurs around $n \approx 250$? This likely reflects a transition between levels of the memory hierarchy. Such bumps are not uncommon in today's complex computer systems. Cache performance in data structure design should be an important but not overriding consideration. The asymptotic gains due to adjacency lists more than trumped any impact of the cache.

Three main lessons can be taken away from our experience developing *Combinatorica*:

- *To make a program run faster, just wait* – Sophisticated hardware eventually trickles down to everybody. We observe a speedup of more than 200-fold for the original version of *Combinatorica* as a consequence of 15 years of hardware evolution. In this context, the further speedups we obtained from upgrading the package become particularly dramatic.

- *Asymptotics eventually do matter* – It was my mistake not to anticipate future developments in technology. While no one has a crystal ball, it is fairly safe to say that future computers will have more memory and run faster than today's. This gives the edge to asymptotically more efficient algorithms/data structures, even if their performance is close on today's instances. If the implementation complexity is not substantially greater, play it safe and go with the better algorithm.

- *Constant factors can matter* – With the growing importance of the study of networks, Wolfram Research has recently moved basic graph data structures into the core of *Mathematica*. This permits them to be written in a compiled instead of interpreted language, speeding all operations by about a factor of 10 over *Combinatorica*.

Speeding up a computation by a factor of 10 is often very important, taking it from a week down to a day, or a year down to a month. This book focuses largely on asymptotic complexity, because we seek to teach fundamental principles. But constants can matter in practice.

7.4   War Story: Getting the Graph

"It takes five minutes just to *read* the data. We will *never* have time to make it do something interesting."

The young graduate student was bright and eager, but green to the power of data structures. She would soon come to appreciate their power.

As described in a previous war story (see Section 3.6 (page 89)), we were experimenting with algorithms to extract triangular strips for the fast rendering of triangulated surfaces. The task of finding a small number of strips that cover each triangle in a mesh can be modeled as a graph problem. The graph has a vertex for every *triangle* of the mesh, with an edge between every pair of vertices representing adjacent triangles. This *dual graph* representation (see

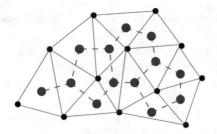

Figure 7.8: The dual graph (dashed lines) of a triangulation

Figure 7.8) captures all information needed to partition the triangulation into triangle strips.

The first step in crafting a program that constructs a good set of strips was to build the dual graph of the triangulation. This I sent the student off to do. A few days later, she came back and announced that it took her machine over five minutes to construct the dual graph of a few thousand triangles.

"Nonsense!" I proclaimed. "You must be doing something very wasteful in building the graph. What format is the data in?"

"Well, it starts out with a list of the three-dimensional coordinates of the vertices used in the model and then follows with a list of triangles. Each triangle is described by a list of three indices into the vertex coordinates. Here is a small example,"

```
VERTICES 4
0.000000 240.000000 0.000000
204.000000 240.000000 0.000000
204.000000 0.000000 0.000000
0.000000 0.000000 0.000000
TRIANGLES 2
0 1 3
1 2 3
```

"I see. So the first triangle uses all but the third point, since all the indices start from zero. The two triangles must share an edge formed by points 1 and 3."

"Yeah, that's right," she confirmed.

"OK. Now tell me how you built your dual graph from this file."

"Well, the geometric position of the points doesn't affect the structure of the graph, so I can ignore it. My dual graph is going to have as many vertices as the number of triangles. I set up an adjacency list data structure with that many vertices. As I read in each triangle, I compare it to each of the others to check whether it has two end points in common. If it does, then I add an edge from the new triangle to this one."

I started to sputter. "But *that's* your problem right there! You are comparing each triangle against every other triangle, so that constructing the dual graph will be quadratic in the number of triangles. Reading the input graph should take linear time!"

"I'm not comparing every triangle against *every* other triangle. On average, it only tests against half or a third of the triangles."

"Swell. But that still leaves us with an $O(n^2)$ algorithm. That is much too slow."

She stood her ground. "Well, don't just complain. Help me fix it!"

Fair enough. I started to think. We needed some quick method to screen away most of the triangles that would not be adjacent to the new triangle (i, j, k). What we really needed was a separate list of all the triangles that go through each of the points i, j, and k. By Euler's formula for planar graphs, the average point is incident to fewer than six triangles. This would compare each new triangle against fewer than twenty others, instead of most of them.

"We are going to need a data structure consisting of an array with one element for every vertex in the original data set. This element is going to be a list of all the triangles that pass through that vertex. When we read in a new triangle, we will look up the three relevant lists in the array and compare each of these against the new triangle. Actually, only two of the three lists need be tested, since any adjacent triangles will share two points in common. We will add an edge to our graph for every triangle pair sharing two vertices. Finally, we will add our new triangle to each of the three affected lists, so they will be updated for the next triangle read."

She thought about this for a while and smiled. "Got it, Chief. I'll let you know what happens."

The next day she reported that the graph could be built in seconds, even for much larger models. From here, she went on to build a successful program for extracting triangle strips, as reported in Section 3.6 (page 89).

Take-Home Lesson: Even elementary problems like initializing data structures can prove to be bottlenecks in algorithm development. Programs working with large amounts of data must run in linear or near-linear time. Such tight performance demands leave no room to be sloppy. Once you focus on the need for linear-time performance, an appropriate algorithm or heuristic can usually be found to do the job.

7.5   Traversing a Graph

Perhaps the most fundamental graph problem is to visit every edge and vertex in a graph in a systematic way. Indeed, all the basic bookkeeping operations on graphs (such as printing or copying graphs, and converting between alternative representations) are applications of graph traversal.

Mazes are naturally represented by graphs, where each graph vertex denotes a junction of the maze, and each graph edge denotes a passageway in the maze. Thus, any graph traversal algorithm must be powerful enough to get us out of an arbitrary maze. For *efficiency*, we must make sure we don't get trapped in the maze and visit the same place repeatedly. For *correctness*, we must do the traversal in a systematic way to guarantee that we find a way out of the maze. Our search must take us through every edge and vertex in the graph.

The key idea behind graph traversal is to mark each vertex when we first visit it and keep track of what we have not yet completely explored. Bread crumbs and unraveled threads have been used to mark visited places in fairy-tale mazes, but we will rely on Boolean flags or enumerated types.

Each vertex will exist in one of three states:

- *Undiscovered* – the vertex is in its initial, virgin state.

- *Discovered* – the vertex has been found, but we have not yet checked out all its incident edges.

- *Processed* – the vertex after we have visited all of its incident edges.

Obviously, a vertex cannot be *processed* until after we discover it, so the state of each vertex progresses from *undiscovered* to *discovered* to *processed* over the course of the traversal.

We must also maintain a structure containing the vertices that we have discovered but not yet completely processed. Initially, only the single start vertex is considered to be discovered. To completely explore a vertex v, we must evaluate each edge leaving v. If an edge goes to an undiscovered vertex x, we mark x *discovered* and add it to the list of work to do in the future. If an edge goes to a *processed* vertex, we ignore that vertex, because further contemplation will tell us nothing new about the graph. Similarly, we can ignore any edge going to a *discovered* but not *processed* vertex, because that destination already resides on the list of vertices to process.

Each undirected edge will be considered exactly twice, once when each of its endpoints is explored. Directed edges will be considered only once, when exploring the source vertex. Every edge and vertex in the connected component must eventually be visited. Why? Suppose that there exists a vertex u that remains unvisited, whose neighbor v *was* visited. This neighbor v will eventually be explored, after which we will certainly visit u. Thus, we must find everything that is there to be found.

I describe the mechanics of these traversal algorithms and the significance of the traversal order below.

7.6 Breadth-First Search

The basic breadth-first search algorithm is given below. At some point during the course of a traversal, every node in the graph changes state from *undiscovered* to *discovered*. In a breadth-first search of an undirected graph, we assign a direction to each edge, from the discoverer u to the discovered v. We thus denote u to be the parent of v. Since each node has exactly one parent, except for the root, this defines a tree on the vertices of the graph. This tree, illustrated in

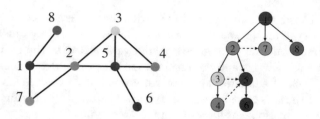

Figure 7.9: An undirected graph and its breadth-first search tree. Dashed lines, which are not part of the tree, show graph edges that go to discovered or processed vertices.

Figure 7.9, defines a shortest path from the root to every other node in the tree. This property makes breadth-first search very useful in shortest path problems.

BFS(G, s)
    Initialize each vertex $u \in V[G]$ so
        $state[u]$ = "undiscovered"
        $p[u] = nil$, i.e. no parent is in the BFS tree
    $state[s]$ = "discovered"
    $Q = \{s\}$
    while $Q \neq \emptyset$ do
        u = dequeue[Q]
        process vertex u if desired
        for each vertex v that is adjacent to u
            process edge (u, v) if desired
            if $state[v]$ = "undiscovered" then
                $state[v]$ = "discovered"
                $p[v] = u$
                enqueue[Q, v]
        $state[u]$ = "processed"

The graph edges that do not appear in the breadth-first search tree also have special properties. For undirected graphs, non-tree edges can point only to vertices on the same level as the parent vertex, or to vertices on the level directly below the parent. These properties follow easily from the fact that each path in the tree must be a shortest path in the graph. For a directed graph, a back-pointing edge (u, v) can exist whenever v lies closer to the root than u does.

Implementation

Our breadth-first search implementation `bfs` uses two Boolean arrays to maintain our knowledge about each vertex in the graph. A vertex is `discovered` the first time we visit it. A vertex is considered `processed` after we have traversed all outgoing edges from it. Thus, each vertex passes from `undiscovered` to `discovered` to `processed` over the course of the search. This information could have been maintained using one enumerated type variable, but we used two Boolean variables instead.

```
bool processed[MAXV+1];     /* which vertices have been processed */
bool discovered[MAXV+1];    /* which vertices have been found */
int parent[MAXV+1];         /* discovery relation */
```

Each vertex is initialized as undiscovered:

```
void initialize_search(graph *g) {
    int i;                      /* counter */

    time = 0;

    for (i = 0; i <= g->nvertices; i++) {
        processed[i] = false;
        discovered[i] = false;
        parent[i] = -1;
    }
}
```

Once a vertex is discovered, it is placed on a queue. Since we process these vertices in first-in, first-out (FIFO) order, the oldest vertices are expanded first, which are exactly those closest to the root:

```
void bfs(graph *g, int start) {
    queue q;            /* queue of vertices to visit */
    int v;              /* current vertex */
    int y;              /* successor vertex */
    edgenode *p;        /* temporary pointer */

    init_queue(&q);
    enqueue(&q, start);
    discovered[start] = true;

    while (!empty_queue(&q)) {
        v = dequeue(&q);
        process_vertex_early(v);
        processed[v] = true;
        p = g->edges[v];
        while (p != NULL) {
            y = p->y;
            if ((!processed[y]) || g->directed) {
                process_edge(v, y);
```

```
        }
        if (!discovered[y]) {
            enqueue(&q,y);
            discovered[y] = true;
            parent[y] = v;
        }
        p = p->next;
    }
    process_vertex_late(v);
  }
}
```

7.6.1   Exploiting Traversal

The exact behavior of `bfs` depends upon the functions `process_vertex_early()`, `process_vertex_late()`, and `process_edge()`. Through these functions, we can customize what the traversal does as it makes its one official visit to each edge and each vertex. Initially, we will do all vertex processing on entry, so `process_vertex_late()` returns without action:

```
void process_vertex_late(int v) {

}
```

By setting the active functions to

```
void process_vertex_early(int v) {
    printf("processed vertex %d\n", v);
}
```

```
void process_edge(int x, int y) {
    printf("processed edge (%d,%d)\n", x, y);
}
```

we print each vertex and edge exactly once. If we instead set `process_edge` to

```
void process_edge(int x, int y) {
    nedges = nedges + 1;
}
```

we get an accurate count of the number of edges. Different algorithms perform different actions on vertices or edges as they are encountered. These functions give us the freedom to easily customize these actions.

7.6.2 Finding Paths

The `parent` array that is filled over the course of `bfs()` is very useful for finding interesting paths through a graph. The vertex that first discovered vertex i is defined as the `parent[i]`. Every vertex is discovered once during the course of traversal, so every node has a parent, except for the start node. This parent relation defines a tree of discovery, with the start node as the root of the tree.

Because vertices are discovered in order of increasing distance from the root, this tree has a very important property. The unique tree path from the root to each node $x \in V$ uses the smallest number of edges (or equivalently, intermediate nodes) possible on any root-to-x path in the graph.

We can reconstruct this path by following the chain of ancestors from x to the root. Note that we have to work backward. We cannot find the path from the root to x, because this does not agree with the direction of the parent pointers. Instead, we must find the path from x to the root. Since this is the reverse of how we normally want the path, we can either (1) store it and then explicitly reverse it using a stack, or (2) let recursion reverse it for us, as follows:

```
void find_path(int start, int end, int parents[]) {
    if ((start == end) || (end == -1)) {
        printf("\n%d", start);
    } else {
        find_path(start, parents[end], parents);
        printf(" %d", end);
    }
}
```

On our breadth-first search graph example (Figure 7.9) our algorithm generated the following parent relation:

vertex	1	2	3	4	5	6	7	8
parent	−1	1	2	3	2	5	1	1

For the shortest path from 1 to 6, this parent relation yields the path $\{1, 2, 5, 6\}$.

There are two points to remember when using breadth-first search to find a shortest path from x to y: First, the shortest path tree is only useful if BFS was performed with x as the root of the search. Second, BFS gives the shortest path only if the graph is unweighted. We will present algorithms for finding shortest paths in weighted graphs in Section 8.3.1 (page 258).

7.7   Applications of Breadth-First Search

Many elementary graph algorithms perform one or two traversals of the graph, while doing something along the way. Properly implemented using adjacency lists, any such algorithm is destined to be linear, since BFS runs in $O(n + m)$

time for both directed and undirected graphs. This is optimal, since this is as fast as one can ever hope to just *read* an n-vertex, m-edge graph.

The trick is seeing when such traversal approaches are destined to work. I present several examples below.

7.7.1   Connected Components

We say that a graph is *connected* if there is a path between any two vertices. Every person can reach every other person through a chain of links if the friendship graph is connected.

A *connected component* of an undirected graph is a maximal set of vertices such that there is a path between every pair of vertices. The components are separate "pieces" of the graph such that there is no connection between the pieces. If we envision tribes in remote parts of the world that have not yet been encountered, each such tribe would form a separate connected component in the friendship graph. A remote hermit, or extremely uncongenial fellow, would represent a connected component of one vertex.

An amazing number of seemingly complicated problems reduce to finding or counting connected components. For example, deciding whether a puzzle such as Rubik's cube or the 15-puzzle can be solved from any position is really asking whether the graph of possible configurations is connected.

Connected components can be found using breadth-first search, since the vertex order does not matter. We begin by performing a search starting from an arbitrary vertex. Anything we discover during this search must be part of the same connected component. We then repeat the search from any undiscovered vertex (if one exists) to define the next component, and so on until all vertices have been found:

```
void connected_components(graph *g) {
    int c;                   /* component number */
    int i;                   /* counter */

    initialize_search(g);

    c = 0;
    for (i = 1; i <= g->nvertices; i++) {
        if (!discovered[i]) {
            c = c + 1;
            printf("Component %d:", c);
            bfs(g, i);
            printf("\n");
        }
    }
}
```

```
void process_vertex_early(int v) {        /* vertex to process */
    printf(" %d", v);
}



void process_edge(int x, int y) {

}
```

Observe how we increment a counter c denoting the current component number with each call to bfs. Alternatively, we could have explicitly bound each vertex to its component number (instead of printing the vertices in each component) by changing the action of process_vertex.

There are two distinct notions of connectivity for directed graphs, leading to algorithms for finding both weakly connected and strongly connected components. Both of these can be found in $O(n + m)$ time, as discussed in Section 18.1 (page 542).

7.7.2   Two-Coloring Graphs

The *vertex-coloring* problem seeks to assign a label (or color) to each vertex of a graph such that no edge links any two vertices of the same color. We can easily avoid all conflicts by assigning each vertex a unique color. However, the goal is to use as few colors as possible. Vertex coloring problems often arise in scheduling applications, such as register allocation in compilers. See Section 19.7 (page 604) for a full treatment of vertex-coloring algorithms and applications.

A graph is *bipartite* if it can be colored without conflicts while using only two colors. Bipartite graphs are important because they arise naturally in many applications. Consider the "mutually interested-in" graph in a heterosexual world, where people consider only those of opposing gender. In this simple model, gender would define a two-coloring of the graph.

But how can we find an appropriate two-coloring of such a graph, thus separating men from women? Suppose we declare by fiat that the starting vertex is "male." All vertices adjacent to this man must be "female," provided the graph is indeed bipartite.

We can augment breadth-first search so that whenever we discover a new vertex, we color it the opposite of its parent. We check whether any non-tree edge links two vertices of the same color. Such a conflict means that the graph cannot be two-colored. If the process terminates without conflicts, we have constructed a proper two-coloring.

```
void twocolor(graph *g) {
    int i;      /* counter */

    for (i = 1; i <= (g->nvertices); i++) {
        color[i] = UNCOLORED;
    }

    bipartite = true;

    initialize_search(g);

    for (i = 1; i <= (g->nvertices); i++) {
        if (!discovered[i]) {
            color[i] = WHITE;
            bfs(g, i);
        }
    }
}


void process_edge(int x, int y) {
    if (color[x] == color[y]) {
        bipartite = false;
        printf("Warning: not bipartite, due to (%d,%d)\n", x, y);
    }

    color[y] = complement(color[x]);
}


int complement(int color) {
    if (color == WHITE) {
        return(BLACK);
    }

    if (color == BLACK) {
        return(WHITE);
    }

    return(UNCOLORED);
}
```

We can assign the first vertex in any connected component to be whatever color/gender we wish. BFS can separate men from women, but we can't tell which gender corresponds to which color just by using the graph structure. Also, bipartite graphs require distinct and binary categorical attributes, so they don't model the real-world variation in sexual preferences and gender identity.

> *Take-Home Lesson:* Breadth-first and depth-first search provide mechanisms to visit each edge and vertex of the graph. They prove the basis of most simple, efficient graph algorithms.

7.8 Depth-First Search

There are two primary graph traversal algorithms: *breadth-first search* (BFS) and *depth-first search* (DFS). For certain problems, it makes absolutely no difference which you use, but in others the distinction is crucial.

The difference between BFS and DFS lies in the order in which they explore vertices. This order depends completely upon the container data structure used to store the *discovered* but not *processed* vertices.

- *Queue* – By storing the vertices in a first-in, first-out (FIFO) queue, we explore the oldest unexplored vertices first. Our explorations thus radiate out slowly from the starting vertex, defining a breadth-first search.

- *Stack* – By storing the vertices in a last-in, first-out (LIFO) stack, we explore the vertices by forging steadily along along a path, visiting a new neighbor if one is available, and backing up only when we are surrounded by previously discovered vertices. Our explorations thus quickly wander away from the starting point, defining a depth-first search.

Our implementation of `dfs` maintains a notion of traversal *time* for each vertex. Our `time` clock ticks each time we enter or exit a vertex. We keep track of the *entry* and *exit* times for each vertex.

Depth-first search has a neat recursive implementation, which eliminates the need to explicitly use a stack:

$\text{DFS}(G, u)$
  $state[u] = \text{``discovered''}$
  process vertex u if desired
  $time = time + 1$
  $entry[u] = time$
  for each vertex v that is adjacent to u
    process edge (u, v) if desired
    if $state[v] = \text{``undiscovered''}$ then
      $p[v] = u$
      $\text{DFS}(G, v)$
  $state[u] = \text{``processed''}$
  $exit[u] = time$
  $time = time + 1$

The time intervals have interesting and useful properties with respect to depth-first search:

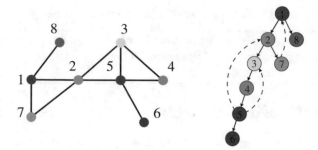

Figure 7.10: An undirected graph and its depth-first search tree. Dashed lines, which are not part of the tree, indicate back edges.

- *Who is an ancestor?* – Suppose that x is an ancestor of y in the DFS tree. This implies that we must enter x before y, since there is no way we can be born before our own parent or grandparent! We also must exit y before we exit x, because the mechanics of DFS ensure we cannot exit x until after we have backed up from the search of all its descendants. Thus, the time interval of y must be properly nested within the interval of ancestor x.

- *How many descendants?* – The difference between the exit and entry times for v tells us how many descendants v has in the DFS tree. The clock gets incremented on each vertex entry and vertex exit, so half the time difference denotes the number of descendants of v.

We will use these entry and exit times in several applications of depth-first search, particularly topological sorting and biconnected/strongly connected components. We may need to take separate actions on each entry and exit, thus motivating distinct `process_vertex_early` and `process_vertex_late` routines called from `dfs`. For the DFS tree example presented in Figure 7.10, the parent of each vertex with its entry and exit times are:

vertex	1	2	3	4	5	6	7	8
parent	−1	1	2	3	4	5	2	1
entry	1	2	3	4	5	6	11	14
exit	16	13	10	9	8	7	12	15

The other important property of a depth-first search is that it partitions the edges of an undirected graph into exactly two classes: *tree edges* and *back edges*. The tree edges discover new vertices, and are those encoded in the `parent` relation. Back edges are those whose other endpoint is an ancestor of the vertex being expanded, so they point back into the tree.

An amazing property of depth-first search is that all edges fall into one of these two classes. Why can't an edge go to a sibling or cousin node, instead of an ancestor? All nodes reachable from a given vertex v are expanded before we finish with the traversal from v, so such topologies are impossible for undirected graphs. This edge classification proves fundamental to the correctness of DFS-based algorithms.

Implementation

Depth-first search can be thought of as breadth-first search, but using a stack instead of a queue to store unfinished vertices. The beauty of implementing `dfs` recursively is that recursion eliminates the need to keep an explicit stack:

```
void dfs(graph *g, int v) {
    edgenode *p;            /* temporary pointer */
    int y;                  /* successor vertex */

    if (finished) {
        return;             /* allow for search termination */
    }
    discovered[v] = true;
    time = time + 1;
    entry_time[v] = time;

    process_vertex_early(v);

    p = g->edges[v];
    while (p != NULL) {
        y = p->y;
        if (!discovered[y]) {
            parent[y] = v;
            process_edge(v, y);
            dfs(g, y);
        } else if (((!processed[y]) && (parent[v] != y)) || (g->directed)) {
            process_edge(v, y);
        }

        if (finished) {
            return;
        }
        p = p->next;
    }

    process_vertex_late(v);
    time = time + 1;
    exit_time[v] = time;
    processed[v] = true;
}
```

Depth-first search uses essentially the same idea as *backtracking*, which we study in Section 9.1 (page 281). Both involve exhaustively searching all possibilities by advancing if it is possible, and backing up only when there is no remaining unexplored possibility for further advance. Both are most easily understood as recursive algorithms.

> *Take-Home Lesson:*   DFS organizes vertices by entry/exit times, and edges into tree and back edges. This organization is what gives DFS its real power.

7.9   Applications of Depth-First Search

As algorithm design paradigms go, depth-first search isn't particularly intimidating. It is surprisingly *subtle*, however, meaning that its correctness requires getting details right.

The correctness of a DFS-based algorithm depends upon specifics of exactly *when* we process the edges and vertices. We can process vertex v either before we have traversed any outgoing edge from v (`process_vertex_early()`), or after we have finished processing all of them (`process_vertex_late()`). Sometimes we will take special actions at both times, say `process_vertex_early()` to initialize a vertex-specific data structure, which will be modified on edge-processing operations and then analyzed afterwards using `process_vertex_late()`.

In undirected graphs, each edge (x, y) sits in the adjacency lists of vertex x and y. There are thus two potential times to process each edge (x, y), namely when exploring x and when exploring y. The labeling of edges as tree edges or back edges occurs the first time the edge is explored. This first time we see an edge is usually a logical time to do edge-specific processing. Sometimes, we may want to take different action the second time we see an edge.

But when we encounter edge (x, y) from x, how can we tell if we have previously traversed the edge from y? The issue is easy if vertex y is undiscovered: (x, y) becomes a tree edge so this must be the first time. The issue is also easy if y has been completely processed: we explored the edge when we explored y so this must be the second time. But what if y is an ancestor of x, and thus in a discovered state? Careful reflection will convince you that this must be our first traversal *unless* y is the immediate ancestor of x—that is, (y, x) is a tree edge. This can be established by testing if `y == parent[x]`.

I find that the subtlety of depth-first search-based algorithms kicks me in the head whenever I try to implement one.[2] I encourage you to analyze these implementations carefully to see where the problematic cases arise and why.

7.9.1   Finding Cycles

Back edges are the key to finding a cycle in an undirected graph. If there is no back edge, all edges are tree edges, and no cycle exists in a tree. But *any* back

---

[2]Indeed, the most horrifying errors in the previous edition of this book came in this section.

Figure 7.11: An articulation vertex is the weakest point in the graph.

edge going from x to an ancestor y creates a cycle with the tree path from y to x. Such a cycle is easy to find using `dfs`:

```
void process_edge(int x, int y) {
    if (parent[y] != x) {     /* found back edge! */
        printf("Cycle from %d to %d:", y, x);
        find_path(y, x, parent);
        finished = true;
    }
}
```

The correctness of this cycle detection algorithm depends upon processing each undirected edge exactly once. Otherwise, a spurious two-vertex cycle (x, y, x) could be composed from the two traversals of any single undirected edge. We use the `finished` flag to terminate after finding the first cycle. Without it we would waste time discovering a new cycle with every back edge before stopping; a complete graph has $\Theta(n^2)$ such cycles.

7.9.2 Articulation Vertices

Suppose you are a vandal seeking to disrupt the telephone trunk line network. Which station in Figure 7.11 should you blow up to cause the maximum amount of damage? Observe that there is a single point of failure—a single vertex whose deletion disconnects a connected component of the graph. Such a vertex v is called an *articulation vertex* or *cut-node*. Any graph that contains an articulation vertex is inherently fragile, because deleting v causes a loss of connectivity between other nodes.

I presented a breadth-first search-based connected components algorithm in Section 7.7.1 (page 218). In general, the *connectivity* of a graph is the smallest number of vertices whose deletion will disconnect the graph. The connectivity is 1 if the graph has an articulation vertex. More robust graphs without such a vertex are said to be *biconnected*. Connectivity will be further discussed in Section 18.8 (page 568).

Testing for articulation vertices by brute force is easy. Temporarily delete each candidate vertex v, and then do a BFS or DFS traversal of the remaining graph to establish whether it is still connected. The total time for n such traversals is $O(n(m + n))$. There is a clever linear-time algorithm, however, that tests all the vertices of a connected graph using a single depth-first search.

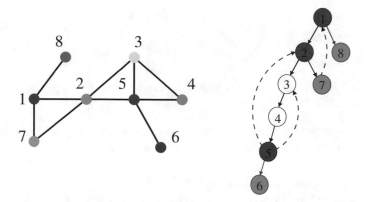

Figure 7.12: DFS tree of a graph containing three articulation vertices (namely 1, 2, and 5). Back edges keep vertices 3 and 4 from being cut-nodes, while green vertices 6, 7, and 8 escape by being leaves of the DFS tree. Red edges (1, 8) and (5, 6) are bridges whose deletion disconnect the graph.

What might the depth-first search tree tell us about articulation vertices? This tree connects all the vertices of a connected component of the graph. If the DFS tree represented the entirety of the graph, all internal (non-leaf) nodes would be articulation vertices, since deleting any one of them would separate a leaf from the root. But blowing up a leaf (shown in green in Figure 7.12) would not disconnect the tree, because it connects no one but itself to the main trunk.

The root of the search tree is a special case. If it has only one child, it functions as a leaf. But if the root has two or more children, its deletion disconnects them, making the root an articulation vertex.

General graphs are more complex than trees. But a depth-first search of a general graph partitions the edges into tree edges and back edges. Think of these back edges as security cables linking a vertex back to one of its ancestors. The security cable from x back to y ensures that none of the vertices on the tree path between x and y can be articulation vertices. Delete any of these vertices, and the security cable will still hold all of them to the rest of the tree.

Finding articulation vertices requires keeping track of the extent to which back edges (i.e., security cables) link chunks of the DFS tree back to ancestor nodes. Let `reachable_ancestor[v]` denote the earliest reachable ancestor of vertex v, meaning the oldest ancestor of v that we can reach from a descendant of v by using a back edge. Initially, `reachable_ancestor[v] = v`:

```
int reachable_ancestor[MAXV+1];  /* earliest reachable ancestor of v */
int tree_out_degree[MAXV+1];     /* DFS tree outdegree of v */
```

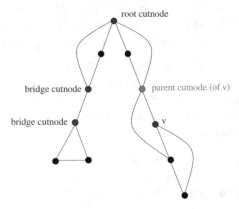

Figure 7.13: The three cases of articulation vertices: root, bridge, and parent cut-nodes.

```
void process_vertex_early(int v) {
    reachable_ancestor[v] = v;
}
```

We update `reachable_ancestor[v]` whenever we encounter a back edge that takes us to an earlier ancestor than we have previously seen. The relative age/rank of our ancestors can be determined from their `entry_time`'s:

```
void process_edge(int x, int y) {
    int class;      /* edge class */

    class = edge_classification(x, y);

    if (class == TREE) {
        tree_out_degree[x] = tree_out_degree[x] + 1;
    }

    if ((class == BACK) && (parent[x] != y)) {
        if (entry_time[y] < entry_time[reachable_ancestor[x]]) {
            reachable_ancestor[x] = y;
        }
    }
}
```

The key issue is determining how the reachability relation impacts whether vertex v is an articulation vertex. There are three cases, illustrated in Figure 7.13 and discussed below. Note that these cases are not mutually exclusive. A single vertex v might be an articulation vertex for multiple reasons:

- *Root cut-nodes* – If the root of the DFS tree has two or more children, it must be an articulation vertex. No edges from the subtree of the second child can possibly connect to the subtree of the first child.

- *Bridge cut-nodes* – If the earliest reachable vertex from v is v, then deleting the single edge $(parent[v], v)$ disconnects the graph. Clearly $parent[v]$ must be an articulation vertex, since it cuts v from the graph. Vertex v is also an articulation vertex unless it is a leaf of the DFS tree. For any leaf, nothing falls off when you cut it.

- *Parent cut-nodes* – If the earliest reachable vertex from v is the parent of v, then deleting the parent must sever v from the tree unless the parent is the root. This is always the case for the deeper vertex of a bridge, unless it is a leaf.

The routine below systematically evaluates each of these three conditions as we back up from the vertex after traversing all outgoing edges. We use `entry_time[v]` to represent the age of vertex v. The reachability time `time_v` calculated below denotes the oldest vertex that can be reached using back edges. Getting back to an ancestor above v rules out the possibility of v being a cut-node:

```c
void process_vertex_late(int v) {
    bool root;               /* is parent[v] the root of the DFS tree? */
    int time_v;              /* earliest reachable time for v */
    int time_parent;         /* earliest reachable time for parent[v] */

    if (parent[v] == -1) {   /* test if v is the root */
        if (tree_out_degree[v] > 1) {
            printf("root articulation vertex: %d \n",v);
        }
        return;
    }

    root = (parent[parent[v]] == -1);        /* is parent[v] the root? */

    if (!root) {
        if (reachable_ancestor[v] == parent[v]) {
            printf("parent articulation vertex: %d \n", parent[v]);
        }

        if (reachable_ancestor[v] == v) {
            printf("bridge articulation vertex: %d \n",parent[v]);

            if (tree_out_degree[v] > 0) {    /* is v is not a leaf? */
                printf("bridge articulation vertex: %d \n", v);
```

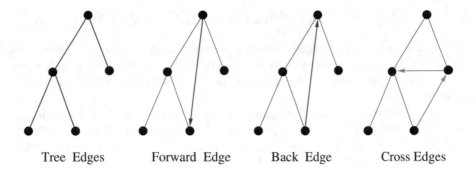

Figure 7.14: The possible edge cases for graph traversal. Forward edges and cross edges can occur in DFS *only* on directed graphs.

```
            }
        }
    }

    time_v = entry_time[reachable_ancestor[v]];
    time_parent = entry_time[reachable_ancestor[parent[v]]];

    if (time_v < time_parent) {
        reachable_ancestor[parent[v]] = reachable_ancestor[v];
    }
}
```

The last lines of this routine govern when we back up from a node's highest reachable ancestor to its parent, namely whenever it is higher than the parent's earliest ancestor to date.

We can alternatively talk about vulnerability in terms of edge failures instead of vertex failures. Perhaps our vandal would find it easier to cut a cable instead of blowing up a switching station. A single edge whose deletion disconnects the graph is called a *bridge*; any graph without such an edge is said to be *edge-biconnected*.

Identifying whether a given edge (x, y) is a bridge is easily done in linear time, by deleting the edge and testing whether the resulting graph is connected. In fact all bridges can be identified in the same $O(n+m)$ time using DFS. Edge (x, y) is a bridge if (1) it is a tree edge, and (2) no back edge connects from y or below to x or above. This can be computed with a appropriate modification to the process_late_vertex function.

7.10   Depth-First Search on Directed Graphs

Depth-first search on an undirected graph proves useful because it organizes the edges of the graph in a very precise way. Over the course of a DFS from a given source vertex, each edge will be assigned one of potentially four labels, as shown in Figure 7.14.

When traversing *undirected* graphs, every edge is either in the depth-first search tree or will be a back edge to an ancestor in the tree. It is important to understand why. Might we encounter a "forward edge" (x, y), directed toward a descendant vertex? No, because in this case, we would have first traversed (x, y) while exploring y, making it a back edge. Might we encounter a "cross edge" (x, y), linking two unrelated vertices? Again no, because we would have first discovered this edge when we explored y, making it a tree edge.

But for *directed graphs*, depth-first search labelings can take on a wider range of possibilities. Indeed, all four of the edge cases in Figure 7.14 can occur in traversing directed graphs. This classification still proves useful in organizing algorithms on directed graphs, because we typically take a different action on edges from each different class.

The correct labeling of each edge can be readily determined from the state, discovery time, and parent of each vertex, as encoded in the following function:

```c
int edge_classification(int x, int y) {
    if (parent[y] == x) {
        return(TREE);
    }

    if (discovered[y] && !processed[y]) {
        return(BACK);
    }

    if (processed[y] && (entry_time[y]>entry_time[x])) {
        return(FORWARD);
    }

    if (processed[y] && (entry_time[y]<entry_time[x])) {
        return(CROSS);
    }

    printf("Warning: self loop (%d,%d)\n", x, y);

    return -1;
}
```

Just as with BFS, this implementation of the depth-first search algorithm includes places to optionally process each vertex and edge—say to copy them, print them, or count them. Both DFS and BFS will traverse all edges in the same

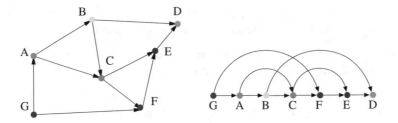

Figure 7.15: A DAG with only one topological sort (G, A, B, C, F, E, D)

connected component as the starting point. Both must start with a vertex in each component to traverse a disconnected graph. The only important difference between them is the way they organize and label the edges.

I encourage the reader to convince themselves of the correctness of the four conditions above. What I said earlier about the subtlety of depth-first search goes double for directed graphs.

7.10.1 Topological Sorting

Topological sorting is the most important operation on directed acyclic graphs (DAGs). It orders the vertices on a line such that all directed edges go from left to right. Such an ordering cannot exist if the graph contains a directed cycle, because there is no way you can keep moving right on a line and still return back to where you started from!

Each DAG has at least one topological sort. The importance of topological sorting is that it gives us an ordering so we can process each vertex before any of its successors. Suppose the directed edges represented precedence constraints, such that edge (x, y) means job x must be done before job y. Any topological sort then defines a feasible schedule. Indeed, there can be many such orderings for a given DAG.

But the applications go deeper. Suppose we seek the shortest (or longest) path from x to y in a DAG. No vertex v appearing after y in the topological order can possibly contribute to any such path, because there will be no way to get from v back to y. We can appropriately process all the vertices from left to right in topological order, considering the impact of their outgoing edges, and know that we will have looked at everything we need before we need it. Topological sorting proves very useful in essentially any algorithmic problem on DAGs, as discussed in the catalog in Section 18.2 (page 546).

Topological sorting can be performed efficiently using depth-first search. A directed graph is a DAG iff no back edges are encountered. Labeling the vertices in the reverse order that they are marked *processed* defines a topological sort of a DAG. Why? Consider what happens to each directed edge (x, y) as we encounter it exploring vertex x:

- If y is currently *undiscovered*, then we start a DFS of y before we can continue with x. Thus, y must be marked *processed* before x is, so x appears before y in the topological order, as it must.

- If y is *discovered* but not *processed*, then (x, y) is a back edge, which is impossible in a DAG because it creates a cycle.

- If y is *processed*, then it will have been so labeled before x. Therefore, x appears before y in the topological order, as it must.

Study the following implementation:

```
void process_vertex_late(int v) {
    push(&sorted, v);
}


void process_edge(int x, int y) {
    int class;      /* edge class */

    class = edge_classification(x, y);

    if (class == BACK) {
        printf("Warning: directed cycle found, not a DAG\n");
    }
}


void topsort(graph *g) {
    int i;      /* counter */

    init_stack(&sorted);

    for (i = 1; i <= g->nvertices; i++) {
        if (!discovered[i]) {
            dfs(g, i);
        }
    }
    print_stack(&sorted);      /* report topological order */
}
```

We push each vertex onto a stack as soon as we have evaluated all outgoing edges. The top vertex on the stack always has no incoming edges from any vertex on the stack. Repeatedly popping them off yields a topological ordering.

7.10.2   Strongly Connected Components

A directed graph is *strongly connected* if there is a directed path between any two vertices. Road networks had better be strongly connected: otherwise there will be places you can drive to but not drive home from without violating one-way signs.

It is straightforward to use graph traversal to test whether a graph $G = (V, E)$ is strongly connected in linear time. The graph is strongly connected iff from any vertex v in G (1) all vertices are reachable from v and (2) all vertices can reach v. To test if condition (1) holds, we can do a BFS or DFS traversal from v to establish whether all vertices get discovered. If so, all must be reachable from v.

To test if there are paths from every vertex to v, we construct the *transpose* graph $G^T = (V, E')$, which has the same vertex and edge set as G but with all edges reversed—that is, directed edge $(x, y) \in E$ iff $(y, x) \in E'$.

```
graph *transpose(graph *g) {
    graph *gt;       /* transpose of graph g */
    int x;           /* counter */
    edgenode *p;     /* temporary pointer */

    gt = (graph *) malloc(sizeof(graph));
    initialize_graph(gt, true);            /* initialize directed graph */
    gt->nvertices = g->nvertices;

    for (x = 1; x <= g->nvertices; x++) {
        p = g->edges[x];
        while (p != NULL) {
            insert_edge(gt, p->y, x, true);
            p = p->next;
        }
    }

    return(gt);
}
```

Any path from v to z in G^T corresponds to a path from z to v in G. By doing a second DFS, this one from v in G^T, we identify all vertices that have paths *to* v in G.

All directed graphs can be partitioned into *strongly connected components*, such that a directed path exists between every pair of vertices in the component, as shown in Figure 7.16 (left). The set of such components can be determined using a more subtle variation of this double DFS approach:

```
void strong_components(graph *g) {
    graph *gt;         /* transpose of graph g */
    int i;             /* counter */
    int v;             /* vertex in component */
```

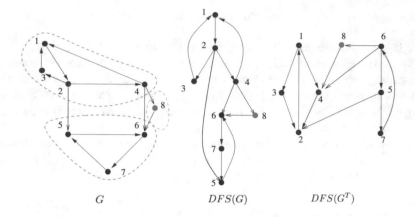

Figure 7.16: The strongly connected components of a graph G (left), with its associated DFS tree (center). The reverse of its DFS finishing order from vertex 1 is $[3, 5, 7, 6, 8, 4, 2, 1]$, which defines the vertex order for the second traversal of the transpose G^T (right).

```
init_stack(&dfs1order);
initialize_search(g);
for (i = 1; i <= (g->nvertices); i++) {
    if (!discovered[i]) {
        dfs(g, i);
    }
}

gt = transpose(g);
initialize_search(gt);

components_found = 0;
while (!empty_stack(&dfs1order)) {
    v = pop(&dfs1order);
    if (!discovered[v]) {
        components_found ++;
        printf("Component %d:", components_found);
        dfs2(gt, v);
        printf("\n");
    }
}
}
```

The first traversal pushes the vertices on a stack in the reverse order they were processed, just as with topological sort in Section 7.10.1 (page 231). The connection makes sense: DAGs are directed graphs where each vertex forms its own strongly connected component. On a DAG, the top vertex on the stack will be one that cannot reach any other vertex. The bookkeeping here is identical to topological sort:

```
void process_vertex_late(int v) {
    push(&dfs1order,v);
}
```

The second traversal, on the transposed graph, behaves like the connected component algorithm of Section 7.7.1 (page 218), except we consider starting vertices in the order they appear on the stack. Each traversal from v will discover all reachable vertices from the transpose G^T, meaning the vertices that have paths to v in G. These reachable vertices define the strongly connected component of v, because they represent the least reachable vertices in G:

```
void process_vertex_early2(int v) {
    printf(" %d", v);
}
```

The correctness of this is subtle. Observe that first DFS places vertices on the stack in groups based on reachability from successive starting vertices in the original directed graph G. Thus, the vertices in the top group have the property that *none* were reachable from *any* earlier group vertex. The second traversal in G^T, starting from the last vertex v of G, finds all the reachable vertices from v in G^T that themselves reach v, meaning they define a strongly connected component.

Chapter Notes

Our treatment of graph traversal represents an expanded version of material from chapter 9 of Skiena and Revilla [SR03]. The *Combinatorica* graph library discussed in the war story is best described in the old [Ski90] and new editions [PS03] of the associated book. Accessible introductions to the science of social networks include Barabasi [Bar03], Easley and Kleinberg [EK10], and Watts [Wat04]. Interest in graph theory has surged with the emergence of the multi-disciplinary field of network science, see the introductory textbooks by Barabasi [B+16] and Newman [New18].

7.11   Exercises

Simulating Graph Algorithms

7-1. *[3]* For the following weighted graphs G_1 (left) and G_2 (right):

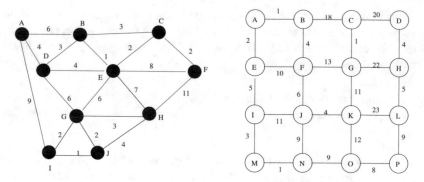

(a) Report the order of the vertices encountered on a breadth-first search starting from vertex A. Break all ties by picking the vertices in alphabetical
order (i.e. A before Z).

(b) Report the order of the vertices encountered on a depth-first search starting
from vertex A. Break all ties by picking the vertices in alphabetical order
(i.e. A before Z).

7-2. *[3]* Do a topological sort of the following graph G:

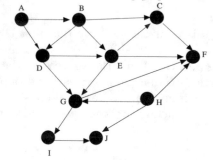

Traversal

7-3. *[3]* Prove that there is a unique path between any pair of vertices in a tree.

7-4. *[3]* Prove that in a breadth-first search on a undirected graph G, every edge is
either a tree edge or a cross edge, where x is neither an ancestor nor descendant
of y in cross edge (x, y).

7-5. *[3]* Give a linear algorithm to compute the chromatic number of graphs where
each vertex has degree at most 2. Any bipartite graph has a chromatic number
of 2. Must such graphs be bipartite?

7-6. *[3]* You are given a connected, undirected graph G with n vertices and m edges.
Give an $O(n + m)$ algorithm to identify an edge you can remove from G while
still leaving G connected, if one exists. Can you reduce the running time to
$O(n)$?

7-7. *[5]* In breadth-first and depth-first search, an undiscovered node is marked *discovered* when it is first encountered, and marked *processed* when it has been
completely searched. At any given moment, several nodes might be simultaneously in the *discovered* state.

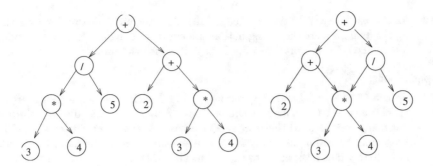

Figure 7.17: Expression 2+3*4+(3*4)/5 as a tree and a DAG

(a) Describe a graph on n vertices and a particular starting vertex v such that $\Theta(n)$ nodes are simultaneously in the *discovered* state during a breadth-first search starting from v.

(b) Describe a graph on n vertices and a particular starting vertex v such that $\Theta(n)$ nodes are simultaneously in the *discovered* state during a depth-first search starting from v.

(c) Describe a graph on n vertices and a particular starting vertex v such that at some point $\Theta(n)$ nodes remain *undiscovered*, while $\Theta(n)$ nodes have been *processed* during a depth-first search starting from v. (Hint: there may also be *discovered* nodes.)

7-8. *[4]* Given pre-order and in-order traversals of a binary tree (discussed in Section 3.4.1), is it possible to reconstruct the tree? If so, sketch an algorithm to do it. If not, give a counterexample. Repeat the problem if you are given the pre-order and post-order traversals.

7-9. *[3]* Present correct and efficient algorithms to convert an undirected graph G between the following graph data structures. Give the time complexity of each algorithm, assuming n vertices and m edges.

   (a) Convert from an adjacency matrix to adjacency lists.

   (b) Convert from an adjacency list representation to an incidence matrix. An incidence matrix M has a row for each vertex and a column for each edge, such that $M[i, j] = 1$ if vertex i is part of edge j, otherwise $M[i, j] = 0$.

   (c) Convert from an incidence matrix to adjacency lists.

7-10. *[3]* Suppose an arithmetic expression is given as a tree. Each leaf is an integer and each internal node is one of the standard arithmetical operations $(+, -, *, /)$. For example, the expression 2+3*4+(3*4)/5 is represented by the tree in Figure 7.17(a). Give an $O(n)$ algorithm for evaluating such an expression, where there are n nodes in the tree.

7-11. *[5]* Suppose an arithmetic expression is given as a DAG (directed acyclic graph) with common subexpressions removed. Each leaf is an integer and each internal node is one of the standard arithmetical operations $(+, -, *, /)$. For example, the expression 2+3*4+(3*4)/5 is represented by the DAG in Figure 7.17(b). Give an $O(n+m)$ algorithm for evaluating such a DAG, where there are n nodes and m edges in the DAG. (Hint: modify an algorithm for the tree case to achieve the desired efficiency.)

7-12. *[8]* The war story of Section 7.4 (page 210) describes an algorithm for constructing the dual graph of the triangulation efficiently, although it does not guarantee
linear time. Give a worst-case linear algorithm for the problem.

Applications

7-13. *[3]* The Chutes and Ladders game has a board with n cells where you seek to
travel from cell 1 to cell n. To move, a player throws a six-sided dice to determine
how many cells forward they move. This board also contains chutes and ladders
that connect certain pairs of cells. A player who lands on the mouth of a chute
immediately falls back down to the cell at the other end. A player who lands on
the base of a ladder immediately travels up to the cell at the top of the ladder.
Suppose you have rigged the dice to give you full control of the number for each
roll. Give an efficient algorithm to find the minimum number of dice throws to
win.

7-14. *[3]* Plum blossom poles are a Kung Fu training technique, consisting of n large
posts partially sunk into the ground, with each pole p_i at position (x_i, y_i). Students practice martial arts techniques by stepping from the top of one pole to
the top of another pole. In order to keep balance, each step must be more than
d meters but less than $2d$ meters. Give an efficient algorithm to find a safe path
from pole p_s to p_t if it exists.

7-15. *[5]* You are planning the seating arrangement for a wedding given a list of guests,
V. For each guest g you have a list of all other guests who are on bad terms
with them. Feelings are reciprocal: if h is on bad terms with g, then g is on bad
terms with h. Your goal is to arrange the seating such that no pair of guests
sitting at the same table are on bad terms with each other. There will be only
two tables at the wedding. Give an efficient algorithm to find an acceptable
seating arrangement if one exists.

Algorithm Design

7-16. *[5]* The *square* of a directed graph $G = (V, E)$ is the graph $G^2 = (V, E^2)$ such
that $(u, w) \in E^2$ iff there exists $v \in V$ such that $(u, v) \in E$ and $(v, w) \in E$; that
is, there is a path of exactly two edges from u to w.

Give efficient algorithms for both adjacency lists and matrices.

7-17. *[5]* A *vertex cover* of a graph $G = (V, E)$ is a subset of vertices V' such that
each edge in E is incident to at least one vertex of V'.

  (a) Give an efficient algorithm to find a minimum-size vertex cover if G is a
       tree.

  (b) Let $G = (V, E)$ be a tree such that the weight of each vertex is equal to the
       degree of that vertex. Give an efficient algorithm to find a minimum-weight
       vertex cover of G.

  (c) Let $G = (V, E)$ be a tree with arbitrary weights associated with the ver-
       tices. Give an efficient algorithm to find a minimum-weight vertex cover
       of G.

7-18. *[3]* A *vertex cover* of a graph $G = (V, E)$ is a subset of vertices V' such that each
edge in E is incident to at least one vertex of V'. Delete all the leaves from any
depth-first search tree of G. Must the remaining vertices form a vertex cover of
G? Give a proof or a counterexample.

7-19. *[5]* An *independent set* of an undirected graph $G = (V, E)$ is a set of vertices U such that no edge in E is incident to two vertices of U.

    (a) Give an efficient algorithm to find a maximum-size independent set if G is a tree.

    (b) Let $G = (V, E)$ be a tree with weights associated with the vertices such that the weight of each vertex is equal to the degree of that vertex. Give an efficient algorithm to find a maximum-weight independent set of G.

    (c) Let $G = (V, E)$ be a tree with arbitrary weights associated with the vertices. Give an efficient algorithm to find a maximum-weight independent set of G.

7-20. *[5]* A *vertex cover* of a graph $G = (V, E)$ is a subset of vertices V' such that each edge in E is incident on at least one vertex of V'. An *independent set* of graph $G = (V, E)$ is a subset of vertices $V' \in V$ such that no edge in E contains both vertices from V'.

An *independent vertex cover* is a subset of vertices that is both an independent set and a vertex cover of G. Give an efficient algorithm for testing whether G contains an independent vertex cover. What classical graph problem does this reduce to?

7-21. *[5]* Consider the problem of determining whether a given undirected graph $G = (V, E)$ contains a *triangle*, that is, a cycle of length 3.

    (a) Give an $O(|V|^3)$ algorithm to find a triangle if one exists.

    (b) Improve your algorithm to run in time $O(|V| \cdot |E|)$. You may assume $|V| \leq |E|$.

Observe that these bounds give you time to convert between the adjacency matrix and adjacency list representations of G.

7-22. *[5]* Consider a set of movies $M_1, M_2, \ldots, M_k$. There is a set of customers, each one of which indicates the two movies they would like to see this weekend. Movies are shown on Saturday evening and Sunday evening. Multiple movies may be screened at the same time.

You must decide which movies should be televised on Saturday and which on Sunday, so that every customer gets to see the two movies they desire. Is there a schedule where each movie is shown at most once? Design an efficient algorithm to find such a schedule if one exists.

7-23. *[5]* The *diameter* of a tree $T = (V, E)$ is given by

$$\max_{u,v \in V} \delta(u, v)$$

(where $\delta(u, v)$ is the number of edges on the path from u to v). Describe an efficient algorithm to compute the diameter of a tree, and show the correctness and analyze the running time of your algorithm.

7-24. *[5]* Given an undirected graph G with n vertices and m edges, and an integer k, give an $O(m + n)$ algorithm that finds the maximum induced subgraph F of G such that each vertex in F has degree $\geq k$, or prove that no such graph exists. Graph $F - (U, R)$ is an induced subgraph of graph $G - (V, E)$ if its vertex set U is a subset of the vertex set V of G, and R consists of all edges of G whose endpoints are in U.

7-25. *[6]* Let v and w be two vertices in an unweighted directed graph $G = (V, E)$. Design a linear-time algorithm to find the *number* of different shortest paths (not necessarily vertex disjoint) between v and w.

7-26. *[6]* Design a linear-time algorithm to eliminate each vertex v of degree 2 from a graph by replacing edges (u, v) and (v, w) by an edge (u, w). It should also eliminate multiple copies of edges by replacing them with a single edge. Note that removing multiple copies of an edge may create a new vertex of degree 2, which has to be removed, and that removing a vertex of degree 2 may create multiple edges, which also must be removed.

Directed Graphs

7-27. *[3]* The *reverse* of a directed graph $G = (V, E)$ is another directed graph $G^R = (V, E^R)$ on the same vertex set, but with all edges reversed; that is, $E^R = \{(v, u) : (u, v) \in E\}$. Give an $O(n + m)$ algorithm for computing the reverse of an n-vertex m-edge graph in adjacency list format.

7-28. *[5]* Your job is to arrange n ill-behaved children in a straight line, facing front. You are given a list of m statements of the form "i hates j." If i hates j, then you do not want to put i somewhere behind j, because then i is capable of throwing something at j.

   (a) Give an algorithm that orders the line (or says that it is not possible) in $O(m + n)$ time.

   (b) Suppose instead you want to arrange the children in rows such that if i hates j, then i must be in a lower numbered row than j. Give an efficient algorithm to find the minimum number of rows needed, if it is possible.

7-29. *[3]* A particular academic program has n required courses, certain pairs of which have prerequisite relations so that (x, y) means you must take course x before y. How would you analyze the prerequisite pairs to make sure it is possible for people to complete the program?

7-30. *[5]* Gotcha-solitaire is a game on a deck with n distinct cards (all face up) and m gotcha pairs (i, j) such that card i must be played sometime before card j. You play by sequentially choosing cards, and win if you pick up the entire deck without violating any gotcha pair constraints. Give an efficient algorithm to find a winning pickup order if one exists.

7-31. *[5]* You are given a list of n words each of length k in a language you don't know, although you are told that words are sorted in lexicographic (alphabetical) order. Reconstruct the order of the α alphabet letters (characters) in that language.

   For example, if the strings are $\{QQZ, QZZ, XQZ, XQX, XXX\}$, the character order must be Q before Z before X.

   (a) Give an algorithm to efficiently reconstruct this character order. (Hint: use a graph structure, where each node represents one letter.)

   (b) What is its running time, as a function of n, k, and α?

7-32. *[3]* A *weakly connected component* in a directed graph is a connected component ignoring the direction of the edges. Adding a single directed edge to a directed graph can reduce the number of weakly connected components, but by at most how many components? What about the number of strongly connected components?

7-33. *[5]* Design a linear-time algorithm that, given an undirected graph G and a particular edge e in it, determines whether G has a cycle containing e.

7-34. *[5]* An *arborescence* of a directed graph G is a rooted tree such that there is a directed path from the root to every other vertex in the graph. Give an efficient and correct algorithm to test whether G contains an arborescence, and its time complexity.

7-35. *[5]* A *mother* vertex in a directed graph $G = (V, E)$ is a vertex v such that all other vertices G can be reached by a directed path from v.

   (a) Give an $O(n + m)$ algorithm to test whether a given vertex v is a mother of G, where $n = |V|$ and $m = |E|$.

   (b) Give an $O(n + m)$ algorithm to test whether graph G contains a mother vertex.

7-36. *[8]* Let G be a directed graph. We say that G is *k-cyclic* if every (not necessarily simple) cycle in G contains at most k distinct nodes. Give a linear-time algorithm to determine if a directed graph G is k-cyclic, where G and k are given as inputs. Justify the correctness and running time of your algorithm.

7-37. *[9]* A *tournament* is a directed graph formed by taking the complete undirected graph and assigning arbitrary directions on the edges—that is, a graph $G = (V, E)$ such that for all $u, v \in V$, exactly one of (u, v) or (v, u) is in E. Show that every tournament has a Hamiltonian path—that is, a path that visits every vertex exactly once. Give an algorithm to find this path.

Articulation Vertices

7-38. *[5]* An articulation vertex of a connected graph G is a vertex whose deletion disconnects G. Let G be a graph with n vertices and m edges. Give a simple $O(n+m)$ algorithm for finding a vertex of G that is *not* an articulation vertex— that is, whose deletion does not disconnect G.

7-39. *[5]* Following up on the previous problem, give an $O(n+m)$ algorithm that finds a deletion order for the n vertices such that no deletion disconnects the graph. (Hint: think DFS/BFS.)

7-40. *[3]* Suppose G is a connected undirected graph. An edge e whose removal disconnects the graph is called a *bridge*. Must every bridge e be an edge in a depth-first search tree of G? Give a proof or a counterexample.

7-41. *[5]* A city that only has two-way streets has decided to change them all into one-way streets. They want to ensure that the new network is strongly connected so everyone can legally drive anywhere in the city and back.

   (a) Let G be the original undirected graph. Prove that there is a way to properly orient/direct the edges of G provided G does not contain a bridge.

   (b) Give an efficient algorithm to orient the edges of a bridgeless graph G so the result is strongly connected.

Interview Problems

7-42. *[3]* Which data structures are used in depth-first and breath-first search?

7-43. *[4]* Write a function to traverse binary search tree and return the ith node in sorted order.

LeetCode

7-1. `https://leetcode.com/problems/minimum-height-trees/`

7-2. `https://leetcode.com/problems/redundant-connection/`

7-3. `https://leetcode.com/problems/course-schedule/`

HackerRank

7-1. `https://www.hackerrank.com/challenges/bfsshortreach/`

7-2. `https://www.hackerrank.com/challenges/dfs-edges/`

7-3. `https://www.hackerrank.com/challenges/even-tree/`

Programming Challenges

These programming challenge problems with robot judging are available at `https://onlinejudge.org`:

7-1. "Bicoloring"—Chapter 9, problem 10004.

7-2. "Playing with Wheels"—Chapter 9, problem 10067.

7-3. "The Tourist Guide"—Chapter 9, problem 10099.

7-4. "Edit Step Ladders"—Chapter 9, problem 10029.

7-5. "Tower of Cubes"—Chapter 9, problem 10051.

Chapter 8

Weighted Graph Algorithms

The data structures and traversal algorithms of Chapter 7 provide the basic building blocks for any computation on graphs. However, all the algorithms presented there dealt with unweighted graphs—in other words, graphs where each edge has identical value or weight.

There is an alternate universe of problems for *weighted graphs*. The edges of road networks are naturally bound to numerical values such as construction cost, traversal time, length, or speed limit. Identifying the shortest path in such graphs proves more complicated than breadth-first search in unweighted graphs, but opens the door to a wide range of applications.

The graph data structure from Chapter 7 quietly supported edge-weighted graphs, but here this is made explicit. Our adjacency list structure again consists of an array of linked lists, such that the outgoing edges from vertex x appear in the list `edges[x]`:

```
typedef struct {
    edgenode *edges[MAXV+1];   /* adjacency info */
    int degree[MAXV+1];        /* outdegree of each vertex */
    int nvertices;             /* number of vertices in the graph */
    int nedges;                /* number of edges in the graph */
    int directed;              /* is the graph directed? */
} graph;
```

Each `edgenode` is a record containing three fields, the first describing the second endpoint of the edge (`y`), the second enabling us to annotate the edge with a weight (`weight`), and the third pointing to the next edge in the list (`next`):

S. S. Skiena, *The Algorithm Design Manual*, Texts in Computer Science, https://doi.org/10.1007/978-3-030-54256-6_8

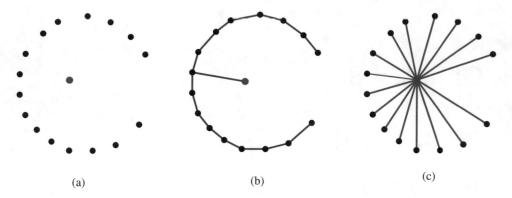

(a)   (b)   (c)

Figure 8.1: (a) The distances between points define a complete weighted graph, (b) its minimum spanning tree, and (c) the shortest path from center tree.

```
typedef struct edgenode {
    int y;                    /* adjacency info */
    int weight;               /* edge weight, if any */
    struct edgenode *next;    /* next edge in list */
} edgenode;
```

We now describe several sophisticated algorithms for weighted graphs that use this data structure, including minimum spanning trees, shortest paths, and maximum flows. That all of these optimization problems can be solved efficiently is a feat quite worthy of our respect. Recall that no such algorithm exists for the first weighted graph problem we encountered, namely the traveling salesman problem.

8.1 Minimum Spanning Trees

A *spanning tree* of a connected graph $G = (V, E)$ is a subset of edges from E forming a tree connecting all vertices of V. For edge-weighted graphs, we are particularly interested in the *minimum spanning tree*—the spanning tree whose sum of edge weights is as small as possible.

Minimum spanning trees are the answer whenever we need to connect a set of points (representing cities, homes, junctions, or other locations) cheaply using the smallest amount of roadway, wire, or pipe. *Any* tree is the smallest possible connected graph in terms of number of edges, but the minimum spanning tree is the smallest connected graph in terms of edge weight. In geometric problems, the point set $p_1, \ldots, p_n$ defines a complete graph, with edge (v_i, v_j) assigned a weight equal to the distance from p_i to p_j. An example of a geometric minimum spanning tree is illustrated in Figure 8.1. Additional applications of minimum spanning trees are discussed in Section 18.3 (page 549).

A minimum spanning tree minimizes the total edge weight over all possible spanning trees. However, there can be more than one minimum spanning tree of a given graph. Indeed, all spanning trees of an unweighted (or equally weighted) graph G are minimum spanning trees, since each contains exactly $n - 1$ equal-weight edges. Such a spanning tree can be found using either DFS or BFS. Finding a minimum spanning tree is more difficult for general weighted graphs. But two different algorithms are presented below, both demonstrating the optimality of specific greedy heuristics.

8.1.1 Prim's Algorithm

Prim's minimum spanning tree algorithm starts from one vertex and grows the rest of the tree one edge at a time until all vertices are included.

Greedy algorithms make the decision of what to do next by selecting the best local option from all available choices without regard to the global structure. Since we seek the tree of minimum weight, the natural greedy algorithm for minimum spanning tree (MST) repeatedly selects the smallest weight edge that will enlarge the number of vertices in the tree.

> Prim-MST(G)
> Select an arbitrary vertex s to start the tree T_{prim} from.
> While (there are still non-tree vertices)
>   Find the minimum-weight edge between a tree and non-tree vertex
>   Add the selected edge and vertex to the tree T_{prim}.

Prim's algorithm clearly creates a spanning tree, because no cycle can be introduced by adding edges between tree and non-tree vertices. But why should it be of minimum weight over *all* spanning trees? We have seen ample evidence of other natural greedy heuristics that do not yield a global optimum. Therefore, we must be particularly careful to demonstrate any such claim.

We use proof by contradiction. Suppose that there existed a graph G for which Prim's algorithm did not return a minimum spanning tree. Since we are building the tree incrementally, this means that there must have been some particular instant where we went wrong. Before we inserted edge (x, y), T_{prim} consisted of a set of edges that was a subtree of some minimum spanning tree T_{min}, but choosing edge (x, y) fatally took us away from any possible minimum spanning tree (see Figure 8.2(a)).

But how could we have gone wrong? There must be a path p from x to y in T_{min}, as shown in Figure 8.2(b). This path must use an edge (v_1, v_2), where v_1 is already in T_{prim}, but v_2 is not. This edge (v_1, v_2) must have weight at least that of (x, y), or else Prim's algorithm would have selected it before (x, y) when it had the chance. Inserting (x, y) and deleting (v_1, v_2) from T_{min} leaves a spanning tree no larger than before, meaning that Prim's algorithm could not have made a fatal mistake in selecting edge (x, y). Therefore, by contradiction, Prim's algorithm must construct a minimum spanning tree.

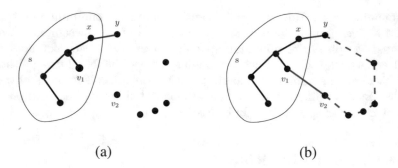

(a)   (b)

Figure 8.2: Does Prim's algorithm go bad? No, because picking edge (x, y) before (v_1, v_2) implies that $weight(v_1, v_2) \geq weight(x, y)$.

Implementation

Prim's algorithm grows the minimum spanning tree in stages, starting from a given vertex. At each iteration, we add one new vertex into the spanning tree. A greedy algorithm suffices for correctness: we always add the lowest-weight edge linking a vertex in the tree to a vertex on the outside. The simplest implementation of this idea would assign to each vertex a Boolean variable denoting whether it is already in the tree (the array **intree** in the code below), and then search all edges at each iteration to find the minimum-weight edge with exactly one **intree** vertex.

Our implementation is somewhat smarter. It keeps track of the cheapest edge linking every non-tree vertex in the tree. The cheapest such edge over all remaining non-tree vertices gets added in the next iteration. We must update the costs of getting to the non-tree vertices after each insertion. However, since the most recently inserted vertex is the only change in the tree, all possible edge-weight updates must come from its outgoing edges:

```
int prim(graph *g, int start) {
    int i;                      /* counter */
    edgenode *p;                /* temporary pointer */
    bool intree[MAXV+1];        /* is the vertex in the tree yet? */
    int distance[MAXV+1];       /* cost of adding to tree */
    int v;                      /* current vertex to process */
    int w;                      /* candidate next vertex */
    int dist;                   /* cheapest cost to enlarge tree */
    int weight = 0;             /* tree weight */

    for (i = 1; i <= g->nvertices; i++) {
        intree[i] = false;
        distance[i] = MAXINT;
        parent[i] = -1;
```

```
    }

    distance[start] = 0;
    v = start;

    while (!intree[v]) {
        intree[v] = true;
        if (v != start) {
            printf("edge (%d,%d) in tree \n",parent[v],v);
            weight = weight + dist;
        }
        p = g->edges[v];
        while (p != NULL) {
            w = p->y;
            if ((distance[w] > p->weight) && (!intree[w])) {
                distance[w] = p->weight;
                parent[w] = v;
            }
            p = p->next;
        }

        dist = MAXINT;
        for (i = 1; i <= g->nvertices; i++) {
            if ((!intree[i]) && (dist > distance[i])) {
                dist = distance[i];
                v = i;
            }
        }
    }

    return(weight);
}
```

Analysis

Prim's algorithm is correct, but how efficient is it? This depends on which data structures are used to implement it. In the pseudocode, Prim's algorithm makes n iterations sweeping through all m edges on each iteration—yielding an $O(mn)$ algorithm.

But our implementation avoids the need to test all m edges on each pass. It only considers the $\leq n$ cheapest known edges represented in the **parent** array and the $\leq n$ edges out of a new tree vertex v to update **parent**. By maintaining a Boolean flag along with each vertex to denote whether it is in the tree, we test whether the current edge joins a tree with a non-tree vertex in constant time.

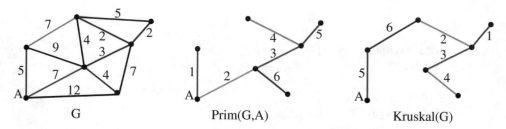

Figure 8.3: A graph G (left) with minimum spanning trees produced by Prim's (center) and Kruskal's (right) algorithms. The numbers and edge colors on the trees denote the order of insertion; ties are broken arbitrarily.

The result is an $O(n^2)$ implementation of Prim's algorithm, and a good illustration of the power of data structures to speed up algorithms. In fact, more sophisticated priority-queue data structures lead to an $O(m + n \lg n)$ implementation, by making it faster to find the minimum-cost edge to expand the tree at each iteration.

The minimum spanning tree itself can be reconstructed in two different ways. The simplest method would be to augment this procedure with statements that print the edges as they are found, and totals the weight of all selected edges to get the cost. Alternatively, the tree topology is encoded by the **parent** array, so it completely describes edges in the minimum spanning tree.

8.1.2   Kruskal's Algorithm

Kruskal's algorithm is an alternative approach to finding minimum spanning trees that proves more efficient on sparse graphs. Like Prim's, Kruskal's algorithm is greedy. Unlike Prim's, it does not start with a particular vertex. As shown in Figure 8.3, Kruskal might produce a different spanning tree than Prim's algorithm, although both will have the same weight.

Kruskal's algorithm builds up connected components of vertices, culminating in the complete minimum spanning tree. Initially, each vertex forms its own separate component in the tree-to-be. The algorithm repeatedly considers the lightest remaining edge and tests whether its two endpoints lie within the same connected component. If so, this edge will be discarded, because adding it would create a cycle. If the endpoints lie in different components, we insert the edge and merge the two components into one. Since each connected component always is a tree, we never need to explicitly test for cycles.

```
Kruskal-MST(G)
    Put the edges into a priority queue ordered by increasing weight.
    count = 0
    while (count < n − 1) do
        get next edge (v, w)
        if (component (v) ≠ component(w))
```

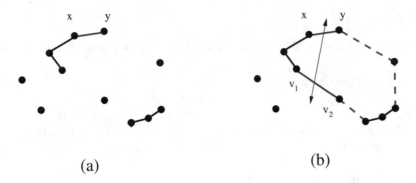

(a) (b)

Figure 8.4: Could Kruskal's algorithm go bad after selecting red edge (x, y) (on left)? No, because edge (v_1, v_2), inserted later, must be heavier than (x, y) (on right).

> increment *count*
> add (v, w) to $T_{kruskal}$
> merge component(v) and component(w)

This algorithm adds $n - 1$ edges without creating a cycle, so it must create a spanning tree for any connected graph. But why does this have to be a *minimum* spanning tree? Suppose it wasn't. As with the correctness proof of Prim's algorithm, there must be a particular graph G on which it fails. In particular, there must an edge (x, y) in G whose insertion first prevented $T_{kruskal}$ from being a minimum spanning tree T_{min}. Inserting this edge (x, y) into T_{min} will create a cycle with the path from x to y, as shown in Figure 8.4. Since x and y were in different components at the time of inserting (x, y), at least one edge (say (v_1, v_2)) on this path must have been evaluated by Kruskal's algorithm at a later time than (x, y). But this means that $weight(v_1, v_2) \geq weight(x, y)$, so exchanging the two edges yields a tree of weight at most T_{min}. Thus, we could not have made a fatal mistake in selecting (x, y), and the correctness follows.

What is the time complexity of Kruskal's algorithm? Sorting the m edges takes $O(m \lg m)$ time. The `while` loop makes at most m iterations, each testing the connectivity of two trees plus an edge. In the most simple-minded implementation, this can be done by breadth-first or depth-first search in the sparse partial tree graph with at most n edges and n vertices, thus yielding an $O(mn)$ algorithm.

However, a faster implementation results if we can implement the component test in faster than $O(n)$ time. In fact, a clever data structure called *union–find*, can support such queries in $O(\lg n)$ time, and it is discussed in Section 8.1.3 (page 250). With this data structure, Kruskal's algorithm runs in $O(m \lg m)$ time, which is faster than Prim's for sparse graphs. Observe again the impact that the right data structure can have when implementing a straightforward algorithm.

Implementation

The implementation of the main routine follows directly from the pseudocode:

```
int kruskal(graph *g) {
    int i;                      /* counter */
    union_find s;               /* union-find data structure */
    edge_pair e[MAXV+1];        /* array of edges data structure */
    int weight=0;               /* cost of the minimum spanning tree */

    union_find_init(&s, g->nvertices);

    to_edge_array(g, e);
    qsort(&e,g->nedges, sizeof(edge_pair), &weight_compare);

    for (i = 0; i < (g->nedges); i++) {
        if (!same_component(&s, e[i].x, e[i].y)) {
            printf("edge (%d,%d) in MST\n", e[i].x, e[i].y);
            weight = weight + e[i].weight;
            union_sets(&s, e[i].x, e[i].y);
        }
    }

    return(weight);
}
```

8.1.3   The Union–Find Data Structure

A *set partition* parcels out the elements of some universal set (say the integers 1 to n) into a collection of disjoint subsets, where each element is in exactly one subset. Set partitions naturally arise in graph problems such as connected components (each vertex is in exactly one connected component) and vertex coloring (a vertex may be white or black in a bipartite graph, but not both or neither). Section 17.6 (page 524) presents algorithms for generating set partitions and related objects.

The connected components in a graph can be represented as a set partition. For Kruskal's algorithm to run efficiently, we need a data structure that efficiently supports the following operations:

- *Same component(v_1, v_2)* – Do vertices v_1 and v_2 occur in the same connected component of the current graph?

- *Merge components(C_1, C_2)* – Merge the given pair of connected components into one component in response to the insertion of an edge between them.

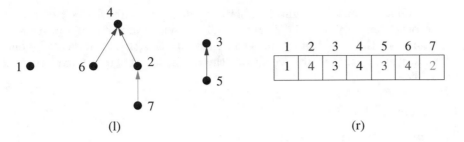

Figure 8.5: Union–find example—the structure represented as a forest of trees (left), and an array of parent pointers (right).

The two obvious data structures for this task each support only one of these operations efficiently. Explicitly labeling each element with its component number enables the *same component* test to be performed in constant time, but updating the component numbers after a merger would require linear time. Alternatively, we can treat the merge components operation as inserting an edge in a graph, but then we must run a full graph traversal to identify the connected components on demand.

The *union–find* data structure represents each subset as a "backwards" tree, with pointers from a node to its parent. Each node of this tree contains a set element, and the *name* of the set is taken from the key at the root, as shown in Figure 8.5. For reasons that will become clear, we also keep track of the number of elements in the subtree rooted in each vertex v:

```
typedef struct {
    int p[SET_SIZE+1];        /* parent element */
    int size[SET_SIZE+1];     /* number of elements in subtree i */
    int n;                    /* number of elements in set */
} union_find;
```

We implement our desired component operations in terms of two simpler operations, *union* and *find*:

- *Find(i)* – Find the root of the tree containing element i, by walking up the parent pointers until there is nowhere to go. Return the label of the root.

- *Union(i,j)* – Link the root of one of the trees (say containing i) to the root of the tree containing the other (say j) so *find(i)* now equals *find(j)*.

We seek to minimize the time it takes to execute the *worst possible* sequence of unions and finds. Tree structures can be very unbalanced, so we must limit the height of our trees. Our most obvious means of control is the choice of which of the two component roots becomes the root of the merged component on each *union*.

To minimize the tree height, it is better to make the smaller tree the subtree of the bigger one. Why? The heights of all the nodes in the root subtree stay the same, but the height of the nodes merged into this tree all increase by one. Thus, merging in the smaller tree leaves the height unchanged on the larger set of vertices.

Implementation

The implementation details are as follows:

```
void union_find_init(union_find *s, int n) {
    int i;      /* counter */

    for (i = 1; i <= n; i++) {
        s->p[i] = i;
        s->size[i] = 1;
    }
    s->n = n;
}


int find(union_find *s, int x) {
    if (s->p[x] == x) {
        return(x);
    }
    return(find(s, s->p[x]));
}


void union_sets(union_find *s, int s1, int s2) {
    int r1, r2;     /* roots of sets */

    r1 = find(s, s1);
    r2 = find(s, s2);

    if (r1 == r2) {
        return;     /* already in same set */
    }

    if (s->size[r1] >= s->size[r2]) {
        s->size[r1] = s->size[r1] + s->size[r2];
        s->p[r2] = r1;
    } else {
        s->size[r2] = s->size[r1] + s->size[r2];
        s->p[r1] = r2;
    }
}
```

```
bool same_component(union_find *s, int s1, int s2) {
    return (find(s, s1) == find(s, s2));
}
```

Analysis

On each union, the tree with fewer nodes becomes the child. But how tall can such a tree get as a function of the number of nodes in it? Consider the smallest possible tree of height h. Single-node trees have height 1. The smallest tree of height 2 has two nodes: it is made from the union of two single-node trees. Merging in more single-node trees won't further increase the height, because they just become children of the rooted tree of height 2. Only when we merge two height 2 trees together can we get a tree of height 3, now with at least four nodes.

See the pattern? We must double the number of nodes in the tree to get an extra unit of height. How many doublings can we do before we use up all n nodes? At most $\lg n$ doublings can be performed. Thus, we can do both unions and finds in $O(\log n)$, fast enough to make Kruskal's algorithm efficient on sparse graphs. In fact, union–find can be done even faster, as discussed in Section 15.5 (page 456).

8.1.4 Variations on Minimum Spanning Trees

The algorithms that construct minimum spanning trees can also be used to solve several closely related problems:

- *Maximum spanning trees* – Suppose an evil telephone company is contracted to connect a bunch of houses together, such that they will be paid a price proportional to the amount of wire they install. Naturally, they will seek to build the most expensive possible spanning tree! The *maximum spanning tree* of any graph can be found by simply negating the weights of all edges and running Prim's or Kruskal's algorithm. The most negative spanning tree in the negated graph is the maximum spanning tree in the original.

  Most graph algorithms do not adapt so easily to negative numbers. Indeed, shortest path algorithms have trouble with negative weights, and certainly do *not* generate the longest possible path using this weight negation technique.

- *Minimum product spanning trees* – Suppose we seek the spanning tree that minimizes the product of edge weights, assuming all edge weights are positive. Since $\lg(a \cdot b) = \lg(a) + \lg(b)$, the minimum spanning tree on a graph whose edge weights are replaced with their logarithms gives the minimum product spanning tree on the original graph.

- *Minimum bottleneck spanning tree* – Sometimes we seek a spanning tree that minimizes the maximum edge weight over all possible trees. In fact, every minimum spanning tree has this property. The proof follows directly from the correctness of Kruskal's algorithm.

  Such bottleneck spanning trees have interesting applications when the edge weights are interpreted as costs, capacities, or strengths. A less efficient but conceptually simpler way to solve such problems might be to delete all "heavy" edges from the graph and ask whether the result is still connected. These kinds of tests can be done with BFS or DFS.

The minimum spanning tree of a graph is unique if all m edge weights in the graph are distinct. Otherwise the order in which Prim's/Kruskal's algorithm breaks ties determines which minimum spanning tree is returned.

There are two important variants of a minimum spanning tree that are *not* solvable with the techniques presented in this section:

- *Steiner tree* – Suppose we want to wire a bunch of houses together, but have the freedom to add extra intermediate vertices to serve as a shared junction. This problem is known as a *minimum Steiner tree*, and is discussed in the catalog in Section 19.10.

- *Low-degree spanning tree* – Alternatively, what if we want to find the minimum spanning tree where the highest degree of a node in the tree is small? The lowest max-degree tree possible would be a simple path, consisting of $n - 2$ nodes of degree 2 and two endpoints of degree 1. Such a path that visits each vertex once is called a *Hamiltonian path*, and is discussed in the catalog in Section 19.5.

8.2   War Story: Nothing but Nets

I'd been tipped off about a small printed circuit board testing company in need of some algorithmic consulting. And so I found myself inside a nondescript building in a nondescript industrial park, talking with the president of Integri-Test and one of his lead technical people.

"We're leaders in robotic printed circuit board testing devices. Our customers have very high reliability requirements for their PC boards. They must check that each and every board has no wire breaks *before* filling it with components. This means testing that each and every pair of points on the board that are supposed to be connected *are* connected."

"How do you do the testing?" I asked.

"We have a robot with two arms, each with electric probes. The arms simultaneously contact both of the points to test whether two points are properly connected. If they are properly connected, then the probes will complete a circuit. For each net, we hold one arm fixed at one point and move the other to cover the rest of the points."

"Wait!" I cried. "What is a net?"

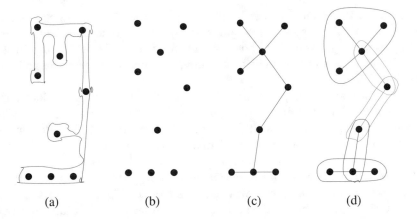

(a)   (b)   (c)   (d)

Figure 8.6: An example net showing (a) the metal connection layer, (b) the contact points, (c) their minimum spanning tree, and (d) the points partitioned into clusters.

"Circuit boards have certain sets of points that are all connected together with a metal layer. This is what we mean by a net. Often a net is just a direct connection between two points. But sometimes a net can have 100 to 200 points, such as all the connections to power or ground."

"I see. So you have a list of all the connections between pairs of points on the circuit board, and you want to trace out these wires."

He shook his head. "Not quite. The input for our testing program consists only of the net contact points, as shown in Figure 8.6(b). We don't know where the actual wires are, but we don't have to. All we must do is verify that all the points in a net are connected together. We do this by putting the left robot arm on the left-most point in the net, and then have the right arm move around to test its connectivity with all the other points in the net. If it is, they must all be connected to each other."

I thought for a moment. "OK. So your right arm has to visit all the other points in the net. How do you choose the order to visit them?"

The technical guy spoke up. "Well, we sort the points from left to right and then go in that order. Is that a good thing to do?"

"Have you ever heard of the traveling salesman problem?" I asked.

He was an electrical engineer, not a computer scientist. "No, what's that?"

"Traveling salesman is the name of the problem that you are trying to solve. Given a set of points to visit, how do you best order them to minimize travel time? Algorithms for the traveling salesman problem have been extensively studied. For small nets, you will be able to find the optimal tour by doing an exhaustive search. For big nets, there are heuristics that will get you close to the optimal tour." I would have pointed them to Section 19.4 (page 594) if I'd had this book handy.

The president scribbled down some notes and then frowned. "Fine. Maybe you can order the points in a net better for us. But that's not our real problem. When you watch our robot in action, the right arm sometimes has to run all the way to the right side of the board on a given net, while the left arm just sits there. It seems we would benefit by breaking nets into smaller pieces to balance things out."

I sat down and thought. The left and right arms each have interlocking TSPs to solve. The left arm would move between the left-most points of each net, while the right arm visits all the other points in each net. By breaking each net into smaller nets we would avoid making the right arm cross all the way across the board. Further, a lot of little nets meant there would be more points in the left TSP, so each left-arm movement was likely to be short as well.

"You are right. We should win if we can break big nets into small nets. We want the nets to be small, both in the number of points and in physical area. But we must ensure that by validating the connectivity of each small net, we will have confirmed that the big net is connected. One point in common between two little nets is sufficient to show that the bigger net formed by their union is connected, because current can flow between any pair of points."

We thus had to break each net into overlapping pieces, where each piece was small. This is a clustering problem. Minimum spanning trees are often used for clustering, as discussed in Section 18.3 (page 549). In fact, that was the answer! We could find the minimum spanning tree of the net points and break it into small clusters whenever a spanning tree edge got too long. As shown in Figure 8.6(d), each cluster would share exactly one point in common with another cluster, with connectivity ensured because we are covering the edges of a spanning tree. The shape of the clusters will reflect the points in the net. If the points lay along a line across the board, the minimum spanning tree would be a path, and the clusters would be pairs of points. If the points all fell in a tight region, there would be one nice fat cluster for the right arm to scoot around.

So I explained the idea of constructing the minimum spanning tree of a graph. The boss listened, scribbled more notes, and frowned again.

"I like your clustering idea. But minimum spanning trees are defined on graphs. All you've got are points. Where do the weights of the edges come from?"

"Oh, we can think of it as a complete graph, where every pair of points is connected. The weight of each edge will be the distance between the two points. Or is it. . . ?"

I went back to thinking. The edge cost should reflect the travel time between two points. While distance is related to travel time, it isn't necessarily the same thing.

"Hey. I have a question about your robot. Does it take the same amount of time to move the arm left–right as it does up–down?"

They thought a minute. "Sure it does. We use the same type of motor to control horizontal and vertical movements. Since these two motors are independent, we can simultaneously move each arm both horizontally and vertically."

"So the time to move both one foot left *and* one foot up is exactly the same as just moving one foot left? This means that the weight for each edge should *not* be the Euclidean distance between the two points, but instead the biggest difference between either the x- or y-coordinates. This is something we call the L_∞ metric, but we can capture it by changing the edge weights in the graph. Anything else funny about your robots?" I asked.

"Well, it takes some time for the arm to come up to full speed. I guess we should factor in its acceleration and deceleration time."

"Darn right. The more accurately you can model the time your arm takes to move between two points, the better our solution will be. But now we have a very clean formulation. Let's code it up and let's see how well it works!"

They were somewhat skeptical about whether this approach would do any good. But a few weeks later they called me back and reported that the new algorithm reduced the distance traveled by about 30% over their previous approach, at a cost of a little extra computation. But their testing machine cost $200,000 a pop compared to a lousy $2,000 for a computer, so this was an excellent tradeoff, particularly since the algorithm need only be run once when testing repeated instances of a particular board design.

The key idea here was modeling the job in terms of classical algorithmic graph problems. I smelled TSP the instant they started talking about minimizing robot motion. Once I realized that they were implicitly using a star-shaped spanning tree to ensure connectivity, it was natural to ask whether the minimum spanning tree would perform any better. This idea led to clustering, and thus partitioning each net into smaller nets. Finally, by carefully designing our distance metric to accurately model the costs of the robot, we could incorporate complicated properties (such as acceleration) without changing our fundamental graph model or algorithm design.

> *Take-Home Lesson:* Most applications of graphs can be reduced to standard graph properties where well-known algorithms can be used. These include minimum spanning trees, shortest paths, and other problems presented in the catalog.

8.3 Shortest Paths

A *path* is a sequence of edges connecting two vertices. There are typically an enormous number of possible paths connecting two nodes in any given road or social network. The path that minimizes the sum of edge weights, that is, the *shortest path*, is likely to be the most interesting, reflecting the fastest travel path or the closest kinship between the nodes.

A shortest path from s to t in an unweighted graph can be identified using a breadth-first search from s. The minimum-link path is recorded in the breadth-first search tree, and hence provides the shortest path when all edges have equal weight.

But BFS does *not* suffice to find shortest paths in weighted graphs. The shortest weighted path might require a large number of edges, just as the fastest route from home to office may involve complicated backroad shortcuts, as shown in Figure 8.7.

This section will present two distinct algorithms for finding the shortest paths in weighted graphs.

8.3.1   Dijkstra's Algorithm

Dijkstra's algorithm is the method of choice for finding shortest paths in an edge- and/or vertex-weighted graph. Starting from a particular vertex s, it finds the shortest path from s to all other vertices in the graph, including your desired destination t.

Suppose the shortest path from s to t in graph G passes through a particular intermediate vertex x. Clearly, the best s-to-t path must contain the shortest path from s to x as its prefix, because if it doesn't we can improve the path by starting with the shorter s-to-x prefix. Thus, we must compute the shortest path from s to x before we find the path from s to t.

Dijkstra's algorithm proceeds in a series of rounds, where each round establishes the shortest path from s to *some* new vertex. Specifically, x is the vertex that minimizes $dist(s, v_i) + w(v_i, x)$ over all unfinished vertices v_i. Here $w(a, b)$ denotes the weight of the edge from vertex a to vertex b, and $dist(a, b)$ is the length of the shortest path between them.

This suggests a dynamic programming-like strategy. The shortest path from s to itself is trivial, so $dist(s, s) = 0$.[1] If (s, y) is the lightest edge incident to s, then $dist(s, y) = w(s, y)$. Once we determine the shortest path to a node x, we check all the outgoing edges of x to see whether there is a shorter path from s through x to some unknown vertex.

ShortestPath-Dijkstra(G, s, t)
    $known = \{s\}$
    for each vertex v in G, $dist[v] = \infty$
    $dist[s] = 0$
    for each edge (s, v), $dist[v] = w(s, v)$
    $last = s$
    while ($last \neq t$)
        select v_{next}, the unknown vertex minimizing $dist[v]$
        for each edge (v_{next}, x), $dist[x] = \min[dist[x], dist[v_{next}] + w(v_{next}, x)]$
        $last = v_{next}$
        $known = known \cup \{v_{next}\}$

The basic idea is very similar to Prim's algorithm. In each iteration, we add exactly one vertex to the tree of vertices for which we *know* the shortest path from s. As in Prim's, we keep track of the best path seen to date for all vertices outside the tree, and insert them in order of increasing cost.

---

[1] Actually, this is true only when the graph does not contain negative weight edges, which is why we assume that all edges are of positive weight in the discussion that follows.

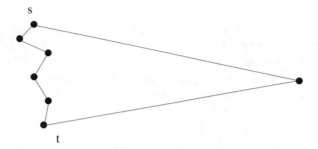

Figure 8.7: The shortest path from s to t might pass through many intermediate vertices rather than use the fewest possible edges.

In fact, the *only* difference between Dijkstra's and Prim's algorithms is how they rate the desirability of each outside vertex. In the minimum spanning tree algorithm, we sought to minimize the weight of the next potential tree edge. In shortest path, we want to identify the closest outside vertex (in shortest-path distance) to s. This desirability is a function of both the new edge weight *and* the distance from s to the tree vertex it is adjacent to.

Implementation

The pseudocode above obscures just how similar the two algorithms are. Below, we give an implementation of Dijkstra's algorithm based on changing exactly four lines from our Prim's implementation—one of which is simply the name of the function!

```
int dijkstra(graph *g, int start) {
    int i;                      /* counter */
    edgenode *p;                /* temporary pointer */
    bool intree[MAXV+1];        /* is the vertex in the tree yet? */
    int distance[MAXV+1];       /* cost of adding to tree */
    int v;                      /* current vertex to process */
    int w;                      /* candidate next vertex */
    int dist;                   /* cheapest cost to enlarge tree */
    int weight = 0;             /* tree weight */

    for (i = 1; i <= g->nvertices; i++) {
        intree[i] = false;
        distance[i] = MAXINT;
        parent[i] = -1;
    }

    distance[start] = 0;
    v = start;
```

```
        while (!intree[v]) {
            intree[v] = true;
            if (v != start) {
                printf("edge (%d,%d) in tree \n",parent[v],v);
                weight = weight + dist;
            }
            p = g->edges[v];
            while (p != NULL) {
                w = p->y;
                if (distance[w] > (distance[v]+p->weight)) { /* CHANGED */
                    distance[w] = distance[v]+p->weight;      /* CHANGED */
                    parent[w] = v;                            /* CHANGED */
                }
                p = p->next;
            }

            dist = MAXINT;
            for (i = 1; i <= g->nvertices; i++) {
                if ((!intree[i]) && (dist > distance[i])) {
                    dist = distance[i];
                    v = i;
                }
            }
        }

        return(weight);
}
```

This algorithm defines a shortest-path spanning tree rooted in s. For unweighted graphs, this would be the breadth-first search tree, but in general it provides the shortest path from s to all other vertices, not just t.

Analysis

What is the running time of Dijkstra's algorithm? As implemented here, the complexity is $O(n^2)$, exactly the same running time as a proper version of Prim's algorithm. This is because, except for the extension condition, it *is* exactly the same algorithm as Prim's.

The length of the shortest path from start to a given vertex t is exactly the value of distance[t]. How do we use dijkstra to find the actual path? We follow the backward parent pointers from t until we hit start (or -1 if no such path exists), exactly as was done in the BFS/DFS find_path() routine of Section 7.6.2 (page 217).

Dijkstra works correctly only on graphs without negative-cost edges. The reason is that during the execution we may encounter an edge with weight so negative that it changes the cheapest way to get from s to some other vertex already in the tree. Indeed, the most cost-effective way to get from your house to your next-door neighbor would be to repeatedly cycle through the lobby of any bank offering you enough free money to make the detour worthwhile. Unless that bank limits its reward to one per customer, you might so benefit by making an unlimited number of trips through the lobby that you would *never* actually reach your destination!

Fortunately, most applications don't have negative weights, making this discussion largely academic. Floyd's algorithm, discussed below, works correctly with negative-cost edges provided there are no negative cost *cycles*, which grossly distort the shortest-path structure.

Stop and Think: Shortest Path with Node Costs

Problem: Suppose we are given a directed graph whose weights are on the vertices instead of the edges. Thus, the cost of a path from x to y is the sum of the weights of all vertices on the path. Give an efficient algorithm for finding shortest paths on vertex-weighted graphs.

Solution: A natural idea would be to adapt the algorithm we have for edge-weighted graphs (Dijkstra's) to the new vertex-weighted domain. It should be clear that this will work. We replace any reference to the weight of any directed edge (x, y) with the weight of the destination vertex y. This can be looked up as needed from an array of vertex weights.

However, my preferred approach would leave Dijkstra's algorithm intact and instead concentrate on constructing an edge-weighted graph on which Dijkstra's algorithm will give the desired answer. Set the weight of each directed edge (i, j) in the input graph to the cost of vertex j. Dijkstra's algorithm now does the job. Try to *design graphs, not algorithms*, as I will encourage in Section 8.7.

This technique can be extended to a variety of different domains, including when there are costs on both vertices and edges. ∎

8.3.2 All-Pairs Shortest Path

Suppose you want to find the "center" vertex in a graph—the one that minimizes the longest or average distance to all the other nodes. This might be the best place to start a new business. Or perhaps you need to know a graph's *diameter*—the largest shortest-path distance over all pairs of vertices. This might correspond to the longest possible time it can take to deliver a letter or network packet. These and other applications require computing the shortest path between all pairs of vertices in a given graph.

We could solve *all-pairs shortest path* by calling Dijkstra's algorithm from each of the n possible starting vertices. But Floyd's all-pairs shortest-path algorithm is a slick way to construct this $n \times n$ distance matrix from the original weight matrix of the graph.

Floyd's algorithm is best employed on an adjacency matrix data structure, which is no extravagance since we must store all n^2 pairwise distances anyway. Our `adjacency_matrix` type allocates space for the largest possible matrix, and keeps track of how many vertices are in the graph:

```
typedef struct {
    int weight[MAXV+1][MAXV+1];   /* adjacency/weight info */
    int nvertices;                /* number of vertices in graph */
} adjacency_matrix;
```

The critical issue in an adjacency matrix implementation is how we denote the edges absent from the graph. A common convention for unweighted graphs denotes graph edges by 1 and non-edges by 0. This gives exactly the wrong interpretation if the numbers denote edge weights, because the non-edges get interpreted as a free ride between vertices. Instead, we should initialize each non-edge to `MAXINT`. This way we can both test whether it is present and automatically ignore it in shortest-path computations.

There are several ways to characterize the shortest path between two nodes in a graph. The Floyd–Warshall algorithm starts by numbering the vertices of the graph from 1 to n. We use these numbers not to label the vertices, but to order them. Define $W[i, j]^k$ to be the length of the shortest path from i to j using only vertices numbered from $1, 2, ..., k$ as possible intermediate vertices.

What does this mean? When $k = 0$, we are allowed no intermediate vertices, so the only allowed paths are the original edges in the graph. The initial all-pairs shortest-path matrix thus consists of the initial adjacency matrix. We will perform n iterations, where the kth iteration allows only the first k vertices as possible intermediate steps on the path between each pair of vertices x and y.

With each iteration, we allow a richer set of possible shortest paths by adding a new vertex as a possible intermediary. The kth vertex helps only if there is a shortest path that goes through k, so

$$W[i, j]^k = \min(W[i, j]^{k-1}, W[i, k]^{k-1} + W[k, j]^{k-1})$$

The correctness of this is somewhat subtle, and I encourage you to convince yourself of it. Indeed, it is a great example of dynamic programming, the algorithmic paradigm that is the focus of Chapter 10. But there is nothing subtle about how simple the implementation is:

```
void floyd(adjacency_matrix *g) {
    int i, j;            /* dimension counters */
    int k;               /* intermediate vertex counter */
    int through_k;       /* distance through vertex k */

    for (k = 1; k <= g->nvertices; k++) {
        for (i = 1; i <= g->nvertices; i++) {
            for (j = 1; j <= g->nvertices; j++) {
                through_k = g->weight[i][k]+g->weight[k][j];
                if (through_k < g->weight[i][j]) {
                    g->weight[i][j] = through_k;
                }
            }
        }
    }
}
```

The Floyd–Warshall all-pairs shortest-path algorithm runs in $O(n^3)$ time, which is asymptotically no better than n calls to Dijkstra's algorithm. However, the loops are so tight and the program so short that it runs better in practice. It is notable as one of the rare graph algorithms that work better on adjacency matrices than adjacency lists.

The output of Floyd's algorithm, as it is written, does not enable one to reconstruct the actual shortest path between any given pair of vertices. These paths can be recovered if we retain a parent matrix P containing our choice of the last intermediate vertex used for each vertex pair (x, y). Say this value is k. The shortest path from x to y is the concatenation of the shortest path from x to k with the shortest path from k to y, which can be reconstructed recursively given the matrix P. Note, however, that most all-pairs applications only need the resulting distance matrix. These are the jobs that Floyd's algorithm was designed for.

8.3.3 Transitive Closure

Floyd's algorithm has another important application, that of computing *transitive closure*. We are often interested in which vertices in a directed graph are reachable from a given node. As an example, consider the *blackmail graph*, where there is a directed edge (i, j) if person i has sensitive-enough private information on person j so that i can get j to do whatever they want. You wish to hire one of these n people to be your personal representative. Who has the most power in terms of blackmail potential?

A simplistic answer would be the vertex of highest out-degree, but an even better representative would be the person who has blackmail chains leading to the most other parties. Steve might only be able to blackmail Miguel directly, but if Miguel can blackmail everyone else then Steve is the person you want to hire.

The vertices reachable from any single node can be computed using breadth-first or depth-first search. But the complete set of relationships can be found using an all-pairs shortest path. If the shortest path from i to j remains `MAXINT` after running Floyd's algorithm, you can be sure that no directed path exists from i to j. Any vertex pair of weight less than `MAXINT` must be reachable, both in the graph-theoretic and blackmail senses of the word.

Transitive closure is discussed in more detail in the catalog in Section 18.5.

8.4   War Story: Dialing for Documents

I was part of a group visiting Periphonics, then an industry leader in building telephone voice-response systems. These are more advanced versions of the *Press 1 for more options, Press 2 if you didn't press 1* telephone systems that blight everyone's lives. The tour guide was so enthusiastic about the joy of using their product it set off the crustiest member of our delegation.

"Like typing, my pupik!" came a voice from the rear of our group. "I *hate* typing on a telephone. Whenever I call my brokerage house to get stock quotes some machine tells me to type in the three letter code. To make things worse, I have to hit two buttons to type in one letter, in order to distinguish between the three letters printed on each key of the telephone. I hit the 2 key and it says Press 1 for A, Press 2 for B, Press 3 for C. Pain in the neck if you ask me."

"Maybe you don't have to hit two keys for each letter!" I chimed in. "Maybe the system could figure out the correct letter from context!"

"There isn't a whole lot of context when you type in three letters of stock market code."

"Sure, but there would be plenty of context if we typed in English sentences. I'll bet that we could reconstruct English text correctly if it was typed in a telephone at one keystroke per letter."

The guy from Periphonics gave me a disinterested look, then continued the tour. But when I got back to the office, I decided to give it a try.

Not all letters are equally likely to be typed on a telephone. In fact, not all letters *can* be typed, since Q and Z are not labeled on a standard American telephone. Therefore, we adopted the convention that Q, Z, and "space" all sat on the * key. We could take advantage of the uneven distribution of letter frequencies to help us decode the text. For example, if you hit the 3 key while typing English, you more likely meant to type an E than either a D or F. Our first attempt to predict the typed text used the frequencies of three characters (trigrams) in a window of the text. But the results were not good. The trigram statistics did a decent job of translating it into gibberish, but a terrible job of transcribing English.

One reason was clear. This algorithm knew nothing about English words. If we coupled it with a dictionary, we might be onto something. But two words in the dictionary are often represented by the exact same string of phone codes. For an extreme example, the code string "22737" collides with eleven distinct English words, including *cases, cares, cards, capes, caper,* and *bases.* For our

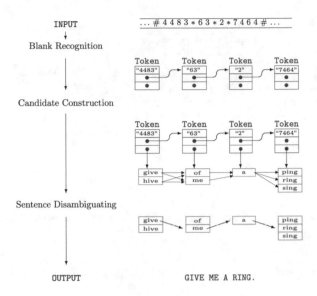

Figure 8.8: The phases of the telephone code reconstruction process.

next attempt, we reported the unambiguous characters of any words that collided in the dictionary, and used trigrams to fill in the rest of the characters.

This also did a terrible job. Most words appearing in the text came from ambiguous codes mapping to more than one vocabulary word. Somehow, we had to distinguish between the different dictionary words that got hashed to the same code. We could factor in the relative popularity of each word, but this still made too many mistakes.

At this point, I started working with Harald Rau on the project, who proved to be a great collaborator. First, he was a bright and persistent graduate student. Second, as a native German speaker, he believed every lie I told him about English grammar. Harald built up a phone code reconstruction program along the lines of Figure 8.8. It worked on the input one sentence at a time, identifying dictionary words that matched each code string. The key problem was how to incorporate grammatical constraints.

"We can get good word-use frequencies and grammatical information from a big text database called the Brown Corpus. It contains thousands of typical English sentences, each parsed according to parts of speech. But how do we factor it all in?" Harald asked.

"Let's think about it as a graph problem," I suggested.

"*Graph problem?* What graph problem? Where is there even a graph?"

"Think of a sentence as a series of tokens, each representing a word in the sentence. Each token has a list of words from the dictionary that match it. How can we choose which one is right? Each possible sentence interpretation can be thought of as a path in a graph. Each vertex of this graph is one word from the complete set of possible word choices. There will be an edge from each possible choice for the ith word to each possible choice for the $(i+1)$st

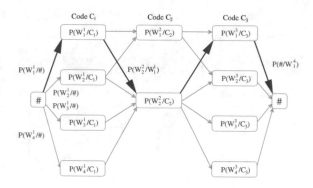

Figure 8.9: The minimum-cost path defines the best interpretation for a sentence.

word. The cheapest path across this graph defines the best interpretation of the sentence."

"But all the paths look the same. They have the same number of edges. Now I see! We have to add weight to the edges to make the paths different."

"Exactly! The cost of an edge will reflect how likely it is that we will travel through the given pair of words. Perhaps we can count how often that pair of words occurred together in previous texts. Or we can weigh them by the part of speech of each word. Maybe nouns don't like to be next to nouns as much as they like being next to verbs."

"It will be hard to keep track of word-pair statistics, since there are so many possible pairs. But we certainly know the frequency of each word. How can we factor that into things?"

"We can pay a cost for walking through a particular vertex that depends upon the frequency of the word. Our best sentence will be given by the shortest path across the graph."

"But how do we figure out the relative weights of these factors?"

"Try what seems natural to you and then we can experiment with it."

Harald implemented this shortest-path algorithm. With proper grammatical and statistical constraints, the system performed great. Look at the Gettysburg Address, with all the reconstruction errors highlighted:

FOURSCORE AND SEVEN YEARS AGO OUR FATHERS BROUGHT FORTH
ON THIS CONTINENT A NEW NATION CONCEIVED IN LIBERTY AND DED-
ICATED TO THE PROPOSITION THAT ALL MEN ARE CREATED EQUAL.
NOW WE ARE ENGAGED IN A GREAT CIVIL WAR TESTING WHETHER
THAT NATION OR ANY NATION SO CONCEIVED AND SO DEDICATED CAN
LONG ENDURE. WE ARE MET ON A GREAT BATTLEFIELD OF THAT **WAS**.
WE HAVE COME TO DEDICATE A PORTION OF THAT FIELD AS A FINAL
SERVING PLACE FOR THOSE WHO HERE **HAVE** THEIR LIVES THAT THE
NATION MIGHT LIVE. IT IS ALTOGETHER FITTING AND PROPER THAT
WE SHOULD DO THIS. BUT IN A LARGER SENSE WE CAN NOT DEDICATE
WE CAN NOT CONSECRATE WE CAN NOT HALLOW THIS GROUND. THE
BRAVE MEN LIVING AND DEAD WHO STRUGGLED HERE HAVE CONSE-
CRATED IT FAR ABOVE OUR POOR POWER TO ADD OR DETRACT. THE
WORLD WILL LITTLE NOTE NOR LONG REMEMBER WHAT WE SAY HERE
BUT IT CAN NEVER FORGET WHAT THEY DID HERE. IT IS FOR US THE

While we still made a few mistakes, we typically guessed about 99% of all characters correctly. The results were clearly good enough for many applications. Periphonics certainly thought so, for they licensed our program to incorporate into their products. The reconstruction time was faster than anyone can type text in on a phone keypad.

The constraints for many pattern recognition problems can be naturally formulated as shortest-path problems in graphs. There is a particularly convenient dynamic programming solution for these problems (the Viterbi algorithm) that is widely used in speech and handwriting recognition systems. Despite the fancy name, the Viterbi algorithm is basically solving a shortest-path problem on a DAG. Hunting for a graph formulation to solve your problem is often the right idea.

8.5 Network Flows and Bipartite Matching

An edge-weighted graph can be interpreted as a network of pipes, where the weight of an edge determines the *capacity* of the pipe. Capacities can be thought of as a function of the cross-sectional area of the pipe. A wide pipe might be able to carry 10 units of flow, that is, the amount of material in a given time, whereas a narrower pipe can only carry 5 units. The *network flow problem* asks for the maximum amount of flow that can be sent from vertices s to t in a given weighted graph G while respecting the maximum capacities of each pipe.

8.5.1 Bipartite Matching

While the network flow problem is of independent interest, its primary importance lies in solving other important graph problems. A classic example is bipartite matching. A *matching* in a graph $G = (V, E)$ is a subset of edges $E' \subset E$ such that no two edges of E' share a vertex. A matching pairs off certain vertices such that every vertex is in at most one such pair, as shown in Figure 8.10.

Graph G is *bipartite* or *two-colorable* if the vertices can be divided into two sets, L and R, such that all edges in G have one vertex in L and one vertex in R. Many naturally defined graphs are bipartite. For example, one class of vertices may represent jobs to be done and the remaining vertices represent people who can potentially do them. The existence of edge (j, p) means that job j can be done by person p. Or let certain vertices represent boys and certain

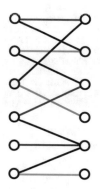 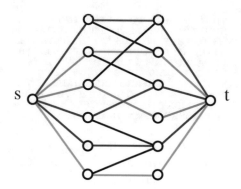

Figure 8.10: Bipartite graph with a maximum matching (left). The corresponding network flow instance highlighting the maximum s–t flow (right).

vertices represent girls, with edges representing compatible pairs. Matchings in these graphs have natural interpretations as job assignments or as traditional marriages, and are the focus of Section 18.6 (page 562).

The maximum cardinality bipartite matching can be readily found using network flow. Create a *source* node s that is connected to every vertex in L by an edge of weight 1. Create a *sink* node t and connect it to every vertex in R by an edge of weight 1. Finally, assign each edge in the central bipartite graph G a weight of 1. Now, the maximum possible flow from s to t defines the largest matching in G. Certainly we can find a flow as large as the matching, by using the matching edges and their source-to-sink connections. Further, there can be no other solution that achieves greater flow, because we can't possibly get more than one flow unit through any given vertex.

8.5.2   Computing Network Flows

Traditional network flow algorithms are based on the idea of *augmenting paths*: finding a path of positive capacity from s to t and adding it to the flow. It can be shown that the flow through a network is optimal iff it contains no augmenting path. Since each augmentation increases the flow, by repeating the process until no such path remains we must eventually find the global maximum.

The key structure is the *residual flow graph*, denoted as $R(G, f)$, where G is the input graph whose weights are the capacities, and f is array of flows through G. The directed, edge-weighted graph $R(G, f)$ contains the same vertices as G. For each edge (i, j) in G with capacity $c(i, j)$ and flow $f(i, j)$, $R(G, f)$ may contain two edges:

(i) an edge (i, j) with weight $c(i, j) - f(i, j)$, if $c(i, j) - f(i, j) > 0$ and

(ii) an edge (j, i) with weight $f(i, j)$, if $f(i, j) > 0$.

The weight of the edge (i, j) in the residual graph gives the exact amount of extra flow that can be pushed from i to j. A path in the residual flow graph

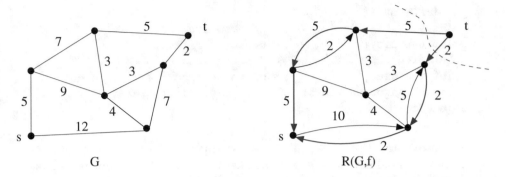

Figure 8.11: Maximum s–t flow in a graph G (on left) showing the associated residual graph $R(G, f)$ and minimum s–t cut (dotted line near t) on right. The undirected edges in $R(G, f)$ have zero flow, so they have residual capacity in both directions.

from s to t implies that more flow can be pushed from s to t. The smallest edge weight on this path defines the amount of extra flow that can be pushed along it.

Figure 8.11 illustrates this idea. The maximum s–t flow in graph G is 7. Such a flow is revealed by the two directed t to s paths in the residual graph $R(G)$, of flows 2 and 5 respectively. These flows completely saturate the capacity of the two edges incident to vertex t, so no augmenting path remains. Thus, the flow is optimal. A set of edges whose deletion separates s from t (like the two edges incident to t) is called an s–t cut. Clearly, no s to t flow can exceed the weight of the minimum such cut. In fact, a flow equal to the minimum cut is always possible.

> *Take-Home Lesson:* The maximum flow from s to t always equals the weight of the minimum s–t cut. Thus, flow algorithms can be used to solve general edge and vertex connectivity problems in graphs.

Implementation

We cannot do full justice to the theory of network flows here. However, it is instructive to see how augmenting paths can be identified and the optimal flow computed.

For each edge in the residual flow graph, we must keep track of both the amount of flow currently going through the edge, as well as its remaining *residual capacity*. Thus, we must modify our edge structure to accommodate the extra fields:

```
typedef struct {
    int v;                      /* neighboring vertex */
    int capacity;               /* capacity of edge */
    int flow;                   /* flow through edge */
    int residual;               /* residual capacity of edge */
    struct edgenode *next;      /* next edge in list */
} edgenode;
```

We use a breadth-first search to look for any path from source to sink that increases the total flow, and use it to augment the total. We terminate with the optimal flow when no such *augmenting* path exists.

```
void netflow(flow_graph *g, int source, int sink) {
    int volume;      /* weight of the augmenting path */

    add_residual_edges(g);

    initialize_search(g);
    bfs(g, source);

    volume = path_volume(g, source, sink);

    while (volume > 0) {
        augment_path(g, source, sink, volume);
        initialize_search(g);
        bfs(g, source);
        volume = path_volume(g, source, sink);
    }
}
```

Any augmenting path from source to sink increases the flow, so we can use bfs to find such a path. We only consider network edges that have remaining capacity or, in other words, positive residual flow. The predicate below helps bfs distinguish between saturated and unsaturated edges:

```
bool valid_edge(edgenode *e) {
    return (e->residual > 0);
}
```

Augmenting a path transfers the maximum possible volume from the residual capacity into positive flow. This amount is limited by the path edge with the smallest amount of residual capacity, just as the rate at which traffic can flow is limited by the most congested point.

```
int path_volume(flow_graph *g, int start, int end) {
    edgenode *e;      /* edge in question */

    if (parent[end] == -1) {
        return(0);
    }

    e = find_edge(g, parent[end], end);

    if (start == parent[end]) {
        return(e->residual);
    } else {
        return(min(path_volume(g, start, parent[end]), e->residual));
    }
}
```

Recall that `bfs` uses the **parent** array to record the discoverer of each vertex on the traversal, enabling us to reconstruct the shortest path back to the root from any vertex. The edges of this tree are vertex pairs, not the actual edges in the graph data structure on which the search was performed. The call `find_edge(g,x,y)` returns a pointer to the record encoding edge (x, y) in graph g, necessary to obtain its residual capacity. The `find_edge` routine can find this pointer by scanning the adjacency list of x (`g->edges[x]`), or (even better) from an appropriate table lookup data structure.

Sending an additional unit of flow along directed edge (i, j) reduces the residual capacity of edge (i, j) but *increases* the residual capacity of edge (j, i). Thus, the act of augmenting a path requires modifying both forward and reverse edges for each link on the path.

```
void augment_path(flow_graph *g, int start, int end, int volume) {
    edgenode *e;      /* edge in question */

    if (start == end) {
        return;
    }

    e = find_edge(g, parent[end], end);
    e->flow += volume;
    e->residual -= volume;

    e = find_edge(g, end, parent[end]);
    e->residual += volume;

    augment_path(g, start, parent[end], volume);
}
```

Initializing the flow graph requires creating directed flow edges (i, j) and (j, i) for each network edge $e = (i, j)$. Initial flows are all set to 0. The initial residual flow of (i, j) is set to the capacity of e, while the initial residual flow of (j, i) is set to 0.

Analysis

The augmenting path algorithm above eventually converges to the optimal solution. However, each augmenting path may add only a little to the total flow, so, in principle, the algorithm might take an arbitrarily long time to converge.

However, Edmonds and Karp [EK72] proved that always selecting a *shortest* unweighted augmenting path guarantees that $O(n^3)$ augmentations suffice for optimization. In fact, the Edmonds–Karp algorithm is what is implemented above, since a breadth-first search from the source is used to find the next augmenting path.

8.6 Randomized Min-Cut

Clever randomized algorithms have been developed for many different types of problems. We have so far seen randomized algorithms for sorting (quicksort), searching (hashing), string matching (Rabin–Karp), and number-theoretic (primality testing) problems. Here we expand this list to graph algorithms.

The minimum-cut problem in graphs seeks to partition the vertices of graph G into sets V_1 and V_2 so that the smallest possible number of edges (x, y) span across these two sets, meaning $x \in V_1$ and $y \in V_2$. Identifying the minimum cut often arises in network reliability analysis: what is the smallest failure set whose deletion will disconnect the graph? The minimum-cut problem is discussed in greater detail in Section 18.8. The graph shown there has a minimum-cut set size of 2, while the graph in Figure 8.12 (left) can be disconnected with just one edge deletion.

Suppose the minimum cut C in G is of size k, meaning that k edge deletions are necessary to disconnect G. Each vertex v must therefore be connected to at least k other vertices, because if not there would be a smaller cut-set disconnecting v from the rest of the graph. This implies that G must contain at least $kn/2$ edges, where n is the number of vertices, because each edge contributes one to the degree of exactly two vertices.

A *contraction* operation for edge (x, y) collapses vertices x and y into a single merged vertex called (say) xy. Any edge of the form (x, z) or (y, z) gets replaced by (xy, z). The upshot is that the number of vertices shrinks by one on an edge contraction. The number of edges stays the same, although a self-loop (xy, xy) replaces (x, y), and two copies of edge (xy, z) are created if both (x, z) and (y, z) were in G before the contraction.

What happens to the size of the minimum cut after contracting (x, y) in G? Each contraction reduces the space of possible V_1, V_2 partitions, since the new vertex xy cannot ever be subdivided. The critical observation is that the

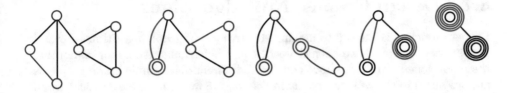

Figure 8.12: If we get lucky, a sequence of random edge contractions does not increase the size of the smallest cut set.

minimum-cut size is unchanged *unless* we contract one of the k edges of the optimal cut. If we did contract one of these cut edges, the minimum-cut size of the resulting graph might grow, because the best partition is no longer available.

This suggests the following randomized algorithm. Pick a random edge of G and contract it. Repeat a total of $n-2$ times, until we are left with a two-vertex graph with multiple parallel edges between them. These edges describe a cut in the graph, although it might not be the smallest possible cut of G. We could repeat this entire procedure r times, and report the smallest cut we ever see as our proposed minimum cut. Properly implemented, this contraction series for one given graph can be implemented in $O(nm)$ time, resulting in a Monte Carlo algorithm with $O(rmn)$ running time, but no guarantee of an optimal solution.

What are the chances of success on any given iteration? Consider the initial graph. A contraction of a random edge e preserves the minimum cut C provided e is not one of the k cut edges. Since G has at least $kn/2$ edges, the probability p_i of a successful ith edge contraction is:

$$p_i \geq 1 - \frac{k}{k(n-i+1)/2} = 1 - \frac{2}{n-i+1} = \frac{n-i-1}{n-i+1}$$

The odds on success for all but the last few contractions in a large graph are strongly in our favor.

To end up with a minimum cut C for a particular run, we must succeed on every one of our $n-2$ contractions, which occurs with probability

$$\prod_{i=1}^{n-2} p_i = \prod_{i=1}^{n-2} \frac{n-i-1}{n-i+1} = \left(\frac{n-2}{n}\right)\left(\frac{n-3}{n-1}\right)\left(\frac{n-4}{n-2}\right)\cdots\left(\frac{3}{5}\right)\left(\frac{2}{4}\right)\left(\frac{1}{3}\right) = \frac{2}{n(n-1)}$$

The product cancels magically, and leaves a success probability of $\Theta(1/n^2)$. That isn't very large, but if we run $r = n^2 \log n$ times it becomes very likely we will stumble upon the minimum cut at least once.

Take-Home Lesson: The key to success in any randomized algorithm is setting up a situation where we can bound our probability of success. The analysis can be tricky, but the resulting algorithms are often quite simple, as they are here. After all, complicated randomized algorithms likely become too difficult to analyze.

8.7   Design Graphs, Not Algorithms

Proper modeling is the key to making effective use of graph algorithms. Several properties of graphs have been defined, and efficient algorithms for computing them developed. All told, about two dozen different algorithmic graph problems are presented in the catalog, mostly in Sections 18 and 19. These classical graph problems provide a framework for modeling most applications.

The secret is learning to design graphs, not algorithms. We have already seen a few instances of this idea:

- The *maximum* spanning tree can be found by negating the edge weights of the input graph G and using a *minimum* spanning tree algorithm on the result. The spanning tree of $-G$ that has the most negative weight will define the maximum-weight tree in G.

- To solve bipartite matching, we constructed a special network flow graph such that the maximum flow corresponds to a matching having the largest number of edges.

The applications below demonstrate the power of proper modeling. Each arose in a real-world application, and each can be modeled as a graph problem. Some of the modelings are quite clever, but they illustrate the versatility of graphs in representing relationships. As you read a problem, try to devise an appropriate graph representation before peeking to see how it was done.

Stop and Think: The Pink Panther's Passport to Peril

Problem: I'm looking for an algorithm to design natural routes for video-game characters to follow through an obstacle-filled room. How should I do it?

Solution: Presumably the desired route should look like a path that an intelligent being would choose. Since intelligent beings are either lazy or efficient, this should be modeled as a shortest-path problem.

But what is the graph? One approach might be to lay a grid of points in the room. Create a vertex for each grid point that is a valid place for the character to stand, one that does not lie within an obstacle. Construct an edge between any pair of nearby vertices, weighted proportionally to the distance between them. Although direct geometric methods are known for shortest paths (see Section 18.4 (page 554)), it is easier to model this discretely as a graph. ∎

Stop and Think: Ordering the Sequence

Problem: A DNA sequencing project generates experimental data consisting of small fragments. For each given fragment f, we know certain other fragments are forced to lie to the left of f, and certain other fragments are forced to be on

f's right. How can we find a consistent ordering of the fragments from left to right that satisfies all the constraints?

Solution: Create a directed graph, where each fragment is assigned a unique vertex. Insert a directed edge (l, f) from any fragment l that is forced to be to the left of f, and a directed edge (f, r) to any fragment r forced to be to the right of f. We seek an ordering of the vertices such that all the edges go from left to right. This is a *topological sort* of the resulting directed acyclic graph. The graph must be acyclic, because cycles would make finding a consistent ordering impossible. ∎

Stop and Think: Bucketing Rectangles

Problem: In my graphics work I must solve the following problem. Given an arbitrary set of rectangles in the plane, how can I distribute them into a minimum number of buckets such that no rectangles in any given bucket intersect one another? In other words, there cannot be any overlapping area between two rectangles in the same bucket.

Solution: We formulate a graph where each vertex represents a rectangle, and there is an edge if two rectangles intersect. Each bucket corresponds to an *independent set* (see Section 19.2 (page 589)) of rectangles, so there is no overlap between any two. A *vertex coloring* (see Section 19.7 (page 604)) of a graph is a partition of the vertices into independent sets, so minimizing the number of colors is exactly what the problem is asking for. ∎

Stop and Think: Names in Collision

Problem: In porting code from Unix to DOS, I have to shorten several hundred file names down to at most eight characters each. I can't just use the first eight characters from each name, because "filename1" and "filename2" would be assigned the exact same name. How can I meaningfully shorten the names while ensuring that they do not collide?

Solution: Construct a bipartite graph with vertices corresponding to each original file name f_i for $1 \leq i \leq n$, as well as a collection of acceptable shortenings for each name $f_{i1}, \ldots, f_{ik}$. Add an edge between each original and shortened name. We now seek a set of n edges that have no vertices in common, so each file name is mapped to a distinct acceptable substitute. *Bipartite matching* is exactly this problem of finding an independent set of edges in a graph. ∎

Stop and Think: Separate the Text

Problem: We need a way to separate the lines of text in the optical character-recognition system that we are building. Although there is some white space between the lines, problems like noise and the tilt of the page make it hard to find. How can we do line segmentation?

Solution: Consider the following graph formulation. Treat each pixel in the image as a vertex in the graph, with an edge between two neighboring pixels. The weight of this edge should be proportional to how dark the pixels are. A segmentation between two lines is a path in this graph from the left to right side of the page. We seek a relatively straight path that avoids as much blackness as possible. This suggests that the *shortest path* in the pixel graph will likely find a good line segmentation. ▌

> *Take-Home Lesson:* Designing novel graph algorithms is very hard, so don't do it. Instead, try to design graphs that enable you to use classical algorithms to model your problem.

Chapter Notes

Network flows are an advanced algorithmic technique, and recognizing whether a particular problem can be solved by network flow requires experience. I point the reader to books by Williamson [Wil19] and Cook and Cunningham [CC97] for more detailed treatments of the subject.

The augmenting path method for network flows is due to Ford and Fulkerson [FF62]. Edmonds and Karp [EK72] proved that always selecting a *shortest* geodesic augmenting path guarantees that $O(n^3)$ augmentations suffice for optimization.

The phone code reconstruction system that was the subject of the war story is described in more technical detail in Rau and Skiena [RS96].

8.8   Exercises

Simulating Graph Algorithms

8-1. *[3]* For the graphs in Problem 7-1:

(a) Draw the spanning forest after every iteration of the main loop in Kruskal's algorithm.

(b) Draw the spanning forest after every iteration of the main loop in Prim's algorithm.

(c) Find the shortest-path spanning tree rooted in A.

(d) Compute the maximum flow from A to H.

Minimum Spanning Trees

8-2. *[3]* Is the path between two vertices in a minimum spanning tree necessarily a shortest path between the two vertices in the full graph? Give a proof or a counterexample.

8-3. *[3]* Assume that all edges in the graph have distinct edge weights (i.e., no pair of edges have the same weight). Is the path between a pair of vertices in a minimum spanning tree necessarily a shortest path between the two vertices in the full graph? Give a proof or a counterexample.

8-4. *[3]* Can Prim's and Kruskal's algorithms yield different minimum spanning trees? Explain why or why not.

8-5. *[3]* Does either Prim's or Kruskal's algorithm work if there are negative edge weights? Explain why or why not.

8-6. *[3]* (a) Assume that all edges in the graph have distinct edge weights (i.e., no pair of edges have the same weight). Is the *minimum spanning tree* of this graph unique? Give a proof or a counterexample.

(b) Again, assume that all edges in the graph have distinct edge weights (i.e. no pair of edges have the same weight). Is the *shortest-path spanning tree* of this graph unique? Give a proof or a counterexample.

8-7. *[5]* Suppose we are *given* the minimum spanning tree T of a given graph G (with n vertices and m edges) and a new edge $e = (u, v)$ of weight w that we will add to G. Give an efficient algorithm to find the minimum spanning tree of the graph $G + e$. Your algorithm should run in $O(n)$ time to receive full credit.

8-8. *[5]* (a) Let T be a minimum spanning tree of a weighted graph G. Construct a new graph G' by adding a weight of k to every edge of G. Do the edges of T form a minimum spanning tree of G'? Prove the statement or give a counterexample.

(b) Let $P = \{s, \ldots, t\}$ describe a shortest path between vertices s and t of a weighted graph G. Construct a new graph G' by adding a weight of k to every edge of G. Does P describe a shortest path from s to t in G'? Prove the statement or give a counterexample.

8-9. *[5]* Devise and analyze an algorithm that takes a weighted graph G and finds the smallest change in the cost to a non-minimum spanning tree edge that would cause a change in the minimum spanning tree of G. Your algorithm must be correct and run in polynomial time.

8-10. *[4]* Consider the problem of finding a minimum-weight connected subset T of edges from a weighted connected graph G. The weight of T is the sum of all the edge weights in T.

   (a) Why is this problem not just the minimum spanning tree problem? (Hint: think negative weight edges.)

   (b) Give an efficient algorithm to compute the minimum-weight connected subset T.

8-11. *[5]* Let $T = (V, E')$ be a minimum spanning tree of a given graph $G = (V, E)$ with positive edge weights. Now suppose the weight of a particular edge $e \in E$ is modified from $w(e)$ to a new value $\hat{w}(e)$. We seek to update the minimum spanning tree T to reflect this change without recomputing the entire tree from scratch. For each of the following four cases, give a linear-time algorithm to update the tree:

(a) $e \notin E'$ and $\hat{w}(e) > w(e)$

(b) $e \notin E'$ and $\hat{w}(e) < w(e)$

(c) $e \in E'$ and $\hat{w}(e) < w(e)$

(d) $e \in E'$ and $\hat{w}(e) > w(e)$

8-12. *[4]* Let $G = (V, E)$ be an undirected graph. A set $F \subseteq E$ of edges is called a *feedback-edge set* if every cycle of G has at least one edge in F.

(a) Suppose that G is unweighted. Design an efficient algorithm to find a minimum-size feedback-edge set.

(b) Suppose that G is a weighted undirected graph with positive edge weights. Design an efficient algorithm to find a minimum-weight feedback-edge set.

Union–Find

8-13. *[5]* Devise an efficient data structure to handle the following operations on a weighted directed graph:

(a) Merge two given components.

(b) Locate which component contains a given vertex v.

(c) Retrieve a minimum edge from a given component.

8-14. *[5]* Design a data structure that enables a sequence of m *union* and *find* operations on a universal set of n elements, consisting of a sequence of all *unions* followed by a sequence of all *finds*, to be performed in time $O(m + n)$.

Shortest Paths

8-15. *[3]* The *single-destination shortest-path* problem for a directed graph seeks the shortest path *from* every vertex to a specified vertex v. Give an efficient algorithm to solve the single-destination shortest-path problem.

8-16. *[3]* Let $G = (V, E)$ be an undirected weighted graph, and let T be the shortest-path spanning tree rooted at a vertex v. Suppose now that all the edge weights in G are increased by a constant number k. Is T still the shortest-path spanning tree from v?

8-17. *[3]* (a) Give an example of a weighted connected graph $G = (V, E)$ and a vertex v, such that the minimum spanning tree of G is the same as the shortest-path spanning tree rooted at v.

(b) Give an example of a weighted connected directed graph $G = (V, E)$ and a vertex v, such that the minimum spanning tree of G is very different from the shortest-path spanning tree rooted at v.

(c) Can the two trees be completely disjoint?

8-18. *[3]* Either prove the following or give a counterexample:

(a) Is the path between a pair of vertices in a minimum spanning tree of an undirected graph necessarily the shortest (minimum-weight) path?

(b) Suppose that the minimum spanning tree of the graph is unique. Is the path between a pair of vertices in a minimum spanning tree of an undirected graph necessarily the shortest (minimum-weight) path?

8-19. *[3]* Give an efficient algorithm to find the shortest path from x to y in an undirected weighted graph $G = (V, E)$ with positive edge weights, subject to the constraint that this path must pass through a particular vertex z.

8-20. *[5]* In certain graph problems, vertices can have weights instead of or in addition to the weights of edges. Let C_v be the cost of vertex v, and $C_{(x,y)}$ the cost of the edge (x, y). This problem is concerned with finding the cheapest path between vertices a and b in a graph G. The cost of a path is the sum of the costs of the edges and vertices encountered on the path.

  (a) Suppose that each edge in the graph has a weight of zero (while non-edges have a cost of ∞). Assume that $C_v = 1$ for all vertices $1 \le v \le n$ (i.e., all vertices have the same cost). Give an *efficient* algorithm to find the cheapest path from a to b and its time complexity.

  (b) Now suppose that the vertex costs are not constant (but are all positive) and the edge costs remain as above. Give an *efficient* algorithm to find the cheapest path from a to b and its time complexity.

  (c) Now suppose that both the edge and vertex costs are not constant (but are all positive). Give an *efficient* algorithm to find the cheapest path from a to b and its time complexity.

8-21. *[5]* Give an $O(n^3)$ algorithm that takes an n-vertex directed graph G with positive edge lengths, and returns the length of the shortest cycle in the graph. This length is ∞ in the case of an acyclic graph.

8-22. *[5]* A highway network is represented by a weighted graph G, with edges corresponding to roads and vertices corresponding to road intersections. Each road is labeled with the maximum possible height of vehicles that can pass through the road. Give an efficient algorithm to compute the maximum possible height of vehicles that can successfully travel from s to t. What is the runtime of your algorithm?

8-23. *[5]* You are given a directed graph G with possibly negative weighted edges, in which the shortest path between any two vertices is guaranteed to have at most k edges. Give an algorithm that finds the shortest path between two vertices u and v in $O(k \cdot (n + m))$ time.

8-24. *[5]* Can we solve the single-source *longest*-path problem by changing *minimum* to *maximum* in Dijkstra's algorithm? If so, then prove your algorithm correct. If not, then provide a counterexample.

8-25. *[5]* Let $G = (V, E)$ be a weighted acyclic directed graph with possibly negative edge weights. Design a linear-time algorithm to solve the single-source shortest-path problem from a given source v.

8-26. *[5]* Let $G = (V, E)$ be a directed weighted graph such that all the weights are positive. Let v and w be two vertices in G and $k \le |V|$ be an integer. Design an algorithm to find the shortest path from v to w that contains exactly k edges. Note that the path need not be simple.

8-27. *[5]* *Arbitrage* is the use of discrepancies in currency-exchange rates to make a profit. For example, there may be a small window of time during which 1 U.S. dollar buys 0.75 British pounds, 1 British pound buys 2 Australian dollars, and 1 Australian dollar buys 0.70 U.S. dollars. At such a time, a smart trader can trade one U.S. dollar and end up with $0.75 \times 2 \times 0.7 = 1.05$ U.S. dollars—a profit of 5%. Suppose that there are n currencies $c_1, ..., c_n$ and an $n \times n$ table R of exchange rates, such that one unit of currency c_i buys $R[i, j]$ units of currency c_j. Devise and analyze an algorithm to determine the maximum value of

$$R[c_1, c_{i_1}] \cdot R[c_{i_1}, c_{i_2}] \cdots R[c_{i_{k-1}}, c_{i_k}] \cdot R[c_{i_k}, c_1]$$

(Hint: think all-pairs shortest path.)

Network Flow and Matching

8-28. *[3]* A matching in a graph is a set of disjoint edges—that is, edges that do not have common vertices. Give a linear-time algorithm to find a maximum matching in a tree.

8-29. *[5]* An *edge cover* of an undirected graph $G = (V, E)$ is a set of edges such that each vertex in the graph is incident to at least one edge from the set. Give an efficient algorithm, based on matching, to find the minimum-size edge cover for G.

LeetCode

8-1. https://leetcode.com/problems/cheapest-flights-within-k-stops/

8-2. https://leetcode.com/problems/network-delay-time/

8-3. https://leetcode.com/problems/find-the-city-with-the-smallest-number-of-neighbors-at-a-threshold-distance/

HackerRank

8-1. https://www.hackerrank.com/challenges/kruskalmstrsub/

8-2. https://www.hackerrank.com/challenges/jack-goes-to-rapture/

8-3. https://www.hackerrank.com/challenges/tree-pruning/

Programming Challenges

These programming challenge problems with robot judging are available at https://onlinejudge.org:

8-1. "Freckles"—Chapter 10, problem 10034.

8-2. "Necklace"—Chapter 10, problem 10054.

8-3. "Railroads"—Chapter 10, problem 10039.

8-4. "Tourist Guide"—Chapter 10, problem 10199.

8-5. "The Grand Dinner"—Chapter 10, problem 10249.

Chapter 9

Combinatorial Search

Surprisingly large problems can be solved using exhaustive search techniques, albeit at great computational cost. But for certain applications, it may be worth it. A good example occurs in testing a circuit or a program. You can prove the correctness of the device by trying all possible inputs and verifying that they give the correct answer. Verified correctness is a property to be proud of: just claiming that it works correctly on all the inputs you tried is worth much less.

Modern computers have clock rates of a few *gigahertz*, meaning billions of operations per second. Since doing something interesting takes a few hundred instructions, you can hope to search millions of items per second on contemporary machines.

It is important to realize how big (or how small) one million is. One million permutations means all arrangements of roughly 10 objects, but not more. One million subsets means all combinations of roughly 20 items, but not more. Solving significantly larger problems requires carefully pruning the search space to ensure we look at only the elements that really matter.

This section introduces backtracking as a technique for listing all possible solutions for a combinatorial algorithm problem. I illustrate the power of clever pruning techniques to speed up real search applications. For problems that are too large to contemplate using combinatorial search, heuristic methods like simulated annealing are presented in Chapter 12. Such heuristics are important weapons in any practical algorist's arsenal.

9.1 Backtracking

Backtracking is a systematic way to run through all the possible configurations of a search space. These configurations may represent all possible arrangements of objects (permutations) or all possible ways of building a collection of them (subsets). Other common situations demand enumerating all spanning trees of a graph, all paths between two vertices, or all possible ways to partition vertices into color classes.

© The Editor(s) (if applicable) and The Author(s), under exclusive license to
Springer Nature Switzerland AG 2020
S. S. Skiena, *The Algorithm Design Manual*, Texts in Computer Science,
https://doi.org/10.1007/978-3-030-54256-6_9

What these problems have in common is that we must generate each possible configuration *exactly* once. Avoiding repetitions and missed configurations means that we must define a systematic generation order. We will model our combinatorial search solution as a vector $a = (a_1, a_2, ..., a_n)$, where each element a_i is selected from a finite ordered set S_i. Such a vector might represent an arrangement where a_i contains the ith element of the permutation. Or perhaps a is a Boolean vector representing a given subset S, where a_i is true iff the ith element of the universal set is in S. The solution vector can even represent a sequence of moves in a game or a path in a graph, where a_i contains the ith game move or graph edge in the sequence.

At each step in the backtracking algorithm, we try to extend a given partial solution $a = (a_1, a_2, ..., a_k)$ by adding another element at the end. After this extension, we must test whether what we now have is a complete solution: if so, we should print it or count it. If not, we must check whether the partial solution is still potentially extendable to some complete solution.

Backtracking constructs a tree of partial solutions, where each node represents a partial solution. There is an edge from x to y if node y was created by extending x. This tree of partial solutions provides an alternative way to think about backtracking, for the process of constructing the solutions corresponds exactly to doing a depth-first traversal of the backtrack tree. Viewing backtracking as a depth-first search on an implicit graph yields a natural recursive implementation of the basic algorithm.

> Backtrack-DFS(a, k)
>   if $a = (a_1, a_2, ..., a_k)$ is a solution, report it.
>   else
> $k = k + 1$
>   construct S_k, the set of candidates for position k of a
>   while $S_k \neq \emptyset$ do
> a_k = an element in S_k
> $S_k = S_k - \{a_k\}$
>   Backtrack-DFS(a, k)

Although a breadth-first search could also be used to enumerate solutions, a depth-first search is greatly preferred because it uses much less space. The current state of a search is completely represented by the path from the root to the current depth-first search node. This requires space proportional to the *height* of the tree. In breadth-first search, the queue stores all the nodes at the current level, which is proportional to the *width* of the search tree. For most interesting problems, the width of the tree grows exponentially with its height.

Implementation

Backtracking ensures correctness by enumerating all possibilities. It ensures efficiency by never visiting a state more than once. To help you understand how this works, my generic `backtrack` code is given below:

```
void backtrack(int a[], int k, data input) {
    int c[MAXCANDIDATES];        /* candidates for next position */
    int nc;                      /* next position candidate count */
    int i;                       /* counter */

    if (is_a_solution(a, k, input)) {
        process_solution(a, k,input);
    } else {
        k = k + 1;
        construct_candidates(a, k, input, c, &nc);
        for (i = 0; i < nc; i++) {
            a[k] = c[i];
            make_move(a, k, input);
            backtrack(a, k, input);
            unmake_move(a, k, input);

            if (finished) {
                return;          /* terminate early */
            }
        }
    }
}
```

Study how recursion yields an elegant and easy implementation of the backtracking algorithm. Because a new candidates array c is allocated with each recursive procedure call, the subsets of not-yet-considered extension candidates at each position will not interfere with each other.

The application-specific parts of this algorithm consist of five subroutines:

- is_a_solution(a,k,input) – This Boolean function tests whether the first k elements of vector a form a complete solution for the given problem. The last argument, input, allows us to pass general information into the routine. We can use it to specify n—the size of a target solution. This makes sense when constructing permutations or subsets of n elements, but other data may be relevant when constructing variable-sized objects such as sequences of moves in a game.

- construct_candidates(a,k,input,c,&nc) – This routine fills an array c with the complete set of possible candidates for the kth position of a, given the contents of the first $k - 1$ positions. The number of candidates returned in this array is denoted by nc. Again, input may be used to pass auxiliary information.

- process_solution(a,k,input) – This routine prints, counts, stores, or processes a complete solution once it is constructed.

- make_move(a,k,input) *and* unmake_move(a,k,input) – These routines enable us to modify a data structure in response to the latest move, as well as clean up this data structure if we decide to take back the move. Such a data structure can always be rebuilt from scratch using the solution vector a, but this can be inefficient when each move involves small incremental changes that can easily be undone.

These calls will function as null stubs in all of this section's examples, but will be employed in the Sudoku program of Section 9.4 (page 290).

A global `finished` flag is included to allow for premature termination, which could be set in any application-specific routine.

9.2   Examples of Backtracking

To really understand how backtracking works, you must see how such objects as permutations and subsets can be constructed by defining the right state spaces. Examples of several state spaces are described in the following subsections.

9.2.1   Constructing All Subsets

Designing an appropriate state space to represent combinatorial objects starts by counting how many objects need representing. How many subsets are there of an n-element set, say the integers $\{1, \ldots, n\}$? There are exactly two subsets for $n = 1$, namely $\{\}$ and $\{1\}$. There are four subsets for $n = 2$, and eight subsets for $n = 3$. Each new element doubles the number of possibilities, so there are 2^n subsets of n elements.

Each subset is described by the elements that are contained in it. To construct all 2^n subsets, we set up a Boolean array/vector of n cells, where the value of a_i (true or false) signifies whether the ith item is in the given subset. In the scheme of our general backtrack algorithm, $S_k = (true, false)$ and a is a solution whenever $k = n$. We can now construct all subsets with simple implementations of `is_a_solution()`, `construct_candidates()`, and `process_solution()`.

```
int is_a_solution(int a[], int k, int n) {
    return (k == n);
}

void construct_candidates(int a[], int k, int n, int c[], int *nc) {
    c[0] = true;
    c[1] = false;
    *nc = 2;
}
```

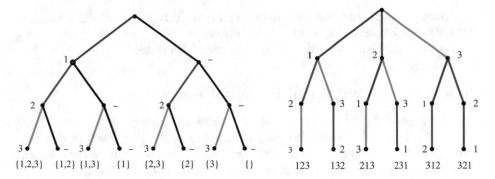

Figure 9.1: Search tree enumerating all subsets (left) and permutations (right) of $\{1, 2, 3\}$. The color of the search tree edges reflects the element being inserted into the partial solution.

```
void process_solution(int a[], int k, int input) {
    int i;     /* counter */

    printf("{");
    for (i = 1; i <= k; i++) {
        if (a[i] == true) {
            printf(" %d", i);
        }
    }

    printf(" }\n");
}
```

Ironically, printing out each subset after constructing it proves to be the most complex of these three routines!

Finally, we must instantiate the call to **backtrack** with the right arguments. Specifically, this means giving a pointer to the empty solution vector, setting $k = 0$ to denote that it is in fact empty, and specifying the number of elements in the universal set:

```
void generate_subsets(int n) {
    int a[NMAX];                    /* solution vector */

    backtrack(a, 0, n);
}
```

In what order will the subsets of $\{1, 2, 3\}$ be generated? It depends on the order of moves as returned from **construct_candidates**. Since *true* always appears before *false*, the subset of all trues is generated first, and the all-false empty set is generated last: $\{123\}$, $\{12\}$, $\{13\}$, $\{1\}$, $\{23\}$, $\{2\}$, $\{3\}$, and $\{\}$.

Trace through this example (shown in Figure 9.1 (left)) carefully to make sure you understand the backtracking procedure. The problem of generating subsets is more thoroughly discussed in Section 17.5 (page 521).

9.2.2 Constructing All Permutations

Counting permutations of $\{1, \ldots, n\}$ is a necessary prerequisite to generating them. There are n distinct choices for the value of the first element of a permutation. Once we have fixed a_1, there are $n - 1$ candidates remaining for the second position, since we can have any value except a_1 in this slot (because repetitions are forbidden in permutations). Repeating this argument yields a total of $n! = \prod_{i=1}^{n} i$ distinct permutations.

This counting argument suggests a suitable representation. Set up an array/vector a of n cells. The set of candidates for the ith position will be all elements that have not appeared in the $(i - 1)$ elements of the partial solution, corresponding to the first $i - 1$ elements of the permutation.

In the scheme of the general backtrack algorithm, $S_k = \{1, \ldots, n\} - \{a_1, \ldots, a_k\}$, and a is a solution whenever $k = n$:

```
void construct_candidates(int a[], int k, int n, int c[], int *nc) {
    int i;                   /* counter */
    bool in_perm[NMAX];      /* what is now in the permutation? */

    for (i = 1; i < NMAX; i++) {
        in_perm[i] = false;
    }

    for (i = 1; i < k;  i++) {
        in_perm[a[i]] = true;
    }

    *nc = 0;
    for (i = 1; i <= n;  i++) {
        if (!in_perm[i]) {
            c[ *nc ] = i;
            *nc = *nc + 1;
        }
    }
}
```

Testing whether i is a candidate for the kth slot in the permutation could be done by iterating through all $k - 1$ elements of a and verifying that none of them matched. However, we prefer to set up a bit-vector data structure (see Section 15.5 (page 456)) to keep track of which elements are in the partial solution. This gives a constant-time legality check.

Completing the job requires specifying `process_solution` and `is_a_solution`, as well as setting the appropriate arguments to `backtrack`. All are essentially the same as for subsets:

```c
void process_solution(int a[], int k, int input) {
    int i;     /* counter */

    for (i = 1; i <= k; i++) {
        printf(" %d", a[i]);
    }
    printf("\n");
}


int is_a_solution(int a[], int k, int n) {
    return (k == n);
}


void generate_permutations(int n) {
    int a[NMAX];                        /* solution vector */

    backtrack(a, 0, n);
}
```

As a consequence of the candidate order, these routines generate permutations in *lexicographic*, or sorted order—that is, 123, 132, 213, 231, 312, and 321, as shown in Figure 9.1 (right). The problem of generating permutations is more thoroughly discussed in Section 17.4 (page 517).

9.2.3 Constructing All Paths in a Graph

In a *simple path* no vertex appears more than once. Enumerating all the simple s to t paths in a given graph is a more complicated problem than just listing permutations or subsets. There is no explicit formula that counts solutions as a function of the number of edges or vertices, because the number of paths depends upon the structure of the graph.

The input data we must pass to `backtrack` to construct the paths consists of the input graph g, the source vertex s, and target vertex t:

```c
typedef struct {
    int s;                        /* source vertex */
    int t;                        /* destination vertex */
    graph g;                      /* graph to find paths in */
} paths_data;
```

The starting point of any path from s to t is always s. Thus, s is the only candidate for the first position and $S_1 = \{s\}$. The possible candidates for the second position are the vertices v such that (s, v) is an edge of the graph, for the path wanders from vertex to vertex using edges to define the legal steps. In general, S_{k+1} consists of the set of vertices adjacent to a_k that have not been used elsewhere in the partial solution a.

```
void construct_candidates(int a[], int k, paths_data *g, int c[],
    int *nc) {
    int i;                      /* counters */
    bool in_sol[NMAX+1];        /* what's already in the solution? */
    edgenode *p;                /* temporary pointer */
    int last;                   /* last vertex on current path */

    for (i = 1; i <= g->g.nvertices; i++) {
        in_sol[i] = false;
    }

    for (i = 0; i < k; i++) {
        in_sol[a[i]] = true;
    }

    if (k == 1) {
        c[0] = g->s;            /* always start from vertex s */
        *nc = 1;
    } else {
        *nc = 0;
        last = a[k-1];
        p = g->g.edges[last];
        while (p != NULL) {
            if (!in_sol[ p->y ]) {
                c[*nc] = p->y;
                *nc= *nc + 1;
            }
            p = p->next;
        }
    }
}
```

We report a successful path whenever $a_k = t$.

```
int is_a_solution(int a[], int k, paths_data *g) {
    return (a[k] == g->t);
}
```

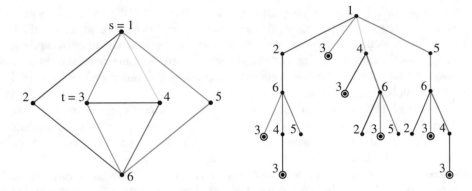

Figure 9.2: Search tree (right) enumerating all simple s–t paths in the given graph (left). The color of a search tree edge reflects the color of the corresponding graph edge.

The number of paths discovered can be counted in process_solution by incrementing a global variable solution_count. The sequence of vertices for each path is stored in the solution vector a, ready to be printed:

```
void process_solution(int a[], int k, paths_data *input) {
    int i;    /* counter */

    solution_count ++;

    printf("{");
    for (i = 1; i <= k; i++) {
        printf(" %d",a[i]);
    }
    printf(" }\n");
}
```

This solution vector must have room for all n vertices, although most paths should be shorter than this. Figure 9.2 shows the search tree giving all paths from the source vertex in a particular graph.

9.3   Search Pruning

Backtracking ensures correctness by enumerating all possibilities. A correct algorithm to find the optimal traveling salesman tour constructs all $n!$ permutations of the n vertices of graph G. For each permutation, we check whether all edges implied by the tour really exist in G and if so add the weights of these edges together. The tour with minimum weight is the solution.

However, it is wasteful to construct all the permutations first and then analyze them later. Suppose our search started from vertex v_1, and it happened that vertex-pair (v_1, v_2) was not an edge in G. The $(n-2)!$ permutations enumerated starting with (v_1, v_2) as its prefix would be a complete waste of effort. Much better would be to stop the search after $[v_1, v_2]$ and then continue from $[v_1, v_3]$. By restricting the set of next elements to reflect only legal moves with respect to the current partial configuration, we significantly reduce the total search complexity.

Pruning is the technique of abandoning a search direction the instant we can establish that a given partial solution cannot be extended into a full solution. For traveling salesman, we seek the cheapest tour that visits all vertices. Suppose that in the course of our search we find a tour t whose cost is C_t. Later, we may have a partial solution a whose edge sum $C_a \geq C_t$. Is there any reason to continue exploring this node? No, because any tour with a as its prefix will have cost greater than tour t, and hence is doomed to be non-optimal. Cutting away such failed partial tours from the search tree as soon as possible can have an enormous impact on running time.

Exploiting symmetry is another avenue for reducing combinatorial search. Pruning away partial solutions equivalent to those previously considered requires recognizing underlying symmetries in the search space. For example, consider the state of our TSP search after we have tried all partial positions beginning with v_1. Does it pay to continue the search with partial solutions beginning with v_2? No. Any tour starting and ending at v_2 can be viewed as a rotation of one starting and ending at v_1, for TSP tours are closed cycles. There are thus only $(n-1)!$ distinct tours on n vertices, not $n!$. By restricting the first element of the tour to v_1, we save a factor of n in time without missing any interesting solutions. Detecting such symmetries can be subtle, but once identified they can usually be easily exploited.

Take-Home Lesson: Combinatorial search, when augmented with tree-pruning techniques, can be used to find the optimal solution for small optimization problems. How small depends upon the specific problem, but typical size limits are somewhere between twenty and a hundred items.

9.4 Sudoku

A Sudoku craze has swept the world. Many newspapers publish daily Sudoku puzzles, and millions of books about Sudoku have been sold. British Airways sent a formal memo forbidding its cabin crews from doing Sudoku during takeoffs and landings. Indeed, I have noticed plenty of Sudoku going on in the back of my algorithms classes during lecture.

What is Sudoku? In its most common form, it consists of a 9×9 grid filled with blanks and the digits 1 to 9. The puzzle is completed when every row, column, and sector (3×3 subproblems corresponding to the nine sectors of a tic-tac-toe puzzle) contain the digits 1 through 9 with no omissions or repetition. Figure 9.3 presents a challenging Sudoku puzzle and its solution.

Backtracking lends itself nicely to the task of solving Sudoku puzzles. We will

							1	2
				3	5			
			6				7	
7						3		
			4			8		
1								
			1	2				
	8						4	
	5					6		

6	7	3	8	9	4	5	1	2
9	1	2	7	3	5	4	8	6
8	4	5	6	1	2	9	7	3
7	9	8	2	6	1	3	5	4
5	2	6	4	7	3	8	9	1
1	3	4	5	8	9	2	6	7
4	6	9	1	2	8	7	3	5
2	8	7	3	5	6	1	4	9
3	5	1	9	4	7	6	2	8

Figure 9.3: A challenging Sudoku puzzle (left) with its completed solution (right).

use Sudoku here to illustrate pruning techniques for combinatorial search. Our state space will be the collection of open squares, each of which must ultimately be filled in with a digit. The candidates for open squares (i, j) are exactly the integers from 1 to 9 that have not yet appeared in row i, column j, or the 3×3 sector containing (i, j). We backtrack as soon as we are out of candidates for a square.

The solution vector `a` supported by `backtrack` only accepts a single integer per position. This is enough to store the contents of a square (1–9) but not the coordinates of the square. Thus, we keep a separate array of `move` positions as part of our `boardtype` data type provided below. The basic data structures we need to support our solution are:

```
#define DIMENSION   9                         /* 9*9 board */
#define NCELLS      DIMENSION*DIMENSION       /* 81 cells in 9-by-9-board */
#define MAXCANDIDATES   DIMENSION+1           /* max digit choices per cell */

bool finished = false;

typedef struct {
    int x, y;       /* row and column coordinates of square */
} point;

typedef struct {
   int m[DIMENSION+1][DIMENSION+1];       /* board contents */
   int freecount;                         /* open square count */
   point move[NCELLS+1];                  /* which cells have we filled? */
} boardtype;
```

Constructing the move candidates for the next position requires first picking which open square we want to fill next (next_square), and then identifying which digits are candidates to fill that square (possible_values). These routines are basically bookkeeping, although the details of how they work can have a substantial impact on performance.

```c
void construct_candidates(int a[], int k, boardtype *board, int c[],
    int *nc) {
    int i;                          /* counter */
    bool possible[DIMENSION+1]; /* which digits fit in this square */

    next_square(&(board->move[k]), board); /* pick square to fill next */

    *nc = 0;

    if ((board->move[k].x < 0) && (board->move[k].y < 0)) {
        return;   /* error condition, no moves possible */
    }

    possible_values(board->move[k], board, possible);
    for (i = 1; i <= DIMENSION; i++) {
        if (possible[i]) {
            c[*nc] = i;
            *nc = *nc + 1;
        }
    }
}
```

We must update our board data structure to reflect the effect of putting a candidate value into a square, as well as remove these changes should we backtrack from this position. These updates are handled by make_move and unmake_move, both of which are called directly from backtrack:

```c
void make_move(int a[], int k, boardtype *board) {
    fill_square(board->move[k], a[k], board);
}

void unmake_move(int a[], int k, boardtype *board) {
    free_square(board->move[k], board);
}
```

One important job for these board update routines is maintaining how many free squares remain on the board. A solution is found when there are no more free squares remaining to be filled. Here, steps is a global variable recording the complexity of our search for Table 9.4:

```
bool is_a_solution(int a[], int k, boardtype *board) {
    steps = steps + 1;              /* count steps for results table */

    return (board->freecount == 0);
}
```

We print the configuration and then turn off the backtrack search after finding a solution by setting the global **finished** flag. This can be done without consequence because "official" Sudoku puzzles are allowed to have only one solution. But there can be non-official Sudoku puzzles with enormous numbers of solutions. The empty puzzle, where initially no digits are specified anywhere, can be filled in exactly 6,670,903,752,021,072,936,960 ways. We ensure we don't see all of them by turning off the search:

```
void process_solution(int a[], int k, boardtype *board) {
    finished = true;
    printf("process solution\n");
    print_board(board);
}
```

This completes the program modulo details of identifying the next open square to fill (**next_square**) and identifying the candidates that might fill it (**possible_values**). Two natural heuristics to select the next square are:

- *Arbitrary square selection* – Pick the first open square we encounter, be it the first, the last, or a random open square. All are equivalent in that there seems to be no reason to believe that one variant will perform better than the others.

- *Most constrained square selection* – Here, we check each open square (i, j) to see how many digits remain possible candidates to fill it—that is, digits that have not already been used in row i, column j, or the sector containing (i, j). We pick the square with the smallest number of candidates.

Although both possibilities work correctly, the second option is much, much better. If there are open squares with only one remaining candidate, the choice is forced. We might as well fill them first, especially since pinning these squares down will help trim the possibilities for other open squares. Of course, we will spend more time selecting each candidate square, but if the puzzle is easy enough we may never have to backtrack at all.

If the most constrained square has two possibilities, we have a 50% chance of guessing right the first time, as opposed to a probability of 1/9 for a completely unconstrained square. Reducing our average number of choices from (say) three per square to two per square is an enormous win, because it multiplies with each position. If we have (say) twenty positions to fill, we must enumerate only $2^{20} = 1,048,576$ solutions. A branching factor of 3 at each of twenty positions requires over 3,000 times as much work!

Pruning condition		Puzzle complexity		
next_square	possible_values	Easy	Medium	Hard
arbitrary	local count	1,904,832	863,305	never finished
arbitrary	look ahead	127	142	12,507,212
most constrained	local count	48	84	1,243,838
most constrained	look ahead	48	65	10,374

Figure 9.4: Sudoku run times (in number of steps) for different pruning strategies.

Our final decision concerns the possible_values we allow for each square. We have two possibilities:

- *Local count* – Our backtrack search works correctly if the routine that generates candidates for board position (i, j) (possible_values) does the obvious thing and allows all digits 1 to 9 that have not appeared in the given row, column, or sector.

- *Look ahead* – But what if our current partial solution has some *other* open square where there are no candidates remaining under the local count criteria? There is no possible way to complete this partial solution into a full Sudoku grid. Thus, there *really* are zero possible moves to consider for (i, j) because of what is happening elsewhere on the board!

  We will discover this obstruction eventually, when we pick this square for expansion, discover it has no moves, and then have to backtrack. But why wait, since all our efforts until then will be wasted? We are *much* better off backtracking immediately and moving on.[1]

Successful pruning requires looking ahead to see when a partial solution is doomed to go nowhere, and backing off as soon as possible.

Figure 9.4 presents the number of calls to is_a_solution for all four backtracking variants on three Sudoku instances of varying complexity:

- The *Easy* board was intended to be easy for a human player. Indeed, my program solved it without any backtracking steps when the most constrained square was selected as the next position.

- The *Medium* board stumped all the contestants at the finals of the World Sudoku Championship in March 2006. But the decent search variants here required only a few backtrack steps to dispatch this problem.

---

[1] This look-ahead condition might have naturally followed from the most-constrained square selection, had it been permitted to select squares with no moves. However, my implementation credited squares that already contained digits as having no moves, thus limiting the next square choices to squares with at least one move.

Figure 9.5: Configurations covering 63 but not 64 squares.

- The *Hard* problem is the board displayed in Figure 9.3, which initially contains only 17 filled squares. This is the fewest specified number of positions of any problem instance known to have a unique solution.

What is considered to be a "hard" problem instance depends upon the given heuristic. Some people find math/theory harder than programming, but others think differently. Heuristic A may well think instance I_1 is easier than I_2, while heuristic B ranks them in the other order.

What can we learn from these experiments? Looking ahead to eliminate dead positions as soon as possible is the best way to prune a search. Without this operation, we could not finish the hardest puzzle and took thousands of times longer on the easier ones than we should have.

Smart square selection had a similar impact, even though it nominally just rearranges the order in which we do the work. But doing more constrained positions first is tantamount to reducing the out-degree of each node in the tree, and each additional position we fix adds constraints that help lower the degree of subsequent selections.

It took the better part of an hour (48:44) to solve the puzzle in Figure 9.3 when I selected an arbitrary square for my next move. Sure, my program was faster in most instances, but Sudoku puzzles are designed to be solved by people using pencils in much less time than this. Making the next move in the most constrained square reduced search time by a factor of over 1,200. Each puzzle we tried can now be solved in seconds—the time it takes to reach for the pencil if you prefer to do it by hand.

This is the power of search pruning. Even simple pruning strategies can suffice to reduce running times from impossible to instantaneous.

9.5 War Story: Covering Chessboards

Every researcher dreams of solving a classical problem—one that has remained open and unsolved for over a century. There is something romantic about communicating across the generations, being part of the evolution of science, and helping to climb another rung up the ladder of human progress. There is also a

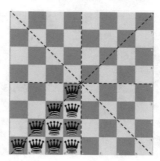

Figure 9.6: The ten unique positions for the queen, with respect to rotational and reflective symmetry.

pleasant sense of smugness that comes from figuring out how to do something that nobody could do before you.

There are several possible reasons why a problem might stay open for such a long period of time. Perhaps it is so difficult and profound as to require a uniquely powerful intellect to solve. A second reason is technological—the ideas or techniques required to solve the problem may not have existed when it was first posed. The final possibility is that no one may have cared about the problem enough in the interim to seriously bother with it. Once, I helped solve a problem that had been open for over a hundred years. Decide for yourself which reason best explains why.

Chess is a game that has fascinated people for thousands of years. In addition, it has inspired many combinatorial problems of independent interest. The combinatorial explosion was first recognized with the legend that the inventor of chess demanded payment of one grain of rice for the first square of the board, and twice as much for the $(i + 1)$st square than the ith square. The king was astonished to learn he had to cough up $\sum_{i=1}^{64} 2^{i-1} = 2^{64} - 1 = 18{,}446{,}744{,}073{,}709{,}551{,}615$ grains of rice. In beheading the inventor, the wise king first established pruning as a technique for dealing with the combinatorial explosion.

In 1849, Kling posed the question of whether all 64 squares on the board could be simultaneously threatened by an arrangement of the eight main pieces on the chess board—the king, queen, two knights, two rooks, and two bishops on oppositely colored squares. Pieces do not threaten the square they sit on. Configurations that simultaneously threaten 63 squares, such as those in Figure 9.5, have been long known, but whether this was the best possible remained an open problem. This problem seemed ripe for solution by exhaustive combinatorial searching, although whether it was solvable depended upon the size of the search space.

How many ways can the eight main chess pieces be positioned on a chessboard? The trivial bound is $64!/(64 - 8)! = 178{,}462{,}987{,}637{,}760 \approx 2 \times 10^{14}$ positions. That's far too many: anything larger than about 10^9 positions would be unreasonable to search on a modest computer in a modest amount of time.

Getting the job done would require significant pruning. Our first idea was to remove symmetries. Accounting for orthogonal and diagonal symmetries left only ten distinct positions for the queen, as shown in Figure 9.6.

Once the queen is placed, there remain $32 \cdot 31$ distinct positions for the bishops, then $61 \cdot 60/2$ for the rooks, $59 \cdot 58/2$ for the knights, and 57 remaining for the king. Such an exhaustive search would test $1{,}770466{,}147{,}200 \approx 1.8 \cdot 10^{12}$ distinct positions—still much too large to try.

We could use backtracking to construct all possible chess boards, but we had to find a way to prune the search space significantly. To prune the search we needed a quick way to detect when there was no way to complete a partially filled-in position to cover all 64 squares. Suppose we had already placed seven pieces on the board, and together they covered all but 10 squares of the board. Say the remaining piece was the king. Can there possibly be a way to place the king so that all squares are threatened? The answer must be no, because the king can threaten at most 8 squares according to the rules of chess. There can be no reason to test any king position. We might win big pruning away such partial configurations.

This pruning strategy required carefully ordering the evaluation of the pieces. Each piece can threaten a certain maximum number of squares: the queen 27, the king/knight 8, the rook 14, and the bishop 13. We would want to insert the pieces in decreasing order of mobility: Q, R_1, R_2, B_1, B_2, K, N_1, N_2. We can prune whenever the number of unthreatened squares exceeds the sum of the maximum coverage of the unplaced pieces. This sum is minimized by using the decreasing order of mobility.

When we implemented a backtrack search using this pruning strategy, it eliminated over 95% of the search space. After optimizing our move generation, our program could search over 1,000 positions per second on a machine of its day. But this was still too slow, for $10^{11}/10^3 = 10^8$ seconds meant 1,000 days! Although we might further tweak the program to speed it up by an order of magnitude or so, what we really needed was to find a way to prune more nodes.

Effective pruning means eliminating large numbers of positions at a single stroke. Our previous attempt was too weak. What if instead of placing up to eight pieces on the board simultaneously, we placed *more* than eight pieces. Obviously, the more pieces we placed simultaneously, the more likely they would threaten all 64 squares. But *if* they didn't cover, no subset of eight distinct pieces from the set could possibly threaten all squares. The potential existed to eliminate a vast number of positions by pruning a single node.

So in our final version, the nodes of our search tree corresponded to chessboards that could have any number of pieces, and more than one piece on a square. For a given board, we distinguished *strong* and *weak* attacks on a square. A strong attack corresponds to the usual notion of a threat in chess. A weak attack ignores any possible blocking effects of intervening pieces. All 64 squares can be weakly attacked with eight pieces, as shown in Figure 9.7.

Our algorithm consisted of two passes. The first pass listed boards where every square was weakly attacked. The second pass filtered the list by considering blocking pieces. A weak attack is much faster to compute (no blocking

Figure 9.7: Weakly covering 64 squares.

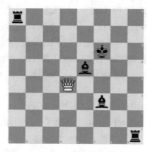

Figure 9.8: Seven pieces suffice when superimposing queen and knight (shown as a white queen).

to worry about), and any strong attack set is always a subset of a weak attack set. The position could be pruned whenever there was a non-weakly threatened square.

This program was efficient enough to complete the search in under a day. It did not find a single position covering all 64 squares with the bishops on opposite colored squares. However, our program showed that it is possible to cover the board with *seven* pieces provided a queen and a knight can occupy the same square, as shown in Figure 9.8.

Take-Home Lesson: Clever pruning can make short work of surprisingly hard combinatorial search problems. Proper pruning will have a greater impact on search time than other factors like data structures or programming language.

9.6   Best-First Search

An important idea to speed up search is to explore your best options before the less promising choices. In the `backtrack` implementation presented above, the search order was determined by the sequence of elements generated by the `construct_candidates` routine. Items near the front of the candidates array were tried before those further back. A good candidate ordering can have a very powerful effect on the time to solve the problem.

The examples so far in this chapter have focused on *existential* search problems, where we look for a single solution (or all solutions) satisfying a given set of constraints. *Optimization problems* seek the solution with the lowest or highest value of some objective function. A simple strategy to deal with optimization problems is to construct all possible solutions, and then report the one that scores best by the optimization criterion. But this can be expensive. Much better would be to generate solutions in order from best to worst, and report the best as soon as we can prove it is the best.

Best-first search, also called *branch and bound*, assigns a cost to every partial solution we have generated. We use a priority queue (named q below) to keep track of these partial solutions by cost, so the most promising partial solution can be easily identified and expanded. As in backtracking, we explore the next partial solution by testing if it is_a_solution and calling process_solution if it is. We identify all ways to expand this partial solution by calling construct_candidates, each of which gets inserted into the priority queue with its associated cost. A generic best-first search, which we apply to the traveling salesman problem (TSP), is implemented as follows:

```
void branch_and_bound (tsp_solution *s, tsp_instance *t) {
    int c[MAXCANDIDATES];      /* candidates for next position */
    int nc;                    /* next position candidate count */
    int i;                     /* counter */

    first_solution(&best_solution,t);
    best_cost = solution_cost(&best_solution, t);
    initialize_solution(s,t);
    extend_solution(s,t,1);
    pq_init(&q);
    pq_insert(&q,s);

    while (top_pq(&q).cost < best_cost) {
        *s = extract_min(&q);
        if (is_a_solution(s, s->n, t)) {
            process_solution(s, s->n, t);
        }
        else {
            construct_candidates(s, (s->n)+1, t, c, &nc);
            for (i=0; i<nc; i++) {
                extend_solution(s,t,c[i]);
                pq_insert(&q,s);
                contract_solution(s,t);
            }
        }
    }
}
```

The `extend_solution` and `contract_solution` routines handle the book-keeping of creating and pricing the partial solutions associated with each new candidate:

```
void extend_solution(tsp_solution *s, tsp_instance *t, int v) {
    s->n++;
    s->p[s->n] = v;
    s->cost = partial_solution_lb(s,t);
}

void contract_solution(tsp_solution *s, tsp_instance *t) {
    s->n--;
    s->cost = partial_solution_lb(s,t);
}
```

What should be the cost of a partial solution? There are $(n-1)!$ circular permutations on n points, so we can represent each tour as an n-element permutation starting with 1 so there are no repetitions. Partial solutions construct a prefix of the tour starting with vertex v_1, so a natural cost function might be the sum of the edge weights on this prefix source. An interesting property of such a cost function is that it serves as a *lower bound* on the cost of any expanded tour, assuming that all edge weights are positive.

But does the first full solution from a best-first search have to be an optimal solution? No, not necessarily. There was certainly no cheaper partial solution available when we pulled it off the priority queue. But extending this partial solution came with a cost, that of the next edge we added to this tour. It is certainly possible that a slightly more costly partial tour might be finishable using a less-expensive next edge, thus producing a better solution.

Thus, to get the global optimal, we must continue to explore the partial solutions coming off the priority queue until they are more expensive than the best solution we already know about. Note that this requires that the cost function for partial solutions be a lower bound on the cost of an optimal solution. Otherwise, there might be something deeper in the queue that would expand to a better solution. That would leave us with no choice but to expand everything on the priority queue completely to be sure we found the right solution.

9.7   The A* Heuristic

Best-first search can take a while, even if our partial cost function is a lower bound on the optimal tour, so we can stop as soon as we have a solution cheaper than the best unexplored partial solution. Consider the partial solutions we will encounter on a search for the optimal traveling salesman tour. Costs increase with the number of edges in the partial solution, so partial solutions with few nodes will always look more promising than longer ones nearer to completion. Even the most awful prefix path on $n/2$ nodes will likely be cheaper than the

	Backtracking			Branch and Bound	
n	all	cost < best	lb < best	cost < best	lb < best
5	24	22	17	11	7
6	120	86	62	28	20
7	720	217	153	51	42
8	5,040	669	443	111	85
9	40,320	2,509	1,619	354	264
10	362,880	5,042	3,025	655	475
11	3,628,800	12,695	6,391	848	705

Figure 9.9: Number of complete TSP solutions evaluated by different search variants. The A* heuristic employed with branch and bound did best, substantially better than backtracking.

optimal solution on all n nodes, meaning that we must expand all partial solutions until their prefix cost is greater than the cost of the best full tour. This will be horribly expensive to work through.

The *A* heuristic* (pronounced "A-star") is an elaboration on the branch-and-bound search presented above, where at each iteration we expanded the best (cheapest) partial solution that we have found so far. The idea is to use a lower bound on the cost of all possible partial solution extensions that is stronger than just the cost of the current partial tour. This will make promising partial solutions look more interesting than those that have the fewest vertices.

How can we lower bound the cost of the full tour, which contains n edges, from a partial solution with k vertices (and thus $k - 1$ edges)? We know it will eventually get $n - k + 1$ additional edges. If `minlb` is a lower bound on the cost of any edge, specifically the distance between the two closest points, adding $(n - k + 1) \times$ `minlb` gives a cost lower bound that is much more realistic for the partial solution:

```
double partial_solution_cost(tsp_solution *s, tsp_instance *t) {
    int i;                 /* counter */
    double cost = 0.0;     /* cost of solution */

    for (i = 1; i < (s->n); i++) {
        cost = cost + distance(s, i, i + 1, t);
    }

    return(cost);
}

double partial_solution_lb(tsp_solution *s, tsp_instance *t) {
    return(partial_solution_cost(s,t) + (t->n - s->n + 1) * minlb);
}
```

Figure 9.9 presents the number of full solution cost evaluations in finding the optimal TSP tour for several search variants. Brute-force backtracking without pruning requires $(n-1)!$ such calls, but we do much better when we prune on partial costs—and even better when we prune using the full lower bound. But branch and bound and A* do even better here.

Note that the number of full solutions encountered is a gross underestimate of the total work done on the search, which includes even partial solutions that got pruned just one move before the end of the tour. But Figure 9.9 does capture the fact that best-first search might have to look at a substantially smaller part of the search tree than backtracking, even with the same pruning criteria.

Best-first search is sort of like breadth-first search. A disadvantage of BFS over DFS is the space required. A backtracking/DFS tree uses memory proportional to the height of the tree, but a best-first/BFS tree requires maintaining all partial solutions, more akin to the width of the tree.

The resulting size of the priority queue for best-first search is a real problem. Consider the TSP experiments above. For $n = 11$, the queue size got to 202,063 compared to a stack size of just 11 for backtracking. Space will kill you quicker than time. To get an answer from a slow program you just have to be patient enough, but a program that crashes because of lack of memory will not give an answer no matter how long you wait.

> *Take-Home Lesson:*   The promise of a given partial solution is not just its cost, but also includes the potential cost of the remainder of the solution. A tight solution cost estimate which is still a lower bound makes best-first search much more efficient.

The A* heuristic proves useful in a variety of different problems, most notably finding shortest paths from s to t in a graph. Recall that Dijkstra's algorithm for shortest path starts from s and with each iteration adds a new vertex to which it knows the shortest path. When the graph describes a road network on the surface of the earth, this known region should expand like a growing disk around s.

But that means that half the growth is in a direction away from t, thus moving farther from the goal. A best-first search, with the as-the-crow-flies straight line distance from each in-tree vertex v to t added to the in-tree distance from s to v, gives a lower bound on the driving distance from s to t, favoring growth in the right direction. The existence of such heuristics for shortest path computations explains how online mapping services can supply you with the route home so quickly.

Chapter Notes

My treatment of backtracking here is partially based on my book *Programming Challenges* [SR03]. In particular, the `backtrack` routine presented here is a generalization of the version in chapter 8 of [SR03]. Look there for my solution to the famous *eight queens problem*, which seeks all chessboard configurations of eight mutually non-attacking queens on an 8×8 board.

More details on our combinatorial search for optimal chessboard-covering positions appear in Robison et al. [RHS89].

9.8 Exercises

Permutations

9-1. *[3]* A *derangement* is a permutation p of $\{1, \ldots, n\}$ such that no item is in its proper position, that is, $p_i \neq i$ for all $1 \leq i \leq n$. Write an efficient backtracking program with pruning that constructs all the derangements of n items.

9-2. *[4]* *Multisets* are allowed to have repeated elements. A multiset of n items may thus have fewer than $n!$ distinct permutations. For example, $\{1, 1, 2, 2\}$ has only six distinct permutations: $[1, 1, 2, 2]$, $[1, 2, 1, 2]$, $[1, 2, 2, 1]$, $[2, 1, 1, 2]$, $[2, 1, 2, 1]$, and $[2, 2, 1, 1]$. Design and implement an efficient algorithm for constructing all permutations of a multiset.

9-3. *[5]* For a given a positive integer n, find all permutations of the $2n$ elements of the multiset $S = \{1, 1, 2, 2, 3, 3, \ldots, n, n\}$ such that for each integer from 1 to n the number of intervening elements between its two appearances is equal to value of the element. For example, when $n = 3$ the two possible solutions are $[3, 1, 2, 1, 3, 2]$ and $[2, 3, 1, 2, 1, 3]$.

9-4. *[5]* Design and implement an algorithm for testing whether two graphs are isomorphic. The graph isomorphism problem is discussed in Section 19.9 (page 610). With proper pruning, graphs on hundreds of vertices can be tested in a reasonable time.

9-5. *[5]* The set $\{1, 2, 3, \ldots, n\}$ contains a total of $n!$ distinct permutations. By listing and labeling all of the permutations in ascending lexicographic order, we get the following sequence for $n = 3$:

$$[123, 132, 213, 231, 312, 321]$$

Give an efficient algorithm that returns the kth of $n!$ permutations in this sequence, for inputs n and k. For efficiency it should *not* construct the first $k - 1$ permutations in the process.

Backtracking

9-6. *[5]* Generate all structurally distinct binary search trees that store values $1 \ldots n$, for a given value of n.

9-7. *[5]* Implement an algorithm to print all valid (meaning properly opened and closed) sequences of n pairs of parentheses.

9-8. *[5]* Generate all possible topological orderings of a given DAG.

9-9. *[5]* Given a specified total t and a multiset S of n integers, find all distinct subsets from S whose elements add up to t. For example, if $t = 4$ and $S = \{4, 3, 2, 2, 1, 1\}$, then there are four different sums that equal t: 4, $3 + 1$, $2 + 2$, and $2 + 1 + 1$. A number can be used within a sum up to the number of times it appears in S, and a single number counts as a sum.

9-10. *[8]* Design and implement an algorithm for solving the subgraph isomorphism problem—given graphs G and H, does there exist a subgraph H' of H such that G is isomorphic to H'? Report how your program performs on such special cases of subgraph isomorphism as Hamiltonian cycle, clique, independent set, and graph isomorphism.

9-11. *[5]* A *team assignment* of $n = 2k$ players is a partitioning of them into two teams with exactly k people per team. For example, if the players are named $\{A, B, C, D\}$, there are three distinct ways to partition them into two equal teams: $\{\{A, B\}, \{C, D\}\}$, $\{\{A, C\}, \{B, D\}\}$, and $\{\{A, D\}, \{B, C\}\}$. (a) List the 10 possible team assignments for $n = 6$ players. (b) Give an efficient backtracking algorithm to construct all possible team assignments. Be sure to avoid repeating any solution.

9-12. *[5]* Given an alphabet Σ, a set of forbidden strings S, and a target length n, give an algorithm to construct a string of length n on Σ without any element of S as a substring. For $\Sigma = \{0, 1\}$, $S = \{01, 10\}$, and $n = 4$, the two possible solutions are 0000 and 1111. For $S = \{0, 11\}$ and $n = 4$, no such string exists.

9-13. *[5]* In the k-partition problem, we need to partition a multiset of positive integers into k disjoint subsets that have equal sum. Design and implement an algorithm for solving the k-partition problem.

9-14. *[5]* You are given a weighted directed graph G with n vertices and m edges. The *mean weight* of a cycle is the sum of its edge weights divided by the number of its edges. Find a cycle in G of minimum mean weight.

9-15. *[8]* In the turnpike reconstruction problem, you are given a multiset D of $n(n - 1)/2$ distances. The problem is to place n points on the line such that their pairwise distances are D. For example, the distances $D = \{1, 2, 3, 4, 5, 6\}$ can be obtained by placing the second point 1 unit from the first, the third point 3 from the second, and the fourth point 2 from the third. Design and implement an efficient algorithm to find all solutions to the turnpike reconstruction problem. Exploit additive constraints when possible to accelerate the search. With proper pruning, problems with hundreds of points can be solved in reasonable time.

Games and Puzzles

9-16. *[5]* Anagrams are rearrangements of the letters of a word or phrase into a different word or phrase. Sometimes the results are quite striking. For example, "MANY VOTED BUSH RETIRED" is an anagram of "TUESDAY NOVEMBER THIRD," which correctly predicted the result of the 1992 US presidential election. Design and implement an algorithm for finding anagrams using combinatorial search and a dictionary.

9-17. *[5]* Construct all sequences of moves that a knight on an $n \times n$ chessboard can make where the knight visits every square only once.

9-18. *[5]* A Boggle board is an $n \times m$ grid of characters. For a given board, we seek to find all possible words that can be formed by a sequence of adjacent characters on the board, without repetition. For example, the board:

e	t	h	t
n	d	t	i
a	i	h	n
r	h	u	b

contains words like *tide*, *dent*, *raid*, and *hide*. Design an algorithm to construct the most words for a given board B consistent with a dictionary D.

9-19. *[5]* A Babbage square is a grid of words that reads the same across as it does down. Given a k-letter word w and a dictionary of n words, find all Babbage squares starting with that word. For example, two squares for the word *hair* are:

h	a	i	r	h	a	i	r
a	i	d	e	a	l	t	o
i	d	l	e	i	t	e	m
r	e	e	f	r	o	m	b

9-20. *[5]* Show that you can solve any given Sudoku puzzle by finding the minimum vertex coloring of a specific, appropriately constructed $9 \times 9 + 9$ vertex graph.

Combinatorial Optimization

For problems 9-21 to 9-27, implement a combinatorial search program to solve it for small instances. How well does your program perform in practice?

9-21. *[5]* Design and implement an algorithm for solving the bandwidth minimization problem discussed in Section 16.2 (page 470).

9-22. *[5]* Design and implement an algorithm for solving the maximum satisfiability problem discussed in Section 17.10 (page 537).

9-23. *[5]* Design and implement an algorithm for solving the maximum clique problem discussed in Section 19.1 (page 586).

9-24. *[5]* Design and implement an algorithm for solving the minimum vertex coloring problem discussed in Section 19.7 (page 604).

9-25. *[5]* Design and implement an algorithm for solving the minimum edge coloring problem discussed in Section 19.8 (page 608).

9-26. *[5]* Design and implement an algorithm for solving the minimum feedback vertex set problem discussed in Section 19.11 (page 618).

9-27. *[5]* Design and implement an algorithm for solving the set cover problem discussed in Section 21.1 (page 678).

Interview Problems

9-28. *[4]* Write a function to find all permutations of the letters in a given string.

9-29. *[4]* Implement an efficient algorithm for listing all k-element subsets of n items.

9-30. *[5]* An anagram is a rearrangement of the letters in a given string into a sequence of dictionary words, like *Steven Skiena* into *Vainest Knees*. Propose an algorithm to construct all the anagrams of a given string.

9-31. *[5]* Telephone keypads have letters on each numerical key. Write a program that generates all possible words resulting from translating a given digit sequence (e.g. 145345) into letters.

9-32. *[7]* You start with an empty room and a group of n people waiting outside. At each step, you may either admit one person into the room, or let one out. Can you arrange a sequence of 2^n steps, so that every possible combination of people is achieved exactly once?

9-33. *[4]* Use a random number generator (rng04) that generates numbers from $\{0, 1, 2, 3, 4\}$ with equal probability to write a random number generator that generates numbers from 0 to 7 (rng07) with equal probability. What is the expected number of calls to rng04 per call of rng07?

LeetCode

9-1. `https://leetcode.com/problems/subsets/`

9-2. `https://leetcode.com/problems/remove-invalid-parentheses/`

9-3. `https://leetcode.com/problems/word-search/`

HackerRank

9-1. `https://www.hackerrank.com/challenges/sudoku/`

9-2. `https://www.hackerrank.com/challenges/crossword-puzzle/`

Programming Challenges

These programming challenge problems with robot judging are available at `https://onlinejudge.org`:

9-1. "Little Bishops"—Chapter 8, problem 861.

9-2. "15-Puzzle Problem"—Chapter 8, problem 10181.

9-3. "Tug of War"—Chapter 8, problem 10032.

9-4. "Color Hash"—Chapter 8, problem 704.

Chapter 10

Dynamic Programming

The most challenging algorithmic problems involve optimization, where we seek to find a solution that maximizes or minimizes an objective function. Traveling salesman is a classic optimization problem, where we seek the tour visiting all vertices of a graph at minimum total cost. But as shown in Chapter 1, it is easy to propose TSP "algorithms" that generate reasonable-looking solutions but do not *always* produce the minimum cost tour.

Algorithms for optimization problems require proof that they *always* return the best possible solution. Greedy algorithms that make the best local decision at each step are typically efficient, but usually do not guarantee global optimality. Exhaustive search algorithms that try all possibilities and select the best always produce the optimum result, but usually at a prohibitive cost in terms of time complexity.

Dynamic programming combines the best of both worlds. It gives us a way to design custom algorithms that systematically search all possibilities (thus guaranteeing correctness) while storing intermediate results to avoid recomputing (thus providing efficiency). By storing the *consequences* of all possible decisions and using this information in a systematic way, the total amount of work is minimized.

After you understand it, dynamic programming is probably the easiest algorithm design technique to apply in practice. In fact, I find that dynamic programming algorithms are often easier to reinvent than to try to look up. That said, *until* you understand dynamic programming, it seems like magic. You have to figure out the trick before you can use it.

Dynamic programming is a technique for efficiently implementing a recursive algorithm by storing partial results. It requires seeing that a naive recursive algorithm computes the same subproblems over and over and over again. In such a situation, storing the answer for each subproblem in a table to look up instead of recompute can lead to an efficient algorithm. Dynamic programming starts with a recursive algorithm or definition. Only after we have a correct recursive algorithm can we worry about speeding it up by using a results matrix.

Dynamic programming is generally the right method for optimization prob-

S. S. Skiena, *The Algorithm Design Manual*, Texts in Computer Science,
https://doi.org/10.1007/978-3-030-54256-6_10

lems on combinatorial objects that have an inherent *left-to-right* order among components. Left-to-right objects include character strings, rooted trees, polygons, and integer sequences. Dynamic programming is best learned by carefully studying examples until things start to click. I present several war stories where dynamic programming played the decisive role to demonstrate its utility in practice.

10.1　Caching vs. Computation

Dynamic programming is essentially a tradeoff of space for time. Repeatedly computing a given quantity can become a drag on performance. If so, we are better off storing the results of the initial computation and looking them up instead of recomputing them.

The tradeoff between space and time exploited in dynamic programming is best illustrated when evaluating recurrence relations such as the Fibonacci numbers. We look at three different programs for computing them below.

10.1.1　Fibonacci Numbers by Recursion

The Fibonacci numbers were defined by the Italian mathematician Fibonacci in the thirteenth century to model the growth of rabbit populations. Rabbits breed, well, like rabbits. Fibonacci surmised that the number of pairs of rabbits born in a given month is equal to the number of pairs of rabbits born in each of the two previous months, starting from one pair of rabbits at the start. Thus, the number of rabbits born in the nth month is defined by the recurrence relation:

$$F_n = F_{n-1} + F_{n-2}$$

with basis cases $F_0 = 0$ and $F_1 = 1$. Thus, $F_2 = 1$, $F_3 = 2$, and the series continues $3, 5, 8, 13, 21, 34, 55, 89, 144, \ldots$. As it turns out, Fibonacci's formula didn't do a great job of counting rabbits, but it does have a host of interesting properties and applications.

That they are defined by a recursive formula makes it easy to write a recursive program to compute the nth Fibonacci number. A recursive function written in C looks like this:

```c
long fib_r(int n) {
    if (n == 0) {
        return(0);
    }

    if (n == 1) {
        return(1);
    }

    return(fib_r(n-1) + fib_r(n-2));
}
```

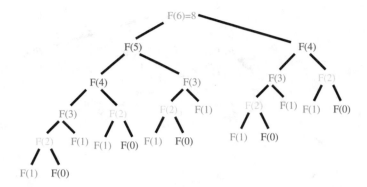

Figure 10.1: The recursion tree for computing Fibonacci numbers.

The course of execution for this recursive algorithm is illustrated by its *recursion tree*, as illustrated in Figure 10.1. This tree is evaluated in a depth-first fashion, as are all recursive algorithms. I encourage you to trace this example by hand to refresh your knowledge of recursion.

Note that $F(4)$ is computed on both sides of the recursion tree, and $F(2)$ is computed no less than five times in this small example. The weight of all this redundancy becomes clear when you run the program. It took 4 minutes and 40 seconds for this program to compute $F(50)$ on my laptop. You might well do it faster by hand using the algorithm below.

How much time does the recursive algorithm take to compute $F(n)$? Since $F_{n+1}/F_n \approx \phi = (1+\sqrt{5})/2 \approx 1.61803$, this means that $F_n > 1.6^n$ for sufficiently large n. Since our recursion tree has only 0 and 1 as leaves, summing them up to get such a large number means we must have at least 1.6^n leaves or procedure calls. This humble little program takes exponential time to run!

10.1.2 Fibonacci Numbers by Caching

In fact, we can do much better. We can explicitly store (or *cache*) the results of each Fibonacci computation $F(k)$ in a table data structure indexed by the parameter k—a technique also known as *memoization*. The key to implement the recursive algorithm efficiently is to explicitly check whether we already know a particular value before trying to compute it:

```
#define MAXN      92      /* largest n for which F(n) fits in a long */
#define UNKNOWN  -1      /* contents denote an empty cell */
long f[MAXN+1];           /* array for caching fib values */
```

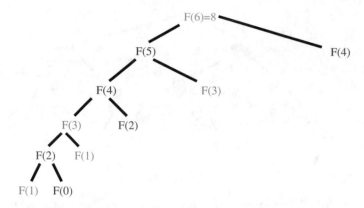

Figure 10.2: The recursion tree for computing Fibonacci numbers with caching.

```
long fib_c(int n) {
    if (f[n] == UNKNOWN) {
        f[n] = fib_c(n-1) + fib_c(n-2);
    }

    return(f[n]);
}


long fib_c_driver(int n) {
    int i;        /* counter */

    f[0] = 0;
    f[1] = 1;

    for (i = 2; i <= n; i++) {
        f[i] = UNKNOWN;
    }

    return(fib_c(n));
}
```

To compute $F(n)$, we call fib_c_driver(n). This initializes our cache to the two values we initially know ($F(0)$ and $F(1)$) as well as the UNKNOWN flag for all the rest that we don't. It then calls a look-before-crossing-the-street version of the recursive algorithm.

This cached version runs instantly up to the largest value that can fit in a long integer. The new recursion tree (Figure 10.2) explains why. There is no meaningful branching, because only the left-side calls do computation. The right-side calls find what they are looking for in the cache and immediately return.

What is the running time of this algorithm? The recursion tree provides more of a clue than looking at the code. In fact, it computes $F(n)$ in linear time (in other words, $O(n)$ time) because the recursive function `fib_c(k)` is called at most twice for each value $0 \leq k \leq n - 1$.

This general method of explicitly caching (or *tabling*) results from recursive calls to avoid recomputation provides a simple way to get *most* of the benefits of full dynamic programming. It is thus worth a careful look. In principle, such caching can be employed on any recursive algorithm. However, storing partial results would have done absolutely no good for such recursive algorithms as *quicksort*, *backtracking*, and *depth-first search* because all the recursive calls made in these algorithms have distinct *parameter values*. It doesn't pay to store something you will use once and never refer to again.

Caching makes sense only when the space of distinct parameter values is modest enough that we can afford the cost of storage. Since the argument to the recursive function `fib_c(k)` is an integer between 0 and n, there are only $O(n)$ values to cache. A linear amount of space for an exponential amount of time is an excellent tradeoff. But as we shall see, we can do even better by eliminating the recursion completely.

Take-Home Lesson: Explicit caching of the results of recursive calls provides *most* of the benefits of dynamic programming, usually including the same running time as the more elegant full solution. If you prefer doing extra programming to more subtle thinking, I guess you can stop here.

10.1.3 Fibonacci Numbers by Dynamic Programming

We can calculate F_n in linear time more easily by explicitly specifying the order of evaluation of the recurrence relation:

```
long fib_dp(int n) {
    int i;                      /* counter */
    long f[MAXN+1];             /* array for caching values */

    f[0] = 0;
    f[1] = 1;

    for (i = 2; i <= n; i++) {
        f[i] = f[i-1] + f[i-2];
    }

    return(f[n]);
}
```

Observe that we have removed all recursive calls! We evaluate the Fibonacci numbers from smallest to biggest and store all the results, so we *know* that we have F_{i-1} and F_{i-2} ready whenever we need to compute F_i. The linearity of this algorithm is now obvious. Each of the n values is simply computed as the sum of two integers, in $O(n)$ total time and space.

More careful study shows that we do not need to store all the intermediate values for the entire period of execution. Because the recurrence depends on two arguments, we only need to retain the last two values we have seen:

```
long fib_ultimate(int n)
{
        int i;                       /* counter */
        long back2=0, back1=1;       /* last two values of f[n] */
        long next;                   /* placeholder for sum */

        if (n == 0) return (0);

        for (i=2; i<n; i++) {
                next = back1+back2;
                back2 = back1;
                back1 = next;
        }
        return(back1+back2);
}
```

This analysis reduces the storage demands to constant space with no asymptotic degradation in running time.

10.1.4   Binomial Coefficients

We now show how to compute *binomial coefficients* as another example of how to eliminate recursion by specifying the order of evaluation. The binomial coefficients are the most important class of counting numbers, where $\binom{n}{k}$ counts the number of ways to choose k things out of n possibilities.

How do you compute binomial coefficients? First, $\binom{n}{k} = \frac{n!}{k!\,(n-k)!}$, so in principle you can compute them straight from factorials. However, this method has a serious drawback. Intermediate calculations can easily cause arithmetic overflow, even when the final coefficient fits comfortably within an integer.

A more stable way to compute binomial coefficients is using the recurrence relation implicit in the construction of Pascal's triangle:

```
                        1
                     1     1
                  1     2     1
               1     3     3     1
            1     4     6     4     1
         1     5    10    10     5     1
```

n / k	0	1	2	3	4	5
0	A					
1	B	G				
2	C	1	H			
3	D	2	3	I		
4	E	4	5	6	J	
5	F	7	8	9	10	K

n / k	0	1	2	3	4	5
0	1					
1	1	1				
2	1	2	1			
3	1	3	3	1		
4	1	4	6	4	1	
5	1	5	10	10	5	1

Figure 10.3: Evaluation order for `binomial_coefficient` at $M[5,4]$ (left). The initialization conditions are labeled A–K and recurrence evaluations labeled 1–10. The matrix contents after evaluation are shown on the right.

Each number is the sum of the two numbers directly above it. The recurrence relation implicit in this is

$$\binom{n}{k} = \binom{n-1}{k-1} + \binom{n-1}{k}$$

Why does this work? Consider whether the nth element appears in one of the $\binom{n}{k}$ subsets having k elements. If it does, we can complete the subset by picking $k-1$ other items from the remaining $n-1$. If it does not, we must pick all k items from the remaining $n-1$. There is no overlap between these cases, and all possibilities are included, so the sum counts all k-element subsets.

No recurrence is complete without basis cases. What binomial coefficient values do we know without computing them? The left term of the sum eventually drives us down to $\binom{m}{0}$. How many ways are there to choose zero things from a set? Exactly one, the empty set. If this is not convincing, then it is equally good to accept $\binom{m}{1} = m$ as the basis case. The right term of the sum runs us up to $\binom{m}{m}$. How many ways are there to choose m things from a m-element set? Exactly one—the complete set. Together, these basis cases and the recurrence define the binomial coefficients on all interesting values.

Figure 10.3 demonstrates a proper evaluation order for the recurrence. The initialized cells are marked A–K, denoting the order in which they were assigned values. Each remaining cell is assigned the sum of the cell directly above it and the cell immediately above and to the left. The triangle of cells marked 1–10 denote the evaluation order in computing $\binom{5}{4} = 5$ using the following code:

```
long binomial_coefficient(int n, int k) {
    int i, j;                         /* counters */
    long bc[MAXN+1][MAXN+1];          /* binomial coefficient table */

    for (i = 0; i <= n; i++) {
        bc[i][0] = 1;
    }
```

```
for (j = 0; j <= n; j++) {
    bc[j][j] = 1;
}

for (i = 2; i <= n; i++) {
    for (j = 1; j < i; j++) {
        bc[i][j] = bc[i-1][j-1] + bc[i-1][j];
    }
}

return(bc[n][k]);
}
```

Study this function carefully to make sure you see how we did it. The rest of this chapter will focus more on formulating and analyzing the appropriate recurrence than the mechanics of table manipulation demonstrated here.

10.2 Approximate String Matching

Searching for patterns in text strings is a problem of unquestionable importance. Back in Section 6.7 (page 188) I presented algorithms for *exact* string matching— finding where the pattern string P occurs as a substring of the text string T. But life is often not that simple. Words in either the text or pattern can be mispelled (sic), robbing us of exact similarity. Evolutionary changes in genomic sequences or language usage mean that we often search with archaic patterns in mind: "Thou shalt not kill" morphs over time into "You should not murder."

How can we search for the substring closest to a given pattern, to compensate for spelling errors? To deal with inexact string matching, we must first define a cost function telling us how far apart two strings are. A reasonable distance measure reflects the number of *changes* that must be made to convert one string to another. There are three natural types of changes:

- *Substitution* – Replace a single character in pattern P with a different character, such as changing *shot* to *spot*.

- *Insertion* – Insert a single character into pattern P to help it match text T, such as changing *ago* to *agog*.

- *Deletion* – Delete a single character from pattern P to help it match text T, such as changing *hour* to *our*.

Properly posing the question of string similarity requires us to set the cost of each such transform operation. Assigning each operation an equal cost of 1 defines the *edit distance* between two strings. Approximate string matching arises in many applications, as detailed in Section 21.4 (page 688).

Figure 10.4: In a single string edit operation, the last character must be either matched/substituted, inserted, or deleted.

Approximate string matching seems like a difficult problem, because we must decide exactly where to best perform a complicated sequence of insert/delete operations in pattern and text. To solve it, let's think about the problem in reverse. What information would we need to select the final operation correctly? What can happen to the last character in the matching for each string?

10.2.1 Edit Distance by Recursion

We can define a recursive algorithm using the observation that the last character in the string must either be matched, substituted, inserted, or deleted. There is no other possible choice, as shown in Figure 10.4. Chopping off the characters involved in this last edit operation leaves a pair of smaller strings. Let i and j be the indices of the last character of the relevant prefix of P and T, respectively. There are three pairs of shorter strings after the last operation, corresponding to the strings after a match/substitution, insertion, or deletion. *If* we knew the cost of editing these three pairs of smaller strings, we could decide which option leads to the best solution and choose that option accordingly. We *can* learn this cost through the magic of recursion.

More precisely, let $D[i, j]$ be the minimum number of differences between the substrings $P_1 P_2 \ldots P_i$ and $T_1 T_2 \ldots T_j$. $D[i, j]$ is the *minimum* of the three possible ways to extend smaller strings:

- If $(P_i = T_j)$, then $D[i-1, j-1]$, else $D[i-1, j-1] + 1$. This means we either match or substitute the ith and jth characters, depending upon whether these tail characters are the same. More generally, the cost of a single character substitution can be returned by a function $match(P_i, T_j)$.

- $D[i, j-1] + 1$. This means that there is an extra character in the text to account for, so we do not advance the pattern pointer and we pay the cost of an insertion. More generally, the cost of a single character insertion can be returned by a function $indel(T_j)$.

- $D[i-1, j] + 1$. This means that there is an extra character in the pattern to remove, so we do not advance the text pointer and we pay the cost of a deletion. More generally, the cost of a single character deletion can be returned by a function $indel(P_i)$.

```
#define MATCH        0        /* enumerated type symbol for match */
#define INSERT       1        /* enumerated type symbol for insert */
#define DELETE       2        /* enumerated type symbol for delete */


int string_compare_r(char *s, char *t, int i, int j) {
    int k;             /* counter */
    int opt[3];        /* cost of the three options */
    int lowest_cost;   /* lowest cost */

    if (i == 0) {      /* indel is the cost of an insertion or deletion */
        return(j * indel(' '));
    }

    if (j == 0) {
        return(i * indel(' '));
    }

                       /* match is the cost of a match/substitution */
    opt[MATCH]  = string_compare_r(s,t,i-1,j-1) + match(s[i],t[j]);
    opt[INSERT] = string_compare_r(s,t,i,j-1) + indel(t[j]);
    opt[DELETE] = string_compare_r(s,t,i-1,j) + indel(s[i]);

    lowest_cost = opt[MATCH];
    for (k = INSERT; k <= DELETE; k++) {
        if (opt[k] < lowest_cost) {
            lowest_cost = opt[k];
        }
    }

    return(lowest_cost);
}
```

This program is absolutely correct—convince yourself. It also turns out to be impossibly slow. Running on my computer, the computation takes several seconds to compare two 11-character strings, and disappears into Never-Never Land on anything longer.

Why is the algorithm so slow? It takes exponential time because it re-computes values again and again and again. At every position in the string, the recursion branches three ways, meaning it grows at a rate of at least 3^n—indeed, even faster since most of the calls reduce only one of the two indices, not both of them.

10.2.2 Edit Distance by Dynamic Programming

So, how can we make this algorithm practical? The important observation is that most of these recursive calls compute things that have been previously computed. How do we know? There can only be $|P| \cdot |T|$ possible unique recursive calls, since there are only that many distinct (i, j) pairs to serve as the argument parameters of the recursive calls. By storing the values for each of these (i, j) pairs in a table, we can look them up as needed and avoid recomputing them.

A table-based, dynamic programming implementation of this algorithm is given below. The table is a two-dimensional matrix m where each of the $|P| \cdot |T|$ cells contains the cost of the optimal solution to a subproblem, as well as a parent field explaining how we got to this location:

```
typedef struct {
    int cost;          /* cost of reaching this cell */
    int parent;        /* parent cell */
} cell;


cell m[MAXLEN+1][MAXLEN+1];    /* dynamic programming table */
```

Our dynamic programming implementation has three differences from the recursive version. First, it gets its intermediate values using table lookup instead of recursive calls. Second, it updates the **parent** field of each cell, which will enable us to reconstruct the edit sequence later. Third, it is implemented using a more general **goal_cell()** function instead of just returning **m[|P|][|T|].cost**. This will enable us to apply this routine to a wider class of problems.

Be aware that we adhere to special string and index conventions in the routine below. In particular, we assume that each string has been padded with an initial blank character, so the first real character of string **s** sits in **s[1]**. Why did we do this? It enables us to keep the matrix indices in sync with those of the strings for clarity. Recall that we must dedicate the zeroth row and column of **m** to store the boundary values matching the empty prefix. Alternatively, we could have left the input strings intact and adjusted the indices accordingly.

```
int string_compare(char *s, char *t, cell m[MAXLEN+1][MAXLEN+1]) {
    int i, j, k;       /* counters */
    int opt[3];        /* cost of the three options */

    for (i = 0; i <= MAXLEN; i++) {
        row_init(i, m);
        column_init(i, m);
    }
```

```
for (i = 1; i < strlen(s); i++) {
    for (j = 1; j < strlen(t); j++) {
        opt[MATCH] = m[i-1][j-1].cost + match(s[i], t[j]);
        opt[INSERT] = m[i][j-1].cost + indel(t[j]);
        opt[DELETE] = m[i-1][j].cost + indel(s[i]);

        m[i][j].cost = opt[MATCH];
        m[i][j].parent = MATCH;
        for (k = INSERT; k <= DELETE; k++) {
            if (opt[k] < m[i][j].cost) {
                m[i][j].cost = opt[k];
                m[i][j].parent = k;
            }
        }
    }
}

goal_cell(s, t, &i, &j);
return(m[i][j].cost);
}
```

To determine the value of cell (i, j), we need to have three values sitting and waiting for us in matrix m—namely, the cells $m(i - 1, j - 1)$, $m(i, j - 1)$, and $m(i - 1, j)$. Any evaluation order with this property will do, including the row-major order used in this program.[1] The two nested loops do in fact evaluate m for every pair of string prefixes, one row at a time. Recall that the strings are padded such that s[1] and t[1] hold the first character of each input string, so the lengths (strlen) of the padded strings are one character greater than those of the input strings.

As an example, we show the cost matrix for turning $P = $ "thou shalt" into $T = $ "you should" in five moves in Figure 10.5. I encourage you to evaluate this example matrix by hand, to nail down exactly how dynamic programming works.

10.2.3 Reconstructing the Path

The string comparison function returns the cost of the optimal alignment, but not the alignment itself. Knowing you can convert "thou shalt" to "you should" in only five moves is dandy, but what is the sequence of editing operations that does it?

The possible solutions to a given dynamic programming problem are described by paths through the dynamic programming matrix, starting from the

---

[1]Suppose we create a graph with a vertex for every matrix cell, and a directed edge (x, y), when the value of cell x is needed to compute the value of cell y. Any topological sort on the resulting DAG (why must it be a DAG?) defines an acceptable evaluation order.

T		y	o	u	-	s	h	o	u	l	d
P	pos	0	1	2	3	4	5	6	7	8	9	10
:		**0**	1	2	3	4	5	6	7	8	9	10
t:	1	**1**	1	2	3	4	5	6	7	8	9	10
h:	2	2	**2**	2	3	4	5	5	6	7	8	9
o:	3	3	3	**2**	3	4	5	6	5	6	7	8
u:	4	4	4	3	**2**	3	4	5	6	5	6	7
-:	5	5	5	4	3	**2**	3	4	5	6	6	7
s:	6	6	6	5	4	3	**2**	3	4	5	6	7
h:	7	7	7	6	5	4	3	**2**	**3**	4	5	6
a:	8	8	8	7	6	5	4	3	3	**4**	5	6
l:	9	9	9	8	7	6	5	4	4	4	**4**	5
t:	10	10	10	9	8	7	6	5	5	5	5	**5**

Figure 10.5: Example of a dynamic programming matrix for editing distance computation, with the underlined entries appearing on the optimal alignment path. Blue values denote insertions, green values deletions, and red values match/substitution.

initial configuration (the pair of empty strings $(0,0)$) down to the final goal state (the pair of full strings $(|P|,|T|)$). The key to building the solution is reconstructing the decisions made at every step along the optimal path that leads to the goal state. These decisions have been recorded in the `parent` field of each array cell.

Reconstructing these decisions is done by walking backward from the goal state, following the `parent` pointer back to an earlier cell. We repeat this process until we arrive back at the initial cell, analogous to how we reconstructed the path found by BFS or Dijkstra's algorithm. The `parent` field for `m[i][j]` tells us whether the operation at (i, j) was MATCH, INSERT, or DELETE. Tracing back through the parent matrix in Figure 10.6 yields the edit sequence DSMMMMMISMS from "thou_shalt" to "you_should"—meaning delete the first "t"; replace the "h" with "y"; match the next five characters before inserting an "o"; replace "a" with "u"; and finally replace the "t" with a "d".

Walking backward reconstructs the solution in reverse order. However, clever use of recursion can do the reversing for us:

```
void reconstruct_path(char *s, char *t, int i, int j,
                        cell m[MAXLEN+1][MAXLEN+1]) {
    if (m[i][j].parent == -1) {
        return;
    }

    if (m[i][j].parent == MATCH) {
        reconstruct_path(s, t, i-1, j-1, m);
```

P	T pos	0	y 1	o 2	u 3	- 4	s 5	h 6	o 7	u 8	l 9	d 10
	0	-1	1	1	1	1	1	1	1	1	1	1
t:	1	2	0	0	0	0	0	0	0	0	0	0
h:	2	2	0	0	0	0	0	0	1	1	1	1
o:	3	2	0	0	0	0	0	0	0	1	1	1
u:	4	2	0	2	0	1	1	1	1	0	1	1
-:	5	2	0	2	2	0	1	1	1	1	0	0
s:	6	2	0	2	2	2	0	1	1	1	1	0
h:	7	2	0	2	2	2	2	0	1	1	1	1
a:	8	2	0	2	2	2	2	2	0	0	0	0
l:	9	2	0	2	2	2	2	2	0	0	0	1
t:	10	2	0	2	2	2	2	2	0	0	0	0

Figure 10.6: Parent matrix for edit distance computation, with the optimal alignment path underlined to highlight. Again, blue values denote insertions, green values deletions, and red values match/substitution.

```
        match_out(s, t, i, j);
        return;
    }

    if (m[i][j].parent == INSERT) {
        reconstruct_path(s, t, i, j-1, m);
        insert_out(t, j);
        return;
    }

    if (m[i][j].parent == DELETE) {
        reconstruct_path(s, t, i-1, j, m);
        delete_out(s, i);
        return;
    }
}
```

For many problems, including edit distance, the solution can be reconstructed from the cost matrix without explicitly retaining the last-move array. In edit distance, the trick is working backward from the costs of the three possible ancestor cells and corresponding string characters to reconstruct the move that took you to the current cell at the given cost. But it is cleaner and easier to explicitly store the moves.

10.2.4 Varieties of Edit Distance

The `string_compare` and path reconstruction routines reference several functions that we have not yet defined. These fall into four categories:

- *Table initialization* – The functions `row_init` and `column_init` initialize the zeroth row and column of the dynamic programming table, respectively. For the string edit distance problem, cells $(i, 0)$ and $(0, i)$ correspond to matching length-i strings against the empty string. This requires exactly i insertions/deletions, so the definition of these functions is clear:

```
row_init(int i)                  column_init(int i)
{                                {
  m[0][i].cost = i;                m[i][0].cost = i;
  if (i>0)                         if (i>0)
    m[0][i].parent = INSERT;         m[i][0].parent = DELETE;
  else                             else
    m[0][i].parent = -1;             m[i][0].parent = -1;
}                                }
```

- *Penalty costs* – The functions `match(c,d)` and `indel(c)` present the costs for transforming character c to d and inserting/deleting character c. For standard edit distance, `match` should cost 0 if the characters are identical, and 1 otherwise; while `indel` returns 1 regardless of what the argument is. But application-specific cost functions can be employed, perhaps with substitution more forgiving for characters located near each other on standard keyboard layouts or those that sound or look similar.

```
int match(char c, char d)        int indel(char c)
{                                {
  if (c == d) return(0);           return(1);
  else return(1);                }
}
```

- *Goal cell identification* – The function `goal_cell` returns the indices of the cell marking the endpoint of the solution. For edit distance, this is always defined by the length of the two input strings. However, other applications we will soon encounter do not have fixed goal locations.

```
void goal_cell(char *s, char *t, int *i, int *j) {
    *i = strlen(s) - 1;
    *j = strlen(t) - 1;
}
```

- *Traceback actions* – The functions `match_out`, `insert_out`, and `delete_out` perform the appropriate actions for each edit operation during traceback.

For edit distance, this might mean printing out the name of the operation or character involved, as determined by the needs of the application.

```
insert_out(char *t, int j)          match_out(char *s, char *t,
{                                                  int i, int j)
    printf("I");                    {
}                                       if (s[i]==t[j]) printf("M");
                                        else printf("S");
delete_out(char *s, int i)          }
{
    printf("D");
}
```

All of these functions are quite simple for edit distance computation. However, we must confess it is difficult to get the boundary conditions and index manipulations correct. Although dynamic programming algorithms are easy to design once you understand the technique, getting the details right requires clear thinking and thorough testing.

This may seem like a lot of infrastructure to develop for such a simple algorithm. However, several important problems can be solved as special cases of edit distance using only minor changes to some of these stub functions:

- *Substring matching* – Suppose we want to find where a short pattern P best occurs within a long text T—say searching for "Skiena" in all its misspellings (Skienna, Skena, Skina, ...) within a long file. Plugging this search into our original edit distance function will achieve little sensitivity, since the vast majority of any edit cost will consist of deleting all that is not "Skiena" from the body of the text. Indeed, matching any scattered $...S...k...i...e...n...a...$ and deleting the rest will yield an optimal solution.

  We want an edit distance search where the cost of starting the match is independent of the position in the text, so that we are not prejudiced against a match that starts in the middle of the text. Now the goal state is not necessarily at the end of both strings, but the cheapest place to match the entire pattern somewhere in the text. Modifying these two functions gives us the correct solution:

```
void row_init(int i, cell m[MAXLEN+1][MAXLEN+1]) {
    m[0][i].cost = 0;          /* NOTE CHANGE */
    m[0][i].parent = -1;       /* NOTE CHANGE */
}
```

```
void goal_cell(char *s, char *t, int *i, int *j) {
    int k;     /* counter */

    *i = strlen(s) - 1;
    *j = 0;

    for (k = 1; k < strlen(t); k++) {
        if (m[*i][k].cost < m[*i][*j].cost) {
            *j = k;
        }
    }
}
```

- *Longest common subsequence* – Perhaps we are interested in finding the longest scattered string of characters included within both strings, without changing their relative order. Indeed, this problem will be discussed in Section 21.8. Do Democrats and Republicans have anything in common? Certainly! The *longest common subsequence* (LCS) between "democrats" and "republicans" is *ecas*.

  A common subsequence is defined by all the identical-character matches in an edit trace. To maximize the number of such matches, we must prevent substitution of non-identical characters. With substitution forbidden, the only way to get rid of the non-common subsequence will be through insertion and deletion. The minimum cost alignment has the fewest such "in-dels," so it must preserve the longest common substring. We get the alignment we want by changing the match-cost function to make substitutions expensive:

```
int match(char c, char d) {
    if (c == d) {
        return(0);
    }
    return(MAXLEN);
}
```

  Actually, it suffices to make the substitution penalty greater than that of an insertion plus a deletion for substitution to lose any allure as a possible edit operation.

- *Maximum monotone subsequence* – A numerical sequence is *monotonically increasing* if the ith element is at least as big as the $(i-1)$st element. The *maximum monotone subsequence* problem seeks to delete the fewest number of elements from an input string S to leave a monotonically increasing subsequence. A maximum monotone subsequence of 243517698 is 23568.

In fact, this is just a longest common subsequence problem, where the second string is the elements of S sorted in increasing order: 123456789. Any common sequence of these two must (a) represent characters in proper order in S, and (b) use only characters with increasing position in the collating sequence—so the longest one does the job. Of course, this approach can be modified to give the longest decreasing sequence simply by reversing the sorted order.

As you can see, our edit distance routine can be made to do many amazing things easily. The trick is observing that your problem is just a special case of approximate string matching.

The alert reader may notice that it is unnecessary to keep all $O(mn)$ cells to compute the cost of an alignment. If we evaluate the recurrence by filling in the columns of the matrix from left to right, we will never need more than two columns of cells to store what is necessary to complete the computation. Thus, $O(m)$ space is sufficient to evaluate the recurrence without changing the time complexity. This is good, but unfortunately we cannot reconstruct the alignment without the full matrix.

Saving space in dynamic programming is very important. Since memory on any computer is limited, using $O(nm)$ space proves more of a bottleneck than $O(nm)$ time. Fortunately, there is a clever divide-and-conquer algorithm that computes the actual alignment in the same $O(nm)$ time but only $O(m)$ space. It is discussed in Section 21.4 (page 688).

10.3 Longest Increasing Subsequence

There are three steps involved in solving a problem by dynamic programming:

1. Formulate the answer you want as a recurrence relation or recursive algorithm.

2. Show that the number of different parameter values taken on by your recurrence is bounded by a (hopefully small) polynomial.

3. Specify an evaluation order for the recurrence so the partial results you need are always available when you need them.

To see how this is done, let's see how we would develop an algorithm to find the longest monotonically increasing subsequence within a sequence of n numbers. Truth be told, this was described as a special case of edit distance in Section 10.2.4 (page 323), where it was called *maximum monotone subsequence*. Still, it is instructive to work it out from scratch. Indeed, dynamic programming algorithms are often easier to reinvent than look up.

We distinguish an increasing sequence from a *run*, where the elements must be physical neighbors of each other. The selected elements of both must be sorted in increasing order from left to right. For example, consider the sequence

$$S = (2, 4, 3, 5, 1, 7, 6, 9, 8)$$

The longest increasing subsequence of S is of length 5: for example, $(2,3,5,6,8)$. In fact, there are eight of this length (can you enumerate them?). There are four increasing runs of length 2: $(2, 4)$, $(3, 5)$, $(1, 7)$, and $(6, 9)$.

Finding the longest increasing *run* in a numerical sequence is straightforward. Indeed, you should be able to easily devise a linear-time algorithm. But finding the longest increasing subsequence is considerably trickier. How can we identify which scattered elements to skip?

To apply dynamic programming, we need to design a recurrence relation for the length of the longest sequence. To find the right recurrence, ask what information about the first $n-1$ elements of $S = (s_1, \ldots, s_n)$ would enable you to find the answer for the entire sequence:

- The length L of the longest increasing sequence in $(s_1, s_2, \ldots, s_{n-1})$ seems a useful thing to know. In fact, this will be the length of the longest increasing sequence in S, unless s_n extends some increasing sequence of the same length.

  Unfortunately, this length L is not enough information to complete the full solution. Suppose I told you that the longest increasing sequence in $(s_1, s_2, \ldots, s_{n-1})$ was of length 5 and that $s_n = 8$. Will the length of the longest increasing subsequence of S be 5 or 6? It depends on whether the length-5 sequence ended with a value < 8.

- We need to know the length of the longest sequence that s_n will extend. To be certain we know this, we really need the length of the longest sequence ending at *every* possible value s_i.

This provides the idea around which to build a recurrence. Define L_i to be the length of the longest sequence ending with s_i. The longest increasing sequence containing s_n will be formed by appending it to the longest increasing sequence to the left of n that ends on a number smaller than s_n. The following recurrence computes L_i:

$$L_i \;=\; 1 + \max_{\substack{0 \le j < i \\ s_j < s_i}} L_j,$$

$$L_0 \;=\; 0$$

These values define the length of the longest increasing sequence ending at each sequence element. The length of the longest increasing subsequence of S is given by $L = \max_{1 \le i \le n} L_i$, since the winning sequence must end somewhere. Here is the table associated with our previous example:

Index i	1	2	3	4	5	6	7	8	9
Sequence s_i	2	4	3	5	1	7	6	9	8
Length L_i	1	2	2	3	1	4	4	5	5
Predecessor p_i	–	1	1	2	–	4	4	6	6

What auxiliary information will we need to store to reconstruct the actual sequence instead of its length? For each element s_i, we will store its *predecessor*— the index p_i of the element that appears immediately before s_i in a longest increasing sequence ending at s_i. Since all of these pointers go towards the left, it is a simple matter to start from the last value of the longest sequence and follow the pointers back so as to reconstruct the other items in the sequence.

What is the time complexity of this algorithm? Each one of the n values of L_i is computed by comparing s_i against the $i - 1 \leq n$ values to the left of it, so this analysis gives a total of $O(n^2)$ time. In fact, by using dictionary data structures in a clever way, we can evaluate this recurrence in $O(n \lg n)$ time. However, the simple recurrence would be easy to program and therefore is a good place to start.

> *Take-Home Lesson:*   Once you understand dynamic programming, it can be easier to work out such algorithms from scratch than to try to look them up.

10.4   War Story: Text Compression for Bar Codes

Ynjiun waved his laser wand over the torn and crumpled fragments of a bar code label. The system hesitated for a few seconds, then responded with a pleasant *blip* sound. He smiled at me in triumph. "Virtually indestructible."

I was visiting the research laboratories of Symbol Technologies (now Zebra), the world's leading manufacturer of bar code scanning equipment. Although we take bar codes for granted, there is a surprising amount of technology behind them. Bar codes exist because conventional optical character recognition (OCR) systems are not sufficiently reliable for inventory operations. The bar code symbology familiar to us on each box of cereal, pack of gum, or can of soup encodes a ten-digit number with enough error correction that it is virtually impossible to scan the wrong number, even if the can is upside-down or dented. Occasionally, the cashier won't be able to get a label to scan at all, but once you hear that *blip* you know it was read correctly.

The ten-digit capacity of conventional bar code labels provides room enough to only store a single ID number in a label. Thus, any application of supermarket bar codes must have a database mapping (say) 11141-47011 to a particular brand and size of soy sauce. The holy grail of the bar code world had long been the development of higher-capacity bar code symbologies that can store entire documents, yet still be read reliably.

"PDF-417 is our new, two-dimensional bar code symbology," Ynjiun explained. A sample label is shown in Figure 10.7. Although you may be more familiar with QR codes, PDF-417 is now a well accepted standard. Indeed, the back of every New York State drivers license contains the criminal record of its owner, elegantly rendered in PDF-417.

"How much data can you fit in a typical 1-inch label?" I asked him.

"It depends upon the level of error correction we use, but about 1,000 bytes. That's enough for a small text file or image," he said.

Figure 10.7: A two-dimensional barcode label of the Gettysburg Address using PDF-417.

"Interesting. You should use some data compression technique to maximize the amount of text you can store in a label." See Section 21.5 (page 693) for a discussion of standard data compression algorithms.

"We do incorporate a data compaction method," he explained. "We understand the different types of files our customers will want to make labels for. Some files will be all in uppercase letters, while others will use mixed-case letters and numbers. We provide four different text modes in our code, each with a different subset of alphanumeric characters available. We can describe each character using only 5 bits as long as we stay within a mode. To switch modes, we issue a mode switch command first (taking an extra 5 bits) and then code for the new character."

"I see. So you designed the mode character sets to minimize the number of mode switch operations on typical text files." The modes are illustrated in Figure 10.8.

"Right. We put all the digits in one mode and all the punctuation characters in another. We also included both mode *shift* and mode *latch* commands. We can *shift* into a new mode just for the next character, perhaps to produce a punctuation mark. Or we can *latch* permanently into a different mode, if we are at the start of a run of several characters from there, like a phone number."

"Wow!" I said. "With all of this mode switching going on, there must be many different ways to encode any given text as a label. How do you find the smallest such encoding?"

"We use a greedy algorithm. We look a few characters ahead and then decide which mode we would be best off in. It works fairly well."

I pressed him on this. "How do you know it works fairly well? There might be significantly better encodings that you are simply not finding."

"I guess I don't know. But it's probably NP-complete to find the optimal coding." Ynjiun's voice trailed off. "Isn't it?"

I started to think. Every encoding starts in a given mode and consists of a sequence of intermixed character codes and mode shift/latch operations. From any given position in the text, we can either output the next character code (assuming it is available in our current mode) or decide to shift. As we moved from left to right through the text, our current state would be completely reflected by our current character position and current mode. For a given position/mode

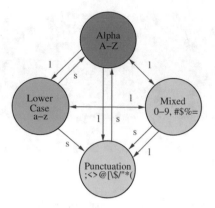

Figure 10.8: Mode switching in PDF-417.

pair, we would have been interested in the cheapest way of getting there, over all possible encodings. . . .

My eyes lit up so bright they cast shadows on the walls.

"The optimal encoding for any given text in PDF-417 can be found using dynamic programming. For each possible mode $1 \leq m \leq 4$, and each character position $1 \leq i \leq n$, we fill a matrix $M[i, m]$ with the cost of the cheapest encoding of the first i characters ending in mode m. Our next move from each mode/position is either match, shift, or latch, so there are only a few possible operations to consider at each position."

Basically,

$$M[i, j] = \min_{1 \leq m \leq 4} \left(M[i - 1, m] + c(S_i, m, j) \right)$$

where $c(S_i, m, j)$ is the cost of encoding character S_i and switching from mode m to mode j. The cheapest possible encoding results from tracing back from $M[n, m]$, where m is the value of k that minimizes $M[n, k]$. Each of the $4n$ cells can be filled in constant time, so it takes time linear in the length of the string to find the optimal encoding.

Ynjiun was skeptical, but he encouraged us to implement an optimal encoder. A few complications arose due to weirdnesses of PDF-417 mode switching, but my student Yaw-Ling Lin rose to the challenge. Symbol compared our encoder to theirs on 13,000 labels and concluded that dynamic programming gave an 8% tighter encoding on average. This was significant, because no one wants to waste 8% of their potential storage capacity, particularly in an environment where the capacity is only a few hundred bytes. Of course, an 8% *average* improvement meant that it did much better than that on certain labels, and it never did worse than the original encoder. While our encoder took slightly longer to run than the greedy encoder, this was not significant, because the bottleneck would be the time needed to print the label.

Our observed impact of replacing a heuristic solution with the global optimum is probably typical of most applications. Unless you really botch up your heuristic, you should get a decent solution. Replacing it with an optimal result, however, usually gives a modest but noticeable improvement, which can have pleasing consequences for your application.

10.5   Unordered Partition or Subset Sum

The *knapsack* or *subset sum* problem asks whether there exists a subset S' of an input multiset of n positive integers $S = \{s_1, \ldots, s_n\}$ whose elements add up a given target k. Think of a backpacker trying to completely fill a knapsack of capacity k with possible selections from set S. Applications of this important problem are discussed in greater detail in Section 16.10.

Dynamic programming works best on linearly ordered items, so we can consider them from left to right. The ordering of items in S from s_1 to s_n provides such an arrangement. To formulate a recurrence relation, we need to determine what information we need on items s_1 to s_{n-1} in order to decide what to do about s_n.

Here is the idea. Either the nth integer s_n is part of a subset adding up to k, or it is not. If it is, then there must be a way to make a subset of the first $n - 1$ elements of S adding up to $k - s_n$, so the last element can finish the job. If not, there may well be a solution that does not use s_n. Together this defines the recurrence:

$$T_{n,k} = T_{n-1,k} \lor T_{n-1,k-s_n}$$

This gives an $O(nk)$ algorithm to decide whether target k is realizable:

```
bool sum[MAXN+1][MAXSUM+1];      /* table of realizable sums */
int parent[MAXN+1][MAXSUM+1];    /* table of parent pointers */

bool subset_sum(int s[], int n, int k) {
    int i, j;                    /* counters */

    sum[0][0] = true;
    parent[0][0] = NIL;

    for (i = 1; i <= k; i++) {
        sum[0][i] = false;
        parent[0][i] = NIL;
    }

    for (i = 1; i <= n; i++) {    /* build table */
        for (j = 0; j <= k; j++) {
            sum[i][j] = sum[i-1][j];
            parent[i][j] = NIL;

            if ((j >= s[i-1]) && (sum[i-1][j-s[i-1]]==true)) {
```

```
                sum[i][j] = true;
                parent[i][j] = j-s[i-1];
        }

    }

}


    return(sum[n][k]);
}
```

The **parent** table encodes the actual subset of numbers totaling to k. An appropriate subset exists whenever **sum[n][k]==true**, but it does not use s_n as an element when **parent[n][k]==NIL**. Instead, we walk up the matrix until we find an interesting parent, and follow the corresponding pointer:

```
void report_subset(int n, int k) {
    if (k == 0) {
        return;
    }

    if (parent[n][k] == NIL) {
        report_subset(n-1,k);
    }
    else {
        report_subset(n-1,parent[n][k]);
            printf(" %d ",k-parent[n][k]);
    }
}
```

Below is an example showing the **sum** table for input set $S = \{1, 2, 4, 8\}$ and target $k = 11$. The true in the lower right corner signals that the sum is realizable. Because S here represents all the powers of twos, and every target integer can be written in binary, the entire bottom row consists of trues:

i	s_i	0	1	2	3	4	5	6	7	8	9	10	11
0	0	T	F	F	F	F	F	F	F	F	F	F	F
1	1	T	T	F	F	F	F	F	F	F	F	F	F
2	2	T	T	T	T	F	F	F	F	F	F	F	F
3	4	T	T	T	T	T	T	T	T	F	F	F	F
4	8	T	T	T	T	T	T	T	T	T	T	T	T

Below is the corresponding **parents** array, encoding the solution $1 + 2 + 8 = 11$. The 3 in the lower right corner reflects that $11 - 8 = 3$. The red bolded cells represent those encountered on the walk back to recover the solution.

i	s_i	0	1	2	3	4	5	6	7	8	9	10	11
0	0	**-1**	-1	-1	-1	-1	-1	-1	-1	-1	-1	-1	-1
1	1	-1	**0**	-1	-1	-1	-1	-1	-1	-1	-1	-1	-1
2	2	-1	-1	**0**	**1**	-1	-1	-1	-1	-1	-1	-1	-1
3	4	-1	-1	-1	**-1**	**0**	**1**	**2**	**3**	-1	-1	-1	-1
4	8	-1	-1	-1	-1	-1	-1	-1	-1	**0**	**1**	**2**	**3**

The alert reader might wonder how we can have an $O(nk)$ algorithm for subset sum when subset sum in an NP-complete problem? Isn't this polynomial in n and k? Did we just prove that $P = NP$?

Unfortunately, no. Note that the target number k can be specified using $O(\log k)$ bits, meaning that this algorithm runs in time exponential in the *size* of the input, which is $O(n \log k)$. This is the same reason why factoring integer N by explicitly testing all $\sqrt{N}$ candidates for smallest factor is not polynomial, because the running time is exponential in the $O(\log N)$ bits of the input.

Another way to see the problem is to consider what happens to the algorithm when we take a specific problem instance and multiply each integer by 1,000,000. Such a transform would not have affected the running time of sorting or minimum spanning tree, or any other algorithm we have seen so far in this book. But it would slow down our dynamic programming algorithm by a factor of 1,000,000, and require a million times as much space for storing the table. The range of the numbers matters in the subset sum problem, which becomes hard for large integers.

10.6   War Story: The Balance of Power

One of the many (presumably too many) uncharitable suspicions I hold is that most electrical engineering (EE) students today would not know how to build a radio. The reason for this is that the EE students I encounter study electrical and *computer* engineering, focusing on computer architecture and embedded systems that involve as much software as hardware. When a natural disaster comes, these guys are not going to be very concerned about restoring the operation of my favorite AM radio station.

Thus, it was a relief when an EE professor and his students came to me with an honest EE problem, about optimizing the performance of the power grid.

"Alternating current (AC) power systems transmit electricity on each of three different phases. Call them A, B, and C. The system works best when the loads on each phase are roughly equal." he explained.

"I guess loads are the machines needing power, right?" I asked insightfully.

"Yeah, think of every house on the street as being a load. Each house will get assigned one of the three phases as its source of power."

"Presumably they connect every third house A, B, C, A, B, C as they wire up the street to balance the load."

"Something like that," the EE professor confirmed. "But not all houses use the same amount of power, and it is even worse in industrial areas. One company might just turn on the lights when another runs an arc furnace. After we measure the loads people are actually using, we would like to move some to different phases to balance the loads."

Now I saw the algorithmic problem. "So given a set of numbers representing the various loads, you want to assign them phases A, B, and C so the load is balanced as well as possible, right?"

"Yeah. Can you give me a fast algorithm to do this?," he asked.

This seemed clear enough to me. It smelled like an integer partition problem, namely the subset sum problem of the previous section where the target $k = (\sum_{i=1}^{n} s_n)/2$. The most balanced possible partition occurs when the sum of elements in the selected subset (here k) equals the sum of the elements left behind (here $\sum_{i=1}^{n} s_n - k$).

The generalization of the problem to partition into three subsets instead of two was straightforward, but it wasn't going to get any easier to solve. Adding a single new item $s_{n+1} = k$ and asking for a partitioning of S into three equal weight subsets requires solving an integer partition on the original elements.

I broke the bad news gently. "Integer partition is an NP-complete problem, and three-phase balancing is just as hard as it is. There is no polynomial-time algorithm for your problem."

They got up and started to leave. But then I remembered the dynamic programming algorithm for subset sum described in Section 10.5 (page 329). Why couldn't this be extended to three phases? Indeed, define the function $C[n, A, B]$ for a given set of loads S, where $C[n, w_A, w_B]$ is true if there is a way to partition the first n loads of S such that the weight on phase A is w_A and the weight on phase B is w_B. Note that there is no need to explicitly keep track of the weight on phase C, because $w_C = \sum_{i=1}^{n} s_i - w_A - w_B$. Then we get the following recurrence, defined by which subset we put the nth load on:

$$C[n, w_A, w_B] = C[n-1, w_A - s_n, w_B] \vee C[n-1, w_A, w_B - s_n] \vee C[n-1, w_A, w_B]$$

This took constant time per cell to update, but there were nk^2 cells to update, where k is the maximum amount of power we are willing to consider on any single phase. Thus, we could optimally balance the phases in $O(nk^2)$ time.

This pleased them immensely, and they set to work to implement the algorithm. But I had one question before they went off, which I purposely directed to one of the computer engineering students. "Why is it that AC power has three phases?"

"Uh, maybe impedance matching and, uh, complex numbers?" he fumphered. His advisor shot him a dirty look, as I felt the warm glow of reassurance.

But that computer engineering student could code, and that was what mattered here. He quickly implemented the dynamic programming algorithm and performed experiments on representative problems, reported in [WSR13].

Our dynamic programming algorithm always produced at least as good a solution as several heuristics, and usually better. This is no surprise, since we always produced an optimal solution and they didn't. Our dynamic program had a running time that grew quadratically in the range of the loads, which could be a problem, but binning the loads by (say) $\lfloor s_i/10 \rfloor$ would reduce the running time by a factor of 100 and produce solutions that were still pretty good for the original problem.

Dynamic programming really proved its worth when our electrical engineers got interested in more ambitious objective functions. It is not a cost-free operation to change which phase a load is on, and so they wanted to find a relatively balanced load assignment which minimized the number of changes required to achieve it. This is essentially the same recurrence, storing the cheapest cost to realize each state instead of just a flag indicating that you could reach it:

$$C[n, w_A, w_B] = \min(C[n-1, w_A - s_n, w_B] + 1,$$
$$C[n-1, w_A, w_B - s_n] + 1,$$
$$C[n-1, w_A, w_B])$$

They then got greedy, and wanted the lowest cost solution that never got seriously unbalanced at any point on the line. A globally balanced solution might choose to fill the total load on A before any loads on B or C, and that this would be bad. But the same recurrence above *still* does the job, provided we set $C[n, w_A, w_B] = \infty$ whenever the loads at this state are deemed too unbalanced to be desirable.

That is the power of dynamic programming. Once you can reduce your state space to a small enough size, you can optimize just about anything. Just walk through each possible state and score it appropriately.

10.7   The Ordered Partition Problem

Suppose that three workers are given the task of scanning through a shelf of books in search of a given piece of information. To get the job done fairly and efficiently, the books are to be partitioned among the three workers. To avoid the need to rearrange the books or separate them into piles, it is simplest to divide the shelf into three regions and assign each region to one worker.

But what is the fairest way to divide up the shelf? If all books are the same length, the job is pretty easy. Just partition the books into equal-sized regions,

$$100\ 100\ 100 \mid 100\ 100\ 100 \mid 100\ 100\ 100$$

so that everyone has 300 pages to deal with.

But what if the books are not the same length? Suppose we used the same partition when the book sizes looked like this:

$$100\ 200\ 300 \mid 400\ 500\ 600 \mid 700\ 800\ 900$$

I would volunteer to take the first section, with only 600 pages to scan, instead of the last one, with 2,400 pages. The fairest possible partition for this shelf would be

$$100\ 200\ 300\ 400\ 500 \mid 600\ 700 \mid 800\ 900$$

where the largest job is only 1,700 pages.

In general, we have the following problem:

Problem: Integer Partition without Rearrangement

Input: An arrangement S of non-negative numbers $s_1, \ldots, s_n$ and an integer k.

Output: Partition S into k or fewer ranges, to minimize the maximum sum over all the ranges, without reordering any of the numbers.

This so-called *ordered partition* problem arises often in parallel processing. We seek to balance the work done across processors to minimize the total elapsed running time. The bottleneck in this computation will be the processor assigned the most work. Indeed, the war story of Section 5.8 (page 161) revolves around a botched solution to the very problem discussed here.

Stop for a few minutes and try to find an algorithm to solve the linear partition problem.

A novice algorist might suggest a heuristic as the most natural approach to solving the partition problem, perhaps by computing the average weight of a partition, $\sum_{i=1}^{n} s_i/k$, and then trying to insert the dividers to come close to this average. However, such heuristic methods are doomed to fail on certain inputs because they do not systematically evaluate all possibilities.

Instead, consider a recursive, exhaustive search approach to solving this problem. Notice that the kth partition starts right after the $(k-1)$st divider. Where can we place this last divider? Between the ith and $(i+1)$st elements for some i, where $1 \leq i \leq n$. What is the cost after this insertion? The total cost will be the larger of two quantities:

- the cost of the last partition $\sum_{j=i+1}^{n} s_j$, and

- the cost of the largest partition formed to the left of the last divider.

What is the size of this left partition? To minimize our total, we must use the $k-2$ remaining dividers to partition the elements $s_1, \ldots, s_i$ as equally as possible. *This is a smaller instance of the same problem, and hence can be solved recursively!*

Therefore, define $M[n, k]$ to be the minimum possible cost over all partitionings of $s_1, \ldots, s_n$ into k ranges, where the cost of a partition is the largest sum of elements in one of its parts. This function can be evaluated:

$$M[n, k] = \min_{i=1}^{n} \left(\max\left(M[i, k-1], \sum_{j=i+1}^{n} s_j\right) \right)$$

We also need to specify the boundary conditions of the recurrence relation. These boundary conditions resolve the smallest possible values for each of the arguments of the recurrence. For this problem, the smallest reasonable value of the first argument is $n = 1$, meaning that the first partition consists of a single element. We can't create a first partition smaller than s_1 regardless of how

M	k				D	k		
s	1	2	3		s	1	2	3
1	1	1	1		1	–	–	–
1	2	1	1		1	–	1	1
1	3	2	1		1	–	1	2
1	4	2	2		1	–	2	2
1	5	3	2		1	–	2	3
1	6	3	2		1	–	3	4
1	7	4	3		1	–	3	4
1	8	4	3		1	–	4	5
1	9	5	3		1	–	4	6

M	k				D	k		
s	1	2	3		s	1	2	3
1	1	1	1		1	–	–	–
2	3	2	2		2	–	1	1
3	6	3	3		3	–	2	2
4	10	6	4		4	–	3	3
5	15	9	6		5	–	3	4
6	21	11	9		6	–	4	5
7	28	15	11		7	–	5	6
8	36	21	15		8	–	5	6
9	45	24	17		9	–	6	7

Figure 10.9: Dynamic programming matrices M and D for two instances of the ordered partition problem. Partitioning $(1,1,1,1,1,1,1,1,1)$ into $((1,1,1),(1,1,1),(1,1,1))$ (left) and $(1,2,3,4,5,6,7,8,9)$ into $((1,2,3,4,5),(6,7),(8,9))$ (right). Prefix sum entries appear in red and the optimal solution divider positions in blue.

many dividers are used. The smallest reasonable value of the second argument is $k = 1$, implying that we do not partition S at all. In summary:

$$M[1,k] = s_1, \text{ for all } k > 0$$

$$M[n,1] = \sum_{i=1}^{n} s_i$$

How long does it take to compute this when we store the partial results? There are a total of $k \cdot n$ cells in the table. How much time does it take to compute the values of $M[n',k']$ for $1 \leq n' \leq n$, $1 \leq k' \leq k$? Calculating this quantity using the general recurrence involves finding the minimum of n' quantities, each of which is the larger of two numbers: a table lookup and the sum of at most n' elements (taking $O(n')$ time). If filling each of kn boxes takes at most n^2 time per box, the total recurrence can be computed in $O(kn^3)$ time.

The evaluation order computes the smaller values before the bigger values, so that each evaluation has what it needs waiting for it. Full details are provided in the following implementation:

```
void partition(int s[], int n, int k) {
    int p[MAXN+1];              /* prefix sums array */
    int m[MAXN+1][MAXK+1];      /* DP table for values */
    int d[MAXN+1][MAXK+1];      /* DP table for dividers */
    int cost;                   /* test split cost */
    int i,j,x;                  /* counters */

    p[0] = 0;                   /* construct prefix sums */
    for (i = 1; i <= n; i++) {
        p[i] = p[i-1] + s[i];
    }
```

```
for (i = 1; i <= n; i++) {
    m[i][1] = p[i];    /* initialize boundaries */
}

for (j = 1; j <= k; j++) {
    m[1][j] = s[1];
}

for (i = 2; i <= n; i++) {    /* evaluate main recurrence */
    for (j = 2; j <= k; j++) {
        m[i][j] = MAXINT;
        for (x = 1; x <= (i-1); x++) {
            cost = max(m[x][j-1], p[i]-p[x]);
            if (m[i][j] > cost) {
                m[i][j] = cost;
                d[i][j] = x;
            }
        }
    }
}
reconstruct_partition(s, d, n, k);    /* print book partition */
}
```

This implementation above, in fact, runs faster than advertised. Our original analysis assumed that it took $O(n^2)$ time to update each cell of the matrix. This is because we selected the best of up to n possible points to place the divider, each of which requires the sum of up to n possible terms. In fact, it is easy to avoid the need to compute these sums by storing the n prefix sums $p_i = \sum_{k=1}^{i} s_k$, since $\sum_{k=i}^{j} s_k = p_j - p_{i-1}$. This enables us to evaluate the recurrence in linear time per cell, yielding an $O(kn^2)$ algorithm. These prefix sums also appear as the initialization values for $k = 1$, and are shown in the dynamic programming matrices of Figure 10.9.

By studying the recurrence relation and the dynamic programming matrices of these two examples, you should be able to convince yourself that the final value of $M[n, k]$ will be the cost of the largest range in the optimal partition. But for most applications, we need the actual partition that does the job. Without it, all we are left with is a coupon with a great price on an out-of-stock item.

The second matrix, D, is used to reconstruct the optimal partition. Whenever we update the value of $M[i, j]$, we record which divider position was used to achieve this value. We reconstruct the path used to get the optimal solution by working backwards from $D[n, k]$, and add a divider at each specified position. This backwards walking is best achieved by a recursive subroutine:

```
void reconstruct_partition(int s[],int d[MAXN+1][MAXK+1], int n, int k) {
    if (k == 1) {
        print_books(s, 1, n);
    } else {
        reconstruct_partition(s, d, d[n][k], k-1);
        print_books(s, d[n][k]+1, n);
    }
}


void print_books(int s[], int start, int end) {
    int i;      /* counter */

    printf("\{");
    for (i = start; i <= end; i++) {
        printf(" %d ", s[i]);
    }
    printf("}\n");
}
```

10.8 Parsing Context-Free Grammars

Compilers identify whether a particular program is a legal expression in a particular programming language, and reward you with syntax errors if it is not. This requires a precise description of the language syntax, typically given by a *context-free grammar*, as shown in Figure 10.10(l). Each *rule* or *production* of the grammar defines an interpretation for the named symbol on the left side of the rule as a sequence of symbols on the right side of the rule. The right side can be a combination of *nonterminals* (themselves defined by rules) or *terminal* symbols defined simply as strings, such as *the*, *a*, *cat*, *milk*, and *drank*.

Parsing a given text sequence S as per a given context-free grammar G is the algorithmic problem of constructing a *parse tree* of rule substitutions defining S as a single nonterminal symbol of G. Figure 10.10(right) presents the parse tree of a simple sentence using our sample grammar.

Parsing seemed like a horribly complicated subject when I took a compilers course as a graduate student. But, more recently a friend easily explained it to me over lunch. The difference is that I understand dynamic programming much better now than when I was a student.

We assume that the sequence S has length n while the grammar G itself is of constant size. This is fair, because the grammar defining a particular programming language (say C or Java) is of fixed length regardless of the size of the program we seek to compile.

Further, we assume that the definitions of each rule are in *Chomsky normal form*, like the example of Figure 10.10. This means that the right sides of every rule consists of either (a) exactly two nonterminals, for example, $X \rightarrow YZ$, or

sentence ::= noun–phrase
        verb–phrase
noun–phrase ::= article noun
verb–phrase ::= verb noun–phrase
article ::= *the, a*
noun ::= *cat, milk*
verb ::= *drank*

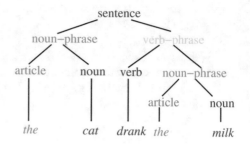

Figure 10.10: A context-free grammar (on left) with an associated parse tree (right)

(b) exactly one terminal symbol, $X \to \alpha$. Any context-free grammar can be easily and mechanically transformed into Chomsky normal form by repeatedly shortening long right-hand sides at the cost of adding extra nonterminals and productions. Thus, there is no loss of generality with this assumption.

So how can we efficiently parse S using a context-free grammar where each interesting rule produces two nonterminals? The key observation is that the rule applied at the root of the parse tree (say $X \to YZ$) splits S at some position i such that the left part, $S_1 \cdots S_i$, must be *generated* by nonterminal Y, and the right part $(S_{i+1} \cdots S_n)$ generated by Z.

This suggests a dynamic programming algorithm, where we keep track of all nonterminals generated by each contiguous subsequence of S. Define $M[i, j, X]$ to be a Boolean function that is true iff subsequence $S_i \cdots S_j$ is generated by nonterminal X. This is true if there exists a production $X \to YZ$ and breaking point k between i and j such that the left part generates Y and the right part Z. In other words, for $i < j$ we have

$$M[i, j, X] = \bigvee_{(X \to YZ) \in G} \left(\bigvee_{k=i}^{j-1} M[i, k, Y] \wedge M[k+1, j, Z] \right)$$

where $\vee$ denotes the logical *or* over all productions and split positions, and $\wedge$ denotes the logical *and* of two Boolean values.

The terminal symbols define the boundary conditions of the recurrence. In particular, $M[i, i, X]$ is true iff there exists a production $X \to \alpha$ such that $S_i = \alpha$.

What is the complexity of this algorithm? The size of our state-space is $O(n^2)$, as there are $n(n + 1)/2$ subsequences defined by (i, j) pairs with $i \geq j$. Multiplying this by the number of nonterminals, which is finite because the grammar was defined to be of constant size, has no impact on the Big Oh. Evaluating $M[i, j, X]$ requires testing all intermediate values k where $i \leq k < j$, so it takes $O(n)$ in the worst case to evaluate each of the $O(n^2)$ cells. This yields an $O(n^3)$ or cubic-time algorithm for parsing.

Stop and Think: Parsimonious Parserization

Problem: Programs often contain trivial syntax errors that prevent them from compiling. Given a context-free grammar G and input sequence S, find the smallest number of character substitutions you must make to S so that the resulting sequence is accepted by G.

Solution: This problem seemed extremely difficult when I first encountered it. But on reflection, it is just a very general version of edit distance, addressed naturally by dynamic programming. Parsing first sounded difficult, too, but fell to the same technique. Indeed, we can solve the combined problem by generalizing the recurrence relation we used for simple parsing.

Define $M'[i, j, X]$ to be an *integer* function that reports the minimum number of changes to subsequence $S_i \cdots S_j$ so it can be generated by nonterminal X. This symbol will be generated by some production $X \rightarrow YZ$. Some of the changes to S may be to the left of the breaking point and some to the right, but all we care about is minimizing the sum. In other words, for $i < j$ we have

$$M'[i, j, X] = \min_{(X \rightarrow YZ) \in G} \left(\min_{k=i}^{j-1} M'[i, k, Y] + M'[k+1, j, Z] \right)$$

The boundary conditions also change mildly. If there exists a production $X \rightarrow \alpha$, the cost of matching at position i depends on the contents of S_i. If $S_i = \alpha$, $M'[i, i, X] = 0$. Otherwise, we can pay one substitution to change S_i to α, so $M'[i, i, X] = 1$ if $S_i \neq \alpha$. If the grammar does not have a production of the form $X \rightarrow \alpha$, there is no way to substitute a single character string into something generating X, so $M'[i, i, X] = \infty$ for all i. ∎

Take-Home Lesson: For optimization problems on left-to-right objects, such as characters in a string, elements of a permutation, points around a polygon, or leaves in a search tree, dynamic programming likely leads to an efficient algorithm to find the optimal solution.

10.9 Limitations of Dynamic Programming: TSP

Dynamic programming doesn't always work. It is important to see why it can fail, to help avoid traps leading to incorrect or inefficient algorithms.

Our algorithmic poster child will once again be the traveling salesman problem, where we seek the shortest tour visiting all the cities in a graph. We will limit attention here to an interesting special case:

Problem: Longest Simple Path

Input: A weighted graph $G = (V, E)$, with specified start and end vertices s and t.

Output: What is the most expensive path from s to t that does not visit any vertex more than once?

This problem differs from TSP in two quite unimportant ways. First, it asks for a path instead of a closed tour. This difference isn't substantial: we get a closed tour simply by including the edge (t, s). Second, it asks for the most expensive path instead of the least expensive tour. Again this difference isn't very significant: it encourages us to visit as many vertices as possible (ideally all), just as in TSP. The critical word in the problem statement is *simple*, meaning we are not allowed to visit any vertex more than once.

For *unweighted* graphs (where each edge has cost 1), the longest possible simple path from s to t is of weight $n - 1$. Finding such *Hamiltonian paths* (if they exist) is an important graph problem, discussed in Section 19.5 (page 598).

10.9.1   When is Dynamic Programming Correct?

Dynamic programming algorithms are only as correct as the recurrence relations they are based on. Suppose we define $LP[i, j]$ to be the length of the longest simple path from i to j. Note that the longest simple path from i to j has to visit some vertex x right before reaching j. Thus, the last edge visited must be of the form (x, j). This suggests the following recurrence relation to compute the length of the longest path, where $c(x, j)$ is the cost/weight of edge (x, j):

$$LP[i, j] = \max_{\substack{x \in V \\ (x,j) \in E}} LP[i, x] + c(x, j)$$

This idea seems reasonable, but can you see the problem? I see at least two of them.

First, this recurrence does nothing to enforce simplicity. How do we know that vertex j has not appeared previously on the longest simple path from i to x? If it did, then adding the edge (x, j) will create a cycle. To prevent this, we must define a recursive function that explicitly remembers where we have been. Perhaps we could define $LP'[i, j, k]$ to denote the length of the longest path from i to j avoiding vertex k? This would be a step in the right direction, but still won't lead to a viable recurrence.

The second problem concerns evaluation order. What can you evaluate first? Because there is no left-to-right or smaller-to-bigger ordering of the vertices on the graph, it is not clear what the *smaller* subprograms are. Without such an ordering, we get stuck in an infinite loop as soon as we try to do anything.

Dynamic programming can be applied to any problem that obeys the *principle of optimality*. Roughly stated, this means that partial solutions can be optimally extended given the *state* after the partial solution, instead of the specifics of the partial solution itself. For example, in deciding whether to extend an approximate string matching by a substitution, insertion, or deletion, we did not need to know the sequence of operations that had been performed to date. In fact, there may be several different edit sequences that achieve a cost of C on the first p characters of pattern P and t characters of string T. Future decisions are made based on the *consequences* of previous decisions, not the actual decisions themselves.

Problems do not satisfy the principle of optimality when the specifics of the operations matter, as opposed to just their cost. Such would be the case with a special form of edit distance where we are not allowed to use combinations of operations in certain particular orders. Properly formulated, however, many combinatorial problems respect the principle of optimality.

10.9.2 When is Dynamic Programming Efficient?

The running time of any dynamic programming algorithm is a function of two things: (1) the number of partial solutions we must keep track of, and (2) how long it takes to evaluate each partial solution. The first issue—namely the size of the state space—is usually the more pressing concern.

In all of the examples we have seen, the partial solutions are completely described by specifying the possible stopping *places* in the input. This is because the combinatorial objects being worked on (typically strings and numerical sequences) have an implicit order defined upon their elements. This order cannot be scrambled without completely changing the problem. Once the order is fixed, there are relatively few possible stopping places or *states*, so we get efficient algorithms.

When the objects are not firmly ordered, however, we likely have an exponential number of possible partial solutions. Suppose the state of our partial longest simple path solution is the entire path P taken from the start to end vertex. Thus, $LP[i, j, P_{ij}]$ denotes the cost of longest simple path from i to j, where P_{ij} is the sequence of intermediate vertices between i and j on this path. The following recurrence relation works correctly to compute this, where $P + x$ denotes appending x to the end of P:

$$LP[i, j, P_{ij}] = \max_{\substack{j \notin P_{ix} \\ (x,j) \in E \\ P_{ij} = P_{ix} + j}} LP[i, x, P_{ix}] + c(x, j)$$

This formulation is correct, but how efficient is it? The path P_{ij} consists of an ordered sequence of up to $n - 3$ vertices, so there can be up to $(n - 3)!$ such paths! Indeed, this algorithm is really using combinatorial search (like backtracking) to construct all the possible intermediate paths. In fact, the max here is somewhat misleading, as there can only be one value of P_{ij} to construct the state $LP[i, j, P_{ij}]$.

We can do something better with this idea, however. Let $LP'[i, j, S_{ij}]$ denote the longest simple path from i to j, where where S_{ij} is the *set* of the intermediate vertices on this path. Thus, if $S_{ij} = \{a, b, c, i, j\}$, there are exactly six paths consistent with S_{ij}: *iabcj, iacbj, ibacj, ibcaj, icabj,* and *icbaj*. This state space has at most 2^n elements, and is thus smaller than the enumeration of all the paths. Further, this function can be evaluated using the following recurrence relation:

$$LP'[i, j, S_{ij}] = \max_{\substack{j \notin S_{ix} \\ (x,j) \in E \\ S_{ij} = S_{ix} \cup \{j\}}} LP'[i, x, S_{ix}] + c(x, j)$$

where $S \cup \{x\}$ denotes unioning S with x.

The longest simple path from i to j can then be found by maximizing over all possible intermediate vertex subsets:

$$LP[i, j] = \max_S LP'[i, j, S]$$

There are only 2^n subsets of n vertices, so this is a big improvement over enumerating all $n!$ tours. Indeed, this method can be used to solve TSPs for up to thirty vertices or more, where $n = 20$ would be impossible using the $O(n!)$ algorithm. Still, dynamic programming proves most effective on well-ordered objects.

Take-Home Lesson: Without an inherent left-to-right ordering on the objects, dynamic programming is usually doomed to require exponential space and time.

10.10   War Story: What's Past is Prolog

"But our heuristic works very, very well in practice." My colleague was simultaneously boasting and crying for help.

Unification is the basic computational mechanism in logic programming languages like Prolog. A Prolog program consists of a set of rules, where each rule has a head and an associated action whenever the rule head matches or unifies with the current computation.

An execution of a Prolog program starts by specifying a goal, say $p(a, X, Y)$, where a is a constant and X and Y are variables. The system then systematically matches the head of the goal with the head of each of the rules that can be *unified* with the goal. Unification means binding the variables with the constants, if it is possible to match them. For the nonsense program below, $p(X, Y, a)$ unifies with either of the first two rules, since X and Y can be bound to match the extra characters. The goal $p(X, X, a)$ would only match the first rule, since the variable bound to the first and second positions must be the same.

$$p(a, a, a) := h(a);$$
$$p(b, a, a) := h(a) * h(b);$$
$$p(c, b, b) := h(b) + h(c);$$
$$p(d, b, b) := h(d) + h(b);$$

"In order to speed up unification, we want to preprocess the set of rule heads so that we can quickly determine which rules match a given goal. We must organize the rules in a trie data structure for fast unification."

Tries are extremely useful data structures in working with strings, as discussed in Section 15.3 (page 448). Every leaf of the trie represents one string. Each node on the path from root to leaf is labeled with exactly one character of the string, with the ith node of the path corresponding to the string's ith character.

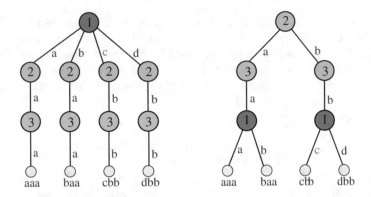

Figure 10.11: Two different tries for the same set of Prolog rule heads, where the trie on the right has four less edges.

"I agree. A trie is a natural way to represent your rule heads. Building a trie on a set of strings of characters is straightforward: just insert the strings starting from the root. So what is your problem?" I asked.

"The efficiency of our unification algorithm depends very much on minimizing the number of edges in the trie. Since we know all the rules in advance, we have the freedom to reorder the character positions in the rules. Instead of the root node always representing the first argument in the rule, we can choose to have it represent the third argument. We would like to use this freedom to build a minimum-size trie for a set of rules."

He showed me the example in Figure 10.11. A trie constructed according to the original string position order $(1, 2, 3)$ uses a total of 12 edges. However, by permuting the character order to $(2, 3, 1)$ on both sides, we could obtain a trie with only 8 edges.

"Interesting..." I started to reply before he cut me off again.

"There's one other constraint. We must keep the leaves of the trie ordered, so that the leaves of the underlying tree go left to right in the same order as the rules appear on the page. The order of rules in Prolog programs is very, very important. If you change the order of the rules, the program returns different results."

Then came my mission.

"We have a greedy heuristic for building good, but not optimal, tries that picks as the root the character position that minimizes the degree of the root. In other words, it picks the character position that has the smallest number of distinct characters in it. This heuristic works very, very well in practice. But we need you to prove that finding the best trie is NP-complete so our paper is, well, complete."

I agreed to try to prove the hardness of the problem, and chased him from my office. The problem did seem to involve some nontrivial combinatorial optimization to build the minimal tree, but I couldn't see how to factor the left-to-right order of the rules into a hardness proof. In fact, I couldn't think of any NP-complete problem that had such a left-to-right ordering constraint. After all, if a given set of n rules contained a character position in common to all the rules, this character position must be probed first in any minimum-size tree. Since the rules were ordered, each node in the subtree must represent the root of a run of consecutive rules. Thus, there were only $\binom{n}{2}$ possible nodes to choose from for this tree....

Bingo! That settled it.

The next day I went back to my colleague and told him. "I can't prove that your problem is NP-complete. But how would you feel about an efficient dynamic programming algorithm to find the best possible trie!" It was a pleasure watching his frown change to a smile as the realization took hold. An efficient algorithm to compute what you need is infinitely better than a proof saying you can't do it!

My recurrence looked something like this. Suppose that we are given n ordered rule heads $s_1, \ldots, s_n$, each with m arguments. Probing at the pth position, $1 \le p \le m$, partitions the rule heads into runs $R_1, \ldots, R_r$, where each rule in a given run $R_x = s_i, \ldots, s_j$ has the same character value as $s_i[p]$. The rules in each run must be consecutive, so there are only $\binom{n}{2}$ possible runs to worry about. The cost of probing at position p is the cost of finishing the trees formed by each created run, plus one edge per tree to link it to probe p:

$$C[i,j] = \min_{p=1}^{m} \left(\sum_{k=1}^{r} (C[i_k, j_k] + 1) \right)$$

A graduate student immediately set to work implementing this algorithm to compare with their heuristic. On many inputs, the optimal and greedy algorithms constructed the exact same trie. However, for some examples, dynamic programming gave a 20% performance improvement over greedy—that is, 20% better than very, very well in practice. The run time spent in doing the dynamic programming was a bit larger than with greedy, but in compiler optimization you are always happy to trade off a little extra compilation time for better execution time in the performance of your program. Is a 20% improvement worth this effort? That depends upon the situation. How useful would you find a 20% increase in your salary?

The fact that the rules had to remain ordered was the crucial property that we exploited in the dynamic programming solution. Indeed, without it I was able to prove that the problem *was* NP-complete with arbitrary rule orderings, something we put in the paper to make it complete.

Take-Home Lesson: The global optimum (found perhaps using dynamic programming) is often noticeably better than the solution found by typical heuristics. How important this improvement is depends on your application, but it can never hurt.

Chapter Notes

Bellman [Bel58] is credited with inventing the technique of dynamic programming. The edit distance algorithm is originally due to Wagner and Fischer [WF74]. A faster algorithm for the book partition problem appears in Khanna et al. [KMS97].

Techniques such as dynamic programming and backtracking can be used to generate worst-case efficient (although still non-polynomial) algorithms for many NP-complete problems. See Downey and Fellows [DF12] and Woeginger [Woe03] for nice surveys of such techniques.

More details about the war stories in this chapter are available in published papers. See Dawson et al. [DRR+95] for more on the Prolog trie minimization problem. Our algorithm for phase-balancing power loads from Section 10.6 (page 331) is reported in Wang et al. [WSR13]. Two-dimensional bar codes, presented in Section 10.4 (page 326), were developed largely through the efforts of Theo Pavlidis and Ynjiun Wang at Stony Brook [PSW92].

The dynamic programming algorithm presented for parsing is known as the *CKY* algorithm after its three independent inventors (Cocke, Kasami, and Younger). See [You67]. The generalization of parsing to edit distance is due to Aho and Peterson [AP72].

10.11   Exercises

Elementary Recurrences

10-1. *[3]* Up to k steps in a single bound! A child is running up a staircase with n steps and can hop between 1 and k steps at a time. Design an algorithm to count how many possible ways the child can run up the stairs, as a function of n and k. What is the running time of your algorithm?

10-2. *[3]* Imagine you are a professional thief who plans to rob houses along a street of n homes. You know the loot at house i is worth m_i, for $1 \leq i \leq n$, but you cannot rob neighboring houses because their connected security systems will automatically contact the police if two adjacent houses are broken into. Give an efficient algorithm to determine the maximum amount of money you can steal without alerting the police.

10-3. *[5]* Basketball games are a sequence of 2-point shots, 3-point shots, and 1-point free throws. Give an algorithm that computes how many possible mixes (1s,2s,3s) of scoring add up to a given n. For $n = 5$ there are four possible solutions: $(5, 0, 0)$, $(2, 0, 1)$, $(1, 2, 0)$, and $(0, 1, 1)$.

10-4. *[5]* Basketball games are a sequence of 2-point shots, 3-point shots, and 1-point free throws. Give an algorithm that computes how many possible scoring sequences add up to a given n. For $n = 5$ there are thirteen possible sequences, including 1-2-1-1, 3-2, and 1-1-1-1-1.

10-5 *[5]* Given an $s \times t$ grid filled with non negative numbers, find a path from top left to bottom right that minimizes the sum of all numbers along its path. You can only move either down or right at any point in time.

(a) Give a solution based on Dijkstra's algorithm. What is its time complexity as a function of s and t?

(b) Give a solution based on dynamic programming. What is its time complexity as a function of s and t?

Edit Distance

10-6. *[3]* Typists often make transposition errors exchanging neighboring characters, such as typing "setve" for "steve." This requires two substitutions to fix under the conventional definition of edit distance.

Incorporate a swap operation into our edit distance function, so that such neighboring transposition errors can be fixed at the cost of one operation.

10-7. *[4]* Suppose you are given three strings of characters: X, Y, and Z, where $|X| = n$, $|Y| = m$, and $|Z| = n + m$. Z is said to be a *shuffle* of X and Y iff Z can be formed by interleaving the characters from X and Y in a way that maintains the left-to-right ordering of the characters from each string.

  (a) Show that *cchocohilaptes* is a shuffle of *chocolate* and *chips*, but *chocochilatspe* is not.

  (b) Give an efficient dynamic programming algorithm that determines whether Z is a shuffle of X and Y. (Hint: the values of the dynamic programming matrix you construct should be Boolean, not numeric.)

10-8. *[4]* The longest common *substring* (not subsequence) of two strings X and Y is the longest string that appears as a run of consecutive letters in both strings. For example, the longest common substring of *photograph* and *tomography* is *ograph*.

  (a) Let $n = |X|$ and $m = |Y|$. Give a $\Theta(nm)$ dynamic programming algorithm for longest common substring based on the longest common subsequence/edit distance algorithm.

  (b) Give a simpler $\Theta(nm)$ algorithm that does not rely on dynamic programming.

10-9. *[6]* The *longest common subsequence (LCS)* of two sequences T and P is the longest sequence L such that L is a subsequence of both T and P. The *shortest common supersequence (SCS)* of T and P is the smallest sequence L such that both T and P are a subsequence of L.

  (a) Give efficient algorithms to find the LCS and SCS of two given sequences.

  (b) Let $d(T, P)$ be the minimum edit distance between T and P when no substitutions are allowed (i.e., the only changes are character insertion and deletion). Prove that $d(T, P) = |SCS(T, P)| - |LCS(T, P)|$ where $|SCS(T, P)|$ ($|LCS(T, P)|$) is the size of the shortest SCS (longest LCS) of T and P.

10-10. *[5]* Suppose you are given n poker chips stacked in two stacks, where the edges of all chips can be seen. Each chip is one of three colors. A turn consists of choosing a color and removing all chips of that color from the tops of the stacks. The goal is to minimize the number of turns until the chips are gone.

For example, consider the stacks $(RRGG, GBBB)$. Playing red, green, and then blue suffices to clear the stacks in three moves. Give an $O(n^2)$ dynamic programming algorithm to find the best strategy for a given pair of chip piles.

Greedy Algorithms

10-11. *[4]* Let $P_1, P_2, \ldots, P_n$ be n programs to be stored on a disk with capacity D megabytes. Program P_i requires s_i megabytes of storage. We cannot store them all because $D < \sum_{i=1}^{n} s_i$

(a) Does a greedy algorithm that selects programs in order of non-decreasing s_i maximize the number of programs held on the disk? Prove or give a counter-example.

(b) Does a greedy algorithm that selects programs in order of non-increasing s_i use as much of the capacity of the disk as possible? Prove or give a counter-example.

10-12. *[5]* Coins in the United States are minted with denominations of 1, 5, 10, 25, and 50 cents. Now consider a country whose coins are minted with denominations of $\{d_1, \ldots, d_k\}$ units. We seek an algorithm to make change of n units using the minimum number of this country's coins.

(a) The greedy algorithm repeatedly selects the biggest coin no bigger than the amount to be changed and repeats until it is zero. Show that the greedy algorithm does not always use the minimum number of coins in a country whose denominations are $\{1, 6, 10\}$.

(b) Give an efficient algorithm that correctly determines the minimum number of coins needed to make change of n units using denominations $\{d_1, \ldots, d_k\}$. Analyze its running time.

10-13. *[5]* In the United States, coins are minted with denominations of 1, 5, 10, 25, and 50 cents. Now consider a country whose coins are minted with denominations of $\{d_1, \ldots, d_k\}$ units. We want to count how many distinct ways $C(n)$ there are to make change of n units. For example, in a country whose denominations are $\{1, 6, 10\}$, $C(5) = 1$, $C(6)$ to $C(9) = 2$, $C(10) = 3$, and $C(12) = 4$.

(a) How many ways are there to make change of 20 units from $\{1, 6, 10\}$?

(b) Give an efficient algorithm to compute $C(n)$, and analyze its complexity. (Hint: think in terms of computing $C(n, d)$, the number of ways to make change of n units with highest denomination d. Be careful to avoid overcounting.)

10-14. *[6]* In the *single-processor scheduling problem*, we are given a set of n jobs J. Each job i has a processing time t_i, and a deadline d_i. A feasible schedule is a permutation of the jobs such that when the jobs are performed in that order, every job is finished before its deadline. The greedy algorithm for single-processor scheduling selects the job with the earliest deadline first.

Show that if a feasible schedule exists, then the schedule produced by this greedy algorithm is feasible.

Number Problems

10-15. *[3]* You are given a rod of length n inches and a table of prices obtainable for rod-pieces of size n or smaller. Give an efficient algorithm to find the maximum value obtainable by cutting up the rod and selling the pieces. For example, if $n = 8$ and the values of different pieces are:

length	1	2	3	4	5	6	7	8
price	1	5	8	9	10	17	17	20

then the maximum obtainable value is 22, by cutting into pieces of lengths 2 and 6.

10-16. *[5]* Your boss has written an arithmetic expression of n terms to compute your annual bonus, but permits you to parenthesize it however you wish. Give an efficient algorithm to design the parenthesization to maximize the value. For the expression:

$$6 + 2 \times 0 - 4$$

there exist parenthesizations with values ranging from -32 to 2.

10-17. *[5]* Given a positive integer n, find an efficient algorithm to compute the smallest number of perfect squares (e.g. $1, 4, 9, 16, \ldots$) that sum to n. What is the running time of your algorithm?

10-18. *[5]* Given an array A of n integers, find an efficient algorithm to compute the largest sum of a continuous run. For $A = [-3, 2, 7, -3, 4, -2, 0, 1]$, the largest such sum is 10, from the second through fifth positions.

10-19. *[5]* Two drivers have to divide up m suitcases between them, where the weight of the ith suitcase is w_i. Give an efficient algorithm to divide up the loads so the two drivers carry equal weight, if possible.

10-20. *[6]* The *knapsack problem* is as follows: given a set of integers $S = \{s_1, s_2, \ldots, s_n\}$, and a given target number T, find a subset of S that adds up exactly to T. For example, within $S = \{1, 2, 5, 9, 10\}$ there is a subset that adds up to $T = 22$ but not $T = 23$.

Give a dynamic programming algorithm for knapsack that runs in $O(nT)$ time.

10-21. *[6]* The *integer partition* takes a set of positive integers $S = \{s_1, \ldots, s_n\}$ and seeks a subset $I \subset S$ such that

$$\sum_{i \in I} s_i = \sum_{i \notin I} s_i$$

Let $\sum_{i \in S} s_i = M$. Give an $O(nM)$ dynamic programming algorithm to solve the integer partition problem.

10-22. *[5]* Assume that there are n numbers (some possibly negative) on a circle, and we wish to find the maximum contiguous sum along an arc of the circle. Give an efficient algorithm for solving this problem.

10-23. *[5]* A certain string processing language allows the programmer to break a string into two pieces. It costs n units of time to break a string of n characters into two pieces, since this involves copying the old string. A programmer wants to break a string into many pieces, and the order in which the breaks are made can affect the total amount of time used. For example, suppose we wish to break a 20-character string after characters 3, 8, and 10. If the breaks are made in left-to-right order, then the first break costs 20 units of time, the second break costs 17 units of time, and the third break costs 12 units of time, for a total of 49 units. If the breaks are made in right-to-left order, the first break costs 20 units of time, the second break costs 10 units of time, and the third break costs 8 units of time, for a total of only 38 units.

Give a dynamic programming algorithm that takes a list of character positions after which to break and determines the cheapest break cost in $O(n^3)$ time.

10-24. *[5]* Consider the following data compression technique. We have a table of m text strings, each at most k in length. We want to encode a data string D of length n using as few text strings as possible. For example, if our table contains *(a,ba,abab,b)* and the data string is *bababbaababa*, the best way to encode it is *(b,abab,ba,abab,a)*—a total of five code words. Give an $O(nmk)$ algorithm to find the length of the best encoding. You may assume that every string has at least one encoding in terms of the table.

10-25. *[5]* The traditional world chess championship is a match of 24 games. The current champion retains the title in case the match is a tie. Each game ends in a win, loss, or draw (tie) where wins count as 1, losses as 0, and draws as 1/2. The players take turns playing white and black. White plays first and so has an advantage. The champion plays white in the first game. The champ has probabilities w_w, w_d, and w_l of winning, drawing, and losing playing white, and has probabilities b_w, b_d, and b_l of winning, drawing, and losing playing black.

(a) Write a recurrence for the probability that the champion retains the title. Assume that there are g games left to play in the match and that the champion needs to get i points (which may be a multiple of 1/2).

(b) Based on your recurrence, give a dynamic programming algorithm to calculate the champion's probability of retaining the title.

(c) Analyze its running time for an n game match.

10-26. *[8]* Eggs break when dropped from great enough height. Specifically, there must be a floor f in any sufficiently tall building such that an egg dropped from the fth floor breaks, but one dropped from the $(f-1)$st floor will not. If the egg always breaks, then $f = 1$. If the egg never breaks, then $f = n + 1$.

You seek to find the critical floor f using an n-floor building. The only operation you can perform is to drop an egg off some floor and see what happens. You start out with k eggs, and seek to make as few drops as possible. Broken eggs cannot be reused. Let $E(k, n)$ be the minimum number of egg drops that will always suffice.

(a) Show that $E(1, n) = n$.

(b) Show that $E(k, n) = \Theta(n^{\frac{1}{k}})$.

(c) Find a recurrence for $E(k, n)$. What is the running time of the dynamic program to find $E(k, n)$?

Graph Problems

10-27. *[4]* Consider a city whose streets are defined by an $X \times Y$ grid. We are interested in walking from the upper left-hand corner of the grid to the lower right-hand corner.

Unfortunately, the city has bad neighborhoods, whose intersections we do not want to walk in. We are given an $X \times Y$ matrix *bad*, where *bad[i,j]* = *"yes"* iff the intersection between streets i and j is in a neighborhood to avoid.

(a) Give an example of the contents of *bad* such that there is no path across the grid avoiding bad neighborhoods.

(b) Give an $O(XY)$ algorithm to find a path across the grid that avoids bad neighborhoods.

(c) Give an $O(XY)$ algorithm to find the *shortest* path across the grid that avoids bad neighborhoods. You may assume that all blocks are of equal length. For partial credit, give an $O(X^2 Y^2)$ algorithm.

10-28. *[5]* Consider the same situation as the previous problem. We have a city whose streets are defined by an $X \times Y$ grid. We are interested in walking from the upper left-hand corner of the grid to the lower right-hand corner. We are given an $X \times Y$ matrix *bad*, where *bad[i,j]* = *"yes"* iff the intersection between streets i and j is somewhere we want to avoid.

If there were no bad neighborhoods to contend with, the shortest path across the grid would have length $(X - 1) + (Y - 1)$ blocks, and indeed there would be many such paths across the grid. Each path would consist of only rightward and downward moves.

Give an algorithm that takes the array *bad* and returns the *number* of safe paths of length $X + Y - 2$. For full credit, your algorithm must run in $O(XY)$.

10-29. *[5]* You seek to create a stack out of n boxes, where box i has width w_i, height h_i, and depth d_i. The boxes cannot be rotated, and can only be stacked on top of one another when each box in the stack is strictly larger than the box above it in width, height, and depth. Give an efficient algorithm to construct the tallest possible stack, where the height is the sum of the heights of each box in the stack.

Design Problems

10-30. *[4]* Consider the problem of storing n books on shelves in a library. The order of the books is fixed by the cataloging system and so cannot be rearranged. Therefore, we can speak of a book b_i, where $1 \leq i \leq n$, that has a thickness t_i and height h_i. The length of each bookshelf at this library is L.

Suppose all the books have the same height h (i.e., $h = h_i$ for all i) and the shelves are all separated by a distance greater than h, so any book fits on any shelf. The greedy algorithm would fill the first shelf with as many books as we can until we get the smallest i such that b_i does not fit, and then repeat with subsequent shelves. Show that the greedy algorithm always finds the book placement that uses the minimum number of shelves, and analyze its time complexity.

10-31. *[6]* This is a generalization of the previous problem. Now consider the case where the height of the books is not constant, but we have the freedom to adjust the height of each shelf to that of the tallest book on the shelf. Here the cost of a particular layout is the sum of the heights of the largest book on each shelf.

- Give an example to show that the greedy algorithm of stuffing each shelf as full as possible does not always give the minimum overall height.

- Give an algorithm for this problem, and analyze its time complexity. (Hint: use dynamic programming.)

10-32. *[5]* Consider a linear keyboard of lowercase letters and numbers, where the leftmost 26 keys are the letters A–Z in order, followed by the digits 0–9 in order, followed by the 30 punctuation characters in a prescribed order, and ended on a blank. Assume you start with your left index finger on the "A" and your right index finger on the blank.

Give a dynamic programming algorithm that finds the most efficient way to type a given text of length n, in terms of minimizing total movement of the fingers involved. For the text $ABABABAB\ldots ABAB$, this would involve shifting both fingers all the way to the left side of the keyboard. Analyze the complexity of your algorithm as a function of n and k, the number of keys on the keyboard.

10-33. *[5]* You have come back from the future with an array G, where $G[i]$ tells you the price of Google stock i days from now, for $1 \leq i \leq n$. You seek to use this information to maximize your profit, but are only permitted to complete at most one transaction (i.e. either buy one or sell one share of the stock) per day. Design an efficient algorithm to construct the buy–sell sequence to maximize your profit. Note that you cannot sell a share unless you currently own one.

10-34. *[8]* You are given a string of n characters $S = s_1 \ldots s_n$, which you believe to be a compressed text document in which all spaces have been removed, like **itwasthebestoftimes**.

(a) You seek to reconstruct the document using a dictionary, which is available in the form of a Boolean function $dict(w)$, where $dict(w)$ is true iff string w is a valid word in the language. Give an $O(n^2)$ algorithm to determine whether string S can be reconstituted as a sequence of valid words, assuming calls to $dict(w)$ take unit time.

(b) Now assume you are given the dictionary as a set of m words each of length at most l. Give an efficient algorithm to determine whether string S can be reconstituted as a sequence of valid words, and its running time.

10-35. *[8]* Consider the following two-player game, where you seek to get the biggest score. You start with an n-digit integer N. With each move, you get to take either the first digit or the last digit from what is left of N, and add that to your score, with your opponent then doing the same thing to the now smaller number. You continue taking turns removing digits until none are left. Give an efficient algorithm that finds the best possible score that the first player can get for a given digit string N, assuming the second player is as smart as can be.

10-36. *[6]* Given an array of n real numbers, consider the problem of finding the maximum sum in any contiguous subarray of the input. For example, in the array

$$[31, -41, 59, 26, -53, 58, 97, -93, -23, 84]$$

the maximum is achieved by summing the third through seventh elements, where $59 + 26 + (-53) + 58 + 97 - 187$. When all numbers are positive, the entire array is the answer, while when all numbers are negative, the empty array maximizes the total at 0.

- Give a simple and clear $\Theta(n^2)$-time algorithm to find the maximum contiguous subarray.

- Now give a $\Theta(n)$-time dynamic programming algorithm for this problem. To get partial credit, you may instead give a *correct* $O(n \log n)$ divide-and-conquer algorithm.

10-37. *[7]* Consider the problem of examining a string $x = x_1 x_2 \ldots x_n$ from an alphabet of k symbols, and a multiplication table over this alphabet. Decide whether or not it is possible to parenthesize x in such a way that the value of the resulting expression is a, where a belongs to the alphabet. The multiplication table is neither commutative or associative, so the order of multiplication matters.

	a	b	c
a	a	c	c
b	a	a	b
c	c	c	c

For example, consider the above multiplication table and the string *bbbba*. Parenthesizing it $(b(bb))(ba)$ gives a, but $((((bb)b)b)a)$ gives c.

Give an algorithm, with time polynomial in n and k, to decide whether such a parenthesization exists for a given string, multiplication table, and goal symbol.

10-38. *[6]* Let α and β be constants. Assume that it costs α to go left in a binary search tree, and β to go right. Devise an algorithm that builds a tree with optimal expected query cost, given keys $k_1, \ldots, k_n$ and the probabilities that each will be searched $p_1, \ldots, p_n$.

Interview Problems

10-39. *[5]* Given a set of coin denominations, find the minimum number of coins to make a certain amount of change.

10-40. *[5]* You are given an array of n numbers, each of which may be positive, negative, or zero. Give an efficient algorithm to identify the index positions i and j to obtain the maximum sum of the ith through jth numbers.

10-41. *[7]* Observe that when you cut a character out of a magazine, the character on the reverse side of the page is also removed. Give an algorithm to determine whether you can generate a given string by pasting cutouts from a given magazine. Assume that you are given a function that will identify the character and its position on the reverse side of the page for any given character position.

LeetCode

10-1. `https://leetcode.com/problems/binary-tree-cameras/`

10-2. `https://leetcode.com/problems/edit-distance/`

10-3. `https://leetcode.com/problems/maximum-product-of-splitted-binary-tree/`

HackerRank

10-1. https://www.hackerrank.com/challenges/ctci-recursive-staircase/

10-2. https://www.hackerrank.com/challenges/coin-change/

10-3. https://www.hackerrank.com/challenges/longest-increasing-subsequent/

Programming Challenges

These programming challenge problems with robot judging are available at https://onlinejudge.org:

10-1. "Is Bigger Smarter?"—Chapter 11, problem 10131.

10-2. "Weights and Measures"—Chapter 11, problem 10154.

10-3. "Unidirectional TSP"—Chapter 11, problem 116.

10-4. "Cutting Sticks"—Chapter 11, problem 10003.

10-5. "Ferry Loading"—Chapter 11, problem 10261.

Chapter 11

NP-Completeness

I will now introduce techniques for proving that *no* efficient algorithm can exist for a given problem. The practical reader is probably squirming at the notion of proving anything, and will be particularly alarmed at the idea of investing time to prove that something does not exist. Why are you better off knowing that something you don't know how to do in fact can't be done at all?

The truth is that the theory of NP-completeness is an immensely useful tool for the algorithm designer, even though all it provides are negative results. The theory of NP-completeness enables us to focus our efforts more productively, by revealing when the search for an efficient algorithm is doomed to failure. Whenever one tries and *fails* to show a problem is hard, that suggests there may well be an efficient algorithm to solve it. Two war stories in Chapter 10 described happy results springing from bogus claims of hardness.

The theory of NP-completeness also enables us to identify which properties make a particular problem hard. This can provide direction to model it in different ways, or exploit more benevolent characteristics of the problem. Developing a sense for which problems are hard is an important skill for algorithm designers, and only comes from hands-on experience with proving hardness.

The fundamental concept we will use here is *reduction*, showing that two problems are really equivalent. We illustrate this idea through a series of reductions, each of which either yields an efficient algorithm for one problem or an argument that no such algorithm can exist for the other. We also provide a brief introduction to the complexity-theoretic aspects of NP-completeness, one of the most fundamental notions in computer science.

11.1 Problems and Reductions

We have encountered several problems in this book where we couldn't find any efficient algorithm. The theory of NP-completeness provides the tools needed to show that these problems are all, on some level, really the same problem.

The key idea to demonstrating the hardness of a problem is that of a *re-*

S. S. Skiena, *The Algorithm Design Manual*, Texts in Computer Science,
https://doi.org/10.1007/978-3-030-54256-6_11

duction, or translation, between two problems. The following allegory of NP-completeness may help explain the idea. A bunch of kids take turns fighting each other in the school yard to prove how "tough" they are. Adam beats up Bill, who then beats up Dwayne. So who if any among them qualifies as "tough?" The truth is that there is no way to know without an external standard. If I tell you that the action takes place in a kindergarten school yard, then the fight results don't mean very much. But suppose instead that I tell you Dwayne was in fact Dwayne "The Rock" Johnson, certified tough guy. You have to be impressed—both Adam and Bill must be at least as tough as he is. In this telling, each fight represents a reduction, and Dwayne Johnson takes on the role of *satisfiability*—a certifiably hard problem.

Reductions are algorithms that convert one problem into another. To describe them, we must be somewhat rigorous in our definitions. An algorithmic *problem* is a general question, with parameters for input and conditions on what constitutes a satisfactory answer or solution. An *instance* is a problem with the input parameters specified. The difference can be made clear by an example:

Problem: The Traveling Salesman Problem (TSP)
Input: A weighted graph G.
Output: Which tour $(v_1, v_2, ..., v_n)$ minimizes $\sum_{i=1}^{n-1} d[v_i, v_{i+1}] + d[v_n, v_1]$?

Any weighted graph defines an instance of TSP. Each particular *instance* has at least one minimum cost tour. The general traveling salesman *problem* asks for an algorithm to find the optimal tour for any possible instance.

11.1.1   The Key Idea

Now consider two algorithmic problems, called *Bandersnatch* and *Bo-billy*. Suppose that I gave you the following reduction/algorithm to solve the *Bandersnatch* problem:

> Bandersnatch(G)
>   Translate the input G to an instance Y of the Bo-billy problem.
>   Call the subroutine Bo-billy to solve instance Y.
>   Return the answer of Bo-billy(Y) as the answer to Bandersnatch(G).

This algorithm will *correctly* solve the Bandersnatch problem provided that the translation to Bo-billy always preserves the correctness of the answer. In other words, provided that the translation has the property that for any instance G,

$$\text{Bandersnatch}(G) = \text{Bo-billy}(Y)$$

A translation of instances from one type of problem to instances of another such that the answers are preserved is what we mean by a *reduction*.

Now suppose this reduction translates instance G to Y in $O(P(n))$ time. There are two possible implications:

- *If* my Bo-billy subroutine ran in $O(P'(n))$, this yields an algorithm to solve the Bandersnatch problem in $O(P(n) + P'(n))$, by translating the problem and then executing the Bo-billy subroutine to solve it.

- *If* I know that $\Omega(P'(n))$ is a lower bound on computing Bandersnatch, meaning there definitely cannot exist a faster algorithm to solve it, then $\Omega(P'(n) - P(n))$ *must* be a lower bound to compute Bo-billy. Why? If I *could* solve Bo-billy faster than this, the above reduction would violate my lower bound on solving Bandersnatch. Because this is impossible, there can be no way to solve Bo-billy any faster than claimed.

Essentially, this reduction shows that Bo-billy is no easier than Bandersnatch. Therefore, if Bandersnatch is hard this means Bo-billy must also be hard. We will illustrate this point by giving a variety of problem reductions in this chapter.

> *Take-Home Lesson:* Reductions are a way to show that two problems are essentially identical. A fast algorithm (or the lack of one) for one of the problems implies a fast algorithm (or the lack of one) for the other.

11.1.2 Decision Problems

Reductions translate between problems so that their answers are identical in every problem instance. Problems differ in the *range* or *type* of possible answers. The traveling salesman problem returns a permutation of vertices as the answer, while other types of problems may return strings or numbers as answers, perhaps restricted to positive numbers or integers.

The simplest interesting class of problems have answers restricted to true and false. These are called *decision problems*. It proves convenient to reduce/translate answers between decision problems because both only allow true and false as possible answers.

Fortunately, most interesting optimization problems can be phrased as decision problems that capture the essence of the computation. For example, the traveling salesman decision problem is defined as:

Problem: The Traveling Salesman Decision Problem (TSDP)
Input: A weighted graph G and integer k.
Output: Does there exist a TSP tour with cost $\leq k$?

This decision version captures the heart of the traveling salesman problem, in that if you had a fast algorithm for the decision problem, you could do a binary search with different values of k and quickly home in on the cost of the optimal TSP solution. With just a bit more cleverness, you could reconstruct the actual tour permutation using a fast solution to the decision problem.

From now on I will generally talk about decision problems, because they prove easier to work with and still capture the power of the theory of NP-completeness.

11.2   Reductions for Algorithms

Reductions are an honorable way to generate new algorithms from old ones. Whenever we can translate the input for a problem we *want to solve* into input for a problem we *know how to solve*, we can compose the translation and the solution into an algorithm to deal with our problem.

In this section, we look at several reductions that lead to efficient algorithms. To solve problem A, we translate/reduce the A instance to an instance of B, and then solve this instance using an efficient algorithm for problem B. The overall running time is the time needed to perform the reduction plus that to solve the B instance.

11.2.1   Closest Pair

The *closest-pair* problem asks to find the pair of numbers within a set S that have the smallest difference between them. For example, the closest pair in $S = \{10, 4, 8, 3, 12\}$ is $(3, 4)$. We can make it a decision problem by asking if this value is less than some threshold:

Input: A set S of n numbers, and threshold t.
Output: Is there a pair $s_i, s_j \in S$ such that $|s_i - s_j| \leq t$?

Finding the closest pair is a simple application of sorting, since the closest pair must be neighbors after sorting. This gives the following algorithm:

> CloseEnoughPair(S,t)
>   Sort S.
>   Is $\min_{1 \leq i < n} |s_{i+1} - s_i| \leq t$?

There are several things to note about this simple reduction:

- The decision version captured what is interesting about the general problem, meaning it is no easier than finding the actual closest pair.

- The complexity of this algorithm depends upon the complexity of sorting. Use an $O(n \log n)$ algorithm to sort, and it takes $O(n \log n + n)$ to find the closest pair.

- This reduction and the fact that there is an $\Omega(n \log n)$ lower bound on sorting *does not* prove that the close-enough pair problem must take $\Omega(n \log n)$ time in the worst case. Perhaps this is just a slow algorithm for close-enough pair, and there is a faster approach that avoids sorting?

- On the other hand, *if* we knew that a close-enough pair required $\Omega(n \log n)$ time to solve in the worst case, this reduction would suffice to prove that sorting couldn't be solved any faster than $\Omega(n \log n)$ because that would imply a faster algorithm for close-enough pair.

11.2.2 Longest Increasing Subsequence

Recall Chapter 10, where dynamic programming was used to solve a variety of problems, including string edit distance. To review:

Problem: Edit Distance
Input: Integer or character sequences S and T; penalty costs for each insertion (c_{ins}), deletion (c_{del}), and substitution (c_{sub}).
Output: What is the cost of the least expensive sequence of operations that transforms S to T?

It was shown that many other problems can be solved using edit distance. But these algorithms can often be viewed as reductions. Consider:

Problem: Longest Increasing Subsequence (LIS)
Input: An integer or character sequence S.
Output: What is the length of the longest sequence of positions $p_1, \ldots, p_m$ such that $p_i < p_{i+1}$ and $S_{p_i} < S_{p_{i+1}}$?

In Section 10.3 (page 324) I demonstrated that longest increasing subsequence can be solved as a special case of edit distance:

$$\text{LongestIncreasingSubsequence}(S)$$
$$T = \text{Sort}(S)$$
$$c_{ins} = c_{del} = 1$$
$$c_{sub} = \infty$$
$$\text{Return } (|S| - \text{EditDistance}(S,T,c_{ins},c_{del},c_{sub})/2)$$

Why does this work? By constructing the second sequence T as the elements of S sorted in increasing order, we ensure that any common subsequence must be an increasing subsequence. If we are never allowed to do any substitutions (because $c_{sub} = \infty$), the optimal alignment of S and T finds the longest common subsequence between them and removes everything else. For example, transforming $S = cab$ to abc costs two, namely inserting and deleting the unmatched c. The length of S minus half this cost gives the length of the LIS.

What are the implications of this reduction? The reduction takes $O(n \log n)$ time because of the cost of sorting. Because edit distance takes time $O(|S| \cdot |T|)$, this gives a quadratic algorithm to find the longest increasing subsequence of S. In fact, there exists a faster $O(n \log n)$ algorithm for LIS using clever data structures, while edit distance is known to be quadratic in the worst case. Hence, our reduction gives us a simple but not optimal polynomial-time algorithm.

11.2.3 Least Common Multiple

The *least common multiple* (lcm) and *greatest common divisor* (gcd) problems arise often in working with integers. We say b *divides* a (written $b \mid a$) if there exists an integer d such that $a = bd$. Then:

Problem: Least Common Multiple (lcm)
Input: Two positive integers x and y.
Output: Return the smallest positive integer m such that m is a multiple of x and also a multiple of y.

Problem: Greatest Common Divisor (gcd)
Input: Two positive integers x and y.
Output: Return the largest integer d such that d divides both x and y.

For example, $lcm(24, 36) = 72$ and $gcd(24, 36) = 12$. Both problems can be solved easily after reducing x and y to their prime factorizations, but no efficient algorithm is known for factoring integers (see Section 16.8 (page 490)). Fortunately, Euclid's algorithm gives an efficient way to solve greatest common divisor without factoring. It is a recursive algorithm that rests on two observations. First,

if $(b \mid a)$, then $gcd(a, b) = b$.

This should be pretty clear. if b divides a, then $a = bk$ for some integer k, and thus $gcd(bk, b) = b$. Second,

If $a = bt + r$ for integers t and r, then $gcd(a, b) = gcd(b, r)$.

Then, for $a \geq b$, Euclid's algorithm repeatedly replaces (a, b) by $(b, a \bmod b)$ until $b = 0$. Its worst-case running time is $O(\log b)$.

Since $x \cdot y$ is a multiple of both x and y, $lcm(x, y) \leq xy$. The only way that there can be a smaller common multiple is if there is some non-trivial factor shared between x and y. This observation, coupled with Euclid's algorithm, provides an efficient way to compute least common multiple, namely:

LeastCommonMultiple(x,y)
      Return $(xy/gcd(x, y))$.

This reduction gives us a nice way to reuse Euclid's efforts for lcm.

11.2.4   Convex Hull (*)

My final example of a reduction from an "easy" problem (meaning one that can be solved in polynomial time) involves finding convex hulls of point sets. A polygon is *convex* if the straight line segment drawn between any two points inside the polygon P lies completely within the polygon. This is the case when P contains no notches or *concavities*, so convex polygons are nicely shaped. The convex hull provides a very useful way to provide structure to a point set. Applications are presented in Section 20.2 (page 626).

Problem: Convex Hull
Input: A set S of n points in the plane.
Output: Find the smallest convex polygon containing all the points of S.

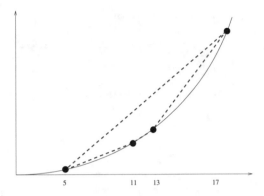

Figure 11.1: Reducing sorting to convex hull by mapping points to a parabola

I will now show how to transform an instance of sorting (say $\{13, 5, 11, 17\}$) to an instance of the convex hull problem. This means we must translate each number to a point in the plane. We do so by mapping x to (x, x^2). Why? This maps each integer to a point on the parabola $y = x^2$, as shown in Figure 11.1. Since the region above this parabola is convex, every point must be on the convex hull. Furthermore, since neighboring points on the convex hull have neighboring x values, the convex hull returns the points sorted by the x-coordinate—that is, the original numbers. Creating and reading off these points takes $O(n)$ time:

> Sort(S)
> > For each $i \in S$, create point (i, i^2).
> > Call subroutine convex-hull on this point set.
> > From the left-most point in the hull,
> > > read off the points from left to right.

What does this mean? Recall the sorting lower bound of $\Omega(n \log n)$. If we could compute convex hull in better than $n \log n$, this reduction would imply that we could sort faster than $\Omega(n \log n)$, which violates our lower bound. Thus, convex hull must take $\Omega(n \log n)$ as well! Observe that any $O(n \log n)$ convex hull algorithm also gives us a complicated but correct $O(n \log n)$ sorting algorithm when coupled with this reduction.

11.3 Elementary Hardness Reductions

The reductions in Section 11.2 (page 358) demonstrate transformations between pairs of problems for which efficient algorithms exist. However, we are mainly concerned with using reductions to prove hardness, by showing that *Bo-billy* is at least as hard as *Bandersnatch*.

For now, I want you to trust me when I say that *Hamiltonian cycle* and *vertex cover* are hard problems. The entire picture (presented in Figure 11.2) will become clear by the end of the chapter.

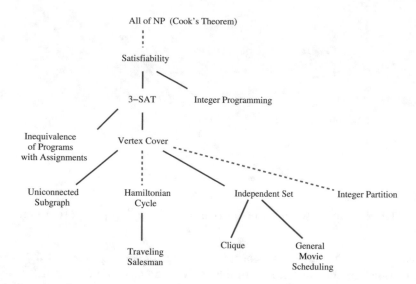

Figure 11.2: A portion of the reduction tree for NP-complete problems. Blue lines denote the reductions presented in this chapter.

11.3.1   Hamiltonian Cycle

The Hamiltonian cycle problem is one of the most famous in graph theory. It seeks a tour that visits each vertex of a given graph exactly once. Hamiltonian cycle has a long history and many applications, as discussed in Section 19.5. Formally, it is defined as:

Problem: Hamiltonian Cycle
Input: An unweighted graph G.
Output: Does there exist a simple tour that visits each vertex of G without repetition?

Hamiltonian cycle has some obvious similarity to the traveling salesman problem. Both problems seek a tour that visits each vertex exactly once. There are also differences between the two problems. TSP works on weighted graphs, while Hamiltonian cycle works on unweighted graphs. The following reduction from Hamiltonian cycle to traveling salesman shows that the similarities are greater than the differences:

HamiltonianCycle($G = (V, E)$)
    Construct a complete weighted graph $G' = (V', E')$ where $V' = V$.
    $n = |V|$
    for $i = 1$ to n do
        for $j = 1$ to n do
            if $(i, j) \in E$ then $w(i, j) = 1$ else $w(i, j) = 2$
    Return the answer to Traveling-Salesman-Decision-Problem(G', n).

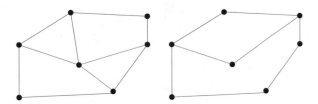

Figure 11.3: Graphs with (left) and without (right) a Hamiltonian cycle.

The actual reduction is quite simple, with the translation from unweighted to weighted graph designed to ensure that the answers of the two problems will be identical. If the graph G has a Hamiltonian cycle $(v_1, \ldots, v_n)$, then this very same tour will correspond to n edges in E' each of weight 1: this defines a TSP tour in G' of weight exactly n. If G does not have a Hamiltonian cycle, then every tour in G' must contain at least one weight 2 edge, so there cannot be a TSP tour of weight n.

This reduction is truth preserving and fast, running in $\Theta(n^2)$ time. A fast algorithm for TSP would imply a fast algorithm for Hamiltonian cycle, while a hardness proof for Hamiltonian cycle would imply that TSP is hard. Since the latter is the case, this reduction shows that TSP is hard, at least as hard as Hamiltonian cycle.

11.3.2 Independent Set and Vertex Cover

The vertex cover problem, discussed more thoroughly in Section 19.3 (page 591), asks for a small set of vertices that touch every edge in a graph. More formally:

Problem: Vertex Cover
Input: A graph $G = (V, E)$ and integer $k \leq |V|$.
Output: Is there a subset S of at most k vertices such that every $e \in E$ contains at least one vertex in S?

It is trivial to find *a* vertex cover of a graph: consider the cover that consists of *all* the vertices. More tricky is to cover the edges using as small a set of vertices as possible. For the graph in Figure 11.4, four of the eight vertices are sufficient to cover.

A set of vertices S of graph G is *independent* if there are no edges (x, y) where both $x \in S$ and $y \in S$. This means there are no edges between any two vertices in an independent set. Again, finding *an* independent set is trivial: just take any single vertex. As discussed in Section 19.2 (page 589), independent set arises in facility location problems. The maximum independent set decision problem is defined:

Problem: Independent Set
Input: A graph G and integer $k \leq |V|$.
Output: Does there exist a set of k independent vertices in G?

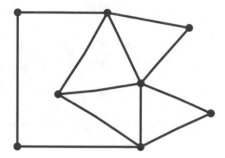

Figure 11.4: Red vertices form a vertex cover of G, so the blue vertices must define an independent set.

Both vertex cover and independent set are problems that revolve around finding special subsets of vertices: the first with representatives of every edge, the second with no edges. If S is a vertex cover of G, then the remaining vertices $V - S$ must form an independent set, for if there was an edge (x, y) that had both vertices in $V - S$, then S could not have been a vertex cover. This gives us a reduction between the two problems:

$$\text{VertexCover}(G, k)$$
$$G' = G$$
$$k' = |V| - k$$
$$\text{Return the answer to IndependentSet}(G', k')$$

Again, a simple reduction shows that one problem is at least as hard as the other. Notice how translation occurs without any knowledge of the answer: we transform the *input*, not the solution. This reduction shows that the hardness of vertex cover implies that independent set must also be hard. It is easy to reverse the roles of the two problems in this particular reduction, thus proving that the two problems are equally hard.

Stop and Think: Hardness of General Movie Scheduling

Problem: Recall the movie scheduling problem, discussed in Section 1.2 (page 8). There, each possible movie project came with a single time interval during which filming took place. We sought the largest possible subset of movie projects such that no two conflicting projects (meaning both requiring the actor at the same time) were selected.

The general problem allows movie projects to have discontinuous schedules. For example, Project A running both January–March and May–June does not intersect Project B running in April and August, but *does* collide with Project C running from June to July.

Prove that the *general* movie scheduling problem is NP-complete, with a reduction from independent set.

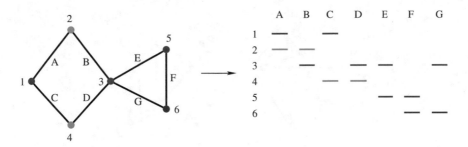

Figure 11.5: Reduction from independent set to generalized movie scheduling, with numbered vertices and lettered edges.

Problem: General Movie Scheduling Decision Problem
Input: A set $I = \{I_1, \ldots, I_n\}$ of n *sets of* intervals on the line, integer k.
Output: Can a subset of at least k mutually non-overlapping interval *sets* be selected from I?

Solution: To prove a problem hard, we first need to establish which is Bandersnatch and which is Bo-billy. Here we need to show how to translate *all* independent set instances into instances of general movie scheduling—meaning sets of disjoint line intervals. Thus, independent set is Bandersnatch and general movie scheduling is Bo-billy.

What is the correspondence between the two problems? Both problems involve selecting the largest subsets possible—of vertices and movies, respectively. This suggests we must translate vertices into movies. Furthermore, both require the selected elements not to interfere, by sharing an edge or overlapping an interval, respectively.

My construction is as follows. Create an interval on the line for each of the m edges of the graph. The movie associated with each vertex will contain the intervals for the edges adjacent with it, as shown in Figure 11.5.

  IndependentSet(G, k)
   $I = \emptyset$
   For the ith edge (x, y), $1 \le i \le m$
    Add interval $[i, i + 0.5]$ to movie x's interval set I_x in I
    Add interval $[i, i + 0.5]$ to movie y's interval set I_y in I
   Return the answer to GeneralMovieScheduling(I, k)

Each pair of vertices sharing an edge (forbidden to be in an independent set) defines a pair of movies sharing a time interval (forbidden to be in the actor's schedule). Thus, the largest satisfying subsets for both problems are the same, and a fast algorithm for solving general movie scheduling gives us a fast algorithm for solving independent set. Thus, general movie scheduling must be at least as hard as independent set, and hence NP-complete. ∎

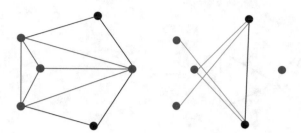

Figure 11.6: A small graph with a four-vertex clique (left), with the corresponding independent set in black forming a two-vertex clique in the graph's complement (right).

11.3.3   Clique

A social clique is a group of mutual friends who all hang around together. Everyone knows everybody. A graph-theoretic *clique* is a complete subgraph, where each vertex pair has an edge between them. Cliques are the densest possible subgraphs:

Problem: Maximum Clique
Input: A graph $G = (V, E)$ and integer $k \leq |V|$.
Output: Does the graph contain a clique of k vertices, meaning is there a subset of vertices S where $|S| = k$ such that every pair of vertices in S defines an edge of G?

The graph in Figure 11.6 contains a clique of four blue vertices. Within the friendship graph, we would expect to see large cliques corresponding to families, workplaces, neighborhoods, religious organizations, and schools. Applications of clique are further discussed in Section 19.1 (page 586).

In the independent set problem, we looked for a subset S with no edges between two vertices of S. This contrasts with clique, where we insist that there *always* be an edge between two vertices. A reduction between these problems follows by reversing the roles of edges and non-edges—an operation known as *complementing* the graph:

> IndependentSet(G, k)
>   Construct a graph $G' = (V', E')$ where $V' = V$, and
>   For all $(i, j) \notin E$, add (i, j) to E'
>   Return Clique(G', k)

These last two reductions provide a chain linking three different problems together. The hardness of clique is implied by the hardness of independent set, which is implied by the hardness of vertex cover. By constructing reductions in a chain, we link together pairs of problems in implications of hardness. Our work is complete once all these chains begin with a single "Dwayne Johnson" problem that is accepted as hard. Satisfiability is the problem that will serve as the first link in this chain.

11.4 Satisfiability

To demonstrate the hardness of problems by using reductions, we must start from a single problem that is absolutely, certifiably, undeniably hard to compute. The mother of all NP-complete problems is a logic problem named *satisfiability*:

Problem: Satisfiability (SAT)
Input: A set of Boolean variables V and a set of logic clauses C over V.
Output: Does there exist a satisfying truth assignment for C—in other words, a way to set each of the variables $\{v_1, \ldots, v_n\}$ either true or false so that every clause contains at least one true literal?

This can be made clear with two examples. Consider $C = \{\{v_1, \bar{v}_2\}, \{\bar{v}_1, v_2\}\}$ over the Boolean variables $V = \{v_1, v_2\}$. We use $\bar{v}_i$ to denote the complement of the variable v_i, because $\bar{v}_i$ means "not v_i." We get credit for satisfying a particular clause containing v_i if $v_i = $ true, or a clause containing $\bar{v}_i$ if $v_i = $ false. Therefore, satisfying a particular set of clauses involves making a series of n true or false decisions, trying to find the right truth assignment to satisfy all of them. The example clause set $C = \{\{v_1, \bar{v}_2\}, \{\bar{v}_1, v_2\}\}$ corresponds to the logical expression

$$(v_1 \vee \bar{v}_2) \wedge (\bar{v}_1 \vee v_2)$$

and can be satisfied either by setting $v_1 = v_2 = $ true or $v_1 = v_2 = $ false.

However, consider the set of clauses $\{\{v_1, v_2\}, \{v_1, \bar{v}_2\}, \{\bar{v}_1\}\}$. Here there can be no satisfying assignment, because v_1 *must* be false to satisfy the third clause, which means that v_2 *must* be false to satisfy the second clause, which then leaves the first clause unsatisfiable. Although you try, and you try, and you try, you can't get no satisfaction.

For a combination of social and technical reasons, it is well accepted that satisfiability is a hard problem; one for which no worst-case polynomial-time algorithm exists. Literally every top-notch algorithm expert in the world (and countless lesser lights) has directly or indirectly tried to come up with a fast algorithm to test whether any given set of clauses is satisfiable. All have failed. Furthermore, many strange and impossible-to-believe things in the field of computational complexity have been shown to be true if there exists a fast satisfiability algorithm. Proving something is as hard as satisfiability means that it is hard. See Section 17.10 (page 537) for more on the satisfiability problem and its applications.

11.4.1 3-Satisfiability

Satisfiability's role as the first NP-complete problem implies that the problem is hard to solve in the worst case. But certain special-case instances of the problem are not necessarily so tough. Suppose that each clause contains exactly one literal, say $\{v_i\}$ or $\{\bar{v}_j\}$. There is only one way to set the literal so as to satisfy such a clause: clearly v_i had better be set true and v_j set false to have any hope of satisfying the full set of clauses. Only when we have two clauses that directly contradict each other, such as $C = \{\{v_1\}, \{\overline{v}_1\}\}$, will the set not be satisfiable.

Since it is so easy to determine whether clause sets with exactly one literal per clause are satisfiable, we are interested in slightly larger classes. How many literals per clause do you need to turn the problem from polynomial to hard? This transition occurs when each clause contains three literals, that is,

Problem: 3-Satisfiability (3-SAT)
Input: A collection of clauses C where each clause contains exactly 3 literals, over a set of Boolean variables V.
Output: Is there a truth assignment to V such that each clause is satisfied?

Since 3-SAT is a restricted case of satisfiability, the hardness of 3-SAT would imply that general satisfiability is hard. The converse isn't true, since the hardness of general satisfiability could conceivably depend upon having long clauses. But we can show the hardness of 3-SAT using a reduction that translates every instance of satisfiability into an instance of 3-SAT without changing whether it is satisfiable.

This reduction transforms each clause independently based on its *length*, by adding new clauses and Boolean variables along the way. Suppose clause C_i contained k literals:

- $k = 1$, say $C_i = \{z_1\}$ – We create two new variables v_1, v_2 and four new 3-literal clauses: $\{v_1, v_2, z_1\}, \{v_1, \overline{v}_2, z_1\}, \{\overline{v}_1, v_2, z_1\}, \{\overline{v}_1, \overline{v}_2, z_1\}$. Observe that the only way that all four of these clauses can be simultaneously satisfied is if $z_1 = \text{true}$, which means the original C_i will have been satisfied.

- $k = 2$, say $C_i = \{z_1, z_2\}$ – We create one new variable v_1 and two new clauses: $\{v_1, z_1, z_2\}, \{\overline{v}_1, z_1, z_2\}$. Again, the only way to satisfy both of these clauses is to have at least one of z_1 and z_2 be true, thus satisfying C_i.

- $k = 3$, say $C_i = \{z_1, z_2, z_3\}$ – We copy C_i into the 3-SAT instance unchanged: $\{z_1, z_2, z_3\}$.

- $k > 3$, say $C_i = \{z_1, z_2, ..., z_k\}$ – Here we create $k-3$ new variables and $k-2$ new clauses in a chain, where $C_{i,1} = \{z_1, z_2, \overline{v}_{i,1}\}$, $C_{i,j} = \{v_{i,j-1}, z_{j+1}, \overline{v}_{i,j}\}$ for $2 \leq j \leq k-3$, and $C_{i,k-2} = \{v_{i,k-3}, z_{k-1}, z_k\}$. This is best illustrated with an example. The clause

$$C_i = \{z_1, z_2, z_3, z_4, z_5, z_6\}$$

gets transformed into the following set of four 3-literal clauses with three new Boolean variables: $v_{i,1}$, $v_{i,2}$, and $v_{i,3}$:

$$\{\{z_1, z_2, \overline{v}_{i,1}\}, \{v_{i,1}, z_3, \overline{v}_{i,2}\}, \{v_{i,2}, z_4, \overline{v}_{i,3}\}, \{v_{i,3}, z_5, z_6\}\}$$

The most complicated case is that of the large clauses. If none of the original literals in C_i are true, then there are not enough new free variables to be able to satisfy all the new subclauses. You can satisfy $C_{i,1}$ by setting $v_{i,1} =$ false, but this forces $v_{i,2} =$ false, and so on until finally $C_{i,k-2}$ cannot be satisfied. However, if any single literal $z_i =$ true, then we have $k - 3$ free variables and $k - 3$ remaining 3-clauses, so we can satisfy all of them.

This transform takes $O(n+c)$ time if there were c clauses and n total literals in the SAT instance. Since any solution to the original SAT problem instance also satisfies the 3-SAT instance we have constructed, and vice versa, the transformed problem is equivalent to the original.

Note that a slight modification to this construction would serve to prove that 4-SAT, 5-SAT, or any $(k \geq 3)$-SAT is also NP-complete. However, this construction breaks down if we try to use it for 2-SAT, since there is no way to stuff anything into the chain of clauses. It turns out that a breadth-first search on an appropriate graph can be used to give a linear-time algorithm for 2-SAT, as discussed in Section 17.10 (page 537).

11.5 Creative Reductions from SAT

Since both satisfiability and 3-SAT are known to be hard, we can use either of them in future reductions. Usually 3-SAT is the better choice, because it is simpler to work with. What follows are a pair of more complicated reductions, designed to serve as examples and also increase our repertoire of known hard problems. Many reductions are quite intricate, because we are essentially programming one problem in the language of a significantly different problem.

One perpetual point of confusion is getting the direction of the reduction right. Recall that we must transform *any* instance of a known NP-complete problem (Bandersnatch) into an instance of the problem we are really interested in (Bo-billy). If we perform the reduction the other way, all we get is a slow way to solve the problem of interest, by using a subroutine that takes exponential time. This always is confusing at first, because it seems backwards. Make sure you understand the direction of reduction now, and think back to this whenever you get confused.

11.5.1 Vertex Cover

Algorithmic graph theory proves to be a fertile ground for hard problems. The prototypical NP-complete graph problem is vertex cover, previously defined in Section 11.3.2 (page 363) as follows:

Problem: Vertex Cover
Input: A graph $G = (V, E)$ and integer $k \leq |V|$.
Output: Is there a subset S of at most k vertices such that every $e \in E$ has at least one vertex in S?

Demonstrating the hardness of vertex cover proves more difficult than the previous reductions we have seen, because the structure of the two relevant problems seems very different. A reduction from 3-SAT to vertex cover must

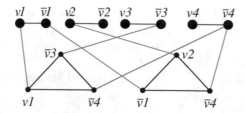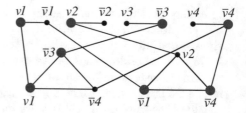

Figure 11.7: Reducing 3-SAT instance $\{\{v_1, \bar{v}_3, \bar{v}_4\}, \{\bar{v}_1, v_2, \bar{v}_4\}\}$ to vertex cover (left). The red vertices (on right) define a minimum vertex cover, and hence the red variable vertices on top define a satisfying truth assignment.

construct a graph G and bound k from the variables and clauses of the satisfiability instance.

Here is a way to do it. First, we translate the variables of the 3-SAT problem. For each Boolean variable v_i, we create two vertices v_i and $\bar{v}_i$ connected by an edge. At least n of these $2n$ vertices will be needed just to cover these edges, because we need at least one vertex per pair.

Second, we translate the clauses of the 3-SAT problem. For each of the c clauses, we create three new vertices: one for each literal in the clause. These three vertices will be connected so as to form a triangle for each clause. At least two vertices per triangle must be included in any vertex cover of these triangles, for a total of $2c$ cover vertices.

Finally, we will connect these two sets of components together. Each literal in a vertex "gadget" is connected to vertices in the clause gadgets (triangles) that share the given literal. From a 3-SAT instance with n variables and c clauses, this constructs a graph with $2n + 3c$ vertices. The complete reduction for the 3-SAT problem $\{\{v_1, \bar{v}_3, \bar{v}_4\}, \{\bar{v}_1, v_2, \bar{v}_4\}\}$ is shown in Figure 11.7.

This graph has been very carefully designed to have a vertex cover of size $n+2c$ iff the original expression is satisfiable. By the earlier analysis, any vertex cover must contain at least $n+2c$ vertices. To show that our reduction is correct, we must demonstrate that:

- *Every satisfying truth assignment gives a vertex cover of size $n + 2c$ –* Given a satisfying truth assignment for the clauses, select the n vertices from the vertex gadgets that correspond to true literals to be members of the vertex cover. Since this defines a satisfying truth assignment, a true literal from each clause must cover at least one of the three cross edges connecting each triangle vertex to a vertex gadget. Therefore, by selecting the other two vertices of each clause triangle, we also pick up all remaining cross edges.

- *Every vertex cover of size $n + 2c$ gives a satisfying truth assignment –* In any vertex cover C of this size, exactly n of the vertices must belong to the vertex gadgets. Let these first stage vertices define the truth assignment, with the remaining $2c$ cover vertices distributed at two per clause-gadget for otherwise a clause-gadget edge would go uncovered. These clause-gadget vertices can cover only two of the three connecting edges per clause. Therefore, if C gives a vertex cover, at least one connecting edge per clause must be covered by first-stage vertices, meaning that the corresponding truth assignment satisfies all clauses.

This proof of the hardness of vertex cover, chained with the clique and independent set reductions of Section 11.3.2 (page 363), gives us a library of hard graph problems that we can use to make future hardness proofs easier.

> *Take-Home Lesson:* A small set of NP-complete problems (3-SAT, vertex cover, integer partition, and Hamiltonian cycle) suffice to prove the hardness of most other hard problems.

11.5.2 Integer Programming

As discussed in Section 16.6 (page 482), integer programming is a fundamental combinatorial optimization problem. It is best thought of as linear programming, with the variables restricted to take only integer (instead of real) values.

Problem: Integer Programming (IP)
Input: A set of integer variables V, a set of linear inequalities over V, a linear maximization function $f(V)$, and an integer B.
Output: Does there exist an assignment of integers to V such that all inequalities are true and $f(V) \geq B$?

Consider the following two examples. Suppose

$$V_1 \geq 1, \quad V_2 \geq 0$$

$$V_1 + V_2 \leq 3$$

$$f(V) : 2V_2, \quad B = 3$$

A solution to this would be $V_1 = 1$, $V_2 = 2$. Note that this respects integrality, and yields an objective value $f(V) = 4 \geq B$. Not all problems have realizable solutions, however. For the following problem:

$$V_1 \geq 1, \quad V_2 \geq 0$$

$$V_1 + V_2 \leq 3$$

$$f(V) : 2V_2, \quad B = 5$$

the maximum possible value of $f(V)$ given the constraints is $2 \times 2 = 4$, so there can be no solution to the associated decision problem.

We show that integer programming is hard using a reduction from general satisfiability. 3-SAT generally makes reductions easier, and would work equally as well here—in an identical manner.

In which direction must the reduction go? We want to prove integer programming is hard, and know that satisfiability is hard. If we could solve satisfiability using integer programming and integer programming were easy, this would mean that satisfiability would be easy. Now the direction should be clear: we must translate satisfiability (Bandersnatch) into integer programming (Bo-billy).

What should the translation look like? Every satisfiability instance contains Boolean (true/false) variables and clauses. Every integer programming instance contains integer variables and constraints. A reasonable idea is to make the integer variables correspond to Boolean variables and use constraints to serve the same role as the clauses do in the original problem.

Our translated integer programming problem will have twice as many variables as the SAT instance—one for each Boolean variable and one for its complement. For each variable v_i in the satisfiability problem, we add the following constraints:

- We restrict each integer programming variable V_i to values of either 0 or 1, by adding constraints $0 \leq V_i \leq 1$ and $0 \leq \overline{V}_i \leq 1$. Coupled with integrality, they correspond to values of true and false.

- We ensure that exactly one of the two integer programming variables associated with a given SAT variable is true, by adding constraints so that $1 \leq V_i + \overline{V}_i \leq 1$.

For each clause $C_i = \{z_1, \ldots, z_k\}$, construct the constraint

$$Z_1 + \ldots + Z_k \geq 1$$

To satisfy this constraint, at least one literal per clause must be set to 1, thus corresponding to a true literal. Satisfying this constraint is therefore equivalent to satisfying the clause.

The maximization function and bound prove relatively unimportant here, because we have already encoded the entire satisfiability instance. By using $f(V) = V_1$ and $B = 0$, we ensure that they will not interfere with any variable assignment satisfying all the inequalities. Clearly, this reduction can be done in polynomial time. To establish that this reduction preserves the answer, we must verify two things:

- *Every SAT solution gives a solution to the IP problem* – In any SAT solution, a true literal corresponds to a 1 in the integer program, since the clause is satisfied. Therefore, the sum in each clause inequality is ≥ 1.

- *Every IP solution gives a solution to the original SAT problem* – All variables must be set to either 0 or 1 in any solution to this integer programming instance. If $V_i = 1$, then set literal $z_i =$ true. If $V_i = 0$, then set literal $z_i =$ false. This is a legal assignment that satisfies all the clauses.

This reduction works both ways, so integer programming must be hard. Notice the following properties, which hold true in general for NP-completeness proofs:

- This reduction preserved the structure of the problem. It did not *solve* the problem, just put it into a different format.

- The possible IP instances resulting from this transformation represent only a small subset of all possible IP instances. But because the instances in this small subset are hard, the more general problem is obviously hard.

- The transformation captures the essence of *why* IP is hard. It has nothing to do with big coefficients or large ranges on the variables, because restricting them all to 0/1 is enough. It has nothing to do with having inequalities having large numbers of variables. Integer programming is hard because satisfying a large set of constraints is hard. A careful study of the properties needed for a reduction can tell us a lot about the problem.

11.6 The Art of Proving Hardness

Proving that problems are hard is a skill. But once you get the hang of it, reductions can be surprisingly straightforward and pleasurable to do. Indeed, the dirty little secret of NP-completeness proofs is that they are usually easier to create than explain, in much the same way that it can be easier to rewrite old code than to understand and modify it.

It takes experience to judge which problems are likely to be hard. The quickest way to gain this experience is through careful study of the catalog. Slightly changing the wording of a problem can make the difference between it being polynomial or NP-complete. Finding the shortest path in a graph is easy, but finding the longest path in a graph is hard. Constructing a tour that visits all the edges once in a graph is easy (Eulerian cycle), but constructing a tour that visits all the vertices once is hard (Hamiltonian cycle).

The first place to look when you suspect a problem might be NP-complete is Garey and Johnson's book *Computers and Intractability* [GJ79], which contains a list of several hundred problems known to be NP-complete. Likely one of these is the problem you are interested in.

Otherwise I offer the following advice to those seeking to prove the hardness of a given problem:

- *Make your source problem as simple (meaning restricted) as possible* – Never try to use the general traveling salesman problem (TSP) as a source problem. Better, use Hamiltonian cycle: TSP where all the weights are restricted 1 or ∞. Even better, use Hamiltonian path instead of cycle, so you never have to worry about closing up the cycle. Best of all, use Hamiltonian path on directed planar graphs where each vertex has total degree 3. All of these problems are equally hard, but the more you can restrict the problem that you are translating from, the less work your reduction has to do.

As another example, never try to use full satisfiability to prove hardness. Start with 3-satisfiability. In fact, you don't even have to use full 3-

satisfiability. Instead, you can use *planar 3-satisfiability*, where there must exist a way to draw the clauses as a graph in the plane such that you can connect all instances of the same literal together without edges crossing. This property tends to be useful in proving the intractability of geometric problems. All these variants are equally hard, and hence NP-completeness reductions using any of them are equally convincing.

- *Make your target problem as hard as possible* – Don't be afraid to add extra constraints or freedoms to make your target problem more general and therefore harder. Perhaps your undirected graph problem can be generalized into a directed graph problem, and can hence only be easier to prove hard. After you have a proof of hardness for the harder problem, you can then go back and try to simplify the target.

- *Select the right source problem for the right reason* – Selecting the right source problem makes a big difference in how difficult it is to prove hardness. This is the first and easiest place to go wrong, although theoretically any NP-complete problem works as well as any other. When trying to prove that a problem is hard, some people fish around through lists of dozens of problems, looking for the best fit. These people are amateurs: odds are they will never recognize the problem they are looking for when they see it.

  I use four (and only four) problems as candidates for my hard source problem. Limiting them to four means that I can know a lot about each one, like which variants of the problems are hard and which are not. My favorite source problems are:

  - *3-SAT*: The old reliable. When none of the three problems below seem appropriate, I go back to the original source.
  - *Integer partition*: This is the one and only choice for problems whose hardness seems to require using large numbers.
  - *Vertex cover*: This is the answer for any graph problem whose hardness depends upon *selection*. Chromatic number, clique, and independent set all involve trying to select the right subset of vertices or edges.
  - *Hamiltonian path*: This is my choice for any graph problem whose hardness depends upon *ordering*. If you are trying to route or schedule something, Hamiltonian path is likely your lever into the problem.

- *Amplify the penalties for making the undesired selection* – Many people are too timid in their thinking about hardness proofs. You want to translate one problem into another, while keeping the problems as close to their original identities as possible. The easiest way to do this is by being bold with your penalties, to punish for deviating from your intended solution. Your thinking should be, "if you select this element, then you must pick up this huge set that blocks you from finding an optimal solution." The sharper the consequences for doing what is undesired, the easier it is to prove the equivalence of the results.

- *Think strategically at a high level, then build gadgets to enforce tactics –* You should be asking yourself questions, like:

  1. How can I force that A or B is chosen but not both?
  2. How can I force that A is taken before B?
  3. How can I clean up the things I did not select?

  Once you know what you want your gadgets to do, you can then worry about how to actually craft them.

- *When you get stuck, switch between looking for an algorithm and a reduction –* Sometimes the reason you cannot prove hardness is that there exists an efficient algorithm to solve your problem! Techniques such as dynamic programming or reducing problems to powerful but polynomial-time graph problems like matching or network flow can yield surprising algorithms. When you can't prove hardness, it pays to stop and try to find an algorithm—just to keep yourself honest.

11.7 War Story: Hard Against the Clock

My class's attention span was running down like sand through an hourglass. Eyes were starting to glaze, even in the front row. Breathing had become soft and regular in the middle of the room. Heads were tilted back and eyes shut in the back.

There were twenty minutes left to go in my lecture on NP-completeness, and I couldn't really blame them. They had already seen several reductions like the ones presented here. But NP-completeness reductions are often easier to create than to explain. They had to watch one being created in order to appreciate how things worked.

I reached for my trusty copy of Garey and Johnson's book [GJ79], which contains a list of over three hundred different known NP-complete problems in an appendix.

"Enough of this!" I announced loudly enough to startle those in the back row. "NP-completeness proofs are sufficiently routine that we can construct them on demand. I need a volunteer with a finger. Can anyone help me?"

A few students in the front held up their hands. A few students in the back held up their fingers. I opted for one from the front row.

"Select a problem at random from the back of this book. I can prove the hardness of any of these problems in the now seventeen minutes remaining in this class. Stick your finger in and read me a problem."

I had definitely gotten their attention. But I could have done that by offering to juggle chain saws. Now I had to deliver results without cutting myself into ribbons.

The student picked out a problem. "OK, prove that *Inequivalence of Programs with Assignments* is hard," she said.

"Huh? I've never heard of that problem before. What is it? Read me the entire problem description so I can write it on the board." The problem was as follows:

Problem: Inequivalence of Programs with Assignments
Input: A finite set X of variables, two programs P_1 and P_2, each a sequence of assignments of the form

$$x_0 \leftarrow \text{ if } (x_1 = x_2) \text{ then } x_3 \text{ else } x_4$$

where the x_i are in X; and a value set V.
Output: Is there an initial assignment of a value from V to each variable in X such that programs P_1 and P_2 yield different final values for some variable in X?

I looked at my watch. Fifteen minutes to go. But everything was now on the table. I was faced with a language problem. The input was two programs with variables, and I had to test whether they always do the same thing.

"First things first. We need to select a source problem for our reduction. Do we start with integer partition? 3-SAT? Vertex cover or Hamiltonian path?"

Since I had an audience, I tried thinking out loud. "Our target is not a graph problem or a numerical problem, so let's start thinking about the old reliable: 3-SAT. There seem to be some similarities. 3-SAT has variables. This thing has variables. To be more like 3-SAT, we could try limiting the variables in this problem so they only take on Boolean values—$V = \{\text{true}, \text{false}\}$. Yes. That seems convenient."

My watch said fourteen minutes left. "So, class, which way does the reduction go? 3-SAT to program or program to 3-SAT?"

The front row correctly murmured, "3-SAT to program."

"Right. So we have to translate our set of clauses into two programs. How can we do that? We might consider trying to split the clauses into two sets and write separate programs for each of them. But how do we split them? I don't see any natural way to do it, because eliminating any single clause from the problem might suddenly make an unsatisfiable formula satisfiable, thus completely changing the answer.

Instead, let's try something else. We can translate all the clauses into one program, and then let the second program be trivial. For example, the second program might ignore the input and always output either only true or only false. This sounds better. *Much* better."

I was still talking out loud to myself, which wasn't that unusual. But I had people listening to me, which was.

"Now, how can we turn a set of clauses into a program? We want to know whether the set of clauses can be satisfied, or in other words if there is an assignment of the variables to make it true. Suppose we constructed a program to evaluate whether $C_1 = \{x_1, \overline{x}_2, x_3\}$ is satisfied."

It took me a few minutes of scratching before I found the right program to simulate a clause. I assumed that I had access to constants for true and false:

$$C_1 = \text{if } (x_1 = \text{true}) \text{ then true else false}$$
$$C_1 = \text{if } (x_2 = \text{false}) \text{ then true else } C_1$$
$$C_1 = \text{if } (x_3 = \text{true}) \text{ then true else } C_1$$

"Great. Now I have a way to evaluate the truth of each clause. I can do the same thing at the end to evaluate whether all the clauses are satisfied:"

$$sat = \text{if } (C_1 = \text{true}) \text{ then true else false}$$
$$sat = \text{if } (C_2 = \text{true}) \text{ then } sat \text{ else false}$$
$$\vdots$$
$$sat = \text{if } (C_c = \text{true}) \text{ then } sat \text{ else false}$$

Now the back of the classroom was getting excited. They were starting to see a ray of hope that class would end on time.

"Great. So now we have a program that can evaluate to be true if and only if there is a way to assign the variables to satisfy the set of clauses. We need a second program to finish the job. What about $sat = $ false? Yes, that is all we need. Our language problem asks whether the two programs always output the same thing, regardless of the possible variable assignments. If the clauses are satisfiable, that means that there must be an assignment of the variables such that the long program would output true. Testing whether the programs are equivalent is exactly the same as asking if the clauses are satisfiable."

I lifted my arms in triumph. "And so, the problem is neat, sweet, and NP-complete." I got the last word out just before the bell rang.

11.8 War Story: And Then I Failed

This exercise of picking a random NP-complete problem from Garey and Johnson's book and proving hardness on demand was so much fun that I have repeated it each time I have taught the algorithms course. Sure enough, I got it eight times in a row. But just as Joe DiMaggio's 56-game hitting streak came to an end, and Google will eventually have a losing quarter financially, the time came for me to get my comeuppance.

The class had voted to see a reduction from the graph theory section of the catalog, and a randomly selected student picked number 30. Problem GT30 turned out to be the following:

Problem: Uniconnected Subgraph
Input: A directed graph $G = (V, A)$, positive integer $k \leq |A|$.
Output: Is there a subset of arcs $A' \subseteq A$ with $|A'| \geq k$ such that $G' = (V, A')$ has at most one directed path between any pair of vertices?

It took a while for me to grok this problem. An undirected version of this would be finding a spanning tree, because that defines exactly one path between any pair of vertices. Adding even a single edge (x, y) to this tree would create a cycle, meaning two distinct paths between x and y.

Any form of directed tree would also be uniconnected. But this problem asks for the *largest* such subgraph. Consider a bipartite-DAG consisting of directed edges (l_i, r_j) all going from a given set of "left" vertices to distinct "right" vertices. No path in this graph consists of more than one edge, yet the graph can contain $\Omega(n^2)$ edges.

"It is a selection problem," I realized after grokking. After all, we had to select the largest possible subset of arcs so that there were no pair of vertices with multiple paths between them. This meant that vertex cover was the problem of choice.

I worked through how the two problems stacked up. Both sought subsets, although vertex cover wanted subsets of vertices and uniconnected subgraph wanted subsets of edges. Vertex cover wanted the smallest possible subset, while unidirected subgraph wanted the largest possible subset. My source problem had undirected edges while my target had directed arcs, so somehow I would have to add edge direction into the reduction.

I had to do something to direct the edges of the vertex cover graph. I could try to replace each undirected edge (x, y) with a single arc, say from y to x. But quite different directed graphs would result depending upon which direction I selected. Finding the "right" orientation of edges might be a hard problem, certainly too hard to use in the translation phase of the reduction.

I realized I could direct the edges so the resulting graph was a DAG. But then, so what? DAGs certainly can have many different directed paths between pairs of vertices.

Alternately, I could try to replace each undirected edge (x, y) with *two* arcs, from x to y and y to x. Now there was no need to chose the right arcs for my reduction, but the graph certainly got complicated. I couldn't see how to force things to prevent vertex pairs from having unwanted multiple paths between them.

Meanwhile, the clock was running and I knew it. A sense of panic set in during the last ten minutes of the class, and I realized I wasn't going to get it this time.

There is no feeling worse for a professor than botching up a lecture. You stand up there flailing away, knowing (1) that the students don't understand what you are saying, but (2) they do understand that you also don't understand what you are saying. The bell rang and the students left the room with faces either sympathetic or smirking.

I promised them a solution for the next class, but somehow I kept getting stuck in the same place each time I thought about it. I even tried to cheat and look up the proof in a journal. But the reference cited by Garey and Johnson was a 30-year old unpublished technical report. It wasn't on the web or in our library.

I dreaded returning to give my next class, the last lecture of the semester. But the night before class the answer came to me in a dream. *"Split the edges,"* the voice said. I awoke with a start and looked at the clock. It was 3:00 AM.

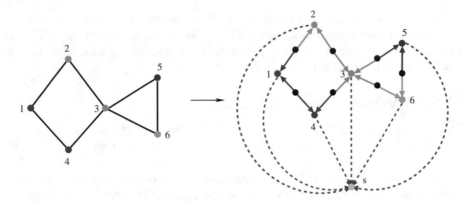

Figure 11.8: Reducing vertex cover to unidirected subgraph, by dividing edges and adding a sink node.

I sat up in bed and scratched out the proof. Suppose I replace each undirected edge (x, y) with a gadget consisting of a new central vertex v_{xy} with arcs going from it to x and y, respectively. This is nice. Now, which vertices are capable of having multiple paths between them? The new vertices have only outgoing edges, so only they can serve as the source of multiple paths. The old vertices have only incoming edges. There is at most one way to get from one of the new source vertices to any of the original vertices of the vertex cover graph, so these could not result in multiple paths.

But now add a sink node s with edges from all the original vertices. There are exactly two paths from each new source vertex to this sink—one through each of the two original vertices it is adjacent to. One of these has to be broken to create a uniconnected subgraph. How could we break it? We could pick one of these two vertices to disconnect from the sink by deleting either arc (x, s) or (y, s) for new vertex v_{xy}. We maximize the size of our subgraph by finding the smallest number of arcs to delete. We must delete the outgoing arc from at least one of the two vertices defining each original edge. *But this is exactly the same as finding the vertex cover in this graph!* The reduction is illustrated in Figure 11.8.

Presenting this proof in class provided some personal vindication, because it validates the principles I teach for proving hardness. Observe that the reduction really wasn't that difficult after all: just split the edges and add a sink node. NP-completeness reductions are often surprisingly simple once you look at them the right way.

11.9   P vs. NP

The theory of NP-completeness rests on a foundation of rigorous but subtle definitions from automata and formal language theory. This terminology is typically confusing to (or misused by) beginners who lack a mastery of these foundations. These details are not really essential to the practical aspects of designing and applying reductions. That said, the question "Is P=NP?" is the most profound open problem in computer science, so any educated algorist should have some idea what the stakes are.

11.9.1   Verification vs. Discovery

The primary issue in P vs. NP is whether *verification* is really an easier task than initial *discovery*. Suppose that while taking an exam you "happen" to notice the answer of the student next to you. Are you now better off? You wouldn't dare turn it in without checking, since an able student like you could answer the question correctly if you spent enough time to solve it. The question here is whether you really can verify the answer faster than you can find it from scratch.

For the NP-complete decision problems we have studied here, the answer *seems* obvious:

- Can you verify that a proposed TSP tour has weight of at most k in a given graph? Sure. Just add up the weights of the edges on the tour and show it is at most k. That is easier than finding the tour from scratch, *isn't it?*

- Can you verify that a given truth assignment represents a solution to a given satisfiability problem? Sure. Just check each clause and make sure it contains at least one true literal from the given truth assignment. That is easier than finding the satisfying assignment from scratch, *isn't it?*

- Can you verify that a given k-vertex subset S is a vertex cover of graph G? Sure. Just traverse each edge (u, v) of G, and check that either u or v is in S. That is easier than finding the vertex cover from scratch, *isn't it?*

At first glance, this seems obvious. The given solutions can be verified in linear time for all three of these problems, while no algorithm faster than exponential brute-force search is known for any of them. The catch is that we have no rigorous lower bound *proof* that prevents the existence of fast algorithms to solve these problems. Perhaps there are in fact polynomial algorithms (say $O(n^{87})$) that we have just been too blind to see yet.

11.9.2   The Classes P and NP

Every well-defined algorithmic problem must have an asymptotically fastest-possible algorithm solving it, as measured in the Big Oh, worst-case sense of fastest.

We can think of the class P as an exclusive club for algorithmic problems where there exists a polynomial-time algorithm to solve it from scratch. Shortest path, minimum spanning tree, and the original movie scheduling problem are all members in good standing of this class P. The P stands for *polynomial time.*

A less exclusive club welcomes all the algorithmic problems whose solutions can be *verified* in polynomial time. As shown above, this club contains traveling salesman, satisfiability, and vertex cover, none of which currently have the credentials to join P. However, all the members of P get a free pass into this less exclusive club. If you can solve a decision problem from scratch in polynomial time, you certainly can verify another solution to it that fast: just check whether you agree on the yes–no answer.

We call this less-exclusive club *NP*. You can think of this as standing for *Not necessarily Polynomial time.*[1]

The $1,000,000 question is whether there are problems in NP that are not members of P. If no such problem exists, the classes must be the same and $P = NP$. If even one such a problem exists, the two classes are different and $P \neq NP$. The opinion of most algorists and complexity theorists is that $P \neq NP$, meaning that some NP problems do not have polynomial-time algorithms, but a much stronger proof than "I can't find a fast enough algorithm" is needed.

11.9.3   Why Satisfiability is Hard

An enormous tree of NP-completeness reductions has been established that entirely rests on the hardness of satisfiability. The portion of this tree demonstrated and/or stated in this chapter is shown in Figure 11.2.

This may seem like a fragile foundation. What would it mean if someone *did* find a polynomial-time algorithm for satisfiability? A fast algorithm for any given NP-complete problem (say traveling salesman) implies fast algorithm for all the problems on the path in the reduction tree between TSP and satisfiability (Hamiltonian cycle, vertex cover, and 3-SAT). But a fast algorithm for satisfiability doesn't immediately yield us anything because the reduction path from SAT to SAT is empty.

Fear not. There exists an extraordinary super-reduction called *Cook's theorem* reducing *all* the problems in NP to satisfiability. Thus, if you prove that satisfiability (or equivalently any single NP-complete problem) is in P, then *all* other problems in NP follow and $P = NP$. Since essentially every problem mentioned in this book is in NP, this would be an enormously powerful and surprising result.

Cook's theorem proves that satisfiability is as hard as any problem in NP. Furthermore, it proves that every NP-complete problem is as hard as any other. Any domino falling (meaning a polynomial-time algorithm to solve just one NP-complete problem) knocks them all down. Our inability to find a fast algorithm for any of these problems is a strong reason for believing that they are all truly hard, meaning $P \neq NP$.

---

[1] In fact, it stands for *non-deterministic polynomial time*. This is in the sense of non-deterministic automata, if you happen to know about such things.

11.9.4   NP-hard vs. NP-complete?

The final technicality we will discuss is the difference between a problem being NP-hard and NP-complete. I tend to be somewhat loose with my terminology, but there is a subtle (usually irrelevant) distinction between the two concepts.

We say that a problem is *NP-hard* if, like satisfiability, it is at least as hard as any problem in NP. We say that a problem is *NP-complete* if it is NP-hard, and also in NP itself. Because NP is such a large class of problems, most NP-hard problems you encounter will actually be in NP and thus NP-complete. The issue can always be settled by giving a (usually simple) verification strategy for the problem. All the NP-hard problems encountered in this book are also NP-complete.

That said, there are some problems that appear to be NP-hard yet are not in NP. These problems might be *even harder* than NP-complete! Two-player games such as chess provide examples of problems that are not in NP. Imagine sitting down to play chess with some know-it-all who is playing white. He pushes his king's pawn up two squares to start the game, and announces "checkmate." The only obvious way to verify that he is right would be to construct the full tree of all your possible moves with his irrefutable replies and demonstrate that you, in fact, cannot win from the current position. This full tree will have a number of nodes exponential in its height, which is the number of moves before you lose playing your most spirited possible defense. Clearly this tree cannot be constructed and analyzed in polynomial time, so the problem is not in NP.

Chapter Notes

The notion of NP-completeness was first developed by Cook [Coo71]. Satisfiability really is a $1,000,000 problem, and the Clay Mathematics Institute has offered such a prize to any person who resolves the $P = NP$ question. See http://www.claymath.org/ for more on the problem and the prize.

Karp [Kar72] showed the importance of Cook's result by providing reductions from satisfiability to more than twenty important algorithmic problems. I recommend Karp's paper for its sheer beauty and economy—he condenses each reduction to three line descriptions showing the problem equivalence. Together, these provided the tools to resolve the complexity of literally hundreds of important problems where no efficient algorithms were known.

The best introduction to the theory of NP-completeness remains Garey and Johnson's book *Computers and Intractability* [GJ79]. It introduces the general theory, including an accessible proof of Cook's theorem [Coo71] that satisfiability is as hard as anything in NP. They also provide an essential reference catalog of more than 300 NP-complete problems, which is a great resource for learning what is known about the most interesting hard problems. The reductions claimed but omitted from this chapter can be found in Garey and Johnson, or textbooks like Cormen et al. [CLRS09].

Factor Man [Gin18] is an exciting novel about a man who discovers a polynomial-time algorithm for satisfiability, and must dodge government agents and assassins for his troubles. I give it two thumbs up. *The Golden Ticket* [For13] is an accessible tour of complexity theory, and the question of $P = NP$.

A few catalog problems exist in a limbo state where it is not yet known whether the problem has a fast algorithm or is NP-complete. The most prominent of these are graph isomorphism (see Section 19.9 (page 610)) and integer factorization (see Section 16.8 (page 490)). That this limbo list is so short is quite a tribute to the state-of-the-art in algorithm design, and the power of the theory of NP-completeness. For almost every important problem we either know a fast algorithm or have a good solid reason why one doesn't exist.

For an alternative and inspiring view of NP-completeness, check out the videos of Erik Demaine's MIT course Algorithmic Lower Bounds: Fun with Hardness Proofs at `http://courses.csail.mit.edu/6.890/fall14/`. The war story problem on unidirected subgraph was originally proven hard in Maheshwari [Mah76].

11.10 Exercises

Transformations and Satisfiability

11-1. *[2]* Give the 3-SAT formula that results from applying the reduction of satisfiability to 3-SAT for the formula:

$$(x \vee y \vee \overline{z} \vee w \vee u \vee \overline{v}) \wedge (\overline{x} \vee \overline{y} \vee z \vee \overline{w} \vee u \vee v) \wedge (x \vee \overline{y} \vee \overline{z} \vee w \vee u \vee \overline{v}) \wedge (x \vee \overline{y})$$

11-2. *[3]* Draw the graph that results from the reduction of 3-SAT to vertex cover for the expression

$$(x \vee \overline{y} \vee z) \wedge (\overline{x} \vee y \vee \overline{z}) \wedge (\overline{x} \vee y \vee z) \wedge (x \vee \overline{y} \vee \overline{x})$$

11-3. *[3]* Prove that 4-SAT is NP-hard.

11-4. *[3]* *Stingy SAT* is the following problem: given a set of clauses (each a disjunction of literals) and an integer k, find a satisfying assignment in which at most k variables are true, if such an assignment exists. Prove that stingy SAT is NP-hard.

11-5. *[3]* The *Double SAT* problem asks whether a given satisfiability problem has **at least two different satisfying assignments**. For example, the problem $\{\{v_1, v_2\}, \{\overline{v_1}, v_2\}, \{\overline{v_1}, \overline{v_2}\}\}$ is satisfiable, but has only one solution ($v_1 = F, v_2 = T$). In contrast, $\{\{v_1, v_2\}, \{\overline{v_1}, \overline{v_2}\}\}$ has exactly two solutions. Show that Double-SAT is NP-hard.

11-6. *[4]* Suppose we are given a subroutine that can solve the traveling salesman decision problem on page 357 in (say) linear time. Give an efficient algorithm to find the actual TSP tour by making a polynomial number of calls to this subroutine.

11-7. *[7]* Implement a SAT to 3-SAT reduction that translates satisfiability instances into equivalent 3-SAT instances.

11-8. *[7]* Design and implement a backtracking algorithm to test whether a set of clause sets is satisfiable. What criteria can you use to prune this search?

11-9. *[8]* Implement the vertex cover to satisfiability reduction, and run the resulting clauses through a satisfiability solver code. Does this seem like a practical way to compute things?

Basic Reductions

11-10. *[4]* An instance of the *set cover* problem consists of a set X of n elements, a family F of subsets of X, and an integer k. The question is, does there exist k subsets from F whose union is X?

For example, if $X = \{1, 2, 3, 4\}$ and $F = \{\{1, 2\}, \{2, 3\}, \{4\}, \{2, 4\}\}$, there does not exist a solution for $k = 2$, but there does for $k = 3$ (for example, $\{1, 2\}, \{2, 3\}, \{4\}$).

Prove that set cover is NP-hard with a reduction from vertex cover.

11-11. *[4]* The *baseball card collector problem* is as follows. Given packets $P_1, \ldots, P_m$, each of which contains a subset of this year's baseball cards, is it possible to collect all the year's cards by buying $\leq k$ packets?

For example, if the players are $\{Aaron, Mays, Ruth, Skiena\}$ and the packets are

$$\{\{Aaron, Mays\}, \{Mays, Ruth\}, \{Skiena\}, \{Mays, Skiena\}\},$$

there does not exist a solution for $k = 2$, but there does for $k = 3$, such as

$$\{Aaron, Mays\}, \{Mays, Ruth\}, \{Skiena\}$$

Prove that the baseball card collector problem is NP-hard using a reduction from vertex cover.

11-12. *[4]* The *low-degree spanning tree problem* is as follows. Given a graph G and an integer k, does G contain a spanning tree such that all vertices in the tree have degree *at most* k (obviously, only tree edges count towards the degree)? For example, in the following graph, there is no spanning tree such that all vertices have a degree at most three.

(a) Prove that the low-degree spanning tree problem is NP-hard with a reduction from Hamiltonian *path*.

(b) Now consider the *high-degree spanning tree problem*, which is as follows. Given a graph G and an integer k, does G contain a spanning tree whose highest degree vertex is *at least* k? In the previous example, there exists a spanning tree with a highest degree of 7. Give an efficient algorithm to solve the high-degree spanning tree problem, and an analysis of its time complexity.

11-13. *[5]* In the *minimum element set cover* problem, we seek a set cover $S \subseteq C$ of a universal set $U = \{1, \ldots, n\}$ such that sum of the sizes of the subsets in S is at most k. (a) Show that $C = \{\{1, 2, 3\}, \{1, 3, 4\}, \{2, 3, 4\}, \{3, 4, 5\}\}$ has a cover of size 6, but none of size 5 because of a repeated element. (b) Prove that this problem is NP-hard. (Hint: set cover remains hard if all subsets are of the same size.)

11-14. *[3]* The *half-Hamiltonian cycle problem* is, given a graph G with n vertices, determine whether G has a simple cycle of length exactly $\lfloor n/2 \rfloor$, where the floor function rounds its input down to the nearest integer. Prove that this problem is NP-hard.

11-15. *[5]* The *3-phase power balance problem* asks for a way to partition a set of n positive integers into three sets A, B, or C such that $\sum_i a_i = \sum_i b_i = \sum_i c_i$. Prove that this problem is NP-hard using a reduction from integer partition or subset sum (see Section 10.5 (page 329)).

11-16. *[4]* Show that the following problem is NP-hard:

Problem: Dense Subgraph

Input: A graph G, and integers k and y.

Output: Does G contain a subgraph of exactly k vertices and at least y edges?

11-17. *[4]* Show that the following problem is NP-hard:

Problem: Clique, No-clique

Input: An undirected graph $G = (V, E)$ and an integer k.

Output: Does G contain both a clique of size k and an independent set of size k?

11-18. *[5]* An *Eulerian cycle* is a tour that visits every edge in a graph exactly once. An *Eulerian subgraph* is a subset of the edges and vertices of a graph that has an Eulerian cycle. Prove that the problem of finding the number of edges in the largest Eulerian subgraph of a graph is NP-hard. (Hint: the Hamiltonian circuit problem is NP-hard even if each vertex in the graph is incident upon exactly three edges.)

11-19. *[5]* Show that the following problem is NP-hard:

Problem: Maximum Common Subgraph

Input: Two graphs $G_1 = (V_1, E_1)$ and $G_2 = (V_2, E_2)$, and a budget b.

Output: Two sets of nodes $S_1 \subseteq V_1$ and $S_2 \subseteq V_2$ whose deletion leaves at least b nodes in each graph, and makes the two graphs identical.

11-20. *[5]* A *strongly independent set* is a subset of vertices S in a graph G such that for any two vertices in S, there is no path of length two in G. Prove that strongly independent set is NP-hard.

11-21. *[5]* A *kite* is a graph on an even number of vertices, say $2n$, in which n of the vertices form a clique and the remaining n vertices are connected in a tail that consists of a path joined to one of the vertices of the clique. Given a graph and a goal g, the *max kite* problem asks for a subgraph that is a kite and contains $2g$ nodes. Prove that *max kite* is NP-hard.

Creative Reductions

11-22. *[5]* Prove that the following problem is NP-hard:

Problem: Hitting Set

Input: A collection C of subsets of a set S, positive integer k.

Output: Does S contain a subset S' such that $|S'| \leq k$ and each subset in C contains at least one element from S'?

11-23. *[5]* Prove that the following problem is NP-hard:

Problem: Knapsack

Input: A set S of n items, such that the ith item has value v_i and weight w_i. Two positive integers: weight limit W and value requirement V.

Output: Does there exist a subset $S' \subseteq S$ such that $\sum_{i \in S'} w_i \leq W$ and $\sum_{i \in S'} v_i \geq V$?   (Hint: start from integer partition.)

11-24. *[5]* Prove that the following problem is NP-hard:

Problem: Hamiltonian Path

Input: A graph G, and vertices s and t.

Output: Does G contain a path that starts from s, ends at t, and visits all vertices without visiting any vertex more than once?   (Hint: start from Hamiltonian cycle.)

11-25. *[5]* Prove that the following problem is NP-hard:

Problem: Longest Path

Input: A graph G and positive integer k.

Output: Does G contain a path that visits at least k different vertices without visiting any vertex more than once?

11-26. *[6]* Prove that the following problem is NP-hard:

Problem: Dominating Set

Input: A graph $G = (V, E)$ and positive integer k.

Output: Is there a subset $V' \subseteq V$ such that $|V'| \leq k$ where for each vertex $x \in V$ either $x \in V'$ or there exists an edge $(x, y) \in E$ such that $y \in V'$.

11-27. *[7]* Prove that the vertex cover problem (does there exist a subset S of k vertices in a graph G such that every edge in G is incident upon at least one vertex in S?) remains NP-hard even when all the vertices in the graph are restricted to have even degrees.

11-28. *[7]* Prove that the following problem is NP-hard:

Problem: Set Packing

Input: A collection C of subsets of a set S, positive integer k.

Output: Does C contain at least k disjoint subsets (i.e., such that no pair of subsets has any elements in common)?

11-29. *[7]* Prove that the following problem is NP-hard:

Problem: Feedback Vertex Set

Input: A directed graph $G = (V, A)$ and positive integer k.

Output: Is there a subset $V' \subseteq V$ such that $|V'| \leq k$, such that deleting the vertices of V' from G leaves a DAG?

11-30. *[8]* Give a reduction from Sudoku to the vertex coloring problem in graphs. Specifically, describe how to take any partially filled Sudoku board and construct a graph that can be colored with nine colors iff the Sudoku board is solvable.

Algorithms for Special Cases

11-31. *[5]* A Hamiltonian path P is a path that visits each vertex exactly once. The problem of testing whether a graph G contains a Hamiltonian path is NP-complete. There does not have to be an edge in G from the ending vertex to the starting vertex of P, unlike in the Hamiltonian cycle problem.

Give an $O(n + m)$-time algorithm to test whether a directed acyclic graph G (a DAG) contains a Hamiltonian path. (Hint: think about topological sorting and DFS.)

11-32. *[3]* Consider the k-clique problem, which is the general clique problem restricted to graphs in which every vertex has degree at most k. Prove that k-clique has an efficient algorithm for any given k, meaning that k is a constant.

11-33. *[8]* The 2-SAT problem is, given a Boolean formula in 2-conjunctive normal form (CNF), to decide whether the formula is satisfiable. 2-SAT is like 3-SAT, except that each clause can have only two literals. For example, the following formula is in 2-CNF:

$$(x_1 \vee x_2) \wedge (\bar{x}_2 \vee x_3) \wedge (x_1 \vee \bar{x}_3)$$

Give a polynomial-time algorithm to solve 2-SAT.

$P = NP$?

11-34. *[4]* Show that the following problems are in NP:

- Does graph G have a simple path (i.e., with no vertex repeated) of length k?

- Is integer n composite (i.e., not prime)?

- Does graph G have a vertex cover of size k?

11-35. *[7]* Until 2002, it was an open question whether the decision problem "Is integer n a composite number, in other words, not prime?" could be computed in time polynomial in the size of its input. Why doesn't the following algorithm suffice to prove it is in P, since it runs in $O(n)$ time?

```
PrimalityTesting(n)
    composite = false
    for i := 2 to n − 1 do
        if (n mod i) = 0 then
            composite = true
```

LeetCode

11-1. `https://leetcode.com/problems/target-sum/`

11-2. `https://leetcode.com/problems/word-break-ii/`

11-3. `https://leetcode.com/problems/number-of-squareful-arrays/`

HackerRank

11-1. `https://www.hackerrank.com/challenges/spies-revised`

11-2. `https://www.hackerrank.com/challenges/brick-tiling/`

11-3. `https://www.hackerrank.com/challenges/tbsp/`

Programming Challenges

These programming challenge problems with robot judging are available at `https://onlinejudge.org`:

11-1. "The Monocycle"—Chapter 12, problem 10047.

11-2. "Dog and Gopher"—Chapter 13, problem 111301.

11-3. "Chocolate Chip Cookies"—Chapter 13, problem 10136.

11-4. "Birthday Cake"—Chapter 13, problem 10167.

These are not particularly relevant to NP-completeness, but are added for completeness.

Chapter 12

Dealing with Hard Problems

For the practical person, demonstrating that a problem is NP-complete is never the end of the line. Presumably, there was a reason why you wanted to solve it in the first place. That application won't go away after you learn there is no polynomial-time algorithm. You still seek a program that solves the problem of interest. All you know is that you won't find one that quickly solves the problem to optimality in the worst case. There are still three possibilities:

- *Algorithms fast in the average case* – Examples of such algorithms include backtracking algorithms with substantial pruning.

- *Heuristics* – Heuristic methods like simulated annealing or greedy approaches can be used to quickly find a solution, albeit with no guarantee that it will be the best one.

- *Approximation algorithms* – The theory of NP-completeness stipulates that it is hard to get the *exact* answer. With clever, problem-specific heuristics, we can get provably *close* to the optimal answer on all possible instances.

This chapter will investigate these possibilities deeper. I include a brief introduction to quantum computing, an exciting technology that is shaking (but not really breaking) the boundaries of what problems are efficiently computable.

12.1 Approximation Algorithms

Approximation algorithms produce solutions with a guarantee attached, namely that the quality of the optimal solution is provably bounded by the quality of your heuristic solution. Thus, no matter what your input instance is and how lucky you are, such an approximation algorithm is destined to produce a

© The Editor(s) (if applicable) and The Author(s), under exclusive license to
Springer Nature Switzerland AG 2020
S. S. Skiena, *The Algorithm Design Manual*, Texts in Computer Science,
https://doi.org/10.1007/978-3-030-54256-6_12

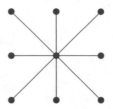

Figure 12.1: Failing to pick the center vertex leads to a terrible vertex cover.

correct answer. Furthermore, provably good approximation algorithms are often conceptually simple, fast, and easy to program.

One thing that is usually not clear, however, is how well the solution from an approximation algorithm compares to what you might get from a heuristic that gives you no guarantees. The answer may be worse, or it could be better. Leaving your money in a bank savings account may guarantee you 3% interest without risk. Still, you likely will do much better investing your money in stocks than leaving it in the bank, even though performance is not guaranteed.

One way to get the best of approximation algorithms and unwashed heuristics is to run both of them on the given problem instance, and pick the solution giving the better result. This way, you will get a solution that comes with a guarantee *and* a second chance to do even better. When it comes to heuristics for hard problems, sometimes you can have it both ways.

12.2   Approximating Vertex Cover

Recall the vertex cover problem, where we seek a small subset S of the vertices of a given graph G such that for every edge (x, y) in G, at least one of x or y is in S. As we have seen, finding the minimum vertex cover of a graph is NP-complete. However, a very simple procedure will always find a cover that is at most twice as large as the optimal cover. It repeatedly selects an uncovered edge, and picks *both* of its vertices for the cover:

```
VertexCover(G = (V, E))
    While (E ≠ ∅) do:
        Select an arbitrary edge (u, v) ∈ E
        Add both u and v to the vertex cover
        Delete all edges from E that are incident to either u or v.
```

It should be apparent that this procedure always produces a vertex cover, since each edge is deleted only after an incident vertex has been added to the cover. More interesting is the claim that the best vertex cover must use at least half as many vertices as this one. Why? Consider only the k edges selected by the algorithm that constitute a matching in the graph. No two of these matching edges can share a vertex, so *any* cover of just these k edges must

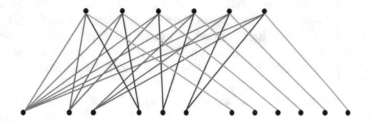

Figure 12.2: A bad example for the greedy heuristic for vertex cover. The optimal cover of this bipartite graph is the row of vertices on top, yet the greedy heuristic will select the vertices from the bottom row from left to right. This example can be enlarged to create an instance where the greedy solution is $\Theta(\log n)$ times larger than the minimum vertex cover.

include at least one vertex per edge, which makes it at least half the size of this $2k$-vertex greedy cover.

There are several interesting things to notice about this algorithm:

- *Although the procedure is simple, it is not stupid* – Many seemingly smarter heuristics can give a far worse performance in the worst case. For example, why not modify the above procedure to select only one of the two vertices for the cover, instead of both? After all, the selected edge will be equally well covered by only one vertex. But consider the star-shaped graph of Figure 12.1. The original heuristic will produce a two-vertex cover, while the single-vertex heuristic *might* return a cover as large as $n - 1$ vertices, should we get unlucky and repeatedly select the leaf instead of the center as the cover vertex we retain.

- *Greedy isn't always the answer* – Perhaps the most natural heuristic for vertex cover would repeatedly select (and then delete) the vertex with highest remaining degree for the vertex cover. After all, this vertex will cover the largest number of possible edges. However, in the case of ties or near ties, this heuristic can go seriously astray. In the worst case, it can yield a cover that is $\Theta(\lg n)$ times optimal, as shown by the example of Figure 12.2.

- *Making a heuristic more complicated does not necessarily make it better* – It is easy to complicate heuristics by adding more special cases or details. For example, the procedure above did not specify which edge should be selected next for the matching. It might seem reasonable to pick the edge whose endpoints have the highest total degree. However, this does not improve the worst-case bound, and just makes it more difficult to analyze.

- *A post-processing cleanup step can't hurt* – The flip side of designing simple heuristics is that they can often be modified to yield better-in-practice solutions without weakening the approximation bound. For example, a post-processing step that deletes any unnecessary vertex from the cover can only improve things in practice, even though it won't help the worst-case bound. And it is fair to repeat the process multiple times with different starting edges and take the best of the resulting runs.

The important property of approximation algorithms is relating the size of the solution produced to a lower bound on the optimal solution. Instead of thinking about how well we might do, we must think about the worst case— that is, how badly the algorithm might perform.

Stop and Think: Leaving Behind a Vertex Cover

Problem: Suppose we do a depth-first search of graph G, naturally building a depth-first search tree T in the process. A leaf node in a tree is any non-root vertex of degree 1. Delete every leaf node from T. Show that (1) the set of all non-leaf nodes of T form a vertex cover of graph G, and (2) that this vertex cover is of size at most twice that of the minimum vertex cover.

Solution: Why must the set of all non-leaf nodes in the DFS tree T form a vertex cover? Recall that the magic property of depth-first search is that it partitions all edges into tree edges and back edges. If a vertex v is a leaf of T, then there is a single tree edge (x, v) containing it, which will be covered by taking non-leaf vertex x. If there are other edges containing v, they must be back edges going to ancestors of v, all of which were selected to be in the cover. So all edges will be covered by the set of non-leaves.

But why is the set of non-leaves at most twice the size of the optimal cover? Start from any leaf v and walk up the tree to the root. Suppose this path is of length k edges, meaning $k + 1$ vertices leaf-to-root. This heuristic will select the k non-leaf vertices for the cover. But the best possible cover for this path requires $\lceil k/2 \rceil$ vertices, so we are always within a factor of at most two times optimal. ∎

12.2.1   A Randomized Vertex Cover Heuristic

Although we proved that our original vertex cover heuristic of selecting arbitrary uncovered edges and adding *both* vertices to the cover yields a factor two approximation algorithm, it feels wrong to grow the cover by two vertices when either one would equally cover the given edge. However, the star-shaped example of Figure 12.1 shows that if we repeatedly pick the wrong (meaning non-center) vertex for each edge, we could end up with a cover of size $n - 1$ instead of 1.

Such a horrible performance requires making the wrong decision $n - 1$ times in a row, which implies either a special talent or horrendous luck. We can make it a matter of luck by choosing the vertex at random:

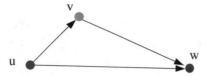

Figure 12.3: The triangle inequality, that $d(u, w) \leq d(u, v) + d(v, w)$, holds for distances defined between geometric points.

VertexCover($G = (V, E)$)
    While ($E \neq \emptyset$) do:
        Select an arbitrary edge $(u, v) \in E$
        Randomly pick either u or v, and add it to the vertex cover
        Delete all edges from E that are incident to the selected vertex.

At the end of this procedure, we will end up with a vertex cover, but how well does its expected size compare to a particular minimum size cover C? Observe that with each edge (u, v) we select, at least one of the two endpoints must appear in the optimal cover C. Thus, at least half the time we get lucky and pick the "right" vertex. At the end of this procedure we will have picked a set $C' \subset C$ of cover vertices plus a set D of vertices from $V - C$ for our cover. We know that $|C'|$ always must be less than or equal to $|C|$. Further, the expected size of D is equal to that of C'. Thus, in expectation $|C'| + |D| \leq 2|C|$, and we get a solution whose size is expected to be at most twice that of optimal.

Randomization is a very powerful tool for developing approximation algorithms. Its role is to make bad special cases go away by making it very unlikely that they will occur. The careful analysis of such probabilities often requires sophisticated efforts, but the heuristics themselves are generally very simple and easy to implement.

12.3 Euclidean TSP

In most natural applications of the traveling salesman problem (TSP), direct routes are inherently shorter than indirect routes. For example, when a graph's edge weights are the straight-line distances between pairs of cities, the shortest path from x to y must always be "as the crow flies."

The edge weights induced by Euclidean geometry satisfy the *triangle inequality*, namely that $d(u, w) \leq d(u, v) + d(v, w)$ for all triples of vertices u, v, and w. The general reasonableness of this condition is demonstrated in Figure 12.3. The cost of airfares is an example of a distance function that *violates* the triangle inequality, because it is often cheaper to fly through an intermediate city than to fly direct to the destination—which is why finding the cheapest fare can be such a pain. But the triangle inequality holds naturally for many problems and applications.

The traveling salesman problem remains hard when the edge weights are defined by Euclidean distances between points. But we can approximate the

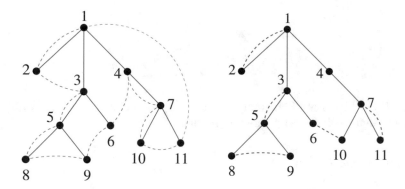

Figure 12.4: A depth-first traversal of a spanning tree, with the shortcut tour (left). The same DFS tree with a minimum weight matching between odd degree vertices, creating an Eulerian graph for the Christofides heuristic (right).

optimal traveling salesman tour on such graphs that obey the triangle inequality using minimum spanning trees. First, observe that the weight of the minimum spanning tree of graph G must be a lower bound on the cost of the optimal TSP tour T of G. Why? Distances are always non-negative, so deleting any edge from tour T leaves a path with total weight no greater than that of T. This path has no cycles, and hence is a tree, which means its weight must be *at least* that of the minimum spanning tree. The weight of the minimum spanning tree thus gives a lower bound on the cost of the optimal TSP tour.

Consider now what happens when performing a depth-first traversal of a spanning tree. We visit each edge twice, once going down the tree when discovering the edge and once more going up after exploring the entire subtree. For example, the depth-first search of Figure 12.4 (left) visits the vertices in order:

$$1, 2, 1, 3, 5, 8, 5, 9, 5, 3, 6, 3, 1, 4, 7, 10, 7, 11, 7, 4, 1$$

This circuit travels along each edge of the minimum spanning tree twice, and hence costs at most twice the optimal tour.

However, many vertices will be repeated on this depth-first search circuit. To remove the extra copies, we can take a direct path to the next unvisited vertex at each step. The shortcut tour for the circuit above is

$$1, 2, 3, 5, 8, 9, 6, 4, 7, 10, 11, 1$$

Because we have replaced a chain of edges by a single direct edge, the triangle inequality ensures that the tour can only get shorter. Thus, this shortcut tour, which can be constructed in $O(n + m)$ time on an n-vertex and m-edge graph G, always has weight at most twice that of the optimal TSP tour of G.

12.3.1 The Christofides Heuristic

There is another way of looking at this minimum spanning tree doubling idea, which will lead to an even better approximation algorithm for TSP. Recall that an *Eulerian cycle* in graph G is a circuit traversing each edge of G exactly once.[1] There is a simple characterization to test when a connected graph contains an Eulerian cycle, namely each vertex must be of even degree. This even-degree condition is obviously necessary, because you must be able to walk out of each vertex exactly the number of times you walk in. But it is also sufficient, and furthermore an Eulerian cycle on any connected, even-degree graph can be easily found in linear time.

We can reinterpret the minimum spanning tree heuristic for TSP in terms of Eulerian cycles. Construct a multigraph M, which consists of *two* copies of each edge of the minimum spanning tree of G. This n-vertex, $2(n-1)$-edge multigraph must be Eulerian, because every vertex has degree twice that of the minimum spanning tree of G. Any Eulerian cycle of M will define a circuit with exactly the same properties as the DFS circuit described above, and hence can be shortcut in the same way to construct a TSP tour with cost at most twice that of the optimal tour.

This suggests that we might find an even better approximation for TSP if we could find a cheaper way to ensure that all vertices are of even degree. Recall (from Section 8.5.1 (page 267)) that a *matching* in a graph $G = (V, E)$ is a subset of edges $E' \subset E$ such that no two edges of E' share a vertex. Adding a set of matching edges to a given graph thus raises the degree of affected vertices by one, turning odd-degree vertices even and even-degree vertices odd, as shown in Figure 12.4 (right).

So let's start by identifying the odd-degree vertices in the minimum spanning tree of G, which are the obstacle preventing us from finding an Eulerian cycle on the minimum spanning tree itself. There must be an even number of odd-degree vertices in any graph. By adding a set of matching edges between these odd-degree vertices, we make the graph Eulerian. The lowest cost *perfect* matching (meaning every vertex must appear in exactly one matching edge) can be computed efficiently, as discussed in Section 18.6 (page 562).

The Christofides heuristic constructs a multigraph M consisting of the minimum spanning tree of G plus the minimum weight set of matching edges between odd-degree vertices in this tree. Thus, M is an Eulerian graph, and contains an Eulerian cycle that can be shortcut to build a TSP tour of weight at most M.

Note that the cost of this matching of just the odd-degree vertices must be a lower bound on the cost of the lowest cost matching of the full graph G, presuming it satisfies the triangle inequality.

Observe in Figure 12.5 that the alternating edges of any TSP tour must define a matching, because each vertex appears only once in the given edge set. These red edges (or blue edges) must cost at least as much as the minimum weight matching of G, and (for the lighter color) weigh at most half that of the TSP tour. The matching edges we added to M thus must have cost at most half that of the optimal TSP tour.

In conclusion, the total weight of M must be at most $(1 + (1/2)) = (3/2)$ times that of the optimal TSP tour, and thus the Christofides heuristic con-

---

[1] Or, if you don't recall this, tour Section 18.7 (page 565) for a refresher.

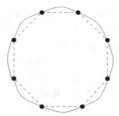

Figure 12.5: Any TSP tour in a graph with an even number of vertices that observes the triangle inequality can be partitioned into red and blue matchings, one of which must be at most half the cost of the tour.

structs a tour of weight at most 3/2 times that of the optimal tour. As with the minimum spanning tree heuristic, the weight lost due to shortcuts might mean the resulting tour is even better than this guarantee. But it can never do worse.

12.4   When Average is Good Enough

In the mythical land of Lake Wobegon, all the children are above average. For certain optimization problems, all (or most) of the solutions are seemingly close to the best possible. Recognizing this yields very simple approximation algorithms with provable guarantees, that can often be refined by the heuristic search strategies we will discuss in Section 12.6 (page 399) into something even better.

12.4.1   Maximum k-SAT

Recall the problem of 3-SAT discussed in Section 11.4.1 (page 367), where we are given a set of three-element logic clauses like

$$v_3 \text{ or } \bar{v}_{17} \text{ or } v_{24}$$

and asked to find an assignment of either true or false to each variable v_i so as to make all the clauses true.

A more general problem is *maximum 3-SAT*, where we seek the Boolean variable assignment that makes the largest number of these clauses true. Asking whether you can satisfy 100% of the clauses is the original 3-SAT problem, so maximum 3-SAT must still be hard. But now it is an optimization problem, so we can think about approximation algorithms for it.

What happens when we flip a coin to decide the value of each variable v_i, and thus construct a completely random truth assignment? What fraction of clauses would we expect to satisfy? Consider the clause above. It will be satisfied unless we pick $v_3 = \text{false}$, $v_{17} = \text{true}$, and $v_{24} = \text{false}$. The probability we get a good assignment for the clause is $1 - (1/2)^3 = 7/8$. Thus, we expect that any random assignment will satisfy (7/8) of the clauses, in other words 87.5% of them.

That seems pretty good for a mindless approach to an NP-complete problem. For a maximum k-SAT instance with m input clauses, we expect to satisfy $m(1 - (1/2)^k)$ of them with any random assignment. From an approximation standpoint, the longer the clauses, the easier it is to get close to the optimum.

12.4.2 Maximum Acyclic Subgraph

Directed acyclic graphs (DAGs) are easier to work with than general directed graphs. Sometimes it is useful to simplify a given graph by deleting a set of edges or vertices that suffice to break all cycles. Such *feedback set* problems are discussed in Section 19.11 (page 618).

Here we consider an interesting problem in this class, where we seek to retain as many edges as possible while breaking all directed cycles:

Problem: Maximum Directed Acyclic Subgraph
Input: A directed graph $G = (V, E)$.
Output: Find the largest possible subset $E' \subseteq E$ such that $G' = (V, E')$ is acyclic.

In fact, there is a very simple algorithm that guarantees you a solution with at least half as many edges as optimum. I encourage you to try to find it now before peeking.

Problem: Construct *any* permutation of the vertices, and interpret it as a left-to-right ordering, akin to topological sorting. Now some of the edges will point from left to right, while the rest point from right to left.

One of these two edge subsets must be at least as large as the other. This means it contains at least half the edges. Furthermore, each of these two edge subsets must be acyclic for the same reason only DAGs can be topologically sorted—you cannot form a cycle by repeatedly moving in one direction. Thus, the larger edge subset must be acyclic, and contain at least half the edges of the optimal solution.

This approximation algorithm *is* simple almost to the point of being stupid. But note that heuristics can make it perform better in practice without losing this guarantee. Perhaps we can try many random permutations, and pick the best. Or we can try to exchange pairs of vertices in the permutations retaining those swaps that throw more edges onto the bigger side.

12.5 Set Cover

The previous sections may encourage a false belief that every problem can be approximated to within a constant factor. Indeed, several catalog problems such as maximum clique cannot be approximated to *any* interesting factor.

Set cover occupies a middle ground between these extremes, having a factor-$\Theta(\lg n)$ approximation algorithm. Set cover is a more general version of the vertex cover problem. As defined in Section 21.1 (page 678):

milestone class	6	5		4					3			2	1	0
uncovered elements	64	51	40	30	25	22	19	16	13	10	7	4	2	1
selected subset size	13	11	10	5	3	3	3	3	3	3	3	2	1	1

Figure 12.6: The coverage process for the greedy heuristic on a particular instance of set cover. The width w is defined by the five subsets in milestone class 4, when the number of uncovered elements gets halved from at least $2^5 - 1$ to at most 2^4.

Problem: Set Cover
Input: A collection of subsets $S = \{S_1, \ldots, S_m\}$ of the universal set $U = \{1, \ldots, n\}$.
Output: What is the smallest subset T of S whose union equals the universal set—i.e., $\cup_{i=1}^{|T|} T_i = U$?

The natural heuristic is greedy. Repeatedly select the subset that covers the largest collection of thus-far uncovered elements, until everything is covered. In pseudocode,

> SetCover(S)
> > While ($U \neq \emptyset$) do:
> > > Identify the subset S_i with the largest intersection with U
> > > Select S_i for the set cover
> > > $U = U - S_i$

One consequence of this selection process is that the number of freshly covered elements defines a non-increasing sequence as the algorithm proceeds. Why? If not, greedy would have picked the more powerful subset earlier if it, in fact, existed.

Thus we can view this heuristic as reducing the number of uncovered elements from n down to zero by progressively smaller amounts. A trace of such an execution is shown in Figure 12.6. An important milestone in such a trace occurs each time the number of remaining uncovered elements reduces past a power of two. Clearly there can be at most $\lceil \lg n \rceil$ such events.

Let w_i denote the number of subsets that were selected by the heuristic to cover elements between milestones $2^{i+1} - 1$ and 2^i. Define the width w to be the maximum w_i, where $0 \leq i \leq \lg n$. In the example of Figure 12.6, the maximum width is given by the five subsets needed to go from $2^5 - 1$ down to 2^4.

Since there are at most $\lg n$ such milestones, the solution produced by the greedy heuristic must contain at most $w \cdot \lg n$ subsets. But I claim that the optimal solution must contain *at least* w subsets, so the heuristic solution is no worse than $\lg n$ times optimal.

Why? Consider the average number of new elements covered as we move between milestones $2^{i+1} - 1$ and 2^i. These 2^i elements require w_i subsets, so the average coverage is $\mu_i = 2^i / w_i$. More to the point, the last/smallest of these subsets can cover at most μ_i subsets. Thus, *no subset exists in S that can cover more than μ_i of the remaining 2^i elements.* So, to finish the job, we need at least $2^i/\mu_i = w_i$ subsets.

The surprising thing here is that there are set cover instances where the greedy heuristic finds solutions that are $\Omega(\lg n)$ times optimal: recall the bad vertex cover instance of Figure 12.2. This logarithmic approximation ratio is an inherent property of the problem/heuristic, not an artifact of weak analysis.

> *Take-Home Lesson:* Approximation algorithms guarantee answers that are always close to the optimal solution. They can provide a practical approach to dealing with NP-complete problems.

12.6 Heuristic Search Methods

Backtracking gave us a method to find the *best* of all possible solutions, as scored by a given objective function. However, any algorithm searching all configurations is doomed to be impossibly expensive on large instances. Heuristic search methods provide an alternate approach to difficult combinatorial optimization problems.

In this section, I will discuss approaches to heuristic search. The bulk of our attention will be devoted to simulated annealing, which I find to be the most reliable method to apply in practice. Heuristic search algorithms have an air of voodoo about them, but how they work and why one method can work better than another follows logically enough if you think them through.

In particular, we will look at three different heuristic search methods: random sampling, gradient descent search, and simulated annealing. The traveling salesman problem will be our ongoing example for comparing heuristics. All three heuristics share two common components:

- *Solution candidate representation* – This is a complete yet concise description of possible solutions for the problem, just like we used for backtracking. For traveling salesman, the solution space consists of $(n-1)!$ elements—namely all possible circular permutations of the vertices. We need a data structure that can represent each element of the solution space. For TSP, the candidate solutions can naturally be represented using an array S of $n-1$ vertices, where S_i defines the $(i+1)$st vertex on the tour starting from v_1.

- *Cost function* – Search methods need a *cost* or *evaluation* function to assess the quality of each possible solution. Our search heuristic identifies the element with the best score—either the highest or lowest depending upon the nature of the problem. For TSP, the cost function for evaluating candidate solutions S just sums up the weights of all edges (S_i, S_{i+1}), where S_0 and S_n both denote v_1.

12.6.1   Random Sampling

The simplest approach to search in a solution space uses random sampling, also known as the *Monte Carlo method*. We repeatedly construct random solutions and evaluate them, stopping as soon as we get a good enough solution, or (more likely) when we get tired of waiting. We report the best solution found over the course of our sampling.

True random sampling requires that we select elements from the solution space *uniformly at random*. This means that each of the elements of the solution space must have an equal probability of being the next candidate selected. Such sampling can be a subtle problem. Algorithms for generating random permutations, subsets, partitions, and graphs are discussed in Sections 17.4 through 17.7.

```c
void random_sampling(tsp_instance *t, int nsamples, tsp_solution *s) {
    tsp_solution s_now;                /* current tsp solution */
    double best_cost;                  /* best cost so far */
    double cost_now;                   /* current cost */
    int i;                             /* counter */

    initialize_solution(t->n, &s_now);
    best_cost = solution_cost(&s_now, t);
    copy_solution(&s_now, s);

    for (i = 1; i <= nsamples; i++) {
        random_solution(&s_now);
        cost_now = solution_cost(&s_now, t);

        if (cost_now < best_cost) {
            best_cost = cost_now;
            copy_solution(&s_now, s);
        }

        solution_count_update(&s_now, t);
    }
}
```

When might random sampling do well?

- *When there is a large proportion of acceptable solutions* – Finding a piece of hay in a haystack is easy, since almost anything you grab is a straw. When good solutions are plentiful, a random search should find one quickly.

  Finding prime numbers is a domain where a random search proves successful. Generating large random prime numbers for keys is an important aspect of cryptographic systems such as RSA. Roughly one out of every $\ln n$ integers is prime, so only a modest number of random samples need to be taken to discover primes that are several hundred digits long.

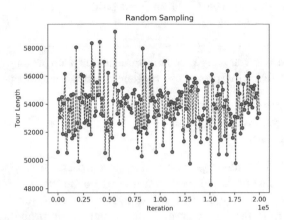

Figure 12.7: Search time/quality tradeoffs for TSP using random sampling. Progress is made infrequently, whenever a new best (here smallest) solution is stumbled upon.

- *When there is no coherence in the solution space* – Random sampling is the right thing to do when there is no sense of when we are getting *closer* to a solution. Suppose you wanted to find one of your friends who has a social security number that ends in 00. There is not much else you can do but tap an arbitrary fellow on the shoulder and ask. No cleverer method will be better than random sampling.

  Consider again the problem of hunting for a large prime number. Primes are scattered quite arbitrarily among the integers. Random sampling is as systematic as anything else would be.

How does random sampling do on TSP? Pretty lousy. The best solution I found after testing 100 million random permutations of a TSP instance with 150 sites was 43,251, which is more than eight times the cost of the optimal tour! The solution space consists almost entirely of mediocre to bad solutions, so quality grows very slowly with the amount of sampling/running time we invest. Figure 12.7 shows the arbitrary up-and-down movements of the generally poor quality solutions encountered using random sampling, so you can get a sense of how the score varied over each iteration.

Most problems we encounter, like TSP, have relatively few good solutions and a highly coherent solution space. More powerful heuristic search algorithms are required to hunt where the needle in the haystack is likely to be.

Stop and Think: Picking the Pair

Problem: We need an efficient and unbiased way to generate random pairs of vertices to perform random vertex swaps. Propose an efficient algorithm to generate elements from the $\binom{n}{2}$ *unordered* pairs on $\{1, \ldots, n\}$ uniformly at random.

Solution: Uniformly generating random structures is a surprisingly subtle problem. Consider the following procedure to generate random unordered pairs:

```
i = random_int(1,n-1);
j = random_int(i+1,n);
```

It is clear that this indeed generates unordered pairs, since $i < j$. Further, it is clear that all $\binom{n}{2}$ unordered pairs can indeed be generated, presuming that random_int generates integers uniformly between its two arguments.

But are they uniform? The answer is no. What is the probability that pair $(1, 2)$ is generated? There is a $1/(n - 1)$ chance of getting the 1, and then a $1/(n - 1)$ chance of getting the 2, which yields $p(1, 2) = 1/(n - 1)^2$. But what is the probability of getting $(n - 1, n)$? Again, there is a $1/(n - 1)$ chance of getting the first number, but now there is only one possible choice for the second candidate! This pair will occur $n - 1$ times more often than $(1, 2)$!

The problem is that fewer pairs start with big numbers than little numbers. We could solve this problem by calculating exactly how many unordered pairs start with i (exactly $(n-i)$) and appropriately bias the probability. The second value could then be selected uniformly at random from $i + 1$ to n.

But instead of working through the math, let's exploit the fact that randomly generating the n^2 *ordered* pairs uniformly is easy. Just pick two integers independently of each other. Ignoring the ordering, by permuting the ordered pair to unordered pair (x, y) where $x < y$, gives us a $2/n^2$ probability of generating each unordered pair of distinct elements. If we happen to generate a pair (x, x), we discard it and try again. We will get unordered pairs uniformly at random in constant expected time by using the following algorithm:

```
do {
    i = random_int(1,n);
    j = random_int(1,n);
    if (i > j) swap(&i,&j);
} while (i==j);
```

∎

12.6.2   Local Search

Now suppose you want to hire an algorithms expert as a consultant to solve your problem. You *could* dial a phone number at random, ask if they are an algorithms expert, and hang up if they say no. After many repetitions you will eventually find one, but it would probably be more efficient to ask the person

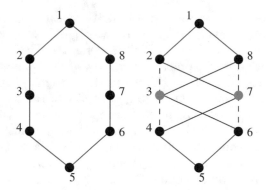

Figure 12.8: Improving a candidate TSP tour by swapping vertices 3 and 7 replaces four old tour edges with four new ones.

on the phone for someone more likely to be an algorithms expert, and call *them* up instead.

A local search scans the *neighborhood* around elements in the solution space. Think of each such candidate solution x as a vertex, with a directed edge (x, y) to every other candidate solution y that is a neighbor of x. Our search proceeds from x to the most promising candidate in x's neighborhood.

We certainly do *not* want to explicitly construct this neighborhood graph for any sizable solution space. Think about TSP, which will have $(n-1)!$ vertices in this graph. We are conducting a heuristic search precisely because we cannot hope to do this many operations in a reasonable amount of time.

Instead, we want a general transition mechanism that takes us to a nearby solution by slightly modifying the current one. Typical transition mechanisms include swapping a random pair of items or changing (inserting or deleting) a single item in the solution.

A reasonable transition mechanism for TSP would be to swap the current tour positions of a random pair of vertices S_i and S_j, as shown in Figure 12.8. This changes up to eight edges on the tour, deleting the four edges currently adjacent to both S_i and S_j, and adding their replacements. The effect of such an incremental change on the quality of the solution can be computed incrementally, so the cost function evaluation takes time proportional to the size of the change (typically constant), which is a big win over being linear in the size of the solution. Even better might be to swap two *edges* on the tour with two others that replace it, since it may be easier to find moves that improve the cost of the tour.

Local search heuristics start from an arbitrary element of the solution space, and then scan the neighborhood looking for a favorable transition to take. In a favorable vertex swap, the four edges we insert are cheaper than the four edges we delete, a computation performed by the transition function. In a greedy *hill-climbing* procedure, we try to find the top of a mountain (or alternately, the lowest point in a ditch) by starting at some arbitrary point and taking any step that leads in the direction we want to travel. We repeat until we have reached a point where all our neighbors lead us in the wrong direction. We are now *King of the Hill*, or for a minimization problem *Dean of the Ditch*.

But unfortunately, we are probably not *King of the Mountain*. Suppose you wake up in a ski lodge, eager to reach the top of the neighboring peak. Your first transition to gain altitude might be to go upstairs to the top of the building. And then you are trapped. To reach the top of the mountain, you must go downstairs and walk outside, but this violates the requirement that each step must increase your score. Hill climbing and closely related heuristics such as greedy search or local search are great at finding local optima quickly, but often fail to find the globally best solution.

```
void hill_climbing(tsp_instance *t, tsp_solution *s) {
    double cost;              /* best cost so far */
    double delta;             /* swap cost */
    int i, j;                 /* counters */
    bool stuck;               /* did I get a better solution? */

    initialize_solution(t->n, s);
    random_solution(s);
    cost = solution_cost(s, t);

    do {
        stuck = true;
        for (i = 1; i < t->n; i++) {
            for (j = i + 1; j <= t->n; j++) {
                delta = transition(s, t, i, j);
                if (delta < 0) {
                    stuck = false;
                    cost = cost + delta;
                } else {
                    transition(s, t, j, i);
                }
                solution_count_update(s, t);
            }
        }
    } while (stuck);
}
```

When does local search do well?

- *When there is great coherence in the solution space* – Hill climbing is at its best when the solution space is *convex*. In other words, it consists of exactly one hill. No matter where you start on the hill, there is always a direction to walk up until you are at the absolute global maximum.

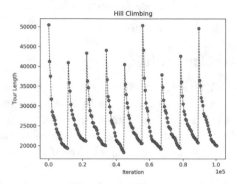

Figure 12.9: Search time/quality tradeoffs for TSP using hill climbing.

Many natural problems have this property. We can think of a binary search as starting in the middle of a search space, where exactly one of the two possible directions we can walk will get us closer to the target key. The simplex algorithm for linear programming (see Section 16.6 (page 482)) is nothing more than hill climbing over the right solution space, yet it guarantees us the optimal solution to any linear programming problem.

- *Whenever the cost of incremental evaluation is much cheaper than global evaluation* – It costs $\Theta(n)$ to evaluate the cost of an arbitrary n-vertex candidate TSP solution, because we must sum up the cost of each edge in the circular permutation describing the tour. Once that is found, however, the cost of the tour after swapping a given pair of vertices can be determined in constant time.

  If we are given a very large value of n and a very small budget of how much time we can spend searching, we are better off using it to do a bunch of incremental evaluations than a few random samples, even if we are looking for a needle in a haystack.

The primary drawback of a local search is that there isn't anything more for us to do after we find the local optimum. Sure, if we have more time we could restart from different random points, but in landscapes of many low hills we are unlikely to stumble on the optimum.

How does local search do on TSP? Much better than random sampling for a similar amount of time. This best local search tour found on our hard 150-site TSP instance had a length of 15,715—improving the quality of our solution by almost a factor of three over random sampling.[2]

This is good, but not great. You would not be happy to learn you are paying twice the taxes than you should be. Figure 12.9 illustrates the trajectory of a local search: repeated streaks from random tours down to decent solutions of

---

[2]This is still more than double the optimal solution cost of 6,828, so the minimum spanning tree approximation of Section 12.3 (page 393) would beat it.

fairly similar quality. We need more powerful methods to get closer to the
optimal solution.

12.6.3   Simulated Annealing

Simulated annealing is a heuristic search procedure that allows occasional transitions leading to more expensive (and hence inferior) solutions. This may not
sound like progress, but it helps keep our search from getting stuck in local
optima. That poor fellow trapped on the second floor of the ski lodge would do
better to break the glass and jump out the window if they really want to reach
the top of the mountain.

The inspiration for simulated annealing comes from the physical process of
cooling molten materials down to the solid state. In thermodynamic theory,
the energy state of a system is described by the energy state of each particle constituting it. But a particle's energy state jumps about randomly, with
such transitions governed by the temperature of the system. In particular,
the backwards transition from a low energy (high quality) state e_i to higher
energy (lower quality) state e_j at temperature T is accepted with probability
$P(e_i, e_j, T)$, where

$$P(e_i, e_j, T) = e^{(e_i - e_j)/(k_B T)}$$

Here k_B is a positive constant—called Boltzmann's constant, used to tune the
desired frequency of backwards moves.

What does this formula mean? Transitioning from a low energy state to
higher energy state, $e_i - e_j < 0$, implies that the exponent is negative. Note
that $0 \leq e^{-x} = 1/e^x \leq 1$ for any positive x. This makes it a probability, one
that gets smaller as $|e_i - e_j|$ gets larger. There is thus a non-zero probability
of accepting a transition into a high-energy (lower quality) state. Small jumps
are much more likely than big ones. The higher the temperature T is, the more
likely such energy jumps will occur.

> Simulated-Annealing()
>   Create initial solution s
>   Initialize temperature T
>   repeat
>   for $i = 1$ to *iteration-length* do
>   Randomly select a neighbor of s to be s_i
>   If $(C(s) \geq C(s_i))$ then $s = s_i$
>   else if $(e^{(C(s) - C(s_i))/(k_B \cdot T)} > random[0, 1))$ then $s = s_i$
>   Reduce temperature T
>   until (no change in $C(s)$)
>   Return s

What relevance does this have for combinatorial optimization? A physical
system, as it cools, seeks to reach a minimum-energy state. Minimizing the total
energy is a combinatorial optimization problem for any set of discrete particles.

Through random transitions generated according to the given probability distribution, we can mimic the physics to solve arbitrary combinatorial optimization problems.

Take-Home Lesson: Don't worry about this molten metal business. Simulated annealing is effective because it spends much more of its time working on good elements of the solution space than on bad ones, and because it avoids getting trapped in local optimum.

As with a local search, the problem representation includes both a representation of the solution space and an easily computable cost function $C(s)$ measuring the quality of a given solution. The new component is the *cooling schedule*, whose parameters govern how likely we are to accept a bad transition as a function of time.

At the beginning of the search, we are eager to use randomness to explore the search space widely, so the probability of accepting a bad transition should be high. As the search progresses, we seek to limit transitions to local improvements and optimizations. This cooling schedule can be regulated by the following parameters:

- *Initial system temperature* – Typically $T_1 = 1$.

- *Temperature decrement function* – Typically $T_i = \alpha \cdot T_{i-1}$, where $0.8 \le \alpha \le 0.99$. This implies an exponential decay in the temperature, as opposed to a linear decay.

- *Number of iterations between temperature change* – Typically, 1,000 iterations or so might be permitted before lowering the temperature. Also, it generally pays to stay at a given temperature for multiple rounds so long as we are making progress there.

- *Acceptance criteria* – A typical criterion is to accept any good transition, and also accept a bad transition whenever

$$e^{\frac{C(s_{i-1})-C(s_i)}{k_B T}} > r,$$

  where r is a random number $0 \le r < 1$. The "Boltzmann" constant k_B scales this cost function so that almost all transitions are accepted at the starting temperature.

- *Stop criteria* – Typically, when the value of the current solution has not changed or improved within the last iteration or so, the search is terminated and the current solution reported.

Creating the proper cooling schedule is a trial-and-error process of mucking with constants and seeing what happens. It probably pays to start from an existing implementation of simulated annealing, so experiment with my full implementation at www.algorist.com.

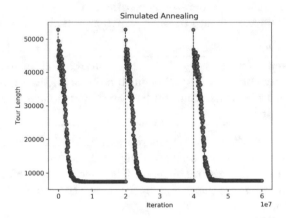

Figure 12.10: Search time/quality tradeoffs for TSP using simulated annealing.

Compare the time/quality profiles of our three heuristics. Simulated annealing does best of all. Figure 12.10 shows three runs from three different random initializations, each looking like a dying heartbeat as it converges to a minima. Because they don't get stuck in a local optimum, all three runs lead to much better solutions than the best hill-climbing result. Further, the rapid plunges toward optimum show that it takes relatively few iterations to score most of the improvement.

After ten million iterations simulated annealing gave us a solution of cost 7,212—only 10.4% over the optimum. Even better solutions are available to those willing to wait a bit longer. Letting it run for one billion iterations (taking only 5 minutes, 21 seconds on my laptop) got the score down to 6,850, just 4.9% over the optimum.

In expert hands, the best problem-specific heuristics for TSP will slightly outperform simulated annealing. But here the simulated annealing solution works admirably. It is my heuristic method of choice for optimization problems.

Implementation

The implementation follows the pseudocode quite closely:

```
void anneal(tsp_instance *t, tsp_solution *s) {
    int x, y;                       /* pair of items to swap */
    int i, j;                       /* counters */
    bool accept_win, accept_loss;   /* conditions to accept transition */
    double temperature;             /* the current system temp */
    double current_value;           /* value of current state */
    double start_value;             /* value at start of loop */
    double delta;                   /* value after swap */
    double exponent;                /* exponent for energy funct */

    temperature = INITIAL_TEMPERATURE;

    initialize_solution(t->n, s);
    current_value = solution_cost(s, t);

    for (i = 1; i <= COOLING_STEPS; i++) {
        temperature *= COOLING_FRACTION;

        start_value = current_value;

        for (j = 1; j <= STEPS_PER_TEMP; j++) {
            /* pick indices of elements to swap */
            x = random_int(1, t->n);
            y = random_int(1, t->n);

            delta = transition(s, t, x, y);
            accept_win = (delta < 0);       /* did swap reduce cost? */

            exponent = (-delta / current_value) / (K * temperature);
            accept_loss = (exp(exponent) > random_float(0,1));

            if (accept_win || accept_loss) {
                current_value += delta;
            } else {
                transition(s, t, x, y);     /* reverse transition */
            }
            solution_count_update(s, t);
        }

        if (current_value < start_value) {  /* rerun at this temp */
            temperature /= COOLING_FRACTION;
        }
    }
}
```

12.6.4   Applications of Simulated Annealing

We provide several examples to demonstrate how careful modeling of the state representation and cost function can lead to elegant simulated annealing solutions for real combinatorial search problems.

Maximum Cut

The *maximum cut* problem seeks to partition the vertices of a weighted graph G into sets V_1 and V_2 to maximize the weight (or number) of edges with one vertex in each set. For graphs that specify an electronic circuit, the maximum cut in the graph defines the largest amount of simultaneous data communication that can take place in the circuit. As discussed in Section 19.6 (page 601), maximum cut is NP-complete.

How can we formulate maximum cut for simulated annealing? The solution space consists of all 2^{n-1} possible vertex partitions. We save a factor of two over all vertex subsets by fixing vertex v_1 to be on the left side of the partition. The subset of vertices accompanying it can be represented using a bit vector. The cost of a solution is the sum of the weights cut in the current configuration. A natural transition mechanism selects one vertex at random and moves it across the partition simply by flipping the corresponding bit in the bit vector. The change in the cost function will be the weight of its old neighbors minus the weight of its new neighbors. This can be computed in time proportional to the degree of the vertex.

This kind of simple, natural modeling represents the right type of heuristic to seek in practice.

Independent Set

An *independent set* of a graph G is a subset of vertices S such that there is no edge with both endpoints in S. The maximum independent set of a graph is the largest vertex set that induces an empty (i.e. edgeless) subgraph. Finding large independent sets arises in dispersion problems associated with facility location and coding theory, as discussed in Section 19.2 (page 589).

The natural state space for a simulated annealing solution would consist of all 2^n possible subsets of the vertices, represented as a bit vector. As with maximum cut, a simple transition mechanism would add or delete one vertex from S.

One natural objective function for subset S might be 0 if the S-induced subgraph contains an edge, and $|S|$ if it is indeed an independent set. Such a function would ensures that we work towards an independent set at all times. However, this condition is so strict that we are liable to move in only a narrow portion of the total search space. More flexibility and quicker objective function computations can result from allowing non-empty graphs at the early stages of cooling. This can be obtained with an objective function like $C(S) = |S| - \lambda \cdot m_S/T$, where λ is a constant, T is the temperature, and m_S is the number of edges in the subgraph induced by S. This objective likes large subsets with few edges, and the dependence of $C(S)$ on T ensures that the search will eventually drive the edges out as the system cools.

Circuit Board Placement

In designing printed circuit boards, we are faced with the problem of positioning modules (typically, integrated circuits) appropriately on the board. Desired criteria in a layout may include (1) minimizing the area or optimizing the aspect ratio of the board so that it properly fits within the allotted space, and (2) minimizing the total or longest wire length in connecting the components. Circuit board placement is representative of the type of messy, multicriterion optimization problems for which simulated annealing is ideally suited.

Formally, we are given a collection of rectangular modules $r_1, \ldots, r_n$, each with associated dimensions $h_i \times l_i$. Further, for each pair of modules r_i, r_j, we are given the number of wires w_{ij} that must connect the two modules. We seek a placement of the rectangles that minimizes area and wire length, subject to the constraint that no two rectangles overlap each other.

The state space for this problem must describe the positions of each rectangle on the board. To make this discrete, these rectangles can be restricted to lie on vertices of an integer grid. Reasonable transition mechanisms include moving one rectangle to a different location, or swapping the position of two rectangles. A natural cost function might be:

$$C(S) = \lambda_{area}(S_{height} \cdot S_{width}) + \sum_{i=1}^{n} \sum_{j=1}^{n} (\lambda_{wire} \cdot w_{ij} \cdot d_{ij} + \lambda_{overlap}(r_i \cap r_j))$$

where λ_{area}, λ_{wire}, and $\lambda_{overlap}$ are weights governing the impact of these components on the cost function. Presumably, $\lambda_{overlap}$ should be a decreasing function of temperature, so after gross placement it adjusts the rectangle positions to not overlap.

> *Take-Home Lesson:* Simulated annealing is a simple but effective technique for efficiently obtaining good but not optimal solutions to combinatorial search problems.

12.7 War Story: Only it is Not a Radio

"Think of it as a radio," he chuckled. "Only it is not a radio."

I'd been whisked by corporate jet to the research center of a large but very secretive company located somewhere east of California. They were so paranoid that I never did get to see the object we were working on, but the people who brought me in did a great job of abstracting the problem.

The issue concerned a manufacturing technique known as *selective assembly*. Eli Whitney helped kick start the Industrial Revolution through his system of *interchangeable parts*. He carefully specified the manufacturing tolerances on each part in his machine so that the parts were *interchangeable*, meaning that

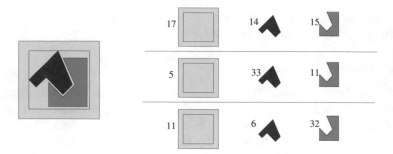

Figure 12.11: Part assignments for three not-radios, such that each had at most fifty defect points.

any legal cog-widget could be used to replace any other legal cog-widget. This greatly sped up the process of manufacturing, because the workers could just put parts together instead of having to stop to file down rough edges and the like. It made replacing broken parts a snap. This was a very good thing.

Unfortunately, it also resulted in large piles of cog-widgets that were slightly outside the manufacturing tolerance, and thus had to be discarded. Another clever fellow then observed that maybe one of these defective cog-widgets could be used when all the *other* parts in the given assembly *exceeded* their required manufacturing tolerances. Good plus bad could well equal good enough. This is the idea of *selective assembly*.

"Each not-radio is made up of n different types of not-radio parts," he told me. For the ith part type (say the right flange gasket), we have a pile of s_i instances of this part type. Each part (flange gasket) comes with a measure of how much it deviates from perfection. We need to match up the parts so as to create the greatest number of working not-radios as possible."

The situation is illustrated in Figure 12.11. Each not-radio consists of three parts, and the sum of the defects in any functional not-radio must total at most fifty. By cleverly balancing the good and bad parts in each machine, we can use all the parts and make three working not-radios.

I thought about the problem. The simplest procedure would take the best part for each part type, make a not-radio out of them, and repeat until the not-radio didn't play (or do whatever a not-radio is supposed to do). But this would create a small number of not-radios drastically varying in quality, whereas they wanted as many decent not-radios as possible.

The goal was to match up good parts and bad parts so the total amount of badness wasn't so bad. Indeed, the problem sounded related to *matching* in graphs (see Section 18.6 (page 562)). Suppose we build a graph where the vertices were the part instances, and add an edge for all two part instances that were within the total error tolerance. In graph matching, we seek the largest number of edges such that no vertex appears more than once in the matching. This is analogous to the largest number of two-part assemblies we can form from the given set of parts.

"I can solve your problem using bipartite matching," I announced, "provided not-radios are each made of only two parts."

There was silence. Then they all started laughing at me. "*Everyone* knows not-radios have more than two parts," they said, shaking their heads.

That spelled the end of this algorithmic approach. Extending to more than two parts turned the problem into matching on hypergraphs[3]—a problem that is NP-complete. Further, it might take exponential time in the number of part types just to build the graph, since we had to explicitly construct each possible hyperedge/assembly.

I went back to the drawing board. They wanted to put parts into assemblies so that no assembly would have more total defects than allowed. Described that way, it sounded like a packing problem. In the *bin packing* problem (see Section 20.9 (page 652)), we are given a collection of items of different sizes and asked to store them using the smallest possible number of bins, each of which has a fixed capacity of size k. Here, the assemblies represented the bins, each of which could absorb total defect $\leq k$. The items to pack were the individual parts, whose size would reflect its quality of manufacture.

It wasn't pure bin packing, however, because parts came in different types, and the task imposed constraints on the allowable contents of each bin. Creating the maximum number of not-radios meant that we sought a packing that maximized the number of bins that contained exactly one part for each of the m different parts types.

Bin packing is NP-complete, but is a natural candidate for a heuristic search approach. The solution space consists of assignments of parts to bins. We initially assign a random part of each type for each bin to provide a starting configuration for the search.

The local neighborhood operation involves moving parts around from one bin to another. We might move one part at a time, but more effective was *swapping* parts of the same type between two randomly chosen bins. In such a swap, both bins remain complete not-radios, hopefully with better error tolerance than before. Thus, our swap operator required three random integers—one to select the appropriate part type (from 1 to m) and two more to select the assembly bins involved (between 1 and b).

The key decision was the cost function to use. They supplied the hard limit k on the total defect level for each *individual* assembly. But what was the best way to score a *set* of assemblies? We could just return the number of acceptable complete assemblies as our score—an integer from 1 to b. Although this was indeed what we wanted to optimize, it would not be sensitive enough to detect when we were making partial progress towards a solution. Suppose one of our swaps succeeded in bringing one of the non-functional assemblies much closer to the not-radio limit k. That would be a better starting point for further progress than the original, and should be favored.

My final cost function was as follows. I gave one point for every working

---

[3]A *hypergraph* is made up of edges that can contain more than two vertices each. They can be thought of as general collections of subsets of vertices/elements.

assembly, and a significantly smaller credit for each non-working assembly based on how close to the threshold k it was. The score for a nonworking assembly decreased exponentially based on how much it was over k. The optimizer would thus seek to maximize the number of working assemblies, and then try to drive down the number of defects in another assembly that was close to the limit.

I implemented this algorithm, and then ran the search on the test case they provided. It was an instance taken directly from the factory floor. It turns out that not-radios contain $m = 8$ important parts types. Some parts types are more expensive than others, and so they have fewer candidates available to consider. The most constrained parts type had only eight representatives, so there could be at most eight possible assemblies from this given mix.

I watched as simulated annealing chugged and bubbled on this problem instance. The number of completed assemblies instantly climbed (one, two, three, four) before progress started to slow a bit. Then came five and six in a hiccup, with a pause before assembly seven came triumphantly together. But try as it might, the program could not put eight not-radios together before I lost interest in watching.

I called and tried to admit defeat, but they wouldn't hear of it. It turned out that the best the factory had managed after extensive efforts was only *six* working not-radios, so my result represented a significant improvement!

12.8   War Story: Annealing Arrays

The war story of Section 3.9 (page 98) reported how we used advanced data structures to simulate a new method for sequencing DNA. Our method, interactive sequencing by hybridization (SBH), required building arrays of specific oligonucleotides on demand.

A biochemist at Oxford University got interested in our technique, and moreover he had in his laboratory the equipment we needed to test it out. The Southern Array Maker, manufactured by Beckman Instruments, prepared discrete oligonucleotide sequences in 64 parallel rows across a polypropylene substrate. The device constructs arrays by appending single characters to each cell along specific rows and columns of arrays. Figure 12.12 shows how to construct an array of all $2^4 = 16$ purine (A or G) 4-mers by building the prefixes along four rows and the suffixes along four columns. This technology provided an ideal environment for testing the feasibility of interactive SBH in a laboratory, because with proper programming it gave a way to fabricate a wide variety of oligonucleotide arrays on demand.

However, we had to provide the proper programming. Fabricating complicated arrays required solving a difficult combinatorial problem. We were given as input a set of n strings (representing oligonucleotides) to fabricate in an $m \times m$ array (where $m = 64$ on the Southern apparatus). We had to produce a schedule of row and column commands to realize the set of strings S. We proved that the problem of designing dense arrays was NP-complete, but that didn't really matter. My student Ricky Bradley and I had to solve it anyway.

	suffix			
prefix	AA	AG	GA	GG
AA	$AAAA$	$AAAG$	$AAGA$	$AAGG$
AG	$AGAA$	$AGAG$	$AGGA$	$AGGG$
GA	$GAAA$	$GAAG$	$GAGA$	$GAGG$
GG	$GGAA$	$GGAG$	$GGGA$	$GGGG$

Figure 12.12: A prefix–suffix array of all purine 4-mers.

"We are going to have to use a heuristic," I told him. "So how can we model this problem?"

"Well, each string can be partitioned into prefix and suffix pairs that realize it. For example, the string ACC can be realized in four different ways: prefix '' and suffix ACC, prefix A and suffix CC, prefix AC and suffix C, or prefix ACC and suffix ''. We seek the smallest set of prefixes and suffixes that together realize all the given strings," Ricky said.

"Good. This gives us a natural representation for simulated annealing. The state space will consist of all possible subsets of prefixes and suffixes. The natural transitions between states might include inserting or deleting strings from our subsets, or swapping a pair in or out."

"What's a good cost function?" he asked.

"Well, we need as small an array as possible that covers all the strings. How about taking the maximum of the number of rows (prefixes) or columns (suffixes) used in our array, plus the number of strings from S that are not yet covered. Try it and let's see what happens."

Ricky went off and implemented a simulated annealing program along these lines. It printed out the state of the solution each time a transition was accepted and was fun to watch. The program quickly kicked out unnecessary prefixes and suffixes, and the array began shrinking rapidly in size. But after several hundred iterations, progress started to slow. A transition would knock out an unnecessary suffix, wait a while, then add a different suffix back again. After a few thousand iterations, no real improvement was happening.

"The program doesn't seem to recognize when it is making progress. The evaluation function only gives credit for minimizing the larger of the two dimensions. Why not add a term to give some credit to the other dimension."

Ricky changed the evaluation function, and we tried again. This time, the program did not hesitate to improve the shorter dimension. Indeed, our arrays started to turn into skinny rectangles instead of squares.

"OK. Let's add another term to the evaluation function to give it points for being roughly square."

Ricky tried again. Now the arrays were the right shape, and progress was in the right direction. But the progress was still slow.

"Too many of the insertion moves don't affect many strings. Maybe we should skew the random selections so that the important prefix/suffixes get

Figure 12.13: Compression of the HIV array by simulated annealing—after 0, 500, 1,000, and 5,750 iterations.

picked more often."

Ricky tried again. Now it converged faster, but sometimes it still got stuck. We changed the cooling schedule. It did better, but was it doing well? Without a lower bound knowing how close we were to optimal, it couldn't really tell how good our solution was. We tweaked and tweaked until our program stopped improving.

Our final solution refined the initial array by applying the following random moves:

- *Swap* – swap a prefix/suffix on the array with one that isn't.

- *Add* – add a random prefix/suffix to the array.

- *Delete* – delete a random prefix/suffix from the array.

- *Useful add* – add the prefix/suffix with the highest usefulness to the array.

- *Useful delete* – delete the prefix/suffix with the lowest usefulness from the array.

- *String add* – randomly select a string not on the array, and add the most useful prefix and/or suffix to cover this string.

We used a standard cooling schedule, with an exponentially decreasing temperature (dependent upon the problem size) and a temperature-dependent Boltzmann criterion for accepting states that have higher costs. Our final cost function was defined as

$$cost = 2 \times max + min + \frac{(max - min)^2}{4} + 4(str_{total} - str_{in})$$

where max is the size of the maximum chip dimension, min is the size of the minimum chip dimension, $str_{total} = |S|$, and str_{in} is the number of strings from S currently on the chip.

How well did we do? Figure 12.13 shows the convergence of an array consisting of the 5,716 unique 7-mers of the HIV virus. Figure 12.13 shows snapshots of the state of the chip at four points during the annealing process, after 0, 500, 1,000, and finally 5,750 iterations. Red pixels represent the first occurrence of an HIV 7-mer. The final chip size here is 130×132—quite an improvement over the initial size of 192×192. It took about fifteen minutes of computation to complete the optimization, which was perfectly acceptable for the application.

But how well did we do? Since simulated annealing is only a heuristic, we really don't know how close to optimal our solution is. I think we did pretty well, but can't really be sure. Simulated annealing is a good way to handle complex optimization problems. However, to get the best results, expect to spend more time tweaking and refining your program than you did in writing it in the first place. This is dirty work, but sometimes you have to do it.

12.9 Genetic Algorithms and Other Heuristics

Many heuristic search methods have been proposed for combinatorial optimization problems. Like simulated annealing, many of these techniques rely on analogies to real-world processes, including *genetic algorithms*, *neural networks*, and *ant colony optimization*.

The intuition behind these methods is highly appealing, but skeptics decry them as voodoo optimization techniques that rely more on superficial analogies to nature than producing superior computational results on real problems compared to other methods.

The question isn't whether you can get decent answers for many problems given enough effort using these techniques. Clearly you can. The real question is whether they lead to *better* solutions with *less implementation complexity* or *greater efficiency* than the other methods we have discussed.

In general, I don't believe that they do. But in the spirit of free inquiry, I introduce genetic algorithms, which is the most popular of these methods. See the Chapter Notes section for more detailed readings.

Genetic Algorithms

Genetic algorithms draw their inspiration from evolution and natural selection. Through the process of natural selection, organisms adapt to optimize their chances for survival in a given environment. Random mutations occur in an organism's genetic description, which then get passed on to its offspring. Should a mutation prove helpful, these children are more likely to survive and reproduce. Should it prove harmful, they won't, and so the bad trait will die with them.

Genetic algorithms maintain a "population" of solution candidates for the given problem. Elements are drawn at random from this population and allowed to "reproduce" by combining aspects of the two-parent solutions. The probability that an element is chosen to reproduce is based on its "fitness," essentially the quality of the solution it represents. Unfit elements are removed from the population, to be replaced by a successful-solution offspring.

The idea behind genetic algorithms is extremely appealing. However, they just don't seem to work as well on practical combinatorial optimization problems as simulated annealing does. There are two primary reasons for this. First, it is quite unnatural to model applications in terms of genetic operators like mutation and crossover on bit strings. The pseudo-biology adds another level of complexity between you and your problem. Second, genetic algorithms take a very long time on non-trivial problems. The crossover and mutation operations typically make no use of problem-specific structure, so most transitions lead to inferior solutions, and convergence is slow. Indeed, the analogy with evolution—where significant progress require millions of years—can be quite appropriate.

I will not discuss genetic algorithms further, except to discourage you from considering them for your applications. However, pointers to implementations of genetic algorithms are provided in Section 16.5 (page 478) if you really insist on playing with them.

Take-Home Lesson: I have *never* encountered any problem where genetic algorithms seemed to me the right way to attack it. Further, I have *never* seen any computational results reported using genetic algorithms that favorably impressed me. Stick to simulated annealing for your heuristic search voodoo needs.

12.10   Quantum Computing

We live in an era where the random access machine (RAM) model of computation introduced in Section 2.1 is being augmented by a new class of computing devices. These devices are powered by the principles of quantum mechanics, which ascribes seemingly impossible properties to how systems of atoms behave. *Quantum computers* exploit these properties to perform certain types of computations with algorithmic efficiencies asymptotically faster than conventional machines.

Quantum mechanics is well known for being so completely unintuitive that no one really understands it. Superposition! Quantum weirdness! Entanglement! Schrödinger's cat! Collapsing wave functions! *Gaa!!* I must make clear that there is absolutely no controversy about the rules of how quantum computers will behave. People in the know agree on the properties of quantum mechanics and the theoretical powers of these machines. You don't *have* to understand why a law exists in order to follow it to the letter. Research in quantum computing revolves around developing technologies to implement large and reliable quantum systems, and devising new algorithms to exploit this power.

I presume that you the reader may well never have taken a physics course, and are likely rusty with linear algebra and complex numbers, so I will try to dispense with such matters. My goal here is to show why these machines have great potential power, and to provide some insight as how this can be exploited to yield asymptotically faster algorithms for certain problems. My approach is to make up a new model of a "quantum" computer. This model isn't really correct, but hopefully provides insight as to what makes these machines exciting. I provide a taste of how three of the most famous quantum algorithms work. Finally, I make some predictions on what the future holds for quantum computing, and confess some of my model's lies in Section 12.10.5.

12.10.1 Properties of "Quantum" Computers

Consider a conventional deterministic computer with n bits of memory labeled $b_0, \ldots, b_{n-1}$. There are exactly $N = 2^n$ possible states this computer can exist in, because each bit can be set to either 0 or 1. The ith state of the machine corresponds to the string of bits corresponding to the binary representation of integer i. Each instruction that is executed changes the state of the machine, flipping a specific set of bits.

We can think of a conventional deterministic computer as maintaining a probability distribution of the machine's current state. At any moment, the probability $p(i)$ of being in state i equals zero for $2^n - 1$ of the possible states, with $p(j) = 1$ if the computer happens to be in state j. Yes, this is a probability distribution over states, but not a very interesting one.

Quantum computers come with n *qubits* of memory, $q_0, \ldots, q_{n-1}$. Thus, there are also $N = 2^n$ possible bit patterns associated with this computer, but the actual state it is in at any moment is a probability distribution. Each of the 2^n bit patterns has an associated probability, where $p(i)$ is the probability that when the machine is read, it would report being in state i. This probability distribution is much richer than with conventional deterministic machines: there can be a non-zero probability of being in all N states at any given time. Being able to manipulate probability distribution in parallel for all $N = 2^n$ states is the real win of quantum computing. As is true of any probability distribution, all these probabilities must sum to one, so $\sum_{i=0}^{N-1} p(i) = 1$.

"Quantum" computers support the following operations:

- *Initialize-state(Q,n,D)* – Initialize the probability distribution of the n qubits of machine Q as per the description D. Obviously this would take $\Theta(2^n)$ time if D was given as an explicit list of the desired probability of each state. We thus seek general descriptions that are smaller, say $O(n)$ in size, like "set each of the $N = 2^n$ states to be of equal probability, so $p(i) = 1/2^n$." The time of the *Initialize-state* operation is $O(|D|)$, not $O(N)$.

- *Quantum-gate(Q,c)* – Change the probability distribution of machine Q according to a quantum gate condition c. Quantum gates are logic operations akin to *and* or *or*, changing the probabilities of states according to the current contents of (say) qubits q_x and q_y. The time of this operation is proportional to the number of qubits involved with condition c, but typically is $O(1)$.

- *Jack(Q,c)* – Increase the probabilities of all states defined by condition c. For example, we might want to jack up the probabilities of all states

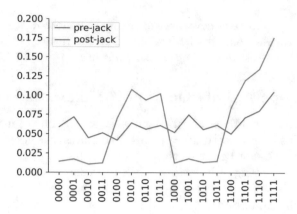

Figure 12.14: Jacking the probability of all states where qubit $q_2 = 1$.

where $c = $ "qubit $q_2 = 1$", as shown in Figure 12.14. This takes time proportional to the number of qubits in condition c, but typically is $O(1)$.

That this can be done in constant time should be recognized as surprising. Even if the condition raises the probability of just one state i, in order to keep the sum totaling to one the probabilities of all $2^n - 1$ other states must be lowered. That this can be done in constant time is one of the strange properties of "quantum" physics.

- *Sample(Q)* – Select exactly one of the 2^n states at random as per the current probability distribution of machine Q, and return the values of the n qubits $q_0, \ldots, q_{n-1}$. This takes $O(n)$ time to report the state of the machine.

"Quantum" algorithms are sequences of these powerful operations, perhaps augmented with control logic from conventional computers.

12.10.2   Grover's Algorithm for Database Search

The first algorithm we will consider is for database search, or more generally function inversion. Suppose we think of the unique binary string defining each of the $N = 2^n$ states as a key, with each state i associated with a value $v(i)$. If these values take m bits to store, we can think of our database as consisting of 2^n unsorted strings with non-zero probability each $n + m$ qubits long, with the value $v(i)$ stored in the m highest-order qubits.

We can create such a system Q using the *Initialize-state(Q,n,D)* instruction, with an appropriate condition D. To any $n + m$ qubit string of the form i concatenated to $v(i)$ we assign a probability of $1/2^n$. All of the $2^m - 1$ other $n+m$ qubit strings with prefix i are assigned probability zero. For a given m-bit search string S, we seek to return an $n + m$ qubit string such that $S = q_n \ldots q_{n+m-1}$.

The instruction set described above does not give us a print statement, other than *Sample(Q)*. To make it likely that this sample returns what we want, we must jack up the probability of all strings with the right value qubits. This suggests an algorithm like the following:

Search(Q,S)
    Repeat
        Jack(Q, "all strings where $S = q_n \ldots q_{n+m-1}$")
        Until probability of success is high enough
        Return the first n bits of *Sample(Q)*

Each such *Jack* operation takes constant time, which is fast. But it increases the probabilities at a slow-enough rate that $\Theta(\sqrt{N})$ rounds are necessary and sufficient to make success likely. Thus, this algorithm returns the appropriate string in $\Theta(\sqrt{N})$, a big win over the $\Theta(N)$ complexity of sequential search.

Solving Satisfiability?

An interesting application of Grover's database searching algorithm is an algorithm for solving satisfiability, the mother of all NP-complete problems discussed in Section 11.4. As we have seen, an n-qubit quantum system represents all $N = 2^n$ binary strings of length n. By thinking of each 1 as true and 0 as false, each such string defines a truth assignment on n Boolean variables.

Now let's add an $(n + 1)$st qubit to the system, to store whether the ith string satisfies a given Boolean logic formula F. Testing whether a given truth assignment satisfies a given set of clauses is easy: explicitly test whether each clause contains at least one true literal. Such testing can be done in parallel using a sequence of quantum gates. If F is a 3-SAT formula with k clauses, this testing can be done making a pass over each clause, using a total of roughly $3k$ quantum gates. We set qubit $q_n = 1$ if $q_0, \ldots, q_{n-1}$ is a satisfying truth assignment, and $q_n = 0$ if not. This means that all strings s corresponding to incorrect values of F get set to $p(s) = 0$.

Now suppose we perform a *Search(Q,($q_n == 1$))* operation, returning quantum state $\{q_0, \ldots, q_{n-1}\}$. With high probability, this is a satisfying truth assignment, so this gives us an efficient quantum algorithm to solve satisfiability! Grover's search algorithm runs in $O(\sqrt{N})$ time, where $N = 2^n$. Since $\sqrt{N} = \sqrt{2^n} = (\sqrt{2})^n$, this runs in $O(1.414^n)$ vs. the naive bound, which is a big improvement. For $n = 100$, this cuts the number of steps from 1.27×10^{30} to 1.13×10^{15}. But $(\sqrt{2})^n$ still grows exponentially, so this is not a polynomial-time algorithm. This quantum algorithm is a big win over brute-force search, but it is not enough to have the effect of $P = NP$.

> *Take-Home Lesson:* Despite their powers, **quantum computers cannot solve NP-complete problems in polynomial time**. Of course, the world changes if $P = NP$, but presumably $P \neq NP$. We believe that the class of problems that can be solved in polynomial time on a quantum computer (called BQP) does not contain NP with roughly the same confidence that we believe $P \neq NP$.

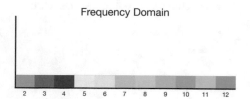
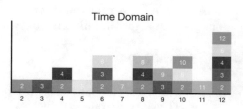

Figure 12.15: Transforming the integers from the frequency to the time domain computes the number of factors for each integer.

12.10.3   The Faster "Fourier Transform"

The fast Fourier Transform (FFT) is the most important algorithm in signal processing, converting an N-element numerical time series to an equivalent representation as the sum of N periodic functions of different frequencies. Many filtering and compression algorithms on signals reduce to eliminating high and/or low frequency components from the Fourier transform of the signal, as discussed in Section 16.11.

Computing Fourier transforms is most simply done using an $O(N^2)$ algorithm, with each of the N elements produced as the sum of N terms. The FFT is a divide-and-conquer algorithm for evaluating this convolution in $O(N \log N)$, which we discussed in Section 5.9. The FFT can be also implemented as a circuit with $\log M$ stages, where each stage involves M independent, parallel multiplications.

It happens that each of these stages of the FFT circuit can be implemented using $\log M$ quantum gates. Thus, the Fourier transform of the $N = 2^n$ states of an n-qubit system can be solved on a "quantum" computer in $O((\log N)^2) = O(n^2)$ time. This is an exponential-time improvement over the FFT!

But there is a catch. We now have an n-qubit quantum system Q, where the Fourier coefficient associated with each input element $0 \leq a_i \leq N - 1$ is represented as the probability of string i in Q. There is no way to get even one of these 2^n coefficients out of machine Q. All we can do is call $Sample(Q)$ and get the index of a (presumably) large coefficient, selected with probability proportional to its magnitude.

This is a very restricted type of "Fourier transform," returning just the index of what is likely to have a large coefficient. But it lays the groundwork for perhaps the most famous quantum algorithm, Shor's algorithm for integer factorization.

12.10.4   Shor's Algorithm for Integer Factorization

There is an interesting connection between periodic functions (meaning they repeat at fixed intervals) and the integers that have a given integer k as a factor, or equivalently as a divisor. It should be clear that these integers must

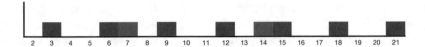

Figure 12.16: Sampling from multiples of the factors of M (here 21) is unlikely to directly yield a factor, but the greatest common divisor of multiple samples is almost certain to give a factor.

occur k positions apart on a number line. What are the integers that have 7 as factor? They are $7, 14, 21, 28, \ldots$, clearly a periodic function with a period of $k = 7$.

The FFT enables us to convert a series in the "time domain" (here consisting of all multiples of 7 less than N) into the "frequency domain," with a non-zero value for the seventh Fourier coefficient signifying a repeating function of period seven. And vice versa. As Figure 12.15 shows, an n-qubit system Q with each state i initialized to a probability $p(i) = 1/2^n$ represents the period of every possible factor i. Taking an FFT of Q takes us into the time domain, yielding the number of factors for every integer $0 \leq i \leq N - 1$.

Now suppose we are interested in factoring a given integer $M < N$. Through quantum magic we can set up system Q in a time domain where exactly the integers that are multiples of factors of M have large probabilities, as shown in Figure 12.16. Whenever we sample from this system, we get a random multiple of a factor of M. For $M = 77 = 7 \times 11$, we might get samples like 33, 42, and 55. These seem helpful, but note that none of these is actually a factor of M.

The *greatest common divisor* $gcd(x, y)$ is the largest integer d such that d divides x and d also divides y. Fast algorithms exist to compute $gcd(x, y)$, as discussed in Section 11.2.3. Whenever the greatest common divisor is greater than 1, we have an excellent candidate for a factor of M. Observe that $gcd(33, 55) = 11$.

The complete Shor's algorithm for factoring, in pseudocode, is as follows:

Factor(M)
    Set up an n-qubit quantum system Q, where $N = 2^n$ and $M < N$.
    Initialize Q so that $p(i) = 1/2^n$ for all $0 \leq i \leq N - 1$.
    Repeat
        Jack(Q, "all i such that $(gcd(i, M) > 1)$*")*
    Until the probabilities of all terms relatively prime to M are very small.
    FFT(Q).
    For $j = 1$ to n
        $S_j = Sample(Q)$
        If $((d = GCD(S_j, S_k)) > 1)$ and (d divides M)), for some $k < j$
            Return(d) as a factor of M
    Otherwise report no factor was found

Each of these operations takes time proportional to n, not $M = \Theta(2^n)$, so this is an exponential-time improvement over divide-and-test factoring. No polynomial-time algorithm is known for integer factoring on conventional machines, but neither is it NP-complete. Thus, no complexity-theoretic assumptions are violated by having a fast algorithm for integer factorization.

12.10.5   Prospects for Quantum Computing

What are the prospects for quantum computing? I write this in the year of normal vision (2020), and developments are happening quickly. My vision is not necessarily better than anyone else's, but I will make some educated guesses:

- *Quantum computing is a real thing, and is gonna happen* – One develops a reasonably trustworthy bullsh*t detector after watching technology hype-cycles for forty years, and quantum computing now passes my sniff test by a safe margin. I see very smart people excited by the prospects of the field, clear and steady technological progress, and substantial investment by big companies and other important players. At this point, I would be surprised if it fizzles out completely.

- *Quantum computing is unlikely to impact the problems considered in this book* – The value of your hard-won algorithms expertise gained from reading my book will hold up just fine in the quantum computing era. I see it as a technology with specialized applications, akin to the way the fastest supercomputers are seen primarily in scientific computing instead of industry. With the exception of factoring integers, there is nothing in this book that I see as potentially better done on quantum computers.

  The fastest technology does not necessarily take over the world. The highest achievable data transmission rates involve giant aircraft packed with DVDs or an even denser storage media. Still, no one has figured out a way to exploit this technology commercially. Similarly, quantum computing is not necessarily a good fit with most of traditional computing.

- *The big wins are likely to be in problems computer scientists don't really care about* – It is not yet clear what the killer app for quantum computing will be, but the most promising applications seem to involve simulating quantum systems. This is a big deal in chemistry and material science, and may well lead to amazing revolutions in drug design and engineering. But it is unclear to what degree computer scientists will lead the fight in this revolution.

We shall see. I look forward to writing the fourth edition of this book, perhaps in 2035, to learn how well these predictions held up.

You should be aware that the "quantum" computing model I describe here differs from real quantum computers in several important ways, although I believe it basically captures the flavor of how they work. Still, in real quantum computers:

- *The role of probabilities are played by complex numbers* – Probabilities are real numbers between 0 and 1 that add up to 1 over all elements of the probability space. Quantum probabilities are complex numbers whose squares lie between 0 and 1 that add up to 1 over all elements of the probability space. Recall that the FFT algorithm described in Section 5.9 works on complex numbers, which is the source of its power.

- *Reading the state of a quantum system destroys it* – When we randomly sample from the state of a quantum system, we lose all information about the remaining $2^n - 1$ states. Thus, we cannot repeatedly sample from the distribution, as we did above. But we *can* recreate the system from scratch as many times as we need, and sample each of these to get the effect of repeated sampling.

  The key hurdle of quantum computing is how to extract the answer we want, because this measurement yields only a tiny bit of the information inherent in Q. If some but not all of the qubits in Q are measured, then the remaining qubits also get "measured" in that their state collapses accordingly. This is the real source of magic in quantum computing, and is referred to as *entanglement*.

- *Real quantum systems breakdown (or decohere) easily* – Manipulating individual atoms to do complex things is not child's play. Quantum computers are generally run at extremely low temperatures and in shielded environments to get them to hang together as long as possible. With current technologies this isn't very long, limiting the complexity of algorithms that can be run and mandating the development of error-correction technologies for quantum systems.

- *Initializing quantum states and the powers of quantum gates are somewhat different than described above* – I have played fast and loose with exactly how you can initialize quantum states and what operations you can do with them. Quantum gates are essentially unitary matrices, multiplication by which changes the probabilities of Q. These operations are well defined by the properties of quantum mechanics, but the details matter here.

Chapter Notes

Kirkpatrick et al.'s original paper on simulated annealing [KGV83] included an application to VLSI module placement problems. The applications from Section 12.6.4 (page 410) are based on material from Aarts and Korst [AK89]. There is a class of quantum computers manufactured by D-Wave that aspire to quantum annealing to solve optimization problems, but the jury is still out as to whether this is an important technology.

The heuristic TSP solutions presented here employ vertex-swap as the local neighborhood operation. In fact, edge-swap is a more powerful operation. Each edge-swap changes two edges in the tour at most, as opposed to at most four edges with a vertex-swap. This improves the possibility of a local improvement. However, more sophisticated data structures are necessary to efficiently maintain the order of the resulting tour. See Fredman et al. [FJMO93].

The different heuristic search techniques are ably presented in Aarts and Lenstra [AL97], which I strongly recommend for those interested in learning more about heuristic searches. Their coverage includes *tabu search*, a variant of simulated annealing that uses extra data structures to avoid transitions to recently visited states. Ant colony optimization is discussed in Dorigo and Stutzle [DT04]. Livnat and Papadimitriou [LP16] propose a theory for why genetic algorithms are generally lousy: the purpose of sexual reproduction is to create diverse populations, not highly optimized individuals. Still, see Michalewicz and Fogel [MF00] for a more favorable view of genetic algorithms and the like.

Our work using simulated annealing to compress DNA arrays was reported in Bradley and Skiena [BS97]. See Pugh [Pug86] and Coullard et al. [CGJ98] for more on selective assembly.

If my introduction to quantum computing whetted your interest, I would encourage you to look at more definitive sources. Books with interesting treatments of quantum computing include [Aar13, Ber19, DPV08], with Yanofsky and Mannucci [YM08] particularly gentle and accessible. Scott Aaronson's blog `https://www.scottaaronson.com/blog/` is fascinating reading, covering the latest in quantum computing algorithms, as well as the broader world of complexity theory.

12.11   Exercises

Special Cases of Hard Problems

12-1. *[5]* Dominos are tiles represented by integer pairs (x_i, y_i), where each of the values x_i and y_i are integers between 1 and n. Let S be a sequence of m integer pairs $[(x_1, y_1), (x_2, y_2), ..., (x_m, y_m)]$. The goal of the game is to create long chains $[(x_{i1}, y_{i1}), (x_{i2}, y_{i2}), ..., (x_{it}, y_{it})]$ such that $y_{ij} = x_{i(j+1)}$. Dominos can be flipped, so (x_i, y_i) equivalent to (y_i, x_i). For $S = [(1, 3), (4, 2), (3, 5), (2, 3), (3, 8)]$, the longest domino sequences include $[(4, 2), (2, 3), (3, 8)]$ and $[(1, 3), (3, 2), (2, 4)]$.

(a) Prove that finding the longest domino chain is NP-complete.

(b) Give an efficient algorithm to find the longest domino chain where the numbers increase along the chain. For S above, the longest such chains are $[(1, 3), (3, 5)]$ and $[(2, 3), (3, 5)]$.

12-2. *[5]* Let $G = (V, E)$ be a graph and x and y be two distinct vertices of G. Each vertex v contains a given number of tokens $t(v)$ that you can collect if you visit v.

(a) Prove that it is NP-complete to find the path from x to y where you can collect the greatest possible number of tokens.

(b) Give an efficient algorithm if G is a directed acyclic graph (DAG).

12-3. *[8]* The *Hamiltonian completion problem* takes a given graph G and seeks an algorithm to add the smallest number of edges to G so that it contains a Hamiltonian cycle. This problem is NP-complete for general graphs; however, it has an efficient algorithm if G is a tree. Give an efficient and *provably correct* algorithm to add the minimum number of possible edges to tree T so that T plus these edges is Hamiltonian.

Approximation Algorithms

12-4. *[4]* In the *maximum satisfiability problem*, we seek a truth assignment that satisfies as many clauses as possible. Give an heuristic that always satisfies at least half as many clauses as the optimal solution.

12-5. *[5]* Consider the following heuristic for vertex cover. Construct a DFS tree of the graph, and delete all the leaves from this tree. What remains must be a vertex cover of the graph. Prove that the size of this cover is at most twice as large as optimal.

12-6. *[5]* The *maximum cut* problem for a graph $G = (V, E)$ seeks to partition the vertices V into disjoint sets A and B so as to maximize the number of edges $(a, b) \in E$ such that $a \in A$ and $b \in B$. Consider the following heuristic for maximum cut. First assign v_1 to A and v_2 to B. For each remaining vertex, assign it to the side that adds the most edges to the cut. Prove that this cut is at least half as large as the optimal cut.

12-7. *[5]* In the *bin-packing problem*, we are given n objects with weights $w_1, w_2, ..., w_n$, respectively. Our goal is to find the smallest number of bins that will hold the n objects, where each bin has a capacity of at most one kilogram.

The *first-fit heuristic* considers the objects in the order in which they are given. For each object, place it into the first bin that has room for it. If no such bin exists, start a new bin. Prove that this heuristic uses at most twice as many bins as the optimal solution.

12-8. *[5]* For the first-fit heuristic described just above, give an example where the packing it finds uses at least 5/3 times as many bins as optimal.

12-9. *[5]* Given an undirected graph $G = (V, E)$ in which each node has degree $\leq d$, show how to efficiently find an independent set whose size is at least $1/(d + 1)$ times that of the largest independent set.

12-10. *[5]* A *vertex coloring* of graph $G = (V, E)$ is an assignment of colors to vertices of V such that each edge (x, y) implies that vertices x and y are assigned different colors. Give an algorithm for vertex coloring G using at most $\Delta + 1$ colors, where Δ is the maximum vertex degree of G.

12-11. *[5]* Show that you can solve any given Sudoku puzzle by finding the minimum vertex coloring of a specific, appropriately constructed $(9 \times 9) + 9$ vertex graph.

Combinatorial Optimization

For each of the problems below, design and implement a simulated annealing heuristic to get reasonable solutions. How well does your program perform in practice?

12-12. *[5]* Design and implement a heuristic for the bandwidth minimization problem discussed in Section 16.2 (page 470).

12-13. *[5]* Design and implement a heuristic for the maximum satisfiability problem discussed in Section 17.10 (page 537).

12-14. *[5]* Design and implement a heuristic for the maximum clique problem discussed in Section 19.1 (page 586).

12-15. *[5]* Design and implement a heuristic for the minimum vertex coloring problem discussed in Section 19.7 (page 604).

12-16. *[5]* Design and implement a heuristic for the minimum edge coloring problem discussed in Section 19.8 (page 608).

12-17. *[5]* Design and implement a heuristic for the minimum feedback vertex set problem discussed in Section 19.11 (page 618).

12-18. *[5]* Design and implement a heuristic for the set cover problem discussed in Section 21.1 (page 678).

"Quantum" Computing

12-19. *[5]* Consider an n qubit "quantum" system Q, where each of the $N = 2^n$ states start out with equal probability $p(i) = 1/2^n$. Say the $Jack(Q, 0^n)$ operation doubles the probability of the state where all qubits are zero. How many calls to this $Jack$ operation are necessary until the probability of sampling this null state becomes $\geq 1/2$?

12-20. *[5]* For the satisfiability problem, construct (a) an instance on n variables that has exactly one solution, and (b) an instance on n variables that has exactly 2^n different solutions.

12-21. *[3]* Consider the first ten multiples of 11, namely 11, 22, ... 110. Pick two of them (x and y) at random. What is the probability that $gcd(x, y) = 11$?

12-22. *[8]* IBM quantum computing (`https://www.ibm.com/quantum-computing/`) offers the opportunity to program a quantum computing simulator. Take a look at an example quantum computing program and run it to see what happens.

LeetCode

12-1. `https://leetcode.com/problems/split-array-with-same-average/`

12-2. `https://leetcode.com/problems/smallest-sufficient-team/`

12-3. `https://leetcode.com/problems/longest-palindromic-substring/`

HackerRank

12-1. `https://www.hackerrank.com/challenges/mancala6/`

12-2. `https://www.hackerrank.com/challenges/sams-puzzle/`

12-3. `https://www.hackerrank.com/challenges/walking-the-approximate longest-path/`

Programming Challenges

These programming challenge problems with robot judging are available at `https://onlinejudge.org`:

12-1. "Euclid Problem"—Chapter 7, problem 10104.

12-2. "Chainsaw Massacre"—Chapter 14, problem 10043.

12-3. "Hotter Colder"—Chapter 14, problem 10084.

12-4. "Useless Tile Packers"—Chapter 14, problem 10065.

Chapter 13

How to Design Algorithms

Designing the right algorithm for a given application is a major creative act—that of taking a problem and pulling a solution out of the air. The space of choices you can make in algorithm design is enormous, leaving you plenty of freedom to hang yourself.

This book has been designed to make you a better algorithm designer. The techniques presented in Part I provide the basic ideas underlying all combinatorial algorithms. The problem catalog of Part II will help you with modeling your application, and inform you what is known about the relevant problems. However, being a successful algorithm designer requires more than book knowledge. It requires a certain attitude—the right problem-solving approach. It is difficult to teach this mindset in a book, yet getting it is essential to becoming a successful algorithm designer.

The key to algorithm design (or any other problem-solving task) is to proceed by asking yourself questions to guide your thought process. "What if we do this? What if we do that?" Should you get stuck on the problem, the best thing to do is move onto the next question. In any group brainstorming session, the most useful person in the room is the one who keeps asking "Why can't we do it this way?"; not the nitpicker who keeps telling them why. Because he or she will eventually stumble on an approach that can't be shot down.

Towards this end, I provide a sequence of questions designed to guide your search for the right algorithm for your problem. To use it effectively, you must not only ask the questions, but answer them. The key is working through the answers carefully by writing them down in a log. The correct answer to "Can I do it this way?" is never "no," but "no, because. . . ." By clearly articulating your reasoning as to why something doesn't work, you can check whether you have glossed over a possibility that you didn't think hard enough about. It is amazing how often the reason you can't find a convincing explanation for something is because your conclusion is wrong.

The distinction between *strategy* and *tactics* is important to keep aware of during any design process. Strategy represents the quest for the big picture—the framework around which we construct our path to the goal. Tactics are used to win the minor battles we must fight along the way. In problem-solving, it is important to repeatedly check whether you are thinking on the right level. If you do not have a global strategy of how to attack your problem, it is pointless to worry about the tactics. An example of a strategic question is "Can I model my application as a graph algorithm problem?" A tactical question might be "Should I use an adjacency list or adjacency matrix data structure to represent my graph?" Of course, such tactical decisions are critical to the ultimate quality of the solution, but they can be properly evaluated only in light of a successful strategy.

Too many people freeze up in their thinking when faced with a design problem. After reading or hearing the problem, they sit down and realize that they *don't know what to do next*. Avoid this fate. Follow the sequence of questions I provide below and in most of the catalog problem sections. I will try to *tell* you what to do next.

Obviously, the more experience you have with algorithm design techniques such as dynamic programming, graph algorithms, intractability, and data structures, the more successful you will be at working through the list of questions. Part I of this book has been designed to strengthen this technical background. However, it pays to work through these questions regardless of how strong your technical skills are. The earliest and most important questions on the list focus on obtaining a detailed understanding of your problem and do not require any specific expertise.

This list of questions was inspired by a passage in *The Right Stuff* [Wol79]—a wonderful book about the US space program. It concerned the radio transmissions from test pilots just before their planes crashed. One might have expected that they would panic, so ground control would hear the pilot yelling "Ahhhhhhhhhhh—," terminated only by the sound of smacking into a mountain. Instead, the pilots ran through a list of what their possible actions could be. "I've tried the flaps. I've checked the engine. Still got two wings. I've reset the—." They had the right stuff. Because of this, they sometimes managed to miss the mountain.

I hope this book has provided you with the right stuff to be a successful algorithm designer. And that prevents you from crashing along the way.

1. Do I really understand the problem?

    (a) What exactly does the input consist of?

    (b) What exactly are the desired results or output?

    (c) Can I construct an input example small enough to solve by hand? What happens when I try to solve it?

    (d) How important is it to my application that I always find the optimal answer? Might I settle for something close to the best answer?

    (e) How large is a typical instance of my problem? Will I be working on 10 items? 1,000 items? 1,000,000 items? More?

(f) How important is speed in my application? Must the problem be solved within one second? One minute? One hour? One day?

(g) How much time and effort can I invest in implementation? Will I be limited to simple algorithms that can be coded up in a day, or do I have the freedom to experiment with several approaches and see which one is best?

(h) Am I trying to solve a numerical problem? A graph problem? A geometric problem? A string problem? A set problem? Which formulation seems easiest?

2. Can I find a simple algorithm or heuristic for my problem?

(a) Will brute force solve my problem *correctly* by searching through all subsets or arrangements and picking the best one?

   i. If so, why am I sure that this algorithm always gives the correct answer?

   ii. How do I measure the quality of a solution once I construct it?

   iii. Does this simple, slow solution run in polynomial or exponential time? Is my problem small enough that a brute-force solution will suffice?

   iv. Am I certain that my problem is sufficiently well defined to actually *have* a correct solution?

(b) Can I solve my problem by repeatedly trying some simple rule, like picking the biggest item first? The smallest item first? A random item first?

   i. If so, on what types of inputs does this heuristic work well? Do these correspond to the data that might arise in my application?

   ii. On what inputs does this heuristic work badly? If no such examples can be found, can I show that it always works well?

   iii. How fast does my heuristic come up with an answer? Does it have a simple implementation?

3. Is my problem in the catalog of algorithmic problems in the back of this book?

(a) What is known about the problem? Is there an available implementation that I can use?

(b) Did I look in the right place for my problem? Did I browse through all the pictures? Did I look in the index under all possible keywords?

(c) Are there relevant resources available on the World Wide Web? Did I do a Google Scholar search? Did I go to the page associated with this book: www.algorist.com?

4. Are there special cases of the problem that I know how to solve?

(a) Can I solve the problem efficiently when I ignore some of the input parameters?

(b) Does the problem become easier to solve when some of the input parameters are set to trivial values, such as 0 or 1?

(c) How can I simplify the problem to the point where I *can* solve it efficiently? Why can't this special-case algorithm be generalized to a wider class of inputs?

(d) Is my problem a special case of a more general problem in the catalog?

5. Which of the standard algorithm design paradigms are most relevant to my problem?

(a) Is there a set of items that can be sorted by size or some key? Does this sorted order make it easier to find the answer?

(b) Is there a way to split the problem into two smaller problems, perhaps by doing a binary search? How about partitioning the elements into big and small, or left and right? Does this suggest a divide-and-conquer algorithm?

(c) Does the set of input objects have a natural left-to-right order among its components, like the characters in a string, elements of a permutation, or the leaves of a rooted tree? Could I use dynamic programming to exploit this order?

(d) Are there certain operations being done repeatedly, such as searching, or finding the largest/smallest element? Can I use a data structure to speed up these queries? Perhaps a dictionary/hash table or a heap/priority queue?

(e) Can I use random sampling to select which object to pick next? What about constructing many random configurations and picking the best one? Can I use a heuristic search technique like simulated annealing to zoom in on a good solution?

(f) Can I formulate my problem as a linear program? How about an integer program?

(g) Does my problem resemble satisfiability, the traveling salesman problem, or some other NP-complete problem? Might it be NP-complete and thus not have an efficient algorithm? Is it in the problem list in the back of Garey and Johnson [GJ79]?

6. Am I still stumped?

(a) Am I willing to spend money to hire an expert (like the author) to tell me what to do? If so, check out the professional consulting services mentioned in Section 22.4 (page 718).

(b) Go back to the beginning and work through these questions again. Did any of my answers change during my latest trip through the list?

Problem-solving is not a science, but part art and part skill. It is one of the skills most worth developing. My favorite book on problem-solving remains Pólya's *How to Solve It* [Pol57], which features a catalog of problem-solving techniques that is fascinating to browse through.

13.1   Preparing for Tech Company Interviews

I gather that many of you reading this book have been inspired more by a fear of technical job interviews than an inherent love of algorithms. Hopefully you are enjoying your reading—and coming to understand the beauty and power of algorithmic thinking. But this section is devoted to some brief suggestions to help you when it comes time to interview at tech companies.

First, with respect to algorithm design people know what they know, and hence cramming the night before an interview won't help much. I think you can learn something valuable about algorithm design with one serious week spent with this book, and even more with additional time. The material I cover here is useful and well worth knowing, so you are not cramming for an exam whose contents you will forget in 24 hours. Algorithm design techniques tend to stick with you after you learn them, so it pays to put in the time with my book.

Algorithm design problems tend to creep into the interview process in two ways: preliminary coding tests and blackboard design problems. The major tech companies attract so many applications that the first round of screening is often mechanical: can you solve some programming challenge problem on an interview site like HackerRank (`https://www.hackerrank.com/`)? These programming challenge problems test coding speed and correctness, and are generally used to screen out less promising candidates.

That said, performance on these programming challenge problems improves with practice. If you are a college student, try to get involved with your school's ACM International Collegiate Programming Contest (ICPC) team. Each team consists of three students working together to solve five to ten programming challenge problems within five hours. These problems are often algorithmic and usually interesting. There is much to be gained even if you don't make the regional championships.

For self-study, I recommend solving some coding problems on judging sites like HackerRank and LeetCode (`https://leetcode.com/`). Indeed, I suggest appropriate coding challenges on each of these sites in the exercises at the end of each chapter. Start simple and build up your speed, and do it to have fun. But figure out whether your weak spot is in the correctness of boundary cases or errors in your algorithm itself, and then work to improve. I humbly recommend my book *Programming Challenges* [SR03], which is designed as a training manual for such programming problems. If you like this book, you may well benefit from that one as well.

After you get past the preliminary screening, you will be granted a video or on-site interview. Here you may well be asked to solve some algorithm design problems on a whiteboard at the prompting of your interviewer. These will generally be like the exercises at the end of each chapter. Some have been designated *interview questions* because they have been rumored to be used by certain tech companies. But all of my exercises are good for self-study and interview preparation.

What should you do at the whiteboard to look like you know what you are talking about? First, I encourage you to ask enough clarifying questions to make sure you understand exactly what the problem is. You are likely to get few points for correctly solving the wrong problem. I strongly encourage you to first present a simple, slow, and correct algorithm before trying to get fancy. After that you can, and should, see whether you can do better. Usually the questioners want to see how you think, and are less concerned with the actual final algorithm than seeing an active thought process.

Students of mine who take these interviews often report that they are given incorrect solutions by their interviewers! Getting a job at a nice company does not turn you into an algorithms expert by osmosis. Often the interviewers are just asking questions that other people asked them, so don't be intimidated. Do the best you can, and it will likely be good enough.

Finally, I have a confession to make. For several years, I served as Chief Scientist at a startup company named General Sentiment, and interviewed all the developers we hired. Most of them knew me as the author of this book, and feared they were in for a grilling. But I never asked them a single algorithm puzzle question. We needed developers who could wrangle a complicated distributed system, not solve puzzles. I asked a lot of questions about the biggest program they had worked on and what they did with it, in order to get a sense of the sophistication of what they could handle and what they liked to do. I am very proud of the excellent people we hired at General Sentiment, all of whom have gone on to even more exciting places.

Of course I strongly encourage that *other* companies continue to base their screening on algorithm questions. The more companies that do this, the more copies of this book I am going to sell.

Good luck to you on your job quest, and may what you learn from this book help you with your job, professional growth, and career. Follow me on Twitter at @StevenSkiena!

Part II

The Hitchhiker's Guide to Algorithms

Chapter 14

A Catalog of Algorithmic Problems

This is a catalog of algorithmic problems that arise commonly in practice. It describes what is known about them, and gives suggestions about how best to proceed if the problem arises in your application.

What is the best way to use this catalog? First, think about your problem. If you recall the name, look up the catalog entry in the index or table of contents. Read through the entire entry, since it contains pointers to other relevant problems. Leaf through the catalog, looking at the pictures and problem names to see if anything strikes a chord. Don't be afraid to use the index, for every problem in the book is listed there under several possible keywords and applications.

The catalog entries contain a variety of different types of information that were never really collected in one place before. Different fields in each entry present information of practical and historical interest.

To make this catalog more accessible, I introduce each problem with a pair of graphics representing the problem instance or input on the left and the result of solving the problem on the right. Considerable thought has been invested in creating stylized examples that illustrate desired behaviors, more than just definitions. For example, the minimum spanning tree figures illustrate how points can be clustered using minimum spanning trees. I hope that you will be able to flip through the pictures and identify problems that might be relevant to you. These pictures are augmented with more formal problem descriptions to eliminate the ambiguity inherent in a purely pictorial representation.

Once you have identified your problem, the discussion section tells you what you should do about it. I describe applications where the problem is likely to arise, and any special issues with associated data. I also discuss the kind of results you might reasonably hope for and, most importantly, what you should do to get them. For each problem, a quick-and-dirty solution is outlined, with pointers to more powerful algorithms to try if the first attempt is not sufficient.

S. S. Skiena, *The Algorithm Design Manual*, Texts in Computer Science, https://doi.org/10.1007/978-3-030-54256-6_14

Available software implementations are discussed in the implementation field of each catalog entry. Many of these routines are quite good, and can perhaps be plugged directly into your application. Others may be inadequate for production use, but they can provide a good model for your own implementation. In general, implementations are listed in order of descending usefulness, but I will explicitly recommend the best one available for each problem if a clear winner exists. More detailed information for many of these implementations appears in Chapter 22. Just about all of the implementations are available at the website associated with this book: www.algorist.com.

Finally, in deliberately smaller print, the history of each problem will be discussed, and results of primarily theoretical interest presented. I have attempted to report the best results known for each problem, and point out empirical comparisons of algorithms or survey articles if they exist. These should interest students, researchers, and practitioners who need to know whether anything better is possible.

Caveats

This is a catalog of algorithmic problems. It is not a cookbook. It cannot be, because there are too many recipes and too many possible variations on what people want to eat. My goal is to point you in the right direction so that you can solve your own problems. I try to identify the issues you will encounter along the way. In particular:

- For each problem, I suggest algorithms and directions to attack it. These recommendations are based on my experience, and are aimed toward what I see as typical applications. I feel it is better to make concrete recommendations for the masses than to try to cover all possible situations. If you don't agree with my advice, you don't have to follow it. But try to understand my reasoning so you can articulate why your needs violate my assumptions.

- The implementations I recommend are not necessarily complete solutions to your problem. Some programs are useful only as models for you to write your own codes. Others are embedded in large systems and so might be too painful to extract and run on their own. All of them contain bugs. Many are quite serious, so beware.

- Please respect the licensing conditions for any implementations you use commercially. Many of these codes are not open source, and most have license restrictions. See Section 22.1 for a further discussion of this issue.

- I would be interested in hearing about your experiences with my recommendations, both positive and negative. I would be especially interested in learning about any other implementations that you know about.

Chapter 15

Data Structures

Data structures are not so much algorithms as they are the fundamental constructs around which you build your application. Becoming fluent in what the standard data structures can do is essential to get full value from them.

This puts data structures slightly out of sync with the rest of the catalog. Perhaps the most useful aspect of it will be the pointers to various implementations and data structure libraries. Many of these data structures are non-trivial to implement well, so the programs I point to will be useful as models even if they do not do exactly what you need. Certain fundamental data structures, like kd-trees and suffix trees, are not as well known as they should be. Hopefully, this catalog will serve to better publicize them.

There are a large number of books on elementary data structures available. My favorites include:

- *Sedgewick* [SW11] – This comprehensive introduction to algorithms and data structures stands out for the clever and beautiful images of algorithms in action. It comes in C, C++, and Java editions.

- *Weiss* [Wei11] – A nice text, emphasizing data structures more than algorithms. It comes in Java, C, C++, and Ada editions.

- *Goodrich and Tamassia* [GTG14] – The Java edition makes particularly good use of the author's Java Data Structures Library (JDSL).

- *Brass* [Bra08] – This is a good treatment of more advanced data structures than those covered in other texts, with implementations in C++.

The *Handbook of Data Structures and Applications* [MS18] provides a comprehensive and up-to-date survey of research in data structures. The student who took only an elementary course in data structures is likely to be surprised and impressed by the volume of recent work on the subject.

© The Editor(s) (if applicable) and The Author(s), under exclusive license to
Springer Nature Switzerland AG 2020
S. S. Skiena, *The Algorithm Design Manual*, Texts in Computer Science,
https://doi.org/10.1007/978-3-030-54256-6_15

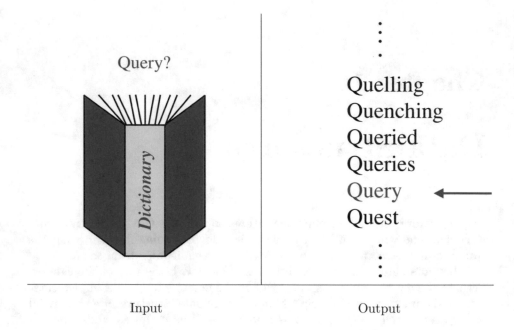

15.1   Dictionaries

Input description: A set of n records, each identified by one or more keys.

Problem description: Build and maintain a data structure to efficiently locate, insert, and delete the record associated with any query key q.

Discussion: The abstract data type "dictionary" is one of the most important structures in computer science. Dozens of data structures have been proposed for implementing dictionaries, including hash tables, skip lists, and balanced/unbalanced binary search trees. This means that choosing the best one can be tricky. *However, in practice, it is more important to avoid using a bad data structure than to identify the single best option available.*

An essential piece of advice is to carefully isolate the implementation of your dictionary data structure from its interface. Use explicit calls to methods or subroutines that initialize, search, and modify the data structure, rather than embed them within the code. This leads to a much cleaner program, but it also makes it easy to experiment with different implementations to see how they perform. Do not obsess about the costs of procedure calls inherent in such an abstraction. If your application is so time-critical that such overhead is meaningful, then it is even more essential that you experiment to find the right dictionary implementation.

To choose the right data structure for your dictionary, answer the following questions:

- *How many items will you have in your data structure?* – Will you know this number in advance? Are you looking at a problem small enough that a simple data structure will suffice? Or one so large that we must worry about running out of memory or virtual memory performance?

- *Do you know the relative frequencies of insert, delete, and search operations?* – Static data structures (like sorted arrays) suffice in applications when there will be no modifications to the data structure after it is first constructed. *Semi-dynamic* data structures, which support insertion but not deletion, can have significantly simpler implementations than fully dynamic ones.

- *Will the access pattern for keys be uniform and random?* – Search queries exhibit a skewed access distribution in many applications, meaning that certain elements are much more popular than others. Further, queries often have a sense of temporal locality, meaning elements are likely to be repeatedly accessed in clusters instead of at fairly regular intervals. Data structures such as splay trees can take advantage of a skewed and clustered universe.

- *Is it critical that each individual operation be fast, or only that the total amount of work done over the entire program be minimized?* – When response time is critical, such as in a program controlling a heart–lung machine, you can't wait too long between steps. But when you are doing a lot of queries over the database, such as identifying all criminals who happen to be politicians, it is not critical that you pick out any particular legislator quickly. Just get them all with the minimum total effort.

An object-oriented generation has emerged that is no more likely to write their own container class than to fix the engine in their car. This is good: default containers should work just fine for most applications. Still, it is sometimes valuable to know exactly what you have under the hood:

- *Unsorted linked lists or arrays* – For small data sets, an unsorted array is probably the easiest data structure to maintain. Linked structures can have terrible cache performance compared with sleek, compact arrays. However, once your dictionary becomes larger than (say) fifty to a hundred items, the linear search time will kill you for either lists or arrays.

  A particularly interesting and useful variant is the *self-organizing list*. Whenever a key is accessed or inserted, we move it to the head of the list. Thus, the key will be near the front if it is accessed again in the near future, and so require only a short search to find it. Most applications exhibit both uneven access frequencies and locality of reference, so the average time for a successful search in a self-organizing list is typically much better than in a sorted or unsorted list. Self-organizing data structures can be built from arrays as well as linked lists and trees.

- *Sorted linked lists or arrays* – Maintaining a sorted linked list is usually not worth the effort unless you are trying to eliminate duplicates, since we cannot perform binary searches in such a data structure. A sorted array will be appropriate iff there are not many insertions or deletions.

- *Hash tables* – For applications involving a moderate-to-large number of keys, a hash table is probably the right way to go. We use a function that maps keys (be they strings, numbers, or whatever) to integers between 0 and $m - 1$. We maintain an array of m buckets, each typically implemented using an unsorted linked list. The hash function immediately identifies which bucket contains a given key. Presuming the hash function spreads the keys out nicely through a sufficiently large hash table, each bucket should contain very few items, thus making linear search acceptable. Insertion and deletion from a hash table reduce to insertion and deletion from the bucket/list. Section 3.7 (page 93) provides a more detailed discussion of hashing and its applications.

A well-tuned hash table will outperform a sorted array in most applications. However, several design decisions go into creating good hash tables:

- *How do I deal with collisions?* Open addressing can lead to more concise tables with better cache performance than bucketing, but performance will be more brittle if the load factor (ratio of occupancy to capacity) of the hash table gets too high.

- *How big should the table be?* With bucketing, m should be about the same as the maximum number of items you expect to put in the table. With open addressing, make the table 30% to 50% larger. Selecting m to be a prime number minimizes the dangers of a bad hash function.

- *What hash function should I use?* For strings, something like

$$H(S) = \alpha^{|S|} + \sum_{i=0}^{|S|-1} \alpha^{|S|-(i+1)} \times char(s_i) \ (\text{mod } m)$$

should work, where α is the size of the alphabet and *char(x)* is the function that maps each character x to its character code. Use Horner's rule (or precompute values of α^x) to implement this hash function computation efficiently, as discussed in Section 16.9 (page 493). A variant of this hash function (discussed in Section 6.7 (page 188)) has the nifty property that hash codes of successive k-character windows of a string can be computed in constant time, instead of $O(k)$.

When evaluating a hash function/table implementation, print statistics on the distribution of keys per bucket to see how uniform it *really* is. Odds are the first hash function you try will not prove to be the best. Botching up the hash function is an excellent way to slow down any application.

- *Binary search trees* – Binary search trees are elegant data structures that support fast insertions, deletions, and queries reviewed in Section 3.4 (page 81). The primary distinction between different types of trees is whether they are explicitly rebalanced after insertion or deletion, and how this rebalancing is done. In simple *random search trees*, we insert each node at the leaf position where we can find it, with no rebalancing. Although such trees perform well under random insertions, most applications are not random. Indeed, unbalanced search trees constructed by inserting keys in sorted order are a disaster, degenerating to a linked list.

  Balanced search trees use local *rotation* operations for restructuring, moving more distant nodes closer to the root while maintaining the in-order search structure of the tree. Among balanced search trees, AVL and 2/3 trees are now passé, and *red–black trees* seem to be most popular. A particularly interesting self-organizing data structure is the *splay tree*, which uses rotations to move any accessed key to the root. Frequently used or recently accessed nodes thus sit near the top of the tree, allowing faster searches.

  Bottom line: Which tree is best for your application? Probably the one of which you have the best implementation. Which flavor of balanced tree is likely not as important as the skill of the programmer who coded it.

- *B-trees* – For data sets so large that they will not fit in main memory your best bet will be some flavor of a B-tree. The search time of a data structure grows by several orders of magnitude once it is stored outside of main memory. With modern cache architectures, similar effects can happen on a smaller scale, because cache is much faster than RAM.

  The idea behind a B-tree is to collapse several levels of a binary search tree into a single large node, so that we can make the equivalent of several search steps before another disk access is needed. B-trees can access enormous numbers of keys using only a few disk accesses. To get the full benefit from using a B-tree, it is important to understand how the secondary storage device and virtual memory systems interact—in particular, constants such as page size and virtual/real address space. *Cache-oblivious algorithms* (described below) can mitigate such concerns.

  Even for modest-sized data sets, unexpectedly poor performance of a data structure may result from excessive swapping, so listen to your disk to help decide whether you should be using a B-tree.

- *Skip lists* – These are somewhat of a cult data structure. A hierarchy of sorted linked lists is maintained, where a coin is flipped for each element to decide whether it gets copied into the next highest list. This implies roughly $\lg n$ lists, each roughly half as large as the one above it. Search starts in the smallest list. The search key lies in an interval between two elements, which is then explored in the next larger list. Each searched interval contains an expected constant number of elements per list, for a total expected $O(\lg n)$ query time. The primary benefits of skip lists are ease of analysis and implementation relative to balanced trees.

Implementations: Modern programming languages provide libraries offering complete and efficient container implementations. The C++ *Standard Template Library* (STL) is now provided with most compilers. See Josuttis [Jos12], Meyers [Mey01], and Musser [MDS01] for more detailed guides to using STL and the C++ standard library. *Java Collections* (JC), a small library of data structures, is included in the `java.util` package of the Java standard edition.

LEDA (see Section 22.1.1 (page 713)) provides a complete collection of dictionary data structures in C++, including hashing, perfect hashing, B-trees, red-black trees, random search trees, and skip lists. Experiments reported in Mehlhorn and Naher [MN99] declared hashing to be the best dictionary choice, with skip lists and 2–4 trees (a special case of B-trees) as the most efficient tree-like structures.

Notes: Knuth [Knu97a] provides the most detailed analysis and exposition on fundamental dictionary data structures, but misses certain modern data structures as red–black and splay trees. Spending some time with his books is an important rite of passage for all computer science students.

The Handbook of Data Structures and Applications [MS18] provides up-to-date surveys on all aspects of dictionary data structures. Other surveys include Mehlhorn and Tsakalidis [MT90b] and Gonnet and Baeza-Yates [GBY91]. Good textbook expositions on dictionary data structures include Sedgewick [Sed98], Weiss [Wei11], and Goodrich et al. [GTG14]. I defer to all these sources to avoid giving original references for each of the data structures described above.

The 1996 DIMACS implementation challenge focused on elementary data structures, including dictionaries. See Goldwasser et al. [GJM02]. Data sets, and codes are accessible from `http://dimacs.rutgers.edu/Challenges`.

The cost of transferring data back and forth between levels of the memory hierarchy (RAM-to-cache or disk-to-RAM) dominates the cost of actual computation for many problems. Each data transfer moves one block of size b, so efficient algorithms seek to minimize the number of block transfers. The complexity of fundamental algorithm and data structure problems on such an external memory model has been extensively studied by Vitter [Vit01]. *Cache-oblivious* data structures offer performance guarantees under such a model without explicit knowledge of the block-size parameter b. Hence, good performance can be obtained on any machine without architecture-specific tuning. See [ABF05, Dem02] for excellent surveys on cache-oblivious data structures.

Splay trees and other modern data structures have been studied using *amortized analysis*, where we bound the total amount of time used by any sequence of operations. In an amortized analysis, a single operation can be very expensive, but only because we have already benefited from enough cheap operations to pay off the higher cost. A data structure realizing an amortized complexity of $O(f(n))$ is less desirable than one whose worst-case complexity is $O(f(n))$ (since a very bad operation might still occur) but better than one with an average-case complexity $O(f(n))$, since the amortized bound will achieve this average on any input.

Related problems: Sorting (see page 506), searching (see page 510).

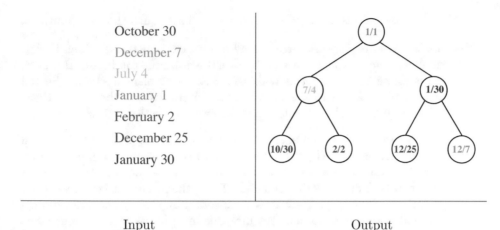

| Input | Output |

15.2 Priority Queues

Input description: A set of records with totally ordered keys.

Problem description: Build and maintain a data structure for providing quick access to the *smallest* or *largest* key in the set.

Discussion: Priority queues are useful data structures in simulations, particularly for maintaining a set of future events ordered by time. They are called "priority" queues because you retrieve items not according to the insertion time (as in a stack or queue), nor by a key match (as in a dictionary), but by highest priority of retrieval.

If your application will perform no insertions after the initial query, there is no need for an explicit priority queue. Simply sort the records by priority and proceed from top to bottom, maintaining a pointer to the last record retrieved. This situation occurs in Kruskal's minimum spanning tree algorithm, or when simulating a completely scripted set of events.

However, you will need a real priority queue when mixing queries with insertions and deletions. The following questions will help select the right one:

- *What other operations do you need?* – Will you be searching for arbitrary keys, or just searching for the smallest? Will you be deleting arbitrary elements from the data, or just repeatedly deleting the top or smallest element?

- *Do you know the maximum data structure size in advance?* – The issue here is whether you can pre-allocate space for the data structure.

- *Might you raise or lower the priority of elements already in the queue?* – Changing the priority of elements requires that we are able to retrieve elements from the queue based on their key, in addition to finding the largest element.

Your choices are between the following basic priority queue implementations:

- *Sorted array or list* – A sorted array is very efficient to both identify the smallest element and "delete" it by decrementing the top index. However, maintaining the total order makes inserting new elements slow. Sorted arrays are suitable when there will be no insertions into the queue. Basic priority queue implementations are reviewed in Section 3.5 (page 87).

- *Binary heaps* – This simple, elegant data structure supports both insertion and extract-min in $O(\lg n)$ time each. Heaps maintain an implicit binary tree structure in an array, such that the key of the root of any subtree is less than that of all its descendants. Thus, the minimum key always sits at the top of the heap. New keys are inserted by placing them at an open leaf and percolating the element upwards until it sits at its proper place in the partial order. An implementation of binary heap construction and retrieval in C appears in Section 4.3.1 (page 116).

  Binary heaps are the right answer when you know an upper bound on the number of items in your priority queue, since you must specify the array size at creation time. Even this constraint can be mitigated by using dynamic arrays (see Section 3.1.1 (page 70)).

- *Bounded-height priority queue* – This array-based data structure permits constant-time insertion and find-min operations whenever the range of possible key values is limited. Suppose we know that all key values will be integers between 1 and n. We can set up an array of n linked lists, such that the ith list serves as a bucket containing all items with key i. We will maintain a *top* pointer to the smallest non-empty list. To insert an item with key k into the queue, add it to the kth bucket and set $top = \min(top, k)$. To extract the minimum, report the first item from bucket *top*, delete it, and move *top* down if the bucket has become empty.

  Bounded-height priority queues are very useful to maintain the vertices of a graph sorted by degree, which is a fundamental operation in graph algorithms. Still, they are not as widely known as they should be. They are usually the right priority queue for any small, discrete range of keys.

- *Binary search trees* – Binary search trees make effective priority queues, since the smallest element is always the left-most leaf, while the largest element is always the right-most leaf. The min (max) is found by simply tracing down left (right) pointers until the next pointer is nil. Binary tree heaps prove most appropriate when you also need other dictionary operations, or if you have an unbounded key range and do not know the maximum priority queue size in advance.

- *Fibonacci and pairing heaps* – These complicated priority queues are designed to speed up *decrease-key* operations, where the priority of an item already in the priority queue is reduced. This arises, for example, in shortest path computations when we discover a shorter route to a vertex

v than previously established. Properly implemented and used, they lead to better performance on very large computations.

Implementations: Modern programming languages provide libraries offering complete and efficient priority queue implementations. The *Java Collections* `PriorityQueue` class is included in the java.util package of Java Standard Edition. Member functions `push`, `top`, and `pop` of the C++ *Standard Template Library* (STL) `priority_queue` template mirror heap operations `insert`, `findmax`, and `deletemax`. See Meyers [Mey01] and Musser [MDS01] for more detailed guides to using STL.

LEDA (see Section 22.1.1 (page 713)) provides a complete collection of priority queues in C++, including Fibonacci heaps, pairing heaps, van Emde Boas trees, and bounded-height priority queues. Experiments reported in Mehlhorn and Naher [MN99] identified simple binary heaps as quite competitive in most applications, with pairing heaps beating Fibonacci heaps in head-to-head tests. Sanders [San00] did extensive experiments demonstrating that his sequence heap, based on k-way merging, was roughly twice as fast as a well-implemented binary heap.

Notes: *The Handbook of Data Structures and Applications* [MS18] provides several up-to-date surveys on all aspects of priority queues. Empirical comparisons between priority queue data structures include [CGS99, GBY91, Jon86, LL96, San00].

Double-ended priority queues extend the basic heap operations to simultaneously support both find-min and find-max. See Sahni [Sah05] for a survey of four different implementations of double-ended priority queues.

Bounded-height priority queues are useful data structures in practice, but do not promise good worst-case performance for unbounded key ranges. However, van Emde Boas priority queues [vEBKZ77] support $O(\lg \lg n)$ insertion, deletion, search, max, and min operations where each key is an element from 1 to n.

Fibonacci heaps [FT87, BLT12] support insert and decrease-key operations in constant amortized time, with $O(\lg n)$ amortized time extract-min and delete operations. The constant-time decrease-key operation leads to faster implementations of classical algorithms for shortest paths, weighted bipartite matching, and minimum spanning tree. In practice, Fibonacci heaps are difficult to implement and have large constant factors associated with them. However, pairing heaps appear to realize the same bounds with less overhead. Experiments with pairing and other heaps are reported in [LST14, SV87].

Heaps define a partial order that can be built using a linear number of comparisons. The familiar linear-time merging algorithm for heap construction is due to Floyd [Flo64]. In the worst case, $1.625n$ comparisons suffice, see [GM86]. Further, $1.5n - O(\lg n)$ comparisons are necessary for heap construction, see [CC92].

Related problems: Dictionaries (see page 440), sorting (see page 506), shortest path (see page 554).

XYZXYZ$

YZXYZ$

ZXYZ$

XYZ$

YZ$

Z$

$

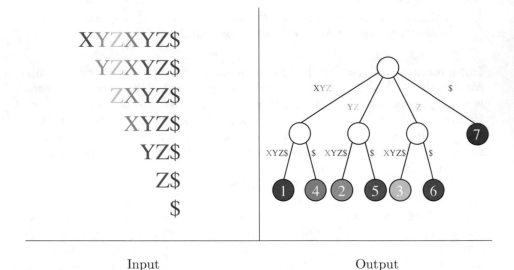

Input   Output

15.3   Suffix Trees and Arrays

Input description: A reference string S.

Problem description: Build a data structure for quickly finding all places where an arbitrary query string q occurs in S.

Discussion: Suffix trees and arrays are phenomenally useful data structures for solving string problems elegantly and efficiently. The proper use of suffix trees often speeds up string processing algorithms from $O(n^2)$ to linear time. Indeed, suffix trees were the hero of the war story reported in Section 3.9 (page 98).

In its simplest instantiation, a suffix tree is simply a *trie* of the n suffixes of an n-character string S. A trie is a tree structure, where each edge represents one character, and the root represents the null string. Each path from the root represents a string, described by the characters labeling the edges traversed. Every finite set of words defines a distinct trie, and two words with common prefixes branch off from each other at the first distinguishing character. Each leaf denotes the end of a string. Figure 15.1 illustrates a simple trie.

Tries are useful in testing whether a given query string q is in the set of strings. We traverse the trie from the root, along branches defined by successive characters of q. If a branch does not exist in the trie, then q cannot be in the set of strings. Otherwise we find the query string in $|q|$ character comparisons *regardless* of how many other strings are in the trie. Tries are very simple to build (repeatedly insert new strings) and very fast to search (just walk down), although they can be expensive in terms of memory.

A *suffix tree* is simply a trie of all proper suffixes of S. The suffix tree enables

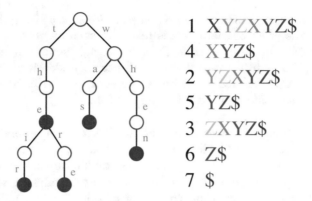

<table>
1 XYZXYZ$
4 XYZ$
2 YZXYZ$
5 YZ$
3 ZXYZ$
6 Z$
7 $
</table>

Figure 15.1: A trie on strings *the, their, there, was,* and *when* (on left). The suffix array of *XYZXYZ$* (on right).

you to test whether q is a substring of S, because any substring of S is the prefix of some suffix (got it?). The search time is again linear in $|q|$.

The catch is that constructing a full suffix tree in this manner can require $O(n^2)$ time and, even worse, $O(n^2)$ space, since the average length of the n suffixes is $n/2$. But if we are clever, linear space suffices to represent a full suffix tree. Observe that most of the nodes in a trie-based suffix tree occur on simple unbranching paths between nodes of outdegree ≥ 2 in the tree. Each simple path corresponds to a substring of the original string. If we store the original string in an array, we can represent any such collapsed path by the starting and ending array indices representing the substring. Thus, each of the tree edges is labeled using only two integers, so we have all the information of the full suffix tree in only $O(n)$ space. The output figure for this section displays a collapsed suffix tree in all its glory.

Even better, there exist $O(n)$ algorithms to construct this collapsed tree, by making clever use of pointers to minimize construction time. These additional pointers can also be used to speed up many applications of suffix trees.

But what can you do with suffix trees? Consider the following applications:

- *Find all occurrences of q as a substring of S* – Just as with a trie, we can walk from the root to the node n_q associated with q. The positions of all occurrences of q in S are represented by the descendants of n_q, which can be identified using a depth-first search from n_q. With a collapsed suffix tree, it takes $O(|q| + k)$ time to find the k occurrences of q in S.

- *Longest substring common to a set of strings* – Build a single collapsed suffix tree containing all suffixes of all strings, with each leaf labeled with its original string. In the course of doing a depth-first search on this tree, we mark each node with both the length of its common prefix and the number of distinct strings that are children of it. From this information, the best node can be selected in linear time.

- *Find the longest palindrome in S* – A *palindrome* is a string that reads the same if the order of characters is reversed, such as *madam*. To find the longest palindrome in a string S, build a single suffix tree containing all suffixes of S and the reversal of S, with each leaf identified by its starting position. A palindrome is defined by any node in this tree that has forward and reversed children from the same position.

Since linear-time suffix tree construction algorithms are non-trivial, I recommend using an existing implementation. But another good option is to use suffix arrays, which do most of what suffix trees do, but are easier to implement.

A suffix array is in principle just an array that contains all the n suffixes of S in sorted order. Thus, a binary search of this array for string q suffices to locate the prefix of a suffix that matches q, permitting an efficient substring search in $O(\lg n)$ string comparisons. With the addition of an index specifying the common prefix length of all bounding suffixes, only $\lg n + |q|$ *character* comparisons need be performed on any query, since we can identify the next character position that must be tested in the binary search. For example, if the lower range of the search is *cowabunga* and the upper range is *cowslip*, all keys in between must share the same first three letters, so only the fourth character of any intermediate key needs to be tested against q.

In practice, suffix arrays are typically as fast or faster to search than suffix trees. They also use much less memory, typically by a factor of four. Each suffix is represented completely by its unique starting position (from 1 to n) and can be read off as needed using a single reference copy of the input string.

Some care must be taken to construct suffix arrays efficiently, because there are $O(n^2)$ characters in the strings being sorted. One solution is to first build a suffix *tree*, then perform an in-order traversal of it to read the strings off in sorted order! However, recent breakthroughs have led to space/time efficient algorithms for constructing suffix arrays directly.

Implementations: There now exist a wealth of suffix array implementations available. Indeed, all of the recent linear-time construction algorithms have been implemented and benchmarked [PST07]. Schürmann and Stoye [SS07] provide an excellent C implementation at `https://bibiserv.cebitec.uni-bielefeld.de/bpr/`.

No less than eight different C/C++ implementations of compressed text indexes appear at the *Pizza&Chili corpus* `http://pizzachili.dcc.uchile.cl/`. These data structures go to great lengths to minimize space usage, typically compressing the input string to near the empirical entropy while still achieving excellent query times!

Suffix tree implementations are also readily available. A `SuffixTree` class is provided in BioJava (`http://www.biojava.org/`)—an open source project providing a Java framework for processing biological data. `Libstree` is a C implementation of Ukkonen's algorithm, available at `http://www.icir.org/christian/libstree/`. Strmat is a collection of C programs implementing exact pattern matching algorithms in association with Gusfield [Gus97], including an implementation of suffix trees. It is available at `https://web.cs.ucdavis.edu/~gusfield/strmat.html`.

Notes: Tries were first proposed by Fredkin [Fre62], the name coming from the central letters of the word "retrieval." A survey of basic trie data structures with extensive references appears in Gonnet and Baeza-Yates [GBY91].

Efficient algorithms for suffix tree construction are due to Weiner [Wei73], Mc-Creight [McC76], and Ukkonen [Ukk92]. Good expositions on these algorithms include Crochmore and Rytter [CR03] and Gusfield [Gus97]. The interesting forty-year history of suffix trees is recounted in Apostolico et al. [ACFC$^+$16].

Suffix arrays were invented by Manber and Myers [MM93], although an equivalent idea called *Pat trees* due to Gonnet and Baeza-Yates appears in [GBY91]. Three teams independently emerged with linear-time suffix array algorithms in 2003 [KSPP03, KA03, KSB06], and progress has continued rapidly. See Puglisi et al. [PST07] for a survey covering all these developments.

Recent work has resulted in the development of compressed full text indexes that offer essentially all the power of suffix trees/arrays in a data structure whose size is proportional to the *compressed* text string. Makinen and Navarro [MN07] survey these remarkable data structures.

The power of suffix trees can be further augmented by using a data structure to compute the *least common ancestor* (LCA) of any pair of nodes x, y in a tree in constant time, after linear-time preprocessing of the tree. The original data structure due to Harel and Tarjan [HT84], has been progressively simplified by Schieber and Vishkin [SV88] and later Bender and Farach [BF00]. Expositions include Gusfield [Gus97]. The least common ancestor of two nodes in a suffix tree or trie defines the node representing the longest common prefix of the two associated strings. That we can answer such queries in constant time is amazing, and proves useful as a building block for many other algorithms.

Related problems: String matching (see page 685), text compression (see page 693), longest common substring (see page 706).

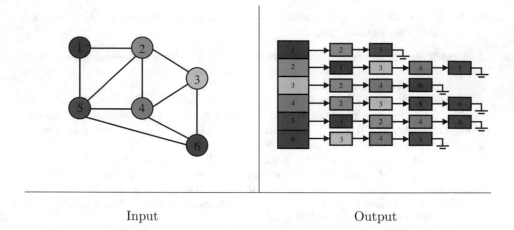

Input   Output

15.4   Graph Data Structures

Input description: A graph G.

Problem description: Represent the graph G using a flexible, efficient data structure.

Discussion: The two basic data structures for representing graphs are *adjacency matrices* and *adjacency lists*. Full descriptions of both these data structures appear in Section 7.2 (page 203), along with an implementation of adjacency lists. In general, for most things, adjacency lists are the way to go.

The issues in deciding which data structure to use include:

- *How big will your graph be?* – How many vertices will it have, both typically and in the worst case? Ditto for the number of edges. Graphs with 1,000 vertices imply adjacency matrices with 1,000,000 entries. This is within the boundary of reality. But adjacency matrices make sense only for small or very dense graphs.

- *How dense will your graph be?* – If your graph is very dense, meaning that a large fraction of the vertex pairs define edges, there is probably no compelling reason to use adjacency lists. You are doomed to use $\Theta(n^2)$ space anyway. Indeed, for complete graphs, matrices will be more concise due to the elimination of pointers.

- *Which algorithms will you be implementing?* – Certain algorithms are more natural on adjacency matrices, such as all-pairs shortest path, but most DFS-based algorithms favor adjacency lists. Adjacency matrices win for algorithms that repeatedly ask, "Is (i, j) in G?" However, most graph algorithms can be designed to eliminate such queries—and if this is all you are doing you would be better off using a hash table of edges.

- *Will you modify the graph over the course of the computation?* – Efficient *static graph* implementations can be used when no edge insertion/deletion operations will be done following initial construction. Indeed, more common than modifying the topology of the graph is modifying the *attributes* of a vertex or edge of the graph, such as size, weight, label, or color. Attributes are best handled as extra fields in the vertex or edge records of adjacency lists.

- *Will your graph be a persistent, online structure?* – Data structures and databases are two different things. People use databases to support commercial-strength applications that must maintain access to large amounts of data. I trust that Facebook is not storing its friendship graph in a memory-resident adjacency list. Graph databases like Neo4j are useful for representing networks in a persistent, online fashion.

Building a good general purpose graph type is a substantial project. I thus suggest that you check out existing implementations (particularly LEDA) before hacking up your own. Note that it costs only time linear in the size of the larger data structure to convert between adjacency matrices and adjacency lists. This conversion is unlikely to be the bottleneck in any application, so you might use both data structures if you have the space to store them. This usually isn't necessary, but could prove simplest if you are confused about the alternatives.

Planar graphs are those that can be drawn in the plane so no two edges cross. Graphs arising in many applications are planar by definition, such as maps of countries. Others are planar by happenstance, like trees. Planar graphs are always sparse, since any n-vertex planar graph can have at most $3n - 6$ edges. They should thus be represented using adjacency lists. If the planar drawing (or *embedding*) of the graph is fundamental to what is being computed, planar graphs are best represented geometrically. See Section 18.12 (page 581) for algorithms for constructing planar embeddings from graphs.

Hypergraphs are generalized graphs where each edge may link subsets of more than two vertices. Suppose we want to encode which representative is on what congressional committee. The vertices of our hypergraph would be the individual congressmen, while each hyperedge would represent one committee. Such arbitrary collections of subsets of a set are naturally thought of as hypergraphs.

Two basic data structures for hypergraphs are:

- *Incidence matrices*, which are analogous to adjacency matrices. They require $n \times m$ space, where m is the number of hyperedges. Each row corresponds to a vertex, and each column to an edge, with a non-zero entry in $M[i, j]$ iff vertex i is incident to edge j. Traditional graphs have exactly two non-zero entries in each column. The degree of each vertex governs the number of non-zero entries in each row.

- *Bipartite incidence structures*, which are analogous to adjacency lists, and thus suited for sparse hypergraphs. We create a vertex of the incidence structure associated for every edge and vertex of the hypergraph, and add an edge (i, j) in the incidence structure whenever vertex i of the hypergraph appears in edge j of the hypergraph. Adjacency lists should be used to represent this incidence structure. Drawing the associated bipartite graph provides a natural way to visualize the hypergraph.

Special efforts must be taken to represent very large graphs efficiently. However, interesting problems have been solved on graphs with millions of edges and vertices. The first step is to make your data structure as lean as possible, by packing your adjacency matrix in a bit vector (see Section 15.5 (page 456)) or removing unnecessary pointers from your adjacency list representation. For example, in static graphs (which do not support edge insertions or deletions) each edge list can be replaced by a packed array of vertex identifiers, eliminating pointers and thus potentially saving half the space.

If your graph is extremely large, it may become necessary to switch to a hierarchical representation, where the vertices are clustered into subgraphs that are compressed into single vertices. Two approaches exist to construct such a hierarchical decomposition. The first breaks the graph into components in a natural or application-specific way. For example, a network of roads and cities suggests a natural decomposition—partition the map into districts, towns, counties, and states. Alternatively, you can run a graph partition algorithm as discussed in Section 19.6 (page 601). A natural decomposition will likely do a better job than some naive heuristic for an NP-complete problem. If your graph is really unmanageably large, you cannot afford to do a very good job of algorithmically partitioning it. First verify that standard data structures fail on your problem before attempting such heroic measures.

Implementations: LEDA (see Section 22.1.1 (page 713)) is a commercial product that provides the best graph data type currently implemented in C++. Study the methods it provides for graph manipulation, so as to see how the right level of abstract graph type makes implementing algorithms clean and easy.

The C++ Boost Graph Library [SLL02] (`http://www.boost.org/libs/graph`) is more readily available. Implementations of adjacency lists, matrices, and edge lists are included, along with a reasonable library of basic graph algorithms. Its interface and components are generic in the same sense as the C++ standard template library (STL).

Neo4j (`https://neo4j.com/`) is a widely used graph database, where the J stands for Java. Needham and Hodler [NH19] present examples of graph algorithms in Neo4j. *JUNG* (`http://jung.sourceforge.net/`) is a Java graph library particularly popular in the social networks community. *JGraphT* (`https://jgrapht.org/`) is a more recent development with similar functionality.

The Stanford Graphbase (see Section 22.1.7 (page 715)) provides a simple but flexible graph data structure in CWEB, a literate version of the C language. It is instructive to see what Knuth does and does not place in his basic data structure, although I recommend other implementations as a better basis for further development.

My (biased) preferences in C language graph types include the libraries from this book, as well as my book *Programming Challenges* [SR03]. See Section 22.1.9 (page 716) for details. Simple graph data structures in Mathematica are provided by *Combinatorica* [PS03], with a library of algorithms and display routines. See Section 22.1.8 (page 716).

Notes: The advantages of adjacency list data structures for graphs became apparent with the linear-time algorithms of Hopcroft and Tarjan [HT73b, Tar72]. The basic adjacency list and matrix data structures are presented in essentially all books on algorithms or data structures, including [CLRS09, AHU83, Tar83]. Hypergraphs are presented in Berge [Ber89].

The improved efficiency of static graph types was revealed by Naher and Zlotowski [NZ02], who sped up certain LEDA graph algorithms by a factor of four by simply switching to a more compact graph structure.

Matrix representations of graphs can exploit the power of linear algebra in problems ranging from shortest paths to partitioning. Laplacians and other matrix structures are presented in Bapat [Bap10]. An interesting question concerns minimizing the number of bits needed to represent arbitrary graphs on n vertices, particularly if certain operations must be supported efficiently. Such issues are surveyed in van Leeuwen [vL90b].

Dynamic graph algorithms [EGI98] are data structures that maintain quick access to an invariant (such as minimum spanning tree or connectivity) under edge insertion and deletion. *Sparsification* [EGIN97] is a general approach to constructing dynamic graph algorithms. Jeff Westbrook, a pioneer in dynamic graph algorithms [Wes89], went on to become a writer for the TV show *The Simpsons*.

Hierarchically defined graphs often arise in VLSI design problems, because designers make extensive use of cell libraries [Len90]. Algorithms specifically for hierarchically defined graphs include planarity testing [Len89], connectivity [LW88], and minimum spanning trees [Len87a].

Related problems: Set data structures (see page 456), graph partition (see page 601).

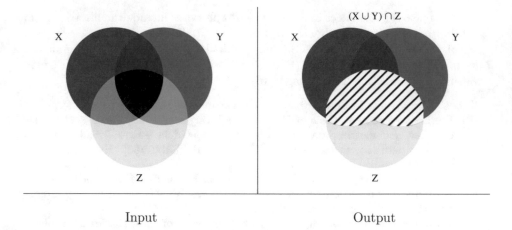

Input   Output

15.5   Set Data Structures

Input description: A universe of items $U = \{u_1, \ldots, u_n\}$ on which is defined a collection of subsets $S = \{S_1, \ldots, S_m\}$.

Problem description: Represent each subset so as to efficiently (1) test whether $u_i \in S_j$, (2) compute the union or intersection of S_i and S_j, and (3) insert or delete members of S.

Discussion: In mathematical terms, a set is an unordered collection of objects drawn from a fixed universal set. However, it is usually useful to represent each set in a single *canonical order*, typically sorted, to speed up or simplify various operations. Sorted order turns the problem of finding the union or intersection of two subsets into a linear-time operation—just sweep from left to right and see what you are missing. It makes possible element searching in sublinear time. Finally, printing the elements of a set in a canonical order paradoxically reminds us that order doesn't really matter.

We distinguish sets from two other types of objects: dictionaries and strings. A collection of objects *not* drawn from a fixed-size universal set is best thought of as a *dictionary*, discussed in Section 15.1 (page 440). Strings are structures where order matters, meaning that $\{A, B, C\}$ is not the same as $\{B, C, A\}$. Section 15.3 and Chapter 21 discuss data structures and algorithms for strings, respectively.

Multisets permit elements to have more than one occurrence. Data structures for sets can generally be extended to multisets by maintaining a count field, or a linked list of equivalent entries for each element.

When every subset contains exactly two elements, they can be thought of as edges in a graph whose vertices represent the universal set. A system of subsets with no restrictions on the cardinality of its members is called a *hypergraph*. It is worth considering whether your problem has a graph-theoretical analogy, like connected components or shortest path in a graph/hypergraph.

Your primary alternatives for representing arbitrary subsets are:

- *Bit vectors* – An n-bit vector or array can represent any subset S on a universal set U containing n items. Bit i is set to 1 if $i \in S$, and 0 if not. Because only one bit is needed per element, bit vectors can be very space efficient for surprisingly large values of $|U|$. Element insertion and deletion simply flips the appropriate bit. Intersection and union are done by "and-ing" or "or-ing" the bits together. The primary drawback of a bit vector is its performance on sparse subsets. For example, it takes $O(n)$ time to explicitly identify all the members of a sparse (or even empty) subset S.

- *Containers or dictionaries* – A subset S can also be represented using a linked list, array, or dictionary containing exactly the elements in S. No notion of a fixed universal set is needed for such a data structure. Dictionaries can be more space and time efficient for sparse subsets than bit vectors, and easier to work with. For efficient union and intersection operations, it pays to keep the elements in each subset sorted, so a linear-time merge of both subsets identifies all duplicates.

- *Bloom filters* – We can emulate a bit vector in the absence of a fixed universal set by hashing each subset element to an integer from 0 to $n-1$ and setting the corresponding bit. Thus, bit $H(e)$ is set to 1 if $e \in S$. Collisions leave some possibility for error under this scheme, however, because a different key might have hashed to the same position.

  A *Bloom filter* uses several (say k) different hash functions $H_1, \ldots H_k$, and sets all k bits $H_i(e)$ upon insertion of key e. Now e can be in S only if all k bits are 1. The probability of false positives can be made arbitrarily low by increasing the number of hash functions k and the table size n.

  This hashing-based data structure is much more space-efficient than dictionaries, for static subset applications that can tolerate a small probability of error. Many can. For instance, spell checkers that occasionally leave some rare random string uncorrected would prove no great tragedy. Bloom filters are more fully described in Section 6.4.

Many applications involve collections of subsets that are pairwise disjoint, meaning that each element occurs in exactly one subset. For example, consider maintaining the connected components of a graph or the party affiliations of politicians. Each vertex/scoundrel appears in exactly one component/party. Such a system of subsets is called a *set partition*. Algorithms for generating partitions of a given set are provided in Section 17.6 (page 524).

The primary issue with set partition data structures is maintaining changes over time, perhaps as edges get added or party members defect. Typical queries include "which set is a particular item in?" and "are two items in the same set?" as we modify the set by (1) changing one item, (2) merging or unioning two sets, or (3) breaking a set apart. Your basic options are:

- *Collection of containers* – Representing each subset in its own container or dictionary permits fast access to all their elements, which facilitates union and intersection operations. The cost comes in membership testing, since we must search each subset data structure independently until we find our target.

- *Generalized bit vector* – Let the ith element of an array denote the number/name of the subset that contains it. Set identification queries and single element modifications can be performed in constant time. However, operations like performing the union of two subsets take time proportional to the size of the universe, since each element in the two subsets must be identified and have its name changed.

- *Dictionary with a subset attribute* – Similarly, each item in a binary tree can be associated with a field that records the name of its subset. Set identification queries and single element modifications can be performed in the time it takes to search the dictionary. But union/intersection operations are again slow. The need to perform such union operations quickly provides the motivation for the ...

- *Union–find data structure* – We represent a subset using a rooted tree where each node points to its *parent* instead of its children. The name of each subset will be the name of the item at the root. To find the subset name for a given element, keep traversing up the parent pointers until you hit the root. Unioning two subsets is also easy. Assign the root of one of two trees to point to the other, so now *all* elements have the same root and hence the same subset name.

  Implementation details have a big impact on asymptotic performance here. Always selecting the larger (or taller) tree as the root in a merger guarantees logarithmic height trees, as with our implementation in Section 8.1.3 (page 250). Shrinking the path traced after each find, by explicitly pointing each path node directly to the root, is called *path compression* and reduces the tree to almost constant height. Union–find is a fast, simple data structure that every programmer should know.

Implementations: Modern programming languages provide libraries offering complete and efficient set implementations. The C++ *Standard Template Library* (STL) provides `set` and `multiset` containers. LEDA (see Section 22.1.1 (page 713)) provides efficient dictionary data structures, sparse arrays, and union–find data structures to maintain set partitions, all in C++. *Java Collections* (JC) contains `HashSet` and `TreeSet` containers and is included in the `java.util` package of Java Standard Edition.

An implementation of union–find underlies any implementation of Kruskal's minimum spanning tree algorithm. For this reason, all the graph libraries of Section 15.4 (page 452) presumably contain an implementation. Minimum spanning tree codes are described in Section 18.3 (page 549).

The computer algebra system *REDUCE* (`http://www.reduce-algebra.com/`) contains `SETS`, a package supporting set-theoretic operations on both explicit and implicit (symbolic) sets. Other computer algebra systems may support similar functionality.

Notes: Optimal algorithms for set operations such as intersection and union were presented in Reingold [Rei72]. Raman [Ram05] provides an excellent survey on data structures for a variety of different set operations. Bloom filters are ably surveyed in Broder and Mitzenmacher [BM05], with experimental results presented in [PSS07]. The cuckoo filter [FAKM14] is an improved Bloom filter variant offering better space/time performance and support for deletions.

Certain balanced tree data structures support merge/meld/link/cut operations, which permit fast ways to union and intersect disjoint subsets. See Tarjan [Tar83] for a nice presentation of such structures. Jacobson [Jac89] augmented the bit-vector data structure to support select operations (where is the ith 1 bit?) efficiently in both time and space.

Galil and Italiano [GI91] survey data structures for disjoint set union. The upper bound of $O(m\alpha(m, n))$ on m union–find operations on an n-element set is due to Tarjan [Tar75], as is a matching lower bound on a restricted model of computation [Tar79]. The inverse Ackerman function $\alpha(m, n)$ grows notoriously slowly, so this performance is close to linear. An interesting connection between the worst-case of union–find and the length of Davenport–Schinzel sequences—a combinatorial structure that arises in computational geometry—is established in Sharir and Agarwal [SA95].

The *power set* of a set S is the collection of all $2^{|S|}$ subsets of S. Explicit manipulation of power sets quickly becomes intractable due to their size. Implicit representations of power sets in symbolic form becomes necessary for non-trivial computations. See [BCGR92] for algorithms on, and computational experience with, symbolic power set representations.

Related problems: Generating subsets (see page 521), generating partitions (see page 524), set cover (see page 678), minimum spanning tree (see page 549).

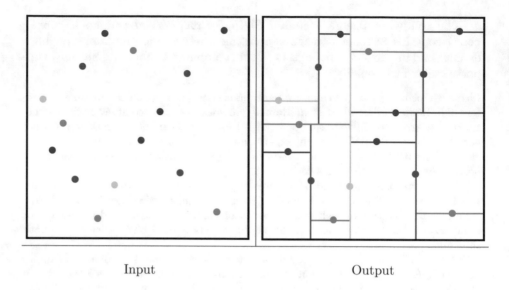

Input   Output

15.6   Kd-Trees

Input description: A set S of n points (or more complicated geometric objects) in k dimensions.

Problem description: Construct a tree that partitions space by half-planes such that each object is contained in its own box-shaped region.

Discussion: Kd-trees and related spatial data structures hierarchically decompose space into a small number of cells, each containing only a few representatives from an input set of points. This provides a fast way to access an object by position. We traverse down the hierarchy to find the smallest cell containing it, and then scan through the objects in this cell until we find what we want.

Typical algorithms construct kd-trees by partitioning point sets. Each node in the tree is defined by a plane cutting through one of the dimensions. Ideally, this plane partitions the subset of points into equal-sized left/right (or up/down) subsets. These children are again partitioned into equal halves, using planes through a different dimension. Partitioning stops after $\lg n$ levels, with every point in its own leaf cell.

The cutting planes along any path from the root to another node defines a unique box-shaped region of space. Each subsequent plane cuts this box into two boxes. Each box-shaped region is defined by $2k$ planes, where k is the number of dimensions. Indeed, the "kd" in *kd-tree* is short for "k-dimensional." We maintain the region of interest defined by the intersection of these half-spaces as we move down the tree.

Flavors of kd-trees differ in exactly how the splitting plane is selected, but options include:

- *Cycling through the dimensions* – partition first on d_1, then $d_2, \ldots, d_k$ before cycling back to d_1.

- *Cutting along the largest dimension* – select the partition dimension that makes the resulting boxes as square or cube-like as possible. Selecting a plane to partition the points equally does not always put the splitter in the middle of the box-shaped region, because all the points may happen to lie in the left side of the box.

- *Quadtrees or octtrees* – Instead of partitioning with single planes, use all axis-parallel planes that pass through a given partition point. In two dimensions, this creates four child cells. In three dimensions, we get eight child cells. Quadtrees seem particularly popular on image data, where leaf cells imply that all pixels in the regions have the same color.

- *Random projection trees* – Here the cutting plane for a node is defined by a random slope/direction (or more generally, a $d - 1$ dimensional plane through the origin). We identify a line/plane perpendicular to this direction that evenly partitions the points into two halves: such a line must exist and can be easily found. Then recur. These trees must have logarithmic height, yet can follow the points better than kd-trees when low dimensional structures occur in higher dimensionsional data.

- *BSP-trees – Binary space partitions* use general (i.e.not just axis-parallel) cutting planes to carve up space into cells such that each cell ends up containing only one object (say a polygon). Such partitions are not possible using only axis-parallel cuts for certain sets of objects. The downside is that such polyhedral cell boundaries are more complicated to work with than boxes.

- *Ball trees* – A ball tree is a hierarchical data structure on points, where each node is associated with a ball, defined by its center and radius. Each node's ball is the smallest that contains the balls of its children. Unlike kd-trees, this means that sibling balls can intersect, and do not partition the entire space. But they are very good for nearest-neighbor search in high dimensions, where kd-trees break down.

Ideally, our partitions will split both the space (ensuring fat, regular regions) and the set of points (ensuring a log height tree) evenly, but doing both simultaneously can be impossible on a given input. The advantages of fat cells become clear in many applications of kd-trees:

- *Point location* – To identify which cell a query point q lies in, start at the root and test which side of the partition plane contains q. By repeating this process on the appropriate child node, we travel down the tree to find the leaf cell containing q in time proportional to its height. See Section 20.7 (page 644) for more on point location.

- *Nearest-neighbor search* – To find the point in S closest to a given query point q, we perform point location to find the cell c containing q. Since this cell is associated with some point p, we can compute the distance $d(p, q)$ from p to q.

  Point p is likely close to q, but it might not be the absolute closest neighbor. Why? Suppose q lies at the right boundary of its cell. Then q's nearest neighbor might well lie towards the left of the boundary of the neighboring cell. Thus, we must traverse all cells that lie within a distance of $d(p, q)$ of q and verify that none of them contain any closer points. In trees with nice, fat cells, very few cells should need to be tested. See Section 20.5 (page 637) for more on nearest-neighbor search.

- *Range search* – Which points lie within a query box or region? Starting from the root, check whether the query region intersects (or contains) the cell defining the current node. If it does, check the children; if not, none of the leaf cells below this node can possibly be of interest. We quickly prune away irrelevant portions of the space. Section 20.6 (page 641) focuses on range search.

- *Partial key search* – Suppose we want to find a point p in S, but we do not have full information about p. Say we are looking for someone of age 59 and height 5 feet 8 inches but of unknown weight in a kd-tree with dimensions of age, weight, and height. Starting from the root, we can identify the correct descendant for all but the weight dimension. To be sure we find the right point, we must search *both children* of these nodes. The more fields we know the better, but such partial key search can be substantially faster than checking all points against the key.

Kd-trees are most useful for a small to moderate number of dimensions, say from two up to maybe twenty dimensions. They lose effectiveness as the dimensionality increases, primarily because the ratio of the volume of a unit sphere in k-dimensions shrinks exponentially compared to the unit cube. Thus, exponentially many cells will have to be searched within a given radius of a query point, say for nearest-neighbor search. Also, the number of neighbors for any cell grows to $2k$ and eventually becomes unmanageable.

The bottom line is you should try to avoid working in high-dimensional spaces, perhaps by discarding (or projecting away) the least important dimensions.

Implementations: *KDTREE 2* contains C++ and Fortran 95 implementations of *kd*-trees for efficient nearest-neighbor search in many dimensions. See `http://arxiv.org/abs/physics/0408067`. Ball trees are included as part of the popular Python package scikit-learn, again for nearest-neighbor search in high dimensional data.

Samet's spatial index demos (`http://donar.umiacs.umd.edu/quadtree/`) provide a series of Java applets illustrating many variants of *kd*-trees, in association with his book [Sam06].

The 1999 DIMACS implementation challenge focused on data structures for nearest-neighbor search [GJM02]. Data sets and codes are accessible from http://dimacs.rutgers.edu/Challenges.

Notes: Samet [Sam06] is the best reference on kd-trees and other spatial data structures. All major (and many minor) variants are developed in substantial detail. Samet's shorter survey [Sam05] is also available. Bentley [Ben75] is generally credited with developing kd-trees, although they have the murky history associated with most folk data structures.

The performance of spatial data structures degrades with high dimensionality. Balltrees [Omo89] and random projection trees [DF08] are examples of data structures designed to address this issue.

Projecting high-dimensional spaces onto a random lower-dimensional hyperplane has recently emerged as a simple but powerful method for dimensionality reduction. Both theoretical [IMS18] and empirical [BM01] results indicate that this method preserves distances quite nicely.

Algorithms that quickly produce a point provably close to the query point are an important aspect higher-dimensional nearest-neighbor search. See [ML14] for recent experimental results on finding high-dimensional near neighbors. Locality sensitive hashing methods such as Andoni and Indyk [AI06] are very popular and effective. Another approach by Arya et al. [AMN+98] builds a sparse weighted-graph structure from the data set, with the nearest neighbor found by starting at a random point and walking greedily in the graph towards the query point. The closest point found over several random trials is declared the winner. Similar methods hold promise for other problems in high-dimensional spaces.

Related problems: Nearest-neighbor search (see page 637), point location (see page 644), range search (see page 641).

Chapter 16

Numerical Problems

If most problems you encounter are numerical in nature, there is an excellent chance that you are reading the wrong book. *Numerical Recipes* [PFTV07] gives a terrific overview of the fundamental problems in numerical computing, including linear algebra, numerical integration, statistics, and differential equations. Different flavors of the book include source code for all the algorithms in C++, Fortran, and even Pascal. Their coverage is somewhat skimpier on the combinatorial/numerical problems considered in this section, but you should be aware of it. Check them out at http://numerical.recipes/.

Numerical computation has been of increasing importance because of machine learning, which heavily relies on linear algebra and unconstrained optimization. But note that numerical algorithms tend to be different beasts than combinatorial algorithms for at least two reasons:

- *Issues of precision and error* – Numerical algorithms typically perform repeated floating-point computations, which accumulate error at each operation until, eventually, the results are meaningless. My favorite example concerns the Vancouver Stock Exchange, which over a 22-month period accumulated enough round-off error to reduce its index from the correct value of 1098.982 to 574.081 [MV99].

  A simple and dependable way to test for round-off errors in numerical programs is to run them both at single and double precision, and then think hard if there is a substantial disagreement.

- *Extensive libraries of codes* – Large, high-quality libraries of numerical algorithms have existed since the 1960s, which is still not yet the case for combinatorial algorithms. There are several reasons for this, including (1) the early emergence of Fortran as a standard for numerical computation, (2) the nature of numerical computations to be recognizably independent modules instead of embedded in large applications, and (3) the existence of large scientific communities needing general numerical libraries.

  Regardless of why, you should exploit this software base. There is probably

S. S. Skiena, *The Algorithm Design Manual*, Texts in Computer Science,
https://doi.org/10.1007/978-3-030-54256-6_16

no reason to implement algorithms for any of the problems in this section, as opposed to using existing codes. Searching Netlib (see Section 22.1.4) is an excellent place to start.

Many scientists and engineers have ideas about algorithms that culturally derive from the simple control and data structures of numerical methods. In contrast, computer scientists grow up programming with pointers and recursion, and hence are comfortable with the more sophisticated data structures required for combinatorial algorithms. Both sides can and should learn from each other, since many problems can be modeled either numerically or combinatorially.

There is a vast literature on numerical algorithms. In addition to *Numerical Recipes*, recommended books include:

- *Chapra and Canale* [CC16] – The contemporary market leader in numerical analysis texts.

- *Mak* [Mak02] – This enjoyable text introduces Java to the world of numerical computation, and vice versa. Source code is provided.

- *Hamming* [Ham87] – This oldie but goodie provides a clear and lucid treatment of fundamental methods in numerical computation. It is available in a low-priced Dover Publications edition.

- *Skeel and Keiper* [SK00] – A readable and interesting treatment of basic numerical methods, avoiding overly detailed algorithm descriptions through its use of the computer algebra system Mathematica. I like it.

- *Cheney and Kincaid* [CK12] – A traditional Fortran-based numerical analysis text, with discussions of optimization and Monte Carlo methods in addition to such standard topics as root-finding, numerical integration, linear systems, splines, and differential equations.

- *Buchanan and Turner* [BT92] – Thorough language-independent treatment of all standard topics, including parallel algorithms. It is the most comprehensive of the texts described here.

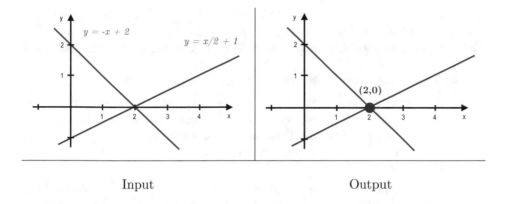

Input   Output

16.1 Solving Linear Equations

Input description: An $m \times n$ matrix A and an $m \times 1$ vector b, together representing m linear equations on n variables.

Problem description: What is the vector x such that $A \cdot x = b$?

Discussion: The need to solve linear systems arises in an estimated 75% of all scientific computing problems [DB74]. For example, applying Kirchhoff's laws to analyze electrical circuits generates a system of equations—the solution of which predicts the current through each branch of the circuit. Many machine learning algorithms reduce to solving linear systems, including linear regression and singular value decomposition (SVD). Even finding the point of intersection between two or more lines reduces to solving a small linear system.

Not all systems of equations have solutions. Just try to solve $2x + 3y = 5$ and $2x + 3y = 6$. Some systems of equations have multiple solutions, such as $2x + 3y = 5$ and $4x + 6y = 10$. Such *degenerate* systems of equations are called *singular*, and they can be recognized by testing whether the determinant of the coefficient matrix is zero.

Solving linear systems is a problem of such scientific and commercial importance that excellent codes are readily available. There is never a good reason to implement your own solver, even though the basic algorithm (Gaussian elimination) is one you learned in high school. This is especially true when working with large systems.

Gaussian elimination is based on the observation that the solution to a system of linear equations is invariant under scaling (if $x = y$, then $2x = 2y$) and adding equations (the solution to $x = y$ and $w = z$ is the same as the solution to $x = y$ and $x + w = y + z$). Gaussian elimination scales and adds equations to eliminate each variable from all but one equation, leaving the system in such a state that the solution can be directly read off from the resulting equations.

The time complexity of Gaussian elimination on an $n \times n$ system of equations is $O(n^3)$, because we add a scaled copy of the n-term ith row to each of the $n-1$

other equations to clear the ith (of n) variable. But on this problem, constants matter. Algorithms that only partially reduce the coefficient matrix and then back substitute to get the answer use 50% fewer floating-point operations than the naive algorithm.

Issues to worry about include:

- *Are round-off errors and numerical stability affecting my solution?* – Gaussian elimination would be quite straightforward to implement except for round-off errors. These accumulate with each row operation and quickly wreak havoc on the solution, particularly with matrices that are *almost* singular.

  To eliminate the danger of numerical errors, it pays to substitute the solution back into each of the original equations and test how close they are to the desired value. *Iterative* techniques for solving linear systems, like the Jacobi and Gauss-Seidel methods, refine initial solutions to obtain more accurate answers. Good linear systems packages will include such routines.

  The key to minimizing round-off errors in Gaussian elimination is selecting the right equations and variables to pivot on, and to scale the equations to eliminate large coefficients. This is an art as much as a science, which is why you should use a well-crafted library routine as described below.

- *Which routine in the library should I use?* – Selecting the right code is also somewhat of an art. If you are taking your advice from this book, start with a general linear system solver and hope it will suffice for your needs. But search through the manual for more efficient procedures solving special types of linear systems. If your matrix happens to be one of these special types, the solution time can reduce from cubic to quadratic or even linear.

- *Is my system sparse?* – A key to recognizing that you have a special-case linear system is establishing how many matrix elements you really need to describe A. If there are only a few non-zero elements, your matrix is *sparse* and you are in luck. If these few non-zero elements are clustered near the diagonal, your matrix is *banded* and you are in even more luck. Algorithms for reducing the bandwidth of a matrix are discussed in Section 16.2. Many other regular patterns of sparse matrices can also be exploited, so consult the manual of your solver or a good book on numerical analysis for details.

- *Will I be solving many systems using the same coefficient matrix?* – In applications such as least-squares curve fitting, people often solve $A \cdot x = b$ repeatedly with different b vectors. We can preprocess A to make this easier. The lower-upper or *LU-decomposition* of A creates lower- and upper-triangular matrices L and U such that $L \cdot U = A$. We can use this decomposition to solve $A \cdot x = b$, since

$$A \cdot x = (L \cdot U) \cdot x = L \cdot (U \cdot x) = b$$

This is efficient because back substitution solves a triangular system of equations in quadratic time. Solving $L \cdot y = b$ and then $U \cdot x = y$ gives the solution x using two $O(n^2)$ steps instead of one $O(n^3)$ step, after the LU-decomposition was computed in $O(n^3)$ time.

The problem of solving linear systems is equivalent to that of matrix inversion, since $Ax = B \leftrightarrow A^{-1}Ax = A^{-1}B$, where $I = A^{-1}A$ is the identity matrix. Avoid it, however, because matrix inversion proves to be three times slower than Gaussian elimination. LU-decomposition is very useful to invert matrices as well as compute determinants (see Section 16.4 (page 475)).

Implementations: The library of choice for solving linear systems is apparently LAPACK—a descendant of LINPACK [DMBS79]. Both of these Fortran codes, as well as many others, are available from Netlib (`https://www.netlib.org/`). Variants of LAPACK exist for other languages, like CLAPACK (C) and LAPACK++ (C++). The *Template Numerical Toolkit* is an interface to such routines in C++, and is available at `http://math.nist.gov/tnt/`.

JScience provides an extensive linear algebra package (including determinants) as part of its comprehensive scientific computing library. *JAMA* is another matrix package written in Java. Links to both and many related libraries are available at `http://math.nist.gov/javanumerics/`.

Numerical Recipes [PFTV07] (`http://numerical.recipes/`) provides guidance and routines for solving linear systems. Lack of confidence in dealing with numerical procedures is the most compelling reason to use these ahead of the free codes.

Notes: Golub and van Loan [GL96] is the standard reference on algorithms for linear systems. Good expositions on algorithms for Gaussian elimination and LU-decomposition include [CLRS09] and a host of numerical analysis texts [BT92, CK12, SK00]. Data structures for linear systems are surveyed in [PT05].

Parallel algorithms for linear systems are discussed in [Gal90, GO14, HNP91, KSV97]. Solving linear systems is one of the most important applications where parallel architectures are used widely in practice.

Matrix inversion and (hence) linear systems solving can be done in matrix multiplication time using Strassen's algorithm plus a reduction. Good expositions on the equivalence of these problems include [AHU74, CLRS09].

The HHL quantum computing algorithm has been proposed to solve $n \times n$ linear systems in $O(\log n)$ time [HHL09], which would be an exponential-time speedup over what is possible on conventional computers. This would be a real game-changer but there are many caveats, clearly explained by Aaronson [Aar15].

Related problems: Matrix multiplication (see page 472), determinant/permanent (see page 475).

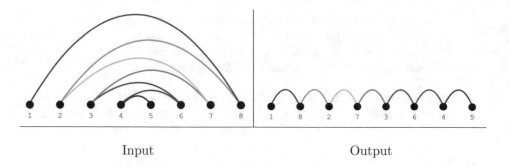

Input   Output

16.2   Bandwidth Reduction

Input description: A graph $G = (V, E)$, representing an $n \times n$ matrix M of zero and non-zero elements.

Problem description: Which permutation p of the vertices minimizes the length of the longest edge—i.e., minimizes $\max_{(i,j) \in E} |p(i) - p(j)|$?

Discussion: Bandwidth reduction lurks as a hidden but important problem for both graphs and matrices. Applied to matrices, bandwidth reduction permutes the rows and columns of a sparse matrix to minimize the distance b of the furthest non-zero entry from the center diagonal. This is important in solving linear systems, because Gaussian elimination can be performed in $O(nb^2)$ on matrices of bandwidth b: a big win over the general $O(n^3)$ algorithm if $b \ll n$.

Bandwidth minimization on graphs arises in more subtle ways. Arranging n circuit components in a row to minimize the length of the longest wire (and hence time delay) is a bandwidth problem, where each vertex of graph G corresponds to a circuit component and there is an edge for every wire linking two components. More general formulations such as rectangular circuit layouts inherit the same hardness and classes of heuristics from the linear versions.

The *bandwidth* problem seeks a linear ordering of the vertices, which minimizes the length of the longest edge. But there are other variations of the problem. In *linear arrangement*, we seek to minimize the sum of the lengths of the edges. This has application to circuit layout, where we seek to position the chips to minimize the total wire length. In *profile minimization*, we seek to minimize the sum of one-way distances, for each vertex v the length of the longest edge whose other vertex is to the left of v.

Unfortunately, all these variants of bandwidth minimization are NP-complete. It stays NP-complete even if the input graph is a tree whose maximum vertex degree is 3, which is about as strong a condition as I have seen on any problem. Therefore, our only options are brute-force search and heuristics.

Fortunately, ad hoc heuristics have been well studied and production-quality implementations of the best heuristics are available. These are based on performing a breadth-first search from a given vertex v, where v is placed at the left-most point of the ordering. All of the vertices that are distance 1 from v are placed to its immediate right, followed by all the vertices at distance 2, and so forth until all vertices in G are accounted for. The popular heuristics differ according to how many different start vertices are considered, and how equidistant vertices are ordered among themselves. Breaking ties with low-degree vertices over to the left generally seems to be a good idea.

Implementations of the most popular heuristics—the Cuthill–McKee and Gibbs–Poole–Stockmeyer algorithms—are discussed below. The worst case time of the Gibbs–Poole–Stockmeyer algorithm is $O(n^3)$, but its performance in practice is close to linear.

Brute-force search programs can find the exact minimum bandwidth, by backtracking through the set of all $n!$ possible permutations of vertices. Considerable pruning can be achieved to reduce the search space by starting with a good heuristic bandwidth solution and alternately adding vertices to the left- and rightmost open slots in the partial permutation.

Implementations: Del Corso and Manzini's [CM99] code for exact solutions to bandwidth problems is available at `https://people.unipmn.it/manzini/bandmin`. Caprara and Salazar-González [CSG05] developed improved methods based on integer programming. Their branch-and-bound implementation in C is available at the Algorithms Repository.

Fortran language implementations of both the Cuthill-McKee algorithm [CM69] and the Gibbs-Poole-Stockmeyer algorithm [GPS76] are available from Netlib. Empirical evaluations of these and other algorithms on a test suite of thirty matrices [Eve79], show Gibbs–Poole–Stockmeyer to be the consistent winner. Petit [Pet03] performed an extensive experimental study on heuristics for the minimum linear arrangement problem. His codes and data are available at `http://www.lsi.upc.edu/~jpetit/MinLA/Experiments/`.

Notes: Diaz et al. [DPS02] provide an excellent survey on algorithms for bandwidth and related graph layout problems. See [CCDG82] for graph-theoretic and algorithmic results on bandwidth minimization, up to 1981. Ad hoc heuristics have been widely studied—a tribute to its importance in numerical computation. Everstine [Eve79] cites no less than 49 different bandwidth reduction algorithms!

The hardness of the bandwidth problem was first established by Papadimitriou [Pap76b], and its hardness on trees of maximum degree 3 in [GGJK78]. There are algorithms that run in polynomial time for fixed bandwidth k [Sax80]. Approximation algorithms offering a polylogarithmic guarantee exist for the general problem [BKRV00, FL07], beyond which the problem is hard to approximate [DFU11].

Related problems: Solving linear equations (see page 467), topological sorting (see page 546).

$$\begin{bmatrix} 0 & 0 & 1 & 0 \\ 1 & 0 & 0 & 0 \\ 0 & 0 & 0 & 1 \\ 0 & 1 & 0 & 0 \end{bmatrix} \begin{bmatrix} 0 & 1 \\ 2 & 3 \\ 4 & 5 \\ 6 & 7 \end{bmatrix} \qquad \begin{bmatrix} 4 & 5 \\ 0 & 1 \\ 6 & 7 \\ 2 & 3 \end{bmatrix}$$

Input	Output

16.3 Matrix Multiplication

Input description: An $x \times y$ matrix A and a $y \times z$ matrix B.

Problem description: Compute the $x \times z$ matrix $A \times B$.

Discussion: Matrix multiplication is a fundamental problem in linear algebra. Its main significance for combinatorial algorithms is its equivalence to many other problems, including transitive closure/reduction, parsing, solving linear systems, and matrix inversion. A faster algorithm for matrix multiplication implies faster algorithms for all of these problems. Matrix multiplication arises in its own right in computing the results of such coordinate transformations as scaling, rotation, and translation for robotics and computer graphics.

Matrix multiplication is often used for data rearrangement problems instead of hard-coded logic. Multiplying by the identity matrix does nothing at all. But consider the input/output figures above. Observe that multiplying by the permutation matrix (on left) rearranged the rows of the output matrix. Efficient sparse matrix multiplication libraries can be astonishingly fast on such operations.

The following tight algorithm computes the product of $x \times y$ matrix A and $y \times z$ matrix B, and runs in $O(xyz)$. Remember to first initialize $M[i, j]$ to 0 for all $1 \le i \le x$ and $i \le j \le z$:

> for $i = 1$ to x do
>   for $j = 1$ to z
>   for $k = 1$ to y
> $M[i, j] = M[i, j] + A[i, k] \cdot B[k, j]$

An implementation in C appears in Section 2.5.4 (page 45). This straightforward algorithm would *seem* to be tough to beat in practice. That said, observe that the three loops can be arbitrarily permuted without changing the resulting answer. Such a permutation will change the memory access patterns and thus how effectively the cache is used. One can expect a 10–20% variation in run time among the six possible implementations, but probably not confidently predict the winner (typically ikj) without running it on your machine with your particular matrices.

When multiplying two bandwidth-b matrices, a speedup to $O(xbz)$ is possible. Zero values cannot contribute to the product, and all non-zero elements of A and B must lie within b positions of their main diagonals.

Asymptotically faster algorithms for matrix multiplication exist, using clever divide-and-conquer recurrences. However, these prove difficult to program, require very large matrices, and are less numerically stable to boot. The most famous of these is Strassen's $O(n^{2.81})$ algorithm. Empirical results (discussed below) disagree on the exact crossover point where Strassen's algorithm beats the simple cubic algorithm, but it is in the ballpark of $n \approx 100$.

There is a better way to save computation when you are multiplying a chain of more than two matrices together. Recall that multiplying an $x \times y$ matrix by a $y \times z$ matrix creates an $x \times z$ matrix. Thus, multiplying a chain of matrices from left to right might create large intermediate matrices, each taking a lot of time to compute. Matrix multiplication is not commutative, but it is associative, so we can parenthesize the chain in whatever manner we deem best without changing the final product. A standard dynamic programming algorithm can be used to construct the optimal parenthesization. Whether it pays to do this optimization will depend upon whether your matrix dimensions are sufficiently irregular and your chain multiplied often enough to justify it. Note that we are optimizing over the sizes of the dimensions in the chain, not the matrices themselves. No improvement is possible when all your matrices have the same dimensions.

Matrix multiplication has a particularly interesting interpretation in counting the number of paths between two vertices in a graph. Let A be the adjacency matrix of a graph G, meaning $A[i, j] = 1$ if there is an edge between i and j. Otherwise, $A[i, j] = 0$. Now consider the square of this matrix, $A^2 = A \times A$. If $A^2[i, j] \geq 1$, there must be a vertex k such that $A[i, k] = A[k, j] = 1$, so i to k to j is a path of length 2 in G. More generally, $A^k[i, j]$ counts the number of paths of length exactly k between i and j. This count includes non-simple paths, where vertices are repeated, such as i to k to i to j.

Implementations: D'Alberto and Nicolau [DN07] have engineered a very efficient matrix multiplication code, which switches from Strassen's to the cubic algorithm at the optimal point. It is available at http://www.fastmmw.com. Earlier experiments put the crossover point where Strassen's algorithm beats the cubic algorithm at about $n = 128$ [BLS91, CR76].

An $O(n^3)$ algorithm will likely be your best bet unless your matrices are very large. The linear algebra library of choice is LAPACK, a descendant of LINPACK [DMBS79], which includes several routines for matrix multiplication. These Fortran codes are available from Netlib.

Notes: Winograd's algorithm [Win68] for fast matrix multiplication reduces the number of multiplications by a factor of two over the straightforward algorithm. This can be a win despite the additional bookkeeping required [DN09].

In my opinion, the history of theoretical algorithm design began when Strassen [Str69] published his $O(n^{2.81})$-time matrix multiplication algorithm. For the first time, improving an algorithm in the asymptotic sense became a respected goal in its own right. Progressive improvements to Strassen's algorithm have gotten progressively less practical. The current best result for matrix multiplication is Williams' [Wil12] $O(n^{2.3727})$ algorithm, beating the previous champion (Coppersmith and Winograd [CW90]) by a factor of $n^{0.003}$. The conjecture is that $\Theta(n^2)$ suffices.

The problem of Boolean matrix multiplication can be reduced to that of general matrix multiplication [CLRS09]. The four-Russians algorithm for Boolean matrix multiplication [ADKF70] uses preprocessing to construct all subsets of $\lg n$ rows for fast retrieval in performing the actual multiplication, yielding a complexity of $O(n^3/\lg n)$. Additional preprocessing can improve this to $O(n^3/\lg^2 n)$ [Ryt85].

Engineering efficient matrix multiplication algorithms requires careful management of cache memory. See [BDN01, HUW02] for studies on these issues.

The inverse of multiplying matrices is factoring them, reducing M to A and B such that $M = A \cdot B$. LU-decomposition is an example of exact matrix factorization, but there is now increasing interest in low dimensional factorization of feature matrices for data science and machine learning [KBV09].

The interest in the squares of graphs goes beyond counting paths. Fleischner [Fle74] proved that the square of any biconnected graph has a Hamiltonian cycle. See [LS95] for results on finding the square roots of graphs—that is, finding A given A^2.

Good expositions of the matrix-chain algorithm include [BvG99, CLRS09], where it is given as a standard textbook example of dynamic programming.

Related problems: Solving linear equations (see page 467), shortest path (see page 554).

$$
\begin{vmatrix} a & b & c \\ d & e & f \\ g & h & i \end{vmatrix}
$$

$$
a \begin{vmatrix} e & f \\ h & i \end{vmatrix} - b \begin{vmatrix} d & f \\ g & i \end{vmatrix} + c \begin{vmatrix} d & e \\ g & h \end{vmatrix}
$$

$$
a(ei - fh) - b(di - fg) + c(dh - eg)
$$

$$
aei - bfg + cdh - ceg - bdi - afh
$$

Input	Output

16.4   Determinants and Permanents

Input description: An $n \times n$ matrix M.

Problem description: What is the determinant $|M|$ or permanent $perm(M)$ of the matrix M?

Discussion: Determinants of matrices provide a clean and useful abstraction that can be used to solve a variety of linear algebra problems:

- Testing whether a matrix is *singular*, meaning that the matrix does not have an inverse. A matrix M is singular iff $|M| = 0$.

- Testing whether a set of d points lies on a plane in fewer than d dimensions. If so, the system of equations they define is singular, so $|M| = 0$.

- Testing whether a point lies to the left or right of a line or plane. This problem reduces to evaluating whether the sign of a determinant is positive or negative, as discussed in Section 20.1 (page 622).

- Computing the area or volume of a triangle, tetrahedron, or other simplicial complex. These quantities are a function of the magnitude of the determinant, also discussed in Section 20.1 (page 622).

The determinant of a matrix M is defined as a sum over all $n!$ possible permutations π_i of the n columns of M:

$$
|M| = \sum_{i=1}^{n!} (-1)^{sign(\pi_i)} \prod_{j=1}^{n} M[j, \pi_j]
$$

where $sign(\pi_i)$ denotes the number of pairs of elements out of order (called *inversions*) in permutation π_i.

Directly implementing this definition yields an $O(n!)$ algorithm, as does the cofactor expansion method I learned in high school. Better algorithms to evaluate determinants are based on LU-decomposition, discussed in Section 16.1 (page 467). The determinant of M is simply the product of the diagonal elements of the LU-decomposition of M, which can be found in $O(n^3)$ time.

The *permanent* is a closely related function that arises in combinatorial problems. For example, the permanent of the adjacency matrix of a graph G counts the number of perfect matchings in G. The permanent of a matrix M is defined by

$$perm(M) = \sum_{i=1}^{n!} \prod_{j=1}^{n} M[j, \pi_j]$$

differing from the determinant only in that all products are positive.

Surprisingly, it is NP-hard to compute the permanent, even though the determinant can easily be computed in $O(n^3)$ time. The fundamental difference is that $det(AB) = det(A) \times det(B)$, while $perm(AB) \neq perm(A) \times perm(B)$. There are permanent algorithms running in $O(n^2 2^n)$ time that are considerably faster than the $O(n!)$ definition. Thus, finding the permanent of a 20×20 matrix is not out of the realm of possibility.

Implementations: The linear algebra package LINPACK contains a variety of Fortran routines for computing determinants, optimized for different data types and matrix structures. *JScience* provides an extensive linear algebra package (including determinants) as part of its comprehensive scientific computing library. *JAMA* is another matrix package written in Java. Links to both and many related libraries are available from http://math.nist.gov/javanumerics/.

Nijenhuis and Wilf [NW78] provide an efficient Fortran routine to compute the permanent of a matrix. See Section 22.1.9 (page 716). Cash [Cas95] provides a C routine to compute the permanent, motivated by the Kekulé structure count of computational chemistry.

Two different codes for approximating the permanent are provided by Barvinok. The first, based on [BS07], provides codes for approximating the permanent and a Hafnian of a matrix, as well as the number of spanning forests in a graph. See http://www.math.lsa.umich.edu/~barvinok/manual.html. The second, based on [SB01], can provide estimates of the permanent of 200×200 matrices in seconds. See http://www.math.lsa.umich.edu/~barvinok/code.html.

Notes: Cramer's rule reduces the problems of matrix inversion and solving linear systems to that of computing determinants. However, algorithms based on LU-decomposition are faster. See [BM53] for an exposition on Cramer's rule.

Determinants can be computed in $o(n^3)$ time using fast matrix multiplication, as shown in [AHU83]. Section 16.3 (page 472) discusses such algorithms. A fast algorithm for computing the sign of the determinant—an important problem in performing robust geometric computations—is due to Clarkson [Cla92].

The problem of computing the permanent was shown to be #P-complete by Valiant [Val79], where #P is the class of problems solvable on a "counting" machine in polynomial time. A counting machine returns the number of distinct solutions to a problem. Counting the number of Hamiltonian cycles in a graph is a #P-complete problem that is trivially NP-hard (and presumably harder), since a count greater than zero proves that the graph is Hamiltonian. But counting problems can be #P-complete even if the corresponding decision problem can be solved in polynomial time, as shown by the permanent and perfect matchings.

Minc [Min78] is the primary reference on permanents. A variant of an $O(n^2 2^n)$-time algorithm due to Ryser for computing the permanent is presented in [NW78].

Probabilistic algorithms have been developed for estimating the permanent, culminating in a fully polynomial randomized approximation scheme that provides an arbitrary close approximation in time that depends polynomially upon the input matrix and the desired error [JSV04].

Related problems: Solving linear systems (see page 467), matching (see page 562), geometric primitives (see page 622).

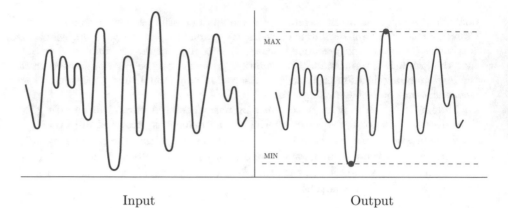

Input   Output

16.5   Constrained/Unconstrained Optimization

Input description: A function $f(x_1, \ldots, x_n)$.

Problem description: Which point $p = (p_1, \ldots, p_n)$ maximizes (or minimizes) the function f?

Discussion: Of the seventy-five problems in this catalog, this is the one whose importance has grown most dramatically since the previous edition of my book. Convex and non-convex optimization are the algorithmic problems most associated with machine learning, from linear regression to deep learning.

Optimization arises whenever you have an objective function that must be tuned for optimal performance. Suppose you are building a program to identify good stocks to invest in. You have certain financial data available to analyze—such as the price–earnings ratio, the interest rate, and the stock price—all as a function of time t. The key question is how much weight we should give to each of these factors, where these weights correspond to coefficients of a formula:

$$\text{stock-goodness}(t) = c_1 \times \text{price}(t) + c_2 \times \text{interest}(t) + c_3 \times \text{PE-ratio}(t)$$

We seek the numerical values c_1, c_2, c_3 whose stock-goodness function does the best job of evaluating stocks from past data. Similar issues arise in tuning evaluation functions for any pattern recognition or machine learning task.

Unconstrained optimization problems also arise in scientific computation. Physical systems from protein molecules to galaxies of stars naturally seek to minimize their "energy" or "potential function." Programs that simulate nature thus often define potential functions assigning a score to each possible object configuration, and then select the configuration that minimizes this potential.

Global optimization problems tend to be hard, and there are many ways to go about them. Ask the following questions to steer yourself in the right direction:

- *Am I doing constrained or unconstrained optimization?* – In unconstrained optimization, there are no limitations on the values of the parameters other than they maximize the value of f. However, many applications impose constraints on these parameters that make certain points illegal, points that might otherwise be the global optimum. For example, companies cannot employ less than zero employees, no matter how much money they think they might save doing so. Constrained optimization problems typically require mathematical programming approaches like linear programming, discussed in Section 16.6 (page 482).

- *Is the function I am trying to optimize described by a formula?* – Sometimes you need to find the maxima or minima of a function presented as an algebraic formula, such as finding the minimum of $f(n) = n^2 - 6n + 2^{n+1}$. If so, the solution is to analytically take its derivative $f'(n)$ and test for which points p' we have $f'(p') = 0$. These points are either local maxima or minima, which can be distinguished by taking a second derivative or just plugging p' back into f and seeing what happens. Be aware that things get more complicated in multivariate functions. Symbolic computation systems such as Mathematica and Maple are quite effective at computing derivatives. Although using computer algebra systems effectively is somewhat of a black art, they are definitely worth a try. And you can always use them to plot a picture of your function, in order to get a better idea of what you are dealing with.

- *Is your function convex?* The main difficulty of global optimization is getting trapped in local optima. Consider the problem of finding the highest point in a mountain range. If there is only one mountain and it is nicely shaped, we can find the top by just walking in whatever direction heads up. However, when there are many false summits, or other mountains in the area, it is difficult to convince ourselves whether we really are at the highest possible point.

  Convex functions have exactly one maxima (or minima), corresponding to a world with a single mountain peak. Gradient descent search analyzes the partial derivatives (slope) to determine the fastest way up (or down) the mountain. The optima is reached when the derivatives at this point are zero. Convex optimization can be used to quickly solve massive problems, even in a large number of dimensions. Consider the task of finding the set of coefficients that minimize fitting error for a linear regression problem in (say) 1,000 input variables or dimensions. The optimal solution (minimal point) here defines 1,000 parameters. If the objective function is convex (as it is in the case of linear regression) gradient descent search will make short work of the problem.

  How do you know whether your function is convex? This is beyond the scope of my book, involving analysis of its derivatives. But trust somebody smart when they tell you a function is convex.

- *How continuous or smooth is my function?* – Even if your function is non-convex, there is a good chance that it is at least smooth. *Smoothness* is the property that points in the local neighborhood of point p should have a value close to that of p. We assume smoothness in any search procedure. If the height at any given point was a completely random value, there would be no algorithm that could hope to find the optima short of sampling every single point.

- *Is it expensive to compute the function at a given point?* – Sometimes we are given a program or subroutine that evaluates f at a given point X, instead of an analytical function. We can request the value at any given point on demand by calling this function, so we can poke around and hunt for the maximum or minimum value.

  Our freedom to search in such a situation depends upon how efficiently we can evaluate $f(X)$. When point evaluation is expensive, the best approach is a simple grid search. Suppose you can afford to test $m \approx s^k$ possible points, for some integer s. Identify a smallest and largest possible value along each of your k dimensions, and then partition each range into s equally spaced values. There are s^k distinct points that can be defined by picking one value from each of these k ranges, points that broadly cover the set of possibilities. Evaluate $f(X)$ on each one of the points, and call the best performing one the "optima." The winner can also be used as a starting point to do a more systematic search.

  Such a situation arises in tuning evaluation functions for games. Suppose that $f(x_1, \ldots, x_n)$ is the board evaluation function in a computer chess program, such that x_1 is how much a pawn is worth, x_2 is how much a bishop is worth, and so forth. To evaluate how good a set of coefficients is as a board evaluator, we must play a bunch of games with it or test it on a library of known positions. Clearly, this is time-consuming, so we must be frugal in the number of evaluations of f we use to optimize the coefficients.

The most efficient algorithms for convex optimization use derivatives and partial derivatives to find local optima, to point out which direction to move from the current point so as to most rapidly increase or decrease the function. Such derivatives can sometimes be computed analytically, or they can be estimated numerically by taking the difference between the values of nearby points. A variety of *steepest descent* and *conjugate gradient* methods to find local optima have been developed—similar in many ways to numerical root-finding algorithms.

For constrained optimization, finding a point that satisfies all the constraints is often the difficult part of the problem. One approach is to use a method for unconstrained optimization, but add a penalty according to how many constraints are violated. This is the idea behind *Lagrangian relaxation*, which you might recall from calculus. Determining the right penalty function is problem-specific, but it often makes sense to vary the penalties as optimization proceeds.

At the end, the penalties should be very high to ensure that all constraints are satisfied.

Simulated annealing is a fairly robust approach to constrained optimization, particularly when we are optimizing over combinatorial structures (permutations, graphs, subsets). Techniques for simulated annealing are described in Section 12.6.3 (page 406).

It is a good idea to try several different methods on any given optimization problem. For this reason, I recommend experimenting with the implementations below before attempting to implement your own method. Clear descriptions of these algorithms are provided in many numerical algorithms books. My favorite is *Numerical Recipes* [PFTV07].

Implementations: The world of constrained/unconstrained optimization is sufficiently complex that several guides have been created to point people to the right codes. Particularly nice is Hans Mittlemann's *Decision Tree for Optimization Software* at `http://plato.asu.edu/guide.html`.

NEOS (Network-Enabled Optimization System) provides a unique service— the opportunity to solve your problem remotely on computers and software at the Wisconsin Institute of Discovery. Linear programming and unconstrained optimization are both supported. Check out `https://neos-server.org` when you need a solution instead of a program.

General purpose simulated annealing implementations are available, and are likely the best place to start experimenting with this technique for constrained optimization. Feel free to try my code from Section 12.6.3 (page 406). Particularly popular is *Adaptive Simulated Annealing (ASA)*, written in C by Lester Ingber and available at `http://asa-caltech.sourceforge.net/`.

Notes: Bertsekas [Ber15], Boyd and Vandenberghe [BV04], and Nesterov [Nes13] provide comprehensive treatments of convex optimization, including methods based on gradient descent. Unconstrained optimization and Lagrangian relaxation are the topics of several books, including [Ber82, PFTV07].

The full objective functions associated with machine learning problems are often linear in the size of the training data, which makes it very expensive to compute partial derivatives for gradient descent. Much better in practice is to estimate the derivatives at the current position using a small random sample of the training data. Such *stochastic gradient descent* algorithms are discussed in [Bot12]. Good books about machine learning include [Bis06, FHT01].

Simulated annealing was devised by Kirkpatrick et al. [KGV83] as a modern variation of the Metropolis algorithm [MRRT53]. Both use Monte Carlo techniques to compute the minimum energy state of a system. Good expositions on all local search variations, including simulated annealing, appear in [AL97].

Genetic algorithms were developed and popularized by Holland [Hol75, Hol92]. Books more partial to genetic algorithms than mine include [LP02, MF00].

Related problems: Linear programming (see page 482), satisfiability (see page 537).

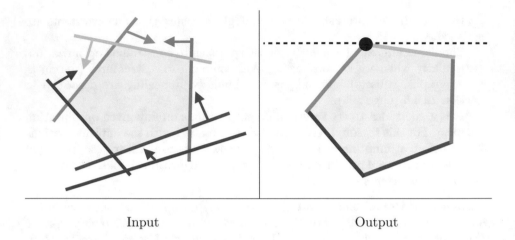

Input   Output

16.6   Linear Programming

Input description: A set S of n linear inequalities on m variables

$$S_i = \sum_{j=1}^{m} c_{ij} \cdot x_j \geq b_i, \ 1 \leq i \leq n$$

and a linear optimization function $f(X) = \sum_{j=1}^{m} c_j \cdot x_j$.

Problem description: Which variable assignment X' maximizes the objective function f while satisfying all inequalities S?

Discussion: Linear programming is the most important problem in mathematical optimization and operations research.   Applications include:

- *Resource allocation* – We seek to invest a given amount of money to maximize our return. Often our possible options, payoffs, and expenses can be expressed as a system of linear inequalities such that we seek to maximize our possible profit given the constraints. Very large linear programming problems are routinely solved by airlines and other corporations.

- *Approximating the solution of inconsistent equations* – A set of n linear equations on m variables x_i, $1 \leq i \leq m$ is over-determined if $n > m$. Such over-determined systems are often *inconsistent*, meaning that there does not exist an assignment to the variables that simultaneously solves all the equations. To find an assignment that best fits the equations, we can replace each variable x_i by $x_i' + \epsilon_i$ and solve the new system as a linear program, minimizing the sum of the error terms.

- *Graph algorithms* – Many graph problems described in this book, including shortest path, bipartite matching, and network flow, can be solved as special cases of linear programming. Most of the rest, including traveling salesman, set cover, and knapsack, can be solved using *integer linear programming*.

The *simplex method* is the standard algorithm for linear programming. Each constraint in a linear programming problem acts like a knife that carves away a region from the space of possible solutions. We seek the point within the remaining region that maximizes (or minimizes) $f(X)$. By appropriately rotating the solution space, this optimal point can always be made to be the highest point in the region. The region (simplex) defined by the intersection of a set of linear constraints is convex, so there is always a higher vertex neighboring any starting point unless we are already at the top. When we cannot find a higher neighbor to walk to, we must have reached the optimal solution.

While the basic simplex algorithm is not particularly complex, there is considerable art to producing an efficient implementation capable of solving large linear programs. Large programs tend to be sparse (meaning that most inequalities use few variables), so sophisticated data structures must be used. There are issues of numerical stability and robustness, as well as choosing the best neighbor to walk to next (the so-called *pivoting rule*). There also exist sophisticated *interior-point* methods, algorithms that cut through the interior of the simplex instead of walking along the outside, and beat simplex in many applications.

The bottom line on linear programming is that you are much better off using an existing LP code than writing your own. Further, you might be better off paying money for it than surfing the web. Linear programming is an algorithmic problem of such economic importance that commercial implementations are superior to free versions.

Issues that arise in linear programming include:

- *Do any variables have integrality constraints?* – It is impossible to send 6.54 airplanes from New York to Washington each business day, even if that value maximizes profit according to your model. Such variables often have natural integrality constraints. A linear program is called an *integer program* if all its variables have integrality constraints, and a *mixed integer program* if some of them do.

  Unfortunately, it is NP-complete to solve integer or mixed programs to optimality. But there are integer programming techniques that work reasonably well in practice. *Cutting plane techniques* solve the problem first as a linear program, and then add extra constraints to enforce integrality around the optimal solution point before solving it again. After sufficiently many iterations, the optimum point of the resulting linear program matches that of the original integer program. As with most exponential-time algorithms, run times for integer programming depend upon the difficulty of the problem instance and are unpredictable.

- *Do I have more variables or constraints?* – Any linear program with m variables and n inequalities can be written as an equivalent *dual* linear program with n variables and m inequalities. This is important to know, because the running time of a given solver might be quite different on the two formulations. In general, linear programs (LPs) with more variables than constraints should be solved directly. If there are more constraints than variables, it is usually better to solve the dual linear program or (equivalently) apply the dual simplex method to the primal LP.

- *What if my optimization function or constraints are not linear?* – In least-squares curve fitting, we seek the line that best approximates a set of points by minimizing the sum of squares of the distance between each point and the line. In formulating this as a mathematical program, the natural objective function is no longer linear, but quadratic. Although fast algorithms exist for least squares fitting, general *quadratic programming* is NP-complete.

  There are three possible courses of action when you must solve a nonlinear program. The best is to try and model it in some other way, as is the case with least-squares fitting. The second is to try to track down special codes for quadratic programming. Finally, you can model your problem as a constrained or unconstrained optimization problem and try to solve it with the codes discussed in Section 16.5 (page 478).

- *What if my model does not match the input format of my LP solver?* – Many linear programming implementations accept models only in so-called *standard form*, where all variables are constrained to be non-negative, the object function must be minimized, and all constraints must be equalities (instead of inequalities).

  Do not fear. There exist standard transformations to map arbitrary LP models into standard form. To convert a maximization problem to a minimization one, just multiply each coefficient of the objective function by -1. The remaining problems can be solved by adding *slack variables* to the model. See any textbook on linear programming for details. Modeling languages such as AMPL can provide a nice interface to your solver and deal with these issues for you.

Implementations: There are at least three reasonable choices in free LP-solvers. `Lp_solve`, written in ANSI C by Michel Berkelaar, can also handle integer and mixed-integer problems. It is available at `http://lpsolve.sourceforge.net/5.5/`, and a substantial user community exists. The simplex solver `CLP` produced under the Computational Infrastructure for Operations Research is available (with other optimization software) at `http://www.coin-or.org/`. Finally, the GNU Linear Programming Kit (GLPK) is intended for solving large-scale linear programming, mixed integer programming (MIP), and other related problems. It is available at `http://www.gnu.org/software/glpk/`.

Benchmark studies [GAD+13, MT12] agree that commercial solvers perform much better than open codes. But they are split as to whether `CLP` or `GLPK` is the better choice. Read the reports for details.

NEOS (Network-Enabled Optimization System) provides a unique service—the opportunity to solve your problem remotely on computers and software at the Wisconsin Institute of Discovery. Linear programming and unconstrained optimization are both supported. Check out `https://neos-server.org` when you need a solution instead of a program.

Notes: The need for optimization via linear programming arose in logistics problems in World War II. The simplex algorithm was invented by George Danzig in 1947 [Dan63]. Klee and Minty [KM72] proved that the simplex algorithm is exponential in worst case, but it is very efficient in practice.

Smoothed analysis measures the complexity of algorithms assuming that their inputs are subject to small amounts of random noise. Carefully constructed worst-case instances for many problems break down under such perturbations. Spielman and Teng [ST04] used smoothed analysis to explain the efficiency of simplex in practice. Kelner and Spielman developed a randomized simplex algorithm running in polynomial time [KS05b].

Khachian's ellipsoid algorithm [Kha79] first proved that linear programming was polynomial in 1979. Karmarkar's algorithm [Kar84] is an interior-point method that has proven to be both a theoretical and practical improvement of the ellipsoid algorithm, as well as a challenge for the simplex method. Good expositions on the simplex and ellipsoid algorithms for linear programming include [Chv83, Gas03, MG07].

Semidefinite programming deals with optimization problems over symmetric positive semidefinite matrix variables, with linear cost function and linear constraints. Important special cases include linear programming and convex quadratic programming with convex quadratic constraints. Semidefinite programming and its applications to combinatorial optimization problems are surveyed in [Goe97, VB96].

Linear programming is P-complete under log-space reductions [DLR79]. This makes it unlikely to have an NC parallel algorithm, where a problem is in NC iff it can be solved on a PRAM in polylogarithmic time using a polynomial number of processors. Any problem that is P-complete under log-space reduction cannot be in NC unless P=NC. See [GHR95] for a thorough exposition of the theory of P-completeness, including an extensive list of P-complete problems.

Related problems: Constrained and unconstrained optimization (see page 478), network flow (see page 571).

HTHTTHTHTHHTTHHHHHTTHTHTT
HTHTHHHTTHHHHTTHHTHHHTTTTT
THTHTTTTHTTHHHHTHHHHTHTHTT
THTTHHHTHTTTHHHHHHHHHTTTT
HTHTHHHHHTHTHTTHHHHTHTTHT
THHTTHHTHHTHHHTHHTTHTTHTT
THTTHTHHHHHHHTHTHTHHHTHHHT
HTTTTTHHTTHTHHHHTTHHTHTTHH
HHHHHTHTHHTTHHHHTTTHTTHTTH
HHTHHHTTHTTTHHHHHTHHHHTTTT

| Input | Output |

16.7 Random Number Generation

Input description: Nothing, or perhaps a seed.

Problem description: Generate a sequence of random integers.

Discussion: Random numbers have a surprising variety of interesting and important applications. They form the foundation of simulated annealing and related heuristic optimization techniques. Discrete event simulations run on streams of random numbers, and are used to model everything from transportation systems to casino poker. Passwords and cryptographic keys are typically generated randomly. Randomized algorithms for graph and geometric problems are revolutionizing these fields, and establishing randomization as one of the fundamental ideas of computer science.

Unfortunately, generating random numbers looks much easier than it really is. Indeed, it is fundamentally impossible to produce truly random numbers on any deterministic device. Von Neumann [Neu63] said it best: "Anyone who considers arithmetical methods of producing random digits is, of course, in a state of sin." The best we can hope for are *pseudorandom* numbers, a stream of numbers that appear as if they were generated randomly.

There can be serious consequences to using a bad random-number generator. In one famous case, a web browser's encryption scheme was broken after the discovery that the seeds of its random-number generator employed too few random bits [GW96]. Simulation accuracy is regularly compromised or invalidated by poor random number generation. This is an area where people shouldn't fool around, but they do. Issues to think about include:

- *Should my program produce the same "random" numbers each time it runs?* – A poker game that deals you the exact same hand every time you play quickly loses interest. One common solution uses the lower-order bits of the machine clock as the *seed* or starting point for a stream of random numbers, so that each time the program runs it does something different.

Such methods are adequate for games, but not for serious simulations. There are liable to be periodicities in the distribution of random numbers whenever calls are made in a loop. Also, debugging is seriously complicated when program results are not repeatable. Should your program crash, you can't go back and discover why. A possible compromise is to use a deterministic pseudorandom-number generator, but write the current seed to a file between runs. During debugging, this file can be overwritten with a fixed initial value of the seed.

- *How good is my compiler's built-in random number generator?* – If you need uniformly generated random numbers, and are not betting the farm on the accuracy of your simulation, my recommendation is simply to use what your compiler provides. Your best opportunity to mess things up will be with a bad choice of the initial seed, so read the manual for its recommendations.

  If you *are* going to bet the farm on the results of your simulation, you had better test your random number generator. Be aware that it is difficult to eyeball the results and decide whether the output is really random. This is because people have very skewed ideas of how random sources should behave, and often see patterns that don't really exist. Several different tests should be used to evaluate a random number generator, and the statistical significance of the results established. The National Institute of Standards and Technology (NIST) has developed test suites for evaluating random number generators, discussed below.

- *What if I must implement my own random-number generator?* – The standard algorithm of choice is the *linear congruential generator*. It is fast, simple, and (if instantiated with the right constants) gives reasonable pseudorandom numbers. The nth random number R_n is a function of the $(n-1)$st random number:

$$R_n = (aR_{n-1} + c) \mod m$$

Linear congruential generators work the same way roulette wheels do. The long path of the ball around and around the wheel (captured by $aR_{n-1} + c$) ends in one of a relatively small number of bins, the choice of which is extremely sensitive to the length of the path (captured by the mod m-truncation).

A substantial theory has been developed to properly select the constants a, c, m, and R_0. The period length is largely a function of the modulus m, which is typically constrained by the word length of your machine.

Note that the stream of numbers produced by a linear congruential generator starts to cycle the instant the first number repeats. Computers are fast enough to make 2^{32} calls to a random-number generator in a few minutes. Thus, any 32-bit linear congruential generator is in danger of cycling, motivating generators with significantly longer periods.

- *What if I don't want such large random numbers?* – The linear congruential generator R_n produces a uniformly distributed sequence of integers between 0 and $m - 1$ that can be easily scaled to produce other uniform distributions. To generate real numbers between 0 and 1, use R_n/m. Note that 1 cannot be realized this way, although 0 can. If you need uniformly distributed integers between l and h, use $\lfloor l + (h - l + 1)R_n/m \rfloor$.

- *What if I need non-uniformly distributed random numbers?* – Generating random numbers according to a given non-uniform distribution can be a tricky business. The most reliable way to do this correctly is the acceptance–rejection method. We bound the geometric region we seek to sample from by a box, and then select a random point p from the box. This in-the-box point can be generated by selecting the x- and y-coordinates independently at random. If p lies within the region of interest, we can return it as being selected at random. Otherwise we throw it away and repeat with another random point. Essentially, we throw darts at random and report those that hit the target.

  This method is correct, but can be slow. When the volume of the region of interest is small relative to that of the bounding box, most of our darts will miss the target. Efficient generators for Gaussian and other special distributions are described in the references and implementations below.

  Be cautious about inventing your own technique, because it can be tricky to obtain the right probability distribution. For example, an *incorrect* way to select points uniformly from a circle of radius r would be to generate polar coordinates by selecting an angle from 0 to 2π and a displacement between 0 and r—both uniformly at random. In such a scheme, half the generated points will lie within $r/2$ of the center, when only one-fourth of them should be! This is different enough to seriously distort the results, while being sufficiently subtle that the skew could easily escape detection.

- *How long should I run my Monte Carlo simulation?* – The longer you run a simulation, the more accurately the results *should* approximate the limiting distribution. But this is true only until you exceed the *period*, or cycle length, of your random-number generator. From then on, your sequence of random numbers repeats itself, so longer runs generate no additional information.

  Instead of jacking up the length of a simulation run to the max, it is usually more informative to do many shorter runs (say ten to a hundred) each with different seeds. Then consider the range of results you obtain. The variance provides a healthy measure of the degree to which your results are repeatable. This exercise corrects the natural tendency to see a simulation as giving "the" correct answer.

Implementations: See `https://www.agner.org/random` for an excellent website on random-number generation, including pointers to papers and many implementations of random-number generators.

Parallel simulations make special demands on random-number generators. How can we ensure that random streams are independent on each machine? L'Ecuyer et.al [LSCK02] provides object-oriented generators with a period length of approximately 2^{191}. Implementations in C, C++, and Java are available at `http://www.iro.umontreal.ca/~lecuyer/myftp/streams00/`. Independent streams of random numbers are supported for parallel applications. Another possibility is the *Scalable Parallel Random Number Generators Library (SPRNG)* [MS00], available at `http://sprng.cs.fsu.edu/`.

The National Institute of Standards [BRS+10] has prepared an extensive statistical test suite to validate random number generators. Both the software and the report describing it are available at `https://csrc.nist.gov/projects/random-bit-generation/documentation-and-software`.

True random-number generators extract random bits by observing physical processes. The website `http://www.random.org` makes available random numbers derived from atmospheric noise that pass the NIST statistical tests. This presents an amusing solution if you need a small quantity of random numbers (say to run a lottery), instead of a random-number generator.

Notes: Knuth [Knu97b] provides a thorough treatment of random-number generation, which I heartily recommend. He gives the theory behind several methods, including the middle-square and shift-register methods I have not described, plus a detailed discussion of statistical tests for validating random-number generators.

That said, see [Gen06, L'E12] for more recent developments in random number generation. The Mersenne twister [MN98] is a fast random number generator of period $2^{19937} - 1$. Other modern methods include [Den05, PLM06]. Methods for generating non-uniform random variates are surveyed in [HLD04]. Comparisons of different random-number generators in practice include [PM88].

Tables of random numbers appear in most mathematical handbooks as relics from the days before there was ready access to computers. Most notable is [RC55], which provides one million random digits. For a good laugh, I encourage you to check out the hundreds of reviews of this book on Amazon. You will be impressed just how many wiseguys exist on the Internet.

The deep relationship between randomness and information is explored within the theory of Kolmogorov complexity, which measures the complexity of a string by its compressibility. Truly random strings are incompressible. The string of seemingly random digits of π are not random under this definition, since the entire sequence is defined by a program implementing the series expansion for π. Li and Vitáni [LV97] provide a thorough treatment of the theory of Kolmogorov complexity.

Related problems: Constrained/unconstrained optimization (see page 478), generating permutations (see page 517), generating subsets (see page 521).

		179424673
		2038074743
8338169264555846052842102071		22801763489
	x	
		8338169264555846052842102071

Input   Output

16.8   Factoring and Primality Testing

Input description: An integer n.

Problem description: Is n a prime number? If not, what are its factors?

Discussion: The dual problems of integer factorization and primality testing have surprisingly many applications for a problem long suspected of being only of mathematical interest. The security of the RSA public-key cryptography system (see Section 21.6 (page 697)) is based on the computational intractability of factoring large integers. It is known that hash table performance typically improves when the table size is a prime number. To get this benefit, an initialization routine must identify a prime near the desired table size. Finally, prime numbers are just interesting to play with. This is why the program to generate large primes used to reside in the games directory of UNIX systems.

Factoring and primality testing are closely related problems, although they are quite different algorithmically. There exist algorithms that can demonstrate that an integer is *composite* (i.e. not prime) without actually giving the factors. To convince yourself of the plausibility of this, note that you can demonstrate the compositeness of any non-trivial integer whose last digit is 0, 2, 4, 5, 6, or 8 without doing the actual division.

The simplest algorithm for both of these problems is brute-force trial division. To factor n, compute the remainder of n/i for all $1 < i \leq \sqrt{n}$. The prime factorization of n will contain at least one instance of every i such that $n/i = \lfloor n/i \rfloor$, unless n is prime. Make sure you handle the multiplicities correctly, and account for any primes larger than $\sqrt{n}$.

Such algorithms can be sped up by using a precomputed table of small primes to avoid testing all possible i. Surprisingly many primes can be represented in surprisingly little space by using bit vectors (see Section 15.5 (page 456)). A bit vector of all odd numbers less than 1,000,000 fits in under 64 kilobytes. Even tighter encodings become possible by eliminating all multiples of 3 and other small primes.

Although trial division runs in $O(\sqrt{n})$ time, it is *not* a polynomial-time algorithm. It only takes $\lg n$ bits to represent n, so trial division takes time exponential in the input size. Considerably faster (but still exponential time) factoring algorithms exist, whose correctness depends upon more substantial number theory. The fastest known algorithm, the *number field sieve*, uses randomness to construct a system of congruences—the solution of which usually gives a factor of the integer. Integers with 250 digits (829 bits) have been factored using this method, although such feats require enormous amounts of computation.

Randomized algorithms make it much easier to test whether an integer is prime. Fermat's little theorem states that $a^{n-1} = 1(\bmod\ n)$ for all a not divisible by n, provided n is prime. Suppose we pick a random value $1 \leq a < n$ and compute the residue of $a^{n-1}(\bmod\ n)$. If this residue is not 1, we have just proven that n must be composite. Such randomized primality tests are very efficient. PGP (see Section 21.6 (page 697)) finds 300+ digit primes using hundreds of these tests in minutes, for use as cryptographic keys.

Although the primes are scattered in a seemingly random way throughout the integers, there is some regularity to their distribution. The *prime number theorem* states that the number of primes less than n (commonly denoted by $\pi(n)$) is approximately $n/\ln n$. Further, there never are large gaps between primes, so in general one should expect to examine about $\ln n$ integers to find the first prime larger than n. This distribution, coupled with the fast randomized primality test, explains how PGP can find such large primes so quickly.

Quantum computers are (theoretically!) capable of factoring large integers very fast, indeed exponentially faster than conventional machines. I would not be shocked if quantum computers capable of fast factoring exist before I write the fourth edition of this book. But serious technical challenges remain: a recent proposal to factor RSA-scale (2,048 bit) integers using Shor's algorithm require 20 million qubits because of the need for extensive error correction [GE19]. I've heard RSA factoring described as the "suicide app" for quantum computing: because the moment it succeeds, RSA stops being used and the application goes away.

Implementations: Several general systems for computational number theory are available. PARI is capable of handling complex number-theoretic problems on arbitrary-precision integers (to be precise, limited to 80,807,123 digits on 32-bit machines), as well as reals, rationals, complex numbers, polynomials, and matrices. It is written mainly in C, with assembly code for inner loops on major architectures, and includes more than 200 special predefined mathematical functions. PARI can be used as a library, but it also possesses a calculator mode that gives instant access to all the types and functions. PARI is available at http://pari.math.u-bordeaux.fr/

A Library for doing Number Theory (NTL) is a high-performance, portable C++ library providing data structures and algorithms to manipulate signed, arbitrary length integers and vectors, matrices, and polynomials over the integers and over finite fields. It is available at `http://www.shoup.net/ntl/`.

Finally, MIRACL (Multiprecision Integer and Rational Arithmetic C/C++ Library) implements six different integer factorization algorithms, including the quadratic sieve. It is available at `https://github.com/miracl/MIRACL`.

Notes: Expositions on modern algorithms for factoring and primality testing include Crandall and Pomerance [CP05] and Yan [Yan03]. More general surveys of computational number theory include Bach and Shallit [BS96] and Shoup [Sho09].

Agrawal, Kayal, and Saxena [AKS04] solved a long-standing open problem giving the first polynomial-time deterministic algorithm to test whether an integer is composite. Their algorithm is surprisingly elementary for such an important result, involving a careful analysis of techniques from earlier randomized algorithms. Its existence serves as somewhat of a rebuke to researchers (like me) who shy away from classical open problems due to fear. Dietzfelbinger [Die04] provides a self-contained treatment of this result.

The complexity class *co-NP* is the set of problems for which a polynomial-time algorithm can verify "no" instances given the appropriate certificate. An important problem in computational complexity theory is whether $P = NP \cap$ co-NP. The decision problem "is n a composite number?" used to be the best candidate for a counterexample. By exhibiting the factors of n, it is trivially in NP. It must be co-NP because every prime has a short proof of its primality [Pra75]. The recent proof that composite numbers testing is in P shot down this line of reasoning. For more information on complexity classes, see [AB09, GJ79].

Shor [Sho99] sparked great interest in quantum computing with his algorithm for factoring integers in polynomial time. Factoring $15 = 3 \times 5$ remains the experimental state of the art [MNM$^+$16] as of this writing. Introductions to quantum computing include [Aar13, Ber19].

The Miller–Rabin [Mil76, Rab80] randomized primality testing algorithm eliminates problems with Carmichael numbers, which are composite integers that always satisfy Fermat's theorem. The best algorithms for integer factorization include the quadratic-sieve [Pom84] and the elliptic-curve methods [Len87b].

Mechanical sieving devices provided the fastest way to factor integers surprisingly far into the computing era. See [SWM95] for a fascinating account of one such device, built during World War I. Hand-cranked, it proved the primality of $2^{31} - 1$ in 15 minutes of sieving time.

The integer RSA-129 was factored in eight months using over 1,600 computers. This was particularly noteworthy because in the original RSA paper [RSA78] they had originally predicted such a factorization would take 40 quadrillion years using 1970s technology. The current record for integer factorization is the successful attack on the 250-digit integer RSA-250 in February 2020.

Related problems: Cryptography (see page 697), high precision arithmetic (see page 493).

$$\frac{4957829128749149515150890542586957 8}{7436743693123724272726335813880436 7}$$

$$2\!\!\Big/\!3$$

Input   Output

16.9 Arbitrary-Precision Arithmetic

Input description: Two very large integers, x and y.

Problem description: What is $x + y$, $x - y$, $x \times y$, and x/y?

Discussion: Every programming language rising above basic assembler supports single- and perhaps double-precision integer/real addition, subtraction, multiplication, and division. But what if we wanted to represent the national debt of the United States in pennies? $22.4 trillion worth of pennies requires 16 decimal digits, which is far more than can fit into a 32-bit integer (although it still fits comfortably in 64-bits).

Other applications require *much* larger integers. The RSA algorithm for public-key cryptography recommends integer keys of at least 2,048 bits or equivalently 617 digits to achieve adequate security. Experimenting with number-theoretic conjectures for fun or research requires playing with large numbers. I once solved a minor open problem by performing an exact computation on the integer $\binom{5906}{2953} \approx 9.93285 \times 10^{1775}$, as presented in Section 6.9 (page 191).

What should you do when you need large integers?

- *Am I solving a problem instance requiring large integers, or do I have an embedded application?* – If all you need is the answer to a specific computation with large integers, as in the number theory example above, you should consider using a Python interpreter or a computer algebra system like Maple or Mathematica. These provide arbitrary-precision arithmetic as a default with easy-to-use language interpreters as the front end— together often reducing your problem to a 5-to-10-line program.

  If you have an embedded application requiring high-precision arithmetic instead, you should use an existing arbitrary precision math library. You are likely to get additional functions beyond the four basic operations, for computing things like greatest common divisor and square roots. See the Implementations section for details.

- *Do I need high- or arbitrary-precision arithmetic?* – Is there an upper bound on how big your integers can get, or do you really need unbounded *arbitrary*-precision? This determines whether you can use a fixed-length array to represent your integer instead of a linked-list of digits. The array should be simpler and not prove a constraint in most applications.

- *What base should I do arithmetic in?* – It is perhaps simplest to implement your own high-precision arithmetic package in decimal, thus representing each integer as a string of base-10 digits. However, it is far more computationally efficient to use a higher base, ideally equal to the square root of the largest integer supported fully by hardware arithmetic.

  Why? The higher the base, the fewer digits we need to represent a number. Compare 64 decimal with 1000000 binary. Since hardware addition usually takes one clock cycle independent of the value of the numbers being added, the best performance is achieved by using the highest supported base. The factor limiting us to base $b = \sqrt{maxint}$ is the desire to avoid overflow when multiplying two of these "digits" together.

  The primary complication of using a larger base is that integers must be converted to and from base-10 for input and output. The conversion is easy to perform using the four basic high-precision arithmetical operations.

- *How low-level are you willing to go for fast computation?* – Hardware addition is much faster than a subroutine call, so you take a significant hit on speed using high-precision arithmetic when low-precision arithmetic suffices. High-precision arithmetic is one of few problems in this book where inner loops in assembly language prove the right idea to speed things up. Similarly, using bit-level masking and shift operations instead of arithmetical operations can be a performance win if you really understand the machine integer representation.

The algorithms of choice for the basic arithmetic operations are as follows:

- *Addition* – The basic schoolhouse method of lining up the decimal points and then adding the digits from right to left with "carries" runs in time linear in the number of digits. More sophisticated carry-look-ahead parallel algorithms are available for low-level hardware implementation. They are presumably used in your microprocessor for low-precision addition.

- *Subtraction* – By fooling with the sign bits of the numbers, subtraction can be a special considered case of addition: $(A-(-B)) = (A+B)$. The tricky part of subtraction is performing the "borrow." This can be simplified by always subtracting the smaller number from the larger absolute value and adjusting the signs afterwards, so we can be certain there is always something to borrow from. Computer architects use two's complement representation to simplify subtraction with signed integers.

- *Multiplication* – The repeated addition method takes exponential time on large integers, so stay away from it. The digit-by-digit schoolhouse method is reasonable to program and runs in $O(n^2)$ time to multiply two n-digit integers. On very large integers, Karatsuba's $O(n^{1.59})$ divide-and-conquer algorithm wins. Dan Grayson, author of Mathematica's arbitrary-precision arithmetic, found that the switch-over happened at well under one hundred digits. An even faster algorithm for very large integers is based on Fourier transforms. See Section 16.11 (page 501).

- *Division* – Repeated subtraction takes exponential time, so the easiest reasonable algorithm to use is the long-division method you hated in school. This requires arbitrary-precision multiplication and subtraction as subroutines, as well as trial-and-error, to determine the correct digit at each position of the quotient.

  In fact, integer division can be reduced to integer multiplication, although in a non-trivial way. So if you are implementing asymptotically fast multiplication, you can reuse that effort in long division. See the references below for details.

- *Exponentiation* – We can evaluate a^n using $n-1$ multiplications, by computing $a \times a \times \ldots \times a$. However, a much better divide-and-conquer algorithm is based on the observation that $n = \lfloor n/2 \rfloor + \lceil n/2 \rceil$. If n is even, then $a^n = (a^{n/2})^2$. If n is odd, then $a^n = a(a^{\lfloor n/2 \rfloor})^2$. In either case, we have halved the size of our exponent at the cost of at most two multiplications, so $O(\lg n)$ multiplications suffice to compute the final value:

$$
\begin{aligned}
&\text{function power}(a, n) \\
&\quad \text{if } (n = 0) \text{ return}(1) \\
&\quad x = \text{power}(a, \lfloor n/2 \rfloor) \\
&\quad \text{if } (n \text{ is even}) \text{ then return}(x^2) \\
&\quad\quad \text{else return}(a \times x^2)
\end{aligned}
$$

High- but not arbitrary-precision arithmetic can be conveniently performed using the Chinese remainder theorem and modular arithmetic. The *Chinese remainder theorem* states that every integer between 1 and $P = \prod_{i=1}^{k} p_i$ is uniquely determined by its set of residues mod p_i, where each p_i, p_j are relatively prime integers. Addition, subtraction, and multiplication (but not division) can be supported using such residue systems, with the advantage that large integers can be manipulated without complicated data structures.

Many algorithms on long integers can be directly applied to computations on polynomials. A particularly useful algorithm is Horner's rule for fast polynomial evaluation. When $P(x) = \sum_{i=0}^{n} c_i \cdot x^i$ is blindly evaluated term by term, $O(n^2)$ multiplications will be performed. Much better is observing that $P(x) = c_0 + x(c_1 + x(c_2 + x(c_3 + \ldots)))$, the evaluation of which uses only a linear number of operations.

Implementations: Python plus commercial computer algebra systems like Maple and Mathematica incorporate high-precision arithmetic. This is your best option for a quick, non-embedded application if you have access. The rest of this section focuses on source code available for embedded applications.

The premier C/C++ library for fast, arbitrary-precision is the GNU Multiple Precision Arithmetic Library (GMP), which operates on signed integers, rational numbers, and floating point numbers. It is widely used and well supported, and available at `http://gmplib.org/`.

The `java.math BigInteger` class provides arbitrary-precision analogues to all of Java's primitive integer operators. `BigInteger` provides additional operations for modular arithmetic, GCD calculation, primality testing, prime generation, bit manipulation, and a few other miscellaneous operations.

A lower-performance, less well-tested but more personal implementation of high-precision arithmetic appears in the library from my book *Programming Challenges* [SR03]. See Section 22.1.9 (page 716) for details.

Several general systems for computational number theory are available. Each of these supports operations of arbitrary-precision integers. Information about the PARI and NTL number-theoretic libraries can be found in Section 16.8 (page 490).

Notes: Knuth [Knu97b] is the primary reference on algorithms for all basic arithmetic operations, including implementations of them in the MIX assembly language. Bach and Shallit [BS96] and Shoup [Sho09] provide more recent treatments of computational number theory. Brent and Zimmermann [BZ10] present a modern treatment of the venerable topic of computer arithmetic.

Expositions on the $O(n^{1.59})$-time divide-and-conquer algorithm for multiplication [KO63] include [AHU74, Man89]. An FFT-based algorithm multiplies two n-bit numbers in $O(n \lg n \lg \lg n)$ time and is due to Schönhage and Strassen [SS71]. Expositions include [AHU74, Knu97b]. This was finally improved to $O(n \lg n)$ by Harvey and Van der Hoven [HVDH19, HVDHL16] in 2019. It is remarkable that an asymptotically optimal algorithm for such a fundamental problem as integer multiplication took so long to be discovered. The reduction between integer division and multiplication is presented in [AHU74, Knu97b]. Applications of fast multiplication to other arithmetic operations are presented by Bernstein [Ber04]

Good expositions of algorithms for modular arithmetic and the Chinese remainder theorem include [AHU74, CLRS09]. A good exposition of circuit-level algorithms for elementary arithmetic algorithms is [CLRS09].

Euclid's algorithm for computing the greatest common divisor of two numbers is perhaps the oldest interesting algorithm. Expositions include [CLRS09, Knu97b].

Related problems: Factoring integers (see page 490), cryptography (see page 697).

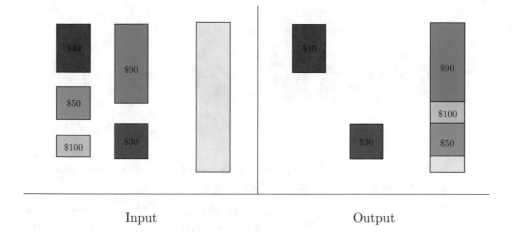

Input	Output

16.10 Knapsack Problem

Input description: A set of items $S = \{1, \ldots, n\}$, where item i has size s_i and value v_i. A knapsack capacity C.

Problem description: Find the subset S' of S that maximizes the value of $\sum_{i \in S'} v_i$, given $\sum_{i \in S'} s_i \leq C$; that is, all the items fit in a knapsack of size C.

Discussion: The knapsack problem arises in resource allocation with financial constraints. How do you select what things to buy given a fixed budget? Everything has a cost and value, so we seek the most value for the given cost. The *knapsack problem* should invoke the image of the backpacker who is constrained by a fixed-size knapsack, and so must fill it with the most useful and portable items.

The most common formulation is the *0/1 knapsack problem*, where each item must either be placed entirely in the knapsack or discarded. Objects cannot be broken up arbitrarily, so it is not fair taking one can of Coke from a six-pack or opening a can to take just a sip. It is this 0/1 property that makes the knapsack problem hard, for a simple greedy algorithm finds the optimal selection when we are allowed to subdivide objects. We compute the "price per pound" for each item, and repeatedly take the most expensive item or the biggest part thereof until the knapsack is full. Unfortunately, this 0/1 constraint is inherent in most applications.

Issues that arise in selecting the best algorithm include:

- *Does every item have the same cost, or perhaps the same size?* – When all items are worth exactly the same, we maximize our total value by taking the greatest number of items. Therefore, the optimal solution is to sort the items in order of increasing size and insert them into the knapsack in this order until no more fit. The problem is similarly solved when all objects

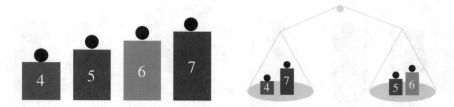

Figure 16.1: Integer partition is a special case of the knapsack problem.

have the same size. Sort by cost, and take the most valuable elements first. These are the easy cases of knapsack.

- *Do all items have the same "price per pound"?* – If so, our problem is equivalent to ignoring the price and just trying to minimize the amount of empty space left in the knapsack. Unfortunately, even this restricted version is NP-complete, so we cannot expect an efficient algorithm that always solves the problem. But don't lose hope, because knapsack proves to be an "easy" hard problem, one that can usually be handled with the algorithms described below.

An important special case of a constant "price-per-pound" knapsack is the *integer partition* problem, presented in cartoon form in Figure 16.1. Here, we seek to partition the elements of S into two sets A and B such that $\sum_{a \in A} a = \sum_{b \in B} b$, or more generally make the difference as small as possible. Integer partition can be thought of as bin packing into two equal-sized bins or knapsack with a capacity of half the total weight, so all three problems are closely related and NP-complete.

The constant "price-per-pound" knapsack problem is often called the *subset sum* problem, because we seek a subset of items that adds up to a specific target number C; that is, the capacity of our knapsack.

- *Are all the sizes relatively small integers?* – When the sizes of the items and the knapsack capacity C are all integers, there exists an efficient dynamic programming algorithm (presented in Section 10.5 (page 329)) that finds the optimal solution in time $O(nC)$ and $O(C)$ space. Whether this works for you depends upon how big C is. It is great for $C \leq 1,000$, but not so great for $C \geq 1,000,000,000$.

The algorithm works as follows: Let S' be a set of items, and let $C[i, S']$ be true if and only if there is a subset of S' whose size adds up exactly to i. Thus, $C[i, \emptyset]$ is false for all $1 \leq i \leq C$. One by one we add a new item s_j to S' and update the affected values of $C[i, S']$. Observe that

$$C[i, S' \cup s_j] = \text{true iff } (C[i, S'] = \text{true }) \text{ or } (C[i - s_j, S'] = true)$$

because we either use s_j in realizing the sum or we don't. We identify all sums that can be realized by performing n sweeps through all C elements— one for each s_j, $1 \leq j \leq n$—and so updating the array. The knapsack solution is given by the index of the true element of largest realizable size. To reconstruct this winning subset, we must also store the name of the item number that turned $C[i]$ from false to true for each $1 \leq i \leq C$ and then scan backwards through the array.

This dynamic programming formulation ignores the values of the items. To generalize the algorithm, now each element of the array stores the value of the best subset to date summing up to i. We update when the sum of the cost of $C[i - s_j, S']$ plus the cost of s_j is better than the previous cost of $C[i, S' \cup s_j]$.

- *What if I have multiple knapsacks?* – When there are multiple knapsacks, your problem might be better thought of as a bin-packing problem. Section 20.9 (page 652) discusses bin-packing/cutting-stock algorithms. That said, algorithms for optimizing over multiple knapsacks are provided in the Implementations section below.

Exact solutions for large capacity knapsacks can be found using integer programming or backtracking. A 0/1 integer variable x_i is used to denote whether item i is present in the optimal subset. We maximize $\sum_{i=1}^{n} x_i \cdot v_i$ given the constraint that $\sum_{i=1}^{n} x_i \cdot s_i \leq C$. Integer programming codes are discussed in Section 16.6 (page 482).

Heuristics must be used when exact solutions prove too costly to compute. The simple greedy heuristic inserts items according to the maximum "price per pound" rule described previously. Often this heuristic solution is close to optimal, but it might prove arbitrarily bad depending upon the problem instance. The "price per pound" rule can also be used to reduce the problem size in exhaustive search-based algorithms by eliminating "cheap but heavy" objects from future consideration.

Another heuristic is based on *scaling*. Dynamic programming works well if the knapsack capacity is a reasonably small integer, say $\leq C_s$. But what if we have a problem with capacity $C > C_s$? We can scale down the sizes of all items by a factor of C/C_s, round the size down to the nearest integer, and then use dynamic programming on the scaled items. Scaling works well in practice, especially when the range of item sizes is not too large.

Implementations: Martello and Toth's collection of Fortran implementations of algorithms for a variety of knapsack problem variants are available at `http://www.or.deis.unibo.it/kp.html`. An electronic copy of the associated book [MT90a] has also been generously made available.

David Pisinger maintains a well-organized collection of C-language codes for knapsack problems and related variants like bin packing and container loading. These are available at `http://www.diku.dk/~pisinger/codes.html`. The strongest code is based on the dynamic programming algorithm of [MPT99].

Algorithm 632 [MT85] of the *Collected Algorithms of the ACM* is a Fortran code for the 0/1 knapsack problem, with the twist that it supports multiple knapsacks. See Section 22.1.4 (page 714).

Notes: Keller, Pferschy, and Pisinger [KPP04] is the most current reference on the knapsack problem and variants. Martello and Toth's book [MT90a] and survey article [MT87] are standard references on the knapsack problem, including both theoretical and experimental results. An excellent exposition on integer programming approaches to knapsack problems appears in [MPT99]. See [MPT00] for a computational study of algorithms for 0/1 knapsack problems.

A polynomial-time approximation scheme (PTAS) is an algorithm that approximates the optimal solution of a problem in time polynomial in both its size and the approximation factor ϵ. This very strong condition implies a smooth tradeoff between running time and approximation quality. Good expositions on polynomial-time approximation schemes [IK75] for knapsack and subset sum includes [BvG99, CLRS09]. Polynomial-time approximation schemes exist even for the case of multiple knapsacks [CK05].

An interesting special variant of the knapsack problem is the *3-sum* problem, where we are given three sets A, B, and C, each with n integers. We seek an $a \in A$, $b \in B$, and $c \in C$ such that $a + b = c$. The best algorithm known for 3-sum is $O(n^2)$, and reductions to 3-sum are often used to suggest that no sub-quadratic algorithm exists for a given problem [GO95].

Vector bin packing is a generalization of knapsack where the knapsack has capacity constraints along d axes (say CPU and memory limits), and each object is defined by a vector of d corresponding demands. Heuristics have been studied for vector bin packing in the context of virtual machine placement [PTUW11].

The first algorithm for generalized public key encryption by Merkle and Hellman [MH78] was based on the hardness of the knapsack problem. See [Sch15] for an exposition.

Related problems: Bin packing (see page 652), integer programming (see page 482).

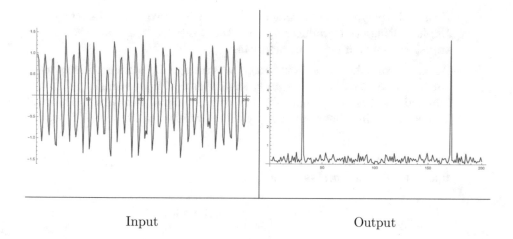

Input   Output

16.11 Discrete Fourier Transform

Input description: A sequence of n real or complex values h_i, $0 \leq i \leq n - 1$, sampled at uniform intervals from a function h.

Problem description: The discrete Fourier transform $H_m = \sum_{k=0}^{n-1} h_k e^{-2\pi i k m/n}$ for $0 \leq m \leq n - 1$.

Discussion: Although computer scientists tend to be fairly ignorant about Fourier transforms, scientists and engineers eat them for breakfast. Functionally, Fourier transforms provide a way to convert samples of a standard time-series into the *frequency domain*. This provides a dual representation of the function, in which certain operations become easier than the original time domain. Applications of Fourier transforms include:

- *Filtering* – Taking the Fourier transform of a function is equivalent to representing it as the sum of sine functions. We can filter an image to remove noise and other artifacts by dropping some of the sine functions, say eliminating undesirable high- and/or low-frequency components. Taking an inverse Fourier transform then gets us back into the time domain. For example, the sharp spike in the plot above represents the period of the single sine function that closely models the input data. The rest is noise.

- *Image compression* – A smoothed, filtered image contains less information than the original, while retaining a similar appearance. By eliminating the coefficients of sine functions that contribute relatively little to the image, we reduce the size of the image at little cost in image fidelity.

- *Convolution and deconvolution* – Fourier transforms can efficiently compute convolutions of two sequences. A *convolution* is the pairwise product of elements from two different sequences, such as in multiplying two

n-variable polynomials f and g or comparing two character strings. Implementing such products directly takes $O(n^2)$, while the fast Fourier transform led to a $O(n \lg n)$ algorithm.

Another example comes from image processing. Because a scanner measures the intensity of an image patch instead of just a single point, the scanned input is always blurred. The original signal can be reconstructed by deconvoluting the input signal with a Gaussian point-spread function.

- *Computing the correlation of functions* – The *correlation function* of two functions $f(t)$ and $g(t)$ is defined by

$$z(t) = \int_{-\infty}^{\infty} f(\tau)g(t+\tau)d\tau$$

  and can be easily computed using Fourier transforms. When functions $f(t)$ and $g(t)$ are similar in shape but one is shifted relative to the other (such as $f(t) = \sin(t)$ and $g(t) = \cos(t)$), the value of $z(t_0)$ will be large at this shift offset t_0. As an application, consider the task of detecting periodicities in our random-number generator. We can generate a large series of random numbers, turn them into a time series (the ith number at time i), and compute the correlation function of this series. Any large spikes will correspond to potential periodicities.

The discrete Fourier transform takes as input n complex numbers h_k, $0 \leq k \leq n - 1$, corresponding to equally spaced points in a time series, and outputs n complex numbers H_k, $0 \leq k \leq n - 1$, each describing a sine function of given frequency. The discrete Fourier transform is defined by

$$H_m = \sum_{k=0}^{n-1} h_k \cdot e^{-2\pi i k m / n} = \sum_{k=0}^{n-1} h_k \left[\cos\left(\frac{2\pi k m}{n}\right) - i \sin\left(\frac{2\pi k m}{n}\right) \right]$$

and the inverse Fourier transform is defined by

$$h_m = \frac{1}{n} \sum_{k=0}^{n-1} H_k \cdot e^{2\pi i k m / n} = \frac{1}{n} \sum_{k=0}^{n-1} H_k \left[\cos\left(\frac{2\pi k m}{n}\right) + i \sin\left(\frac{2\pi k m}{n}\right) \right]$$

which enables us to move easily between h and H.

Since the output of the discrete Fourier transform consists of n numbers, each of which is computed using a formula on n numbers, they can be computed in $O(n^2)$ time. The fast Fourier transform (FFT) is an algorithm that computes the discrete Fourier transform in $O(n \log n)$. This is arguably the most important algorithm known, for it opened the door to modern signal processing. Several different algorithms call themselves FFTs, each based on a divide-and-conquer approach. Essentially, the problem of computing the discrete Fourier transform on n points is reduced to computing two transforms on $n/2$ points each, and then applied recursively.

The FFT usually assumes that n is a power of two. If this is not the case, you are usually better off padding your data with zeros to create $n = 2^k$ elements rather than hunting for a more general code.

Many signal-processing systems have strong real-time constraints, so FFTs are often implemented in hardware, or at least in assembly language tuned to the particular machine. Keep this in mind if the codes below prove too slow.

Implementations: FFTW is a C subroutine library for computing the discrete Fourier transform in one or more dimensions, with arbitrary input size, and supporting both real and complex data. It is the clear choice among freely available FFT codes. Extensive benchmarking proves that FFTW is indeed the "Fastest Fourier Transform in the West." FFTW received the 1999 J. H. Wilkinson Prize for Numerical Software. See http://www.fftw.org/.

FFTPACK is a package of Fortran subprograms for the fast Fourier transform of periodic and other symmetric sequences, written by P. Swartzrauber. It includes complex, real, sine, cosine, and quarter-wave transforms. FFTPACK resides at http://www.netlib.org/fftpack. The GNU Scientific Library for C/C++ provides a re-implementation of FFTPACK. See http://www.gnu.org/software/gsl/.

Notes: Bracewell [Bra99] and Brigham [Bri88] are excellent introductions to Fourier transforms and the FFT. See also the exposition in [PFTV07]. Credit for inventing the fast Fourier transform is usually given to Cooley and Tukey [CT65], but see [Bri88] for a complete history.

A cache-oblivious algorithm for the fast Fourier transform is given in [FLPR99]. This paper first introduced the notion of cache-oblivious algorithms. The FFTW is based on this algorithm. See [FJ05] for more on the design of the FFTW. Faster algorithms are known for the case where we seek only the k largest coefficients of the transform, when $k \ll n$ [HIKP12].

An interesting divide-and-conquer algorithm for polynomial multiplication [KO63] does the job in $O(n^{1.59})$ time and is discussed in [AHU74, Man89]. An FFT-based algorithm that multiplies two n-bit numbers in $O(n \lg n \lg \lg n)$ time is due to Schönhage and Strassen [SS71].

The *quantum Fourier transform* provides an exponential speedup over the classical FFT, working on 2^n amplitudes stored in n qubits using only $O(n^2)$ operations [NC02]. The catch is efficiently getting your desired amplitudes into (or out of) the qubits. This operation is an essential component of Shor's quantum factoring algorithm [Sho99]. For intuition why, consider a set of sine functions with periods $2, 3, 5, 7, 11, \ldots$ in the frequency domain. If n is divisible by any of these, then n must have a peak in the time domain.

It is an open question of whether complex variables are really fundamental to fast algorithms for convolution. Fortunately, fast convolution can generally be used as a black box in applications. Many variants of string matching are based on fast convolution [Ind98].

In recent years, wavelets have been proposed to replace Fourier transforms in filtering. See [BN09] for an introduction to wavelets.

Related problems: Data compression (see page 693), high-precision arithmetic (see page 493).

Chapter 17

Combinatorial Problems

We will now consider several algorithmic problems of a purely combinatorial nature. These include sorting and searching, both among the first non-numerical problems arising on electronic computers. Sorting can be viewed as identifying or imposing a total order on the keys, while searching and selection involve identifying specific keys based on their position in this total order.

The rest of this section deals with combinatorial objects, such as permutations, partitions, subsets, calendars, and schedules. We are particularly interested in algorithms that *rank* and *unrank* combinatorial objects—mapping each distinct object to/from a unique integer. Rank/unrank operations make many other tasks simple, such as generating random objects (pick a random number and unrank) or listing all objects in order (iterate from 1 to n and unrank).

I conclude with the problem of generating graphs. Graph algorithms are more fully presented in subsequent sections of the catalog.

Books on general combinatorial algorithms, in this restricted sense, include:

- *Knuth* – The standard reference on sorting and searching [Knu98]. New material on the generation of permutations, subsets, partitions, and trees comprises the first part of his mythical Volume 4 [Knu11].

- *Ruskey* [Rus03] – He never officially completed it, but this manuscript is a standard reference on generating combinatorial objects. Previews are available if you Google *Ruskey Combinatorial Generation*.

- *Kreher and Stinson* [KS99] – This book on combinatorial generation algorithms, with additional particular focus on algebraic problems such as isomorphism and dealing with symmetry.

- *Pemmaraju and Skiena* [PS03] – This description of *Combinatorica*, a library of over 400 Mathematica functions for generating combinatorial objects and graph theory (see Section 22.1.8), provides a distinctive view of how different algorithms can fit together. Its second author is well qualified to write a manual on algorithm design.

© The Editor(s) (if applicable) and The Author(s), under exclusive license to
Springer Nature Switzerland AG 2020
S. S. Skiena, *The Algorithm Design Manual*, Texts in Computer Science,
https://doi.org/10.1007/978-3-030-54256-6_17

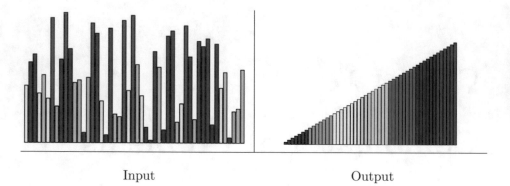

Input   Output

17.1   Sorting

Input description: A set of n items.

Problem description: Arrange the items in increasing order.

Discussion: Sorting is the most fundamental algorithmic problem in computer science. Learning the different sorting algorithms is like learning scales for a musician. Sorting proves to be the first step in solving a host of other algorithm problems, as discussed in Section 4.2 (page 113). Indeed, "when in doubt, sort" is one of the first rules of algorithm design.

Sorting also illustrates most of the standard paradigms of algorithm design. Programmers are generally familiar with several different sorting algorithms, which sows confusion as to which should be used for a given application. The following criteria can help you decide:

- *How many keys will you be sorting?* – For small amounts of data (say $n \leq 100$), it doesn't really matter which quadratic-time algorithm you use. Insertion sort is faster, simpler, and less likely to be buggy than bubblesort. Shellsort is closely related to, but much faster than, insertion sort, although it requires looking up the right insert sequences in Knuth [Knu98].

  But if you have more than one hundred items to sort, it becomes important to use an $O(n \lg n)$-time algorithm like heapsort, quicksort, or mergesort. Partisans may favor one of these algorithms over the others, but it usually doesn't really matter.

  Once you get past (say) 100,000,000 items, it is important to start thinking about external-memory sorting algorithms that minimize disk or network access. Both types of algorithm are discussed below.

- *Will there be duplicate keys?* – The sorted order is completely defined if all items have distinct keys. However, when two items share the same key, something else must determine which one comes first. Sometimes this matters. Ties are often broken by sorting on a secondary key, like the first name or initial when the family names collide.

Ties sometimes get broken by their initial position in the data set. Suppose the 5th and 27th items of the initial data set share the same key. This means the 5th item must appear before the 27th in the final order. A *stable* sorting algorithm preserves the original ordering in case of ties. Most of the quadratic-time sorting algorithms are stable, while many of the $O(n \lg n)$ algorithms are not. If it is important that your sort be stable, it is probably best to explicitly use the initial position as a secondary key in your comparison function instead of trusting the stability of your implementation.

- *What do you know about your data?* – Perhaps you can exploit special knowledge about your data to get it sorted faster or more easily. General sorting is a fast $O(n \lg n)$ operation, so if the time spent sorting *really* is the bottleneck in your application, you are indeed a fortunate person.

  - *Has the data already been partially sorted?* If so, certain algorithms like insertion sort perform better than they otherwise would.

  - *Do you know the distribution of the keys?* If the keys are randomly or uniformly distributed, a *bucket* or *distribution sort* makes sense. Throw the keys into bins based on their first letter, and recur until each bin is small enough to sort by brute force. This is very efficient provided the keys are evenly distributed in key space. Note that bucket sort would perform very badly sorting names on the membership roster of the "Smith Society."

  - *Are your keys very long or hard to compare?* When sorting long text strings, it might pay to use a relatively short prefix (say ten characters) of each key for an initial sort, and then resolve ties using the full key. This is particularly important in external sorting (see below), since you don't want to waste fast memory on the dead weight of irrelevant detail.

    Another idea might be to use radix sort. This always takes time linear in the number of characters in the file, instead of $O(n \lg n)$ times the cost of comparing two keys.

  - *Is the range of possible keys very small?* If you want to sort a subset of (say) $n/2$ distinct integers, each with a value from 1 to n, the fastest algorithm would be to initialize an n-element bit vector, turn on the bits corresponding to keys, then scan from left to right and report the positions with true bits.

- *Should I worry about disk accesses?* – In massive sorting problems, it is not possible to keep all data in memory simultaneously. This problem is called *external sorting*, because the data is maintained on an external storage device. Traditionally, this meant tape drives, and Knuth [Knu98] describes a variety of intricate algorithms for efficiently merging data from different tapes. Today, it usually means virtual memory. Any sorting algorithm will run using virtual memory, but clumsy ones will spend most of their time swapping.

The simplest approach to external sorting loads the data into a B-tree, and then does an in-order traversal of the tree to read the keys off in sorted order. The highest performance sorting programs are based on multiway-mergesort. Files containing portions of the data are sorted using a fast internal sort, and then these sorted runs are merged in stages using 2- or k-way merging. Complicated merging patterns and buffer management based on the properties of the external storage device can be used to optimize performance.

The best general-purpose internal sorting algorithm is quicksort (see Section 4.2 (page 113)), although it requires tuning effort to achieve maximum performance. Indeed, you are much better off using a library function instead of coding it yourself. A poorly written quicksort will likely run more slowly than a poorly written heapsort. If you are determined to implement your own quicksort, use the following heuristics, which make a big difference in practice:

- *Use randomization* – By randomly permuting (see Section 17.4 (page 517)) the keys before sorting, you can eliminate the potential embarrassment of quadratic-time behavior on nearly sorted data.

- *Median of three* – For your pivot element, use the median of the first, last, and middle elements of the array to increase the likelihood of partitioning the array into roughly equal pieces. Experiments suggest using a larger sample on big subarrays and a smaller sample on small ones.

- *Leave small subarrays for insertion sort* – Terminating the quicksort recursion and switching to insertion sort makes sense when the subarrays get small, say fewer than 20 elements. You should experiment to identify the best switch point for your implementation.

- *Do the smaller partition first* – You can minimize run-time memory by processing the smaller partition before the larger one. Since each successive stored call is at most half as large as the previous one, only $O(\lg n)$ stack space is needed.

Before you get started, see Bentley's article on building a faster quicksort [Ben92b].

Implementations: The best freely available sort program is presumably GNU sort, part of the GNU core utilities library. See http://www.gnu.org/software/coreutils/. There are also commercial vendors of high-performance external sorting programs, including Cosort (www.iri.com), Syncsort (www.syncsort.com), and Ordinal Technology (www.ordinal.com).

Modern programming languages provide libraries so you should never need to implement your own sort routine. The C standard library contains `qsort`, a generic implementation of (presumably) quicksort. The C++ *Standard Template Library* (STL) provides both `sort` and `stable_sort` methods. See Josuttis [Jos12] and Meyers [Mey01] for more detailed guides to using STL and the C++ standard library. *Java Collections* (JC), a small library of data structures, is included in the `java.util` package of Java SE. In particular, `SortedMap` and `SortedSet` classes are provided.

High-performance sorting systems distribute the work over many machines. Map-Reduce systems like Hadoop [Whi12] make it relatively easy to implement parallel bucketsort, such as the terabyte champion system reported in [O08]. Efficient sorting algorithms for GPUs are considered by Satish et.al [SHG09].

Numerous websites provide animations of all the basic sorting algorithms, many quite interesting to watch. Indeed, sorting is the canonical problem for algorithm animation. Search on Google and YouTube for "sorting animations" and watch the most popular ones.

Notes: Knuth [Knu98] is the best book that will ever be written on sorting. It is now almost fifty years old, but remains fascinating reading. One area that has developed since Knuth is sorting under presortedness measures, surveyed in [ECW92]. Timsort, a popular candidate for the fastest sorting algorithm in practice, exploits naturally ordered sequences to take $O(n \log p)$ time, where p is the number of sorted runs discovered in the input data [AJNP18].

Expositions on the basic internal sorting algorithms appear in every algorithms textbook. Heapsort was first invented by Williams [Wil64]. Quicksort was invented by Hoare [Hoa62], with careful analysis and implementation by Sedgewick [Sed78]. Von Neumann is credited with having produced the first implementation of mergesort on the EDVAC in 1945. See Knuth [Knu98] for a full discussion of the history of sorting, dating back to the days of punch-card tabulating machines.

The primary competitive forum for high-performance sorting is an annual competition initiated by the late Jim Gray. See `http://sortbenchmark.org/` for current and previous results, which are either inspiring or depressing depending upon how you look at it. The magnitude of progress is inspiring (the million-record instances of the original benchmarks are now too small to bother with) but it is depressing (to me) that systems/memory management issues thoroughly trump the combinatorial/algorithmic aspects of sorting. Modern attempts to engineer high-performance sort programs include work on both cache-conscious [LL99] and cache-oblivious [BFV07] sorting.

Sorting has a well-known $\Omega(n \lg n)$ lower bound under the algebraic decision tree model [BO83]. Determining the exact number of comparisons required for sorting n elements, for small values of n, has generated considerable study. See [Aig88, Raw92] for expositions and Peczarski [Pec04, Pec07] for the latest results.

This lower-bound does not hold under different models of computation. Fredman and Willard [FW93] present an $O(n\sqrt{\lg n})$ algorithm for sorting under a model of computation that permits arithmetic operations on keys. Andersson [And05] surveys algorithms for fast sorting on such non-standard models of computation.

Related problems: Dictionaries (see page 440), searching (see page 510), topological sorting (see page 546).

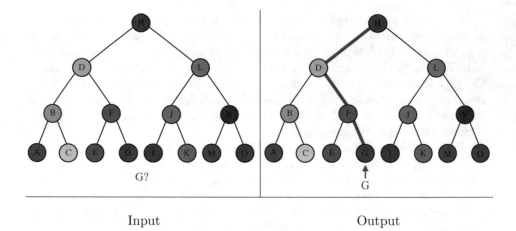

Input   Output

17.2   Searching

Input description: A set of n keys S, and a query key q.

Problem description: Where is q in S?

Discussion: "Searching" is a word that means different things to different people. Searching for the global maximum or minimum of a function is the problem of *unconstrained optimization*, discussed in Section 16.5 (page 478). Chess-playing programs select the best move to play through an exhaustive search of possible candidates using a variation of backtracking (see Section 9.1 (page 281)).

But here we consider the task of searching for a key in a list, array, or tree. Dictionary data structures maintain efficient access to sets of keys under insertion and deletion, and are discussed in Section 15.1 (page 440). Typical dictionaries include binary search trees and hash tables.

I treat searching as a problem distinct from dictionaries because simpler and more efficient solutions can emerge when our primary interest is static searching without insertion/deletion. These little data structures can yield substantial performance improvements when properly employed in an innermost loop. Also, ideas such as binary search and self-organization apply to other problems and well justify our attention.

The two basic approaches we consider are sequential search and binary search. Both are simple, yet have interesting and subtle variations. In *sequential search*, we start from the front of our list/array of keys and compare each successive item against the search key q until we find a match, or reach the end. In *binary search*, we start with a sorted array of keys. To search for q, we compare it to the middle key $S_{n/2}$. If q is before $S_{n/2}$, it must reside in the top half of our set; if not, it must reside in the bottom half of our set. By repeating this process on the correct half, we find q using $\lceil \lg n \rceil$ comparisons. This is a big win over the $n/2$ comparisons we expect with sequential search. See Section 5.1 (page 148) for more on binary search.

A sequential search is the simplest algorithm, and likely to be fastest on up to about twenty elements. Beyond (say) one hundred elements, binary search will clearly be more efficient than sequential search, easily justifying the cost of sorting if there will be multiple queries. But other issues come into play in identifying the proper variant of the algorithm:

- *How much time can you spend programming?* – A binary search is a notoriously tricky algorithm to program correctly. It took seventeen years after its invention until the first *correct* version of a binary search was published! Don't be afraid to start from an implementation described below. Test it completely by writing a driver that searches for every key in the set S as well as between the keys.

- *Are certain items accessed more often than other ones?* – Certain English words (such as "the") are much more likely to occur than others (such as "defenestrate"). We can reduce the number of comparisons in a sequential search by putting the most popular words on the top of the list and the least popular ones at the bottom. Non-uniform access is usually the rule, not the exception. Many real-world distributions are governed by *power laws*. A classic example is word use in English, which is fairly accurately modeled by *Zipf's law*. Under Zipf's law, the ith most frequently accessed key is selected with probability $(i-1)/i$ times the probability of the $(i-1)$st most popular key, for all $1 \leq i \leq n$.

  Knowledge of access frequencies is easy to exploit with sequential search. But the issue is more complicated with binary trees. We want popular keys close to the root (so we hit them quickly) but not at the expense of losing balance and degenerating into sequential search. The answer is to employ a dynamic programming algorithm that builds the *optimal binary search tree*. The critical observation is that every possible root node i partitions the space of keys into those to the left of i and those to the right; each of which should be represented by an optimal binary search tree on a smaller subrange of keys. The root of the optimal tree is selected to minimize the expected search costs of the resulting partition.

- *Might access frequencies change over time?* – Reordering a list or tree to exploit a skewed access pattern requires knowing the access pattern in advance. For many applications, it can be difficult to obtain this information. Better are *self-organizing lists*, where the order of the keys changes in response to the queries. The best self-organizing scheme is move-to-front; that is, we move the most recently searched-for key from its current position to the front of the list. Popular keys keep getting boosted to the front, while unsearched-for keys drift towards the back of the list. There is no need to keep track of the frequency of access; just move the keys on demand. Self-organizing lists also exploit *locality of reference*, since accesses to any given key are likely to occur in clusters. A hot key will be maintained near the top of the list during its cluster of accesses, even if other keys have proven more popular in the past.

Self-organization can extend the useful size range of sequential search, although you should switch to binary search beyond one hundred elements. But consider using *splay trees*, which are self-organizing binary search trees that rotate each searched-for node to the root. They offer excellent amortized performance guarantees.

- *Is the key close by?* – Suppose we know that the target key is to the right of position p, and we think it is nearby. A sequential search is fast if we are correct, but we will be punished severely should we guess wrong. A better idea is to test repeatedly at larger intervals ($p + 1$, $p + 2$, $p + 4$, $p + 8$, $p + 16$, ...) to the right until we find a key to the right of our target. This defines a window containing the target, so now we can proceed with a conventional binary search.

  Such a *one-sided binary search* finds the target at position $p + l$ using at most $2\lceil \lg l \rceil$ comparisons, so it is faster than binary search when $l \ll n$, yet it can never be much worse. One-sided binary search is particularly useful in unbounded search problems, such as in numerical root finding.

- *Is my data structure sitting on external memory?* – Once the number of keys grows *too* large, binary search loses its status as the best search technique. Such a search jumps wildly around the set of keys looking for midpoints to compare, so each comparison requires reading in a new page from external memory. Much better are data structures such as B-trees (see Section 15.1 (page 440)) or van Emde Boas trees (see the Notes section), which cluster the keys into pages to minimize the number of disk accesses per search.

- *Can I guess where the key should be?* – In *interpolation search*, we exploit our understanding of the distribution of keys to guess where to look next. Interpolation search is probably a more accurate description of how we use a telephone book than binary search. Suppose we search for *Washington, George* in a sorted telephone book. We feel safe making our first comparison three-fourths of the way down the list, essentially doing two comparisons for the price of one.

  Although an interpolation search is an appealing idea, I caution against it for three reasons: First, you must work very hard to optimize your search algorithm before you can hope for a speedup over binary search. Second, even if you do beat a binary search, it is unlikely to be by enough to have justified the exercise. Finally, your program will be much less robust and efficient when the distribution changes, such as when your application gets ported to work on French text instead of English.

Implementations: The basic sequential and binary search algorithms are simple enough that you should consider implementing them yourself. That said, the C standard library contains `bsearch`, a generic implementation of (presumably) binary search. The C++ *Standard Template Library* (STL) provides `find` (sequential search) and `binary_search` iterators. *Java Collections* (JC), provides `binarySearch` in the `java.util` package of Java standard edition.

Many data structure textbooks provide extensive and illustrative implementations. Sedgewick (`https://algs4.cs.princeton.edu/code/`) [SW11] and Weiss (`http://www.cs.fiu.edu/~weiss/`) [Wei11] provide implementations of splay trees and other search structures in both C++ and Java.

Notes: *The Handbook of Data Structures and Applications* [MS18] provides up-to-date surveys on all aspects of dictionary data structures. Other surveys include Mehlhorn and Tsakalidis [MT90b] and Gonnet and Baeza-Yates [GBY91]. Knuth [Knu97a] provides a detailed analysis and exposition on all fundamental search algorithms and dictionary data structures, but omits such modern data structures as red–black and splay trees.

The next position probed in linear interpolation search on an array of sorted numbers is given by

$$next = (low - 1) + \lceil \frac{q - S[low - 1]}{S[high + 1] - S[low - 1]} \times (high - low + 1) \rceil$$

where q is the query numerical key and S the sorted numerical array. If the keys are drawn independently from a uniform distribution, the expected search time is $O(\lg \lg n)$ [DJP04, PIA78]. But in practice, they won't be.

Non-uniform access patterns can be exploited in binary search trees by structuring them so that popular keys are located near the root, thus minimizing search time. Dynamic programming can be used to construct such optimal search trees in $O(n \lg n)$ time [Knu98]. Stout and Warren [SW86] provide a slick algorithm to efficiently transform a binary tree to a minimum height (optimally balanced) tree using rotations.

The van Emde Boas layout of a binary tree (or sorted array) offers better external memory performance than conventional binary search, at a cost of greater implementation complexity. See the survey of Arge et al. [ABF05] for more on this and other cache-oblivious data structures.

Grover's algorithm permits searching in an unsorted database in $O(\sqrt{n})$ time on a quantum computer [NC02]. The basic intuition is that quantum computers work on probabilities, initially uniform among n items in superposition. But a single amplitude amplification operation can increase the probability of a target element by a factor of $\sqrt{n}$. After $O(\sqrt{n})$ such amplifications, we become highly likely to sample the target element.

Related problems: Dictionaries (see page 440), sorting (see page 506).

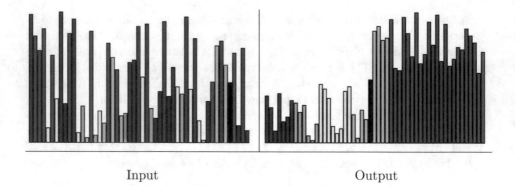

Input   Output

17.3   Median and Selection

Input description: A set of n numbers or keys, and an integer k.

Problem description: Find the key greater than or equal to exactly k of the n keys.

Discussion: Median finding is an essential problem in statistics, where it provides a more robust notion of average than the *mean*. The mean wealth of people who have published research papers on sorting is significantly influenced by the presence of one William Gates [GP79], although his effect on the *median* wealth is merely to cancel out one starving graduate student.

Median finding is a special case of the more general *selection* problem, which seeks the kth element in sorted order. Selection arises in several applications:

- *Filtering outlying elements* – When dealing with noisy data, it is often a good idea to throw out the 10% largest and smallest values. Selection can be used to identify the items defining the 10th and 90th percentiles. The outliers are then filtered away by comparing each item to the selected bounds.

- *Identifying the most promising candidates* – A computer chess program might quickly assess all possible next moves, and then evaluate the top 25% more carefully. This is again selection followed by filtering.

- *Deciles and related divisions* – A useful way to present income distribution in a population charts the salary of the people ranked at regular intervals, say exactly at the 10th percentile, 20th percentile, and so on. Computing these values is simply selection on the appropriate position ranks.

- *Order statistics* – Particularly interesting special cases of selection include finding the smallest element ($k = 1$), the largest element ($k = n$), and the median element ($k = n/2$).

The mean of n numbers can be computed in linear time by summing the elements and dividing by n. But finding the median is a more difficult problem. Algorithms that compute the median can readily be generalized to arbitrary selection. Issues in median finding and selection include:

- *How fast does it have to be?* – The most elementary median-finding algorithm sorts the items in $O(n \lg n)$ time and then returns the item occupying the $(n/2)$nd position. As a plus, you get much more information than just the median, enabling selection for any kth element ($1 \leq k \leq n$) in constant time after the sort. However, there are faster algorithms if all you want is the median.

  In particular, *quick-select* is an $O(n)$ *expected*-time algorithm based on quicksort. Select a random element from the data set as a pivot, and use it to partition the data into sets less than and greater than the pivot. From the sizes of these sets, we know the position of the pivot in the total order, and hence whether the median lies to the left or right of this point. Now we recur on the appropriate subset until it converges on the median. This takes (on average) $O(\lg n)$ iterations, with the cost of each iteration being roughly half that of the previous one. This defines a geometric series that converges to a linear-time algorithm, although if you are very unlucky it takes $\Theta(n^2)$, the same worst-case time as quicksort.

  More complicated algorithms can find the median in worst-case linear time. However, this expected-time algorithm will likely win in practice.

- *What if you only get to see each element once?* – Selection and median finding become expensive on large datasets because they require several passes through external memory. In data-streaming applications, the volume of data is often too large to store, making repeated consideration (and thus exact median finding) impossible. Much better is computing a small summary of the data for future analysis, say approximate deciles of frequency moments (where the kth moment of stream x is defined as $F_k = \sum_i x_i^k$).

  One solution to such a problem is random sampling. Flip a coin for each value to decide whether to save it, with the probability of heads set low enough that you won't overflow your buffer. Likely the median of your samples will be close to that of the underlying data set. Alternatively, you can devote some fraction of memory to retaining (say) decile values of large blocks, and then combine these decile distributions to yield more refined decile bounds.

- *How fast can you find the mode?* – Beyond mean and median lies a third notion of average. The *mode* is defined to be the element that occurs with highest frequency in the data set. The best way to compute the mode sorts the set in $O(n \log n)$ time, which creates runs of identical elements. By doing a linear sweep from left to right on this sorted set, we can count the length of the longest run and hence compute the mode in a total of $O(n \log n)$ time.

The mode can also be found in *expected* linear time using hashing, but no such *worst-case* algorithm is possible. Testing whether there exist two identical elements in a set (a problem called element uniqueness) has an $\Omega(n \log n)$ lower bound. Element uniqueness is equivalent to asking whether the mode occurs more than once. Possibilities exist for improvement, at least theoretically, when the mode is large using fast median computations.

Implementations: The C++ *Standard Template Library* (STL) provides a general selection method (`nth_element`) implemented using the linear expected-time algorithm. See Josuttis [Jos12], Meyers [Mey01], and Musser [MDS01] for more detailed guides to using STL and the C++ standard library.

Notes: The linear expected-time algorithm for median and selection is due to Hoare [Hoa61]. Floyd and Rivest [FR75] provide an algorithm that uses fewer comparisons on average. Good expositions on linear-time selection include [BvG99, CLRS09, Raw92], with [Raw92] being particularly enlightening.

Streaming algorithms have extensive applications to large data sets, and are well surveyed by Muthukrishnan [Mut05] and Cormode [CH09].

A sport of considerable theoretical interest is determining *exactly* how many comparisons are sufficient to find the median of n items. The linear-time algorithm of Blum et al. [BFP+72] proves that $c \cdot n$ comparisons suffice, but we want to know what c is. Dor and Zwick [DZ99] proved that $2.95n$ comparisons suffice to find the median. These algorithms attempt to minimize the number of element comparisons but not the total number of operations, and hence do not lead to faster algorithms in practice. They also hold the current best lower bound of $(2 + \epsilon)$ comparisons for median finding [DZ01].

Tight combinatorial bounds for selection problems are presented in Aigner [Aig88]. An optimal algorithm for computing the mode is given by [DM80].

Related problems: Priority queues (see page 445), sorting (see page 506).

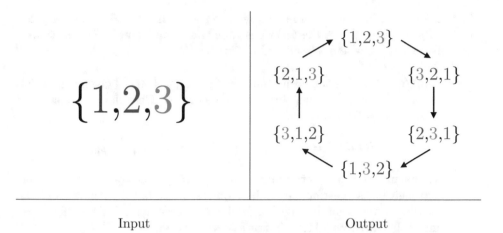

Input   Output

17.4 Generating Permutations

Input description: An integer n.

Problem description: Generate (1) all, or (2) a random, or (3) the next permutation of length n.

Discussion: A permutation describes an arrangement or ordering of items. Many algorithmic problems seek the best way to order a set of objects, including *traveling salesman* (the least-cost order to visit n cities), *bandwidth* (order the vertices of a graph on a line so as to minimize the length of the longest edge), and *graph isomorphism* (order the vertices of one graph so that it is identical to another). Any algorithm that solves such a problem exactly must construct a series of permutations along the way.

There are $n!$ permutations of n items. This grows so quickly that you can't generate all permutations for $n > 15$, because $15! = 1,307,674,368,000$. Numbers like these should cool the ardor of anyone excited by exhaustive search, and help explain the importance of generating random permutations.

Fundamental to any permutation-generation algorithm is the notion of order, the sequence in which the permutations are constructed from first to last. The most natural generation order is *lexicographic*, the sequence they would appear if they were sorted numerically. Lexicographic order for $n = 3$ is $\{1,2,3\}$, $\{1,3,2\}$, $\{2,1,3\}$, $\{2,3,1\}$, $\{3,1,2\}$, and finally $\{3,2,1\}$. Although lexicographic order is aesthetically pleasing, there is often no particular advantage to using it. For example, when searching through a collection of files, it does not matter whether the filenames are encountered in sorted order, so long as you eventually search through all of them. Indeed, non-lexicographic orders lead to faster and simpler permutation generation algorithms.

There are two different paradigms for constructing permutations: ranking/unranking and incremental change methods. The latter are more efficient, but ranking and unranking can be applied to solve a much wider class of problems. The key is to define the functions *rank* and *unrank* on all permutations p and integers n, m, where $|p| = n$ and $0 \leq m \leq n! - 1$.

- *Rank(p)* – What is the position of $p = \{p_1, \ldots, p_n\}$ in the given generation order? A typical ranking function is recursive. Consider the basis case $Rank(\{1\}) = 0$ with

$$Rank(p) = (p_1 - 1) \cdot (|p| - 1)! + Rank(p_2, \ldots, p_{|p|})$$

  We assume that any permutation p is an arrangement of distinct integers 1 trough $|p|$, so getting this right requires relabeling the elements of the smaller permutation to reflect the deleted first element. This is why $\{1, 3\}$ magically turns into $\{1, 2\}$ in the following example:

$$Rank(\{2, 1, 3\}) = (2-1) \cdot 2! + Rank(\{1, 2\}) = 2 + (1-1) \cdot 1! + Rank(\{1\}) = 2$$

- *Unrank(m,n)* – Which permutation is in position m of the $n!$ permutations of n items? A typical unranking function finds the number of times $(n-1)!$ goes into m and proceeds recursively. $Unrank(2, 3)$ tells us that the first element of the permutation must be "2", because $(2-1) \cdot (3-1)! \leq 2$ but $(3-1) \cdot (3-1)! > 2$. Deleting $(2-1) \cdot (3-1)!$ from m leaves the smaller problem $Unrank(0, 2)$. The ranking of 0 corresponds to the total order. The total order on the two remaining elements (since 2 has been used) is $\{1, 3\}$, so $Unrank(2, 3) = \{2, 1, 3\}$.

What the rank and unrank functions actually do does not matter so long as they are inverses of each other. In other words, $p = Unrank(Rank(p), n)$ for all permutations p. You can perform many different tasks once you have ranking and unranking functions for permutations:

- *Sequencing permutations* – To determine the *next* permutation that occurs in order after p, we can $Rank(p)$, add 1, and then $Unrank(p)$. Similarly, the permutation right before p in order is $Unrank(Rank(p)-1, |p|)$. Counting through the integers from 0 to $n!-1$ and unranking them is equivalent to generating all permutations.

- *Generating random permutations* – Selecting a random integer from 0 to $n! - 1$ and then unranking it yields a truly random permutation.

- *Keep track of a set of permutations* – Suppose we are generating random permutations, but want to act only when we encounter one we have never generated before. We can set up a bit vector (see Section 15.5 (page 456)) with $n!$ bits, and set bit i to 1 if permutation $p=Unrank(i,n)$ has already been seen. A similar technique was employed with k-subsets in the Lotto application of Section 1.8 (page 22).

This rank/unrank method is best suited for small values of n, since $n!$ quickly exceeds the capacity of machine integers. Incremental change methods work by defining *next* and *previous* operations that transform one permutation into another, typically by swapping two elements. The tricky part is scheduling the sequence of swaps so that permutations do not repeat until after all $n!$ of them have been generated. The output picture above gives an ordering of the six permutations of $\{1, 2, 3\}$ using a single swap between successive permutations.

Incremental change algorithms for sequencing permutations are elegant but tricky, generally so concise that they can be expressed in a dozen-line program. See the implementation section for pointers to code. Because the incremental change is only a single swap, these algorithms can be extremely fast—on average, constant time independent of the size of the permutation! The secret is to represent permutations using an n-element array to facilitate swaps. In certain applications, only the change between permutations is important. For example, in a brute-force program to search for the optimal TSP tour, the cost of the tour associated with the new permutation will be that of the previous permutation, with the addition and deletion of four edges.

Throughout this discussion, we have assumed that the items we are permuting are all distinguishable. However, should there be duplicates (meaning our set is a *multiset*), you can save considerable time and effort by avoiding identical permutations. For example, there are only ten distinct permutations of $\{1, 1, 2, 2, 2\}$, instead of 120. To avoid repeats, use backtracking and generate the permutations in lexicographic order.

Generating random permutations is an important little problem that people often stumble across, and often botch up. The right way is to use the following two-line, linear-time algorithm often called the Fisher-Yates shuffle. We assume that Random$[i, n]$ generates a random integer between i and n, inclusive:

> for $i = 1$ to n do $a[i] = i$;
> for $i = 1$ to $n - 1$ do swap$[a[i], a[\text{Random}[i, n]]]$;

That this algorithm generates all permutations uniformly at random is not obvious. If you think so, convincingly explain why the following algorithm *does not* generate permutations uniformly:

> for $i = 1$ to n do $a[i] = i$;
> for $i = 1$ to $n - 1$ do swap$[a[i], a[\text{Random}[1, n]]]$;

Such subtleties demonstrate why you must be very careful with random generation algorithms. Indeed, I recommend that you perform reasonably extensive experiments with *any* random generator before trusting it. For example, generate 10,000 random permutations of length 4 and verify that all 24 distinct permutations occur approximately the same number of times. If you know how to test for statistical significance, you are in even better shape.

Implementations: The C++ *Standard Template Library* (STL) provides two functions (`next_permutation` and `prev_permutation`) for sequencing permutations in lexicographic order. C++ routines for generating an astonishing variety of combinatorial objects, including permutations and cyclic permutations, are available at http://www.jjj.de/fxt/.

The *Combinatorial Object Server* (http://combos.org/) developed by Frank Ruskey of the University of Victoria is a unique resource for generating permutations, subsets, partitions, graphs, and other objects. An interactive interface enables you to specify which objects you would like returned to you. Check it out. Implementations in C, Pascal, and Java are available for certain types of objects.

Nijenhuis and Wilf [NW78] is a venerable but still valuable reference on generating combinatorial objects. They provide efficient Fortran implementations of algorithms to construct random permutations and to sequence permutations in minimum-change order. Also included are routines to extract the cycle structure of a permutation. See Section 22.1.9 (page 716) for details.

Combinatorica [PS03] provides Mathematica implementations of algorithms that construct random permutations and sequence permutations in minimum change and lexicographic orders. It also provides a backtracking routine to construct all distinct permutations of a multiset, and supports various permutation group operations. See Section 22.1.8 (page 716).

Notes: The best recent reference on permutation generation is Knuth [Knu11]. Sedgewick's excellent survey on the topic is older [Sed77], but this is not a fast moving field. Good expositions include [KS99, NW78, Rus03].

Fast permutation generation methods make only a single swap between successive permutations. The Johnson–Trotter algorithm [Joh63, Tro62] satisfies an even stronger condition, namely that the two elements being swapped are always adjacent. Simple linear-time ranking and unranking functions for permutations are given by Myrvold and Ruskey [MR01].

Markov chain generation methods construct random objects through random transitions, like swaps. Applying $\Theta(n \log n)$ random swaps to the identity permutation $\{1, 2, \dots, n\}$ suffices to construct a random permutation, as per the coupon collector's problem analysis of Section 6.2.1 (page 180). Sinclair [Sin12] presents the theory of Markov chain generation.

In the days before ready access to computers, books with tables of random permutations [MO63] were used instead of algorithms. The swap-based random permutation algorithm presented above was first described in this book.

Related problems: Random-number generation (see page 486), generating subsets (see page 521), generating partitions (see page 524).

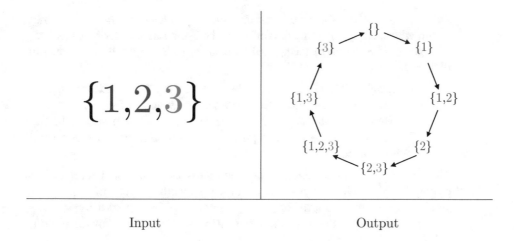

Input   Output

17.5 Generating Subsets

Input description: An integer n.

Problem description: Generate (1) all, or (2) a random, or (3) the next subset of the integers $\{1, \ldots, n\}$.

Discussion: A subset describes a selection of objects, where the order among them does not matter. Many important algorithmic problems search for the best subset of a group of things: *vertex cover* seeks the smallest subset of vertices to touch each edge in a graph, *knapsack* the most profitable subset of items of bounded total size, and *set packing* the smallest subset of subsets that together cover each item exactly once.

There are 2^n distinct subsets of an n-element set, including the empty set as well as the set itself. This grows exponentially, but at a considerably slower rate than the $n!$ permutations of n items. Indeed, a brute-force search through all subsets of 20 elements is easily manageable, because $2^{20} = 1{,}048{,}576$. But since $2^{30} = 1{,}073{,}741{,}824$, you will hit limits for slightly larger values of n.

By definition, the relative order among the elements does not distinguish different subsets. Thus, $\{1, 2, 5\}$ is the same as $\{2, 1, 5\}$. However, it is a very good idea to maintain your subsets in a sorted or *canonical* order to speed up operations such as testing whether or not two subsets are identical.

As with permutations (see Section 17.4 (page 517)), the key to subset generation problems is establishing a numerical sequence among all 2^n subsets. There are three primary alternatives:

- *Lexicographic order* – This means sorted order, and is often the most natural way to generate combinatorial objects. The eight subsets of $\{1, 2, 3\}$ in lexicographic order are $\{\}$, $\{1\}$, $\{1, 2\}$, $\{1, 2, 3\}$, $\{1, 3\}$, $\{2\}$, $\{2, 3\}$, and $\{3\}$. But it is surprisingly difficult to generate subsets in lexicographic order. Unless you have a compelling reason to do so, don't bother.

- *Gray code* – A particularly interesting subset sequence is minimum change order, wherein adjacent subsets differ by the insertion or deletion of exactly one element. Such an ordering, called a *Gray code*, appears in the output picture above.

  Generating subsets in Gray code order can be very fast, because there is a nice recursive construction. Construct a Gray code of $n - 1$ elements G_{n-1}. Reverse a second copy of G_{n-1} and add n to each subset in this copy. Then concatenate them together to create G_n. Study the output example for clarification.

  Since only one element changes between subsets, exhaustive search algorithms built on Gray codes can be quite efficient. A set cover program would only have to update the change in coverage by the addition or deletion of one subset. See the implementation section below for Gray code subset-generation programs.

- *Binary counting* – The simplest approach to subset-generation problems is based on the observation that every subset S' is defined by the items of S that are in S'. We can represent S' by a binary string of n bits, where bit i is 1 iff the ith element of S is in S'. This defines a bijection between the 2^n binary strings of length n, and the 2^n subsets of n items. For $n = 3$, binary counting generates subsets in the following order: {}, {3}, {2}, {2,3}, {1}, {1,3}, {1,2}, {1,2,3}.

  This binary representation is the key to solving all subset generation problems. To generate all subsets in order, simply count from 0 to $2^n - 1$. For each integer, successively mask off each of the bits and compose a subset of exactly the items corresponding to 1 bits. To generate the *next* or *previous* subset, increment or decrement this integer by one. *Unranking* a subset is exactly the masking procedure described above, while *ranking* constructs the binary number with 1's corresponding to items in S and converts it to an integer.

  To generate a random subset, you could try to generate a random integer from 0 to $2^n - 1$ and unrank, although your random number generator might round things off in a way certain subsets can never occur. Much better is to flip a coin n times, with the ith flip deciding whether to include element i in the subset. A coin flip can be robustly simulated by generating a random real or large integer and testing whether it is bigger or smaller than half its range.

Generation problems for two closely related problems arise often in practice:

- *K-subsets* – Instead of constructing all subsets, we may only be interested in the subsets containing exactly k elements. There are $\binom{n}{k}$ such subsets, which is substantially less than 2^n, particularly for small values of k.

  The best way to construct all k-subsets is in lexicographic order. The ranking function is based on the observation that there are $\binom{n-f}{k-1}$ k-subsets whose smallest element is f. Using this, we can determine the smallest element in the mth k-subset of n items, and then proceed recursively for subsequent elements of the subset. See the implementations below for details.

- *Strings* – Generating all subsets is equivalent to generating all 2^n strings of true and false. The same basic techniques apply to generate all or random strings on alphabets of size α, except there will be α^n strings in total.

Implementations: C++ routines for generating an astonishing variety of combinatorial objects, including subsets and k-subsets (combinations), are available in the combinatorics package at http://www.jjj.de/fxt/.

The *Combinatorial Object Server* (http://combos.org/) developed by Frank Ruskey of the University of Victoria is a unique resource for generating permutations, subsets, partitions, graphs, and other objects. An interactive interface enables you to specify which objects you would like returned to you. Check it out. Implementations in C, Pascal, and Java are available for certain types of objects.

Nijenhuis and Wilf [NW78] is an excellent reference on generating combinatorial objects. They provide efficient Fortran implementations of algorithms to construct random subsets, and to sequence subsets in Gray code and lexicographic order. They also provide routines to construct random k-subsets and sequence them in lexicographic order. See Section 22.1.9 (page 716) for details on these programs.

Combinatorica [PS03] provides Mathematica implementations of algorithms to construct random subsets and sequence subsets in Gray code, binary, and lexicographic order. They also provide routines to construct random k-subsets and strings, and sequence them lexicographically. See Section 22.1.8 (page 716) for further information on *Combinatorica*.

Notes: The best reference on subset generation is Knuth [Knu11]. Good expositions include [KS99, NW78, Rus03]. Wilf [Wil89] provides an update of [NW78], including a thorough discussion of modern Gray code generation problems.

Gray codes were first developed [Gra53] to transmit digital information in a robust manner over an analog channel. By assigning the code words in Gray code order, the ith word differs only slightly from the $(i+1)$st, so minor fluctuations in analog signal strength corrupt only a few bits. Gray codes have a particularly nice correspondence to Hamiltonian cycles on the hypercube. Savage [Sav97] gives an excellent survey of Gray codes (minimum change orderings) for a large class of combinatorial objects, including subsets. An interesting new approach to k-subset generation based on rotating bit strings was developed by Ruskey and Williams [RW09].

The popular puzzle *Spinout*®, manufactured by ThinkFun (formerly Binary Arts Corporation), is solved using ideas from Gray codes.

Related problems: Generating permutations (see page 517), generating partitions (see page 524).

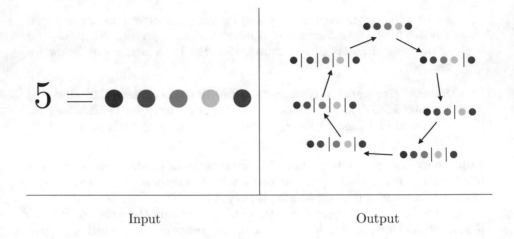

Input   Output

17.6   Generating Partitions

Input description: An integer n.

Problem description: Generate (1) all, or (2) a random, or (3) the next integer or set partitions of length n.

Discussion: Two different types of combinatorial objects are denoted by the word "partition," namely integer partitions and set partitions. They are quite different beasts, but it is good to make both a part of your vocabulary:

- *Integer partitions* are multisets of non-zero integers that add up exactly to n. For example, the seven distinct integer partitions of 5 are $\{5\}$, $\{4,1\}$, $\{3,2\}$, $\{3,1,1\}$, $\{2,2,1\}$, $\{2,1,1,1\}$, and $\{1,1,1,1,1\}$. An interesting application I encountered that required generating integer partitions was a simulation of nuclear fission. When an atom is smashed, the nucleus is broken into a set of smaller clusters. The sum of the particles in the set of clusters must equal the original size n of the nucleus. As such, the integer partitions of n represent all possible ways to smash an atom.

- *Set partitions* divide the elements $\{1, \ldots, n\}$ into non-empty subsets. There are 15 distinct set partitions of $n = 4$: $\{1234\}$, $\{123,4\}$, $\{124,3\}$, $\{12,34\}$, $\{12,3,4\}$, $\{134,2\}$, $\{13,24\}$, $\{13,2,4\}$, $\{14,23\}$, $\{1,234\}$, $\{1,23,4\}$, $\{14,2,3\}$, $\{1,24,3\}$, $\{1,2,34\}$, and $\{1,2,3,4\}$. Algorithm problems returning set partitions as results include *vertex/edge coloring* and *connected components*.

Although the number of integer partitions grows exponentially with n, they do so at a refreshingly slow rate. There are only 627 partitions of $n = 20$. It is even possible to enumerate all integer partitions of $n = 100$, since there are only 190,569,292 of them.

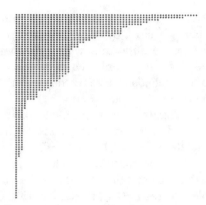

Figure 17.1: The Ferrers diagram of a random partition of n =1,000.

The easiest way to generate integer partitions constructs them in lexico-graphically decreasing order. The first partition is $\{n\}$ itself. The general rule is to subtract 1 from the smallest part that is > 1 and then collect all the 1's so as to match the new smallest part > 1. For example, the partition following $\{4,3,3,3,1,1,1,1\}$ is $\{4,3,3,2,2,2,1\}$, because the five 1's left after $3-1=2$ becomes the smallest part are best packaged as 2,2,1. When the partition is reduced to all 1's, we have completed one pass through the partitions.

This algorithm is sufficiently intricate to program that you should use one of the implementations below. In any case, test to make sure that you get exactly 627 distinct partitions for $n = 20$.

Generating integer partitions uniformly at random is a trickier business than generating random permutations or subsets. This is because selecting the first (i.e. largest) element of the partition has a dramatic effect on the number of possible partitions that can be generated. Observe that there is only one parti-tion of n whose largest part is 1, namely $\{1, 1, \ldots, 1\}$. The number of partitions of n with largest part at most k is given by the recurrence

$$P_{n,k} = P_{n-k,k} + P_{n,k-1}$$

because any such partition either contains a part of size k or it doesn't. The two boundary conditions are $P_{n,1} = 1$ and $P_{x-y,x} = P_{x-y,x-y}$, for all $y \leq x$. The second condition looks funny but it is correct: note that $P_{3,5}$ must be equal to $P_{3,3}$, because there are no partitions of 3 whose biggest part is 4 or 5. This function can be used to select the largest part of your random partition with the right probability and then, by proceeding recursively, eventually construct the entire random partition. Implementations are cited below.

Random partitions tend to have large numbers of fairly small parts, best visualized by a Ferrers diagram as in Figure 17.1. Each row of the diagram corresponds to one part of the partition, sorted by size, with the magnitude of each part represented by that many dots. Such diagrams provide a very good way to think about integer partitions.

One application arises in evaluating the publication record of academic researchers. Research papers get cited by other research papers, and larger citation counts correspond to more important work. A scholar has an *H-index* of h if they have written at least h papers each of which has received at least h citations. The H-index corresponds to size of the central square defined by the Ferrers diagram of an author's citations.

Set partitions can be generated using techniques akin to integer partitions. Each set partition is encoded as a *restricted growth function*, $a_1, \ldots, a_n$, where $a_1 = 0$ and $a_i \leq 1 + \max(a_1, \ldots, a_{i-1})$, for $i = 2, \ldots, n$. Each distinct digit identifies a subset, or *block*, of the partition, while the growth condition ensures that the blocks are sorted into a canonical order based on the smallest element in each block. For example, the restricted growth function $0, 1, 1, 2, 0, 3, 1$ defines the set partition $\{\{1, 5\}, \{2, 3, 7\}, \{4\}, \{6\}\}$.

Since there is a one-to-one equivalence between set partitions and restricted growth functions, we can use lexicographic order on the restricted growth functions to order the partitions. Indeed, the fifteen set partitions of $\{1, 2, 3, 4\}$ listed above are sequenced according to the lexicographic order of their restricted growth function (check it out).

We can use a similar counting strategy to generate random set partitions as we did with integer partitions. The Stirling numbers of the second kind $\{^n_k\}$ count the number of partitions of $\{1, \ldots, n\}$ with exactly k blocks. They are computed using the recurrence

$$\left\{{n \atop k}\right\} = \left\{{n-1 \atop k-1}\right\} + k\left\{{n-1 \atop k}\right\}$$

with the boundary conditions $\left\{{n \atop n}\right\} = \left\{{n \atop 1}\right\} = 1$. The reader is referred to the sections below for more details.

Implementations: C++ routines for generating an astonishing variety of combinatorial objects, including integer partitions and compositions, are available in the combinatorics package at http://www.jjj.de/fxt/. Kreher and Stinson [KS99] generate both integer and set partitions in lexicographic order, including ranking/unranking functions. These implementations in C are available at http://www.math.mtu.edu/~kreher/cages/Src.html.

The *Combinatorial Object Server* (http://combos.org/) developed by Frank Ruskey of the University of Victoria is a unique resource for generating permutations, subsets, partitions, graphs, and other objects. An interactive interface enables you to specify which objects you would like returned to you. Check it out. Implementations in C, Pascal, and Java are available for certain types of objects.

Nijenhuis and Wilf [NW78] remains a valuable resource on generating combinatorial objects. They provide efficient Fortran implementations of algorithms to construct random and sequential integer partitions, set partitions, compositions, and Young tableaux. See Section 22.1.9 (page 716) for details.

Combinatorica [PS03] provides Mathematica implementations of algorithms to construct random and sequential integer partitions, compositions, strings, and Young tableaux, as well as to count and manipulate these objects. See Section 22.1.8 (page 716).

Notes: The best reference on algorithms for generating both integer and set partitions is Knuth [Knu11]. Good expositions include [KS99, NW78, Rus03, PS03]. Andrews [And98] is the primary reference on integer partitions and related topics, with [AE04] his more accessible introduction. Mansour [Man12] is a recent book on set partitions.

Integer and set partitions are both special cases of *multiset partitions*, or set partitions of not necessarily distinct numbers. In particular, the distinct set partitions on the multiset $\{1, 1, 1, \ldots, 1\}$ correspond exactly to integer partitions. Multiset partitions are discussed in Knuth [Knu11].

The (long) history of combinatorial object generation is detailed by Knuth [Knu11]. Particularly interesting are connections between set partitions and a Japanese incense burning game, and the naming of all 52 set partitions for $n = 5$ with distinct chapters from the oldest novel known, *The Tale of Genji*.

The 2015 film *The Man Who Knew Infinity*, about the life of the great Indian mathematician Ramanujan, revolved around his amazing formula to approximately count the number of integer partitions.

Two related combinatorial objects are Young tableaux and integer compositions, although they are less likely to emerge in applications. Generation algorithms for both are presented in [NW78, Rus03, PS03].

Young tableaux are two-dimensional configurations of integers $\{1, \ldots, n\}$ where the number of elements in each row is defined by an integer partition of n. Further, the elements of each row and column are sorted in increasing order, and the rows are left-justified. This notion of shape captures a wide variety of structures as special cases. They have many interesting properties, including the existence of a bijection between pairs of tableaux and permutations.

Compositions represent the possible assignments of n indistinguishable balls to k distinguishable boxes. For example, we can place three balls into two boxes as $\{3,0\}$, $\{2,1\}$, $\{1,2\}$, or $\{0,3\}$. Compositions are most easily constructed sequentially in lexicographic order. To construct them randomly, pick a random $(k-1)$-subset of $n + k - 1$ items using the algorithm of Section 17.5 (page 521), and count the number of unselected items between the selected ones. For example, if $k = 5$ and $n = 10$, the $(5-1)$ subset $\{1,3,7,14\}$ of $1, \ldots, (n+k-1) = 14$ defines the composition $\{0,1,3,6,0\}$, since there are no items to the left of element 1 nor to the right of element 14.

Related problems: Generating permutations (see page 517), generating subsets (see page 521).

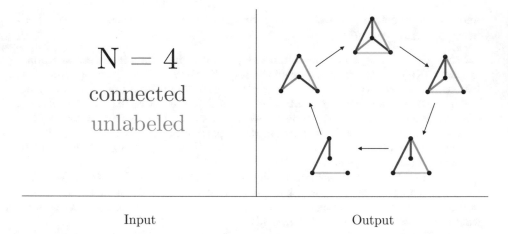

| Input | Output |

17.7   Generating Graphs

Input description: Parameters describing the desired graph, including the number of vertices n, and the number of edges m or edge probability p.

Problem description: Generate (1) all, or (2) a random, or (3) the next graph satisfying the parameters.

Discussion: Graph generation typically arises when constructing test data for programs. Perhaps you have two different programs that solve the same problem, and you want to see which is faster, or make sure that they both give the same (presumably right) answer. Another application is experimental graph theory, verifying whether a particular property is true for all graphs. It is much easier to believe the four-color theorem after you have demonstrated four colorings for all planar graphs on fifteen vertices.

Many factors complicate the problem of generating graphs. First, make sure you know exactly what type of graph you want to generate. Figure 7.2 on page 198 illustrates several important properties of graphs. For purposes of generation, the most important questions are:

- *Do I want labeled or unlabeled graphs?* – The issue here is whether the names of the vertices matter in deciding whether two graphs are the same. In generating *labeled graphs*, we seek to construct all possible labelings of all possible graph topologies. When generating *unlabeled graphs*, we want exactly one representative for each topology and ignore labelings. For example, there are only two connected unlabeled graphs on three vertices—a triangle and a simple path. However, there are four connected labeled graphs on three vertices—one triangle and three 3-vertex paths, each distinguished by the name of their central vertex. In general, labeled graphs are much easier to generate. However, you are likely to get swamped with many isomorphic copies of the same few graphs.

- *Do I want directed or undirected graphs?* – Most natural generation algorithms generate undirected graphs. These can be turned into directed graphs by flipping coins to orient the edges. Any graph can be oriented to be directed and acyclic (i.e., a DAG) by randomly permuting the vertices on a line and aiming each edge from left to right. With all such ideas, careful thought must be given to decide whether you are generating all graphs uniformly at random, and how much this matters to you.

You also must define what you mean by random. There are three primary models of random graphs, all of which generate graphs according to different probability distributions:

- *Random edge generation* – The Erdős–Rényi model is parameterized by a given edge probability p. In this model, a coin is flipped for each pair of vertices x and y to decide whether to add an edge (x, y). All *labeled* graphs will be generated with equal probability when $p = 1/2$, although smaller values of p can be used to construct sparser random graphs.

- *Random edge selection* – The second model is parameterized by the desired number of edges m. It selects m distinct edges uniformly at random. One way to do this is by drawing random (x, y)-pairs and creating an edge if that pair is not already in the graph. An alternative approach constructs the set of $\binom{n}{2}$ possible edges and selects a random m-subset of them, as discussed in Section 17.5 (page 521).

- *Preferential attachment* – Under a rich-get-richer model, newly created edges are more likely to point to high-degree vertices than low-degree ones. Consider new links (edges) being added to the graph of webpages. Under any realistic web generation model, it is much more likely the next link will point to Google than www.algorist.com.[1] Selecting the next neighbor with probability proportional to its degree yields graphs with *power law* properties encountered in many real networks.

Which of these options best models your application? Probably none of them. By definition, random graphs have very little concrete structure. But graphs are used to model relationships, which are often highly structured. Experiments conducted on random graphs, although interesting and easy to perform, often fail to capture the phenomenon that you are looking for.

An alternative to random graphs is "organic" graphs—graphs that reflect the relationships among real-world objects. Many raw sources of relationships are available on the web that can be turned into interesting organic graphs with a little programming and imagination. Consider the graph defined by a set of webpages, with any hyperlink between two pages defining an edge. Or, what about the graph implicit in railroad, subway, or airline networks, with vertices being stations and edges between two stations connected by direct service?

Two classes of graphs have particularly interesting generation algorithms:

---

[1] Please link to us from your homepage to correct this travesty.

- *Trees* – Prüfer codes provide a simple way to rank and unrank *labeled* trees and thus solve all the standard generation problems (see Section 17.4 (page 517)). There are exactly n^{n-2} labeled trees on n vertices, and exactly that many strings of length $n-2$ on the alphabet $\{1, 2, \ldots, n\}$.

  The key to Prüfer's bijection is the observation that every tree has at least two vertices of degree 1. Thus, in any labeled tree the vertex v incident on the leaf with lowest label is well defined. We take v to be S_1, the first character in the code. We then delete the associated leaf and repeat the procedure until only two vertices are left. This defines a unique code S for any given labeled tree that can be used to rank the tree. To go from code to tree, observe that the lowest-labeled leaf will be the smallest integer missing from S, which when paired with S_1 determines the first edge of the tree. The entire tree follows by induction.

  Algorithms for efficiently generating unlabeled rooted trees are discussed in the Implementations section.

- *Fixed degree sequence graphs* – The *degree sequence* of a graph G is an integer partition $p = (p_1, \ldots, p_n)$, where p_i is the degree of the ith highest-degree vertex of G. Each edge contributes to the degree of two vertices, so p is an integer partition of $2m$, where m is the number of edges.

  Not all partitions correspond to degree sequences of graphs. However, there is a recursive procedure that constructs a graph with a given degree sequence if one exists. If a partition is realizable, the highest-degree vertex v_1 can be connected to the next p_1 highest-degree vertices in G, or the vertices corresponding to parts $p_2, \ldots, p_{p_1+1}$. Deleting p_1 and decrementing $p_2, \ldots, p_{p_1+1}$ yields a smaller partition to recur on. The partition will be realized if we terminate without ever creating negative numbers. Since we always connect the highest-degree vertex to other high-degree vertices, it is important to reorder the parts of the partition by size after each iteration.

  Although this construction is deterministic, a semi-random collection of graphs realizing this degree sequence can be generated from G using *edge-flipping* operations. Suppose edges (x, y) and (w, z) are in G, but (x, w) and (y, z) are not. Exchanging these pairs of edges creates a different (not necessarily connected) graph without changing the degrees of any vertex.

Implementations: The Stanford GraphBase [Knu94] is perhaps most useful as an instance generator for constructing graphs to serve as test data for other programs. Because of its machine-independent random-number generators, it provides a way to construct random graphs such that they can be reconstructed elsewhere, thus making them perfect for experimental comparisons of algorithms. See Section 22.1.7 (page 715) for additional information.

Resources for real-world networks include the Network Data Repository (http://networkrepository.com/) and the Stanford Large Network Dataset Collection (https://snap.stanford.edu/data/). Check them out.

Combinatorica [PS03] provides Mathematica generators for such graphs as stars, wheels, complete graphs, random graphs and trees, and graphs with a given degree sequence. Further, it includes operations to construct more interesting graphs from these, including join, product, and line graph.

The *Combinatorial Object Server* (http://combos.org/) developed by Frank Ruskey of the University of Victoria provides routines for generating both free and rooted trees.

The graph isomorphism testing program Nauty (see Section 19.9 (page 610)) includes a suite of programs for generating non-isomorphic graphs, plus special generators for bipartite graphs, digraphs, and multigraphs. They are available at http://users.cecs.anu.edu.au/~bdm/nauty/. Brendan McKay also collects exhaustive catalogs of several families of graphs and trees at http://cs.anu.edu.au/~bdm/data/. The House of Graphs https://hog.grinvin.org/ is a carefully curated set of graphs with interesting properties, designed to break conjectures [BCGM13].

Nijenhuis and Wilf [NW78] provide efficient Fortran routines to enumerate all labeled trees via Prüfer codes and to construct random unlabeled rooted trees. See Section 22.1.9 (page 716). Kreher and Stinson [KS99] generate labeled trees in C, with implementations available at http://www.math.mtu.edu/~kreher/cages/Src.html.

Notes: Extensive literature exists on generating graphs uniformly at random. Surveys include [Gol93, Tin90]. Fast random graph generation on GPUs is demonstrated in [NLKB11]. Closely related to the problem of generating classes of graphs is counting them. Harary and Palmer [HP73] survey results in graphical enumeration.

Knuth [Knu11] is the best recent reference on generating trees. The bijection between $n - 2$ strings and labeled trees is due to Prüfer [Prü18].

Random graph theory is concerned with the properties of random graphs. Threshold laws in random graph theory define the edge density at which properties such as connectedness become highly likely to occur. Expositions on random graph theory include [Bol01, FK15, JLR00].

The preferential attachment model of graphical evolution has emerged relatively recently in the study of networks. See [Bar03, Wat04] for introductions to this exciting field. Methods for generating graphs with prescribed degree sequences are presented in [BD11, VL05].

An integer partition is *graphic* if there exists a simple graph with that degree sequence. Erdős and Gallai [EG60] proved that a degree sequence is graphic if and only if the sequence observes the following condition for each integer $r < n$:

$$\sum_{i=1}^{r} d_i \le r(r-1) + \sum_{i=r+1}^{n} \min(r, d_i)$$

Related problems: Generating permutations (see page 517), graph isomorphism (see page 610).

December 21, 2012? (Gregorian)	5773 Teveth 8 (Hebrew)
	1434 Safar 7 (Islamic)
	1934 Agrahayana 30 (Indian Civil)
	13.0.0.0 (Mayan Long Count)

| Input | Output |

17.8   Calendrical Calculations

Input description: A particular calendar date d: month, day, and year.

Problem description: Which day of the week did d fall on according to the given calendar system?

Discussion: Business applications often need to perform calendrical calculations. Perhaps we want to display a calendar of a specified month and year. Maybe we need to compute what day of the week or year some event occurs, such as figuring out on which date a 180-day futures contract comes due. The importance of correct calendrical calculations was perhaps best revealed by the furor over the "Millennium bug"—the year 2000 crisis in legacy programs that allocated only two digits for storing the year.

More complicated questions arise in international applications, because different nations and ethnic groups use different calendar systems. Some, like the Gregorian calendar used in most of the world, are based on the Sun, while others, like the Hebrew calendar, are lunar calendars. How would you tell today's date according to the Chinese or Islamic calendars?

Calendrical calculations differ from other problems in this book because calendars are historical objects, not mathematical ones. The algorithmic issues here revolve around the rules of the calendrical system and implementing them correctly, rather than designing efficient computational shortcuts.

The basic approach underlying calendar systems is to start from a particular reference date (called the *epoch*) and count up from there. The particular rules for wrapping the count into months and years is what distinguishes one system from another. Implementing a calendar requires two functions: (1) given a date, return the integer number of days that have elapsed since the epoch, and (2) given an integer n, return the calendar date exactly n days from epoch. These are analogous to the ranking and unranking rules for combinatorial objects such as permutations (see Section 17.4 (page 517)).

That the solar year is not an integer number of days long is the major source of complications in calendar systems. To keep a calendar's dates in sync with the seasons, leap days must be added at both regular and irregular intervals. One solar year is 365 days and 5:49:12 hours long, so adding a leap day every four years leaves an extra 10 minutes and 48 seconds unaccounted for every year.

The original Julian calendar (from Julius Caesar) ignored these extra minutes, which had accumulated to ten days by 1582. Pope Gregory XIII then proposed the Gregorian calendar used today, by deleting these ten days and eliminating leap days in years that are multiples of 100 but not 400. Supposedly, riots ensued because the masses feared their lives were being shortened by ten days. Outside the Catholic church, resistance to change slowed the reforms. The deletion of days did not occur in England and America until September 1752, and not until 1927 in Turkey.

The rules for most calendrical systems are sufficiently complicated and pointless that you should lift code from a reliable place rather than attempt to write your own. I identify suitable implementations below.

There are a variety of "impress your friends" algorithms that enable you to compute in your head on what day of the week a particular date occurred. Such algorithms often fail to work reliably outside the given century, and certainly should be avoided for computer implementation.

Implementations: Readily available calendar libraries exist in both C++ and Java. The Boost time-data library provides a reliable implementation of the Gregorian calendar in C++. See `https://www.boost.org/doc/libs/1_70_0/doc/html/date_time.html`. The class GregorianCalendar derived from the abstract superclass Calendar in the package `java.util` implements the Gregorian calendar in Java. Either of these will likely suffice for most applications.

Dershowitz and Reingold provide a uniform algorithmic presentation [RD18] for a variety of different calendar systems, including the Gregorian, ISO, Chinese, Hindu, Islamic, and Hebrew calendars, as well as other calendars of historical interest. *Calendrical* is an implementation of these calendars in Common Lisp, Java, and Mathematica, with routines to convert dates between calendars, day of the week computations, and the determination of secular and religious holidays. *Calendrical* is likely to be the most comprehensive and reliable calendrical routines available. See their website at `http://calendarists.com`.

C and Java implementations of international calendars of unknown reliability are readily available at GitHub (`https://github.com/`). Search for "Gregorian calendar" to avoid the mass of datebook implementations.

Notes: A comprehensive discussion of calendrical computation algorithms appear in the papers of Dershowitz and Reingold [DR90, RDC93], which have been superseded by their book [RD18] that outlines algorithms for no less than twenty-five international and historical calendars. Three hundred years of calendars representing tabulations for all dates from 1900 to 2200 appear in [DR02].

Related problems: Arbitrary-precision arithmetic (see page 493), generating permutations (see page 517).

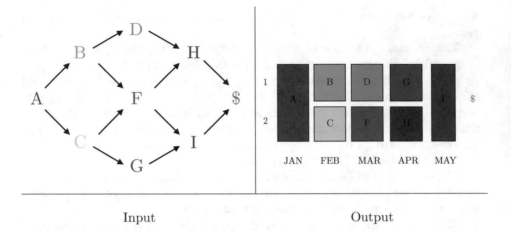

Input   Output

17.9   Job Scheduling

Input description: A directed acyclic graph $G = (V, E)$, with vertices representing jobs, and edge (u, v) implies task u must be done before task v.

Problem description: Which schedule of tasks completes the job using the minimum amount of time or processors?

Discussion: Devising the best schedule to satisfy a set of constraints is fundamental to many applications. Mapping tasks to processors is a critical aspect of any parallel-processing system. Poor scheduling can leave many machines sitting idle while one bottleneck task is performed. Assigning people-to-jobs, meetings-to-rooms, or courses-to-exam periods are all scheduling problems.

Scheduling problems differ widely in the nature of the constraints that must be satisfied and the type of schedule desired. Several other catalog problems have connections to variants of scheduling:

- *Topological sort* constructs a schedule consistent with precedence constraints in a DAG. See Section 18.2 (page 546).

- *Bipartite matching* assigns jobs to workers with appropriate skills for them. See Section 18.6 (page 562).

- *Vertex/edge coloring* assigns jobs to time slots such that no two interfering jobs are assigned the same slot. See Sections 19.7 and 19.8.

- *Traveling salesman* identifies the most efficient delivery route to visit a given set of locations. See Section 19.4 (page 594).

- *Eulerian cycle* defines the most efficient route for a snowplow or mailman to traverse a given set of edges. See Section 18.7 (page 565).

Here the focus is on precedence-constrained scheduling problems for directed acyclic graphs. Suppose you have broken a big job into a number of smaller tasks. For each task, you know how long it should take to finish. Further, for each pair of tasks A and B you know whether it is essential that A be completed before B. The fewer constraints we must enforce, the tighter our schedule can be. These constraints must define a directed acyclic graph—acyclic because a cycle in the precedence constraints represents a Catch-22 situation that can never be resolved.

Several problems on these task networks are of interest:

- *Critical path* – The longest path from the start vertex to the completion vertex defines the *critical path*. This is important to know, for the only conceivable way to shorten the minimum possible completion time is to reduce the length of a task on each critical path. The critical (longest) path in a DAG can be determined in $O(n + m)$ time using dynamic programming.

- *Minimum completion time* – What is the fastest we can get this job completed while respecting precedence constraints, assuming that we have an unlimited number of workers? Were there *no* precedence constraints, each task could be worked on independently by its own worker, and the total time would be that of the longest single task. Were there such strict precedence constraints that each task must follow the completion of its immediate predecessor, the minimum completion time would be obtained by summing up the times for each task.

  The minimum completion time for a DAG can be computed in $O(n + m)$ time, because it is defined by the critical path. To get such a schedule, consider the jobs in topological order, and start each job on a new processor the moment its latest prerequisite completes.

- *What is the tradeoff between number of workers and completion time?* – What we really are interested in is how best to complete the schedule with our given number of workers. Unfortunately, this and most similar problems are NP-complete.

Real scheduling applications often present constraints that may be difficult or impossible to model using these techniques, such as keeping Joe and Bob apart so they won't kill each other. There are two reasonable ways to deal with such problems. First, we can ignore such esoteric constraints until the end, and then modify the schedule to account for them. Alternately, you can formulate your scheduling problem in all its complexity via linear-integer programming (see Section 16.6 (page 482)). I recommend first trying something simple to see how it works before ramping up the complexity.

Another fundamental scheduling problem takes a set of jobs without precedence constraints and assigns them to identical machines to minimize the total elapsed time. Consider a print shop with k copying machines and a stack of jobs to finish today. Such tasks are called *job-shop scheduling*, and can be modeled as bin packing (see Section 20.9 (page 652)). Each job is associated with the time it will take to complete, and each machine is represented by a bin with space equal to the number of hours in a day.

More sophisticated variations of job-shop scheduling provide each task with allowable start and required finishing times. Effective heuristics are known, based on sorting the tasks by size and finishing time. We refer the reader to the references for more information. Note that these scheduling problems become hard only when the tasks cannot be broken up onto multiple machines or interrupted (preempted) and then rescheduled. You should exploit these degrees of freedom if your application allows them.

Implementations: JOBSHOP is a collection of C programs for job-shop scheduling created for a computational study by Applegate and Cook [AC91]. They are available at `http://www.math.uwaterloo.ca/~bico//jobshop/`.

UniTime (`https://www.unitime.org/` is a comprehensive educational scheduling system that supports developing course/exam timetables, and scheduling students to individual classes. It is distributed under an open source license.

LEKIN is a flexible job-shop scheduling system designed for educational use [Pin16]. It supports single machine, parallel machines, flow-shop, flexible flow-shop, job-shop, and flexible job-shop scheduling, and is available at `http://www.stern.nyu.edu/om/software/lekin`.

For commercial scheduling applications, ILOG CP has been reflective of the state-of-the-art for over 20 years [LRSV18]. See `https://www.ibm.com/analytics/cplex-cp-optimizer`. A restricted free version is available.

Notes: The literature on scheduling algorithms is vast. Brucker [Bru07] and Pinedo [Pin16] provide comprehensive overviews of the field. The *Handbook of Scheduling* [LA04] provides a collection of surveys on all aspects of scheduling. Real-time scheduling for computing systems is treated by Buttazzo [But11].

A well-defined taxonomy covers thousands of job-shop scheduling variants, which classifies each problem $\alpha|\beta|\gamma$ according to (α) the machine environment, (β) details of processing characteristics and constraints, and (γ) the objectives to be minimized. Surveys of results include [Bru07, CPW98, LLK83, Pin16].

Gantt charts provide visual representations of job-shop scheduling solutions, where the x-axis represents time and rows represent distinct machines. The output figure above illustrates a Gantt chart, where each scheduled job is represented as a horizontal block identifying its start-time, duration, and server. Project precedence-constrained scheduling techniques are often called PERT/CPM, for *Program Evaluation and Review Technique/Critical Path Method*. Gantt charts and PERT/CPM appear in most textbooks on operations research, including [Pin16].

Timetabling is a term often used in discussion of classroom and related scheduling problems. PATAT (for *Practice and Theory of Automated Timetabling*) is a bi-annual conference reporting new results in the field. See `https://patatconference.org/`.

Related problems: Topological sorting (see page 546), matching (see page 562), vertex coloring (see page 604), edge coloring (see page 608), bin packing (see page 652).

$$(\; x_1 \; \text{ or } \; x_2 \; \text{ or } \; \overline{x_3} \;) \quad \Big| \quad (\; x_1 \; \text{ or } \; x_2 \; \text{ or } \; \overline{x_3} \;)$$

$$(\; x_1 \; \text{ or } \; \overline{x_2} \; \text{ or } \; x_3 \;) \quad \Big| \quad (\; x_1 \; \text{ or } \; \overline{x_2} \; \text{ or } \; x_3 \;)$$

$$(\; \overline{x_1} \; \text{ or } \; \overline{x_2} \; \text{ or } \; \overline{x_3} \;) \quad \Big| \quad (\; \overline{x_1} \; \text{ or } \; \overline{x_2} \; \text{ or } \; \overline{x_3} \;)$$

$$(\; \overline{x_1} \; \text{ or } \; x_2 \; \text{ or } \; x_3 \;) \quad \Big| \quad (\; \overline{x_1} \; \text{ or } \; x_2 \; \text{ or } \; x_3 \;)$$

Input	Output

17.10   Satisfiability

Input description: A set of clauses in conjunctive normal form.

Problem description: Is there a truth assignment to the Boolean variables such that every clause is simultaneously satisfied?

Discussion: Satisfiability (SAT) arises whenever we seek a configuration or object that must be consistent with (i.e. satisfy) a set of logical constraints. A representative application is in verifying that a given hardware or software system design works correctly on all inputs. Suppose that logical formula $S(X)$ denotes the specified result on input variables $X = x_1, \ldots, x_n$, while a different formula $C(X)$ denotes the Boolean logic of a circuit for computing $S(X)$. This circuit is correct unless there exists an $\bar{X}$ such that $S(\bar{X}) \neq C(\bar{X})$.

Satisfiability is *the* original NP-complete problem. Despite its applications to constraint satisfaction, logic, and automatic theorem proving, it is most important theoretically as the root problem from which all other NP-completeness proofs originate. So much engineering has gone into today's best SAT solvers that they represent a reasonable starting point whenever one needs to solve an NP-complete problem *exactly*. That said, employing heuristics that give good but non-optimal solutions is usually the better approach for dealing with NP-complete problems.

Issues in satisfiability testing include:

- *Is your formula the "AND of ORs" or the "OR of ANDs"?* – In satisfiability, the constraints are specified as a logical formula. There are two primary ways of expressing logical formulas—conjunctive normal form (CNF) and disjunctive normal form (DNF). In CNF formulas, we must satisfy all clauses ("AND"), where each clause is constructed by or-ing *literals* (terms of the form v_i or $\bar{v}_i$, the latter denoting "not v_i"), like

$$(v_1 \text{ or } \bar{v}_2) \text{ and } (v_2 \text{ or } v_3)$$

With DNF formulas, it suffices to satisfy any single clause, because they will be or-ed together. The formula above can be written in DNF as

$$(\bar{v}_1 \text{ and } \bar{v}_2 \text{ and } v_3) \text{ or } (\bar{v}_1 \text{ and } v_2 \text{ and } \bar{v}_3) \text{ or }$$
$$(\bar{v}_1 \text{ and } v_2 \text{ and } v_3) \text{ or } (v_1 \text{ and } \bar{v}_2 \text{ and } v_3)$$

Solving DNF-satisfiability is trivial, because every DNF formula can be satisfied unless *all* clauses contain both a literal and its complement (negation). But CNF-satisfiability is NP-complete. This seems paradoxical, because we can use De Morgan's laws to convert CNF formula into equivalent DNF formula, and vice versa. The catch is that an exponential number of terms might be required for the translation, so that the translation cannot be constructed in polynomial time.

- *How big are your clauses?* – k-SAT is a special case of satisfiability, where each clause contains at most k literals. The problem of 1-SAT is trivial, since we must set true every literal appearing in any clause. The problem of 2-SAT is not so trivial, but can still be solved in linear time. This is interesting, because certain problems can be modeled as 2-SAT using a little cleverness. The good times end once clauses contain three literals each (i.e., 3-SAT), for 3-SAT is NP-complete.

- *Does it suffice to satisfy* **most** *of the clauses?* – If you are determined to solve a SAT problem exactly, there is not much you can do except backtracking algorithms like the Davis–Putnam procedure. In the worst case 2^m truth assignments must be tested, but fortunately there are many ways to prune the search. Although satisfiability is NP-complete, how hard it is in practice depends upon the particular problem instance. Naturally defined "random" instances are often surprisingly easy to solve, and in fact it is non-trivial to generate instances that are truly hard.

  Still, we might benefit by relaxing the problem so that the goal becomes satisfying as many clauses as possible. Optimization techniques such as simulated annealing can then be put to work to refine random or heuristic solutions. Indeed, any random truth assignment to the variables will satisfy each k-SAT clause with probability $1 - (1/2)^k$, so our first attempt is likely to satisfy most of the clauses. Finishing off the job is the hard part. Finding an assignment that satisfies the maximum number of clauses is NP-complete even for non-satisfiable instances.

When faced with a problem of unknown complexity, proving it NP-complete can be an important first step. If you think your problem might be hard, skim through Garey and Johnson [GJ79] looking for your problem. If you don't find it, I recommend that you put the book away and try to prove hardness from first principles, using the basic problems of 3-SAT, vertex cover, independent set, integer partition, clique, and Hamiltonian cycle. Chapter 11 focuses on strategies for proving hardness.

Implementations: Recent years have seen tremendous progress in the performance of satisfiability solvers. An annual SAT competition identifies the top performing solvers in each of several categories of instances. The source code for all these solvers and more are available from the competition webpage (http://www.satcompetition.org/).

SAT *Live!* (http://www.satlive.org/) is the most up-to-date source for papers, programs, and test sets for satisfiability and related logic optimization problems.

Notes: The most comprehensive overview of satisfiability testing in practice is Kautz, et al. [KSBD07]. The Davis-Putnam-Logemann-Loveland (DPLL) algorithm is a backtracking algorithm introduced in 1962 for solving satisfiability problems. Local search techniques work better on certain classes of problems that are difficult for DPLL solvers. See [BHvM09, GKSS08, KS07] for surveys of the field of satisfiability testing.

The first part of the second part of Knuth's Volume 4 is on satisfiability algorithms, published independently as a fascicle [Knu15]. He demonstrates a variety of fascinating applications of satisfiability (including the game of life) and careful treatment of backtracking-based search procedures for solving them.

An algorithm for solving 3-SAT in worst-case $O^*(1.4802^n)$ appears in [DGH$^+$02]. Efficient (but non-polynomial) algorithms for NP-complete problems are surveyed in [Woe03].

The primary reference on NP-completeness is [GJ79], featuring a list of roughly 400 NP-complete problems. The book remains an extremely useful reference; it is perhaps the book I reach for most often. Good expositions of Cook's theorem [Coo71], where satisfiability is proven hard, include [CLRS09, GJ79, KT06]. The importance of Cook's result became clear in Karp's paper [Kar72], showing the hardness of more than twenty different combinatorial problems.

A linear-time algorithm for 2-SAT appears in [APT79]. See [WW95] for an interesting application of 2-SAT to map labeling. The best heuristic known approximates maximum 2-SAT to within a factor of 1.0741 [FG95].

Related problems: Constrained optimization (see page 478), traveling salesman problem (see page 594).

Chapter 18

Graph Problems: Polynomial Time

Algorithmic graph problems constitute roughly one third of the material in this catalog. Indeed, several problems from other sections could have been formulated equally well in terms of graphs, such as bandwidth minimization and finite-state automata optimization. Finding the right name of a given graph-theoretic invariant or problem is one of the primary skills of a good algorist. Indeed, the catalog will tell you exactly how to proceed, just as soon as you figure out your particular problem's name.

In this section, we will deal with problems that have efficient algorithms to solve them, with running times that grow polynomially with the size of the graph. There is often more than one way to model a given application, so it makes sense to look here before proceeding on to the harder formulations.

Graphs are often best understood as drawings. Many interesting graph properties follow from the nature of a type of drawing, such as planar graphs. We also discuss algorithms for drawing graphs, trees, and planar graphs.

Many advanced graph algorithms are difficult to program, but good implementations are available if you know where to look. The best general sources include LEDA [MN99] and the Boost Graph Library [SLL02]. However, better special-purpose codes exist for many problems.

See Atallah [AB17], Thulasiraman [TABN16], and van Leeuwen [vL90a] for up-to-date surveys on all areas of graph algorithms. Books of interest include:

- *Sedgewick* [SW11] – The graph algorithms volume of this algorithms text provides a comprehensive but gentle introduction to the field.

- *Ahuja, Magnanti, and Orlin* [AMO93] – While purporting to be a book on network flows, it covers the gamut of graph algorithms with an emphasis on operations research. Strongly recommended.

- *Even* [Eve11] – A respected advanced text on graph algorithms, with a particularly thorough treatment of planarity-testing algorithms.

© The Editor(s) (if applicable) and The Author(s), under exclusive license to Springer Nature Switzerland AG 2020
S. S. Skiena, *The Algorithm Design Manual*, Texts in Computer Science, https://doi.org/10.1007/978-3-030-54256-6_18

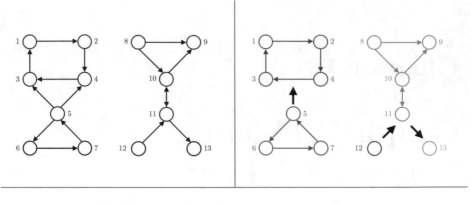

| Input | Output |

18.1   Connected Components

Input description: A directed or undirected graph G.

Problem description: Identify the different pieces or components of G, where vertices x and y are in different components when no path exists from x to y in G.

Discussion: The connected components of a graph represent in a gross sense the separate pieces of the graph. Two vertices are in the same component of G if and only if there exists some path between them.

Finding connected components is at the heart of many important graph applications. For example, consider the challenge of identifying natural clusters among a set of items. We represent each item by a vertex, and add an edge between each pair of items deemed "similar." The connected components of this graph correspond to the different classes of items.

Testing whether a graph is connected is an essential preprocessing step for every graph algorithm. Subtle, hard-to-detect bugs often result when an algorithm is run only on one component of a disconnected graph. Connectivity tests are so quick and easy that you should always verify the integrity of your input graph, even when you *know* that it just has to be connected.

Testing the connectivity of any undirected graph is a job for either depth-first or breadth-first search. Which one you choose doesn't really matter. Both traversals initialize a *component-number* field for each vertex to 0, and then start the search for component 1 from vertex v_1. As each vertex is discovered, the value of this field is set to the current component number. After the initial traversal ends, the component number is incremented, and the search begins again from the first vertex whose *component-number* remains 0. Properly implemented using adjacency lists (as is done in Section 7.7.1 (page 218)) this runs in $O(n + m)$ time.

Other notions of connectivity also arise in practice:

- *What if my graph is directed?* – There are two distinct notions of connected components for directed graphs. A directed graph is *strongly connected* if there is a directed path between every pair of vertices. A directed graph is *weakly connected* if it would be connected when ignoring the direction of edges. This distinction can be made clear by considering the network of one- and two-way streets in a given city. The network is strongly connected if it is possible to drive legally between any two places in town. The network is weakly connected when it is possible to legally or *illegally* drive between any pair of positions. The network is *disconnected* if there is no possible way to drive from some a to some b.

  Weakly and strongly connected components define unique partitions of the vertices. The output figure above shows a directed graph consisting of two weakly connected components, and five strongly connected components (also called *blocks* of G).

  Testing whether a directed graph is weakly connected can be done easily in linear time. Simply turn all edges of G into undirected edges and use the DFS-based connected components algorithm described previously. Tests for strong connectivity are somewhat more complicated. The simplest linear-time algorithm performs a depth-first search from any vertex v to demonstrate that the entire graph is reachable from v. We then construct a transpose graph G' where we reverse all the edges of G. A traversal of G' from v suffices to determine whether all vertices of G can reach v. Graph G is strongly connected iff all vertices can reach, and are reachable, from v.

  All the strongly connected components of G can be extracted in linear time using more sophisticated DFS-based algorithms. A generalization of the above "two-DFS" approach is deceptively easy to program, but somewhat subtle to understand exactly why it works:

  1. Perform a DFS, starting from an arbitrary vertex in G, and labeling each vertex in order of its completion (not discovery).
  2. Reverse the direction of each edge in G, yielding G'.
  3. Perform a DFS of G', starting from the highest numbered vertex in G. If this search does not completely traverse G', perform a new search starting from the highest-numbered unvisited vertex.
  4. Each DFS tree created in Step 3 defines a strongly connected component.

  My implementation of this two-pass algorithm appears in Section 7.10.2 (page 232). In either case, it is probably easier to start from an existing implementation than a textbook description.

- *What is the weakest point in my graph/network?* – A chain is only as strong as its weakest link. Losing one or more internal links causes a chain to become disconnected. The *connectivity* of graphs measures the strength of a graph—how many edges or vertices must be removed to disconnect it. Connectivity is an essential invariant for network design and other structural problems.

Algorithmic connectivity problems are discussed in Section 18.8 (page 568). In particular, *biconnected components* are pieces of the graph that result from cutting the edges incident on a single vertex. All biconnected components can be found in linear time using DFS. See Section 7.9.2 (page 225) for an implementation of this algorithm. Vertices whose deletion disconnects the graph belong to more than one biconnected component, whose edges are uniquely partitioned among the components.

- *Is the graph a tree? How can I find a cycle if one exists?* – The problem of cycle identification often arises, particularly with respect to directed graphs. For example, testing if a sequence of conditions can deadlock often reduces to cycle detection. If I am waiting for Fred, and Fred is waiting for Mary, and Mary is waiting for me, there is a cycle and we are all deadlocked.

  For undirected graphs, the analogous problem is tree identification. A tree is, by definition, an undirected, connected graph without any cycles. Depth-first search can be used to test whether a graph is connected. If the graph is connected and has $n - 1$ edges for n vertices, it is a tree.

  Depth-first search can be used to find cycles in both directed and undirected graphs. Whenever we encounter a back edge in our DFS—that is, an edge to an ancestor vertex in the DFS tree—the back edge and the tree together define a directed cycle. Directed graphs without cycles are called DAGs (directed acyclic graphs). Topological sorting (see Section 18.2 (page 546)) is the fundamental operation on DAGs.

Implementations: The graph data structure implementations of Section 15.4 (page 452) all include implementations of BFS/DFS, and hence connectivity testing to at least some extent. The C++ Boost Graph Library [SLL02] (http://www.boost.org/libs/graph/doc) provides implementations of connected components and strongly connected components. LEDA (see Section 22.1.1 (page 713)) provides these plus biconnected and triconnected components, breadth-first and depth-first search, connected components and strongly connected components, all in C++.

With respect to Java, *JUNG* (http://jung.sourceforge.net/) also provides biconnected component algorithms, while JGraphT (https://jgrapht.org/) does strongly connected components.

My (biased) preference for C language implementations of all basic graph connectivity algorithms, including strongly connected components and biconnected components, is the library associated with this book. See Section 22.1.9 (page 716) for details.

Notes: Depth-first search was first used to find paths out of mazes, and dates back to the nineteenth century [Luc91, Tar95]. Breadth-first search was first reported to find the shortest path by Moore in 1957 [Moo59].

Hopcroft and Tarjan [HT73b, Tar72] established depth-first search as a fundamental technique for efficient graph algorithms. Expositions on depth-first and breadth-first search appear in every book discussing graph algorithms, with [CLRS09] perhaps the most thorough description available.

The first linear-time algorithm for strongly connected components is due to Tarjan [Tar72], with expositions including [BvG99, Eve11, Man89]. Another algorithm—simpler to program and slicker—for finding strongly connected components is due to Sharir and Kosaraju. Good expositions of this algorithm appear in [AHU83, CLRS09]. Cheriyan and Mehlhorn [CM96] propose improved algorithms for certain problems on dense graphs, including strongly connected components.

DFS is hard to parallelize. Parallel algorithms for connected components on Map-Reduce and other models of computation are presented in [KLM+14, SRM14].

Random graphs exhibit interesting connectivity properties, such that once the number of edges exceeds a particular (surprisingly small) threshold, the graph is likely to contain one giant connected component and a small number of tiny components. For example, a random graph with only $n \ln 2 = 0.693n$ edges likely contains a connected component with $n/2$ vertices. By analogy, any large social network like Facebook presumably contains one large component with almost everyone except hermits and new users. This phenomena drives much of the notation that there are "six degrees of separation" between any two people in the world. The birth of the large component is covered in every book on the theory of random graphs, including [Bol01, JLR00].

Related problems: Edge-vertex connectivity (see page 568), shortest path (see page 554).

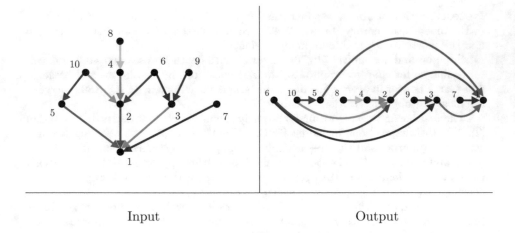

| Input | Output |

18.2   Topological Sorting

Input description: A directed acyclic graph $G = (V, E)$, also known as a *partial order* or *poset*.

Problem description: Find a linear ordering of the vertices of V such that for each edge $(i, j) \in E$, vertex i is to the left of vertex j.

Discussion: Topological sort arises as a subproblem in most algorithms on directed acyclic graphs. Topological sorting orders the vertices and edges of a DAG in a simple and consistent way, and hence plays the same role for DAGs that depth-first search does for general graphs.

Topological sort can be used to schedule jobs under precedence constraints. Suppose we have a set of tasks to do, but certain tasks must be performed before other ones. These precedence constraints form a directed acyclic graph, and any topological sort (also known as a *linear extension*) defines an order to perform these tasks such that each is done only after all of its constraints are satisfied.

Three important facts about topological sorting are:

1. *Only* DAGs can be topologically sorted, because any directed cycle provides an inherent contradiction to a linear order of tasks.

2. *Every* DAG can be topologically sorted, so there always is at least one schedule for any reasonable set of precedence constraints among jobs.

3. DAGs can usually be topologically sorted in many different ways, especially when there are few constraints. Consider n unconstrained jobs. All of the $n!$ permutations of these jobs constitute valid topological orderings.

The conceptually simplest linear-time algorithm for topological sorting performs a depth-first search of the DAG to identify the complete set of *source vertices*, meaning vertices of in-degree zero. At least one such source must exist in any DAG. Source vertices can appear at the front of any schedule without violating any constraints. Deleting all the outgoing edges of these source vertices will create new source vertices, which can then sit comfortably to the immediate right of the first set. We repeat until all vertices are accounted for. With a modest amount of care using the right data structures (adjacency lists and queues), this runs in $O(n + m)$ time.

An alternate algorithm makes use of the observation that ordering the vertices in terms of decreasing DFS finishing time yields a linear extension. An implementation of this algorithm with an argument for correctness is given in Section 7.10.1 (page 231).

Two special considerations with respect to topological sorting are:

- *What if I need all the linear extensions, instead of just one of them?* – Sometimes we need to construct *all* linear extensions of a DAG (say) to identify the best schedule according to a secondary criteria that satisfies all precedence constraints. Beware, because the number of linear extensions typically grows exponentially in the size of the DAG. Even the problem of counting the number of linear extensions is NP-hard.

  Algorithms for listing all linear extensions in a DAG are based on backtracking. They build all possible orderings from left to right, where each one of the in-degree zero vertices is a candidate for the next vertex. The outgoing edges from the selected vertex are deleted before moving on. An optimal algorithm for listing (or counting) linear extensions is discussed below.

  Algorithms to construct random linear extensions start from an arbitrary linear extension. We then repeatedly sample pairs of vertices. These are exchanged if the resulting permutation remains a topological ordering. This results in a uniformly selected linear extension if given enough random samples. See the Notes section for details.

- *What if your graph is not acyclic?* – When a set of constraints contains inherent contradictions, the natural problem becomes removing the smallest set of items that eliminates all inconsistencies. The sets of offending jobs (vertices) or constraints (edges) whose deletion leaves a DAG are known as a *feedback vertex set* or *feedback arc set*, respectively. They are discussed in Section 19.11 (page 618). Unfortunately, both problems are NP-complete.

  Because the DFS-based topological sorting algorithm gets stuck as soon as it identifies a vertex on a directed cycle, we can delete the offending edge or vertex and continue. This quick-and-dirty heuristic will eventually leave a DAG, but might delete more things than necessary. Section 12.4.2 (page 397) describes an approximation algorithm for this problem.

Implementations: Essentially all the graph data structure implementations of Section 15.4 (page 452) include implementations of topological sort. This includes the Boost Graph Library [SLL02] (http://www.boost.org/libs/graph/

doc) and LEDA (see Section 22.1.1 (page 713)) for C++. For Java, check out
JGraphT (`https://jgrapht.org/`).

The *Combinatorial Object Server* (`http://combos.org/`) provides C language programs to generate linear extensions in both lexicographic and Gray
code orders, as well as count them. An interactive interface is also provided.

My (biased) preference for C language implementations of all basic graph algorithms, including topological sorting, is the library associated with this book.
See Section 22.1.9 (page 716) for details.

Notes: Good expositions on topological sorting include [CLRS09, Man89]. No provably I/O-efficient algorithm for topological sorting of graphs in external memory is
known, but Ajwani [ACLZ11] reports his experience engineering a topological sort
for massive graphs. Brightwell and Winkler [BW91] prove that it is #P-complete to
count the number of linear extensions of a partial order, even for posets of height two
[DP18]. The complexity class #P includes NP, so any #P-complete problem must be
NP-hard.

Pruesse and Ruskey [PR86] give an algorithm that generates linear extensions of a
DAG in constant amortized time. Further, each extension differs from its predecessor
by either one or two adjacent transpositions. This algorithm can be used to count
the number of linear extensions $e(G)$ of an n-vertex DAG G in $O(n^2 + e(G))$. The
reverse search technique of Avis and Fukuda [AF96] can also be employed to list linear
extensions. A backtracking program to generate all linear extensions is described in
[KS74].

Huber [Hub06] gives an algorithm to sample linear extensions uniformly at random
from an arbitrary partial order in expected $O(n^3 \lg n)$ time, improving the result of
[BD99].

Related problems: Sorting (see page 506), feedback edge/vertex set (see page
618).

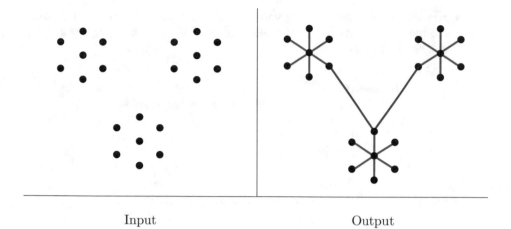

Input   Output

18.3 Minimum Spanning Tree

Input description: A graph $G = (V, E)$ with weighted edges.

Problem description: Find a subset of edges $E' \subset E$ that define a tree of minimum weight on V.

Discussion: The minimum spanning tree (MST) of a graph defines the cheapest subset of edges that connects the graph in a single component. Telephone companies are interested in minimum spanning trees, because the MST of a set of locations defines the wiring scheme that connects the sites using as little wire as possible. MST is the mother of all network design problems.

Minimum spanning trees prove important for several reasons:

- They can be computed quickly and easily, and create a sparse subgraph that reflects a lot about the original graph.

- They provide a way to identify clusters in sets of points. Deleting all long edges from a minimum spanning tree leaves connected components that define natural clusters in the data set. For example, deleting the two red edges from the output figure above leaves three natural clusters.

- They can be used to give approximate solutions to hard problems such as Steiner tree and traveling salesman.

- As an educational tool, MST algorithms provide graphic evidence that greedy algorithms sometimes produce provably optimal solutions.

Three classical algorithms efficiently construct minimum spanning trees. Detailed implementations of two of them (Prim's and Kruskal's) are given with correctness arguments in Section 8.1 (page 244). The third somehow manages to be less well known despite being invented first and (arguably) being both easier to implement and more efficient. The contenders are:

- *Kruskal's algorithm* – Every vertex starts as a separate tree. These trees are merged together by repeatedly adding the lowest cost edge that does not create a cycle.

> Kruskal(G)
>> Sort the edges in order of increasing weight
>> *count* = 0
>> while (*count* < $n - 1$) do
>>> get next edge (v, w)
>>> if (component $(v) \neq$ component(w))
>>>> add to T
>>>> component(v) = component(w)
>>>> count++

The "component(x)?" test can be efficiently implemented using the union–find data structure (Section 15.5 (page 456)) to yield an $O(m \lg m)$ algorithm.

- *Prim's algorithm* – Starts from an arbitrary vertex v to "grow" a tree, repeatedly finding the lowest-cost edge that links some new vertex into this tree. During execution, we label each vertex as either in the tree, in the *fringe* (meaning there exists an edge from a tree vertex), or *unseen* (meaning the vertex is still more than one edge away from the tree).

> Prim(G)
>> Select an arbitrary vertex to start
>> While (there are fringe vertices)
>>> select minimum-weight edge between tree and fringe
>>> add the selected edge and vertex to the tree
>>> update the cost to all affected fringe vertices

This creates a spanning tree for any connected graph, since no cycle can be introduced by adding edges between tree and fringe vertices. That this results in a tree of minimum weight can be proven by contradiction. Prim's algorithm can be implemented in $O(n^2)$ time using simple data structures.

- *Boruvka's algorithm* – This rests on the observation that the lowest-weight edge incident on every vertex must appear in the minimum spanning tree. The union of these edges will result in a spanning forest with at most $n/2$ trees. Now select an edge (x, y) of lowest weight for each of tree T, such that $x \in T$ and $y \notin T$. All of these edges must again appear in a minimum spanning tree, and so again results in a spanning forest with at most half as many trees as before:

    Boruvka(G)
        Initialize spanning forest F to n single-vertex trees
            While (F has more than one tree)
                for each T in F, find the smallest edge from T to $G - T$
                add all selected edges to F, thus merging pairs of trees

The number of trees are at least halved in each round, so we get the MST after at most $\lg n$ iterations, each of which takes linear time. This gives an $O(m \log n)$ algorithm without using any fancy data structures.

MST is only one of several spanning tree problems that arise in practice. The following questions will help you sort your way through them:

- *Are all edges of your graph of identical weight?* – Every spanning tree on n points contains exactly $n - 1$ edges. Thus, if your graph is unweighted, *any* spanning tree must be a minimum spanning tree. Either breadth-first or depth-first search can be used to find a rooted spanning tree in linear time. DFS trees tend to be long and thin, while BFS trees better reflect the distance structure of the graph.

- *Should I use Prim's or Kruskal's algorithm?* – As implemented in Section 8.1 (page 244), Prim's algorithm runs in $O(n^2)$, while Kruskal's algorithm takes $O(m \log m)$ time. Prim's algorithm is thus faster on dense graphs, while Kruskal's is faster on sparse graphs. That said, Prim's algorithm can be implemented in $O(m + n \lg n)$ time using more advanced data structures, so a Prim's implementation using pairing heaps would be the fastest practical choice for both sparse and dense graphs.

- *What if my input consists of points in the plane, instead of a graph?* – Geometric instances, comprising n points in d-dimensions, can be solved by constructing the complete distance graph in $O(n^2)$ and then finding the MST of this complete graph. But for points in two dimensions it proves more efficient to solve the geometric version of the problem directly. First construct the Delaunay triangulation of the points (see Sections 20.3 and 20.4), which gives a graph with $O(n)$ edges containing all the edges of the minimum spanning tree of the point set. Running Kruskal's algorithm on this sparse graph finishes the job in $O(n \lg n)$ time.

- *How can I find a spanning tree that avoids vertices of high degree?* – Another common goal of spanning tree problems is to minimize the maximum degree, typically to minimize the fan out in an interconnection network. Unfortunately, finding a spanning tree of maximum degree 2 is NP-complete, because this is identical to the Hamiltonian path problem. But efficient algorithms are known that can construct spanning trees whose maximum degree is at most one more than required, which should suffice in practice. See the references below.

Implementations: All the graph data structure implementations of Section 15.4 (page 452) include implementations of Prim's and/or Kruskal's algorithms. This includes the Boost Graph Library [SLL02] (http://www.boost.org/libs/graph) and LEDA (see Section 22.1.1 (page 713)) for C++. JGraphT (https://jgrapht.org/) has an extensive graph algorithm library in Java, including Boruvka, Kruskal, and Prims, plus other variants.

Timing experiments on minimum spanning tree algorithms produce contradicting results, suggesting the stakes are really too low to matter. Pascal implementations of Prim's, Kruskal's, and the Cheriton–Tarjan algorithm are provided in [MS91], along with extensive empirical analysis proving that Prim's algorithm with the appropriate priority queue is fastest on most graphs. Kruskal's algorithm proved the fastest of four different MST algorithms in the Stanford GraphBase (see Section 22.1.7 (page 715)).

Combinatorica [PS03] provides Mathematica implementations of Kruskal's MST algorithm and a combinatorial method to efficiently count the number of spanning trees of a graph. See Section 22.1.8 (page 716).

My (biased) preference for C language implementations of all basic graph algorithms, including minimum spanning trees, is the library associated with this book. See Section 22.1.9 (page 716) for details.

Notes: The minimum spanning tree problem dates back to Boruvka's algorithm in 1926, well before Prim's [Pri57] and Kruskal's [Kru56] algorithms. Prim's algorithm was then rediscovered by Dijkstra [Dij59]. See [GH85] for more on the interesting history of MST algorithms. Wu and Chao [WC04] have written a monograph on minimum spanning tree and related problems.

The fastest implementations of Prim's and Kruskal's algorithms use Fibonacci heaps [FT87]. However, pairing heaps have been proposed to realize the same bounds with less overhead. Experiments with pairing heaps are reported in [SV87]. Efficient parallel algorithms for models like Map-Reduce are presented by Andoni [ANOY14].

A simple combination of Boruvka's algorithm with Prim's algorithm yields an $O(m \lg \lg n)$ algorithm. Run Borukva's algorithm for $\lg \lg n$ iterations, yielding a forest of at most $n/\lg n$ trees. Now create a graph G' with one vertex representing each tree in this forest, with the weight of the edge between trees T_i and T_j set to the lightest edge (x, y), where $x \in T_i$ and $y \in T_j$. The MST of G' coupled with the edges selected by Boruvka's algorithm yields the MST of G. Prim's algorithm (implemented with Fibonacci heaps) will take $O(m)$ time on this $n/\lg n$ vertex, m edge graph.

The best theoretical bounds on finding minimum spanning trees tell a complicated story. Karger, Klein, and Tarjan [KKT95] give a linear-time randomized algorithm for MSTs, based again on Borukva's algorithm. Chazelle [Cha00] give a deterministic $O(n\alpha(m, n))$ algorithm, where $\alpha(m, n)$ is the inverse Ackerman function. Pettie and Ramachandran [PR02] give a provably optimal algorithm whose exact running time is (paradoxically) unknown, but lies between $\Omega(n + m)$ and $O(n\alpha(m, n))$.

A *spanner* $S(G)$ of a given graph G is a subgraph that offers an effective compromise between two competing network objectives. To be precise, $S(G)$ must have total weight close to the MST of G while guaranteeing that the shortest path between vertices x and y in $S(G)$ approaches the shortest path in the full graph G. The monograph of Narasimhan and Smid [NS07] provides a complete, up-to-date survey on spanner networks.

The $O(n \log n)$ algorithm for Euclidean MSTs is due to Shamos, and discussed in computational geometry texts such as [dBvKOS08, PS85].

Fürer and Raghavachari [FR94] give an algorithm that constructs a spanning tree whose maximum degree is almost minimized—indeed is at most one more than the lowest-degree spanning tree. The situation is analogous to Vizing's theorem for edge coloring, which also gives an approximation algorithm with an additive factor of one. A recent generalization [SL07] gives a polynomial-time algorithm for finding a spanning tree of maximum degree $\leq k + 1$ whose cost is no more than that of the optimal minimum spanning tree of maximum degree $\leq k$.

Minimum spanning tree algorithms have an interpretation in terms of *matroids*, which are systems of subsets closed under inclusion. The maximum weighted independent set in matroids can be found using a greedy algorithm. The connection between greedy algorithms and matroids was established by Edmonds [Edm71]. Expositions on the theory of matroids include [GM12, Law11, PS98].

Dynamic graph algorithms seek to maintain a graph invariant (such as the MST) efficiently under edge insertion or deletion operations. Holm et al. [HdlT01] gives an efficient, deterministic algorithm to maintain MSTs (and several other invariants) in amortized polylogarithmic time per update.

Algorithms for generating spanning trees in order from minimum to maximum weight are presented in [Gab77]. The complete set of spanning trees of an unweighted graph can be generated in constant amortized time. See Ruskey [Rus03] for an overview of algorithms to generate, rank, and unrank spanning trees.

Related problems: Steiner tree (see page 614), traveling salesman (see page 594).

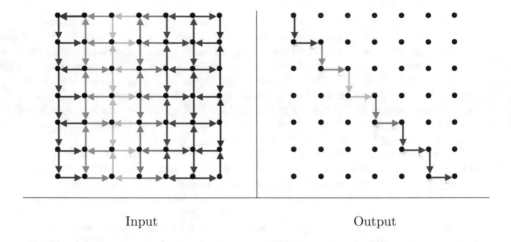

Input	Output

18.4   Shortest Path

Input description: An edge-weighted graph G, with vertices s and t.

Problem description: Find the shortest path from s to t in G.

Discussion: The problem of finding shortest paths in a graph has many applications, some quite surprising:

- Shortest paths often arise in transportation or communications problems, such as finding the best route to drive between Chicago and Phoenix, or the fastest way to direct packets across a network.

- *Segmentation* is the task of partitioning a digitized image into regions containing distinct objects. Region boundaries can be thought of as paths over the image that avoid cutting through object pixels as much as possible. This grid of pixels can be modeled as a graph, with the cost of an edge reflecting the color transitions between neighboring pixels. The shortest path in such a weighted graph defines a promising potential boundary between two regions.

- Recall the dialing for documents war story of Section 8.4 (page 264), where we sought to use some notion of grammatical constraints to identify the right word for each slot in a sentence from a set of possibilities. Such tasks are common in natural language processing and speech recognition systems. We construct a graph whose vertices correspond to these possible word interpretations, with edges between neighboring word-interpretations. If we set the weight of each edge to reflect the likelihood of transition, the shortest path across this graph defines the best interpretation of the sentence.

- In an informative drawing of a graph, the "center" of the graph should appear near the center of the page. A good definition of the graph center is the vertex that minimizes the maximum distance to any other vertex in the graph. Identifying this center point requires knowing the shortest path distance between all pairs of vertices.

The primary algorithm for finding shortest paths is *Dijkstra's algorithm*, which efficiently computes the shortest path from a given starting vertex x to all $n - 1$ other vertices. In each iteration, it identifies a new vertex v for which the shortest path from x to v is known. We maintain a set of vertices S to which we know the shortest path from x, and grow this set by one new vertex in each iteration. In each iteration, we identify the edge (u, v) where $u \in S$ and $v \in V - S$ such that

$$dist(x, v) = \min_{u \in S} (dist(x, u) + weight(u, v))$$

This edge (u, v) gets added to a *shortest path tree*, whose root is x and describes all the shortest paths from x.

An $O(n^2)$ implementation of Dijkstra's algorithm appears in Section 8.3.1 (page 258). Faster times can be achieved using more complicated data structures, as described below. If we just need to know the shortest path from x to y, terminate the algorithm as soon as y enters S.

Dijkstra's algorithm is the right choice to compute single-source shortest path on positively weighted graphs. However, special circumstances sometimes dictate different choices:

- *Is your graph weighted or unweighted?* – If your graph is unweighted, a simple breadth-first search starting from the source vertex will find the shortest path to all other vertices in linear time. Only when edges have different weights will you need more sophisticated algorithms.

- *Does your graph have negative cost weights?* – Dijkstra's algorithm assumes that all edges have positive cost. For graphs with edges of negative weight, you must use the more general (but less efficient) Bellman–Ford algorithm. Graphs with negative cost cycles are an even bigger problem. The shortest x to y path in such a graph is not defined, because we can repeatedly detour from x to the negative cost cycle, making the total cost arbitrarily small.

  Note that adding a fixed amount of weight to make each edge positive *does not* solve the problem. Dijkstra's algorithm will then favor paths using a small number of edges, even if those were not the shortest weighted paths in the original graph.

- *Is your input a set of geometric obstacles, instead of a graph?* – Many applications seek the shortest path between two points in a geometric setting, such as an obstacle-filled room. The most straightforward solution is to convert your problem into a graph of distances to feed to Dijkstra's algorithm. Vertices correspond to the boundary vertices of the obstacles, with edges defined between pairs of vertices that "see" each other.

There are more efficient geometric algorithms that compute the shortest path directly from the arrangement of obstacles. See Section 20.14 (page 667) on motion planning for pointers to such algorithms.

- *Is your graph acyclic—that is, a DAG?* – Shortest paths in directed acyclic graphs can be found in linear time. Perform a topological sort to order the vertices such that all edges go from left to right, starting from source s. The distance from s to itself, $dist(s, s)$, clearly equals 0. We now process the vertices from left to right. Observe that

$$dist(s, j) = \min_{(i,j) \in E} \left(dist(s, i) + weight(i, j) \right)$$

  since we already know the shortest path $dist(s, i)$ for all vertices i to the left of j. Indeed, most dynamic programming problems can be formulated as shortest paths on specific DAGs. This same algorithm (replacing min with max) also suffices to find the *longest path* in a DAG, which proves useful in applications like scheduling (see Section 17.9 (page 534)).

- *Do you need the shortest path between all pairs of points?* – The naive approach to calculate the all-pairs shortest path matrix D (where D_{ij} is the distance from i to j) is to run Dijkstra n times, once with each vertex as the source. The Floyd-Warshall algorithm is a slick $O(n^3)$ dynamic programming algorithm for all-pairs shortest path, which is faster and easier to program than Dijkstra. It works with negative cost edges (but not cycles), and is presented with an implementation in Section 8.3.2 (page 261). Let M denote the edge weight matrix, where $M_{ij} = \infty$ if there is no edge (i, j):

$$D^0 = M$$
  for $k = 1$ to n do
      for $i = 1$ to n do
          for $j = 1$ to n do
              $D_{ij}^k = \min(D_{ij}^{k-1}, D_{ik}^{k-1} + D_{kj}^{k-1})$
  Return D^n

  The key to understanding Floyd's algorithm is that D_{ij}^k denotes "the length of the shortest path from i to j that goes through vertices $1, \ldots, k$ as possible intermediate vertices." Note that $O(n^2)$ space suffices, since we need only keep D^k and D^{k-1} around at time k.

- *How do I find the shortest cycle in a graph?* – One application of all-pairs shortest path is to find the shortest cycle in a graph, called its *girth*. Floyd's algorithm can be used to compute d_{ii} for $1 \leq i \leq n$, which is the length of the shortest way to get from vertex i to i—in other words, the shortest cycle through i.

This *might* be what you want. But the shortest cycle through x is likely to go from x to y back to x, using the same edge twice. A *simple* cycle is one that visits no edge or vertex twice. To find the shortest simple cycle, we compute the lengths of the shortest paths from i to all other vertices, and then explicitly check whether there is an acceptable edge from each vertex back to i.

The problem of finding the *longest* cycle in a graph includes Hamiltonian cycle as a special case (see Section 19.5), so it is NP-complete.

The all-pairs shortest path matrix can be used to compute several useful invariants related to the center of graph G. The *eccentricity* of vertex v is the shortest-path distance to the farthest vertex from v. From the eccentricity come other graph invariants. The *radius* of a graph is the smallest eccentricity of any vertex, while the *center* is the set of vertices whose eccentricity is the radius. The *diameter* of a graph is the maximum eccentricity of any vertex.

Implementations: The highest performance shortest path codes are due to Andrew Goldberg and his collaborators, at `http://www.avglab.com/andrew/soft.html`. In particular, MLB is a C++ short path implementation for non-negative, integer-weighted edges. See [Gol01] for details of the algorithm and its implementation. Its running time is typically only four or five times that of a breadth-first search, and it is capable of handling graphs with millions of vertices. High-performance C implementations of both Dijkstra and Bellman–Ford are also available.

All the C++ and Java graph libraries discussed in Section 15.4 (page 452) include at least an implementation of Dijkstra's algorithm. The C++ Boost Graph Library [SLL02] (`http://www.boost.org/libs/graph`) has a particularly broad collection, including Bellman–Ford's and Johnson's all-pairs shortest-path algorithm. LEDA (see Section 22.1.1 (page 713)) provides good implementations in C++ for all of the shortest-path algorithms we have discussed, including Dijkstra, Bellman–Ford, and Floyd's algorithms. JGraphT (`https://jgrapht.org`) provides both Dijkstra and Bellman–Ford in Java.

Shortest-path algorithms was the subject of the 9th DIMACS Implementation Challenge, held in October 2006. Implementations of efficient algorithms for finding shortest paths were discussed. The papers, instances, and implementations are available at `http://dimacs.rutgers.edu/programs/challenge/`.

Notes: Good expositions on Dijkstra's algorithm [Dij59], the Bellman–Ford algorithm [Bel58, FF62], and Floyd's all-pairs-shortest-path algorithm [Flo62] include [CLRS09]. Nice surveys on shortest path algorithms include [MAR+17, Zwi01]. Geometric shortest-path algorithms are surveyed by Mitchell [PN18].

The fastest algorithm known for single-source shortest-path is Dijkstra's algorithm with Fibonacci heaps, running in $O(m + n \log n)$ time [FT87]. Experimental studies of shortest-path algorithms include [DF79, DGKK79]. However, these experiments were done before Fibonacci heaps were developed. See [CGR99] for a more recent study. Heuristics can be used to enhance the performance of Dijkstra's algorithm in practice. Holzer, et al. [HSWW05] provide a careful experimental study of how four such heuristics interact together.

Online services like Google Maps quickly find at least an approximate shortest path between two points in enormous road networks. This problem differs somewhat from the shortest-path problems here in that (1) preprocessing costs can be amortized over many point-to-point queries, (2) the backbone of high-speed, long-distance highways can reduce the path problem to identifying the best place to get on and off this backbone, and (3) approximate or heuristic solutions suffice in practice.

The A^*-algorithm performs a best-first search for the shortest path coupled with a lower-bound analysis to establish when the best path we have seen is indeed the shortest-path in the graph. Goldberg, Kaplan, and Werneck [GKW06] describe an implementation of A^* capable of answering point-to-point queries in one millisecond on national-scale road networks after two hours of preprocessing. Heuristics for speeding up shortest path algorithms are analyzed in [AFGW10].

Many applications demand multiple short alternative paths, in addition to the optimal path. This motivates the problem of finding the k shortest paths. Variants exist depending upon whether the paths must be simple, or can contain cycles. Eppstein [Epp98] generates an implicit representation of these paths in $O(m + n \log n + k)$ time, from which each path can be reconstructed in $O(n)$ time. Hershberger et al. [HMS03] presents a new algorithm and experimental results.

Fast algorithms for computing the girth are known for both general [IR78] and planar graphs [Dji00].

Related problems: Network flow (see page 571), motion planning (see page 667).

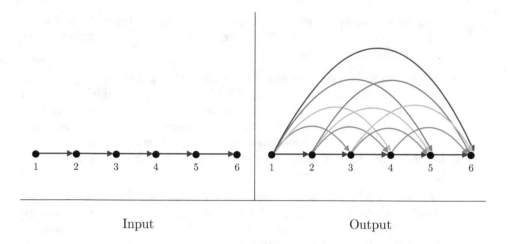

Input   Output

18.5 Transitive Closure and Reduction

Input description: A directed graph $G = (V, E)$.

Problem description: For *transitive closure*, construct a graph G' with edge $(i, j) \in E'$ iff there is a directed path from i to j in G. For *transitive reduction*, construct a graph G' with the smallest number of edges such that a directed path from i to j exists in G' iff there is a directed path from i to j in G.

Discussion: Transitive closure can be thought of as establishing a data structure to efficiently solve reachability queries: "Can I get to y from x?" After constructing a transitive closure matrix M, such queries can be answered in constant time by reporting matrix entry $M[x, y]$.

Transitive closure arises in propagating the consequences of modified attributes of a graph G. Consider the graph underlying a spreadsheet model, whose vertices are cells with an edge from cell i to cell j iff the result of cell j depends on cell i. When a given cell is modified, the values of all cells reachable from it must also be updated. The identity of these cells is revealed by the transitive closure of G. Many database problems reduce to computing transitive closures, for analogous reasons.

There are three basic algorithms for computing transitive closure:

- The simple algorithm performs a breadth-first or depth-first search from each vertex, and keeps track of all vertices encountered. Doing n such traversals gives an $O(n(n + m))$ algorithm, which degenerates to cubic time when the graph is dense. This algorithm is easily implemented, runs well on sparse graphs, and is likely the right answer for your application.

- Warshall's algorithm constructs the transitive closure in $O(n^3)$ time, using a slick approach identical to Floyd's all-pairs shortest-path algorithm of Section 18.4 (page 554). If we are not interested in the length of the resulting paths, we can reduce storage by retaining only one bit per matrix element. Thus, $D_{ij}^k = $ true iff j is reachable from i using only vertices $1, \ldots, k$ as intermediates.

- Matrix multiplication can also be used to solve transitive closure. Let M^1 be the adjacency matrix of graph G. The non-zero matrix entries of $M^2 = M \times M$ identify all length-2 paths in G. Observe that $M^2[i, j] = \sum_x M[i, x] \cdot M[x, j]$, so path (i, x, j) contributes to $M^2[i, j]$. Thus, the union $\cup_i^n M^i$ yields the transitive closure T. Furthermore, this union can be computed using only $O(\lg n)$ matrix operations using the fast exponentiation algorithm in Section 16.9 (page 493).

  This will be faster for large enough n if using Strassen's fast matrix multiplication algorithm, although I for one wouldn't bother trying. Transitive closure is provably as hard as matrix multiplication, so there is little hope for a significantly faster algorithm.

The running time for all three of these methods can be substantially improved on many graphs. Recall that a strongly connected component is a set of vertices for which all pairs are mutually reachable. For example, any directed cycle defines a strongly connected subgraph. All the vertices in any strongly connected component must reach exactly the same subset of G. Thus, we can reduce our problem to finding the transitive closure on a graph of strongly-connected components, that generally has fewer edges and vertices than G. The strongly connected components of G can be computed in linear time (see Section 18.1 (page 542)).

Transitive reduction (also known as *minimum equivalent digraph*) is the inverse operation to transitive closure, namely reducing the number of edges while maintaining identical reachability properties. The transitive closure of G is identical to the transitive closure of the transitive reduction of G. Transitive reduction is space minimization, by eliminating redundant edges from G that do not affect reachability. It also arises in graph drawing, where it is important to eliminate as many implied edges as possible to reduce visual clutter.

Although the transitive closure of G is uniquely defined, a graph may have many different transitive reductions, including G itself. We seek the smallest such reduction, but there are multiple formulations of the problem:

- A quick-and-dirty, linear-time transitive reduction algorithm identifies the strongly connected components of G, replaces each by a simple directed cycle, and then adds the edges bridging the different components. Although this does not always yield the smallest possible reduction, it is likely to be pretty close on many graphs.

  One catch with this heuristic is that it might add edges to the transitive reduction of G that are not in G. Depending on your application, this may or may not be a problem.

- If we are restricted to only using edges from G in our transitive reduction, we must abandon hope of finding the minimum possible reduction. To see why, consider a directed graph consisting of one strongly connected component, where every vertex can reach every other vertex. The smallest possible transitive reduction here will be a simple directed cycle, consisting of exactly n edges. But this is possible if and only if G is Hamiltonian, thus proving that finding the smallest such closure is NP-complete.

A heuristic for edge-preserving transitive reduction successively considers each edge, deleting it if its removal does not change the transitive closure. Implementing this efficiently means minimizing the time spent on reachability tests. Observe that directed edge (i, j) can be eliminated whenever there is another path from i to j avoiding this edge.

- The minimum size reduction using arbitrary pairs of vertices as edges can be found in $O(n^3)$ time. See the references below for details. However, the quick-and-dirty heuristic described above will likely suffice for most applications, being both faster and easier to program.

Implementations: The Boost Graph Library [SLL02] (`http://www.boost.org/libs/graph`) implementations of transitive closure and reduction are particularly well engineered. LEDA (see Section 22.1.1 (page 713)) also provides implementations of both transitive closure and reduction in C++ [MN99]. The extensive graph algorithm library of JGraphT (`https://jgrapht.org/`) contains Java implementations of both algorithms.

Combinatorica [PS03] provides Mathematica implementations of transitive closure and reduction, as well as the display of partial orders requiring transitive reduction. See Section 22.1.8 (page 716).

Notes: Van Leeuwen [vL90a] provides an excellent survey on transitive closure and reduction. The equivalence between matrix multiplication and transitive closure was proven by Fischer and Meyer [FM71], with expositions including [AHU74].

There is a surprising amount of more recent activity on transitive closure, much of it captured in Nuutila [Nuu95]. Penner and Prasanna [PP06] improved the performance of Warshall's algorithm [War62] by roughly a factor of two through a cache-friendly implementation.

The equivalence between transitive closure and reduction was established in [AGU72]. Empirical studies of transitive closure algorithms include [Nuu95, PP06, SD75]. Sparse transitive reductions often contain long paths. *Transitive closure spanners* seek small reductions of diameter at most k [BGJ$^+$12].

Estimating the size of the transitive closure is important in database query optimization. A linear-time algorithm for estimating the size of the closure is given by Cohen [Coh94].

Related problems: Connected components (see page 542), shortest path (see page 554).

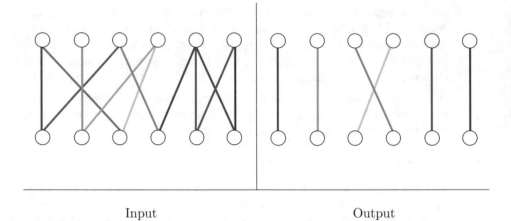

Input   Output

18.6   Matching

Input description: A (weighted) graph $G = (V, E)$.

Problem description: Find the largest set of edges E' from E such that every vertex in V is incident to at most one edge of E'.

Discussion: Suppose you manage a team of workers, each capable of performing a distinct subset of the tasks needed to complete a job.   Who should get assigned to do what? Construct a graph with vertices representing each worker and each task, with edges linking each worker to the tasks they can perform. Each worker is to be assigned one task. The desired assignment is the largest possible set of edges where no employee or job is repeated—that is, a matching.

Matching is a very powerful piece of algorithmic magic, so powerful that it is quite surprising that optimal matchings can be found efficiently, in polynomial time. Applications arise often once you know to look for them.

Marrying off a set of boys to a set of girls such that each couple is happy is another example of a bipartite matching problem. Here the graph contains an edge between any compatible boy and girl. For a synthetic biology application [MPC+06], we needed to shuffle the characters in a string S so as to maximize the number of characters that move. For example, *aaabc* can be rearranged to *bcaaa* so that only the central *a* stays fixed. This is yet another bipartite matching problem, where the boys represent the multiset of alphabet symbols and the girls are the positions in the string (1 to $|S|$). Edges link symbols to all the string positions that originally contained a different symbol.

This basic matching framework can be enhanced in several ways, while remaining essentially the same *assignment* problem:

- *Is your graph bipartite?* – Many matching problems involve bipartite graphs, as in the classic assignment problem of boys to girls. But modern marriages represent a matching on a more general, non-bipartite graph. Efficient algorithms exist for matching on general graphs, although faster and simpler methods work for bipartite graphs.

- *What if employees can be given multiple jobs?* – Natural generalizations of matching include assigning workers more than one task to do, or (equivalently) seeking multiple workers for a given job. Such desires can be modeled by replicating an employee vertex by as many times as we want her to be matched. Indeed, we employed this trick (using multiple occurrences of the same letters) in the string shuffle example above.

- *Is your graph weighted or unweighted?* – The matching applications discussed so far are based on unweighted graphs. We have sought a maximum *cardinality* matching—ideally a *perfect* matching where every vertex is matched to another in the matching.

  But other applications augment each edge with a weight, perhaps reflecting the suitability of a particular employee for a given task, or how much person x likes person y. The problem now becomes constructing a maximum *weight* matching—that is, finding the set of indpendent edges of maximum total cost.

Efficient algorithms for constructing matchings work by finding *augmenting paths* in graphs, which start and end with unmatched edges. An augmenting path P in graph G for a given (partial) matching M is a path of edges that alternate (out-of-M, in-M, ..., out-of-M). We can always enlarge the matching by one edge given such an augmenting path, by replacing the even-numbered edges of P from M with the odd-numbered edges of P. Berge's theorem states that a matching is maximum iff it does not contain any augmenting path. Therefore, we can construct maximum-cardinality matchings by searching for augmenting paths, and stop when none exist.

General graphs prove trickier to match than bipartite ones, because it is possible to have augmenting paths involving odd-length cycles, where the first and last vertices are the same. Such cycles (or blossoms) are impossible in bipartite graphs, which by definition do not contain odd-length cycles.

The standard algorithms for bipartite matching are based on network flow, using a simple transformation to convert a bipartite graph into an equivalent flow graph. Indeed, an implementation of this is given in Section 8.5 (page 267).

Be warned that different approaches are needed to solve weighted matching problems, most notably the matrix-oriented "Hungarian algorithm."

Implementations: High-performance codes for both weighted and unweighted bipartite matching have been developed by Andrew Goldberg and his collaborators. CSA is a weighted bipartite matching code in C based on cost-scaling network flow, developed by Goldberg and Kennedy [GK95]. BIM is a faster unweighted bipartite matching code based on augmenting path methods, developed by Cherkassky, et al. [CGM+98]. Both are available for non-commercial use from http://www.avglab.com/andrew/soft.html.

The First DIMACS Implementation Challenge [JM93] focused on network flows and matching. Several instance generators and implementations for maximum weight and maximum cardinality matching were collected, and can be obtained from `http://dimacs.rutgers.edu/archive/Challenges/`. These include a maximum-cardinality matching solver that implements Gabow's $O(n^3)$ algorithm and a maximum-weighted matching solver, both in C by Edward Rothberg. This is slower but more general than his unweighted solver.

LEDA (see Section 22.1.1 (page 713)) provides efficient implementations in C++ for both maximum-cardinality and maximum-weighted matching, on both bipartite and general graphs. *Blossom IV* [CR99] is an efficient code in C for minimum-weight perfect matching available at `http://www.math.uwaterloo.ca/~bico//software.html`. An $O(mn\alpha(m,n))$ implementation of maximum-cardinality matching in general graphs (`http://www.cs.arizona.edu/~kece/Research/software.html`) is due to Kececioglu and Pecqueur [KP98].

The Stanford GraphBase (see Section 22.1.7 (page 715)) contains an implementation of the Hungarian algorithm for bipartite matching. To provide readily visualized weighted bipartite graphs, Knuth uses a digitized version of the *Mona Lisa* and seeks row/column disjoint pixels of maximum brightness. Matching is also used to construct amusing "domino portraits."

Notes: Lovász and Plummer [LP09] is the definitive reference on matching theory and algorithms. Survey articles on matching algorithms include [Gal86]. Good expositions on network flow algorithms for bipartite matching include [CLRS09, Eve11, Man89], and those on the Hungarian method include [Law11, PS98]. The best algorithm for maximum bipartite matching, due to Hopcroft and Karp [HK73], repeatedly finds the shortest augmenting paths instead of using network flow, and runs in $O(\sqrt{n}m)$. The Hungarian algorithm runs in $O(n(m + n \log n))$ time.

Edmond's algorithm [Edm65] for maximum-cardinality matching is of great historical interest, for provoking questions on what problems can be solved in polynomial time. Expositions on Edmond's algorithm include [Law11, PS98, Tar83]. The best algorithm known for general matching runs in $O(\sqrt{n}m)$ [MV80].

Consider a matching of boys to girls containing edges (B_1, G_1) and (B_2, G_2), where B_1 and G_2 prefer each other to their own spouses. In real life, these two would run off with each other, breaking the marriages. A marriage without any such couples is said to be *stable*. The theory of stable matching is presented in [GI89]. It is a surprising fact that no matter how the boys and girls rate each other, there is always at least one stable marriage. Further, such a marriage can be found in $O(n^2)$ time [GS62]. An important application of stable marriage occurs in the annual matching of medical residents to hospitals.

Online matching problems arise when edge selection must take place without full information about the graph. Such problems arise in Internet advertising, and are ably surveyed in [M+13].

The maximum matching is equal in size to the minimum vertex cover in bipartite graphs. This implies that both the minimum vertex cover problem and maximum independent set problems can be solved in polynomial time on bipartite graphs.

Related problems: Network flow (see page 571). vertex cover (see page 591).

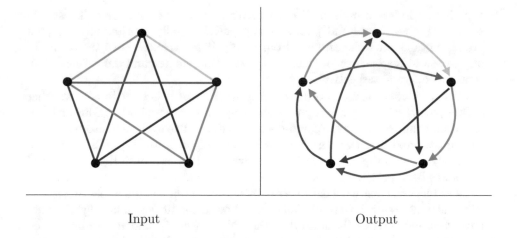

Input   Output

18.7 Eulerian Cycle/Chinese Postman

Input description: A graph $G = (V, E)$.

Problem description: Find the shortest tour visiting every edge of G.

Discussion: Suppose you are charged with designing the daily routes for garbage trucks, snow plows, or postal workers. Every road in the city must be completely traversed at least once in all these applications, to ensure that all deliveries or pickups can be made. For efficiency, you seek to minimize total drive time, or (equivalently) the total distance or number of edges traversed.

Alternatively, consider a human-factors validation of telephone menu systems. Each "Press 4 for more information" option is properly interpreted as a directed edge between two vertices in a graph. Our tester seeks the most efficient way to walk over this graph and visit every link in the system at least once.

Both tasks are variants of the *Eulerian cycle* problem, best characterized by the puzzle that asks children to draw a given figure completely without repeating any edges, or lifting their pencil off the paper. They seek a path or cycle through a graph that visits each edge exactly once.

Well-known conditions exist for determining whether a graph contains an Eulerian cycle:

- An *undirected* graph contains an Eulerian *cycle* iff it is connected, and each vertex is of even degree.

- A *directed* graph contains an Eulerian *cycle* iff it is strongly connected, and each vertex has the same in-degree as out-degree.

For Eulerian *paths*, which cover all edges but do not return to the starting vertex, the degree conditions are weakened. An *undirected* connected graph contains an Eulerian path iff all but two vertices are of even degree. These will serve as the start and end points of any path. Finally, a *directed* connected graph contains an Eulerian path from x to y iff all vertices have the same in-degree as out-degree, except that x has in-degree one less and y one more than their out-degrees, respectively.

These characterizations of Eulerian graphs make it is easy to test whether such a path/cycle exists: verify that the graph is connected using DFS or BFS, and then count the number of odd-degree vertices. The cycle can also be explicitly constructed in linear time by Hierholzer's algorithm. Use DFS to find an arbitrary cycle in the graph. Delete this cycle and repeat, until the entire set of edges has been partitioned into a set of edge-disjoint cycles. Because deleting a cycle reduces the degree of each vertex by an even number, the remaining graph will continue to satisfy the Eulerian degree-bound conditions. These cycles must have common vertices (because the graph is connected), and so can be spliced together in a "figure eight" at any shared vertex. By splicing all extracted cycles together, we get a single circuit containing all of the edges.

An Eulerian cycle, if one exists, solves the motivating snowplow problem, since any tour that visits every edge only once must have minimum length. However, it is unlikely that your road network will satisfy the Eulerian degree conditions. Instead, we need to solve the more general *Chinese postman problem*, which minimizes the length of a cycle that traverses every edge at least once. This minimum cycle will never visit any edge more than twice, so good tours exist for any road network.

The optimal postman tour can be constructed by adding the appropriate edges to the graph G to make it Eulerian. Adding a path between two odd-degree vertices in G makes both of them even-degree, moving G closer to becoming an Eulerian graph.

Finding the best set of shortest paths to add to G reduces to identifying a minimum-weight perfect matching in a special graph G'. For undirected graphs, the vertices of G' correspond to the odd-degree vertices of G, with the weight of edge (i, j) defined by the length of the shortest path from i to j in G. For directed graphs, the vertices of G' correspond to the degree-imbalanced vertices from G. All edges in G' go from out-degree deficient vertices to in-degree deficient ones. Thus, bipartite matching algorithms suffice when G is directed. An optimal cycle can be extracted in linear time, once the graph is Eulerian.

Implementations: Several graph libraries provide implementations of Eulerian cycles, but Chinese postman implementations are rarer. I recommend the implementation of directed Chinese postman by Thimbleby [Thi03]. This Java implementation is available at `http://www.harold.thimbleby.net/cpp/index.html`. JGraphT (`https://jgrapht.org`) provides an implementation of Hierholzer's algorithm in Java.

GOBLIN (`http://goblin2.sourceforge.net/`) is an extensive C++ library dealing with all of the standard graph optimization problems, including Chinese postman for both directed and undirected graphs. LEDA (see Section 22.1.1 (page 713)) provides all the tools for an efficient implementation: Eulerian cycles, matching, and shortest path in both bipartite and general graphs.

Combinatorica [PS03] provides Mathematica implementations of Eulerian cycles and de Bruijn sequences. See Section 22.1.8 (page 716).

Notes: The history of graph theory began in 1736, when Euler first solved the seven bridges of Königsberg problem. Königsberg (now Kaliningrad) is a city on the banks of the Pregel river. In Euler's day there were seven bridges linking the banks and two islands, which can be modeled as a multigraph with seven edges and four vertices. Euler sought a way to walk over each of the bridges exactly once and return home— that is, an Eulerian cycle. Euler proved that such a tour is impossible, since all four of the vertices had odd degrees. The bridges were destroyed in World War II. See [BLW76] for a translation of Euler's original paper and a history of the problem.

Corberán and Laporte [CL13] and Toth and Vigo [TV14] are recent books on vehicle and arc routing problems, including the Chinese postman. Expositions on linear-time algorithms for constructing Eulerian cycles [Ebe88] include [Eve11, Man89]. Fleury's algorithm [Luc91] is a direct and elegant approach to constructing Eulerian cycles. Start walking from any vertex, and erase any edge that has been traversed. The only criterion in picking the next edge is that we avoid using a bridge (an edge whose deletion disconnects the graph), until no other alternative remains.

The Euler's tour technique is an important paradigm in parallel graph algorithms. Many parallel graph algorithms start by finding a spanning tree and then rooting the tree, where the rooting is done using the Euler tour technique. See parallel algorithms texts (e.g., [J92]) for an exposition, and [CB04] for experience in practice. Efficient algorithms exist to count the number of Eulerian cycles in a graph [HP73].

The problem of finding the shortest tour traversing all edges in a graph was introduced by Kwan [Kwa62], hence the name *Chinese* postman. The bipartite matching algorithm for solving Chinese postman is due to Edmonds and Johnson [EJ73]. It works for both directed and undirected graphs, although the problem is NP-complete for mixed graphs [Pap76a]. Mixed graphs contain both directed and undirected edges. Expositions on the Chinese postman algorithm include [Law11].

A *de Bruijn* sequence S of span k on an alphabet Σ of size α is a circular string of length α^k containing all strings of length k as substrings of S, each exactly once. For example, for $k = 3$ and $\Sigma = \{0, 1\}$, the circular string 00011101 contains the following substrings in order: 000, 001, 011, 111, 110, 101, 010, 100. De Bruijn sequences can be thought of as "safe cracker" sequences, describing the shortest sequence of dial turns with α positions sufficient to try out all combinations of length n.

De Bruijn sequences can be constructed by building a directed graph whose vertices represent all α^{n-1} strings of length $n - 1$, with an edge (u, v) iff $u = s_1 s_2 \ldots s_{n-1}$ and $v = s_2 \ldots s_{n-1} s_n$. Any Eulerian cycle on this graph describes a de Bruijn sequence. Expositions on de Bruijn sequences and their construction include [Eve11, PS03].

A cute algorithm for optimizing embroidery patterns based on Eulerian cycles is presented in [AHK$^+$08].

Related problems: Matching (see page 562), Hamiltonian cycle (see page 598).

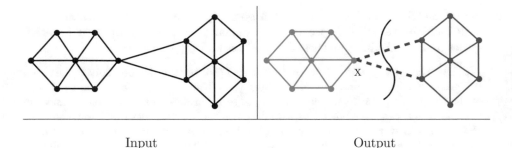

Input   Output

18.8   Edge and Vertex Connectivity

Input description: A graph G. Optionally, a pair of vertices s and t.

Problem description: What is the smallest subset of vertices (or edges) whose deletion will disconnect G, or that will separate s from t?

Discussion: Graph connectivity arises in problems related to network reliability. In the context of communication networks, the vertex connectivity is the smallest number of switching stations that a terrorist must destroy in order to separate the network—that is, prevent two undamaged stations from talking to each other. The edge connectivity is the smallest number of wires that need to be cut to accomplish the same objective. Either one well-placed bomb or snipping the right pair of cables suffices to disconnect the above network.

The edge (vertex) connectivity of a graph G is the smallest number of edge (vertex) deletions sufficient to disconnect G. There is a close relationship between these two quantities. The vertex connectivity is always less than or equal to the edge connectivity, because deleting one vertex from each edge in a cut set disconnects the graph. But smaller vertex subsets may be possible. The minimum vertex degree is an upper bound for both edge and vertex connectivity, because deleting all its neighbors (or cutting the edges to all its neighbors) disconnects a single-vertex component from the rest of the graph.

Several connectivity problems prove to be of interest:

- *Is the graph already disconnected?* – The simplest connectivity problem is testing whether the graph is in fact connected. A depth-first or breadth-first search suffices to identify all connected components in linear time, as discussed in Section 18.1 (page 542). For directed graphs, the issue is whether the graph is *strongly connected*, meaning there is a directed path between every pair of vertices. In a *weakly connected* graph, there may exist paths to nodes from which there is no way to return.

- *Is there a weak link in my graph?* – We say that G is *biconnected* if there is no single vertex whose deletion will disconnect G. Such a weak point is called an *articulation vertex*. A *bridge* is the analogous concept for edges, meaning a single edge whose deletion disconnects G.

The simplest algorithm to identify articulation vertices (or bridges) tries deleting vertices (or edges) one by one, and then uses DFS or BFS to test whether the resulting graph is still connected. More sophisticated linear-time algorithms exist for both problems, based on depth-first search. Indeed, a full implementation is given in Section 7.9.2 (page 225).

- *What if I want to split the graph into equal-sized pieces?* – Often we seek a small cut that breaks the graph into roughly equal-sized pieces. For example, suppose we want to split a big computer program into two maintainable units. We can construct a graph whose vertices represent subroutines. Edges can be added between any two subroutines that interact, namely where one calls the other. We now seek to partition the subroutines into roughly equal-sized sets with few pairs of interacting routines span the divide.

  This is the *graph partition* problem, discussed in Section 19.6 (page 601). Although the problem is NP-complete, reasonable heuristics exist.

- *Are arbitrary cuts OK, or must I separate a given pair of vertices?* – There are two flavors of the general connectivity problem. One asks for the smallest cut-set for the entire graph, the other for the smallest set to separate s from t. Any algorithm for $(s - t)$ connectivity can be used $n - 1$ times to give an algorithm for general connectivity, since vertex v_1 must end up in a different component from at least one of the other $n - 1$ vertices after deleting any cut set.

Edge and vertex connectivity can both be found using network-flow techniques, which interpret a weighted graph as a network of pipes where each edge has a maximum capacity. We seek to maximize the flow between two given vertices of the graph. The maximum flow between v_i and v_j in G is exactly the weight of the smallest set of edges to disconnect v_i from v_j. Thus, the edge connectivity can be found by maximizing the flow between v_1 and each of the $n - 1$ other vertices in an unweighted graph G.

Vertex connectivity is characterized by *Menger's theorem*, which states that a graph is k-connected iff every pair of vertices is joined by at least k vertex-disjoint paths. Network flow can again be used to perform this calculation, because a flow of k between a pair of vertices implies k *edge*-disjoint paths.

To exploit Menger's theorem, we construct a graph G' such that any set of edge-disjoint paths in G' corresponds to vertex-disjoint paths in G. This is done by replacing each vertex v_i of G with two vertices $v_{i,1}$ and $v_{i,2}$, adding an edge $(v_{i,1}, v_{i,2})$ between them in G'. We also replacing every edge $(x, y) \in G$ by the edges (x_0, y_1) and (x_1, y_0) in G'. Thus, two edge-disjoint paths in G' correspond to each vertex-disjoint path in G. As such, the maximum flow in G' gives twice the vertex connectivity of G.

Implementations: MINCUTLIB is a collection of high-performance codes for several different cut algorithms, including both flow and contraction-based methods. They were implemented by Chekuri et al. as part of an excellent experimental study of these algorithms and the heuristics needed to make them run fast [CGK+97]. The codes are available for non-commercial use at `http://www.avglab.com/andrew/soft.html`.

Most of the graph data structure libraries of Section 18.1 (page 542) include routines for connectivity and biconnectivity testing. The C++ Boost Graph Library [SLL02] (`http://www.boost.org/libs/graph/doc`) is distinguished by also including an implementation of edge connectivity testing.

GOBLIN (`http://goblin2.sourceforge.net/`) is an extensive C++ library dealing with all of the standard graph optimization problems, including both edge and vertex connectivity. LEDA (see Section 22.1.1 (page 713)) contains extensive support for both low-level connectivity testing (both biconnected and triconnected components) and edge connectivity/minimum cut in C++.

Notes: Good expositions on the network-flow approach to edge and vertex connectivity include [Eve11, PS03], and the book by Nagamouch and Ibaraki [NI08]. The correctness of these algorithms is based on Menger's theorem [Men27] that connectivity is determined by the number of edge/vertex disjoint paths separating a pair of vertices. The maximum-flow, minimum-cut theorem is due to Ford and Fulkerson [FF62].

The theoretically fastest algorithms for minimum-cut/edge connectivity are based on graph contraction, not network flows. Contracting an edge (x, y) in a graph G merges the two incident vertices into one, removing self-loops but leaving multiedges. Any sequence of such contractions can raise (but not lower) the minimum cut in G, and leaves the cut unchanged if no edge of the cut is contracted. Karger gave a beautiful randomized algorithm for minimum cut, observing that the minimum cut is left unchanged with non-trivial probability over the course of any random series of deletions. See Motwani and Raghavan [MR95] for an excellent treatment of randomized algorithms, including a presentation of Karger's algorithm.

The fastest version of Karger's algorithm runs in $(m \lg^3 n)$ expected time [Kar00]. Slightly faster deterministic algorithms are known [HRW17]. See [CGK+97, HNSS18] for experimental comparisons of algorithms for finding minimum cuts.

Minimum-cut methods have found many applications in computer vision, including image segmentation. Boykov and Kolmogorov [BK04] report on an experimental evaluation of minimum-cut algorithms in this context.

A non-flow-based algorithm for edge k-connectivity in $O(kn^2)$ is due to Matula [Mat87]. Faster k-connectivity algorithms are known for certain small values of k. All 3-connected components of a graph can be generated in linear time [HT73a], while $O(n^2)$ suffices to test 4-connectivity [KR91].

Related problems: Connected components (see page 542), network flow (see page 571), graph partition (see page 601).

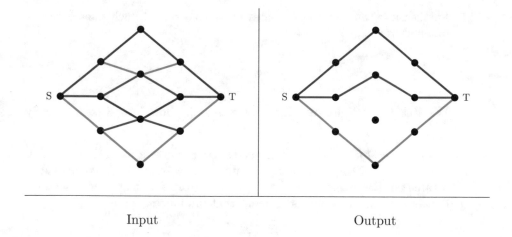

Input   Output

18.9   Network Flow

Input description: A directed graph G, where each edge $e = (i, j)$ has a capacity c_e. A source node s and sink node t.

Problem description: What is the maximum flow you can route from s to t while respecting the capacity constraint of each edge?

Discussion: Applications of network flow go far beyond plumbing. Finding the most cost-effective way to ship goods between a set of factories and a set of stores defines a network-flow problem, as do many resource-allocation problems in communications networks.

But the real power of network flow is that many linear programming problems arising in practice can be modeled by network-flow, including several graph problems that have been discussed in this book: bipartite matching, shortest path, and edge/vertex connectivity. Network-flow algorithms can solve these problems much faster than general-purpose linear programming methods.

The key to exploiting this power is recognizing that your problem can be modeled as network flow. This requires experience and study. My recommendation is that you first construct a linear programming model for your problem and then compare it with linear programs for the two primary classes of network flow problems: *maximum flow* and *minimum-cost flow*.

- *Maximum flow* – Here we seek the heaviest possible flow from s to t, given the edge capacity constraints of G. Let x_{ij} be a variable accounting for the flow from vertex i through directed edge (i, j). The flow through this edge is constrained by its capacity c_{ij}, so

$$0 \leq x_{ij} \leq c_{ij} \text{ for } 1 \leq i, j \leq n$$

Furthermore, an equal flow comes in as goes out at each non-source or non-sink vertex, so

$$\sum_{j=1}^{n} x_{ji} - \sum_{j=1}^{n} x_{ij} = 0 \text{ for all } 1 \leq i \leq n$$

We seek the assignment that maximizes the flow into sink t, namely $\sum_{i=1}^{n} x_{it}$.

- *Minimum cost flow* – Here we have an extra parameter for each edge (i, j), namely the cost (d_{ij}) of sending one unit of flow from i to j. We also have a targeted flow volume f we want to send from s to t at minimum total cost. Hence, we seek the assignment that minimizes

$$\sum_{i=1}^{n} \sum_{j=1}^{n} d_{ij} \cdot x_{ij}$$

subject to the edge and vertex capacity constraints of maximum flow, plus the additional restriction that $\sum_{i=1}^{n} x_{it} = f$.

Special considerations include:

- *What if I have multiple sources and/or sinks?* – No problem. We can modify the network by adding a vertex to serve as a super-source that feeds all the sources, and a super-sink that drains all the sinks.

- *What if all arc capacities are identical, either 0 or 1?* – Faster algorithms exist for 0/1 network flows. See the Notes section for details.

- *What if all my edge costs are identical?* – Use the simpler and faster algorithms for solving maximum flow as opposed to minimum-cost flow. Maximum flow without edge costs arises in many applications, including edge/vertex connectivity and bipartite matching.

- *What if I have multiple types of material moving through the network?* – Every message sent through a telecommunications network has a specific source and destination. Each destination needs to receive *exactly* those calls sent to it, not an equal amount of communication from arbitrary places. This can be modeled as a *multicommodity flow* problem, where each call defines a different commodity and we seek to satisfy all demands without exceeding the total capacity of any edge.

  Linear programming will suffice for multicommodity flow if fractional flows are permitted. Unfortunately, integral multicommodity flow is NP-complete, even for only two commodities.

Network flow algorithms can be complicated, and significant engineering is required to optimize performance. Excellent codes are available and described below. The two primary classes of algorithms are:

- *Augmenting path methods* – These algorithms repeatedly find a path of positive capacity from source to sink and add it to the flow. It can be shown that the flow through a network is optimal if and only if it contains no augmenting path. Since each augmentation adds something to the flow, we eventually reach the maximum. Network-flow algorithms differ in *how* they select the augmenting path. If we are not careful, each augmenting path might add only a little bit to the total flow, and so the algorithm might take a long time to converge.

- *Preflow-push methods* – These algorithms push flows from one vertex to another, initially ignoring the constraint that in-flow must equal out-flow at each vertex. Preflow-push methods prove faster than augmenting-path methods, essentially because multiple paths can be augmented simultaneously. These algorithms are the method of choice and are implemented in the best codes described below.

Implementations: High-performance codes for both maximum flow and minimum cost flow were developed by Andrew Goldberg and his collaborators. The codes `HIPR` and `PRF` [CG94] are provided for maximum flow, with the proviso that `HIPR` is recommended in most cases. For minimum-cost flow, the code of choice is `CS` [Gol97]. Both are written in C and available for non-commercial use from `http://www.avglab.com/andrew/soft.html`.

The C++ Boost Graph Library [SLL02] (`http://www.boost.org/libs/graph`) and Java JGraphT (`https://jgrapht.org`) both provide implementations of several network flow algorithms.

The First DIMACS Implementation Challenge on Network Flows and Matching [JM93] collected several implementations and generators for network flow, which are accessible from `http://dimacs.rutgers.edu/Challenges`. These include: (1) a preflow-push network flow implementation in C by Edward Rothberg, and (2) an implementation of eleven network flow variants in C, including the older Dinic and Karzanov algorithms by Anderson and Setubal.

Notes: Excellent books on network flows and its applications include [AMO93, Wil19], with Goldberg and Tarjan [GT14] a short but nice expository survey. The fundamental maximum-flow, minimum-cut theorem is due to Ford and Fulkerson [FF62]. Expositions on the hardness of multicommodity flow [Ita78] include [Eve11].

Conventional wisdom has long held that network flow should be computable in $O(nm)$ time, finally achieved by Orlin [Orl13]. See [AMO93] for a history of algorithms for the problem. Empirical studies of flow algorithms include [GKK74, Gol97].

Information flows through a network can be modeled as multicommodity flows, with the observation that replicating information at internal nodes can eliminate the need for distinct source-to-sink paths when multiple sinks are interested in the same information. The field of *network coding* [YLCZ05] uses such ideas to achieve information flows at the theoretical limits of the max-flow, min-cut theorem.

Related problems: Linear programming (see page 482), matching (see page 562), connectivity (see page 568).

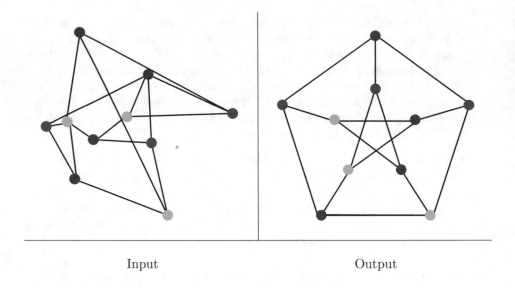

Input   Output

18.10 Drawing Graphs Nicely

Input description: A graph G.

Problem description: Draw a graph G to accurately reflect its structure.

Discussion: Graph drawing is a natural problem, yet it is inherently ill-defined. What exactly is a *nice* drawing? We seek the visualization of a graph that best shows off its structure, so the viewer can understand it. Simultaneously, we want this drawing to look aesthetically pleasing.

Unfortunately, these are "soft" criteria for which it is impossible to design an optimization algorithm. Indeed, there may be many different drawings of a given graph, each of which is most appropriate in a certain context. Three different drawings of the Petersen graph are presented on page 610. Check them out. Which is the "right" one?

Several "hard" criteria can help assess the quality of a drawing:

- *Crossings* – We seek a drawing with as few pairs of crossing edges as possible, because they are distracting.

- *Area* – We seek a drawing that uses as little paper as possible relative to the shortest edge length, because it avoids cramping parts of the graph.

- *Edge length* – We seek a drawing that avoids long edges, because they tend to obscure other features of the drawing.

- *Angular resolution* – We seek a drawing that avoids small angles between two edges incident on a given vertex, because the resulting lines tend to partially or fully overlap.

- *Aspect ratio* – We seek a drawing whose aspect ratio (width/height) reflects the desired output medium (typically a computer screen at 4/3) as closely as possible, because that is the way we will ultimately view it.

Unfortunately, these goals are mutually contradictory, and the problem of finding the best drawing under any non-empty subset of them is presumably NP-complete.

Two final warnings before getting down to business. For graphs without any inherent symmetries or structure, it is likely that no really nice drawing exists. This is especially true for graphs with more than ten to fifteen vertices. The sheer amount of ink needed to draw any large, dense graph will overwhelm even a massive display. A drawing of the complete graph on 100 vertices (K_{100}) contains approximately 5,000 edges. On a $1,000 \times 1,000$ pixel display, this works out to 200 pixels per edge. What can you hope to see, except a black blob in the center of the screen?

Once all this is understood, it must be admitted that graph-drawing algorithms can be quite effective and fun to play with. To help choose the right approach, ask yourself the following questions:

- *Must the edges be straight, or can I have curves and/or bends?* – Straight-line drawing algorithms are relatively simple, but have their limitations. Orthogonal polyline drawings seem to work best to visualize complicated graphs such as circuit designs. *Orthogonal* means that all lines must be drawn either horizontal or vertical, with no intermediate slopes. *Polyline* means that each graph edge is represented by a chain of straight-line segments, connected by vertices or bends.

- *Is there a natural, application-specific drawing?* – If your graph represents a network of cities and roads, you are unlikely to find a better drawing than placing the vertices in the same position as the cities on a map. This basic principle holds for many different applications.

- *Is your graph either planar or a tree?* – If so, use one of the special planar graph or tree-drawing algorithms discussed in Sections 18.11 and 18.12.

- *Is your graph directed?* – Edge direction has a significant impact on the nature of the desired drawing. When drawing directed acyclic graphs (DAGs), it is best that all edges flow in a logical direction—either left–right or top–down.

- *How fast must your algorithm be?* – Your graph drawing algorithm must be very fast if it will be used for interactive update and display. You will be limited to using incremental algorithms, which change the vertex positions only in the immediate neighborhood of any edited vertex. If instead you are printing a pretty picture for extended study, you can afford more time for optimization.

- *Does your graph contain symmetries?* – The output drawing above is attractive because the graph contains symmetries—namely a five-way rotational symmetry. The inherent symmetries in a graph can be identified by computing its *automorphisms*, or self-isomorphisms. Graph isomorphism codes (see Section 19.9 (page 610)) can be readily used to find all automorphisms.

For a quick-and-dirty drawing, I recommend simply spacing the vertices evenly on a circle, and then drawing the edges as straight lines between vertices. Such drawings are easy to program and fast to construct. They have the substantial advantage that no two edges will obscure each other, since no three vertices will be collinear. Such artifacts are hard to avoid once you allow internal vertices into your drawing. An unexpected pleasure with circular drawings is the symmetry they sometimes reveal, because vertices appear in the order they were inserted into the graph. Simulated annealing can be used to permute the circular vertex order to minimize crossings or edge length, and thus significantly improve the drawing.

A good, general purpose graph-drawing heuristic models the graph as a system of springs and then uses energy minimization to space the vertices. Let adjacent vertices attract each other with a force proportional to (say) the logarithm of their separation, while all non-adjacent vertices repel each other with a force proportional to their separation distance. These weights provide incentive for all edges to be short, while spreading the vertices apart. The behavior of such a system can be approximated by determining the force acting on each vertex at a particular time and then moving each vertex a small amount in the appropriate direction. After several iterations, the system will stabilize on a reasonable drawing. The input and output figures above demonstrate the effectiveness of the spring embedding on a particular small graph.

If you need a polyline graph-drawing algorithm, my recommendation is that you study the systems presented below or described in [JM12] to decide whether one of them can do the job. You will have to do a significant amount of work before you can hope to develop a better algorithm.

Drawing your graph opens another can of worms, namely where to place the edge/vertex labels. We seek to place labels very close to the edges or vertices they identify, and yet to position them such that they do not overlap each other or other important graph features. Optimizing label placement can be shown to be an NP-complete problem, but heuristics related to bin packing (see Section 20.9 (page 652)) can be effectively used.

Implementations: GraphViz (http://www.graphviz.org) is a popular and well-supported graph-drawing program developed by Stephen North. It represents edges as spline curves and can construct useful drawings of quite large and complicated graphs. It has sufficed for all of my professional graph-drawing needs over the years.

All of the graph data structure libraries of Section 15.4 (page 452) devote some effort to visualizing graphs. The Boost Graph Library and JGraphT both output graphs in GraphViz's Dot format, instead of reinventing the wheel.

VivaGraph (`https://github.com/anvaka/VivaGraphJS`) and Sigma (`http://sigmajs.org/`) are popular JavaScript packages for interactive display of graphs in browsers.

Graph drawing is a problem where very good commercial products exist, including those from Tom Sawyer Software (`www.tomsawyer.com`) and yFiles (`www.yworks.com`). Pajek [DNMB18] is a package particularly designed for drawing social networks, and available at `http://mrvar.fdv.uni-lj.si/pajek/`. All of these have free trial or non-commercial use downloads.

Combinatorica [PS03] provides Mathematica implementations of several graph-drawing algorithms, including circular, spring, and ranked embeddings. See Section 22.1.8 (page 716) for further information on *Combinatorica*.

Notes: There exists a significant community of researchers in graph drawing, whose annual conference (Graph Drawing and Network Visualization) has run for over 25 years. See more at `http://www.graphdrawing.org/`. Perusing a volume of the proceedings will provide a good view of the state-of-the-art and of what kinds of ideas people are thinking about. The *Handbook of Graph Drawing and Visualization* [Tam13] is the most comprehensive review of the field.

Two excellent books on graph-drawing algorithms are Battista et al. [BETT99] and Kaufmann and Wagner [KW01]. A third book by Jünger and Mutzel [JM12] is organized around systems instead of algorithms, but provides technical details about the drawing methods each system employs. Map-labeling heuristics are described in [BDY06, WW95].

Graph embeddings encode the structural information associated with each vertex as a short vector, providing useful features for machine learning models. Such a two- or three-dimensional embedding can be interpreted as vertex positions for a drawing, although $d = 128$ is more typical for machine learning. Our own DeepWalk [PARS14] is a very popular approach for constructing graph embeddings. See [CPARS18] for a survey of this field. t-SNE [MH08] is a widely used method to project higher dimensional point sets (like these embeddings) down to two dimensions for visualization. Implementations of t-SNE are available at `https://lvdmaaten.github.io/tsne/`.

It is trivial to space n points evenly along the boundary of a circle. However, the problem is considerably more difficult on the surface of a sphere. See Hardin, Sloane, and Smith [HSS07] for extensive tables of such spherical codes for $n \leq 130$.

Related problems: Drawing trees (see page 578), planarity testing (see page 581).

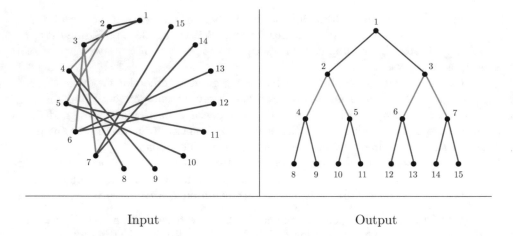

Input	Output

18.11   Drawing Trees

Input description: A tree T, which is a graph without any cycles.

Problem description: Create a nice drawing of the tree T.

Discussion: Many applications require drawing pictures of trees. Tree diagrams are often used to display the hierarchical structure of file system directories. My attempts to Google "tree drawing software" revealed special-purpose applications for visualizing family trees, syntax trees (sentence diagrams), and evolutionary phylogenetic trees all in the top twenty links.

Different aesthetics are associated with each application, making it difficult to generalize. That said, the primary issue in tree drawing is whether you are drawing free or rooted trees:

- *Rooted trees* define a hierarchical order, emanating from a single source node identified as the root. Any drawing should reflect this hierarchical structure, plus any additional application-dependent constraints on the order in which children should appear. For example, family trees are rooted, with sibling nodes typically drawn from left to right in the order of birth.

- *Free trees* do not encode any structure beyond their connection topology. There is no root associated with the minimum spanning tree of a graph, so a hierarchical drawing will be misleading. Such free trees might well inherit their drawing from that of the full underlying graph, such as the map of the cities whose distances define the minimum spanning tree.

Trees are always planar graphs, and hence can and should be drawn so no two edges cross. Any of the planar drawing algorithms discussed in Section 18.12 (page 581) could be used to do so. But such algorithms are overkill, because there are much simpler ways of constructing planar drawings of trees. The spring-embedding heuristics of Section 18.10 (page 574) work well on free trees, but might be too slow for interactive applications.

The most natural tree-drawing algorithms assume rooted trees. However, they can be used equally well with free trees, after selecting one vertex to serve as the root of the drawing. This faux-root can be selected arbitrarily, or, even better, by using a *center* vertex of the tree. A center vertex minimizes the maximum distance to other vertices. This tree center can be identified in linear time, by repeatedly trimming all the leaves until only the center remains.

Your two primary options for drawing rooted trees are *ranked* and *radial* embeddings:

- *Ranked embeddings* – Place the root in the top center of your page, and then partition the page into the root-degree number of top–down strips. Deleting the root creates a set of subtrees, each of which is assigned to its own strip. Draw each subtree recursively, by placing its new root (the vertex adjacent to the old root) in the center of its strip a fixed distance down from the top, with a line from old root to new root. The output figure above is a nice ranked embedding of a balanced binary tree.

  Such ranked embeddings are particularly effective to represent a hierarchy— be it a family tree, data structure, or corporate ladder. The top–down distance illustrates how far each node is from the root. Unfortunately, such repeated subdivision eventually produces very narrow strips, where many vertices get crammed into a small region of the page. Try to adjust the width of each strip to reflect the total number of nodes it will contain, and don't be afraid of expanding into a neighboring region's turf after their shorter subtrees have been completed.

- *Radial embeddings* – Free trees are better drawn using a radial embedding, where the center of the tree is placed in the center of the drawing. The space around this center vertex is divided into angular sectors for each subtree. Although the same problem of cramping will eventually occur, radial embeddings make better use of space than ranked embeddings and appear considerably more natural for free trees.

Implementations: GraphViz (http://www.graphviz.org) is a popular and well-supported graph-drawing program developed by Stephen North. It represents edges as spline curves and can construct useful drawings of quite large and complicated graphs. It has sufficed for all of my professional graph-drawing needs over the years. But all of the tools discussed in Section 18.10 will do something intelligent with trees.

Very good commercial products exist for graph/tree, including those from Tom Sawyer Software (www.tomsawyer.com) and yFiles (www.yworks.com). Tree-diagram oriented products include Lucid (https://www.lucidchart.com) and Visme (https://www.visme.co/tree-diagram-maker/. All of these have free trial or non-commercial use downloads.

Combinatorica [PS03] provides Mathematica implementations of several tree-drawing algorithms, including radial and rooted embeddings. See Section 22.1.8 (page 716) for further information on *Combinatorica*.

Notes: All books and surveys on graph drawing include discussions of tree-drawing algorithms. The *Handbook of Graph Drawing and Visualization* [Tam13] is the most comprehensive review of the field. Two excellent books on graph drawing algorithms are Battista et al. [BETT99] and Kaufmann and Wagner [KW01]. A third book by Jünger and Mutzel [JM12] is organized around systems instead of algorithms, but provides technical detail about the drawing methods each system employs.

A comprehensive resource of tree visualization is `https://treevis.net/`, with an associated survey paper [Sch11]. Treemaps are a popular method for displaying hierarchical data where nodes are represented by rectangles, and the subtrees are nested within their parent [JS91].

Heuristics for tree layout have been studied by several researchers, with Buchheim, et al. [BJL06] reflective of the state of the art. Under certain aesthetic criteria, the problem is NP-complete [SR83].

Certain tree layout algorithms arise from non-drawing applications. The van Emde Boas layout of a binary tree offers better external memory performance than conventional binary search, at a cost of greater complexity. See the survey of Arge et al. [ABF05] for more on this and other cache-oblivious data structures.

Related problems: Drawing graphs (see page 574), planar drawings (see page 581).

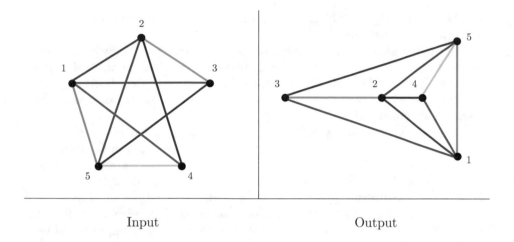

Input   Output

18.12 Planarity Detection and Embedding

Input description: A graph G.

Problem description: Can G be drawn in the plane such that no two edges cross? If so, produce such a drawing.

Discussion: Planar drawings (or *embeddings*) make clear the structure of a given graph by eliminating crossing edges, which can be confused as additional vertices. Graphs defined by road networks or printed circuit board layouts are inherently planar, because they are completely defined by surface structures. Other common graphs are planar by happenstance, like trees.

Planar graphs have nice properties that can be exploited to yield faster algorithms for many problems. The most important fact to know is that every planar graph is *sparse*. Euler's formula shows that $|E| \leq 3|V| - 6$ for every non-trivial planar graph $G = (V, E)$. This means that every planar graph contains a linear number of edges, and further that every planar graph contains a vertex of degree ≤ 5. Every subgraph of a planar graph is planar, so there must always be a sequence of low-degree vertices to delete from G, finally reducing it to the empty graph.

To gain a better appreciation of the subtleties of planar drawings, I encourage the reader to construct a planar (non-crossing) embedding for the graph $K_5 - e$, shown on the input figure above. Don't peek. Then construct such an embedding where all the edges are straight. Finally, add the missing edge to the graph and try to do the same for K_5 itself.

The study of planarity has motivated much of the development of graph theory. It must be confessed, however, that the need for planarity testing arises relatively infrequently in applications. Most graph-drawing systems do not explicitly seek planar embeddings. "Planarity Detection" proved to be among the least frequently hit pages of my Algorithm Repository (www.algorist.com)

[Ski99]. That said, it is still very useful to know how to deal with planar graphs when you encounter them.

It pays to distinguish the problem of planarity testing (does my graph have a planar drawing?) from constructing planar embeddings (create such a drawing), although both can be done in linear time. Many efficient algorithms on planar graphs make no use of the drawing, but instead exploit the low-degree deletion sequence described above.

Algorithms for planarity testing begin by embedding an arbitrary cycle from the graph in the plane and then considering additional paths in G, connecting vertices on this cycle. Whenever two such paths cross, one must be drawn outside the cycle and one inside. When three such paths mutually cross, there is no way to resolve the problem, so the graph cannot be planar. Linear-time algorithms for planarity detection are based on depth-first search. But they are sufficiently complicated that you are wise to seek an existing implementation.

Such path-crossing algorithms can be used to construct a planar embedding by inserting the paths into the drawing one by one. Unfortunately, because they work in an incremental manner, nothing prevents them from inserting many vertices and edges into a small area of the drawing. Such cramping is a major problem, because it leads to ugly drawings that are hard to understand. Better algorithms have been devised that construct *planar-grid embeddings*, where each vertex lies on a $(2n - 4) \times (n - 2)$ grid. Thus, no region can get too cramped and no edge can get too long. Still, the resulting drawings tend not to look as natural as one might wish.

For non-planar graphs, what is often sought is a drawing that minimizes the number of crossings. Unfortunately, computing the crossing number of a graph is NP-complete. Indeed, finding the crossing number for planar graphs with just one additional edge is NP-complete [CM13]. A useful heuristic extracts a large planar subgraph of G, embeds this subgraph, and then inserts the remaining edges one by one to minimize the number of crossings. This won't do much for dense graphs, which are doomed to have many crossings, but it does work well for graphs that are almost planar, such as road networks with overpasses or printed circuit boards with multiple layers. Large planar subgraphs can be found by modifying planarity-testing algorithms so they delete troublemaking edges when encountered.

Implementations: LEDA (see Section 22.1.1 (page 713)) includes linear-time algorithms for both planarity testing and constructing straight-line planar-grid embeddings. Their planarity tester returns an obstructing Kuratowski subgraph (see the Notes section) for any graph deemed non-planar, yielding concrete proof of its non-planarity.

The Open Graph Drawing Framework (`http://www.ogdf.net`), presented in [CGJ+13], is a C++ graph-drawing framework that includes several planarity testing/embedding algorithms, including the PQ-tree algorithm of [CNAO85]. The C++ Boost Graph Library [SLL02] (`http://www.boost.org/libs/graph`) also contains algorithms for planarity detection and embedding.

PIGALE (`http://pigale.sourceforge.net/`) is a C++ graph editor and algorithm library focusing on planar graphs. It contains a variety of algorithms for constructing planar drawings as well as efficient algorithms to test planarity and identify an obstructing subgraph ($K_{3,3}$ or K_5), if one exists.

Greedy randomized adaptive search procedure (GRASP) heuristics to find the largest planar subgraph have been implemented by Ribeiro and Resende [RR99] as Algorithm 797 of the *Collected Algorithms of the ACM* (see Section 22.1.5 (page 715)).

Notes: Kuratowski [Kur30] gave the first characterization of planar graphs, namely that they do not contain a subgraph homeomorphic to $K_{3,3}$ or K_5. Thus, if you are still working on the exercise to embed K_5, now is an appropriate time to give it up. Fáry's theorem [F48] states that every planar graph can be drawn in such a way that each edge is straight.

Hopcroft and Tarjan [HT74] gave the first linear-time algorithm for drawing graphs. Booth and Lueker [BL76] developed an alternate planarity-testing algorithm based on PQ-trees. Simplified planarity-testing algorithms include [MM96, SH99]. Efficient $2n \times n$ planar grid embeddings were first developed by [dFPP90]. The book by Nishizeki and Rahman [NR04] provides a good overview of the spectrum of planar drawing algorithms. Recent surveys on planarity detection include [Pat13, Tam13].

Outerplanar graphs are those that can be drawn such that all vertices lie on the outer face of the drawing. These graphs can be characterized as having no subgraph homeomorphic to $K_{2,3}$, and can be recognized and embedded in linear time.

Generalizations of planarity revolve around embedding graphs in more complicated surfaces than the plane. I encourage the reader to show that both $K_{3,3}$ and K_5 can be embedded without crossings on a bagel or donut. See [GT01] for an introduction to topological graph theory.

Related problems: Graph drawing (see page 574), drawing trees (see page 578).

Chapter 19

Graph Problems: NP-Hard

A cynical view of graph algorithms is that "everything we want to do is hard."
Indeed, all problems in this section are provably NP-complete with the exception
of graph isomorphism—whose complexity status remains an open question. The
theory of NP-completeness demonstrates that either *all* NP-complete problems
have polynomial-time algorithms, or *none* of them do. The former prospect
is sufficiently unlikely that NP-completeness suffices to say that no efficient
algorithm exists to solve the given problem.

Still, do not abandon hope if your problem resides in this chapter. I provide
a recommended attack for every problem, be it combinatorial search, heuristics,
approximation algorithms, or algorithms for restricted instances. Hard problems
require a different methodology to work with than polynomial-time problems,
but with care can usually be dealt with successfully.

The following books will help you deal with NP-complete problems:

- *Garey and Johnson* [GJ79] – This is the classic reference on the theory of
  NP-completeness. Most notably, it contains a concise catalog of over 400
  NP-complete problems, with associated references and comments. Browse
  through this catalog if you suspect your problem might be hard. This is
  the book in my algorithms library that I reach for most often.

- *Crescenzi and Kann* [CK97] – This website (`www.nada.kth.se/~viggo/
  problemlist/`) serves as the "Garey and Johnson" for approximation al-
  gorithms.

- *Williamson and Shmoys* [WS11] – The most comprehensive textbook on
  the theory and design of approximation algorithms.

- *Vazirani* [Vaz04] – A complete treatment of the theory of approximation
  algorithms by a highly regarded researcher in the field.

- *Gonzalez* [Gon18] – This handbook contains current surveys on a variety of
  techniques for dealing with hard problems, both applied and theoretical.

© The Editor(s) (if applicable) and The Author(s), under exclusive license to
Springer Nature Switzerland AG 2020
S. S. Skiena, *The Algorithm Design Manual*, Texts in Computer Science,
https://doi.org/10.1007/978-3-030-54256-6_19

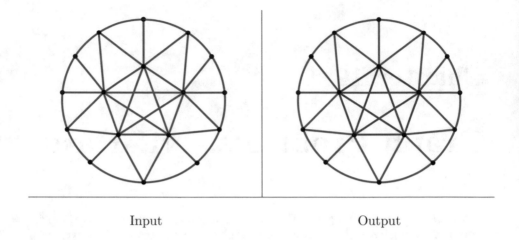

Input   Output

19.1   Clique

Input description: A graph $G = (V, E)$.

Problem description: What is the largest subset S of vertices such that all pairs are connected, that is, for all $x, y \in S$, $(x, y) \in E$?

Discussion: In high school, everybody complained about the "clique"—a group of friends who all hung around together and seemed to dominate everything social. Consider a graph representing the school's social network. Vertices correspond to people, with edges between pairs of people who are friends. Thus, the high school clique defines a (complete subgraph) clique in this friendship graph.

Identifying "clusters" of related objects often reduces to finding large cliques in graphs. An interesting example arose in a program the Internal Revenue Service (IRS) developed to detect organized tax fraud. A popular scam submits large numbers of phony tax returns in the hopes of getting undeserved refunds. But generating large numbers of *different* phony tax returns is hard work. The IRS constructs graphs with vertices corresponding to submitted tax forms and edges between any two forms that appear suspiciously similar. Any large clique in this graph points to fraud.

Since every single edge in a graph represents a clique of two vertices, the challenge lies not in finding *a* clique, but in finding a *large* clique. And it is indeed a challenge, for finding a maximum clique is NP-complete. To make matters worse, it is provably hard to approximate even to within a factor of $n^{1-\epsilon}$. Theoretically, clique is about as hard as a problem in this book can get. So what can we hope to do about it?

- *Will a maximal clique suffice?* – A clique is *maximal* if it cannot be enlarged by adding any additional vertex. A given maximal clique *might* in fact be the largest possible clique, but it probably won't be, and could well be much smaller. To find a hopefully large, maximal clique, sort the vertices from highest degree to lowest degree, put the first vertex in the clique, and then test if subsequent vertices are adjacent to all the clique vertices added thus far. If so, add it to enlarge the clique; if not, continue down the list. By using a bit vector to mark which vertices are currently in the clique, this can be done in $O(n+m)$ time. An alternative approach might incorporate some randomness into the vertex ordering, and accept the largest maximal clique you find after a number of trials.

- *What if I will settle for a large, dense subgraph?* – Insisting on cliques to define clusters in a graph can be risky, because a single missing edge will eliminate a vertex from consideration. Instead, we should seek large *dense* subgraphs—subsets of vertices that contain large numbers of edges between them. Cliques are, by definition, the densest subgraphs possible.

  The largest set of vertices whose induced subgraph has vertex degree $\geq k$ can be found with a simple linear-time algorithm. Delete all the vertices whose degree is less than k. This may reduce the degree of other vertices below k, so they will also have to be deleted. Repeating this process until all remaining vertices have degree $\geq k$ constructs the largest high-degree subgraph. This algorithm can be implemented in $O(n+m)$ time by using adjacency lists and the constant-width priority queue of Section 15.2 (page 445).

- *What if the graph is planar?* – Planar graphs cannot have cliques of a size larger than four, or else they cease to be planar. Since each edge defines a clique of size 2, the only interesting cases are cliques of three and four vertices. Efficient algorithms to find such small cliques consider the vertices from lowest to highest degree. Every planar graph must contain a vertex v of degree at most 5 (see Section 18.12 (page 581)), which has only a constant-sized neighborhood to check exhaustively for the largest clique containing v. We then delete this vertex to leave a smaller planar graph, which contains a different low-degree vertex. Repeat this check-and-delete process until the graph is empty.

If you *really* need to find the largest clique in a graph, an exhaustive search via backtracking provides the only real solution. We search through all k-subsets of the vertices, pruning a subset as soon as it contains a vertex that is not adjacent to all the rest. A simple upper bound on the size of the maximum clique in G is the highest vertex degree plus 1. A better upper bound comes from sorting the vertices in order of decreasing degree. Let j be the largest index such that degree of vertex v_j is at least $j-1$. The largest clique in the graph contains no more than j vertices, since no vertex of degree less than $(j-1)$ can appear in a clique of size j. To speed our search, first delete all such low degree vertices from G.

Heuristics for finding large cliques based on randomized techniques, such as simulated annealing, are likely to work reasonably well.

Implementations: Cliquer is a set of C routines for finding cliques in arbitrary weighted graphs by Patric Östergard. It uses an exact branch-and-bound algorithm, and is available at `http://users.tkk.fi/~pat/cliquer.html`.

Programs for finding cliques and independent sets were sought for the Second DIMACS Implementation Challenge [JT96]. Programs and data from the challenge can be obtained from `http://dimacs.rutgers.edu/archive/Challenges/`. dfmax.c implements a simple-minded branch-and-bound algorithm similar to [CP90]. dmclique.c uses a "semi-exhaustive greedy" scheme for finding large independent sets from [JAMS91].

Kreher and Stinson [KS99] provide branch-and-bound programs in C for finding the maximum clique using a variety of lower-bounds, available at `http://www.math.mtu.edu/~kreher/cages/Src.html`.

GOBLIN (`http://goblin2.sourceforge.net/`) employs branch-and-bound algorithms to find large cliques. They claim to work with graphs as large as 150 to 200 vertices.

Notes: Bomze, et al. [BBPP99] and Wu, et al. [WH15] give comprehensive surveys on the problem of finding maximum cliques. Particularly interesting is the work from the operations research community on branch-and-bound algorithms for finding cliques effectively. More recent experimental results are reported in [JS01].

The proof that clique is NP-complete is due to Karp [Kar72]. His reduction (given in Section 11.3.3 (page 366)) established clique, vertex cover, and independent set as very closely related problems, so heuristics and programs that solve one of them effectively should also produce reasonable solutions for the other two.

The *densest subgraph problem* seeks the subset of vertices whose induced subgraph has the highest average vertex degree. A clique of k vertices is clearly the densest possible subgraph of its size, but larger, less complete subgraphs may achieve higher average degree. This problem is NP-complete, but simple heuristics based on repeatedly deleting the lowest-degree vertex achieve reasonable approximation ratios [AITT00]. See [GKT05] for an interesting application of densest subgraph, namely detecting link spam on the web.

That clique cannot be approximated to within a factor of $n^{1/2-\epsilon}$ unless $P = NP$ (and $n^{1-\epsilon}$ under weaker assumptions) was shown by [Has82]. Picking any single vertex as a clique gives an n-factor approximation to max clique. These hardness results show that no polynomial-time approximation algorithm can do much better than this trivial heuristic.

Related problems: Independent set (see page 589), vertex cover (see page 591).

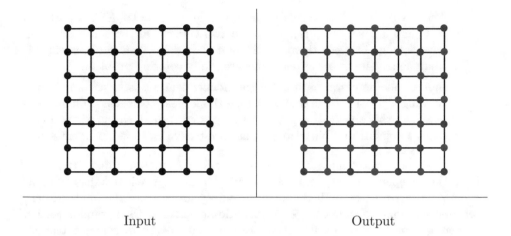

Input   Output

19.2 Independent Set

Input description: A graph $G = (V, E)$.

Problem description: What is the largest subset S of vertices of V such that there is no edge $(x, y) \in E$ where $x \in S$ and $y \in S$?

Discussion: The need to find large independent sets arises in facility dispersion problems, where we seek a set of mutually separated locations. To ensure that no two locations of our new "McAlgorithm" franchise service are so close as to compete with each other, we can create a graph where the vertices represent possible locations, and add edges between any two locations deemed close enough to interfere. The maximum independent set on this graph defines the largest set of franchise locations we can sell without cannibalizing sales.

Independent sets (also known as *stable sets*) avoid conflicts between elements, and hence often arise in coding theory and scheduling problems. For instance, we can define a graph whose vertices represent the set of possible code words, and add edges between any two code words sufficiently similar to be confused due to noise. The maximum independent set of this graph defines the highest capacity code for the given communication channel.

Independent set is closely related to two other NP-complete problems:

- *Clique* – A clique is a subset of pairwise connected vertices, while an independent set is a subset of pairwise non-connected vertices. The *complement* of $G = (V, E)$ is a graph $G' = (V, E')$ where $(i, j) \in E'$ iff (i, j) is not in E. The complement replaces each edge by a non-edge and vice versa. The maximum independent set in G is exactly the maximum clique in G', so the two problems are algorithmically identical. The algorithms and implementations in Section 19.1 (page 586) can thus be used to find the independent set of G'.

- *Vertex coloring* – The vertex coloring of a graph $G = (V, E)$ is a partition of V into a k sets (colors), where no two vertices of the same color can have an edge between them. Each color class defines an independent set. Many independent set applications are really coloring problems.

  Indeed, one heuristic to find a large independent set is to use any vertex coloring algorithm/heuristic, and take the largest color class. One consequence of this observation is that all graphs with small chromatic numbers (such as planar and bipartite graphs) must have large independent sets.

The simplest reasonable heuristic is to find the lowest-degree vertex, add it to the independent set, and then delete it and all vertices adjacent to it. Repeating this process until the graph is empty gives a *maximal* independent set, in that it can't be made larger by just adding vertices. Using randomization or perhaps some degree of exhaustive search might result in somewhat larger independent sets.

The independent set problem is in some sense dual to graph matching. The former asks for a large set of vertices with no edge in common, while the latter asks for a large set of edges with no vertex in common. This suggests trying to rephrase your problem as an efficiently computable matching problem instead of maximum independent set, which is NP-complete.

The maximum independent set of a tree can be found in linear time by (1) stripping off the leaf nodes, (2) adding them to the independent set, (3) deleting all adjacent nodes, and then (4) repeating the first step on the resulting trees until it is empty.

Implementations: Any program for computing the maximum clique in a graph can find maximum independent sets by just complementing the input graph. Therefore, I refer the reader to the clique-finding programs of Section 19.1 (page 586).

GOBLIN (`http://goblin2.sourceforge.net/`) implements a branch-and-bound algorithm for finding independent sets (called stable sets in the manual).

Greedy randomized adaptive search (GRASP) heuristics for independent set have been implemented by Resende, et al. [RFS98] as Algorithm 787 of the *Collected Algorithms of the ACM* (see Section 22.1.5 (page 715)).

Notes: That independent set is NP-complete was shown by Karp [Kar72]. It remains NP-complete for planar cubic graphs [GJ79]. Independent set can be solved efficiently for bipartite graphs [Law11]. This is not trivial—indeed the larger of the "part" of a bipartite graph is not necessarily its maximum independent set.

Finding maximal independent sets is a challenging problem in parallel and distributed models of computation, because concurrent additions might well have edges between them. See [BFS12, Gha16] for representative results.

Related problems: Clique (see page 586), vertex coloring (see page 604), vertex cover (see page 591).

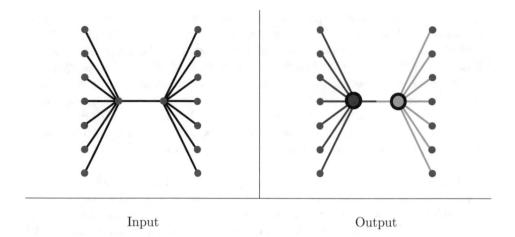

| Input | Output |

19.3 Vertex Cover

Input description: A graph $G = (V, E)$.

Problem description: What is the smallest subset $C \subseteq V$ such that every edge $(x, y) \in E$ contains at least one vertex of C?

Discussion: Vertex cover is a special case of the more general *set cover* problem, which takes as input an arbitrary collection of subsets $S = (S_1, \ldots, S_n)$ over the universal set $U = \{1, \ldots, m\}$. We seek the smallest cover C of subsets from S whose union is U. Set cover often arises in applications associated with buying things sold in fixed lots or assortments. See Section 21.1 (page 678) for a discussion of set cover.

To turn vertex cover into a set cover problem, let universal set U represent the set E of edges from G, and define S_i to be the set of edges incident on vertex i. Selecting subset S_i in the set cover problem is equivalent to selecting vertex i in the vertex cover problem. A set of vertices defines a vertex cover in graph G iff the corresponding subsets of edges define a set cover in this particular instance. Vertex cover instances are simpler than general set cover, because each edge can appear in only two different subsets. Vertex cover is a relative lightweight among NP-complete problems, and can be solved more effectively than general set cover.

Vertex cover and independent set are very closely related problems. Since every edge in E is (by definition) incident on a vertex in any cover S, there cannot be an edge whose endpoints are both in in $V - S$. Thus, $V - S$ must be an independent set. Since minimizing S is the same as maximizing $V - S$, the problems are equivalent, and any independent set solver can be applied to vertex cover as well. Having two ways of looking at your problem can be helpful, since one may appear easier in a given context.

The simplest heuristic for vertex cover selects the vertex with highest degree, adds it to the cover, deletes all adjacent edges, and then repeats until the graph is empty. With the right data structures, this can be done in linear time, and "usually" produces a decent cover. But in the worst case this cover might be $\lg n$ times larger than the optimal cover, on particularly challenging input graphs.

Fortunately, we can always find a vertex cover whose size is at most twice as large as optimal. Find a *maximal* matching M in the graph—a set of edges that share no vertex in common, and that cannot be enlarged by adding edges. Such a maximal matching can be constructed incrementally, by picking an arbitrary edge $e = (x, y)$ in the graph, deleting the two vertices x and y, and repeating until the graph is out of edges.

Taking *both* of the vertices for each edge in this maximal matching gives us a vertex cover. Why is this a cover? Because every edge gets deleted when it is a neighbor to a cover vertex. Why must this cover be at most twice as large as the minimum cover? Because *any* vertex cover must contain *at least* one of the two vertices in each matching edge, just to cover the edges of M.

This heuristic can be tweaked to perform somewhat better in practice, if not in theory. We can select the matching edges so as to "kill off" as many other edges as possible. By starting from the smallest maximal matching we can find, we will minimize the number of pairs of vertices in the vertex cover. Also, some of the vertices from M may not be needed for the cover because all of their incident edges were covered using other matching vertices. We can identify and delete these redundant vertices by making a second pass through our cover.

The vertex cover problem seeks to cover all edges using few vertices. Two other important problems have similar-sounding objectives:

- *Cover all vertices using few vertices* – The *dominating set* problem seeks the smallest set of vertices D such that every vertex in $V - D$ is adjacent to at least one vertex in the dominating set D. Every vertex cover of a connected graph is also a dominating set, but dominating sets can be much smaller. Any single vertex represents the minimum dominating set of complete graph K_n, while $n - 1$ vertices are needed for a vertex cover. Dominating sets tend to arise in communications problems, because they represent the hubs or broadcast centers sufficient to communicate with all sites/users.

  Dominating set problems can also be expressed as instances of set cover (see Section 21.1 (page 678)). Each vertex v_i defines the subset consisting of all vertices it is adjacent to, plus itself. The greedy set cover heuristic running on this instance yields a $\Theta(\lg n)$ approximation to the optimal dominating set.

- *Cover all vertices using few edges* – The *edge cover* problem seeks the smallest set of edges such that each vertex is included in one of the edges. Edge cover can be solved efficiently by finding a maximum cardinality matching (see Section 18.6 (page 562)), and then selecting arbitrary edges to account for the unmatched vertices. It is curious that the dual problems

of edge cover and vertex cover have such different fates: one NP-complete
and the other polynomial.

Implementations: Any program for computing the maximum clique in a graph
can be applied to vertex cover by complementing the input graph and selecting
the vertices that do not appear in the clique. Therefore, I encourage the reader
to check out the clique-finding programs of Section 19.1 (page 586).

JGraphT (`https://jgrapht.org`) is a Java graph library that contains
greedy and 2-approximate heuristics for vertex cover.

Notes: Karp [Kar72] first proved that vertex-cover is NP-complete. A diverse set
of heuristics yield 2-approximation algorithms for vertex cover, including randomized
rounding. Good expositions on these 2-approximation algorithms include [CLRS09,
Pas97, Vaz04, WS11]. The example that the greedy algorithm can be as bad as $\lg n$
times optimal is due to [Joh74] and presented in [PS98]. Experimental studies of vertex
cover heuristics include [ACL12, GMPV06, GW97, RHG07].

Whether there exists a better than 2-factor approximation for vertex cover has
long been one of the major open problems in approximation algorithms. Knot and
Regev [KR08] prove no $(2 - \epsilon)$-approximation exists for vertex cover, assuming the
unique games conjecture. Dinur and Safra [DS05] proved there does not exist a better
than 1.36-factor approximation algorithm, assuming $P \neq NP$.

The primary reference on dominating sets is the monograph of Haynes et al.
[HHS98]. Heuristics for the connected dominating set problem are presented in [GK98].
Dominating set cannot be approximated to better than the $\Omega(\lg n)$ factor [CK97] of
set cover.

Related problems: Independent set (see page 589), set cover (see page 678).

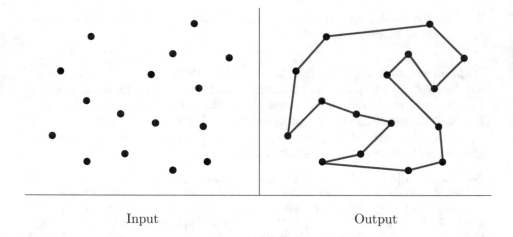

Input   Output

19.4 Traveling Salesman Problem

Input description: A weighted graph G.

Problem description: Find the cycle of minimum cost, visiting each vertex of G exactly once.

Discussion: The traveling salesman problem (TSP) is the most notorious NP-complete problem. This is a function of its general usefulness, and the ease with which it can be explained to the public at large. Imagine a salesperson planning a car trip to visit a set of cities. What is the shortest route that will visit all of them and return home, thus minimizing his or her total driving?

The traveling salesman problem arises in many transportation and routing problems. Optimizing tool paths for manufacturing equipment is a TSP. For example, consider a robot arm assigned to solder all the connections on a printed circuit board. The shortest tour that visits each solder point exactly once defines the most efficient route for the robot.

Several issues arise in solving TSPs:

- *Is the graph unweighted?* – If the graph is unweighted, or is a complete graph where all the edges have one of two possible cost values (long or short), this problem reduces to finding a *Hamiltonian cycle*. See Section 19.5 (page 598) for a discussion.

- *Does your input satisfy the triangle inequality?* – Our sense of how proper distance measures behave is captured by the *triangle inequality*, which states that $d(i,j) \leq d(i,k) + d(k,j)$ for all vertices $i, j, k \in V$. Geometric distances always satisfy the triangle inequality, because the shortest distance between two points is as the crow flies. Commercial air fares do *not* satisfy the triangle inequality, which is why it is so hard to find the cheapest airfare between two points. TSP heuristics work much better on sensible graphs that do obey the triangle inequality.

- *Are you given n points as input or a weighted graph?* – Geometric instances are often easier to work with than a graph representation, because all the pairs of points define a complete graph. Thus, there is never an issue of finding a feasible tour. We can also save space by computing these distances on demand, thus eliminating the need to store an $n \times n$ distance matrix. Geometric instances inherently satisfy the triangle inequality, so we can exploit performance guarantees from certain heuristics. Finally, we can take advantage of geometric data structures like kd-trees to quickly identify close unvisited sites.

- *Can you visit a vertex more than once?* – The restriction that the tour not revisit any vertex is irrelevant in many applications. In air travel, the cheapest way to visit all cities might repeatedly visit an airport hub. Note that this issue never arises when the input observes the triangle inequality.

  TSP with repeated vertices is easily solved by using any conventional TSP code on a new cost matrix D, where $D(i, j)$ is the shortest path distance from i to j. This matrix satisfies the triangle inequality, and can be constructed by solving an all-pairs shortest path (see Section 18.4 (page 554)).

- *Is your distance function symmetric?* – A distance function is *asymmetric* when there exists x, y such that $d(x, y) \neq d(y, x)$. The asymmetric traveling salesman problem (ATSP) is much harder to solve in practice than symmetric (STSP) instances, so try to avoid such pathological distance functions. Be aware that there is a reduction converting ATSP instances to symmetric instances containing twice as many vertices [GP07]. This can be useful, because symmetric solvers are so much better.

- *How important is it to find the optimal tour?* – Heuristic solutions will suffice for most applications. There are two different approaches if you insist on solving your TSP to optimality, however. *Cutting plane methods* model the problem as an integer program, then solve the linear programming relaxation of it. Additional constraints designed to force integrality are then added if the optimal solution is not at an integer point. *Branch-and-bound algorithms* perform a combinatorial search while maintaining careful upper and lower bounds on the cost of a tour. In the hands of professionals, problems with thousands of vertices can be solved. Maybe you can too, if you use the best solver available.

Almost any flavor of TSP is going to be NP-complete, so the right way to proceed is with heuristics. These typically come within a few percent of the optimal solution, which is close enough for engineering work. Literally dozens of heuristics have been proposed for TSP, so the situation can be confusing.

Empirical results in the literature are sometimes contradictory. However, we recommend choosing from among the following heuristics:

- *Minimum spanning trees* – Start by finding the minimum spanning tree of the sites, and then do a depth-first search of the resulting tree. In the course of DFS, we walk over each of the $n-1$ edges exactly twice: once going down to discover a new vertex, and once going up when we backtrack. Now define a tour by ordering the vertices based on when they were discovered. The resulting tour is at most twice the length of the optimal TSP tour, if the graph obeys the triangle inequality. In practice, it is usually better, typically 15% to 20% over optimal. The running time of computing the minimum spanning tree for points in the plane is only $O(n \lg n)$ (see Section 18.3 (page 549)).

- *Incremental insertion methods* – A different class of heuristics starts from a single vertex, and then inserts new points into this partial tour one at a time until the tour is complete. The version of this heuristic that seems to work best is *furthest point* insertion: of all remaining points, insert the point v into a partial tour T such that

$$\max_{v \in V} \min_{i=1}^{|T|} (d(v, v_i) + d(v, v_{i+1}))$$

  The "min" ensures that we insert the vertex in the position that adds the smallest amount of distance to the tour, while the "max" ensures that we pick the worst such vertex first. This seems to work well because it "roughs out" a partial tour first before filling in details. Such tours are typically only 5% to 10% longer than optimal.

- *K-optimal tours* – More powerful are the Kernighan–Lin or *k-opt* heuristics. The method applies local refinements to an initially arbitrary tour in the hopes of improving it. In particular, we delete a subset of k edges from the tour so that the k remaining subchains can be rewired to form a new, hopefully improved, tour. A tour is k-optimal when no subset of k edges can be deleted and rewired to reduce the cost. Two-opting a tour is a fast and effective way to improve any other heuristic. Experiments suggest that 3-optimal tours are usually within a few percent of optimal cost. For $k > 3$, the computation time increases considerably faster than the solution quality. Simulated annealing provides an alternate mechanism to employ edge flips to improve heuristic tours.

Implementations: Concorde is a program for the symmetric traveling salesman problem and related network optimization problems, written in ANSI C. This record-setting program by Applegate, Bixby, Chvatal, and Cook [ABCC07] has obtained the optimal solutions to at least 106 of TSPLIB's 110 instances; the largest of which has 85,900 cities. Concorde is available for academic research use from http://www.math.uwaterloo.ca/tsp. It is the clear choice among available TSP codes. Their website features very interesting material on the history and applications of TSP.

Lodi and Punnen [LP07] put together an excellent survey of available software for solving TSP. Current links to all programs mentioned are maintained at http://or.deis.unibo.it/research_pages/tspsoft.html.

TSPLIB [Rei91] provides the standard collection of hard instances of TSPs that arise in practice. The best-supported version of TSPLIB is available from https://www.iwr.uni-heidelberg.de/groups/comopt/software/TSPLIB95/.

Notes: The book by Applegate et al. [ABCC07] documents the techniques they used in their record-setting TSP solvers, as well as the theory and history behind the problem. Cook also wrote a popular book on TSP [Coo11]. Gutin and Punnen [GP07] now offer the best reference on all aspects and variations of the traveling salesman problem, displacing an older but beloved book by Lawler et al. [LLKS85].

Experimental results on heuristic methods for solving large TSPs include [Ben92a, Rei94, WCL+14]. Typically, it is possible to get within a few percent of optimal with such methods.

The Christofides heuristic [Chr76] is an improvement over the minimum spanning tree heuristic and guarantees a tour whose cost is at most $3/2$ times optimal on Euclidean graphs. It runs in $O(n^3)$, where the bottleneck is the time it takes to find a minimum-weight perfect matching (see Section 18.6 (page 562)). An exciting recent result yields a constant-factor approximation for the more general case of asymmetric TSP [STV17]. The minimum spanning tree heuristic is due to [RSL77].

Polynomial-time approximation schemes for Euclidean TSP have been developed by Arora [Aro98] and Mitchell [Mit99], which offer $1 + \epsilon$ factor approximations in polynomial time for any $\epsilon > 0$. They are of great theoretical interest, although any practical consequences remain to be determined.

The history of progress on optimal TSP solutions is inspiring. In 1954, Dantzig, Fulkerson, and Johnson solved a symmetric TSP instance of 42 US cities [DFJ54]. In 1980, Padberg and Hong solved an instance on 318 vertices [PH80]. Applegate et al. [ABCC07] have solved problems that are almost 300 times larger than this. Some of this increase is due to improved hardware, but most is due to better algorithms. The rate of growth demonstrates that exact solutions to NP-complete problems can be obtained for surprisingly large instances if the stakes are high enough.

For sets of n points in convex position in the plane, the minimum TSP tour is described by its convex hull (see Section 20.2 (page 626)), which can be computed in $O(n \lg n)$ time. Other easy special cases of TSP are known.

Related problems: Hamiltonian cycle (see page 598), minimum spanning tree (see page 549), convex hull (see page 626).

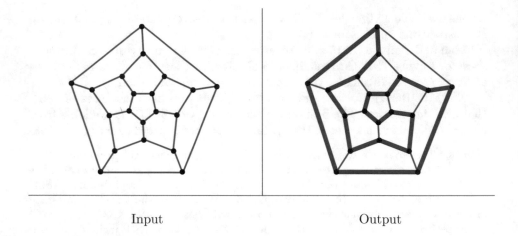

Input   Output

19.5   Hamiltonian Cycle

Input description: A graph $G = (V, E)$.

Problem description: Find a tour of the vertices using only edges from G, such that each vertex is visited exactly once.

Discussion: Hamiltonian cycle or path in a graph G is a special case of the traveling salesman problem on a graph G'—where each edge in G has distance 1 in G' and non-edge vertex pairs are separated by a greater distance. Such a weighted graph has a TSP tour of cost n in G' iff G is Hamiltonian.

Hamiltonian cycles are fundamental structures in graph theory, and a useful way to model a diverse set of phenomena. Sections 17.4–17.6 detail algorithms for generating combinatorial objects like permutations, subsets, and partitions. Minimum change orders (aka *Gray codes*) are the most efficient way of constructing such objects, and can naturally be thought of as defining a Hamiltonian cycle on the appropriate graph.

Closely related is the problem of finding the longest path or cycle in a graph. In precedence-constrained scheduling problems, where directed edge (x, y) implies job x must be completed before job y, the longest path defines the critical path that determines the shortest possible time to complete all the jobs. In signal analysis of circuits, the length of the longest path/cycle defines the time it takes the circuit to settle into a stable state after a perturbation.

The problems of finding longest cycles and paths are both NP-complete, even on very restrictive classes of unweighted graphs. There are several possible lines of attack, however:

- *Is there a serious penalty for visiting vertices more than once?* – Reformulating the Hamiltonian cycle problem as minimizing the total number of vertices visited on a complete tour turns it into an optimization problem, instead of an existential one. This allows possibilities for heuristics and approximation algorithms. Finding a spanning tree of the graph and doing a depth-first search, as discussed in Section 19.4 (page 594), yields a tour with at most $2n$ vertices. Using repeated random trials or simulated annealing might bring the size of this down considerably.

- *Am I seeking the longest path in a directed acyclic graph (DAG)?* – The problem of finding the longest path in a DAG can be solved in linear time using dynamic programming. Conveniently, the algorithm for finding the *shortest* path in a DAG presented in Section 18.4 (page 554) does the job if we replace min with max. DAGs are the most interesting case of the longest-path problem for which efficient algorithms exist.

- *Is my graph dense?* – Sufficiently dense graphs always contain Hamiltonian cycles. Further, the cycles implied by such sufficiency conditions can be efficiently constructed. In particular, any graph where all vertices have degree at least $n/2$ must be Hamiltonian. Stronger sufficient conditions also hold, as discussed in the Notes section.

- *Are you visiting all the vertices or all the edges?* – Verify that you really have a vertex-tour problem and not an edge-tour problem. With a little cleverness, it is sometimes possible to reformulate a Hamiltonian cycle problem in terms of Eulerian cycles, which instead visits every edge in a graph. Perhaps the most famous such instance is the problem of constructing de Bruijn sequences, discussed in Section 18.7 (page 565). The win here is that fast algorithms exist for finding Eulerian cycles and many related variants, while the Hamiltonian cycle problem is NP-complete.

If you really *must* know whether your graph is Hamiltonian, backtracking with pruning is your only possible solution. First check whether your graph is biconnected (see Section 18.8 (page 568)). If not, the graph has an articulation vertex whose deletion will disconnect the graph, and hence cannot be Hamiltonian.

Implementations: The reduction described above (weight 1 for an edge and 2 for a non-edge) turns Hamiltonian cycle into a symmetric TSP problem that obeys the triangle inequality. I therefore refer the reader to the TSP solvers discussed in Section 19.4 (page 594). Foremost among them is `Concorde`, a program for the symmetric traveling salesman problem and related network optimization problems, written in ANSI C. `Concorde` is available for academic research use from `http://www.math.uwaterloo.ca/tsp/concorde`. It is the clear choice among available TSP codes.

An effective program for solving Hamiltonian cycle problems resulted from the masters thesis of Vandegriend [Van98]. Both the code and the thesis are available from `https://webdocs.cs.ualberta.ca/~joe/Theses/vandegriend.html`.

Lodi and Punnen [LP07] put together an excellent survey of available TSP software, including the special case of Hamiltonian cycle. Links to the programs are maintained at `http://or.deis.unibo.it/research_pages/tspsoft.html`. Nijenhuis and Wilf [NW78] provide an efficient routine to enumerate all Hamiltonian cycles of a graph by backtracking. See Section 22.1.9 (page 716).

The football program of the Stanford GraphBase (see Section 22.1.7 (page 715)) uses a stratified greedy algorithm to solve the asymmetric longest-path problem. The goal is to derive a chain of football scores to establish the superiority of one football team over another. After all, if Virginia beat Illinois by 30 points, and Illinois beat Stony Brook by 14 points, then by transitivity Virginia would beat Stony Brook by 44 points if they played, right? We seek the longest simple path in a graph, where the weight of edge (x, y) denotes the number of points by which x beat y.

Notes: Hamiltonian cycles first arose in Euler's study of the knight's tour problem, although they were popularized by Hamilton's "Around the World" game in 1839. See [ABCC07, Coo11, GP07, LLKS85] for comprehensive references on the traveling salesman problem, including discussions on Hamiltonian cycle.

Finding long paths in graphs is very difficult. Although fast algorithms exist to find paths of length $\Theta(\log n)$ in Hamiltonian graphs [KMR97], it is hard to find even a polynomial-factor approximation [BHK04]. Most good texts in graph theory review sufficiency conditions for graphs to be Hamiltonian. My favorite is West [Wes00].

Techniques for solving optimization problems in the laboratory using biological processes have attracted considerable attention. In the original application of these "biocomputing" techniques, Adleman [Adl94] solved a seven-vertex instance of the directed Hamiltonian path problem. Unfortunately, this approach requires an exponential number of molecules, and Avogadro's number implies that such experiments are inconceivable for graphs beyond $n \approx 70$.

Related problems: Eulerian cycle (see page 565), traveling salesman (see page 594).

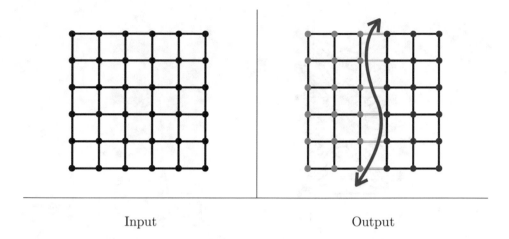

Input   Output

19.6 Graph Partition

Input description: A (weighted) graph $G = (V, E)$ and integers k and m.

Problem description: Partition the vertices into m roughly equal-sized subsets such that the total cost of edges spanning the subsets is at most k.

Discussion: Graph partitioning arises in many divide-and-conquer algorithms, which gain their efficiency by breaking problems into equal-sized pieces whose respective solutions can be combined into a single whole. Minimizing the edges cut in this partition usually simplifies the task of merging.

Graph partition also arises when clustering vertices into logical components. If edges link "similar" pairs of objects, the clusters remaining after partition should reflect coherent groupings. Large graphs are often partitioned into reasonable-sized pieces to improve data locality, or make less cluttered drawings.

Finally, graph partition is a critical step in many parallel algorithms. Consider the finite element method, which is used to compute the physical properties (such as stress and heat transfer) of geometric models. Parallelizing such calculations requires partitioning the models into equal-sized pieces whose interface is small. This is a graph-partitioning problem, since the topology of geometric models are usually represented by graphs.

Several different flavors of graph partitioning arise, depending on the desired objective function:

- *Minimum cut set* – The *smallest* set of edges to cut so as to disconnect a graph can be efficiently found using network flow or randomized algorithms. See Section 18.8 (page 568) for more on connectivity algorithms. This smallest cutset might split off only a single vertex, leaving the resulting partition very unbalanced in size.

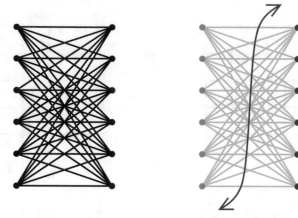

Figure 19.1: The maximum cut of a bipartite graph cuts all the edges.

- *Graph partition* – A better partition criterion seeks a small cut that partitions the vertices into roughly equal-sized pieces. Unfortunately, this version is NP-complete. Fortunately, heuristics work well in practice.

  Certain types of graphs always have small *separators* that partition the vertices into balanced pieces. Every tree contains at least one vertex v whose deletion (or equivalently, the deletion of all edges adjacent to v) partitions it so that no component contains more than half of the original n vertices. These components are not necessarily connected: consider the separating vertex of a star-shaped tree. Such a separating vertex can be found in linear time using depth first-search.

  Each planar graph has a set of $O(\sqrt{n})$ vertices whose deletion leaves no component with more than $2n/3$ vertices. Such separators provide a useful way to decompose geometric models, which are often defined by planar graphs.

- *Maximum cut* – Given an electronic circuit specified by a graph, the *maximum cut* defines the largest amount of data communication that can simultaneously occur in the circuit. The highest speed communications channel should thus span the vertex partition defined by the maximum cut, as shown above. Finding this maximum cut is NP-complete [Kar72], however heuristics similar to those of graph partitioning work well.

The basic approach for dealing with graph partitioning or maximum cut problems constructs an initial partition of the vertices (either randomly or according to some problem-specific strategy), and then sweeps through each vertex v, determining whether the cut improves if we move v over to the other side of the partition. The decision whether to move v can be made in time proportional to its degree, by identifying which side contains more of v's neighbors. Of course, the best side for v may change after its neighbors jump, so multiple iterations are needed before the process converges on a local optimum. Even so, such a local optimum might be arbitrarily far away from the global maximum cut if we are unlucky.

There are many variations of this basic procedure, by changing the order we test the vertices in or moving clusters of vertices simultaneously. Using some form of randomization, particularly simulated annealing, is almost certain to be a good idea. When more than two components are desired, this partitioning heuristic can be applied recursively.

Spectral partitioning methods use sophisticated linear algebra techniques to obtain a good partitioning. The eigenvectors associated with the smallest non-zero eigenvalues of the *Laplacian matrix* of graph G provide excellent features to partition it into highly connected pieces. Spectral methods tend to do a good job of identifying the general shape of a partition, but the results can be cleaned up using local optimization.

Implementations: A very popular code for graph partitioning is METIS (`http://glaros.dtc.umn.edu/gkhome/views/metis`), which has successfully partitioned graphs with over 1,000,000 vertices. Available versions include one variant designed to run on parallel machines and another suitable for partitioning hypergraphs.

Scotch (`http://www.labri.fr/perso/pelegrin/scotch/`) is another well-respected code to consider. Chaco is a widely used graph partitioning code designed to partition graphs for parallel computing applications. It employs several different partitioning algorithms, including both Kernighan–Lin and spectral methods. Chaco is available at `https://cfwebprod.sandia.gov/cfdocs/ CompResearch/templates/insert/softwre.cfm?sw=36`.

The Tenth DIMACS Challenge (`https://www.cc.gatech.edu/dimacs10/`) revolved around the related problems of graph partitioning and graph clustering. Results of the competition are reported in [BMSW13].

Notes: Recent surveys of graph partitioning algorithms include [BS13, BMS⁺16]. The fundamental local improvement heuristics for graph partitioning are the Kernighan–Lin [KL70] and Fiduccia–Mattheyses [FM82] methods. Spectral methods for graph partitioning are discussed in [Chu97, PSL90]. Empirical results on graph partitioning heuristics include [BG95, LR93].

The planar separator theorem and an efficient algorithm for finding such a separator are due to Lipton and Tarjan [LT79, LT80]. For experiences in implementing planar separator algorithms, see [ADGM07, HPS⁺05].

Any random vertex partition will expect to cut half of the edges in the graph, because the probability that the two vertices defining an edge end up on different sides of the partition is 1/2. Goemans and Williamson [GW95] gave an 0.878-factor approximation algorithm for maximum cut, based on semi-definite programming techniques. Tighter analysis of this algorithm was followed by Karloff [Kar96].

Related problems: Edge/vertex connectivity (see page 568), network flow (see page 571).

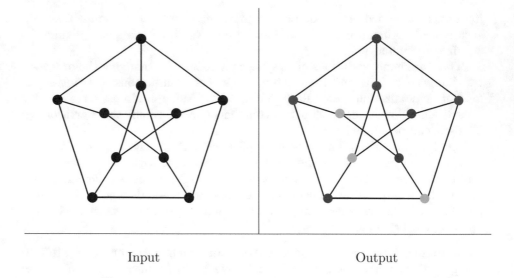

Input   Output

19.7   Vertex Coloring

Input description: A graph $G = (V, E)$.

Problem description: Color the vertices of V using the minimum number of colors such that for all $(i, j) \in E$, vertices i and j have different colors.

Discussion: Vertex coloring arises in scheduling and clustering applications. Register allocation in compiler optimization is the canonical application of coloring. Each variable in a given program fragment has a range of times during which its value must be kept intact: in particular, after it is initialized and before its final use. Any two variables whose life spans intersect cannot be placed in the same register. Construct a graph where each vertex corresponds to a variable, with an edge between any two vertices whose variable life spans intersect. If we color the vertices of this graph, we ensure that no variables assigned the same color will clash. Thus, they all can be assigned to the same register.

Of course, conflicts will never occur if each vertex is colored using a distinct color. But computers have a limited number of registers, so we seek a coloring using the fewest colors. The smallest number of colors sufficient to vertex color a graph is called its *chromatic number*.

Several special cases of interest arise in practice:

- *Can I color the graph using only two colors?* – An important special case is testing whether a graph is *bipartite*, meaning it can be colored using just two different colors. Bipartite graphs arise naturally in such applications as mapping workers to possible jobs. Fast, simple algorithms exist for problems such as matching (see Section 18.6 (page 562)) when restricted to bipartite graphs.

It is easy to test whether a graph G is bipartite. Color the first vertex blue, and then do a depth-first search of the graph. Whenever we discover a new, uncolored vertex, we must color it opposite of its parent because using the same color would cause a clash. If we ever find an edge (x, y) where both x and y have been colored identically, then G cannot be bipartite. Otherwise, the final coloring will be a two-coloring, constructed in $O(n + m)$ time. An implementation of this algorithm is given in Section 7.7.2 (page 219).

- *Is the graph planar, or are all vertices of low degree?* – The famous four-color theorem states that every planar graph can be vertex colored using at most four distinct colors. Efficient algorithms to four-color planar graphs are known, although it is NP-complete to decide whether a given planar graph is three-colorable.

  But there is a very simple algorithm that will vertex color any planar graph using at most six colors. Every planar graph contains a vertex of degree at most five. Delete this vertex v, and recursively color the rest of the graph. Because v has at most five neighbors, it can always be colored using one of the six colors that does not appear as a neighbor. This works because deleting a vertex from a planar graph leaves a planar graph, so it must also have a low-degree vertex to delete. The same idea can be used to color any graph of maximum degree Δ using at most $\Delta + 1$ colors, in $O(n\Delta)$ time.

- *Is this an edge-coloring problem?* – Certain vertex coloring problems can be modeled as *edge coloring*, where we seek to color the edges of a graph G such that edges get different colors if they share a vertex in common. The payoff is that there is an efficient algorithm that always returns a near-optimal edge coloring. Algorithms for edge coloring are the focus of Section 19.8 (page 608).

Computing the chromatic number of a graph is NP-complete. If you need an exact solution you must resort to backtracking, which can be surprisingly effective in coloring certain random graphs. It remains hard to compute a good approximation to the optimal coloring, so expect no guarantees.

Incremental methods prove to be the heuristic of choice for vertex coloring. As in the previously mentioned algorithm for planar graphs, vertices are colored sequentially, with the colors chosen in response to colors already assigned in the vertex's neighborhood. These methods vary in how the next vertex is selected and how it is assigned a color. Experience suggests inserting the vertices in non-increasing order of degree, because high-degree vertices have more color constraints and so are most likely to require an additional color if inserted late. Brèlaz's heuristic [Brè79] dynamically selects the uncolored vertex of highest *color degree* (i.e., adjacent to the most different colors), and colors it with the lowest-numbered unused color.

Incremental methods can be further improved by using *color interchange*. Taking a properly colored graph and exchanging two of the colors (say, painting the red vertices blue and the blue vertices red) leaves a proper vertex coloring. Now suppose we take a properly colored graph and delete all but the red and blue vertices. We can repaint one or more of the resulting connected components, again leaving a proper coloring. After such a recoloring, some vertex v previously adjacent to both red and blue vertices might now be only adjacent to blue vertices, thus freeing v to be colored red.

Color interchange is a win in terms of producing better colorings, at a cost of increased time and implementation complexity. Implementations are described below. Simulated annealing algorithms that incorporate color interchange to move from state to state can be even more effective.

Implementations: Graph coloring has been blessed with two useful web resources. Culberson's graph coloring page, `http://webdocs.cs.ualberta.ca/~joe/Coloring/`, provides an extensive bibliography and programs to generate and solve hard graph coloring instances. Michael Trick's page, `https://mat.tepper.cmu.edu/COLOR/color.html`, provides a nice overview of graph-coloring applications, an annotated bibliography, and a collection of over seventy graph-coloring instances arising in applications such as register allocation and printed circuit board testing. Both include a C language implementation of the DSATUR coloring algorithm.

Programs for the closely related problems of finding cliques and vertex coloring graphs were sought for at the Second DIMACS Implementation Challenge [JT96], held in October 1993. Programs and data from the challenge are accessible from `http://dimacs.rutgers.edu/Challenges`.

The C++ Boost Graph Library [SLL02] (`http://www.boost.org/libs/graph`) and Java JGraphT (`https://jgrapht.org`) both provide implementations of several vertex coloring heuristics. GOBLIN (goblin2.sourceforge.net) implements a branch-and-bound algorithm for vertex coloring.

Nijenhuis and Wilf [NW78] provide an efficient Fortran implementation of chromatic polynomials and vertex coloring by backtracking. See Section 22.1.9 (page 716). *Combinatorica* [PS03] provides Mathematica implementations of bipartite graph testing, heuristic colorings, chromatic polynomials, and vertex coloring by backtracking. See Section 22.1.8 (page 716).

Notes: Recent survey articles on graph coloring include [GHHP13, MT10]. An old but excellent source on vertex coloring heuristics is Syslo et al. [SDK83], which includes experimental results. Classical heuristics for vertex coloring include [Brè79, MMI72, Tur88]. See [GH06, HDD03] for more results.

Wilf [Wil84] proved that backtracking to test whether a random graph has chromatic number k runs in *constant time*, dependent on k but independent of n. This is less impressive than it seems, because only a vanishingly small fraction of such graphs are indeed k-colorable. A number of provably efficient (but still exponential) algorithms for vertex coloring are known. See [Woe03] for a survey.

Paschos [Pas03] reviews what is known about provably good approximation algorithms for vertex coloring. On one hand, it is provably hard to approximate within a polynomial factor [BGS95]. On the other hand, heuristics offer some non-trivial guarantees in terms of various parameters, such as Wigderson's [Wig83] factor of

$n^{1-1/(\chi(G)-1)}$ approximation algorithm, where $\chi(G)$ is the chromatic number of G.

Brook's theorem states that the chromatic number $\chi(G) \leq \Delta(G) + 1$, where $\Delta(G)$ is the maximum degree of a vertex of G. Equality holds only for odd-length cycles (which have chromatic number 3) and complete graphs.

The *four-color problem* is the most famous problem in the history of graph theory, first posed in 1852 and finally settled in 1976 by Appel and Haken using a proof involving extensive computation. Every planar graph can be five-colored using a variation of the color interchange heuristic. Despite the four-color theorem, it is NP-complete to test whether a particular planar graph requires four colors or if three suffice. See [SK86] for an exposition on the history of the four-color problem and the proof. An efficient algorithm to four-color a graph is presented in [RSST96], which more recently has been formally verified [Gon08].

Related problems: Independent set (see page 589), edge coloring (see page 608).

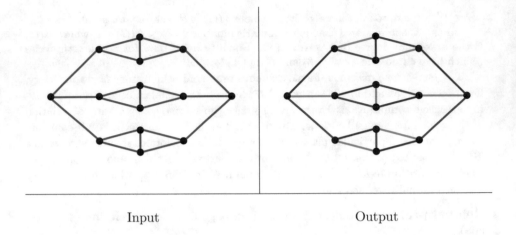

Input   Output

19.8   Edge Coloring

Input description: A graph $G = (V, E)$.

Problem description: What is the smallest set of colors needed to color the edges of G such that no two edges of the same color share a common vertex?

Discussion: Edge coloring arises in scheduling applications, typically associated with minimizing the number of non-interfering rounds needed to complete a given set of tasks. For example, consider a situation where we must schedule a set of two-person interviews, where each interview takes one hour. The meetings could all be scheduled at distinct times to avoid possible conflicts, but it is less wasteful to hold non-conflicting events simultaneously. We construct a graph G whose vertices are people, and whose edges represent the pairs of people who must meet. An edge coloring of G defines the schedule, with the color classes representing the different time periods in the schedule, so that all meetings of the same color happen simultaneously.

The National Football League solves such an edge-coloring problem to make up its schedule each season. The pairs of teams who must play each other are determined by the records of the previous season. Assigning these pairs to weeks of the season is an edge-coloring problem, complicated by secondary constraints like spacing out rematches and ensuring there is a good game every Monday night.

The minimum number of colors needed to edge color a graph is called its *edge-chromatic number* or sometimes *chromatic index*. Even-length cycles can be edge-colored with two colors, while odd-length cycles have an edge-chromatic number of 3.

Edge coloring has a better (if less famous) theorem associated with it than vertex coloring. *Vizing's theorem* states that any graph with a maximum vertex degree of Δ can be edge colored using at most $\Delta + 1$ colors. To put this in perspective, note that *any* edge coloring must have at least Δ colors, because all the edges incident on any vertex must be distinct colors. This is a very tight bound.

The proof of Vizing's theorem is constructive, meaning it can be turned into an $O(nm\Delta)$ algorithm to find an edge-coloring with $\Delta + 1$ colors. Deciding whether we can get away using one less color than this is NP-complete, so it hardly seems worth the effort to try. An implementation of Vizing's theorem is described below.

Edge-coloring on graph G can be converted to the problem of finding a vertex coloring on the *line graph* $L(G)$, which has a vertex of $L(G)$ for each edge of G and an edge of $L(G)$ iff the two edges of G share a common vertex. Line graphs can be constructed in time linear to their size, and then any vertex-coloring code can be employed to color them. That said, it would be disappointing to go the vertex coloring route. Vizing's theorem is our reward for for discovering that we in fact have an edge-coloring problem.

Implementations: The C++ Boost Graph Library [SLL02] (`http://www.boost.org/libs/graph`) has an implementation of Misra and Gries' constructive proof of Vizing's theorem, which runs in $O(nm)$ time. GOBLIN (`http://goblin2.sourceforge.net/`) implements a branch-and-bound algorithm for edge coloring.

See Section 19.7 (page 604) for a larger collection of vertex-coloring codes and heuristics, which can be applied to the line graph of your target graph.

Notes: Stiebitz et al. [SSTF12] is a recent book on edge coloring. Graph-theoretic results on edge coloring are surveyed in [FW77, GT94]. Vizing [Viz64] and Gupta [Gup66] independently proved that any graph can be edge colored using at most $\Delta + 1$ colors. Misra and Gries give a simple constructive proof of this result [MG92]. Despite these tight bounds, it is NP-complete to compute the edge-chromatic number [Hol81]. Bipartite graphs can be edge-colored in polynomial time [Sch98].

Whitney, in introducing line graphs [Whi32], showed that any two connected graphs with isomorphic line graphs are isomorphic, with the exception of K_3 and $K_{1,3}$. It is an interesting exercise to show that the line graph of an Eulerian graph is both Eulerian and Hamiltonian, while the line graph of a Hamiltonian graph is always Hamiltonian.

Related problems: Vertex coloring (see page 604), scheduling (see page 534).

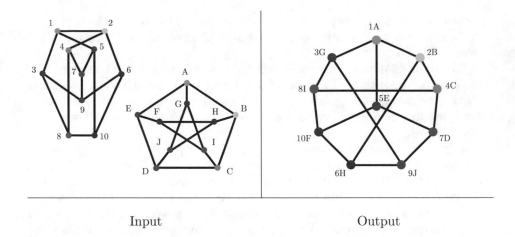

Input   Output

19.9   Graph Isomorphism

Input description: Two graphs, G and H.

Problem description: Find a mapping f from the vertices of G to the vertices of H such that G and H are identical, that is, (x, y) is an edge of G iff $(f(x), f(y))$ is an edge of H.

Discussion: Isomorphism is the problem of testing whether two graphs are really the same. Suppose we are given a collection of graphs, and must perform an expensive operation on each of them. If we can identify which of the graphs are duplicates, we can discard the copies to avoid redundant work.

Many pattern recognition problems can be mapped to graph or subgraph isomorphism. For example, the structure of chemical compounds are naturally described by labeled graphs, with each atom represented by a vertex. Identifying all molecules in a structure database containing a particular functional group is an instance of subgraph isomorphism testing.

What exactly is meant when we say that two graphs are the same? Two labeled graphs $G = (V_g, E_g)$ and $H = (V_h, E_h)$ are *identical* when $(x, y) \in E_g$ iff $(x, y) \in E_h$. In identical graphs, v_i in G corresponds to v_i in H. The more challenging isomorphism problem seeks to find a mapping from the vertices of G to H such that they are identical. The problem of finding this mapping is sometimes called *graph matching*.

Identifying symmetries is another important application of graph isomorphism. A mapping of a graph to itself is called an *automorphism*, and the collection of automorphisms (the automorphism *group*) provides a great deal of information about symmetries in the graph. For example, the complete graph K_n has $n!$ automorphisms (because any mapping will do), while an arbitrary random graph is likely to only have one, because G is always identical to itself.

Several variations of graph isomorphism arise in practice:

- *Is graph G contained in graph H?* – Instead of testing equality, we are often interested in knowing whether a small pattern graph G is a *subgraph* of H. Such problems as clique, independent set, and Hamiltonian cycle are important special cases of subgraph isomorphism.

  There are two distinct graph-theoretic notions of "contained in." *Subgraph isomorphism* asks whether there is a subset of edges and vertices of H that is isomorphic to a smaller graph G. *Induced subgraph isomorphism* asks whether there is a subset of vertices of H whose deletion leaves a subgraph isomorphic to a smaller graph G. For induced subgraph isomorphism, (1) all edges of G must be present in H, and (2) no *non-edges* of G can be present in H. Clique happens to be an instance of both subgraph isomorphism problems, while Hamiltonian cycle is only an example of vanilla subgraph isomorphism.

  Be aware of this distinction in your application. Subgraph isomorphism problems tend to be much harder than graph isomorphism, while induced subgraph problems tend to be even harder than subgraph isomorphism. Some flavor of backtracking is your only viable approach.

- *Are your graphs labeled or unlabeled?* – In many applications, vertices or edges of the graphs are *labeled* with attributes that must be respected when determining isomorphisms. For example, when comparing two bipartite graphs, each with "worker" and "job" vertices, any mapping that equated a job with a worker would make no sense.

  Labels and related constraints can be factored into any backtracking algorithm. Further, such constraints can significantly speed up the search, by creating opportunities for pruning whenever two vertex labels do not match up.

- *Are you testing whether two trees are isomorphic?* – Faster algorithms exist for special cases of graph isomorphism, including trees and planar graphs. Tree isomorphism is a problem that often arises in language pattern matching and parsing applications. A parse tree describes the structure of a text, so two parse trees T_1 and T_2 will be isomorphic when the underlying pair of texts have the same structure.

  Efficient algorithms for tree isomorphism work inward toward the center, starting from the leaves of both trees. Each vertex in T_1 is assigned a label representing the set of vertices in T_2 that might possibly be mapped to it, based on the constraints of labels and vertex degrees. For example, all the leaves in T_1 are potentially equivalent to all leaves of T_2. Working inward, we can partition the next level vertices in T_1 into classes based on how many leafs with matching labels they are adjacent to. Any mismatch means $T_1 \neq T_2$, while completing the process partitions the vertices into equivalence classes defining all isomorphisms. See the references below for more details.

- *How many graphs do you have?* – Many data mining applications search
  for all instances of a particular pattern graph in a big database of graphs,
  like the chemical structure mapping application described above. Such
  databases typically contain a large number of relatively small graphs. This
  puts an onus on indexing the graph database by small substructures (say
  five to ten vertices each), and doing expensive isomorphism tests only
  against those containing the same substructures as the query graph.

No polynomial-time algorithm is known for graph isomorphism, but neither
is it known to be NP-complete. Along with integer factorization (see Section
16.8 (page 490)), it is one of the few important algorithmic problems whose
rough computational complexity is still not known. The conventional wisdom
is that isomorphism lies somewhere between P and NP-complete, assuming P
$\neq$ NP.

Although no worst-case polynomial-time algorithm is known, testing isomorphism is *usually* not very hard in practice. The basic algorithm backtracks
through all of the $n!$ possible relabelings of the vertices of H with the names of
vertices of G, and then tests whether the graphs are identical. Of course, we
prune the search at a given prefix as soon as we detect any mismatch between
edges whose vertices are both in the prefix.

But the real key to efficient isomorphism testing is preprocessing the vertices
into "equivalence classes," partitioning them into sets such that two vertices
in different sets cannot possibly be mapped to each other. The vertices in
each equivalence class must share the same value of every invariant that is
independent of labeling. Possibilities include:

- *Vertex degree* – Two vertices of different degrees can never be identical, or
  mapped to each other. This simple partition can be a big win, but won't
  do much for regular (equal degree) graphs.

- *Shortest path distance* – The all-pairs shortest path matrix (see Section
  18.4 (page 554)) defines a multiset of $n-1$ distances representing the
  distances between v and each of the other vertices. Two vertices belong
  in the same equivalence class only if defining identical distance multisets.

- *Counting length-k paths* – Taking the kth power of the adjacency matrix of
  G yields a matrix $G^k[i,j]$ that counts the number of (non-simple) length-
  k paths from i to j. For each vertex v and each k, this matrix defines
  a multiset of path-counts, which can be used for partitioning as with
  distances above.

By using these invariants, you should be able to partition the vertices of most
graphs into a large number of small equivalence classes. Finishing the job off
with backtracking will then be short work. Each vertex gets assigned the name of
its equivalence class as a label, so we can treat it as a labeled matching problem.
It is harder to detect isomorphisms between highly symmetric graphs than with
random graphs, because of the reduced effectiveness of these equivalence-class
partitioning heuristics.

Implementations: The best isomorphism testing program is `nauty` (No AU-Tomorphisms, Yes?)—a set of very efficient C language procedures for determining the automorphism group of a vertex-colored graph. Nauty also produces a canonical labeling of the graph, to assist in isomorphism testing. It can test most graphs with fewer than a hundred vertices in well under a second. Nauty is available at `http://pallini.di.uniroma1.it/`. The theory behind `nauty` and an affiliate program `Traces` are described in [McK81, MP14].

Valiente [Val02] has made available the implementations of graph/subgraph isomorphism algorithms for both trees and graphs in his book. These C++ implementations run on top of LEDA (see Section 22.1.1 (page 713)), and are available at `http://www.lsi.upc.edu/~valiente/algorithm/`.

Kreher and Stinson [KS99] compute isomorphisms of graphs in addition to more general group-theoretic operations. These implementations in C are available at `http://www.math.mtu.edu/~kreher/cages/Src.html`.

Notes: Graph isomorphism is an important problem in computational complexity theory because of its rare open complexity status. The feel-good algorithm story of 2015 was Lazlo Babai's announcement of a quasi-polynomial (subexponential but super-polynomial) time algorithm for graph isomorphism after a 40-year quest [Bab16].

Monographs on isomorphism detection include [Hof82, KST93]. Valiente [Val02] focuses on algorithms for tree and subgraph isomorphism. Kreher and Stinson [KS99] take a more group-theoretic approach to isomorphism testing. Graph mining systems and algorithms are surveyed in [CH06]. See [FSV01] for performance comparisons between different graph and subgraph isomorphism algorithms.

Polynomial-time algorithms are known for planar graph isomorphism [HW74] and for graphs where the maximum vertex degree is bounded by a constant [Luk80]. The all-pairs shortest path heuristic is due to [SD76], although there exist non-isomorphic graphs that realize the exact same set of distances [BH90]. A linear-time tree isomorphism algorithm for both labeled and unlabeled trees is presented in [AHU74].

A problem is said to be *isomorphism-complete* if it is provably as hard as isomorphism. Bipartite graph isomorphism testing is isomorphism-complete, because any graph can be made bipartite by replacing each edge by two edges connected with a new vertex. The original graphs are isomorphic iff the transformed graphs are.

Related problems: Shortest path (see page 554), string matching (see page 685).

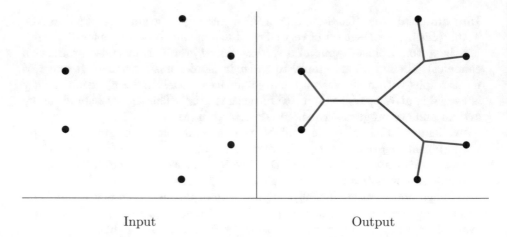

Input   Output

19.10   Steiner Tree

Input description: A graph $G = (V, E)$ and specified subset of vertices $T \subseteq V$. Or set of geometric points T.

Problem description: Find the smallest tree connecting the vertices of T.

Discussion: Steiner trees arise in network design problems, because the minimum Steiner tree describes how to connect a given set of sites using the smallest amount of wire. Analogous problems occur when designing networks of water pipes, heating ducts, or communication devices. Typical Steiner tree problems in electronic circuit design require connecting a set of sites to (say) ground under constraints such as material cost and signal propagation delay.

The Steiner tree problem is distinguished from the minimum spanning tree problem in graphs (see Section 18.3 (page 549)) when $T \neq V$, so we must select which intermediate connection points to use to minimize the cost of the tree. In geometric Steiner tree, the hard part of the problem is defining the positions of the new points we add to shorten the connections. Issues in Steiner tree construction include:

- *How many points must you connect?* – The Steiner tree of just a pair of vertices is the shortest path between them (see Section 18.4 (page 554)). The Steiner tree of all n vertices, when $T = V$, reduces to finding the minimum spanning tree (MST) of G. The general minimum Steiner tree problem is NP-hard despite these special cases, and remains so under a broad range of restrictions.

- *Is the input a set of geometric points?* – The geometric version of Steiner tree works on a set of points as input, typically in the plane, and seeks the lowest weight tree connecting the points. The set of possible intermediate points is *not* given as part of the input but must be deduced from the set of points. These Steiner points must satisfy several geometric properties, which can be used to reduce the set of candidates down to a finite number. For example, every Steiner point will have a degree of exactly 3 in a minimum Steiner tree, and the angles formed between any two of these edges must be exactly 120 degrees.

- *Are there constraints on the edges we define?* – In many wiring problems, all the edges are restricted to being either horizontal or vertical. This geometric version is called the *rectilinear Steiner problem*. A different set of angular and degree conditions apply for rectilinear Steiner trees than for Euclidean trees. In particular, all angles must be multiples of 90 degrees, and each vertex has degree at most 4.

- *Do I really need an optimal tree?* – Certain Steiner tree applications justify investing large amounts of computation to find the best possible Steiner tree. Perhaps our circuit design will be replicated millions of times, or the trenches to hold our pipes cost thousands of dollars per meter to dig. We should then use an exhaustive search technique such as backtracking or branch-and-bound to design the optimal tree.

  There are many opportunities for pruning search based on geometric and graph-theoretic constraints. But Steiner tree remains a hard problem. I recommend experimenting with the implementations described below before attempting your own.

- *What is the meaning of the Steiner vertices?* – A very special type of Steiner tree arises in classification and evolution. *Phylogenic trees* illustrate the relative similarity between different objects or species. Each object represents (typically) a leaf/terminal vertex of the tree, with intermediate vertices representing branching points between classes of objects. For example, an evolutionary tree might have leaf nodes of *human*, *dog*, *snake*, and *lizard*, and internal nodes corresponding to taxa *(animal, mammal, reptile)*. A tree rooted at *animal* with *dog* and *human* classified under *mammal* implies that humans are closer to dogs than to snakes.

  Many phylogenic tree construction algorithms have been developed that differ in what data they attempt to model, and their desired optimization criterion. Each combination of reconstruction algorithm and distance measure is likely to give a different tree, so identifying the "right" method for any given application is somewhat a question of faith. A reasonable procedure is to acquire a standard package of implementations, discussed below, and then see what happens to your data under all of them.

Fortunately, there is a good, efficient heuristic for finding Steiner trees that works well on all versions of the problem. Construct a graph modeling your input, setting the weight of edge (i, j) equal to the distance from point i to point j. The minimum spanning tree of this graph is guaranteed a good approximation for both Euclidean and rectilinear Steiner trees.

The worst case for the minimum spanning tree approximation of the Euclidean Steiner tree is three points forming an equilateral triangle. Any spanning tree will contain two of the sides (for a length of 2), whereas the minimum Steiner tree will connect the three points using an interior point, for a total length of $\sqrt{3}$. This ratio of $\sqrt{3}/2 \approx 0.866$ is always achieved, and the minimum spanning tree is usually within a few percent of the optimal Steiner tree in practice. For rectilinear Steiner trees, the optimal/minimum spanning tree ratio is always $\geq 2/3 \approx 0.667$.

For geometric instances, any suboptimal tree can be refined by inserting a Steiner point whenever the edges incident on a vertex form an angle of less than 120 degrees. Inserting these points and locally readjusting the tree edges can move the solution a little closer towards the optimum. Similar optimizations are possible for rectilinear spanning trees.

Note that we are only interested here in the subtree connecting the terminal vertices. We can trim the minimum spanning tree to retain only the tree edges that lie on the (unique) path between a pair of terminal nodes. The complete set of these can be found in $O(n)$ time by deleting any leaf that is not a terminal node, and then recurring.

An alternative heuristic for graphs is based on finding shortest paths. Start with a tree consisting of the shortest path between two terminals. For each remaining terminal t, find the shortest path to a vertex in the tree and add this path to connect t. The quality of this heuristic depends upon the insertion order of the terminals, but something simple and fairly effective is likely to result.

Implementations: GeoSteiner is a package for solving both Euclidean and rectilinear Steiner tree problems in the plane by Warme et al. [JWWZ18]. It also solves the related problem of minimum spanning trees in hypergraphs, and claims to have solved problems as large as 10,000 points to optimality. It is available from `http://www.geosteiner.com/`. This is almost certainly the best code for geometric Steiner tree instances.

Steiner tree algorithms were the subject of the 11th DIMACS Implementation Challenge, held in December 2014. Implementations of efficient algorithms for finding shortest paths were discussed. The papers, instances, and implementations are available at `http://dimacs.rutgers.edu/programs/challenge/`.

FLUTE (`http://home.eng.iastate.edu/~cnchu/flute.html`) computes rectilinear Steiner trees, emphasizing speed. It contains a user-defined parameter to control the tradeoff between solution quality and run time. GOBLIN (`http://goblin2.sourceforge.net/`) includes both heuristics and search methods for finding Steiner trees in graphs.

The programs PHYLIP (`http://evolution.genetics.washington.edu/phylip.html`) and PAUP (`https://paup.phylosolutions.com/`) are widely used packages for inferring phylogenic trees. Both contain more than twenty different algorithms for constructing phylogenic trees from data. Although many of them are designed to work with molecular sequence data, several general methods accept arbitrary distance matrices as input.

Notes: Monographs on the Steiner tree problem include Hwang, Richards, and Winter [HRW92] and Prömel and Steger [PS02]. Du et al. [DSR00] is a collection of surveys on all aspects of Steiner trees. Empirical results on Steiner tree heuristics include [BC19, SFG82, Vos92].

The Euclidean Steiner problem dates back to Fermat, who asked how to find a point p in the plane minimizing the sum of the distances to three given points. This was solved by Torricelli before 1640. Steiner was apparently one of several mathematicians who worked on the general problem for n points, and was mistakenly credited with originating the problem. An interesting, more detailed history appears in [HRW92].

Gilbert and Pollak [GP68] first conjectured that the ratio of the length of the minimum Steiner tree over the minimum spanning tree is always $\geq \sqrt{3}/2 \approx 0.866$. After twenty years of active research, the Gilbert–Pollak ratio was finally proven by Du and Hwang [DH92]. The Euclidean minimum spanning tree for n points in the plane can be constructed in $O(n \lg n)$ time [PS85].

Arora [Aro98] gave a polynomial-time approximation scheme (PTAS) for Steiner trees in k-dimensional Euclidean space. A 1.55-factor approximation for Steiner trees on graphs is due to Robins and Zelikovsky [RZ05].

The hardness of Steiner tree for Euclidean and rectilinear metrics was established in [GGJ77, GJ77]. Euclidean Steiner tree is not known to be in NP, because of numerical issues in exactly representing the positions of Steiner points.

Analogies can be drawn between minimum Steiner trees and minimum energy configurations in certain physical systems. The case that such analog systems— including the behavior of soap films over wire frames—"solve" the Steiner tree problem is discussed in [DKR10]. Slime molds are also pretty good at building Steiner trees [LSZ+15].

Related problems: Minimum spanning tree (see page 549), shortest path (see page 554).

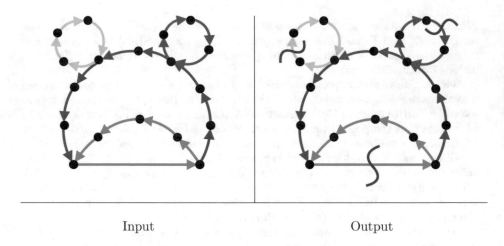

Input   Output

19.11   Feedback Edge/Vertex Set

Input description: A (directed) graph $G = (V, E)$.

Problem description: What is the smallest set of edges E' or vertices V' whose deletion leaves an acyclic graph?

Discussion: Feedback set problems arise because many things are easier to do on directed acyclic graphs (DAGs) than general digraphs. Consider the problem of scheduling jobs with precedence constraints, stating that job A must come before job B. When all constraints are consistent, the resulting graph is a DAG, and topological sort (see Section 18.2 (page 546)) can order the jobs/vertices to respect them. But no such schedule can exist when there are cyclic constraints, such as A before B, B before C, and C before A.

The feedback set identifies a small number of constraints that can be dropped to permit a valid schedule. In the *feedback edge* (or *arc*) set problem, we drop individual precedence constraints. In the *feedback vertex set* problem, we drop entire jobs and all constraints associated with them.

Similar considerations are involved in eliminating race conditions from electronic circuits. This explains why the problem is called "feedback" set. It is also more sensibly known as the *maximum acyclic subgraph problem*.

One final application has to do with ranking tournaments. Suppose we want to order the skills of players at some two-player game, such as chess or tennis. We can construct a directed graph with an arc from x to y whenever x beats y in a game. The higher-ranked player should *usually* beat the lower-ranked player, although upsets do occur. A natural rank ordering is the topological sort resulting after deleting the minimum set of feedback edges (upsets) from the graph.

Issues in feedback set problems include:

- *Do any constraints have to be dropped?* – No changes are needed if the graph is already a DAG, which can be determined via topological sort. One approach to finding a feedback set modifies the topological sorting algorithm to delete whatever edge or vertex is causing the trouble whenever a contradiction is found. Feedback edge set and feedback vertex set are both NP-complete on directed graphs, so heuristic solutions might be much larger than optimal.

- *How can I find a good feedback edge set?* – An effective linear-time heuristic constructs a vertex ordering and then deletes any arc going in the wrong direction. At least half the arcs must go either left–to–right or right–to–left for any vertex order, so take the smaller partition as your feedback set.

  But what is the right vertex order to start with? A good heuristic is to sort the vertices in terms of edge imbalance, namely in-degree minus out-degree. An incremental insertion approach starts by picking an arbitrary vertex v. Any vertex x defined by an in-going edge (x, v) will be placed to the left of v. Any x defined by out-going edge (v, x) will analogously be placed to the right of v. We can now recur on the left and right subsets to complete the vertex order.

- *How can I find a good feedback vertex set?* – The heuristics above yield vertex orders defining (hopefully) few back edges. We seek a small set of vertices that together cover these backedges. This is exactly the vertex cover problem, heuristics for which are discussed in Section 19.3 (page 591).

- *How do I break all cycles in an undirected graph?* – The problem of finding feedback sets in undirected graphs is quite different from digraphs. Trees are undirected graphs without cycles, and every tree on n vertices contains exactly $n-1$ edges. Thus, the smallest feedback edge set of any undirected graph G is $|E| - (n - c)$, where c is the number of connected components of G. The back edges encountered during a depth-first search of G qualify as a minimum feedback edge set.

  The feedback vertex set problem remains NP-complete for undirected graphs, however. A reasonable heuristic uses breadth-first search to identify the shortest cycle in G. Delete one of the vertices in this cycle from G, or all of them if you want a guarantee, and repeat by finding the shortest remaining cycle. This find-and-delete procedure is employed until the graph is acyclic. The optimal feedback vertex set must contain at least one vertex from each of these vertex-disjoint cycles, so the average deleted-cycle length determines just how good our approximation is.

It may pay to refine any of these heuristic solutions using randomization or simulated annealing. To move between states, we can modify the vertex permutation by swapping pairs in order or insert/delete vertices to/from the candidate feedback set.

Implementations: Greedy randomized adaptive search procedure (GRASP) heuristics for both feedback vertex and feedback edge set problems have been implemented by Festa, et al. [FPR01] as Algorithm 815 of the *Collected Algorithms of the ACM* (see Section 22.1.5 (page 715)).

An exact solver for feedback vertex set `https://github.com/wata-orz/fvs` by Iwata and Imanishi [Iwa16] won first place in the Parameterized Algorithms and Computational Experiments Challenge (`https://pacechallenge.wordpress.com/track-b-feedback-vertex-set/`).

GOBLIN (`http://goblin2.sourceforge.net`) includes an approximation heuristic for minimum feedback arc set.

The econ_order program of the Stanford GraphBase (see Section 22.1.7 (page 715)) permutes the rows and columns of a matrix so as to minimize the sum of the numbers below the main diagonal. Using an adjacency matrix as the input and deleting all edges below the main diagonal leaves an acyclic graph.

Notes: See [FPR99] for a survey on the feedback set problem. Expositions of the proof that feedback minimization is hard [Kar72] include [AHU74, Eve11]. Both feedback vertex and edge set remain hard even if no vertex has in-degree or out-degree greater than two [GJ79].

Bafna, et al. [BBF99] gives a 2-factor approximation for feedback vertex set in undirected graphs. Feedback edge sets in directed graphs can be approximated to within a factor of $O(\log n \log \log n)$ [ENSS98]. Heuristics for ranking tournaments are discussed in [LMM$^+$18]. Experiments with heuristics are reported in [Koe05].

Fixed-parameter tractable algorithms are polynomial in the input size n and exponential in the solution size k, say $O(k!n)$. Such an algorithm would be linear so long as k is a constant. Feedback vertex set can be solved with such an algorithm [CCL15]. The book by Downey and Fellows [DF12] offers the best overview of fixed-parameter complexity.

I will confess that I use a feedback edge set approach to grading semester projects in my enormous graduate data science course, where I can't grade all the papers. Teaching assistants and students can generally be trusted to make a binary judgment as to whether paper x is better or worse than paper y. After getting a large number of such judgment pairs, I use a feedback edge set algorithm to remove the smallest number of conflicting judgments, then order the papers using topological sort to assign grades.

An interesting application of feedback arc set to economics is presented in [Knu94]. For each pair A, B of sectors of the economy, we are given how much money flows from A to B. We seek to order the sectors to determine which sectors are primarily producers to other sectors, and which deliver primarily to consumers.

Related problems: Bandwidth reduction (see page 470), topological sorting (see page 546), scheduling (see page 534).

Chapter 20

Computational Geometry

Computational geometry is the algorithmic study of geometric problems. Its emergence coincided with application areas such as computer graphics, computer-aided design/manufacturing, and scientific computing, all of which need geometric computing.

Good books on computational geometry include:

- *de Berg, et al.* [dBvKOS08] – The "three Mark's" book is the best general introduction to the theory of computational geometry and its fundamental algorithms.

- *O'Rourke* [O'R01] – This is the best practical introduction to computational geometry. The emphasis is on careful and correct implementation of geometric algorithms. C and Java code are available from `https://cs.smith.edu/~orourke/books/compgeom.html`.

- *Preparata and Shamos* [PS85] – Although somewhat out of date, this book remains a good general introduction to computational geometry, stressing algorithms for convex hulls, Voronoi diagrams, and intersection detection.

- *Goodman, O'Rourke, and Toth* [TOG18] – This recent collection of survey articles provides a detailed overview of what is known in almost every subfield of discrete and computational geometry.

The leading conference in computational geometry is the ACM Symposium on Computational Geometry, held annually in late May or early June. There is a growing body of implementations of geometric algorithms. We point out specific implementations where applicable in the catalog, but the reader should definitely be aware of `CGAL` (Computational Geometry Algorithms Library)—a comprehensive library of geometric algorithms in C++ produced as a result of a joint European project. Anyone with a serious interest in geometric computing should check it out at `http://www.cgal.org/`.

© The Editor(s) (if applicable) and The Author(s), under exclusive license to
Springer Nature Switzerland AG 2020
S. S. Skiena, *The Algorithm Design Manual*, Texts in Computer Science,
https://doi.org/10.1007/978-3-030-54256-6_20

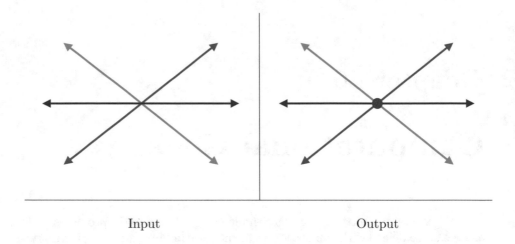

Input   Output

20.1 Robust Geometric Primitives

Input description: A point p and line segment l, or two segments l_1, l_2.

Problem description: Does p lie on, over, or under l? Does l_1 intersect l_2?

Discussion: Implementing basic geometric primitives is a task fraught with peril. Even simple things like returning the intersection point of two lines are more complicated than you may think. What should be returned when two lines are parallel, meaning they don't intersect at all? What if the lines are identical, so the intersection "point" is the entire line? What if one of the lines is horizontal, so that in the course of solving equations for the intersection point you will divide by zero? What if the two lines are *almost* parallel, so that the intersection point is so far from the origin that it causes arithmetic overflows? These issues become even more complicated for intersecting line segments, because additional cases arise that must be watched for and treated separately.

If you are new to implementing geometric algorithms, I suggest that you see O'Rourke's *Computational Geometry in C* [O'R01] for practical advice and complete implementations of basic geometric algorithms and data structures. It will help you avoid many headaches if you follow in his footsteps.

There are two different issues at work here: geometric degeneracy and numerical stability. *Degeneracy* refers to annoying special cases that must be treated in substantially different ways, such as when two lines intersect at more or less than a single point. There are three approaches to dealing with degeneracy:

- *Ignore it* – Make an operating assumption that your program will work correctly only when no three points are collinear, no three lines meet at a point, no intersections happen at the endpoints of line segments, and so on. This is probably the most common approach, and what I recommend for short-term projects where you can live with frequent crashes. The drawback is that interesting data often comes from points sampled on a grid, which tends to be highly degenerate.

- *Fake it* – Randomly or symbolically perturb your data so that it becomes non-degenerate. By moving each of your points a small amount in a random direction, you can break many of the existing degeneracies in the data, hopefully without creating too many new problems. This probably should be the first thing to try once you determine that your program is crashing too often. A problem with random perturbation is that it can change the shape of your data in subtle ways, which may be intolerable for your application. There also exist techniques to "symbolically" perturb your data to remove degeneracies in a consistent manner, but these require serious study to apply correctly.

- *Deal with it* – Geometric applications can be made more robust by writing special code to handle each special case that arises. This can work well if done with care from the beginning, but not so well if kludges are added whenever the system crashes. Expect to expend significant effort if you are determined to do it right.

Geometric computations often involve floating-point arithmetic, which leads to problems with overflows and numerical precision. There are three basic approaches to the issue of *numerical stability*:

- *Integer arithmetic* – By forcing all points of interest to lie on a fixed-size integer grid, you can perform exact comparisons to test whether any two points are equal or two line segments intersect. The cost is that the intersection point of two lines may not be exactly representable as a grid point. This is likely to be the simplest and best method, if you can get away with it.

- *Double-precision reals* – By using double-precision floating point numbers, you can reduce the occurrence of numerical errors. Your best bet might be to keep all data as single-precision reals, and then use double precision for intermediate computations.

- *Arbitrary precision arithmetic* – This is certain to be correct, but also to be slow. Careful analysis can minimize the need for high-precision arithmetic and thus the performance penalty. Still, you should expect high-precision arithmetic to be several orders of magnitude slower than standard floating-point arithmetic.

The best technique to produce robust geometric software is to build your applications around a small set of geometric primitives that handle as much of the low-level geometry as possible. These primitives include:

- *Area of a triangle* – The area $A(t)$ of a triangle $t = (a, b, c)$ is indeed half the base times the height, but computing the length of the base and altitude is messy work with trigonometric functions. Much better is to use the determinant formula for *twice* the area:

$$2 \cdot A(t) = \begin{vmatrix} a_x & a_y & 1 \\ b_x & b_y & 1 \\ c_x & c_y & 1 \end{vmatrix} = a_x b_y - a_y b_x + a_y c_x - a_x c_y + b_x c_y - c_x b_y$$

This formula generalizes to compute $d!$ times the volume of a simplex in d dimensions. Thus, $3! = 6$ times the volume of a tetrahedron $t = (a, b, c, d)$ in three dimensions is

$$6 \cdot A(t) = \begin{vmatrix} a_x & a_y & a_z & 1 \\ b_x & b_y & b_z & 1 \\ c_x & c_y & c_z & 1 \\ d_x & d_y & d_z & 1 \end{vmatrix}$$

These formulae give signed volumes and hence can be negative, so take the absolute value first. See Section 16.4 (page 475) for how to compute determinants.

The conceptually simplest way to compute the area of a polygon (or polyhedron) is to triangulate it and then sum up the area of each triangle. Implementations of a slicker algorithm that avoids triangulation are presented in [O'R01, SR03].

- *Above–below–on test* – Does a given point c lie above, below, or on a given line l? A clean way to deal with this is to represent l as a directed line that passes through point a before point b, and ask whether c lies to the left or right of the directed line l.

This primitive can be implemented using the sign of the triangle area as computed above. If the area of $t(a, b, c) > 0$, then c lies to the left of $\overline{ab}$. If the area of $t(a, b, c) = 0$, then c lies on $\overline{ab}$. Finally, if the area of $t(a, b, c) < 0$, then c lies to the right of $\overline{ab}$. This generalizes naturally to three dimensions, where the sign of the area denotes whether d lies above or below the oriented plane (a, b, c).

- *Line segment intersection* – This above–below primitive can also be used to test whether a line intersects a line segment. It does iff one endpoint of the segment is to the left of the line and the other is to the right. Segment–segment intersection is similar, and I refer you to implementations described below. The question of whether two segments intersect if they only share an endpoint is representative of the problems with degeneracy.

- *In-circle test* – Does point d lie inside or outside the circle defined by points a, b, and c in the plane? This primitive occurs in all Delaunay triangulation algorithms, and can be used as a robust way to do distance comparisons. Assuming that a, b, c are labeled in counterclockwise order around the circle, compute the determinant

$$\text{incircle}(a, b, c, d) = \begin{vmatrix} a_x & a_y & a_x^2 + a_y^2 & 1 \\ b_x & b_y & b_x^2 + b_y^2 & 1 \\ c_x & c_y & c_x^2 + c_y^2 & 1 \\ d_x & d_y & d_x^2 + d_y^2 & 1 \end{vmatrix}$$

In-circle will return 0 if all four points are cocircular, a positive value if d is inside the circle, and negative value if d is outside.

Check out the implementations below before you build your own.

Implementations: CGAL (www.cgal.org) and LEDA (see Section 22.1.1 (page 713)) both provide very complete sets of geometric primitives for planar geometry written in C++. LEDA is easier to learn and to work with, but CGAL is more comprehensive and freely available. Check them out if you are starting a significant geometric application, before you try to write your own.

O'Rourke [O'R01] provides implementations in C of most of the primitives discussed in this section. See http://cs.smith.edu/~jorourke/books/ CompGeom/CompGeom.html. These primitives were implemented primarily for exposition rather than production use, but they should be reliable and appropriate for modest applications.

The *Core Library* (see http://cs.nyu.edu/exact/) provides an API, which supports the Exact Geometric Computation (EGC) approach to numerically robust algorithms. With small changes, any C/C++ program can use it to readily support three levels of accuracy: machine-precision, arbitrary-precision, and guaranteed.

Shewchuk's [She97] robust implementation of basic geometric primitives in C++ is available at http://www.cs.cmu.edu/~quake/robust.html.

Notes: O'Rourke [O'R01] provides an implementation-oriented introduction to computational geometry that stresses robust geometric primitives. It is recommended reading. LEDA [MN99] provides another excellent role model.

Yap [SY18] gives an excellent survey on techniques for achieving robust geometric computation, including an available book draft [MY07]. Kettner, et al. [KMP+04] provides graphic evidence of the troubles that can arise when employing real arithmetic in geometric algorithms for convex hull. Controlled perturbation [MOS11] is a more recent approach for robust computation. Shewchuk [She97] and Fortune and van Wyk [FvW93] present careful studies on the costs of using arbitrary-precision arithmetic for geometric computation. By being careful about when to use it, reasonable efficiency can be maintained while achieving complete robustness.

Related problems: Intersection detection (see page 648), maintaining arrangements (see page 671).

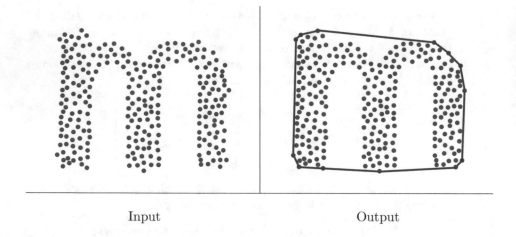

Input   Output

20.2   Convex Hull

Input description: A set S of n points in d-dimensional space.

Problem description: Find the smallest convex polygon (or polyhedron) containing all the points of S.

Discussion: Convex hull is the most important elementary problem in computational geometry, just as sorting is the most important elementary problem in combinatorial algorithms. It arises because constructing the hull captures a rough idea of the shape or extent of a data set.

Convex hull serves as a preprocessing step to many geometric algorithms. For example, consider the problem of finding the diameter of a set of points, meaning the pair of points that lie a maximum distance apart. The diameter must be between two points on the convex hull. The $O(n \lg n)$ algorithm for computing diameter first constructs the convex hull, and then for each hull vertex finds which other hull vertex lies farthest from it. The "rotating-calipers" method can be used to move efficiently from one-diametrically opposed hull vertex pair to the next by always proceeding in a clockwise fashion around the hull.

There are almost as many convex hull algorithms as sorting algorithms. Answer the following questions to help choose between them:

- *How many dimensions are you working with?* – Convex hulls are fast to compute in two and even three dimensions. But certain assumptions valid in lower dimensions break down as the dimensionality increases. For example, any n-vertex polygon in two dimensions has exactly n edges. However, the relationship between the numbers of faces and vertices becomes more complicated even in three dimensions. A cube has eight vertices and six faces, while an octahedron has eight faces and six vertices. This has implications for the data structures that represent hulls—are you just looking for the hull points or do you need the defining polyhedron? Be aware of these complications of high-dimensional spaces if your problem takes you there.

Simple $O(n \log n)$ convex-hull algorithms are available for the important special cases of two and three dimensions, but in higher dimensions things get more complicated. *Gift-wrapping* is the basic approach to construct higher-dimensional convex hulls. Observe that a three-dimensional convex polyhedron is composed of two-dimensional faces, or *facets*, that are connected by one-dimensional lines or *edges*. Each edge joins exactly two facets together. Gift-wrapping starts by finding an initial facet associated with the lowest vertex, and then conducting a breadth-first search from this facet to discover new, additional facets. Each edge e defining the boundary of a facet must be shared with another facet. By iterating through all n points, we can identify which point defines the next facet with e. We "wrap" the points one facet at a time, like bending wrapping paper around an edge until it hits the first point.

The key to efficiency is making sure that each edge is explored only once. Implemented properly in d dimensions, gift-wrapping takes $O(n\phi_{d-1} + \phi_{d-2} \lg \phi_{d-2})$, where ϕ_{d-1} is the number of facets and ϕ_{d-2} is the number of edges in the convex hull. This can be as bad as $O(n^{\lfloor d/2 \rfloor + 1})$ when the convex hull is very complex. Use one of the codes described below rather than roll your own.

- *Is your data given as vertices or half-spaces?* – The problem of finding the intersection of a set of n half-spaces in d dimensions (each containing the origin) is dual to that of computing convex hulls of n points in d dimensions. Thus, the same basic algorithm suffices for both problems. The necessary duality transformation is discussed in Section 20.15 (page 671). The problem of half-plane intersection differs from convex hull when no interior point is given, because infeasible instances arise where the intersection of the half-planes is empty.

- *How many points are likely to be on the hull?* – If your point set was generated "randomly," it is likely that most points lie within the interior of the hull. Planar convex-hull programs can be made more efficient in practice using the observation that the left-most, right-most, top-most, and bottom-most points must all be on the convex hull. This usually gives a set of either three or four distinct hull points, defining a triangle or quadrilateral. Any point inside this region *cannot* be on the convex hull, and so can be discarded in a linear sweep through the points. Ideally, only a few points will then remain to run through the full convex-hull algorithm.

This trick can also be applied beyond two dimensions, although it loses effectiveness as the dimension increases.

- *How do I find the shape of my point set?* – Although convex hulls provide a gross measure of shape, any details associated with the concavities are lost. The convex hull of the "m" from the example input would be indistinguishable from the convex hull of "w." *Alpha-shapes* are a more general structure that can be parameterized so as to retain arbitrarily large concavities. Implementations and references on alpha-shapes are included below.

The *Graham scan* is the most popular convex-hull algorithm in the plane. It starts with one point p known to be on the convex hull, say the point with the lowest x-coordinate, and then sorts the rest of the points in angular order around p. Starting from a partial hull consisting of p and the point with the smallest angle, we proceed counterclockwise adding points. If the angle formed by the new point and the last hull edge is less than 180 degrees, we insert this new point to the hull. If the angle formed by the new point and the last "hull" edge is greater than 180 degrees, then a chain of vertices starting from the last hull edge must be deleted to maintain convexity. The total time is $O(n \lg n)$, because the bottleneck is the cost of sorting the points around p.

This Graham scan procedure can also be used to construct a non-self-intersecting (or *simple*) polygon passing through all the points. Sort the points around v, but instead of testing angles, connect the points in angular order. This gives a polygon without self-intersection, although it typically has many ugly skinny protrusions.

The gift-wrapping algorithm becomes especially simple in two dimensions, since each "facet" becomes an edge, each "edge" becomes a vertex of the polygon, and the "breadth-first search" simply walks around the hull in a clockwise or counterclockwise order. The two-dimensional gift-wrapping (or *Jarvis march*) algorithm runs in $O(nh)$ time, where h is the number of vertices on the convex hull. I recommend sticking with Graham scan unless you know in advance that there are only a few vertices on the hull.

Implementations: The CGAL library (`www.cgal.org`) offers C++ implementations of an extensive variety of convex-hull algorithms for two, three, and arbitrary numbers of dimensions. Alternate C++ implementations of planar convex hulls include LEDA (see Section 22.1.1 (page 713)).

Qhull [BDH97] is a popular low-dimensional, convex-hull code, optimized for two to about eight dimensions. It is written in C and can also construct Delaunay triangulations, Voronoi vertices, furthest-site Voronoi vertices, and half-space intersections. Qhull has been widely used in scientific applications and has a well-maintained homepage at `http://www.qhull.org/`.

O'Rourke [O'R01] provides a robust implementation of the Graham scan in two dimensions and an $O(n^2)$ implementation of an incremental algorithm for convex hulls in three dimensions. C and Java implementations are both available. See Section 22.1.9 (page 717).

Ken Clarkson's higher-dimensional convex-hull code *Hull* also does alpha-shapes, and is available at `http://www.netlib.org/voronoi/hull.html`.

Different codes are needed for enumerating the vertices of intersecting half-spaces in higher dimensions. Avis' *lrs* (`http://cgm.cs.mcgill.ca/~avis/C/lrs.html`) is an arithmetically robust ANSI C implementation of the Avis–Fukuda reverse search algorithm for vertex enumeration/convex-hull problems. Since the polyhedron is implicitly traversed but not explicitly stored in memory, even problems with very large output sizes can sometimes be solved.

Notes: Planar convex hull plays the same role in computational geometry that sorting does in algorithm design. Like sorting, convex hull is a fundamental problem where many different algorithmic approaches lead to interesting or optimal algorithms. Quickhull and mergehull are examples of hull algorithms inspired by sorting algorithms [PS85]. A simple construction involving points on a parabola presented in Section 11.2.4 (page 360) reduces sorting to convex hull, so the information-theoretic lower bound for sorting implies that planar convex hull requires $\Omega(n \lg n)$ time to compute. A stronger lower bound is established in [Yao81].

Good expositions of the Graham scan algorithm [Gra72] and the Jarvis march [Jar73] include [dBvKOS08, CLRS09, O'R01, PS85]. The optimal planar convex-hull algorithm [KS86] takes $O(n \lg h)$ time, where h is the number of hull vertices, and captures the best performance of both Graham scan and gift wrapping. Planar convex hull can be efficiently computed *in-place*, meaning without requiring additional memory in [BIK$^+$04]. Seidel [Sei18] provides an excellent survey of convex hull algorithms and variants, particularly for higher dimensions.

Topology is the study of shape. Edelsbrunner and Harar [EH10] provide an introduction to computational topology. Alpha-hulls, presented in [EKS83], provide a useful notion of the shape of a point set. A generalization to three dimensions, with an implementation, is presented in [EM94].

Reverse-search algorithms for constructing convex hulls are effective in higher dimensions [AF96]. Through a clever lifting-map construction [ES86], the problem of building Voronoi diagrams in d-dimensions can be reduced to constructing convex hulls in $(d+1)$-dimensions. See Section 20.4 (page 634) for more details.

Dynamic algorithms for convex-hull maintenance are data structures that permit inserting and deleting arbitrary points while always representing the current convex hull. Jacob and Brodal [JB19] reduced the cost of such operations to logarithmic amortized time.

Related problems: Sorting (see page 506), Voronoi diagrams (see page 634).

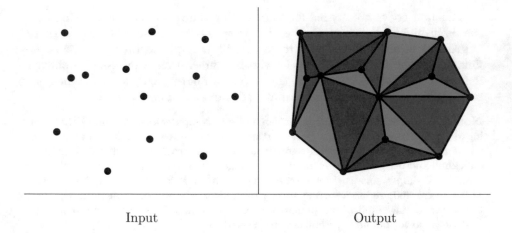

Input   Output

20.3   Triangulation

Input description: A set of points, a polygon, or a polyhedron.

Problem description: Partition the interior of the point set or polyhedron into triangles.

Discussion: A good first step in working with complicated geometric objects is to break them into simple geometric objects. This makes triangulation a fundamental problem in computational geometry, because the simplest geometric objects are triangles in two dimensions. Classical applications of triangulation include finite element analysis and computer graphics.

One interesting application of triangulation is surface interpolation. Suppose that we have sampled the height of a mountain at a number of (x, y, z) points. How can we estimate the height z' at any point $q = (x', y')$ in the plane? We can project the sampled points on the plane, and then triangulate them. This triangulation partitions the plane into triangles, so we can estimate height by interpolating between the heights (z-coordinates) of the three points of the triangle that contain q. Furthermore, this triangulation and the associated height values define a mountain surface suitable for graphics rendering.

A triangulation in the plane is constructed by adding non-intersecting chords between the vertices until no more such chords can be added. Specific issues arising in triangulation include:

- *Are you triangulating a point set or a polygon?* – Often we are given a set of points to triangulate, as in the surface interpolation problem discussed above. This requires first constructing the convex hull of the point set and then carving up the interior into triangles.

  The simplest such $O(n \lg n)$ algorithm first sorts the points by x-coordinate. It then inserts them from left to right as per the convex-hull algorithm of page 111, building the triangulation by adding a chord to each point newly cut off from the hull.

- *Does the shape of the triangles in your triangulation matter?* – There are usually many different ways to partition your input into triangles. Consider a set of n points in convex position in the plane. The simplest way to triangulate them would be to add fan-shaped diagonals from the first point to all $n-1$ other points. But this has the tendency to create skinny triangles.

  Many applications seek to avoid skinny triangles, or equivalently, minimize small angles in the triangulation. The *Delaunay triangulation* of a point set maximizes the minimum angle over all possible triangulations. This isn't exactly what we are looking for, but it is pretty close, and the Delaunay triangulation has enough other interesting properties to make it the quality triangulation of choice. Further, it can be constructed in $O(n \lg n)$ time, using implementations described below.

- *How can I improve the shape of a given triangulation?* – Each internal edge of any triangulation is shared between two triangles. The four vertices defining these two triangles form either (a) a convex quadrilateral, or (b) a triangle with a triangular bite taken out of it. The beauty of the convex case is that exchanging the internal edge with a chord linking the other two vertices yields a different triangulation.

  This gives us a local "edge-flip" operation for changing and possibly improving a given triangulation. Indeed, a Delaunay triangulation can be constructed from any initial triangulation by removing skinny triangles until no locally improving exchange remains.

- *What dimension are we working in?* – Three-dimensional problems are usually harder than two-dimensional problems. The three-dimensional generalization of triangulation involves partitioning the space into four-vertex tetrahedra by adding non-intersecting faces. One important difficulty is that there is no way to tetrahedralize the interior of certain polyhedra without adding extra vertices. Furthermore, it is NP-complete to decide whether such a tetrahedralization exists, so we should not feel afraid to add extra vertices to simplify our problem.

- *What constraints does the input have?* – When we are triangulating a polygon or polyhedra, we can only add chords that do not intersect any of the boundary facets. In general, we may have a set of additional obstacles or constraints that cannot be intersected by inserted chords. The best such triangulation is the *constrained Delaunay triangulation*. Implementations are described below.

- *Are you allowed to add extra points, or move input vertices?* – When the shape of the triangles does matter, it might pay to strategically add a small number of extra "Steiner" points to the data set to facilitate the construction of a triangulation (say) with no small angles. As discussed above, you *must* add Steiner points to triangulate certain polyhedra.

To triangulate a convex polygon in linear time, just pick an arbitrary starting vertex v and insert chords from v to each other vertex in the polygon to form a fan. Only because the polygon is convex can we be confident that no boundary edges of the polygon will be intersected by these chords. The simplest algorithm for general polygon triangulation tests each of the $O(n^2)$ possible chords, and only inserts those that do not intersect a boundary edge or previously inserted chord. There are practical algorithms that run in $O(n \lg n)$ time and theoretically interesting algorithms that run in linear time. See the Implementations and Notes sections for details.

Implementations: Triangle, by Jonathan Shewchuk, is an award-winning C language code that generates Delaunay triangulations, constrained Delaunay triangulations (forced to have certain edges), and quality-conforming Delaunay triangulations (which avoid small angles by inserting extra points). It has been widely used for finite element analysis and is fast and robust. Triangle is the first thing I would try if I needed a two-dimensional triangulation code. It is available at `http://www.cs.cmu.edu/~quake/triangle.html`.

Fortune's Sweep2 is a widely used two-dimensional code for Voronoi diagrams and Delaunay triangulations, written in C. This code may be simpler to work with, if all you need is the Delaunay triangulation of points in the plane. It is based on Fortune's sweep line algorithm [For87] for Voronoi diagrams and is available from Netlib (see Section 22.1.4 (page 714)) at `https://www.netlib.org/voronoi/`.

TetGen (`http://wias-berlin.de/software/tetgen/`) appears to be the software of choice to tetrahedralize three-dimensional polyhedra [Si15]. Both the CGAL (`www.cgal.org`) and LEDA (see Section 22.1.1 (page 713)) libraries offer C++ implementations of an extensive variety of triangulation algorithms for two and three dimensions, including both constrained and furthest site Delaunay triangulations.

Higher-dimensional Delaunay triangulations are a special case of higher-dimensional convex hulls. Qhull [BDH97] is a popular low-dimensional convex hull code, for two to about eight dimensions. It is written in C and can also construct Delaunay triangulations, Voronoi vertices, furthest-site Voronoi vertices, and half-space intersections. Qhull has been widely used in scientific applications and has a well-maintained homepage at `http://www.qhull.org/`. Another choice is Ken Clarkson's higher-dimensional convex-hull code, *Hull*, available at `https://www.netlib.org/voronoi/hull.html`.

Notes: Chazelle [Cha91] gave a linear-time algorithm for triangulating a simple polygon, which was an important theoretical result because triangulation served as the bottleneck for many other geometrical algorithms. Chazelle's algorithm is sufficiently hopeless to implement that it qualifies more as an existence proof, but a simpler randomized algorithm is known [AGR01]. The first $O(n \lg n)$ algorithm for polygon triangulation was given by [GJPT78]. An $O(n \lg \lg n)$ algorithm by Tarjan and van

Wyk [TW88] followed before Chazelle's result. Bern et al. [BSA18] gives a survey on polygon and point-set triangulation.

Books on Delaunay triangulations and quality mesh generation include [AKL13, SDC16]. The *International Meshing Roundtable* is an annual conference for people interested in mesh and grid generation. Excellent surveys on mesh generation include [Ber02, Ede06].

Linear-time algorithms for triangulating monotone polygons have been long known [GJPT78], and are the basis of algorithms for triangulating simple polygons. A polygon is monotone when there exists a direction d such that any line with slope d intersects the polygon in at most two points.

A well-studied class of optimal triangulations seeks to minimize the total length of the chords used. The computational complexity of constructing this *minimum weight triangulation* was resolved when Rote [MR06] proved it NP-complete. Interest has thus shifted to provably good approximation algorithms [RW18]. The minimum weight triangulation of a convex polygon can be found in $O(n^3)$ time using dynamic programming.

Related problems: Voronoi diagrams (see page 634), polygon partitioning (see page 658).

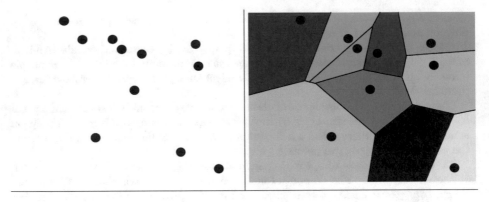

Input   Output

20.4 Voronoi Diagrams

Input description: A set S of points $p_1, \ldots, p_n$.

Problem description: Decompose space into regions such that all points in the region around p_i are closer to p_i than to any other point in S.

Discussion: Voronoi diagrams represent the region of influence around each of a given set of sites S. If these sites represent the locations of McDonald's restaurants, the Voronoi diagram $V(S)$ partitions space into cells around each restaurant. For each person living in a particular cell, the defining McDonald's represents the closest place to get a Big Mac.

Voronoi diagrams have a surprising variety of applications:

- *Nearest-neighbor search* – Finding the nearest neighbor of query point q from a fixed set of points S is simply a matter of determining which cell in $V(S)$ contains q. See Section 20.5 (page 637) for more details.

- *Facility location* – Suppose McDonald's wants to open another restaurant. To minimize interference with existing McDonald's, it should be located as far away from the closest restaurant as possible. This location is always at a vertex of the Voronoi diagram, and hence can be found in a linear-time search through the Voronoi vertices.

- *Largest empty circle* – Suppose you seek a large, contiguous, undeveloped piece of land on which to build a factory. The same condition used to select McDonald's locations is appropriate for other undesirable facilities, namely that they get located as far as possible from any relevant sites of interest. A Voronoi vertex defines the center of the largest empty circle among the points.

- *Path planning* – If the sites of S are the centers of obstacles we seek to avoid, the edges of the Voronoi diagram $V(S)$ define the possible channels that maximize the distance to the obstacles. The "safest" path among the obstacles will thus stick to the edges of the Voronoi diagram.

- *Quality triangulations* – When triangulating a set of points, we often seek nice, fat triangles that avoid small angles and skinny triangles. The *Delaunay triangulation* maximizes the minimum angle over all triangulations, and can be constructed as the dual of the Voronoi diagram. See Section 20.3 (page 630) for details.

Each edge of a Voronoi diagram is a segment of the perpendicular bisector of the line segment formed by the two points in S, because this is the line that partitions the plane between the points. The conceptually simplest method to build a Voronoi diagram is by randomized incremental construction. To add a new site p to the diagram, locate the cell that contains it and add perpendicular bisectors separating p from all sites defining impacted regions. If the sites are inserted in random order, it is likely that only a few regions will be impacted on each insertion.

However, the method of choice is Fortune's sweep line algorithm, especially since robust implementations of it are readily available. The algorithm works by projecting the set of sites in the plane into a set of cones in three dimensions, such that the Voronoi diagram is defined by projecting the cones back onto the plane. Advantages of Fortune's algorithm include that (a) it runs in optimal $\Theta(n \log n)$ time, (b) it is reasonable to implement, and (c) we need not store the entire diagram as we sweep over it.

There is an interesting relationship between convex hulls in $d + 1$ dimensions and Delaunay triangulations (or equivalently Voronoi diagrams) in d-dimensions. After projecting each site in E^d to E^{d+1},

$$(x_1, x_2, \ldots, x_d) \longrightarrow (x_1, x_2, \ldots, x_d, \sum_{i=1}^{d} x_i^2)$$

then taking the convex hull of this $(d + 1)$-dimensional point set, and finally projecting back into d dimensions, we obtain the Delaunay triangulation. Details are given in the Notes section, but this provides the best way to construct Voronoi diagrams in higher dimensions. Programs that compute higher-dimensional convex hulls are presented in Section 20.2 (page 626).

Several important variations of standard Voronoi diagrams arise in practice:

- *Non-Euclidean distance metrics* – Recall that Voronoi diagrams decompose space into regions of influence around each given site. We have assumed that Euclidean distance measures influence, but this is not always appropriate. The time it takes to drive to McDonald's depends upon where the major roads are. Efficient algorithms are known for constructing Voronoi diagrams under a variety of different metrics.

- *Power diagrams* – These structures decompose space into regions of influence around the sites, where the sites are no longer constrained to have all the same power. Imagine a map of radio stations broadcasting at a given frequency. The region of influence around a station depends both on the power of its transmitter and the position/power of its neighbors.

- *K th-order and furthest-site diagrams* – The idea of decomposing space into regions sharing some property can be taken beyond closest-point Voronoi diagrams. All points within a single cell of the kth-order Voronoi diagram share the same set of k nearest neighbors in S. In furthest-site diagrams, any point within a particular region shares the same furthest point in S. Point location (see Section 20.7 (page 644)) on these structures permits fast retrieval of the appropriate points.

Implementations: Fortune's Sweep2 is a widely used two-dimensional code for Voronoi diagrams and Delaunay triangulations, written in C. This code is simple to work with if all you need is the Voronoi diagram. It is based on Fortune's sweep line algorithm [For87] for Voronoi diagrams and is available from Netlib (see Section 22.1.4 (page 714)) at `https://www.netlib.org/voronoi/`.

Both the CGAL (`www.cgal.org`) and LEDA (see Section 22.1.1 (page 713)) libraries offer C++ implementations of a variety of Voronoi diagram and Delaunay triangulation algorithms in two and three dimensions.

Higher-dimensional and furthest-site Voronoi diagrams can be constructed as a special case of higher-dimensional convex hulls. Qhull [BDH97] is a popular low-dimensional convex-hull code, useful for two to about eight dimensions. It is written in C and can also construct Delaunay triangulations, Voronoi vertices, furthest-site Voronoi vertices, and half-space intersections. Qhull has been widely used in scientific applications and has a well-maintained homepage at `http://www.qhull.org/`. Another choice is Ken Clarkson's convex-hull code, *Hull*, available at `https://www.netlib.org/voronoi/hull.html`.

Notes: Voronoi diagrams were studied by Dirichlet in 1850 and are sometimes referred to as *Dirichlet tessellations*. They are named after G. Voronoi, who discussed them in a 1908 paper. In mathematics, concepts get named after the last person to discover them.

Two books [AKL13, OBSC00] offer a complete treatment of Voronoi diagrams and their applications. Fortune [For18] provides an excellent survey on Voronoi diagrams and associated variants such as power diagrams. The first $O(n \lg n)$ algorithm for constructing Voronoi diagrams was based on divide and conquer and is due to Shamos and Hoey [SH75]. Good expositions of both Fortune's sweeping algorithm [For87] and the relationship between Delaunay triangulations and $(d+1)$-dimensional convex hulls [ES86] include [dBvKOS08, O'R01].

In a kth-order Voronoi diagram, we partition the plane such that each point in a region is closest to the same set of k sites. Using the algorithm of [ES86], the complete set of kth-order Voronoi diagrams can be constructed in $O(n^3)$ time. By performing a point location on this structure, the k nearest neighbors of a query point can be found in $O(k + \lg n)$. Expositions on kth-order Voronoi diagrams include [O'R01, PS85].

Related problems: Nearest-neighbor search (see page 637), point location (see page 644), triangulation (see page 630).

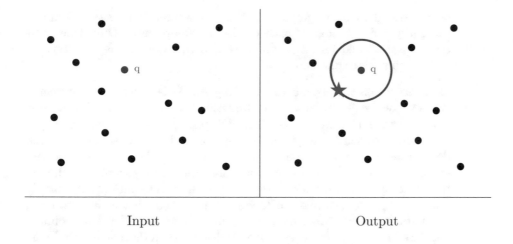

Input   Output

20.5 Nearest-Neighbor Search

Input description: A set S of n points in d dimensions, and query point q.

Problem description: Which point in S is closest to q?

Discussion: The need to quickly find the nearest neighbor of a query point arises in many geometric applications. The classic example involves designing a system to dispatch emergency vehicles to the scene of a fire. As soon as the dispatcher learns the location of the fire, he or she uses a map to identify the firehouse closest to this point so as to minimize transportation delays. This situation occurs in any application mapping customers to service providers.

A nearest-neighbor search is also important in classification. Suppose we are given a collection of numerical data about people (say age, height, weight, years of education, and income level) each of whom has been labeled as either Democrat or Republican. We seek a classifier to predict which way a new voter is likely to vote. Every person in our data set can be represented by a party-labeled point in d-dimensional space. A simple voter classifier can be built by assigning the new point the same party affiliation as its nearest neighbor.

Such nearest-neighbor classifiers are widely used, often in high-dimensional spaces. The vector-quantization method of image compression partitions an image into 8×8 pixel regions. This method uses a predetermined library of several thousand 8×8 pixel tiles and replaces each image region by the most similar library tile. The most similar tile is the point in 64-dimensional space that is closest to the image region in question. Compression is achieved by reporting the 12-bit identifier of the closest library tile instead of the full 64 pixels, at the cost of some minor loss of image fidelity.

Issues arising in nearest-neighbor search include:

- *How many points are you searching?* – When your data set contains only a small number of points (say $n \leq 100$), or if only a few queries are

ever destined to be performed, the simple approach is best. Compare
the query point q against each of the n data points. Only when fast
queries are needed on large numbers of points does it pay to consider
more sophisticated methods.

- *How many dimensions are you working in?* – Nearest-neighbor search
  gets progressively harder as the dimensionality increases. The kd-tree
  data structure, presented in Section 15.6 (page 460), does a very good job
  in moderate-dimensional spaces. But by twenty dimensions, kd-tree search
  degenerates to almost a linear search through the data points. Searches
  in high-dimensional spaces become hard because a sphere of radius r,
  representing all the points with distance $\leq r$ from the center, progressively
  fills up less volume relative to a cube as the dimensionality increases. Thus,
  any data structure based on partitioning points into enclosing volumes
  becomes progressively less effective. Balltrees are a data structure based
  on spheres, which works better in higher dimensions.

  In two dimensions, Voronoi diagrams (see Section 20.4 (page 634)) provide an efficient data structure for nearest-neighbor queries. The Voronoi
  diagram of a point set decomposes the plane into regions such that the
  cell containing data point p consists of all points that are closer to p than
  to any other point in S. Finding the nearest neighbor of query point q reduces to identifying which Voronoi diagram cell contains q, and reporting
  the data point associated with it. Although Voronoi diagrams can be built
  in higher dimensions, their size rapidly grows to the point of unusability.

- *Do you really need the exact nearest neighbor?* – Finding the *absolute*
  nearest neighbor of a point in a very high-dimensional space is hard work.
  Indeed, you probably won't do better than a linear (brute force) search.
  But there are algorithms/heuristics that can give you a reasonably close
  neighbor of your query point fairly quickly.

  One important technique is *dimension reduction*. Projections exist that
  map any set of n points in d-dimensions into a $d' = O(\lg n/\epsilon^2)$-dimensional
  space such that distance to the nearest neighbor in the low-dimensional
  space is within $(1+\epsilon)$ times that of the actual nearest neighbor. Projecting
  the points onto d' random hyperplanes will reliably do the trick.

  Another idea is adding randomness when you search your data structure.
  A kd-tree can be efficiently searched for the cell containing the query
  point q—a cell whose boundary points are good candidates to be close
  neighbors. Now suppose we search for a point q', which is a small random
  perturbation of q. It should land in a different but nearby cell, one of
  whose boundary points might prove to be an even closer neighbor of q.
  Repeating such random queries gives us a way to productively use exactly
  as much computing time as we are willing to spend to improve the answer.

- *Is your data set static or dynamic?* – Will there be occasional insertions
  or deletions of new data points in your application? If these are very

rare events, it might pay to rebuild your data structure from scratch each time. But if they are frequent, select a version of the kd-tree that supports insertions and deletions.

The nearest-neighbor graph on a set S of n points links each vertex to its nearest neighbor. This graph is a subgraph of the Delaunay triangulation, and so can be computed in $O(n \log n)$. This is quite a bargain, since it takes $\Theta(n \log n)$ time just to discover the closest pair of points in S.

As a lower bound, the closest-pair problem in one dimension reduces to sorting. We only need to check the minimum gap between $n - 1$ adjacent pairs after sorting, because the closest pair corresponds to two numbers that lie next to each other. The limiting case occurs when the closest pair lie zero distance apart, meaning that the elements are not unique.

Implementations: *ANN* is a C++ library for both exact and approximate nearest-neighbor searching in arbitrarily high dimensions. It performs well for searches over hundreds of thousands of points in up to about twenty dimensions. It supports all l_p distance norms, including Euclidean and Manhattan distance, and is available at `http://www.cs.umd.edu/~mount/ANN/`. It is the first code I would turn to for nearest-neighbor search.

Annoy (Approximate Nearest Neighbors Oh Yeah) is a C++ library with Python bindings created by Spotify, and made available at `https://github.com/spotify/annoy`. It is based on random projection trees, and designed to support parallel processes sharing the same data. My students swear by the `sklearn.neighbors.BallTree` implementation for nearest-neighbor search in Python.

Samet's spatial index demos (`http://donar.umiacs.umd.edu/quadtree/`) provide a series of Java applets illustrating many variants of *kd*-trees, in association with [Sam06]. *KDTREE 2* contains C++ and Fortran 95 implementations of *kd*-trees for efficient nearest-neighbor search in many dimensions. See `http://arxiv.org/abs/physics/0408067`.

Section 20.4 (page 634) gives a complete collection of Voronoi diagram implementations. In particular, CGAL (`www.cgal.org`) and LEDA (see Section 22.1.1 (page 713)) provide Voronoi diagrams in C++, plus planar point location to make effective use of them for nearest-neighbor search.

Notes: Approximate nearest-neighbor search in high dimensions has been a very active area of research for the past twenty years, with recent results providing a general framework for a broad class of distance metrics [ANN$^+$18]. Andoni and Indyk [AI08, AIR18] ably survey recent results in approximate nearest-neighbor search in high dimensions based on locality sensitive hashing and random projection methods. Both theoretical and empirical results [BM01, ML14, WSSJ14] indicate that these methods preserve distances quite nicely.

The theoretical guarantees underlying Arya and Mount's approximate nearest-neighbor code *ANN* [AM93, AMN$^+$98] are somewhat different. A sparse weighted graph structure is built from the data set, and the nearest neighbor is found by starting at a random point and greedily walking towards the query point in the graph. The closest point found over several random trials is declared the winner. Similar data structures hold promise for other problems in high-dimensional spaces. Nearest-neighbor search was a subject of the Fifth DIMACS challenge, as reported in [GJM02].

Samet [Sam06] is the best reference on kd-trees and other spatial data structures. All major (and many minor) variants are developed in substantial detail. A shorter survey [Sam05] is also available. The technique of using random perturbations of the query point is due to [Pan06].

Good expositions on finding the closest pair of points in the plane [BS76] include [CLRS09, Man89]. These algorithms use a divide-and-conquer approach instead of just selecting from the Delaunay triangulation.

Related problems: Kd-trees (see page 460), Voronoi diagrams (see page 634), range search (see page 641).

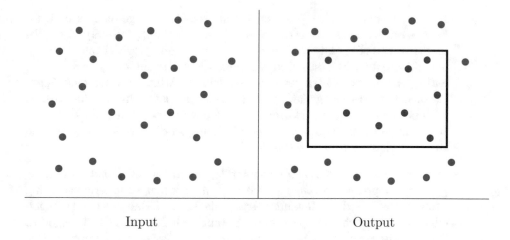

Input   Output

20.6 Range Search

Input description: A set S of n points in d dimensions, and a query region Q.

Problem description: Which points in S lie within Q?

Discussion: Range search problems arise in database and geographic information system (GIS) applications. Data objects with d numerical fields, say people described by their height, weight, and income, can be modeled as points in d-dimensional space. A *range query* asks for all points (or the number of points) in a given region of space. For example, asking for all people with income between $0 and $20,000, height between 6 and 7 feet, and weight between 50 and 150 lbs., defines a box containing people whose wallet and body are both thin.

The difficulty of range search depends on several factors:

- *How many range queries will you perform?* – The simplest approach to range search tests each of the n points against the query polygon Q. This works just fine when the number of queries are small. Algorithms to test whether a point is within a given polygon are presented in Section 20.7 (page 644).

- *What is the shape of your query polygon?* – The easiest regions to query against are *axis-parallel rectangles*, because the inside–outside test reduces to testing whether each coordinate lies within a prescribed range. The output figure above illustrates such an *orthogonal range query*.

  When querying against a non-convex polygon, it pays to partition the polygon into convex pieces or (even better) triangles, and then query the point set against each piece. This works because it is fast to test whether a point lies inside a convex polygon. Algorithms for such convex decompositions are discussed in Section 20.11 (page 658).

- *How many dimensions?* – A general approach to range queries builds a kd-tree on the point set, as discussed in Section 15.6 (page 460). A depth-first traversal of the kd-tree is performed for the query, with each tree node expanded only if its associated rectangle intersects the query region. The entire tree might get traversed for sufficiently large or misaligned query regions, but kd-trees generally lead to an efficient solution. Algorithms with better worst-case performance are known in two dimensions, but kd-trees should work just fine in the plane. In higher dimensions, they provide the only viable solution to the problem.

- *Is your point set static, or might there be insertions/deletions?* – A clever practical approach to range search and many other geometric searching problems is based on Delaunay triangulations. Delaunay triangulation edges connect each point p to nearby points, including its nearest neighbor. To perform a range query, we start by using planar point location (see Section 20.7 (page 644)) to quickly identify a triangle within the region of interest. We then do a depth-first search around a vertex of this triangle, pruning the search whenever it visits a point too distant to have interesting undiscovered neighbors. This should be efficient, because the total number of points visited should be roughly proportional to the number within the query region.

  One nice thing about this approach is that it is relatively easy to employ "edge-flip" operations to fix up a Delaunay triangulation following a point insertion or deletion. See Section 20.3 (page 630) for more details.

- *Can I just count the number of points in a region, or must I identify them?* – It often suffices to count the number of points in a region instead of actually returning them. Harkening back to our introductory example, we may want to know whether there are more thin/poor people or rich/fat ones. The need to find the densest or emptiest region in space often arises, and this can be solved using counting range queries.

  A nice data structure for efficiently answering such aggregate range queries is based on the dominance ordering of the point set. A point x is said to *dominate* point y if y lies both below and to the left of x. Let $D(p)$ be a function that counts the number of points in S that are dominated by p. The number of points m in the orthogonal rectangle defined by $x_{\min} \le x \le x_{\max}$ and $y_{\min} \le y \le y_{\max}$ is given by:

  $$m = D(x_{\max}, y_{\max}) - D(x_{\max}, y_{\min}) - D(x_{\min}, y_{\max}) + D(x_{\min}, y_{\min})$$

  The last additive term corrects for the points for the lower left-hand corner that have been subtracted away twice.

  We can partition the space into n^2 rectangles by drawing a horizontal and vertical line through each of the n points. The set of dominated points will be identical for each point within any rectangle, so the dominance count of the lower left-hand corner of each rectangle can be precomputed, stored, and reported for any query point within it. Range queries reduce to binary search and thus take $O(\lg n)$ time. This data structure takes quadratic space, which is expensive, but the same idea can be adapted to kd-trees to create a more space-efficient search structure.

Implementations: Both CGAL (`www.cgal.org`) and LEDA (see Section 22.1.1 (page 713)) use a dynamic Delaunay triangulation data structure to support circular, triangular, and orthogonal range queries. Both libraries also provide implementations of range tree data structures, which support orthogonal range queries in $O(k + \lg^2 n)$ time where n is the complexity of the subdivision and k is the number of points in the rectangular region.

ANN is a C++ library for both exact and approximate nearest-neighbor searching in arbitrarily high dimensions. It performs well for searches over hundreds of thousands of points in up to about twenty dimensions. It supports fixed-radius, nearest-neighbor queries over all l_p distance norms, which can be used to approximate circular and orthogonal range queries under the l_2 and l_1 norms, respectively. *ANN* is available at `https://www.cs.umd.edu/~mount/ANN/`.

Annoy (Approximate Nearest Neighbors Oh Yeah) is a C++ library with Python bindings created by Spotify, and made available at `https://github.com/spotify/annoy`. It is based on random projection trees, and designed to support parallel processes sharing the same data. Nearest-neighbor search is closely connected to circular range search, since the largest empty circle around a point defines its nearest neighbor.

Notes: Good expositions on data structures with worst-case $O(\lg n + k)$ performance for orthogonal-range searching [Wil85] and *kd*-trees include [dBvKOS08, PS85]. The worst-case performance of *kd*-trees can be very bad: [LW77] describes an instance in two dimensions requiring $O(\sqrt{n})$ time to report that a rectangle is empty. Faster range search and counting structures are available under the word RAM model [CLP11, CW16]. Sun and Blelloch [SB19] present experimental results for both sequential and parallel algorithms for range search.

The problem becomes considerably more difficult for non-orthogonal range queries, where the query region is not an axis-aligned rectangle. For half-plane intersection queries, $O(\lg n)$ time and linear space suffice [CGL85]. For range searching with simplex query regions (such as a triangle in the plane), lower bounds preclude efficient worst-case data structures. See Agrawal [Aga18] for a survey and discussion.

Related problems: Kd-trees (see page 460), point location (see page 644).

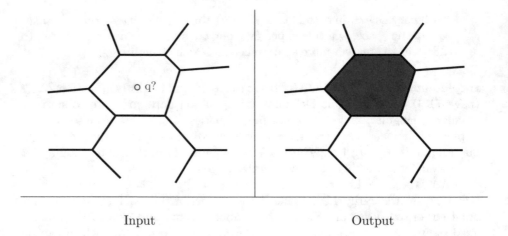

Input   Output

20.7   Point Location

Input description: A decomposition of the plane into polygonal regions, and a query point q.

Problem description: Which region contains the query point q?

Discussion: Point location is often needed as an ingredient to solve larger geometric problems. In a typical police dispatch system, the city will be partitioned into different precincts or districts. Given a map of such regions and a query point (the crime scene), the system must identify which region contains the point. This is exactly the problem of planar point location. Variations include:

- *Is the query point inside or outside of polygon P?* – The simplest version of point location involves only two regions, inside-P and outside-P, and asks which contains a given query point. For polygons with many narrow spirals, this can be surprisingly difficult to tell by inspection. The secret to doing it both by eye or machine is to draw a ray starting from the query point and ending beyond the furthest extent of the polygon. Count the number of times the ray crosses an edge of P. The query point will lie within the polygon iff this number is odd. Whether to count the ray passing through a vertex instead of an edge is evident from context, because we care about how many times we cross the boundary of P. Testing each of the n edges for intersection against the query ray takes $O(n)$ time. Faster algorithms for convex polygons are based on binary search, and take $O(\lg n)$ time.

- *How many queries will be performed?* – We could perform this inside-polygon test separately on each region in a given planar subdivision, but this will be wasteful if many such queries will be performed. It is much better to construct a grid-like or tree-like data structure on top of our subdivision to get us near the correct region quickly. Such search structures are discussed in more detail below.

- *How complicated are the regions of your subdivision?* – More sophisticated inside–outside tests are required when the regions of your subdivision are not convex polygons. But by triangulating all polygonal regions first, each inside–outside test reduces to testing whether a point is in a triangle. Such tests can be made particularly fast and simple, at the minor cost of recording the region name associated with each triangle. An added benefit is that the smaller your regions are, the better grid-like or tree-like superstructures are likely to perform. Care should be taken to avoid long skinny triangles, as discussed in Section 20.3 (page 630).

- *How regularly sized and spaced are your regions?* – When all resulting triangles are roughly the same size and shape, the simplest point location method imposes a regularly spaced $k \times k$ grid of horizontal and vertical lines over the entire subdivision. For each of the k^2 rectangular regions, we maintain a list of all the regions that are at least partially contained within the rectangle. Performing a point location query in such a *grid file* involves a binary search or hash table lookup to identify which rectangle contains query point q, and then searching each region in the associated list to identify the right one.

Such grid files can perform very well, provided that each triangular region overlaps relatively few rectangles (thus minimizing storage space) and each rectangle overlaps only a few triangles (thus minimizing search time). How well it performs depends on the regularity of your subdivision. Some flexibility can be achieved by spacing the horizontal lines irregularly, depending upon where the regions actually lie. The *slab method*, discussed below, is a variation on this idea that guarantees efficient point location at the cost of quadratic space.

- *How many dimensions will you be working in?* – In three or more dimensions, some flavor of kd-tree will almost certainly be the point location method of choice. They may even be the right answer for planar subdivisions that are too irregular for grid files.

Kd-trees, described in Section 15.6 (page 460), decompose space into a hierarchy of rectangular boxes. At each node in the tree, the current box is split into a small number of smaller boxes: typically 2, 4, or 2^d for dimension d. Each leaf box is labeled with the small set of regions at least partially contained in the box. The point location search starts at the root of the tree and traverses down through the child whose box contains the query point q. When the search hits a leaf, we test all relevant regions to see which one contains q. As with grid files, we hope that each leaf contains a small number of regions, and that each region does not cut across too many leaf cells.

- *Am I close to the right cell?* – Walking is a simple point–location technique that might even work well beyond two dimensions. Start from an arbitrary point p in an arbitrary cell, hopefully near to the query point q. Construct the ray from p to q, and identify which face of the cell this hits (a so-called *ray-shooting query*). Such queries take constant time in triangulated arrangements.

  Proceeding to the neighboring cell through this face gets us one step closer to the target. The expected path length will be $O(n^{1/d})$ for sufficiently regular d-dimensional arrangements, although linear in the worst case.

The simplest algorithm that guarantees $O(\lg n)$ worst-case access is the *slab* method, which draws horizontal lines through each vertex, thus defining $n + 1$ "slabs" between the lines. The horizontal slab containing query point q can be found by doing a binary search on the y-coordinate of q. The region containing q within the right slab can be identified by doing a second binary search on the edges that cross the slab. The catch is that a binary search tree must be maintained for each slab, for a worst-case of $O(n^2)$ space. A more space-efficient approach based on building a hierarchy of triangulations over the regions also achieves $O(\lg n)$ search, and is discussed in the Notes section.

Point location methods that are efficient in the worst case tend to require a lot of memory, or are complicated to implement. I identify implementations of such methods below, which are worth experimenting with. But I generally recommend kd-trees for point location applications.

Implementations: Both CGAL (`www.cgal.org`) and LEDA (see Section 22.1.1 (page 713)) provide excellent support for maintaining planar subdivisions in C++. CGAL favors a jump-and-walk strategy, although a worst-case logarithmic search is also provided. LEDA implements $O(\lg n)$ expected-time point location, using partially persistent search trees.

ANN is a C++ library for both exact and approximate nearest-neighbor searching in arbitrarily high dimensions. It can be used to quickly identify a nearby cell boundary point to begin walking from. Check it out at `https://www.cs.umd.edu/~mount/ANN/`.

Arrange is a package for maintaining arrangements of polygons in either the plane or on the sphere. Polygons may be degenerate, and hence represent arrangements of lines. A randomized incremental construction algorithm is used, and efficient point location on the arrangement is supported. *Arrange* is written in C by Michael Goldwasser and is available from `http://euler.slu.edu/~goldwasser/publications/`.

Routines (in C) to test whether a point lies in a simple polygon have been provided by [O'R01, SR03].

Notes: Snoeyink [Sno18] gives an excellent survey of the state of the art in point location, both theoretical and practical. Thorough treatments of deterministic planar-point location data structures are provided by [dBvKOS08, PS85].

Tamassia and Vismara [TV01] use planar point location as a case study of geometric algorithm engineering, in Java. An experimental study of algorithms for planar point location is described in [EKA84]. The winner was a bucketing technique akin to the grid file. Performance of the CGAL point location implementation is reported in [HH09, HKH12].

Kirkpatrick's elegant triangle refinement method [Kir83] builds a hierarchy of triangulations above the actual planar subdivision, such that each triangle on a given level intersects only a constant number of triangles on the following level. Each triangulation is a fraction of the size of the subsequent one, so the total space is linear by summing up a geometric series. Furthermore, the height of the hierarchy is $O(\lg n)$, ensuring fast query times. An alternative algorithm realizing the same time bounds is [EGS86]. The slab method described above is due to [DL76] and is presented in [PS85]. Expositions on the inside-outside test for simple polygons include [O'R01, PS85, SR03].

More recently, there has been interest in dynamic data structures for point location, which support fast incremental updates of the planar subdivision (such as insertions and deletions of edges and vertices) as well as fast point location. Chiang and Tamassia's [CT92] survey is an appropriate place to begin, with updated references in [Sno18]. The best current methods approach logarithmic search and update times [CN18].

Related problems: Kd-trees (see page 460), Voronoi diagrams (see page 634), nearest-neighbor search (see page 637).

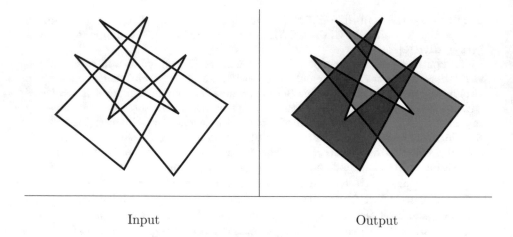

Input   Output

20.8   Intersection Detection

Input description: A set S of lines and line segments $l_1, \ldots, l_n$, or a pair of polygons or polyhedra P_1 and P_2.

Problem description: Which pairs of line segments intersect each other? What is the intersection of P_1 and P_2?

Discussion: Intersection detection is a fundamental geometric primitive with many applications. Picture the virtual-reality simulation of an architectural model for a building. Any illusion of reality vanishes the instant the virtual person walks through a virtual wall. To enforce such physical constraints, any such intersection between polyhedral models must be immediately detected, and the operator notified or constrained.

Another application arises in design rule checking for integrated circuit layout. A minor design defect resulting in two crossing metal strips could short out the chip. But such errors can be detected before fabrication, using programs that find all intersections between line segments.

Issues arising in intersection detection include:

- *Do you want to compute the intersection or just report it?* – We distinguish between intersection detection and computing the actual intersection. Just detecting that an intersection exists can be a substantially easier problem, and often suffices. For the virtual reality application, it might not matter exactly where we hit the wall—just the fact that we hit it.

- *Are you intersecting lines or line segments?* – The big difference here is that any two lines with different slopes will intersect at exactly one point. All the points of intersections can thus be found in $O(n^2)$ time, by comparing each pair of lines. Constructing the arrangement of the lines provides more information than just the intersection points, and is discussed in Section 20.15 (page 671).

Finding all the intersections between n line segments proves considerably more challenging. Even the basic primitive of testing whether two line segments intersect is not as trivial, as discussed in Section 20.1 (page 622). We can explicitly test each line segment pair, and thus find all intersections in $O(n^2)$ time. But faster algorithms exist when there are only a few intersection points.

- *How many intersection points do you expect?* – In integrated circuit design-rule checking, we expect the set of line segments to have few if any intersections. What we seek is an algorithm whose running time is *output sensitive*, taking time proportional to the number of intersection points.

Such output-sensitive algorithms exist for line segment intersection. The fastest algorithm takes $O(n \lg n + k)$ time, where k is the number of intersections. These algorithms are based on the planar sweep line approach.

- *Can you see point x from point y?* – Visibility queries ask whether vertex x has an unobstructed view of vertex y in a room full of obstacles. This can be phrased as a line-segment intersection problem: Does the line segment from x to y intersect any obstacle? Such visibility problems arise in robot motion planning (see Section 20.14) and in hidden-surface elimination for computer graphics.

- *Are the intersecting objects convex?* – Better intersection algorithms exist when the line segments form the boundaries of polygons. The critical issue is whether the polygons are convex. Intersecting a convex n-gon with a convex m-gon can be done in $O(n + m)$ time, using the sweep line algorithm discussed below. This is possible because the intersection of two convex polygons always forms a convex polygon with at most $n + m$ vertices.

However, non-convex polygons are not so well behaved. Consider the intersection of two "combs" generalizing the Picasso-like frontispiece to this section. As shown, the intersection of non-convex polygons may be disconnected and have quadratic size in the worst case.

Intersecting polyhedra is more complicated than polygons, because two polyhedra can intersect even when no edges do. Consider the example of a needle piercing the interior of a face. In general, however, similar issues arise for both polygons and polyhedra.

- *Do the objects move?* – In the walk-through application just described, the room and the objects in it do not change between one scene and the next. But the person moves, and so we must do repeated analysis of the same fixed geometry.

One common technique is to approximate the objects in a scene by simpler objects that enclose them, such as boxes. Whenever two enclosing boxes intersect, then the underlying objects *might* intersect, and so further work is necessary to decide the issue. But it is much more efficient to test whether simple boxes intersect than more complicated objects, so we win when collisions are rare. Many variations on this theme are possible, but this idea leads to large performance improvements for complicated environments.

Planar sweep algorithms can be used to efficiently compute the intersections among a set of line segments, or the intersection/union of two polygons. These algorithms keep track of interesting changes as we sweep a vertical line from left to right over the data. At its left-most position, the line intersects nothing, but we encounter a series of events as it moves to the right:

- *Insertion* – The left-most point of a line segment may be encountered, which now becomes available to intersect some other line segment.

- *Deletion* – The right-most point of a line segment is encountered, which means that we have completely swept over the segment. It can thus be safely deleted, and removed from any further consideration.

- *Intersection* – If we maintain the active line segments that intersect the sweep line as sorted from top to bottom, the next intersection always occurs between neighboring line segments. Following this intersection, these two line segments swap their relative order.

Keeping track of what is going on requires two data structures. The future is maintained by an *event queue*: a priority queue ordered by the x-coordinate of all possible future events of interest: insertion, deletion, and intersection. See Section 15.2 (page 445) for priority queue implementations. The present is represented by the *horizon*—an ordered list of line segments intersecting the current position of the sweep line. The horizon can be maintained using any dictionary data structure, such as a balanced tree.

To compute the intersection or union of polygons, we modify the processing of the three basic event types. This sweep line algorithm becomes quite simple for pairs of convex polygons, because (1) at most four polygon edges intersect the sweep line, so no horizon data structure is needed, and (2) no event-queue sorting is needed, because we can start from the left-most vertex of each polygon and proceed to the right following the polygonal ordering. The details are more complex for general polygon intersection, but the sweep line approach described above still gets the job done.

Implementations: Both LEDA (see Section 22.1.1 (page 713)) and CGAL (www.cgal.org) offer extensive support for line segment and polygonal intersection. In particular, they provide a C++ implementation of the Bentley–Ottmann sweep line algorithm [BO79], finding all k intersection points between n line segments in the plane in $O((n + k) \lg n)$ time.

O'Rourke [O'R01] provides a robust program in C to compute the intersection of two convex polygons. See Section 22.1.9 (page 717).

Finding the mutual intersection of a collection of half-spaces is a special case of the convex hull problem. Qhull [BDH97] is convex hull code of choice for general dimensions. Qhull has been widely used in scientific applications and has a well-maintained homepage at `https://www.qhull.org/`.

Notes: Mount [Mou18] is an excellent survey of algorithms for computing intersections of geometric objects such as line segments, polygons, and polyhedra. Books with chapters discussing such problems include [dBvKOS08, CLRS09, PS85]. Preparata and Shamos [PS85] provide a good exposition on the special case of finding intersections and unions of axis-oriented rectangles—a problem that arises often in integrated circuit design.

An optimal $O(n \lg n + k)$ algorithm for computing line segment intersections is due to Chazelle and Edelsbrunner [CE92]. Simpler, randomized algorithms achieving the same time bound are presented by Mulmuley [Mul94].

Lin et al. [LMK18] survey techniques and software for collision detection. Weller [Wel13] provides a book-length treatment of recent data structures, emphasizing collision detection for haptic feedback. Deformable models, whose shapes change over time, represent another challenge for collision detection [BFB12]. Identifying road junctions from GPS or image data is an interesting intersection detection problem, detailed in [FK10].

Related problems: Maintaining arrangements (see page 671), motion planning (see page 667).

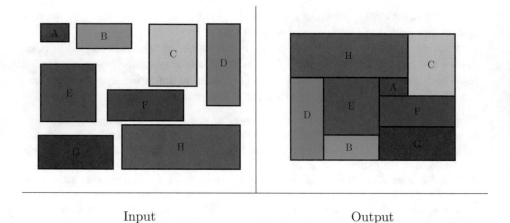

Input   Output

20.9   Bin Packing

Input description: A set of n items with sizes $d_1, \ldots, d_n$. A set of m bins with capacities $c_1, \ldots, c_m$.

Problem description: Store all the items using the fewest number of bins.

Discussion: Bin packing arises in many packaging and manufacturing problems. Suppose that you are manufacturing widgets cut from sheet metal or pants cut from cloth. To minimize cost and waste, we seek to lay out the parts to use as few fixed-size metal sheets or bolts of cloth as possible. Placing each part on a given sheet, in the best location, is a bin-packing variant called the *cutting stock* problem. After our widgets have been successfully manufactured, we face another bin-packing problem—how best to fit the boxes onto trucks so as to minimize the number of vehicles needed to ship everything.

Even the most elementary-sounding bin-packing problems are NP-complete; see the discussion of integer partition in Section 16.10 (page 497). Thus, we are doomed to think in terms of heuristics, instead of worst-case optimal algorithms. Fortunately, relatively simple heuristics tend to work well on most bin-packing problems. Further, many applications have peculiar, problem-specific constraints that would frustrate sophisticated algorithms for bin packing. The following factors will affect the choice of heuristic:

- *What are the shapes and sizes of the objects?* – The character of a bin-packing problem depends greatly on the shapes of the objects to be packed. Solving a standard jigsaw puzzle has a much different flavor than packing squares into a rectangular box. In *one-dimensional bin packing*, each object's size is given simply as an integer. This is equivalent to packing boxes of equal width into a chimney of that width, and makes it a special case of the knapsack problem of Section 16.10 (page 497).

When all the boxes are of identical size and shape, repeatedly filling each row gives a reasonable, but not necessarily optimal, packing. Consider trying to fill a 3×3 square with 2×1 bricks. You can only pack three bricks using one orientation, while four bricks fit using two.

- *Are there constraints on the orientation and placement of objects?* – Many shipping boxes are labeled "this side up" (imposing an orientation on the box) or "do not stack" (requiring them to sit on top of any box pile). Respecting these constraints restricts our flexibility in packing, and hence increases the number of trucks needed to send out certain shipments. Most shippers solve this problem by ignoring the labels. Indeed, your task will be simpler if you don't have to worry about the consequences of them.

- *Is the problem on-line or off-line?* – Do we know the complete set of objects to pack at the beginning of the job (an *off-line* problem)? Or will we get them one at a time and greedily deal with them as they arrive (an *on-line* problem)? This difference is important, because we can do a better job packing when we can take a global view and plan ahead. For example, we can arrange the objects in an order that will facilitate efficient packing, perhaps by sorting them from biggest to smallest.

The standard off-line heuristics for bin packing order the objects by size or shape and then insert them into bins. Typical insertion rules are (1) select the first or left-most bin the object fits in, (2) select the bin with the most room, (3) select the bin that provides the tightest fit, or (4) select a random bin.

Analytical and empirical results suggest that *first-fit decreasing* is the best heuristic. Sort the objects in decreasing order of size, so that the biggest object is first and the smallest last. Insert each object one by one into the first bin that has room for it. If no bin has room, we must start another bin. In the case of one-dimensional bin packing, this can never require more than 22% more bins than necessary, and usually does much better. First-fit decreasing has an intuitive appeal to it, for we pack the bulky objects first and hope that little objects can fill up the remaining cracks.

First-fit decreasing is easily implemented in $O(n \lg n + bn)$ time, where $b \leq \min(n, m)$ is the number of bins actually used. Simply do a linear sweep through the bins to check for space on each insertion. A faster $O(n \lg n)$ implementation is possible by using a binary tree to keep track of the space remaining in each bin.

We can fiddle with the insertion order in such a scheme to deal with problem-specific constraints. For example, it is reasonable to take "do not stack" boxes last (perhaps after artificially lowering the height of the bins to leave some room up top to work with) and to place fixed-orientation boxes at the beginning (so we can use the extra flexibility later to stick boxes on top).

Packing boxes is much easier than packing arbitrary geometric shapes, enough so that one general technique packs each part into its own box, and then packs the boxes. Finding an enclosing rectangle for a polygonal part is easy; just find the upper, lower, left, and right tangents in a given orientation. Finding the orientation that minimizes the area (or volume) of such a box is more difficult, but can be done in both two and three dimensions [O'R85]. See the Implementations section for a fast approximation to minimum enclosing box.

In the case of non-convex parts, considerable useful space can be wasted in the holes created by placing each part in a box.   One solution is to find the *maximum empty rectangle* within each boxed part and use this to contain other parts if it is sufficiently large. More advanced solutions are discussed below.

Implementations: BPPLIB (`http://or.dei.unibo.it/library/bpplib`) is an extensive collection of bin-packing resources, including codes, sample problems, and references [DIM18]. This should be your first stop if you are interested in bin packing or cutting stock problems.

Martello and Toth's collection of Fortran implementations of algorithms for a variety of knapsack problem variants are available at `http://www.or.deis.unibo.it/kp.html`. An electronic copy of [MT90a] has also been generously made available. David Pisinger maintains a well-organized collection of C-language codes for knapsack problems and related variants like bin packing and container loading. These are available at `http://www.diku.dk/~pisinger/codes.html`.

A first step towards packing arbitrary shapes packs each in its own minimum volume box. For a code to find an approximation to the optimal packing, see `https://sarielhp.org/research/papers/00/diameter/diam_prog.html`. This algorithm runs in near-linear time [BH01].

Notes: See [CJCG$^+$13, CKPT17, DIM16, WHS07] for surveys of the extensive literature on bin packing and the cutting stock problem. Keller, Pferschy, and Psinger [KPP04] is a solid reference on the knapsack problem and variants. Experimental results on bin-packing heuristics include [BJLM83, MT87].

Efficient algorithms are known for finding the largest empty rectangle in a polygon [DMR97] and point set [CDL86].

Sphere packing is an important and well-studied special case of bin packing, with applications to error-correcting codes. Particularly notorious was the "Kepler conjecture"—the problem of establishing the densest packing of unit spheres in three dimensions. This conjecture was finally settled by Hales and Ferguson in 1998; see [Szp03] for an exposition. Conway and Sloane [CS93] is the best reference on sphere packing and related problems.

Milenkovic has worked extensively on two-dimensional bin-packing problems for the apparel industry, minimizing the amount of material needed to manufacture pants and other clothing. Reports of this work include [DM97, Mil97].

Related problems: Knapsack problem (see page 497), set packing (see page 682).

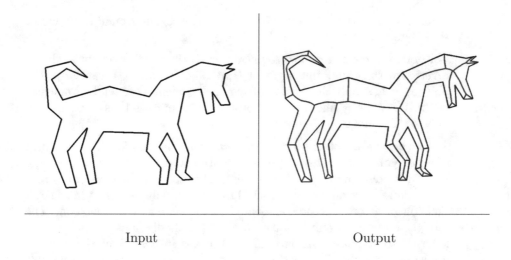

Input   Output

20.10 Medial-Axis Transform

Input description: A polygon or polyhedron P.

Problem description: Find the skeleton of P, the set of points that have more than one closest point on the boundary of P

Discussion: The medial-axis transformation is useful in *thinning* a polygon, or equivalently finding its *skeleton*. The goal is to extract a simple, robust representation of the shape of the polygon. The thinned versions of letters like 'A' and 'B' capture the essence of their shape, and are largely unaffected by changing the thickness of strokes, or adding font-dependent flourishes such as serifs. The skeleton also represents the center of the given shape, a property that leads to other applications, like shape reconstruction and motion planning.

The medial-axis transformation of a polygon is always a tree, making it fairly easy to use dynamic programming to measure the "edit distance" between the skeleton of a known model and the skeleton of an unknown object. Whenever the two skeletons are similar enough, we can classify the unknown object as an instance of our model. This technique has proven useful in computer vision and in optical character recognition. The skeleton of a polygon with holes (like the letters A and B) is not a tree but an embedded planar graph, yet it remains fairly easy to work with.

There are two distinct approaches to computing medial-axis transforms, depending upon whether your input is a geometric point set or a pixel image:

- *Geometric data* – Recall that the Voronoi diagram of a point set S (see Section 20.4 (page 634)) decomposes the plane into regions around each point $s_i \subset S$ such that all points within the region around s_i are closer to s_i than to any other site in S. Similarly, the Voronoi diagram of a set of line segments L decomposes the plane into regions around each line segment $l_i \in L$ such that all points within the region around l_i are closer to l_i than to any other site in L.

Polygons are defined by line segments, such that each segment l_i shares a vertex with neighboring segment l_{i+1}. The medial-axis transform of a polygon P is simply the portion of the line-segment Voronoi diagram that lies within P. Any line-segment Voronoi diagram code thus suffices to do polygon thinning.

The *straight skeleton* is a structure related to the medial-axis transform of a polygon, except that the bisectors are equidistant to the supporting lines of its defining edges. The straight skeleton, medial-axis transform, and Voronoi diagram are all identical for convex polygons. But in general, skeleton bisectors may not be located in the center of the polygon. The straight skeleton is quite similar to a proper medial-axis transform, but is easier to compute. In particular, all edges in a straight skeleton are polygonal. See the Notes section for references with more details on how to compute it.

- *Image data* – Digitized images can be interpreted as points sitting at the lattice points on an integer grid. Thus, we could extract a polygonal description from boundaries in an image and feed it to the geometric algorithms just described. However, the internal vertices of the skeleton will most likely not lie at grid/pixel points. Geometric approaches to image processing problems often flounder, because images are pixel based and not continuous.

  A direct pixel-based approach for constructing a skeleton implements the "brush fire" view of thinning. Imagine a fire burning along all edges of the polygon, racing inward at a constant speed. The skeleton is marked by all points where two or more fires meet. The resulting algorithm traverses all the boundary pixels of the object, identifies those vertices as being in the skeleton, deletes the rest of the boundary, and repeats. This algorithm terminates when all pixels are extreme, leaving an object only one or two pixels thick. When implemented properly, this takes time linear in the number of pixels in the image.

  Algorithms that explicitly manipulate pixels tend to be easy to implement, because they avoid complicated data structures. However, the geometry doesn't work out exactly right in such pixel-based approaches. For example, the skeleton of a polygon is no longer always a tree, or even necessarily connected. Further, the points in the skeleton will be close to but not quite equidistant to two boundary edges. Because you are trying to do continuous geometry in a discrete world, there is no way to solve the problem completely. You just have to live with it.

Implementations: CGAL (www.cgal.org) includes a package for computing the straight skeleton of a polygon P. Associated with it are routines for constructing offset contours defining the polygonal regions within P whose points are at least distance d from the boundary.

VRONI [Hel01] is a robust and efficient program for computing Voronoi diagrams of line segments, points, and arcs in the plane. It can readily compute medial-axis transforms of polygons, because it can construct Voronoi diagrams of arbitrary line segments. *VRONI* has been tested on thousands of synthetic and real-world data sets, some with over a million vertices. For more information, see `http://www.cosy.sbg.ac.at/~held/projects/vroni/vroni.html`. Other programs for constructing Voronoi diagrams are discussed in Section 20.4 (page 634).

Programs that reconstruct or interpolate point clouds often are based on medial-axis transforms. *Cocone* (`http://www.cse.ohio-state.edu/~tamaldey/cocone.html`) constructs an approximate medial-axis transform of the polyhedral surface it interpolates from points in E^3. See [Dey06] for the theory behind *Cocone*. *Powercrust* [ACK01a, ACK01b] constructs a discrete approximation to the medial-axis transform, and then reconstructs the surface from this transform. When the point samples are sufficiently dense, the algorithm is guaranteed to produce a geometrically and topologically correct approximation to the surface. It is available at `https://web.cs.ucdavis.edu/~amenta/powercrust.html`.

Notes: The book by Siddiqi and Pizer [SP08] offers a comprehensive treatment of medial representations and algorithms. Surveys of thinning approaches in image processing and computer graphics, include [LLS92, Ogn93, SBdB16, TDS+16]. The medial-axis transformation was introduced for shape similarity studies in biology [Blu67]. Computational topology is an emerging field for the formal analysis of shape: see the book by Edelsbrunner and Harer [EH10]. Good expositions on the medial-axis transform include [dBvKOS08, O'R01, Pav82].

The medial-axis of a polygon can be computed in $O(n \lg n)$ time for arbitrary n-gons [Lee82], although linear-time algorithms exist for convex polygons [AGSS89]. An $O(n \lg n)$ algorithm for constructing medial-axis transforms in curved regions was given by Kirkpatrick [Kir79].

Straight skeletons were introduced in [AAAG95], with a subquadratic algorithm due to [EE99]. See [LD03] for an interesting application of straight skeletons to defining the roof structures in virtual building models. The input and output figures above were inspired by [dMPF09].

Related problems: Voronoi diagrams (see page 634), Minkowski sum (see page 674).

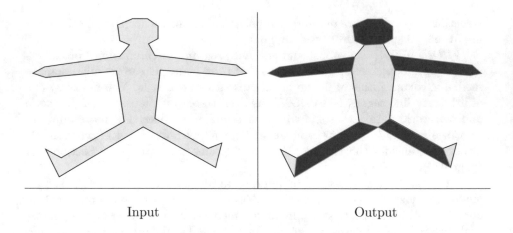

Input   Output

20.11   Polygon Partitioning

Input description: A polygon or polyhedron P.

Problem description: Partition P into a small number of simple (typically convex) pieces.

Discussion: Partitioning is an important pre-processing step for many geometric algorithms, because geometric problems tend to be simpler on convex objects than on non-convex ones. It is often easier to work with a number of convex pieces than with a single non-convex polygon.

Carving a nation into states or counties or districts is a classical problem in polygon partitioning. Typically we seek to balance the population or area of each region, but shape also matters. Gerrymandering is the fine art of drawing districts for the electoral advantage of your particular political party, districts that often assume very complicated shapes to ensure they include all the "right" voters. To prevent the most egregious violations, the law demands that districts be as *compact* as possible, ideally nicely shaped into convex regions.

Several flavors of polygon partitioning arise, depending upon the particular application:

- *Should all the pieces be triangles?* – Triangulation is the mother of all polygon partitioning problems, where we partition the interior of the polygon completely into triangles. Triangles are convex and have only three sides, making them the most elementary possible polygon.

  Every triangulation of an n-vertex polygon contains exactly $n - 2$ triangles. Thus, triangulation cannot be the answer if we seek a small number of convex pieces. A "nice" triangulation is judged by the shape of the triangles, not the count. See Section 20.3 (page 630) for an introduction to triangulation.

- *Do I want to cover or partition my polygon?* – *Partitioning* a polygon means completely dividing the interior into non-overlapping pieces. *Covering* a polygon means that our decomposition is permitted to contain mutually overlapping pieces. Both prove useful in different situations. When decomposing a complicated query polygon for a range search (Section 20.6 (page 641)), we seek a partitioning, so that each point we locate occurs in exactly one piece. But when decomposing a polygon for graphics rendering, a covering suffices, since there is no difficulty with painting a given region twice—provided we use the same color on all regions. We will concentrate here on partitioning, since it is simpler to do right, and any application needing a covering will accept a partitioning. The only drawback is that partitions can be larger than coverings.

- *Am I allowed to add extra vertices?* – A final issue is whether we are allowed to add Steiner vertices to the polygon, either by splitting edges or adding interior points. Otherwise, we are restricted to adding chords between two existing vertices. But adding well-placed vertices may enable us to reduce the number of pieces, at the cost of more complicated algorithms and perhaps messier results.

The Hertel–Mehlhorn heuristic for convex decomposition using diagonals is simple and efficient. It starts from an arbitrary triangulation of the polygon, and then deletes any chord that leaves only convex pieces. A chord deletion results in a non-convex piece only when it creates an internal angle that is greater than 180 degrees. The decision of whether such an angle will result can be made locally from the chords and edges surrounding the deleted chord, in constant time. The result always contains at most four times the minimum number of convex pieces.

I recommend using this heuristic unless it is critical for you to absolutely minimize the number of pieces. By experimenting with different triangulations and various deletion orders, you may be able to obtain somewhat better decompositions.

Dynamic programming can be employed to find the absolute minimum number of diagonals used in a polygon decomposition into convex regions. The simplest implementation, which maintains the number of pieces for all $O(n^2)$ subpolygons split by a chord, runs in $O(n^4)$. Faster algorithms use fancier data structures, running in $O(n + r^2 \min(r^2, n))$ time, where r is the number of reflex vertices. An $O(n^3)$ algorithm that further reduces the number of pieces by adding interior vertices is cited below, although it is complex and presumably difficult to implement.

An alternate decomposition problem partitions polygons into *monotone* pieces. The vertices of a y-monotone polygon can be divided into two chains such that any horizontal line intersects either chain at most once.

Implementations: Many triangulation codes start by finding a trapezoidal or monotone decomposition of polygons. Further, a triangulation is a simple form of convex decomposition. Check out the codes in Section 20.3 (page 630) as a starting point.

CGAL (`www.cgal.org`) contains a polygon-partitioning library that includes (1) the Hertel–Mehlhorn heuristic for partitioning a polygon into convex pieces, (2) finding an optimal convex partitioning using the $O(n^4)$ dynamic programming algorithm, and (3) an $O(n \log n)$ sweep-line heuristic for partitioning into monotone polygons.

A triangulation code of particular relevance here is GEOMPACK—a suite of Fortran 77 and C++ codes for two- and three-dimensional triangulation and convex decomposition problems, available at `http://people.math.sc.edu/Burkardt/cpp_src/geompack/geompack.html`. In particular, it does both Delaunay triangulation and convex decompositions of polygonal and polyhedral regions, as well as arbitrary-dimensional Delaunay triangulations.

Notes: Survey articles on polygon partitioning include [Kei00, OST18]. Keil and Sack [KS85] give an excellent overview of what is known about partitioning and covering polygons. Expositions on the Hertel–Mehlhorn heuristic [HM83] include [O'R01]. The $O(n + r^2 \min(r^2, n))$ dynamic programming algorithm for minimum convex decomposition using diagonals is due to Keil and Snoeyink [KS02]. The $O(r^3 + n)$ algorithm minimizing the number of convex pieces with Steiner points appears in [CD85]. Amato et al. [GALL13, LA06] provide an efficient heuristic for decomposing polygons with holes into "almost convex" polygons in $O(nr)$ time, with later work generalizing this to polyhedra.

Art gallery problems are an interesting topic related to polygon covering, where we seek to position the minimum number of guards in a given polygon such that every point in the interior of the polygon is watched by at least one guard. This corresponds to covering the polygon with a minimum number of star-shaped polygons. O'Rourke [O'R87] is a beautiful book that presents the art gallery problem and its many variations. Although sadly out of print, it is available at `http://cs.smith.edu/~jorourke/books/ArtGalleryTheorems`. The art gallery problem is not obviously in NP because of its dependence on non-integer arithmetic, but hard under an appropriate model of computation [AAM18]. Recent computational results on constructing optimal guard sets are reported in [KBFS12].

Related problems: Triangulation (see page 630), set cover (see page 678).

Input   Output

20.12 Simplifying Polygons

Input description: A polygon or polyhedron p, with n vertices.

Problem description: Find a polygon or polyhedron p' containing only n' vertices, such that the shape of p' is as close as possible to p.

Discussion: Polygon simplification has two primary applications. The first involves cleaning up a noisy representation of a shape, perhaps obtained by scanning a picture of an object. Simplifying the boundary can remove the noise, and reconstruct the original object. The second involves data compression, where we seek to reduce detail on a large and complicated object: yet leave it looking essentially the same. This can be a big win in computer graphics, where the smaller model might be significantly faster to render.

Several issues arise in shape simplification:

- *Do you want the convex hull?* – The simplest simplification is the convex hull of the object's vertices. The convex hull (see Section 20.2 (page 626)) removes all internal concavities from the polygon. If you are simplifying a robot model for motion planning, this is almost certainly a good thing. But using the convex hull would be disastrous in an OCR system, because the concavities of characters provide most of the interesting features. An "X" would be identical to an "I", since both hulls are boxes. Also, taking the convex hull of a convex polygon can do nothing to simplify it further.

- *Am I allowed to insert points, or just delete them?* – The typical goal of simplification is to represent the object as well as possible using a given number of vertices. The simplest approaches do local modifications to the boundary in order to reduce the vertex count. For example, if three consecutive vertices form a small-area triangle or define an extremely large angle, the center vertex can be deleted and replaced with an edge without severely distorting the polygon.

Methods that only delete vertices quickly melt shapes into unrecognizability, however. More robust heuristics move vertices around to cover up the gaps that are created by deletions. Such "split-and-merge" heuristics can do a decent job, although nothing is guaranteed. Better results are likely by using the Douglas–Peucker algorithm, described below.

- *Must the resulting polygon be intersection-free?* – A serious drawback of incremental procedures is that they fail to ensure *simple* polygons, meaning they are without self-intersections. Such "simplified" polygons may have ugly artifacts that cause problems for subsequent routines. If simplicity is important, explicitly test all the line segments of your polygon for pairwise intersections, as discussed in Section 20.8 (page 648).

  An approach to polygon simplification that guarantees a simple approximation involves computing minimum-link paths. The *link distance* of a path between points s and t is the number of straight segments on the path. An as-the-crow-flies path has a link distance of one, while in general the link distance is one more than the number of turns on the path. The link distance between points s and t in a scene with obstacles is defined by the minimum-link distance over all paths from s to t.

  This link distance approach "fattens" the boundary of the polygon by some acceptable error window ϵ (see Section 20.16 (page 674)) in order to construct a channel around the polygon. The minimum-link cycle in this channel represents the simplest polygon that never deviates from the original boundary by more than ϵ. An easy-to-compute approximation to link distance reduces it to breadth-first search, by placing a discrete set of possible turn points within the channel and connecting each pair of mutually visible points by an edge.

- *Are you given an image to clean up, instead of a polygon to simplify?* – The conventional approach to remove noise from a digital image is to take the Fourier transform of the image, filter out the high-frequency elements, and then take the inverse transform to recreate the image. See Section 16.11 (page 501) for details on the fast Fourier transform.

The Douglas–Peucker algorithm for shape simplification starts with a simple approximation and then refines it, instead of starting with a complicated polygon and trying to simplify it. Start by selecting two vertices v_1 and v_2 of polygon P, and propose the degenerate polygon v_1, v_2, v_1 as a simple approximation P'. Now scan through each of the vertices of P, and select the one that is farthest from the corresponding edge of the polygon P'. Inserting this vertex adds the triangle to P' to minimize the maximum deviation from P. Points can be inserted until satisfactory results are achieved. This takes $O(kn)$ time to insert k points when $|P| = n$.

Simplification becomes considerably more difficult in three dimensions. Indeed, it is NP-complete to find the minimum-size surface separating two polyhedra. Higher-dimensional analogies of the planar algorithms discussed here can be used to heuristically simplify polyhedra. See the Notes section.

Implementations: The Douglas–Peucker algorithm is readily implemented. For a C implementation with efficient worst-case performance [HS94], see `https://www.codeproject.com/Articles/1711/A-C-implementation-of-Douglas-Peucker-Line-Approxi`.

QSlim is a quadric-based simplification algorithm that can produce high-quality approximations of triangulated surfaces quite rapidly. It is available at `http://mgarland.org/software/qslim.html`.

Yet another approach to polygonal simplification is based on simplifying and expanding the medial-axis transform of the polygon. The medial-axis transform (see Section 20.10 (page 655)) produces a skeleton of the polygon, which can be trimmed before inverting the transform to yield a simpler polygon. *Cocone* (`http://www.cse.ohio-state.edu/~tamaldey/cocone.html`) constructs an approximate medial-axis transform of the polyhedral surface it interpolates from points in E^3. See [Dey06] for the theory behind *Cocone*. *Powercrust* [ACK01a, ACK01b] constructs a discrete approximation to the medial-axis transform, and then reconstructs the surface from this transform. When the point samples are sufficiently dense, the algorithm is guaranteed to produce a geometrically and topologically correct approximation to the surface. It is available at `https://web.cs.ucdavis.edu/~amenta/powercrust.html`.

CGAL (`www.cgal.org`) provides support for polyline simplification, as well as the most extreme polygon/polyhedral reduction, finding the smallest enclosing circle/sphere.

Notes: The Douglas–Peucker incremental refinement algorithm [DP73] is the basis for most shape simplification schemes, with faster implementations due to [HS94, HS98]. Generalizations include nested polygonal subdivisions [DDS09, XWW11] and area-preserving simplification [BMRS16]. The link distance approach to polygon simplification is presented in [GHMS93]. Shape simplification problems become considerably more complex in three dimensions. Even finding the minimum-vertex convex polyhedron lying between two nested convex polyhedra is NP-complete [DJ92], although approximation algorithms are known [MS95b].

Heckbert and Garland [HG97] survey algorithms for shape simplification. Shape simplification using medial-axis transformations (see Section 20.10) are presented in [TH03].

Testing whether a polygon is simple can be performed in linear time, at least in theory, as a consequence of Chazelle's linear-time triangulation algorithm [Cha91].

Related problems: Thinning (see page 655), convex hull (see page 626).

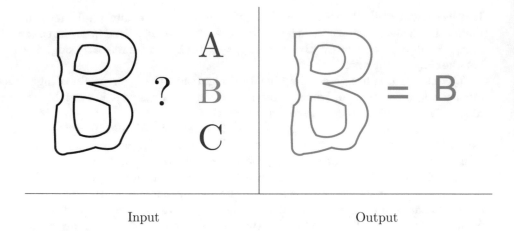

| Input | Output |

20.13   Shape Similarity

Input description: Two polygonal shapes, P_1 and P_2.

Problem description: How similar are P_1 and P_2?

Discussion: Shape similarity is a problem that underlies much of pattern recognition. Consider a system for optical character recognition (OCR). We are given a library of shape models representing letters, and unknown shapes obtained by scanning a page. We seek to identify the unknown shapes by matching each of them to the most similar shape model.

Shape similarity is an inherently ill-defined problem, because what "similar" means is application dependent. Thus, no single algorithmic approach can solve all shape matching problems. Whatever method you select, expect to spend a large chunk of time tweaking it to achieve maximum performance.

Among your possible approaches are:

- *Hamming distance* – Suppose that your two polygons have been properly *registered*, meaning overlaid one on top of the other. The Hamming distance measures the area of symmetric difference between the two polygons—in other words, the area lying within one polygon but not both of them. When two polygons are identical and properly aligned, the Hamming distance is zero. If the polygons differ only in a little noise at the boundary, then the Hamming distance of properly aligned polygons will be small.

  Computing the area of the symmetric difference reduces to finding the intersection or union of two polygons (discussed in Section 20.8 (page 648)) and then computing areas (discussed in Section 20.1). But the difficult problem is finding the right alignment of the two polygons. This overlay problem is simplified in applications such as OCR, because the characters are inherently aligned within lines on the page and are not free to rotate. Efficient algorithms to optimize the overlap of convex polygons without rotation are cited below. Simple but reasonably effective heuristics are based on identifying reference landmarks on each polygon (such as the centroid, bounding box, or extremal vertices) and then matching a subset of these landmarks to define the alignment.

Hamming distance is particularly simple and efficient to compute on bit-mapped images, since after alignment all we do is sum the differences of the corresponding pixels. Although Hamming distance makes sense conceptually and can be simple to implement, it captures only a crude notion of shape and is likely to be ineffective in many applications.

- *Hausdorff distance* – An alternative similarity measure (post-registration) is Hausdorff distance, which identifies the point on P_1 that is the maximum distance from P_2 and returns this distance. The Hausdorff distance is not symmetrical, for the tip of a long but thin protrusion from P_1 can imply a large Hausdorff distance P_1 to P_2, even though every point on P_2 is close to some point on P_1. A fattening of the entire boundary of one of the models (as is liable to happen with boundary noise) by a small amount may substantially increase the Hamming distance yet have little effect on the Hausdorff distance.

  Which is better, Hamming or Hausdorff? It depends upon your application. As with Hamming distance, computing the right alignment between the polygons can be difficult and time-consuming.

- *Comparing Skeletons* – A more powerful approach to shape similarity uses thinning (see Section 20.10 (page 655)) to extract a tree-like skeleton for each object. This skeleton captures many aspects of the original shape. The problem now reduces to comparing the shape of two such skeletons, using such features as the topology of the tree and the lengths/slopes of the edges. This comparison can be modeled as some form of subgraph isomorphism (see Section 19.9 (page 610)), with edges allowed to match whenever their lengths and slopes are sufficiently similar.

- *Machine learning techniques* – A final approach for pattern recognition problems uses machine learning-based techniques such as logistic regression, support vector machines, or deep neural networks. The progress here has been staggering in recent years in applications like face recognition, which works reliably despite differences in pose, orientation, and lighting between template and target images. Such problems feel much harder than the rigid shape similarity discussed earlier in this section.

  Machine learning proves successful when you have a lot of data to train on and no problem-specific ideas of what to do with it. Typically, you first identify a set of easily computed features of the shape, such as area, number of sides, and number of holes—although deep learning methods eliminate the need to do such feature engineering. A black-box program then takes your training data and produces a classification function, which accepts as input your feature vector and returns a measure of what the shape is, or how close it is to a particular shape.

How good are the resulting classifiers? It depends upon the application. Machine learning methods usually take a fair amount of tweaking and tuning to realize their full potential. One problem is interpretability. If you don't know how or why black-box classifiers are making their decisions, you can't know when they will fail. An interesting case was a system built for the military to distinguish between images of cars and tanks. It performed very well on test images but disastrously in the field. Eventually, someone realized that the car images had been filmed on a sunnier day than the tanks, and the program was classifying solely on the presence of clouds in the background of the image!

Implementations: The computational geometry library CGAL (`https://www.cgal.org/`) contains a variety of routines associated with shape detection and matching, including Hausdorff distance computations. An alternate distance metric between polygons can be based on its angle-turning function [ACH+91]. An implementation in C of this turning function metric by Eugene K. Ressler is provided at www.algorist.com.

Several excellent support vector machine classifiers are available. These include Python's scikit-learn (`https://scikit-learn.org/`, SVM^{light} (`http://svmlight.joachims.org/`) and the widely used and well-supported LIBSVM (`https://www.csie.ntu.edu.tw/~cjlin/libsvm/`).

Notes: Veltkamp [Vel01] is an excellent survey on shape matching from a computational geometry perspective. See also the survey by Alt and Guibas [AG00]. General books on pattern classification algorithms include [Che15, DHS00, JD88]. A recent survey of geometric approaches to face recognition is presented in [SBW17]. Goodfellow [GBC16] is the primary reference on deep learning.

The optimal alignment of n and m-vertex convex polygons subject to translation (but not rotation) can be computed in $O((n + m)\log(n + m))$ time [dBDK+98]. An approximation of the optimal overlap under translation and rotation is due to Ahn, et al. [ACP+07].

A linear-time algorithm for computing the Hausdorff distance between two convex polygons is given in Atallah [Ata83], with algorithms for the general case reported in [HK90].

Related problems: Graph isomorphism (see page 610), thinning (see page 655).

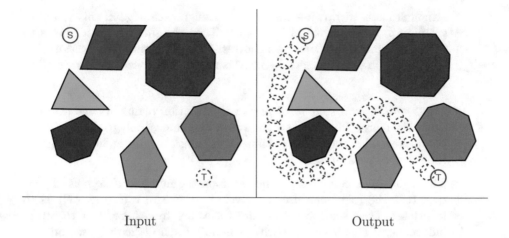

Input   Output

20.14 Motion Planning

Input description: A polygonal-shaped robot starting at position s in a room containing polygonal obstacles, and a goal position t.

Problem description: Find the shortest route taking s to t without intersecting any obstacles.

Discussion: That motion planning is a complex problem is obvious to anyone who has tried to move a large piece of furniture into a small apartment. Plotting paths for mobile robots is the canonical motion-planning application. It also arises in systems for molecular docking. Many drugs are small molecules that act by binding to a given target model. Identifying which binding sites are accessible to a candidate drug is clearly an instance of motion planning.

Finally, motion planning provides a tool for computer animation and virtual reality. Given the set of object models and where they appear in scenes s_1 and s_2, a motion planning algorithm can construct a short sequence of intermediate motions to transform s_1 to s_2. These motions can fill in the intermediate scenes between s_1 and s_2, with such scene interpolation greatly reducing the workload on the animator.

Many factors govern the complexity of motion-planning problems:

- *Is your robot a point?* – For point robots, motion planning reduces to finding the shortest path from s to t around the obstacles. This problem is known as *geometric shortest path*. The most readily implementable approach constructs the *visibility graph* of the polygonal obstacles, plus the points s and t. This visibility graph contains a node for each obstacle vertex, and an edge between two obstacle vertices iff they "see" each other without being blocked by some obstacle edge.

    This visibility graph can be constructed by testing each of the $\binom{n}{2}$ vertex-pair edge candidates for intersection against each of the n obstacle edges, although faster algorithms are known. Assign each edge of this visibility graph with weight equal to its length. Then the shortest path from s to t can be found using Dijkstra's shortest-path algorithm (see Section 18.4 (page 554)) in time bounded by what is required to construct the visibility graph.

- *What motions can your robot perform?* – Motion planning becomes considerably more difficult when the robot becomes a polygon instead of a point. Now all of the corridors that we use must be wide enough to permit the robot to pass through.

The algorithmic complexity depends upon the number of *degrees of freedom* that the robot can use to move. Is it free to rotate as well as to translate? Does the robot have links that are free to bend or to rotate independently, as in an arm with a hand? Each degree of freedom corresponds to a dimension in the search space of possible configurations. Additional freedom makes it more likely that a short path exists from start to goal, although it also makes it harder to find this path.

- *Can you simplify the shape of your robot?* – Motion planning algorithms tend to be complex and time-consuming. Anything you can do to simplify your environment is a win. In particular, consider replacing your robot in an enclosing disk. Any start-to-goal path for this disk defines such a path for the robot inside of it. Furthermore, since any orientation of a disk is equivalent to any other orientation, rotation provides no help in finding a path. All movements can thus be limited to the simpler case of translation.

- *Are motions limited to translation only?* – When rotation is not allowed, the *expanded obstacles* approach can be used to reduce the problem of polygonal motion planning to the previously resolved case of a point robot. Pick a reference point on the robot, and replace each obstacle by its Minkowski sum with the robot polygon (see Section 20.16 (page 674)). This creates a larger, fatter obstacle, defined by the shadow traced as the robot walks a loop around the object while maintaining contact with it. Finding a path from the initial reference position to the goal amidst these fattened obstacles defines a legal path for the polygonal robot in the original environment.

- *Are the obstacles known in advance?* – We have assumed that the robot starts out with a map of its environment. But this can't be true (say) in applications where the obstacles move. There are two strategies for solving motion planning problems without a map. The first approach explores the environment, building a map of what has been seen, and then uses this map to plan a path to the goal. A simpler strategy proceeds like a sightless man with a compass. Walk in the direction towards the goal until progress is blocked by an obstacle, and then trace out a path along the obstacle until the robot is again free to proceed directly towards the goal. Unfortunately, this will fail in environments of sufficient complexity.

The most practical approach to general motion planning involves randomly sampling the *configuration space* of the robot. The configuration space defines the set of legal positions for the robot, using one dimension for each degree of freedom. A planar robot capable of translation and rotation has three degrees of freedom, namely the x- and y-coordinates of a reference point on the robot and the angle θ relative to this point. Certain points in this space represent legal positions, while others intersect obstacles.

Construct a set of legal configuration-space points by random sampling. For each pair of points p_1 and p_2, decide whether there exists a direct, non-intersecting path between them. This defines a graph with vertices for each legal point and edges for each traversable pair. Motion planning now reduces to finding a direct path from the initial/final positions to one or more vertices in the graph, and then solving a shortest-path problem from initial to final points.

There are many ways to enhance this basic technique, such as adding additional vertices to regions of particular interest. Building a road map provides a nice, clean approach to solve problems that would otherwise get very messy.

Implementations: State-of-the-art sampling-based motion planning algorithms are available from the *Open Motion Planning Library* (`https://ompl.kavrakilab.org/`) with hooks to integrate collision checking and visualization. This should presumably be your first stop in any robot or other motion planning project. OMPL is described in [ŞMK12].

The *Motion Planning Toolkit* (MPK) is a C++ library and toolkit for developing single- and multi-robot motion planners. It includes SBL, a fast single-query probabilistic roadmap path planner, and is available at `http://robotics.stanford.edu/~mitul/mpk/`.

The computational geometry library CGAL (`www.cgal.org`) contains many algorithms related to motion planning including visibility graph construction and Minkowski sums. O'Rourke [O'R01] gives a toy implementation of an algorithm to plot motion for a two-jointed robot arm in the plane. See Section 22.1.9 (page 717).

Notes: Latombe's book [Lat91] describes practical approaches to motion planning, including the random sampling method described above. Two other worthy books on motion planning are freely available: by LaValle [LaV06] (`http://planning.cs.uiuc.edu/`) and Laumond [Lau98] (`https://www.laas.fr/~jpl/book.html`) respectively.

Motion planning was originally studied by Schwartz and Sharir as the "piano mover's problem." Their solution constructs the complete free space of robot positions that do not intersect obstacles, and then finds the shortest path within the proper connected component. These free space descriptions are very complicated, involving arrangements of higher-degree algebraic surfaces. The fundamental papers on the piano mover's problem appear in [HSS87], with more recent surveys including [KF11, MLL16, PČY+16, HSS18].

The best general result for this free-space approach to motion planning is due to Canny [Can87], who showed that any problem with d degrees of freedom can be solved in $O(n^d \lg n)$, although faster algorithms exist for special cases of the general motion planning problem. The expanded obstacle approach to motion planning is due to Lozano-Perez and Wesley [LPW79]. The heuristic, sightless man's approach to motion planning discussed above has been studied by Lumelski [LS87].

The time complexity of algorithms based on the free-space approach to motion planning depends intimately on the combinatorial complexity of the arrangement of surfaces defining the free space. Algorithms for maintaining arrangements are presented in Section 20.15 (page 671). Davenport–Schinzel sequences often arise in the analysis of such arrangements. Sharir and Agarwal [SA95] provide a comprehensive treatment of Davenport–Schinzel sequences and their relevance to motion planning.

The visibility graph of n line segments with E pairs of visible vertices can be constructed in $O(n \lg n + E)$ time [GM91, PV96], which is optimal. Hershberger and Suri [HS99] have an $O(n \lg n)$ algorithm for finding shortest paths for point-robots with polygonal obstacles. Chew [Che85] provides an $O(n^2 \lg n)$ for finding shortest paths for a disk-robot in such a scene.

Related problems: Shortest path (see page 554), Minkowski sum (see page 674).

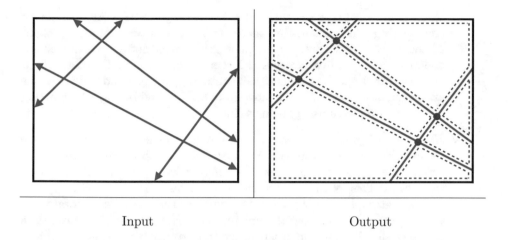

Input   Output

20.15   Maintaining Line Arrangements

Input description: A set of lines $l_1, \ldots, l_n$.

Problem description: What is the decomposition of the plane defined by $l_1, \ldots, l_n$?

Discussion: A fundamental problem in computational geometry is explicitly constructing the regions formed by the intersections of a set of n lines. Many problems reduce to constructing and analyzing such an arrangement of a specific set of lines. Examples include:

- *Degeneracy testing* – Given a set of n lines in the plane, do any three lines pass through the same point? Brute-force testing of all triples takes $O(n^3)$ time. Instead, we can construct the arrangement of the lines and walk over each vertex to explicitly count its degree, all in quadratic time.

- *Satisfying the maximum number of linear constraints* – Suppose that we are given a set of n linear constraints, each of the form $y \leq a_i x + b_i$. Which point in the plane satisfies the largest number of them? Construct the arrangement of the lines. All points in any region or *cell* of this arrangement will satisfy exactly the same set of constraints, so we only need to test one point per cell to find the global maximum.

Thinking of geometric problems in terms of features in an arrangement can be very useful in formulating algorithms. Unfortunately, it must be admitted that arrangements are not as popular in practice as might be supposed. Primarily, this is because of the depth of understanding necessary to apply them correctly. The computational geometry library CGAL provides a general and robust implementation that justifies the effort to figure this out. Issues arising in arrangements include:

- *What is the right way to construct a line arrangement?* – Algorithms for constructing arrangements are incremental. Begin with an arrangement of one or two lines. Subsequent lines get inserted into the arrangement one at a time, yielding larger and larger arrangements. To insert a new line, we start from the left-most cell containing the line and walk over the arrangement to the right. We move from cell to neighboring cell, splitting each into two pieces using the new line.

- *How big will your arrangement be?* – A geometric fact called the *zone theorem* implies that the kth line inserted cuts through k cells of the arrangement, and further that $O(k)$ total edges form the boundary of these cells. This means that we can scan through each edge of every cell we encounter on our insertion walk, confident that only linear total work will be performed inserting the line into the arrangement. The total time to insert all n lines in constructing the full arrangement is thus $O(n^2)$.

- *What do you want to do with your arrangement?* – Given an arrangement and a query point q, we often want to identify which cell of the arrangement contains q. This is the problem of point location, discussed in Section 20.7 (page 644). Given an arrangement of lines or line segments, we are often interested in computing all points of intersection of the lines. The problem of intersection detection is discussed in Section 20.8 (page 648).

- *Does your input consist of points instead of lines?* – Although lines and points seem to be different geometric objects, appearances can be misleading. Through the magic of *duality transformations*, we can turn line L into point p and vice versa:

$$L : y = 2ax - b \leftrightarrow p : (a, b)$$

Duality is important because we can now apply line arrangements to point problems, often with surprising results.

For example, suppose we are given a set of n points, and we want to know whether any three of them all lie on the same line. This sounds similar to the degeneracy testing problem discussed above. But in fact it is *exactly the same*, only with the role of points and lines exchanged. We can dualize our points into lines as above, construct the arrangement, and then search for a vertex with three lines passing through it. The dual of this vertex defines the line on which the three initial vertices lie.

It often becomes useful to traverse each face of an existing arrangement exactly once. Such traversals are called *sweep line algorithms*, and are discussed in some detail in Section 20.8 (page 648). The basic procedure sorts the intersection points by x-coordinate and then walks from left to right while keeping track of all we have seen.

Implementations: CGAL (www.cgal.org) provides a generic and robust package for arrangements of curves (not just lines) in the plane. This should be the starting point for any serious project using arrangements. A recent book by Fogel et al. [FHW12] on CGAL arrangements provides the best starting point.

A robust code for constructing and topologically sweeping an arrangement in C++ is provided at `https://www.cs.tufts.edu/research/geometry/other/sweep/`. An extension of topological sweep to deal with the visibility complex of a collection of pairwise disjoint convex planar sets has been provided in CGAL.

Arrange is a package for maintaining arrangements of polygons in either the plane or on the sphere. Polygons may be degenerate, and hence represent arrangements of lines. A randomized incremental construction algorithm is used, and efficient point location on the arrangement is supported. *Arrange* is written in C by Michael Goldwasser and is available from `http://euler.slu.edu/~goldwasser/publications/`.

Notes: Edelsbrunner [Ede87] provides a comprehensive treatment of the combinatorial theory of arrangements, plus algorithms on arrangements with applications. It is an essential reference for anyone seriously interested in the subject. Recent surveys of combinatorial and algorithmic results include [AS00, HS18]. Good expositions on constructing arrangements include [dBvKOS08, O'R01]. Implementation issues related to arrangements as implemented in CGAL are discussed in [FWH04, HH00].

Arrangements generalize naturally beyond two dimensions. Instead of lines, the space decomposition is defined by planes (or beyond three dimensions, *hyperplanes*). The zone theorem states that any arrangement of n d-dimensional hyperplanes has total complexity $O(n^d)$, and any single hyperplane intersects cells of complexity $O(n^{d-1})$. This provides the justification for incremental construction algorithms for arrangements. Walking around the boundary of each cell to find the next cell that the hyperplane intersects takes time proportional to the number of cells created by inserting the hyperplane.

The history of the zone theorem has become somewhat muddled, because the original proofs were later found to be wrong in higher dimensions. See [ESS93] for a discussion and a correct proof. The theory of Davenport–Schinzel sequences is intimately tied into the study of arrangements, which is presented in [SA95].

The naive algorithm for sweeping an arrangement of lines sorts the n^2 intersection points by x-coordinate and hence requires $O(n^2 \lg n)$ time. The *topological sweep* [EG89, EG91] eliminates the need to sort, and so traverses the arrangement in quadratic time. This algorithm is readily implementable and can be applied to speed up many sweep-line algorithms. See [RSS02] for a robust implementation with experimental results.

Related problems: Intersection detection (see page 648), point location (see page 644).

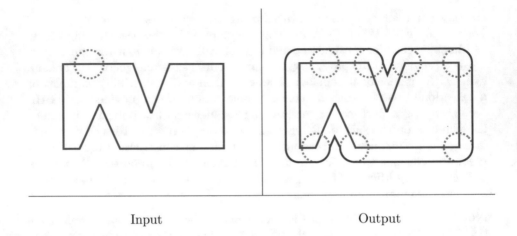

Input   Output

20.16   Minkowski Sum

Input description: Point sets or polygons A and B, containing n and m vertices respectively.

Problem description: What is the convolution of A and B, that is, the Minkowski sum $A + B = \{x + y \mid x \in A, y \in B\}$?

Discussion: Minkowski sums are useful geometric operations that *fatten* up objects in appropriate ways. For example, a popular approach to motion planning for polygonal robots in a room with polygonal obstacles (see Section 20.14 (page 667)) fattens each obstacle by taking the Minkowski sum of them with the shape of the robot. This reduces the problem to the more easily solved case of point robots. Another application is in shape simplification (see Section 20.12 (page 661)). Here we fatten the boundary of an object to create a channel around it, and then let the minimum link path lying within this channel define the simplified shape. Finally, convolving an irregular object with a small circle will smooth out the boundary by eliminating minor nicks and cuts.

The definition of a Minkowski sum assumes that the polygons A and B have been positioned on a coordinate system:

$$A + B = \{x + y \mid x \in A, y \in B\}$$

where $x + y$ is the vector sum of two points. Thinking of this in terms of translation, the Minkowski sum is the union of all translations of A by a point defined within B. Issues arising in computing Minkowski sums include:

- *Are your objects rasterized images or explicit polygons?* – The definition of Minkowski summation suggests a simple algorithm if A and B are rasterized images. Initialize a sufficiently large matrix of pixels by determining the size of the convolution of the bounding boxes of A and B. For each pair of points in A and B, sum up their coordinates and darken the appropriate pixel. These algorithms get more complicated if an explicit polygonal representation of the Minkowski sum is needed.

- *Do you want to fatten your object by a fixed amount?* – The most common fattening operation expands a model M by a given tolerance t, known as *offsetting.* As shown in the figures above, this is accomplished by computing the Minkowski sum of M with a disk of radius t. The basic algorithms still work, although the offset is not a polygon. Its boundary is instead composed of circular arcs and line segments.

- *Are your objects convex?* – The complexity of computing Minkowski sums depends in a serious way on the shape of the polygons. When both A and B are convex, the Minkowski sum can be found in $O(n + m)$ time by tracing the boundary of one polygon with another. If one of them is non-convex, the *size* of the sum (the number of vertices or edges) can be as large as $\Theta(nm)$. Even worse is when both A and B are not convex, where the size can be as large as $\Theta(n^2m^2)$. Minkowski sums of non-convex polygons are often ugly in a majestic sort of way, with holes created or destroyed in surprising fashion.

A straightforward approach to computing the Minkowski sum is based on triangulation and union. First triangulate both polygons, and then compute the Minkowski sum of each triangle of A against every triangle of B. The sum of a triangle against another triangle is an easy-to-compute special case of convex polygons, discussed below. The union of these $O(nm)$ convex polygons will be $A + B$. Algorithms for computing the union of polygons are based on plane sweep, as discussed in Section 20.8 (page 648).

Computing the Minkowski sum of two convex polygons is easier than the general case, because the sum will always be convex. For convex polygons it is easiest to slide A along the boundary of B and compute the sum edge by edge. Partitioning each polygon into a small number of convex pieces (see Section 20.11 (page 658)), and then unioning the Minkowski sum for each pair of pieces, will usually prove more efficient than working with two fully triangulated polygons.

Implementations: The CGAL (`www.cgal.org`) Minkowski sum package provides an efficient and robust code to find the Minkowski sums of two arbitrary polygons, as well as compute both exact and approximate offsets.

An implementation for computing the Minkowski sums of two convex polyhedra in three dimensions is described in [FH06] and available at `https://www.cs.tau.ac.il/~efif/CD/`.

Notes: Good expositions on algorithms for Minkowski sums include [dBvKOS08, O'R01]. The fastest algorithms for various cases of Minkowski sums include [KOS91, Sha87].

The practical efficiency of Minkowski sum in the general case depends upon how the polygons are decomposed into convex pieces. The optimal solution is not necessarily the partition with the fewest number of convex pieces. Agarwal et al. [AFH02] provide a thorough study of decomposition methods for Minkowski sum. Baram et al. [BFH+18] show how to speed up Minkowski sums of polygons with holes through an observation that sufficiently small holes do not affect the shape of the sum.

The combinatorial complexity of the Minkowski sum of two convex polyhedra in three dimensions is completely resolved in [FHW07]. An implementation of Minkowski sum for such polyhedra is described in [FH06].

Related problems: Thinning (see page 655), motion planning (see page 667), simplifying polygons (see page 661).

Chapter 21

Set and String Problems

Sets and strings both represent collections of objects—the difference is whether order matters. Sets are groups of symbols whose order is assumed to carry no significance, while strings are defined by the sequence or arrangement of symbols.

The assumption of a fixed order makes it possible to solve string problems much more efficiently than set problems, through techniques such as dynamic programming and advanced data structures like suffix trees. The interest in and importance of large-scale string processing algorithms have been increasing due to bioinformatics, social media, and other text-processing applications. Good books on string algorithms include:

- *Gusfield* [Gus97] – To my taste, this remains the best introduction to string algorithms. It contains a thorough discussion on suffix trees, with clear and innovative formulations of classical exact string matching algorithms.

- *Crochemore, Hancart, and Lecroq* [CHL07] – A comprehensive treatment of string algorithms, written by a true leader in the field. Translated from the French, but clear and accessible.

- *Navarro and Raffinot* [NR07] – A concise but practical and implementation-oriented treatment of pattern-matching algorithms, with particularly thorough treatment of bit-parallel approaches.

- *Crochemore and Rytter* [CR03] – A survey of specialized topics in string algorithmics emphasizing theory.

Theoreticians working in string algorithmics sometimes refer to their field as *stringology*. The annual Combinatorial Pattern Matching (CPM) conference is the primary venue devoted to both practical and theoretical aspects of string algorithmics and related areas.

S. S. Skiena, *The Algorithm Design Manual*, Texts in Computer Science, https://doi.org/10.1007/978-3-030-54256-6_21

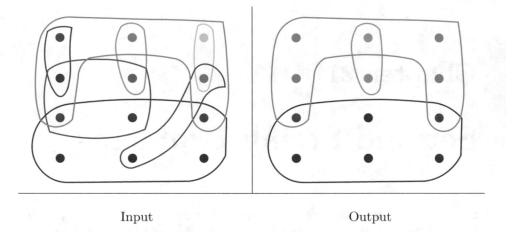

| Input | Output |

21.1   Set Cover

Input description: A collection of subsets $S = \{S_1, \ldots, S_m\}$ of the universal set $U = \{1, \ldots, n\}$.

Problem description: What is the smallest subset T of S whose union equals the universal set, that is,

$$\cup_{i=1}^{|T|} T_i = U \ ?$$

Discussion: Set cover arises when you try to efficiently acquire items that have been packaged in a fixed set of lots. You seek a collection with at least one representative of each item type, while buying as few lots as possible. Finding *a* set cover is easy, because you can always buy one of every possible lot. But identifying a small set cover lets you do the same job for less money. Set cover provides a natural formulation of the Lotto ticket optimization problem discussed in Section 1.8 (page 22). There we seek to buy the smallest number of tickets needed to cover all of a given set of combinations.

Boolean logic minimization is another interesting application of set cover. We are given a specific Boolean function on k variables, which specifies the output of 0 or 1 for all 2^k possible input vectors. We seek the simplest circuit that exactly implements this function. One approach is to find a disjunctive normal form (DNF) formula on the variables and their complements, such as $x_1\bar{x}_2 + \bar{x}_1\bar{x}_2$. We could build one "and" term for each 1 in the input vector, and then "or" them all together. But we can save by factoring out common subsets of variables. Given a set of feasible "and" terms, each of which covers a subset of the vectors we need, we seek to "or" together the smallest number of terms that realize the function. This is exactly the set cover problem.

There are several types of set cover problems you should be aware of:

- *Are you allowed to cover elements more than once?* – The distinction here is between *set cover* and *set packing*, which will be discussed in Section 21.2 (page 682). If you can, take advantage of the freedom to cover elements multiple times because it usually results in a smaller covering.

- *Are your sets derived from the edges or vertices of a graph?* – Set cover is a very general problem, and includes several useful graph problems as special cases. Suppose instead that you seek the smallest set of edges in a graph that will cover each vertex exactly once. You are really looking for a *perfect matching* in the graph (see Section 18.6 (page 562)). Now suppose instead that you seek the smallest set of vertices that cover each edge at least once. This is the *vertex cover* problem, discussed in Section 19.3 (page 591).

  It is instructive to see how to model vertex cover as an instance of set cover. Let the universal set U correspond to the set of edges $\{e_1, \ldots, e_m\}$. Construct n subsets, with S_i consisting of the edges incident on vertex v_i. Although vertex cover is a special case of set cover, you should take advantage of the superior heuristics that exist for the more restricted problem.

- *Do your subsets contain only two elements each?* – You are in luck if all of your subsets have at most two elements each. This special case can be solved efficiently to optimality because it reduces to finding a maximum matching in a graph. Unfortunately, the problem becomes NP-complete as soon as your subsets have three elements each.

- *Do you want to cover elements with sets, or sets with elements?* – In the *hitting set* problem, we seek a small number of items that together represent each subset in a given population. Hitting set is illustrated in Figure 21.1. The input is identical to set cover, but instead we seek the smallest subset of elements $T \subset U$ such that each subset S_i contains at least one element of T. Thus, $S_i \cap T \neq \emptyset$ for all $1 \leq i \leq m$. Suppose we seek a small Congress containing at least one representative of each demographic group. Individuals have multiple identities and thus represent several groups at one time: I am simultaneously male, Jewish, left-handed, and a baby boomer. If each group is defined by a specified subset of people, the minimum hitting set gives the smallest possible politically correct Congress.

  Hitting set is *dual* to set cover, meaning that it is exactly the same problem in disguise. Replace each element of U by a set of the names of the subsets that contain it. Now S and U have exchanged roles, for we seek a set of subsets from U to cover all the elements of S. This is exactly set cover, so we can use any set cover code to solve hitting set problems after performing this simple translation. See Figure 21.1 for an example.

Set cover is at least as hard as vertex cover, so it is also NP-complete. In fact, it is somewhat harder. Approximation algorithms do no worse than twice

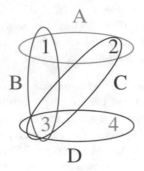
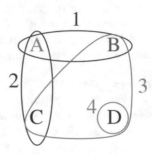

Figure 21.1: A hitting set instance optimally solved by selecting elements 1 and 3, or 2 and 3 (on left). This dual set cover instance of this problem is optimally solved by selecting subsets 1 and 3, or 2 and 3 (on right).

optimal for vertex cover, but the best approximation algorithm for set cover is $\Theta(\lg n)$ times optimal.

Greedy is the most natural and effective heuristic for set cover. Begin by selecting the largest subset for the cover, and then delete all its elements from the universal set. We add the subset containing the largest number of remaining uncovered elements repeatedly until all are covered. This heuristic always gives a set cover using at most $\ln n$ times as many sets as optimal. In practice it usually does a lot better.

The simplest implementation of the greedy heuristic sweeps through the entire input instance of m subsets for each greedy step. However, by using such data structures as linked lists and a bounded-height priority queue (see Section 15.2 (page 445)), the greedy heuristic can be implemented in $O(S)$ time, where $S = \cup_{i=1}^{m} |S_i|$ is the size of the input representation.

It pays to check whether there exist elements that occur in only a few subsets—ideally just one. If so, we should select the biggest subset containing such an element at the very beginning. We must eventually take such a subset, and this will carry along additional elements we might otherwise pay extra to cover if we wait.

Simulated annealing techniques on top of such greedy heuristics is likely to produce somewhat better set covers. Backtracking can be used to guarantee you an optimal solution, but it is often not worth the computational expense.

An often more powerful approach rests on the integer linear programming (ILP) formulation of set cover. Let the 0/1 integer variable s_i denote whether subset S_i is selected for a given cover. Each universal set element $x \in U$ defines a constraint based on all subsets S_i that contain x, namely:

$$\sum_{x \in S_i} s_i \geq 1$$

This ensures that x will be covered by at least one selected subset. The minimum set cover satisfies all of these constraints while minimizing $\sum_i s_i$. This integer program can be easily generalized to weighted set cover (allowing non-uniform costs for different subsets. Relaxing this to a linear program (allowing $0 \leq s_i \leq 1$ instead of constricting each variable to be either 0 or 1) permits efficient and effective heuristics using rounding techniques.

Implementations: Both the greedy heuristic and the integer linear programming formulation above are sufficiently simple in their respective worlds that one has to implement them from scratch.

Pascal implementations of an exhaustive search algorithm for set packing, as well as heuristics for set cover, appear in [SDK83]. See Section 22.1.9 (page 717).

SYMPHONY is a mixed-integer linear programming solver that includes a set partitioning solver. It is available at `https://github.com/coin-or/SYMPHONY`.

Notes: An old but classic survey article on set cover is [BP76], with more recent approximation and complexity analysis surveyed in [Pas97]. See [CFT99, CFT00] for extensive computational studies of integer programming-based set cover heuristics and exact algorithms. An excellent exposition on algorithms and reduction rules for set cover is presented in [SDK83].

Good expositions of the greedy heuristic for set cover include [CLRS09, Hoc96]. An example demonstrating that the greedy heuristic for set cover can be as bad as $\lg n$ is presented in [Joh74, PS98]. This is not a defect of the heuristic. Indeed, it is provably hard to approximate set cover to within an approximation factor better than $(1 - o(1)) \ln n$ [Fei98]. Better results are possible for restricted set cover problems arising from geometric instances, like finding the smallest number of points to stab a given set of circles [AP14, MR10].

Knuth's volume 4A [Knu11] contains a fascinating discussion of Boolean logic optimization reminiscent of set cover.

Related problems: Matching (see page 562), vertex cover (see page 591), set packing (see page 682).

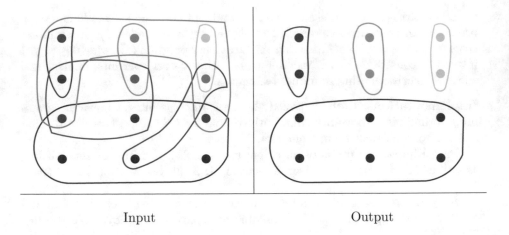

Input   Output

21.2   Set Packing

Input description: A set of subsets $S = \{S_1, \ldots, S_m\}$ of the universal set $U = \{1, \ldots, n\}$.

Problem description: Select a small collection of *mutually disjoint* subsets from S whose union is the universal set.

Discussion: Set packing problems arise in applications where we have strong constraints on what is an allowable partition. The key feature of packing problems (as opposed to covering problems) is that no elements can be covered by more than one selected subset.

Some flavor of this is captured by the independent set problem in graphs, discussed in Section 19.2 (page 589). There we seek a large subset of vertices from graph G such that each edge is adjacent to at most one of the selected vertices. To model this as set packing, let the universal set consist of all edges of G, and subset S_i consist of all edges incident on vertex v_i. Finally, define an additional singleton set for each edge. Any set packing defines a set of vertices with no edge in common—in other words, an independent set. The singleton sets are used to pick up any edges not covered by the selected vertices.

Scheduling airline flight crews is another application of set packing. Each airplane in the fleet needs to have a crew assigned to it, consisting of a pilot, copilot, and navigator. There are constraints on the composition of possible crews, based on their training to fly different types of aircraft, personality conflicts, and work schedules. Given all possible crew and plane combinations, each represented by a subset of items, we need an assignment such that each plane and each person is in exactly one chosen combination. After all, the same person cannot be on two different planes simultaneously, and every plane needs a crew. We need a perfect packing given the subset constraints.

We use set packing here to represent several different problems on sets, all of which are NP-complete:

- *Must every element appear in exactly one selected subset?* – In the *exact cover* problem, we seek a collection of subsets such that each element is covered exactly once. The airplane scheduling problem above has the flavor of exact covering, since every plane and crew has to be employed.

  Unfortunately, exact cover puts us in the same situation as the Hamiltonian cycle on graphs. If we really *must* cover all the elements exactly once, and this existential problem is NP-complete, then all we can do is exponential search. The cost will be prohibitive unless we happen to stumble upon a solution quickly.

- *Does each element have its own singleton set?* – Things will be far better if we can be content with a partial solution, say by including each element of U as a singleton subset of S. Thus, we can expand any set packing into an exact cover by mopping up the unpacked elements of U with singleton sets. Now our problem is reduced to finding a minimum-cardinality set packing, which can be attacked via heuristics.

- *What is the penalty for covering elements twice?* – In set cover (see Section 21.1 (page 678)), there is no penalty for elements existing in many selected subsets. But in exact cover, any such violation is forbidden. For many applications, the truth lies somewhere in between. Such problems can be approached by charging the greedy heuristic more to select a subset that contains previously covered elements.

The right heuristics for set packing are greedy, and similar to those of set cover (see Section 21.1 (page 678)). If we seek a packing with few sets, then we repeatedly select the largest remaining subset, delete all subsets from S that clash with it, and repeat. As usual, augmenting this approach with some exhaustive search or randomization (in the form of simulated annealing) is likely to yield better packings at the cost of additional computation.

A more powerful approach rests on an integer programming formulation akin to that of set cover. Let the integer $0/1$ variable s_i denote whether subset S_i is selected for a given cover. Each universal set element $x \in U$ defines a constraint based on all subsets S_i that contain x, namely:

$$\sum_{x \in S_i} s_i = 1$$

This ensures that x is covered by *exactly* one selected subset. Minimizing or maximizing $\sum_i s_i$ while respecting these constraints enables us to modulate the desired number of sets in the cover.

Implementations: Since set cover is a more popular and more tractable problem than set packing, it might be easier to find an appropriate implementation to solve the cover problem. Such implementations discussed in Section 21.1 (page 678) should be readily modifiable to support certain packing constraints.

Pascal implementations of an exhaustive search algorithm for set packing, as well as heuristics for set cover, appear in [SDK83]. See Section 22.1.9 (page 717) for details on FTP-ing these codes.

SYMPHONY is a mixed-integer linear programming solver that includes a set partitioning solver. It is available at `https://github.com/coin-or/` `SYMPHONY`.

Notes: Survey articles on set packing include [BP76, HP09, Pas97]. A local search heuristic for set packing is presented in [SW13]. Fixed-parameter tractable [FKN^{+}08] and online [EHM^{+}12] versions of set packing have also been studied. Bidding strategies for combinatorial auctions typically reduce to solving set-packing problems, as described in [dVV03].

Set-packing relaxations for integer programs are presented in [BW00]. An excellent exposition on algorithms and reduction rules for set packing is presented in [SDK83], including the airplane scheduling application discussed previously.

Related problems: Independent set (see page 589), set cover (see page 678).

"I often repeat repeat myself,	"I often repeat repeat myself,
I often repeat repeat.	I often repeat repeat.
I often repeat repeat myself,	I often repeat repeat myself,
I often repeat repeat." repeat?	I often repeat repeat."
– Jack Prelutsky, A Pizza the Size of the Sun	– Jack Prelutsky, A Pizza the Size of the Sun

Input	Output

21.3 String Matching

Input description: A text string t of length n. A pattern p of length m.

Problem description: Find an/all instances of pattern p in the text.

Discussion: String matching arises in almost all text processing applications. Every text editor contains a mechanism to search the current document for arbitrary strings. Pattern matching programming languages like Python derive much of their power from their built-in string matching primitives, making it easy to fashion programs that filter and modify text. Spell checkers scan an input text for words appearing in the dictionary and reject any strings that do not match.

String matching is a fundamental algorithmic problem that remains surprisingly active. Several issues arise in identifying the right string matching algorithm for a given application:

- *Are your search patterns and/or texts short?* – If your strings are short and your queries infrequent, the simple $O(mn)$-time search algorithm will suffice. For all possible starting positions $1 \leq i \leq n-m+1$, it tests whether the m characters starting from the ith position of the text are identical to the pattern. An implementation of this algorithm (in C) appears in Section 2.5.3 (page 43).

  For very short patterns (say $m \leq 10$), you can't hope to beat this simple algorithm by much, so you shouldn't try. Further, we expect much better than $O(mn)$ behavior for typical strings, because we advance the pattern the instant we observe a text/pattern mismatch. Indeed, the trivial algorithm *usually* runs in linear time. But the worst case certainly can occur, as with pattern $p = a^m$ and text $t = (a^{m-1}b)^{n/m}$.

- *What about longer texts and patterns?* – String matching can in fact be performed in worst-case linear time. Observe that we need not begin the search from scratch after finding a character mismatch, because the pattern prefix and text must exactly match prior to the point of mismatch. Given a partial match ending at position i, we jump ahead to the first character position in the pattern/text that provides new information about the text in position $i + 1$. The Knuth–Morris–Pratt (KMP) algorithm preprocesses the search pattern to construct such a jump table efficiently. The details are tricky to get correct, but the resulting algorithm is quite short.

- *Do I expect to find the pattern, or not?* – The Boyer–Moore algorithm matches the pattern against the text from right to left, and hence can avoid looking at large chunks of text on a mismatch. Suppose the pattern is *abracadabra*, and the eleventh character of the text is x. This pattern cannot possibly match from any of the first eleven starting positions of the text, and so the next necessary position to test is the twenty-second character. If we get very lucky, only n/m characters need ever be tested. The Boyer–Moore algorithm involves two sets of jump tables in the case of a mismatch: one based on the pattern matched so far, the other on the text character seen in the mismatch.

  Although somewhat more complicated than Knuth–Morris–Pratt, this is worthwhile in practice for patterns of length $m > 10$, unless the pattern is expected to occur many times in the text. Boyer–Moore's worst-case performance is $O(n + rm)$, where r is the number of occurrences of p in t.

- *Will you perform multiple queries on the same text?* – Suppose you are building a program to repeatedly search a particular text database, such as the Bible. Since the text remains fixed, it pays to build a data structure to speed up search queries. The suffix tree and suffix array data structures, discussed in Section 15.3 (page 448), are the right tools for the job.

- *Will you search many texts using the same patterns?* – Suppose you are building a program to screen out dirty words from a text stream. Here, the set of patterns remains stable, while the search texts are free to change. We may need to find all occurrences of any of k different patterns, where k can be quite large.

  Performing a linear-time scan for each pattern yields an $O(k(m+n))$ algorithm. But if k is large, a better solution builds a single finite automaton that recognizes all of these patterns and returns to the appropriate start state on any character mismatch. The Aho–Corasick algorithm builds such an automaton in linear time. Space savings can be achieved by optimizing the pattern recognition automaton, as discussed in Section 21.7 (page 702). This approach was used in the original version of *fgrep*.

  Sometimes multiple patterns are specified not as a list of strings, but concisely as a regular expression. For example, the regular expression $a(a + b + c)^*a$ matches any string on (a, b, c) that begins and ends with a distinct a. The best way to test whether an input string is recognized by a regular expression R constructs the finite automaton equivalent to R and then simulates this machine on the string. Again, see Section 21.7 (page 702) for details on constructing automata from regular expressions.

When the patterns get specified by context-free grammars instead of regular expressions, the problem becomes one of parsing, discussed in Section 10.8 (page 337).

- *What if our text or pattern contains a spelling error?* – The algorithms discussed here work only for exact string matching. If you must allow some tolerance for spelling errors, your problem becomes *approximate string matching*, which is thoroughly discussed in Section 21.4 (page 688).

Implementations: Strmat is a collection of C programs implementing exact pattern matching algorithms in association with [Gus97], including several variants of the KMP and Boyer–Moore algorithms. It is available at `https://www.cs.ucdavis.edu/~gusfield/strmat.html`.

Several versions of the general regular expression pattern matcher (grep) are readily available. GNU *grep* found at `https://directory.fsf.org/project/grep/`, and uses a fast lazy-state deterministic matcher hybridized with a Boyer–Moore search for fixed strings.

The Boost string algorithms library provides C++ routines for basic operations on strings, including search. See `http://www.boost.org/doc/html/string_algo.html`.

Notes: All books on string algorithms contain thorough discussions of exact string matching, including [CHL07, NR07, Gus97]. Good expositions on the Boyer–Moore [BM77] and Knuth-Morris-Pratt algorithms [KMP77] include [BvG99, CLRS09, Man89]. The history of string matching algorithms is somewhat checkered because several published proofs were incorrect or incomplete. See [Gus97] for clarification.

Aho [Aho90] provides a good survey on algorithms for pattern matching in strings, particularly for regular expression patterns. The Aho–Corasick algorithm for multiple patterns is described in [AC75].

Empirical comparisons of string matching algorithms include [DB86, Hor80, Lec95, dVS82, YLDF16]. Which algorithm performs best depends upon the properties of the strings and the size of the alphabet. For long patterns and texts, I recommend that you use the best implementation of Boyer–Moore that you can find. String matching algorithms for GPUs are considered in [LLCC12].

The Rabin–Karp algorithm [KR87] uses a hash function to perform string matching in linear expected time. Its worst-case time remains quadratic, and its performance in practice appears somewhat worse than the character comparison methods described above. This algorithm is presented in Section 6.7 (page 188).

Related problems: Suffix trees (see page 448), approximate string matching (see page 688).

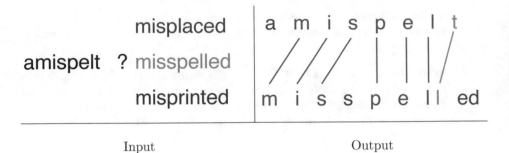

Input	Output

21.4   Approximate String Matching

Input description: A text string t and a pattern string p.

Problem description: What is the minimum-cost way to transform t to p using insertions, deletions, and substitutions?

Discussion: Approximate string matching is important because we live in an error-prone world. Spelling correction programs must identify the closest match for any text string not found in a dictionary. Efficient sequence similarity (homology) searches on large databases of DNA sequences have revolutionized the study of molecular biology. Suppose you are interested in a particular human gene, and discover that it is similar to a particular gene in rats. Likely this new gene does for people what it does for rats, with any differences being the result of genetic mutations during evolution.

I once encountered approximate string matching when evaluating the performance of an optical character-recognition system. We needed to compare the answers produced by our system on a test document with the correct results. To improve our system, we had to identify which letters were getting misidentified. The solution was to do an alignment between the two texts. This same principle is used in file difference programs, which identify the lines that have changed between two versions of a file.

When no changes are permitted, our problem reduces to exact string matching, which is discussed in Section 21.3 (page 685). Here, the discussion is restricted to the problem of matching with errors.

Dynamic programming provides the basic approach to approximate string matching. Let $D[i, j]$ denote the cost of editing the first i characters of the pattern string p into the first j characters of the text t. We must do *something* with the tail characters p_i and t_j. Our only options are matching/substituting one for the other, deleting p_i, or inserting a match for t_j. Thus, $D[i, j]$ is the minimum of the costs of these possibilities:

- If $p_i = t_j$ then $D[i - 1, j - 1]$ else $D[i - 1, j - 1]$ + substitution cost.

- $D[i - 1, j]$ + deletion cost of p_i.

- $D[i, j-1]$ + deletion cost of t_j.

A general implementation in C and more complete discussion appears in Section 10.2 (page 314). Several issues remain before we can make full use of this recurrence:

- *Do I match the pattern against the full text, or against a substring?* – The boundary condition of this recurrence is what distinguishes between algorithms for string matching and substring matching. Suppose we seek to align the full pattern against the full text. Then the cost of $D[i, 0]$ must be that of deleting the first i characters of the pattern, so $D[i, 0] = i$. Similarly, $D[0, j] = j$.

  But now suppose that the pattern can occur anywhere within the text. The proper cost of $D[0, j]$ becomes 0, since there should be no penalty for starting the alignment in the jth position of the text. The cost of $D[i, 0]$ remains i, because the only way to match the first i pattern characters with nothing is to delete them all. The cost of the best substring pattern match against the text will be given by $\min_{k=1}^{n} D[m, k]$.

- *How should I select the substitution and insertion/deletion costs?* – The edit distance algorithm can use different costs for insertion, deletion, and substitution for specific pairs of characters. What costs are most appropriate depends on what you plan to do with the alignment.

  The default choice charges the same for each insertion, deletion, or substitution. Charging a substitution cost of more than insertion + deletion ensures that substitutions will never get performed, because it will always be cheaper to just edit both characters out of their strings. With only insertion and deletion to work with, the problem reduces to *longest common subsequence*, discussed in Section 21.8 (page 706). It often pays to tweak the edit distance costs and study the resulting alignments so you can find the best parameters for the job.

- *How do I find the actual alignment of the strings?* – The recurrence above only yields the cost of the optimal string/pattern alignment, not the sequence of editing operations that achieve it. To obtain such a transcript, we can work backwards from the cell $D[m, n]$ in the complete cost matrix D. We had to come from either $D[m-1, n]$ (pattern deletion/text insertion), $D[m, n-1]$ (text deletion/pattern insertion), or $D[m-1, n-1]$ (substitution/match). The chosen option can be reconstructed from these costs and the given characters p_m and t_n. By repeatedly moving backwards to the previous cell, we can reconstruct the entire alignment. Again, an implementation in C appears in Section 10.2 (page 314).

- *What if the two strings are very similar to each other?* – The dynamic programming algorithm above fills out an $m \times n$ matrix to compute edit distance. But to seek an alignment involving a combination of at most d insertions, deletions, and substitutions, we need only traverse the band of $O(dn)$ cells within a distance d of the central diagonal. If no low-cost alignment exists within this band, then no low-cost alignment can exist in the full cost matrix.

- *Is your pattern short or long?* – A recent approach to string matching exploits the fact that modern computers can do operations on 64-bit words in a single gulp. This is long enough to hold eight 8-bit ASCII characters, providing motivation to design *bit-parallel algorithms* that do more than one comparison with each operation.

  The basic idea is quite clever. Construct a bit-mask B_α for each letter α of the alphabet, such that ith-bit $B_\alpha[i] = 1$ iff the ith character of the pattern is α. Now suppose you have a match bit-vector M_j for position j in the text string, such that $M_j[i] = 1$ iff the first i bits of the pattern exactly match the $(j - i + 1)$st through jth character of the text. We can find *all* the bits of M_{j+1} using just two operations by (1) shifting M_j one bit to the right, and then (2) doing a bitwise AND with B_α, where α is the character in position $j + 1$ of the text.

  The *agrep* program, discussed below, uses such a bit-parallel algorithm generalized to approximate matching. Such algorithms are easy to program and many times faster than dynamic programming.

- *How do I minimize the required storage?* – The quadratic space needed to store the dynamic programming table usually presents a more serious obstacle than its running time. Fortunately, only $O(\min(m, n))$ space is needed to compute $D[m, n]$. We only need to maintain two active rows (or columns) of the matrix to compute the optimal cost. The entire matrix is required only to reconstruct the actual sequence alignment.

  But we can use Hirschberg's clever recursive algorithm to efficiently recover the optimal alignment in linear space. During the first pass of the linear-space algorithm above to compute $D[m, n]$, we can identify which middle-element cell $D[m/2, x]$ was used to optimize $D[m, n]$. This reduces our problem to finding the best paths from $D[1, 1]$ to $D[m/2, x]$ and from $D[m/2, x]$ to $D[m/2, n]$. Both of these can be solved recursively. Each round removes half of the matrix elements of the previous round from consideration, so the total time remains $O(mn)$. This linear-space algorithm proves to be a big win in practice on long strings.

- *Should I score long runs of insertions/deletions differently?* – Many string matching applications look kindly upon alignments where insertions or deletions (*indels*) get bunched in a small number of runs or gaps. Deleting a paragraph from a document should presumably cost less than a similar number of scattered single-character edits, because the word represents a single (albeit substantial) modification.

  String matching with *gap penalties* provides a way to properly account for such changes. Typically, we assign a cost of $A + Bt$ for each indel of t consecutive characters, where A is the cost of starting the gap and B is the per-character deletion cost. If A is large relative to B, the alignment has an incentive to create relatively few runs of deletions.

String matching under such *affine* gap penalties can be done in the same quadratic time as regular edit distance. We use separate recurrences E and F to encode the cost of being in insertion or deletion gap mode respectively, so we only pay the cost of initiating the gap once:

$$V(i,j) = \max(E(i,j), F(i,j), G(i,j))$$
$$G(i,j) = V(i-1, j-1) + match(i,j)$$
$$E(i,j) = \max(E(i, j-1), V(i, j-1) - A) - B$$
$$F(i,j) = \max(F(i-1, j), V(i-1, j) - A) - B$$

With a constant amount of work per cell, this algorithm takes $O(mn)$ time, same as without gap costs.

- *Does similarity mean strings that sound alike?* – Other models of approximate pattern matching become appropriate in certain applications. Particularly interesting is *Soundex*, a hashing scheme that attempts to pair up English words that sound alike. This can be useful in testing whether two names that have been spelled differently are likely to be the same. For example, my last name has been spelled "Skina," "Skinnia," "Schiena," and occasionally "Skiena." All of these hash to the same Soundex code, *S25*.

Soundex drops vowels and silent letters, removes doubled letters, and then assigns the remaining letters numbers from the following classes: *BFPV* gets a 1, *CGJKQSXZ* gets a 2, *DT* gets a 3, *L* gets a 4, *MN* gets a 5, and *R* gets a 6. The characters *HWY* are not assigned a digit. The code starts with the first letter of the name and contains at most three digits. Although this sounds fairly hokey, experience shows that it works reasonably well. Experience indeed: Soundex has been used since the 1920's.

Implementations: Several excellent software tools are available for approximate pattern matching. Manber and Wu's *agrep* [WM92a, WM92b] (approximate general regular expression pattern matcher) is a tool supporting text search with spelling errors. The current version is available at `http://www.tgries.de/agrep/`. Navarro's *nrgrep* [Nav01b] combines bit-parallelism and filtration, resulting in running times that are more constant than *agrep*, although not always faster. It is available at `https://www.dcc.uchile.cl/~gnavarro/software/`.

TRE is a general regular-expression matching library for exact and approximate matching, which is more general than *agrep*. The worst-case complexity is $O(nm^2)$, where m is the list of the regular expressions involved. *TRE* is available at `https://github.com/laurikari/tre/`.

Wikipedia gives programs for computing edit (Levenshtein) distance in a dizzying array of languages (including Ada, C++, Emacs Lisp, JavaScript, Java, PHP, Python, Ruby VB, and C#) Check it out at `https://en.wikibooks.org/wiki/Algorithm_implementation/Strings/Levenshtein_distance`.

Notes: There have been many recent advances in approximate string matching, particularly in bit-parallel algorithms. Navarro and Raffinot [NR07] is the best reference on these techniques, which are also treated in other books on string algorithmics [CHL07, Gus97]. String matching with gap penalties is particularly well treated in [Gus97].

The basic dynamic programming alignment algorithm is attributed to [WF74], although it is apparently folklore. The wide range of applications for approximate string matching was made apparent in Sankoff and Kruskal's book [SK99], which remains a useful historical reference. Surveys on approximate pattern matching include [HD80, Nav01a]. Expositions of Hirschberg's linear-space algorithm [Hir75] include [CR03, Gus97].

Masek and Paterson [MP80] compute the edit distance between m- and n-length strings in time $O(mn/\log(\min\{m, n\}))$ for constant-sized alphabets, using ideas from the four Russians algorithm for Boolean matrix multiplication [ADKF70]. A recent hardness result by Backurs and Indyk [BI15] shows that edit distance cannot be computed in $O(n^{2-\epsilon})$ time without violating the strong exponential time hypothesis (SETH). Another recent breakthrough are subquadratic algorithms that approximate edit distance to within a constant factor [CDG⁺18].

The shortest-path formulation leads to a variety of algorithms that are good when the edit distance is small, including an $O(n \lg n + d^2)$ algorithm due to Myers [Mye86] and an $O(dn)$ algorithm due to Landau and Vishkin [LV88]. Longest increasing subsequence can be done in $O(n \lg n)$ time [HS77], as presented in [Man89].

Bit-parallel algorithms for approximate matching include Myers's [Mye99b] algorithm for approximate matching in $O(mn/w)$ time, where w is the number of bits in the computer word. Experimental studies of bit-parallel algorithms include [FN04, HFN05, NR00].

Soundex was invented and patented by M. K. Odell and R. C. Russell. Expositions on Soundex include [BR95, Knu98]. Metaphone is a modern attempt to improve on Soundex [BR95, Par90]. See [LMS06] for an application of such phonetic hashing techniques to the problem entity name unification.

Related problems: String matching (see page 685), longest common substring (see page 706).

Input   Output

21.5 Text Compression

Input description: A text string S.

Problem description: Create a shorter text string S' such that S can be correctly reconstructed from S'.

Discussion: Secondary storage devices quickly fill up on most computer systems, even though their capacity seems to double every year. Decreasing storage prices only increases interest in data compression, because there is more data to compress than ever before. *Data compression* is the algorithmic problem of finding a space-efficient encoding for a given data file. The rise of computer networks provides a new mission for data compression, that of increasing the effective bandwidth by reducing the number of bits before transmission.

People seem to *like* inventing ad hoc data-compression methods for their particular application. Sometimes these outperform general methods, but usually they don't. Several issues arise in selecting the right compression algorithm:

- *Must we recover the exact input text after compression?* – Lossy vs. *lossless* encoding is the primary issue in data compression. Document storage applications demand lossless encodings, because users become quite disturbed when their data files are silently altered. Fidelity is not the same concern in image or video compression, because small perturbations are imperceptible to the viewer. Significantly greater compression ratios can be obtained using lossy compression, which is why most image/video/audio compression algorithms exploit this freedom.

- *Can I simplify my data before I compress it?* – The most effective way to free space on a disk is to delete the files you don't need. Likewise, any preprocessing you can do to reduce the information content of a file pays off later in better compression. Can we eliminate redundant white space from the file? Might the document be converted entirely to uppercase characters, or have formatting information removed?

A particularly interesting simplification results from applying the *Burrows–Wheeler transform* to the input string. This transform sorts all n cyclic shifts of the n character input, and then reports the last character of each shift. As an example, the cyclic shifts of *abab* are *abab*, *baba*, *abab*, and *baba*. After sorting, these become *abab*, *abab*, *baba*, and *baba*. Reading the last character of each of these strings yields the transform result: *bbaa*.

Provided the last character of the input string is unique (an end-of-string symbol), this transform is perfectly invertible to the original input! The Burrows–Wheeler string is typically 10–15% more compressible than the original text, because repeated words turn into blocks of repeated characters. Further, this transform can be computed in linear time.

- *Does it matter whether the algorithm is patented?* – Certain data compression algorithms have been patented—most notoriously the LZW variation of the Lempel–Ziv algorithm discussed below. Mercifully, this patent has now expired, as has the one covering JPEG. Typically there are unrestricted variations of any compression algorithm that perform about as well as the patented variant, but problems arise if a patented algorithm sneaks into a popular standard.

- *How do I compress image data* – The simplest lossless compression algorithm for image data is *run-length coding*. Here we replace runs of identical pixel values with a single instance of the pixel and an integer giving the length of the run. This works well on binary images with large contiguous regions of similar pixels, like scanned text. But it performs badly on images with many quantization levels and random noise. Correctly selecting (1) the number of bits to allocate to the count field, and (2) the right traversal order to reduce a two-dimensional image into a stream of pixels, has a surprisingly important impact on compression.

  For serious audio/image/video compression applications, I recommend that you use a popular lossy coding method and not fool around with implementing it yourself. JPEG is the standard high-performance image compression method, while MPEG is designed to exploit the frame-to-frame coherence of video.

- *Must compression run in real time?* – Fast decompression is often more important than fast compression. A YouTube video is compressed only once, but decompressed every time someone plays it. In contrast, an operating system that increases effective disk capacity by automatically compressing files will need a symmetric algorithm with fast compression times.

Literally dozens of text compression algorithms are available, but they can be classified into two distinct groups. *Static algorithms*, such as Huffman codes, build a single coding table by analyzing the entire document. *Adaptive algorithms*, such as Lempel–Ziv, build a coding table on the fly that adapts to the local character distribution of the document. Adaptive algorithms usually prove to be the right answer for most problems, but both are interesting:

- *Huffman codes* – Huffman codes replace each alphabet symbol by a variable-length code string. Using eight bits-per-symbol to encode English text is wasteful, because certain characters (such as "e") occur far more frequently than others (such as "q"). Huffman codes assign "e" a short code word, and "q" a longer one to compress the text.

  Huffman codes can be constructed using a greedy algorithm. Sort the symbols in increasing order by frequency. We merge the two least-frequently used symbols x and y into a new symbol xy, whose frequency is the sum of its two child symbols. Replacing x and y by xy leaves a smaller set of symbols. We now repeat this operation $n - 1$ times until all symbols have been merged together. These merging operations define a rooted binary tree, with the original alphabet symbols as leaves. The left or right choices on the root-to-leaf path define the bits of the binary code word for each symbol. Priority queues can efficiently maintain the symbols by frequency during construction, yielding Huffman codes in $O(n \lg n)$ time.

  Huffman codes are popular but have three disadvantages. Two passes must be made over the document on encoding, first to build the coding table, and then to actually encode the text. The coding table must be explicitly stored with the document to decode it, which eats into any space savings on short documents. Finally, Huffman codes only exploit non-uniform symbol distributions, while adaptive algorithms can recognize the higher-order redundancies such as in *0101010101....*

- *Lempel–Ziv algorithms* – Lempel–Ziv algorithms (including the popular LZW variant) compress text by building a coding table on the fly as we read the document. The coding table changes as we move through the text. A clever protocol ensures that the encoder and decoder both work with the exact same code table, so no information is lost.

  Lempel–Ziv algorithms build coding tables of frequent substrings, which can get arbitrarily long. They can thus exploit often-used syllables, words, and phrases to build better encodings. They adapt to local changes in the text distribution, which is important because many documents exhibit significant locality of reference.

  The amazing thing about Lempel–Ziv is how robust it is on different types of data. It is quite difficult to beat it by using your own application-specific compression algorithm. My recommendation is not to try. If you can eliminate application-specific redundancies using a simple preprocessing step, go ahead and do it. But don't waste much time fooling around. You are unlikely to get significantly better text compression than with *gzip* or some other popular program, and you might well do worse.

Implementations: Perhaps the most popular text compression program is *gzip*, which implements a public domain variation of the Lempel–Ziv algorithm. It is distributed under the GNU software license and can be obtained from `https://www.gzip.org`.

There is a natural tradeoff between compression ratio and compression time. Another choice is *bzip2*, which uses the Burrows–Wheeler transform. It produces tighter encodings than *gzip* at somewhat greater cost in running time. Going to the extreme, certain compression algorithms devote enormous run times to squeeze every bit out of a file. Representative programs of this genre are collected at `http://mattmahoney.net/dc/`. Personally, I believe that there is a special place in hell for people who send me files compressed by weird encoders. I always delete such messages, and demand that they resend me a gzip file.

Notes: Many books on data compression are available. Recent and comprehensive books include Sayood [Say17] and Salomon [Sal06], with [SM10] an authoritative reference. Also recommended is the older text by Bell, Cleary, and Witten [BCW90]. Surveys on text compression algorithms include [CL98, KA10].

Good expositions on Huffman codes [Huf52] include [AHU83, CLRS09]. The Lempel–Ziv algorithm and variants are described in [Wel84, ZL78]. The Burrows–Wheeler transform was introduced in [BW94].

The annual IEEE Data Compression Conference (`https://www.cs.brandeis.edu/~dcc/`) is the primary research venue in this field. This is a mature technical area where most current work is shooting for fairly marginal improvements, particularly in the case of text compression. More encouragingly, I note that the conference is held annually at a world-class ski resort in Utah.

Related problems: Shortest common superstring (see page 709), cryptography (see page 697).

The magic words are Squeamish Ossifrage.	I5&AE<&UA9VEC'=0 <F1s"F%R92!3<75E96UI<V V@*3W-S:69R86=E+@K_
Input	Output

21.6 Cryptography

Input description: A plaintext message T or encrypted text E, and key k.

Problem description: Encode T using k giving E, or decode E giving T.

Discussion: Cryptography has grown wildly more important as computer networks increasingly make confidential documents more vulnerable to prying eyes. Cryptography increases security by making messages difficult to read if they fall into the wrong hands. Although the discipline of cryptography is at least two thousand years old, its algorithmic and mathematical foundations have only recently solidified to the point where provably secure cryptosystems can be envisioned.

Cryptographic ideas and applications go far beyond the tasks of "encryption" and "decryption." The field now includes mathematical constructs such as cryptographic hashes, digital signatures, and useful primitive protocols that provide associated security assurances.

There are three classes of cryptosystems everyone should be aware of:

- *Caesar shifts* – The oldest ciphers involve mapping each character of the alphabet to a different letter. The weakest such ciphers rotate the alphabet by some fixed number of characters (say 13), and thus have only 26 possible keys. Better is to use an arbitrary permutation of the letters, giving 26! possible keys. Even so, such systems can be easily attacked by counting the frequency of each symbol and exploiting the fact that "e" occurs more often than "z." While there are variants that will make this more difficult to break, none will be as secure as AES or RSA.

- *Block shuffle ciphers* – This class of algorithms repeatedly shuffles the bits of your text as governed by the key. The classic example of such a cipher is the *Data Encryption Standard* (DES). Although approved as a Federal Information Processing Standard in 1976, *DES* was officially withdrawn as a federal standard in 2005, replaced by the stronger *Advanced Encryption Standard* (AES). A simple variant called *triple DES* permits an effective key length of 112 bits by using three rounds of DES with two 56-bit keys. But it, too, became vulnerable with time, and as of 2018 is depreciated, with usage disallowed after 2023.

But 256-bit AES is still considered very secure. AES libraries are available for all major programming languages, including C/C++, Java, JavaScript, and Python, and form the basis for security in WhatsApp, Facebook Messenger, and many other systems.

- *Public key cryptography* – If you fear bad guys reading your messages, you should also be afraid to tell anyone else the key you need to decrypt them. Public key systems use different keys to encode and decode messages. Since the encoding key is of no help in decoding, it can be made public at no risk to security. This solution to the key distribution problem is literally its key to success.

  RSA is the classic example of a public key cryptosystem, named after its inventors Rivest, Shamir, and Adelman. The security of RSA is based on the relative computational complexities of factoring and primality testing (see Section 16.8 (page 490)).   Encoding is (relatively) fast because it relies on primality testing to construct the key, while the hardness of decryption follows from that of factoring—an assumption rapidly being threatened by advances in quantum computing. As of this writing, the largest integer incontrovertibly factored using Shor's algorithm on a quantum computer appears to be 15, so this is not an imminent threat. But it does bear watching.

  RSA is slow relative to other cryptosystems—roughly 100 to 1,000 times slower than AES. Still, it earns its salt by making public keys possible.

The critical issue in selecting a cryptosystem is identifying your paranoia level. Who are you trying to stop from reading your stuff: your grandmother, local thieves, the Mafia, or the NSA (National Security Agency)?   If you can use an accepted implementation of AES or RSA, you should feel pretty safe against anybody, at least for now. Increasing computer power lays waste to cryptosystems surprisingly quickly, as discussed above. Be sure to use the longest possible keys and keep abreast of algorithmic developments if you are planning a long-term storage of sensitive material.

That said, I will confess that I use DES to encrypt my final exam each semester. It proved more than sufficient the time an ambitious student broke into my office looking for it.   The story would have been different had the NSA been trying to crack it, but it is important to understand that *the most serious security holes are human, not algorithmic.* Ensuring that your password is long enough, hard to guess, and not written down is far more important than obsessing about the encryption algorithm.

Most symmetric key encryption mechanisms are harder to crack than public key ones for the same key size. This means one can get away with much shorter key lengths for symmetric key than for public key encryption. NIST and RSA Labs both provide schedules of recommended key sizes for secure encryption, and as of this writing they recommend 256-bit symmetric keys as equivalent to 15,360-bit asymmetric keys. This difference helps explain why symmetric key algorithms are typically orders of magnitude faster than public key algorithms.

Simple ciphers like the Caesar shift are fun and easy to program. For this reason, it is healthy to use them for applications needing only a casual level of security (such as hiding the punchlines of jokes). Since they are easy to break, they should never be used for serious security applications.

Another thing you should *never* do is try to develop your own novel cryptosystem. The security of AES and RSA is accepted because these systems have survived many years of public scrutiny. In this time, many other cryptosystems have been proposed, proven vulnerable to attack, and then abandoned. This is not a field for amateurs. If you are charged with implementing a cryptosystem, carefully study a respected program such as PGP to see how they handle issues such as key selection and key distribution. Any cryptosystem is as strong as its weakest link.

Certain other problems related to cryptography arise often in practice:

- *How can I validate the integrity of data against random corruption?* – There is often a need to validate that transmitted data is identical to that which has been received. One solution has the receiver transmit the data back to the source, so the original sender can confirm that the two texts are identical. This fails when inverse errors are made in retransmission, but a more serious problem is that such a scheme cuts your available bandwidth in half.

  A more efficient method uses a *checksum*, a hash of the long text down to a large integer. We then transmit the checksum along with the text. The checksum can be recomputed from the text on the receiving end, and bells set off if the computed checksum is not identical to what was received. The simplest checksum scheme just adds up the byte or character values and takes the sum modulo of some constant, say $2^8 = 256$. But an error transposing two or more characters would go undetected under such a scheme, because addition is commutative.

  Cyclic-redundancy check (CRC) – The CRC provides a more powerful method to compute checksums, that is used in most communications systems and internally in computers to validate disk drive transfers. These codes compute the remainder in the ratio of two polynomials, the numerator of which is a function of the input text. The design of these polynomials involves considerable mathematical sophistication, but ensures that all reasonable errors are detected.

- *How can I validate the integrity of data against deliberate corruption?* – CRC is good at detecting random errors, but not malicious changes to a document. *Cryptographic hash functions* such as MD5 and SHA-256 are easy to compute for a given document, but hard to invert. This means that for a particular hash code value x, it is hard to construct a document d such that $H(d) = x$. The property makes them valuable for digital signatures and other applications.

- *How can I prove that a file has not been changed?* – If I send you a contract in electronic form, what is to stop you from editing the file and then claiming that your version is what we had agreed to? I need a way to prove that any modification to a document is fraudulent. *Digital signatures* are a cryptographic way for me to stamp my document as genuine, and play an important role in maintaining the integrity of the blockchain associated with cryptocurrencies like Bitcoin.

I can compute a checksum for any given file, and then encrypt this checksum using my own private key. I send you the file and the encrypted checksum. Sure, you can edit the file, but to fool the judge you must also edit the encrypted checksum such that it can be decrypted to yield the correct checksum. Designing a file that yields the same checksum becomes an insurmountable problem. For full security, we need a trusted third party to authenticate the timestamp and associate the private key with me.

- *How can I restrict access to copyrighted material?* – An important application for cryptography is digital rights management for audio and video. A key issue here is speed of decryption, so it can keep up with data transmission or retrieval in real time. Such *stream ciphers* usually involve efficiently generating a stream of pseudorandom bits, say using a shift-register generator. The exclusive-or of these bits with the data stream gives the encrypted sequence, with the original data recovered by exclusive-oring the result with the same stream of pseudorandom bits.

  High-speed cryptosystems have proven to be relatively easy to break. The state-of-the-art solution to this problem involves erecting laws like the Digital Millennium Copyright Act, which makes it illegal to try to break them.

Implementations: *Nettle* is a comprehensive low-level cryptographic library in C. Cryptographic hash functions include MD5 and SHA-256. Block ciphers include DES, AES, and some more recently developed codes. An implementation of RSA is also provided. *Nettle* is available at `https://www.lysator.liu.se/~nisse/nettle/`. See `http://csrc.nist.gov/groups/ST/toolkit` for related cryptographic resources provided by NIST.

Crypto++ is a large C++ class library of cryptographic schemes, including all that I have mentioned here. It is available at `https://www.cryptopp.com/`.

Many popular open source utilities employ serious cryptography, and serve as good models of current practice. *GnuPG*, an open source version of PGP, is available at `https://www.gnupg.org/`. *OpenSSL*, for authenticating access to computer systems, is available at `https://www.openssl.org/`.

The *Boost CRC Library* provides multiple implementations of cyclic redundancy check algorithms. It is available at `https://www.boost.org/libs/crc/`.

Notes: The *Handbook of Applied Cryptography* [MOV96] provides technical surveys of all aspects of cryptography, and has been generously made available online at `http://www.cacr.math.uwaterloo.ca/hac/`. Schneier [Sch15] provides a thorough overview of different cryptographic algorithms, with [FS03] as perhaps a better introduction.

Kahn [Kah67] presents the fascinating history of cryptography from ancient times to 1967 and is particularly noteworthy in light of the secretive nature of the subject.

Expositions on the RSA algorithm [RSA78] include [CLRS09]. The RSA Laboratories home page http://www.rsa.com/rsalabs/ is very informative. Quantum technologies lead to new secure encryption methods [BB14], and also new ways to break what we already have. Post-quantum cryptography is an area of active research [CJL+16].

Of course, the National Security Agency is the place to go to learn the real state of the art in cryptography. The history of DES is well presented in Schneier [Sch15]. Particularly controversial was the decision by the NSA to limit key length to 56 bits.

MD5 [Riv92] is the hashing function used by PGP to compute digital signatures. Expositions include [Sch15, Sta06]. Problems with the security of MD5 have been exposed [WY05]. The SHA family of hash functions appears more secure, particularly SHA-256 and SHA-512.

Related problems: Factoring and primality testing (see page 490), text compression (see page 693)).

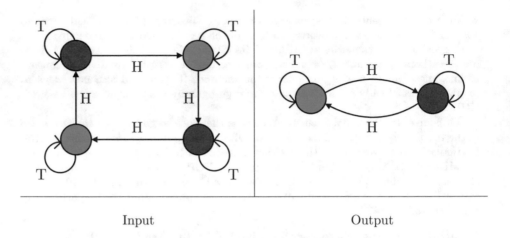

| Input | Output |

21.7   Finite State Machine Minimization

Input description: A deterministic finite automaton M.

Problem description: Create the smallest deterministic finite automaton M' such that M' behaves identically to M.

Discussion: Finite state machines are very useful for specifying and recognizing patterns. Modern programming languages such as Java and Python provide built-in support for *regular expressions*, a particularly natural way of defining automata. Control systems and compilers often use finite state machines to encode the current state and possible actions/transitions. Minimizing the size of these automata reduces both the storage and execution costs of dealing with such machines.

Finite state machines are defined by directed graphs. Each vertex represents a state, and each character-labeled edge defines a transition from one state to another on receipt of the given symbol. The automata shown analyzes a sequence of coin tosses, with dark states signifying that an even number of heads have been observed. Such automata can be represented using any graph data structure (see Section 15.4 (page 452)), or by an $n \times |\Sigma|$ *transition matrix* M where $|\Sigma|$ is the size of the symbol *alphabet* of the automata. Here $M[i, j]$ reports the state transitioned to from state i on receipt of symbol j.

Finite state machines are often used to specify search patterns in the guise of regular expressions, which are patterns formed by and-ing, or-ing, and looping over smaller regular expressions. For example, the regular expression $a(a + b + c)^*a$ matches any string on (a, b, c) that begins and ends with distinct as. The best way to test whether a string s is recognized by a given regular expression R constructs the finite automaton equivalent to R, and then simulates this machine on S. See Section 21.3 (page 685) for alternative approaches to string matching.

We consider three different problems on finite automata:

- *Minimizing deterministic finite state machines* – Transition matrices for finite automata can become prohibitively large for sophisticated machines, thus fueling the need for tighter encodings. The most direct approach is to eliminate redundant states in the automaton. As the example above shows, automata of widely varying sizes can compute the same function.

  Algorithms to minimize the number of states in a deterministic finite automaton (DFA) appear in any book on automata theory. The basic approach partitions the states into gross equivalence classes, and then refines the partition. Initially, the states are partitioned into accepting, rejecting, and non-terminal classes. The transitions from each node now branch to a given class on a given symbol. Whenever two states s, t from the same class C branch to elements of different classes, then C must be partitioned into two subclasses, one containing s, the other containing t.

  This algorithm makes a sweep through all the classes looking for a new partition, and repeats the process from scratch if it finds one. This yields an $O(n^2)$ algorithm for constant-sized alphabets, because at most $n - 1$ sweeps need ever be performed. The final equivalence classes correspond to the states in the minimum automaton. In fact, a more efficient $O(n \log n)$ algorithm is known. Implementations are cited below.

- *Constructing deterministic machines from non-deterministic machines* – DFAs are simple to work with, because the machine is always in exactly one state at any given time. *Non-deterministic finite automata* (NFAs) can be in multiple states at a time, so their current "state" represents a subset of the possible machine states.

  In fact, any NFA can be mechanically converted to an equivalent DFA, which can then be minimized as above. However, converting an NFA to a DFA might cause an exponential blowup in the number of states, which might perversely get eliminated when minimizing the DFA. This exponential blowup makes most NFA minimization problems PSPACE-hard, which is even worse than NP-complete.

  The proofs of equivalence between NFAs, DFAs, and regular expressions are elementary enough to be covered in undergraduate automata theory classes. However, they are surprisingly nasty to code. Implementations are discussed below.

- *Constructing machines from regular expressions* – There are two approaches for translating a regular expression to an equivalent finite automaton. NFAs are easier to construct than deterministic automata, but less efficient to simulate.

  The non-deterministic construction uses ϵ-moves, which are optional transitions that require no input to fire. On reaching a state with an ϵ-move, we must assume that the machine is in both states. It is straightforward to construct an automaton using ϵ-moves, from a depth-first traversal of the parse tree of the regular expression. This machine will have $O(m)$ states, if m is the length of the regular expression. Simulating this machine on a string of length n takes $O(mn)$ time, since we need to consider each state/prefix pair only once.

The deterministic construction starts with the parse tree for the regular expression, observing that each leaf represents an alphabet symbol in the pattern. After recognizing a prefix of the text, we can be left in some subset of these possible positions, which would correspond to a state in the finite automaton. The *derivatives* method builds up this automaton state-by-state as it is needed. Some regular expressions of length m require $O(2^m)$ states in any DFA implementing them, such as $(a+b)^*a(a+b)(a+b)\ldots(a+b)$. There is no way to avoid this exponential space blowup. Fortunately it takes linear time to simulate an input string on any DFA, regardless of the size of the automaton.

Implementations: *Grail+* is a C++ package for symbolic computation with finite automata and regular expressions. Grail enables one to convert between different machine representations and to minimize automata. It can handle large machines defined on large alphabets. All code and documentation are accessible from `http://www.csit.upei.ca/~ccampeanu/Grail/`, as well as pointers to a variety of other automaton packages.

The OpenFst Library (`http://www.openfst.org/`) is a library for constructing, combining, optimizing, and searching weighted finite state transducers (FSTs), which are generalizations of finite state machines where output symbols do not necessary match input symbols. Minimization and conversions to deterministic machines are provided.

JFLAP (Java Formal Languages and Automata Package) is a package of graphical tools for learning the basic concepts of automata theory. Included are functions to convert between DFAs, NFAs, and regular expressions, and minimize the resulting automata. High-level automata are also supported, including context-free languages and Turing machines. *JFLAP* is available at `http://www.jflap.org/`. A related book [RF06] is also available.

Notes: Aho [Aho90] provides a good survey on algorithms for pattern matching, with a particularly clear exposition for regular expression patterns. The technique for regular expression pattern matching with ϵ-moves is due to Thompson [Tho68]. Other expositions on finite automaton pattern matching include [AHU74]. Expositions on finite automata and the theory of computation include [HK11, HMU06, Sip05]. The major annual meeting of interest in this field is the Conference on Implementations and Applications of Automata (CIAA).

Hopcroft [Hop71] gave an optimal $O(n \lg n)$ algorithm for minimizing the number of states in DFAs. The derivatives method of constructing a finite state machine from a regular expression is due to Brzozowski [Brz64] and has been expanded upon in [BS86]. Expositions on the derivatives method includes Conway [Con71]. Recent work on the incremental construction and optimization of automata includes [Wat03]. The problems of compressing a DFA to a minimum NFA [JR93] and testing the equivalence of two non-deterministic finite state machines [SM73] are both PSPACE-complete.

The inefficiencies of algorithmic approaches for regular expression pattern matching have been exploited for denial-of-service attacks on websites. If public-facing code contains a regular expression requiring superlinear computation (in the worst case), an attacker can overload machines by supplying pathological input that takes excessive time to parse. See [DCSL18] for a discussion of such attacks.

Related problems: Satisfiability (see page 537). string matching (see page 685).

AHAAIGDDETNWORTDTS	AHAAIGDDETNWORTDTS
HGITGDDEANWTOSRDSG	HGITGDDEANWTOSRDSG
GTASHIDDENWORDTGAG	GTASHIDDENWORDTGAG
HIGSDDEGNWORGADSTA	HIGSDDEGNWORGADSTA
HAIDAGDENSWORSADTS	HAIDAGDENSWORSADTS

Input   Output

21.8   Longest Common Substring/Subsequence

Input description: A set S of strings $S_1, \ldots, S_n$.

Problem description: What is the longest string S' such that all the characters of S' appear as a substring or subsequence of each S_i, $1 \le i \le n$?

Discussion: Longest common substring/subsequence (LCS) arises whenever we search for similarities across multiple texts. A particularly important application concerns finding a consensus among biological sequences. The genes for building proteins evolve with time, but the regions critical for function must remain intact in order for them to work correctly. The longest common subsequence of a gene in different species provides insight into what has been conserved over time.

The longest common subsequence problem for two strings is a special case of edit distance (see Section 21.4 (page 688)), when substitutions are forbidden and exact character match, insert, and delete are the only allowable edit operations. Under these conditions, the edit distance between P and T is $n+m-2|lcs(P,T)|$, because we can delete the missing characters from P to the $lcs(P,T)$ and then insert the missing characters from T to transform P to T.

Issues arising in LCS include:

- *Are you looking for a common substring?* – When detecting plagiarism, we seek to find the longest phrase shared between two or more documents. Since phrases are strings of consecutive characters, here we want the longest common *substring* between the texts.

  The longest common substring of a set of strings can be identified in linear time, as discussed in Section 15.3 (page 448). The trick is to build a suffix tree containing all the strings, label each leaf with the input string it represents, and then do a depth-first traversal to identify the deepest node with descendants from each input string.

- *Are you looking for a common scattered subsequence?* – For the rest of this section we restrict attention to finding common scattered subsequences. This algorithm is a special case of the dynamic programming edit-distance computation. Indeed, an implementation is given on page 323.

Let $M[i, j]$ denote the number of characters in the longest common subsequence of $S[1], \ldots, S[i]$ and $T[1], \ldots, T[j]$. When $S[i] \neq T[j]$, the last pair of characters cannot match, so $M[i, j] = \max(M[i, j-1], M[i-1, j])$. But if $S[i] = T[j]$, we also have the option to select this character for our substring, so $M[i, j] = \max(M[i-1, j-1] + 1, M[i-1, j], M[i, j-1])$.

This recurrence computes the length of the longest common subsequence in $O(nm)$ time. We can reconstruct the actual sequence by walking backward from $M[n, m]$ to establish which characters were matched along the way.

- *What if the strings are permutations?* – Permutations are strings without repeating characters. Two permutations define n pairs of matching characters, and so the above algorithm runs in $O(n \lg n)$ time. A particularly important case occurs in finding the longest *increasing* subsequence of a numerical sequence p. Sorting the elements of p in increasing order yields a sequence s. The longest common subsequence of p and s gives the longest increasing subsequence.

- *What if there are relatively few sets of matching characters?* – There is a faster algorithm when strings do not contain too many copies of the same character, like the permutations above. Let r be the number of pairs of positions (i, j) such that $S_i = T_j$. In this technique, each of the r pairs defines a point in the plane.

  The complete set of such points can be found in $O(n + m + r)$ time using bucketing techniques. We create a bucket for each alphabet symbol c and each string (S or T), then partition the positions of each character of the string into the appropriate bucket. We then create a point (s, t) from every pair $s \in S_c$ and $t \in T_c$ in the buckets S_c and T_c.

  A common subsequence describes a monotonically non-decreasing path through these points, meaning all moves on this path are up and to the right. The longest such path can be found in $O((n + r) \lg n)$ time. We sort the points in order of increasing x-coordinate, breaking ties in favor of increasing y-coordinate. We insert points in this order, and maintain the minimum terminal y-coordinate of any path going through exactly k points, for $1 \leq k \leq n$. Each new point (p_x, p_y) changes exactly one of these paths, either identifying a new longest subsequence or reducing the y-coordinate of the shortest path whose endpoint lies above p_y.

- *What if we have more than two strings to align?* – The basic dynamic programming algorithm can be generalized to k strings, taking $O(2^k n^k)$ time, where n is the length of the longest string. This algorithm is exponential in the number of strings k, and so it becomes expensive for more than a few strings. This problem is NP-complete, so no better exact algorithm is destined to come along soon.

Many heuristics have been proposed for multiple sequence alignment. They often start by computing the pairwise alignment for each pair of strings. One approach replaces the two most similar sequences with a single merged sequence, and repeats until all alignments have been merged into one. The catch is that two strings often have many different alignments of optimal cost. The "right" alignment to pick depends upon the subsequent sequences to merge, and hence unknowable to the heuristic.

Implementations: Several programs are available for multiple sequence alignment of DNA/protein sequence data. *ClustalW* [THG94] is a popular and well-regarded program for multiple alignment of protein sequences. It is available at `https://www.ebi.ac.uk/Tools/msa/`. Another respectable option is the *MSA* package for multiple sequence alignment [GKS95], which is available at `https://www.ncbi.nlm.nih.gov/CBBresearch/Schaffer/msa.html`.

Any of the dynamic programming-based approximate string matching programs of Section 21.4 (page 688) can be used to find the longest common subsequence of two strings.

Combinatorica [PS03] provides a Mathematica implementation of an algorithm to construct the longest increasing subsequence of a permutation. This algorithm is based on Young tableaux rather than dynamic programming. See Section 22.1.8 (page 716).

Notes: Surveys of algorithmic results on longest common subsequence (LCS) problems include [BHR00, GBY91]. The algorithm for the case where all the characters in each sequence are distinct or infrequent is due to Hunt and Szymanski [HS77], but recently improved [IR09]. Expositions include [Aho90, Man89]. There has been a surprising amount of recent work on this problem, including efficient bit-parallel algorithms for LCS [CIPR01]. Masek and Paterson [MP80] solve the longest common subsequence problem in $O(mn/\log(\min\{m, n\}))$ for constant-sized alphabets, using the four Russians technique. No strongly subquadratic algorithm is possible without refuting the strong exponential time hypothesis [ABW15, BK18].

Construct two random n-character strings on an alphabet of size α. What is the expected length of their LCS? This problem has been extensively studied, with an excellent survey by Dancik [Dan94].

Multiple sequence alignment for computational biology is a large field, with the books of Gusfield [Gus97] and Compeau and Pevzner [CP18] serving as excellent introductions. See [Not02] for a more recent survey. The hardness of multiple sequence alignment follows from that of shortest common subsequence for large sets of strings [Mai78].

We motivated the problem of longest common substring with the application of plagiarism detection. See [SWA03] for the interesting details of how to implement a plagiarism detector for computer programs.

Related problems: Approximate string matching (see page 688), shortest common superstring (see page 709).

ABRACA   **ABRACADABRA**

RACADA   ABRACA

ACADAB   RACADA

CADABR   ACADAB

ADABRA   CADABR

                ADABRA

Input   Output

21.9 Shortest Common Superstring

Input description: A set of strings $S = \{S_1, \ldots, S_m\}$.

Problem description: Find the shortest string S' that contains each string S_i as a substring of S'.

Discussion: Shortest common superstring/supersequence (SCS) arises in a variety of applications. A casino gambling addict once asked me how to reconstruct the pattern of symbols on the wheels of a slot machine. With every spin, the wheel turns to a random position, displaying the selected symbol as well as the symbols immediately before/after it. Given enough observations of the slot machine, the symbol order for each wheel can be determined as the shortest common (circular) superstring of the observed symbol triples.

A more important application of shortest common superstring is data/matrix compression. Suppose we are given a sparse $n \times m$ matrix M, meaning that most elements are zero. We can partition each row into m/k runs of k elements, and construct the shortest common superstring S' of all these runs. We can now represent the matrix by the superstring, plus an $n \times m/k$ array of pointers denoting where each run starts in S'. Any particular element $M[i,j]$ can still be accessed in constant time, but there will be substantial space savings when $|S| \ll mn$.

Perhaps the most compelling application is in DNA sequence assembly. Machines readily sequence fragments of 100 to 1,000 base pairs, or characters of DNA. But the real interest is in sequencing large molecules. Large-scale "shotgun" sequencing clones many copies of the target molecule, breaks them randomly into fragments, sequences the fragments, and then proposes the shortest superstring as the correct sequence.

Finding a superstring of a set of strings is not difficult, since we can simply concatenate them together. Finding the *shortest* such string is what's problematic. Indeed, shortest common superstring is NP-complete for all reasonable classes of strings.

Finding the shortest common superstring can be reduced to the traveling salesman problem (see Section 19.4 (page 594)). Create a directed overlap graph G, where vertex v_i represents string S_i. Assign edge (v_i, v_j) weight equal to the length of S_i minus the overlap of S_j with S_i. Thus, $w(v_i, v_j) = 1$ for $S_i = abc$ and $S_j = bcd$. The minimum weight path visiting all the vertices defines the shortest common superstring. These edge weights are not symmetric: note that $w(v_j, v_i) = 3$ for the example above. Unfortunately, asymmetric TSP problems are much harder to solve in practice than symmetric instances.

The greedy heuristic provides the standard approach to approximating shortest common superstring. Identify which pair of strings have the maximum overlap. Replace them by the merged string, and repeat until only one string remains. This heuristic can be implemented in linear time. The seemingly most time-consuming part is in building the overlap graph. The brute-force approach to finding the maximum overlap of two length-l strings takes $O(l^2)$ time for each of $O(n^2)$ string pairs. But a faster construction is made possible by suffix trees (see Section 15.3 (page 448)). Build a tree containing all suffixes of all strings from S. String S_i overlaps with S_j iff a suffix of S_i matches the prefix of S_j—an event defined by a vertex of the suffix tree. Traversing these vertices in order of distance from the root defines the appropriate merging order.

How well does the greedy heuristic perform? It can certainly be fooled into creating a superstring that is twice as long as optimal. The optimal merging order for strings $c(ab)^k$, $(ba)^k$, and $(ab)^k c$ is left to right. But greedy starts by merging the first and third string, leaving the middle one no overlap possibility. The greedy superstring can never be worse than 3.5 times optimal, and usually will be much better in practice.

Building superstrings becomes difficult when the input contains both positive strings (each of which *must* be a substring of the superstring) and negative strings (each of which is *forbidden* to appear in the final result). Deciding whether *any* such consistent string exists is NP-complete, unless you are allowed to add an extra character to the alphabet to use as a spacer.

Implementations: Several high-performance programs for DNA sequence assembly are available. Such programs correct for sequencing errors, so the final result is not necessarily a superstring of the input reads. At the very least, they will serve as excellent models if you really need a short proper superstring.

CAP3 (Contig Assembly Program) [HM99] and *PCAP* [HWA+03] are the latest in a series of assemblers by Xiaoqiu Huang and his collaborators, which are available from `http://seq.cs.iastate.edu/`. They have been used on mammalian scale assembly projects involving hundreds of millions of bases.

The Celera assembler that sequenced the human genome is now available as open source. See `https://sourceforge.net/projects/wgs-assembler/`.

Notes: The shortest common superstring (SCS) problem and its application to DNA shotgun assembly are ably surveyed in [MKT07, Mye99a, SP15]. Kececioglu and Myers [KM95] report on an algorithm for this more general version of shortest common superstring, where the strings are assumed to have character substitution errors. Their paper is recommended reading to anyone interested in fragment assembly.

Blum et al. [BJL+94] gave the first constant-factor approximation algorithms for shortest common superstring, using a variation of the greedy heuristic. More recent research has beaten this constant down to 2.367 [Muc13, Pal14], progress towards the expected factor-2 result. The best approximation ratio so far proven for the standard greedy heuristic is 3.5 [KS05a]. Fast implementations of such heuristics are described in [Gus94].

Experiments on shortest common superstring heuristics are reported in [RBT04], which suggest that greedy heuristics typically produce solutions within 1.4% of optimal for a reasonable class of inputs. Experiments with genetic algorithm approaches are reported in [ZS04]. Analytical results [YZ99] demonstrate very little compression on the SCS of random sequences, largely because the expected overlap length of any two random strings is small.

Related problems: Suffix trees (see page 448), text compression (see page 693).

Chapter 22

Algorithmic Resources

This chapter briefly describes resources that the working algorithm designer should be familiar with. Although some of this information has appeared elsewhere in the catalog, we collect the most important pointers here for general reference.

22.1 Algorithm Libraries

A good algorithm designer does not reinvent the wheel, and a good programmer does not rewrite code that other people have written. Picasso put it best: "Good artists borrow. Great artists steal."

But a word of caution about stealing. Many of the codes described in this book have been made available only for research or educational use. Commercial use may require a licensing arrangement with the author. I urge you to respect this. Licensing terms from academic institutions are usually quite modest. The recognition that industry is using a particular code is important to the authors, often more important than the money involved. Do the right thing and get a license. Information about terms or whom to contact is usually available embedded within the documentation, or available at the source's website.

Although many of the systems described here may be available by accessing the algorithm repository, www.algorist.com I strongly encourage you to get them from the original sites. First, the version on the original site is *much* more likely to be up-to-date. Second, there are often extra supporting files and documentation that may be of interest to you. Finally, many authors monitor the downloads of their codes, and so you deny them a well-earned thrill if you don't take them from the original site.

22.1.1 LEDA

LEDA, for *Library of Efficient Data types and Algorithms*, is perhaps the best single resource available to support combinatorial computing. LEDA was origi-

© The Editor(s) (if applicable) and The Author(s), under exclusive license to
Springer Nature Switzerland AG 2020
S. S. Skiena, *The Algorithm Design Manual*, Texts in Computer Science,
https://doi.org/10.1007/978-3-030-54256-6_22

nally developed by a group at the Max Planck Institut in Saarbrücken, Germany. LEDA is unique because of (1) the algorithmic sophistication of its developers, and (2) the level of continuity and resources invested in the project.

What LEDA offers is a complete collection of well-implemented C++ data structures and types. Particularly useful is the graph type, which supports all basic operations in an intelligent way. A useful library of graph algorithms is included, which illustrates how cleanly and concisely these algorithms can be implemented using the LEDA data types. Good implementations of the most important data structures supporting dictionaries and priority queues are provided. For more, see the book [MN99].

LEDA is exclusively available from Algorithmic Solutions Software GmbH (`https://www.algorithmic-solutions.com/`). This ensures professional-quality support, and new releases appear often. A free edition contains all the basic data structures, including dictionaries, priority queues, graphs, and numerical types. No source code or advanced algorithms are provided with the free edition. But the licensing fees for the full library are not outlandish, and free trial downloads are available. Check it out.

22.1.2   CGAL

The *Computational Geometry Algorithms Library* or *CGAL* provides efficient and reliable geometric algorithms in C++. It is very comprehensive, offering a rich variety of triangulations, Voronoi diagrams, operations on polygons and polyhedra, line/curve arrangements, alpha-shapes, convex-hull algorithms, and geometric search structures. Many work in three dimensions and beyond.

CGAL (`https://www.cgal.org`) should be the first place to go for serious geometric computing. CGAL is distributed under a dual-license scheme. It can be used together with open source software free of charge, but using CGAL in other contexts requires obtaining a commercial license.

22.1.3   Boost Graph Library

Boost (`www.boost.org`) provides a well-regarded collection of free peer-reviewed portable C++ source libraries, encouraging both commercial and non-commercial use.

The Boost Graph Library [SLL02] (`http://www.boost.org/libs/graph/doc`) is perhaps most relevant for the readers of this book. Implementations of adjacency lists, matrices, and edge lists are included, along with a reasonable library of basic graph algorithms. Its interface and components are generic in the same sense as the C++ Standard Template Library (STL). Other Boost libraries of interest include string/text processing and math/numeric computation.

22.1.4   Netlib

Netlib (`https://netlib.org/`) is an online repository of mathematical software that contains a large number of interesting codes, tables, and papers. Netlib is important because of its breadth and ease of access. Whenever you need a specialized piece of mathematical software, you should look here first.

The Guide to Available Mathematical Software (GAMS), is an indexing service for Netlib and other related software repositories that can help you find what you want. Check it out at `https://gams.nist.gov`. GAMS is a service of the National Institute of Standards and Technology (NIST).

22.1.5   Collected Algorithms of the ACM

An early mechanism for the distribution of useful algorithm implementations was the *Collected Algorithms of the ACM* (*CALGO*). It first appeared in *Communications of the ACM* in 1960, covering such famous algorithms as Floyd's linear-time build heap algorithm. More recently, it has been the province of the *ACM Transactions on Mathematical Software*. Each algorithm/implementation is described in a brief journal article, with the implementation validated and collected. These implementations are maintained at `https://www.acm.org/calgo/` and at Netlib.

Almost 1,000 algorithms have appeared to date. Most of the codes are in Fortran and are relevant to numerical computing, although several interesting combinatorial algorithms have slithered their way into CALGO. Since the implementations have been refereed, they are presumably more reliable than most comparable software.

22.1.6   GitHub and SourceForge

With over 28 million public repositories, the software development platform *GitHub* (`https://github.com`) is the largest host of source code in the world. A GitHub search should be your second stop when seeking implementations of algorithms: go there immediately after you check the relevant catalog entry in this book. All interesting recent codes are likely to be found there, including over a dozen of the systems referred to in the catalog.

SourceForge (`http://sourceforge.net/`) is an older open source software development website, with over 160,000 registered projects. There is still a lot of good stuff to be found, including graph libraries such as *JUNG* and *JGraphT*, optimization engines such as *lpsolve* and *JGAP*, and much more.

22.1.7   The Stanford GraphBase

The Stanford GraphBase is an interesting program for several reasons. First, it was composed as a "literate program," meaning that it was written to be read. If anybody's programs deserve to be read, it is Knuth's, and [Knu94] contains the full source code of the system.

The GraphBase contains implementations of several important combinatorial algorithms, including matching, minimum spanning trees, and Voronoi diagrams, as well as specialized topics like constructing expander graphs and generating combinatorial objects. Finally, it contains programs for several recreational problems, including constructing word ladders (flour–floor–flood–blood–brood–broad–bread) and establishing dominance relations among football teams. Check it out at `http://www-cs-faculty.stanford.edu/~knuth/sgb.html`.

Although the GraphBase is fun to play with, it is not really suited for building general applications on top of. The GraphBase is perhaps most useful as an instance generator for constructing a wide variety of graphs to serve as test data.

22.1.8   Combinatorica

Combinatorica [PS03] is a collection of over 450 algorithms for combinatorics and graph theory written in Mathematica. *Combinatorica* has been widely used for both research and education.

Although (in my totally unbiased opinion) *Combinatorica* is more comprehensive and better integrated than other libraries of combinatorial algorithms, it is also the slowest such system available. *Combinatorica* is best for finding quick solutions to small problems, and (if you can read Mathematica code) as a terse exposition of algorithms for translation into other languages.

Check out `http://www.combinatorica.com` for the latest release and associated resources. As graphs have become more important, much of its functionality has migrated into Mathematica itself. It is also included with the standard Mathematica distribution in *Packages/DiscreteMath/Combinatorica.m*.

22.1.9   Programs from Books

Several books on algorithms include working implementations of the algorithms in a real programming language. Although these implementations are intended primarily for exposition, they can also be useful for computation. Since they are typically small and clean, they can prove the right foundation for simple applications.

The most useful codes of this genre are described below. Most are available from the algorithm repository, www.algorist.com.

- *Programming Challenges* – If you like the C code that appeared in the first half of the text, you should check out the programs I wrote for the book *Programming Challenges* [SR03]. Perhaps most useful are additional examples of dynamic programming, computational geometry routines like convex hull, and a bignum integer arithmetic package. This algorithm library is available at:

  `https://www.cs.stonybrook.edu/~skiena/392/programs/`, and

  `https://github.com/SkienaBooks/Algorithm-Design-Manual-Programs`.

- *Combinatorial Algorithms for Computers and Calculators* – Nijenhuis and Wilf [NW78] specializes in algorithms for constructing basic combinatorial objects such as permutations, subsets, and partitions. Such algorithms are often very short, but they are hard to locate and usually surprisingly subtle. Fortran routines for all of the algorithms are provided, as well as a discussion of the theory behind each of them. Both random and sequential generation algorithms are provided. Descriptions of more recent algorithms for several problems, without code, are provided in [Wil89].

These programs are now available from the algorithm repository website. I tracked them down from Neil Sloane, who had them on a magnetic tape, while the original authors did not!

- *Computational Geometry in C* – O'Rourke [O'R01] is perhaps the best practical introduction to computational geometry, because of its careful and correct C language implementations of all the main algorithms of computational geometry. Fundamental geometric primitives, convex hulls, triangulations, Voronoi diagrams, and motion planning are all included. Although they were implemented primarily for exposition rather than production use, they should be quite reliable. The codes are available from `http://cs.smith.edu/~jorourke/code.html`.

- *Algorithms in C++* – Sedgewick's popular algorithms text [Sed98, SW11] comes in several different language editions, including C, C++, and Java. This book distinguishes itself through its use of algorithm animation and in its broad topic coverage, including numerical, string, and geometric algorithms. Code from the Java edition is available from `https://algs4.cs.princeton.edu/`.

- *Discrete Optimization Algorithms in Pascal* – This collection of 28 programs for solving discrete optimization problems appears in the book by Syslo, Deo, and Kowalik [SDK83]. The package includes programs for integer and linear programming, the knapsack and set cover problems, traveling salesman, vertex coloring, and scheduling, as well as standard network optimization problems. They have been made available from the algorithm repository at www.algorist.com.

22.2 Data Sources

It is often important to have interesting data to feed your algorithms, to serve as test data to ensure correctness or to compare different algorithms for raw speed. Finding good test data can be surprisingly difficult. Here are some pointers:

- *Stanford Network Analysis Project (SNAP)* – This resource couples a general purpose network analysis library that scales to hundreds of millions of nodes with a nice collection of real-world graphs drawn from social networks, citation networks, and communications networks. All are available from `https://snap.stanford.edu/`.

- *TSPLIB* – This well-respected library of test instances for the traveling salesman problem [Rei91] provides the standard collection of hard instances of TSPs. TSPLIB instances are large, real-world graphs, derived from applications such as circuit boards and networks, and are available from `http://www.math.uwaterloo.ca/tsp/data`.

- *Stanford GraphBase* – As discussed in Section 22.1.7, this suite of programs by Knuth provides portable generators for a wide variety of graphs. These include graphs arising from distance matrices, arts, and literature, as well as graphs of more theoretical interest.

22.3   Online Bibliographic Resources

The web is a fantastic resource for people interested in algorithms. What follows is a selective list of the resources that I use most often. All should be in the tool chest of every algorist:

- *Google Scholar* – This free resource (`https://scholar.google.com/`) restricts web searches to things that look like academic papers, making it a sounder search for serious information than a general web search. Particularly useful is the ability to see which papers cite a given paper. This lets you update an old reference to see what has happened since publication, and helps to judge the significance of a particular article.

- *ACM Digital Library* – This collection of bibliographic references provides links to essentially every technical paper ever published in computer science. Check out what is available at `https://portal.acm.org/`.

- *Arxiv* – This preprint server with over 1.6 million papers is where researchers disseminate results before they are formally published—at which time they are out of date. This is the place to go to find the latest research on any topic discussed in this book. Check it out at `https://arxiv.org/`.

22.4   Professional Consulting Services

Algorist Technologies (`http://www.algorist.com`) is a consulting firm that provides its clients with short-term, expert help in algorithm design and implementation. Typically, an Algorist consultant is called in for one to three days worth of intensive discussion and analysis with the client's own development staff. Algorist has built an impressive record of performance improvements with several companies and applications. Email info@algorist.com for more information on our services.

Algorist Technologies
215 West 92nd St. Suite 1F
New York, NY 10025

Chapter 23

Bibliography

[AAAG95]   O. Aichholzer, F. Aurenhammer, D. Alberts, and B. Gartner. A novel type of skeleton for polygons. *J. Universal Computer Science*, 1:752–761, 1995.

[AAM18]   M. Abrahamsen, A. Adamaszek, and T. Miltzown. The art gallery problem is ∃r-complete. In *Proceedings of the 50th Annual ACM SIGACT Symposium on Theory of Computing*, pages 65–73. ACM, 2018.

[Aar13]   Scott Aaronson. *Quantum Computing since Democritus*. Cambridge University Press, 2013.

[Aar15]   Scott Aaronson. Quantum machine learning algorithms: Read the fine print. *Nature Physics*, 11(4):291, 2015.

[AB09]   Sanjeev Arora and Boaz Barak. *Computational Complexity: A Modern Approach*. Cambridge University Press, 2009.

[AB17]   Mikhail J Atallah and Marina Blanton. *Algorithms and Theory of Computation Handbook, Volume 1: General Concepts and Techniques*. Chapman & Hall/CRC, 2017.

[ABCC07]   D. Applegate, R. Bixby, V. Chvatal, and W. Cook. *The Traveling Salesman Problem: A Computational Study*. Princeton University Press, 2007.

[ABF05]   L. Arge, G. Brodal, and R. Fagerberg. Cache-oblivious data structures. In D. Mehta and S. Sahni, editors, *Handbook of Data Structures and Applications*, pages 34:1–34:27. Chapman & Hall/CRC Press, 2005.

[ABW15]   Amir Abboud, Arturs Backurs, and Virginia Vassilevska Williams. Tight hardness results for lcs and other sequence similarity measures. In *2015 IEEE 56th Annual Symposium on Foundations of Computer Science*, pages 59–78. IEEE, 2015.

[AC75]   A. Aho and M. Corasick. Efficient string matching: an aid to bibliographic search. *Communications of the ACM*, 18:333–340, 1975.

[AC91]   D. Applegate and W. Cook. A computational study of the job-shop scheduling problem. *ORSA Journal on Computing*, 3:149–156, 1991.

[ACFC+16]   Alberto Apostolico, Maxime Crochemore, Martin Farach-Colton, Zvi Galil, and Shan Muthukrishnan. Forty years of suffix trees. *Communications of the ACM*, 59(4):66–73, 2016.

© The Editor(s) (if applicable) and The Author(s), under exclusive license to
Springer Nature Switzerland AG 2020
S. S. Skiena, *The Algorithm Design Manual*, Texts in Computer Science,
https://doi.org/10.1007/978-3-030-54256-6

[ACH$^+$91]   E. M. Arkin, L. P. Chew, D. P. Huttenlocher, K. Kedem, and J. S. B. Mitchell.   An efficiently computable metric for comparing polygonal shapes. *IEEE Trans./ PAMI*, 13(3):209–216, 1991.

[ACK01a]   N. Amenta, S. Choi, and R. Kolluri. The power crust. In *Proc. 6th ACM Symp. on Solid Modeling*, pages 249–260, 2001.

[ACK01b]   N. Amenta, S. Choi, and R. Kolluri. The power crust, unions of balls, and the medial axis transform. *Computational Geometry: Theory and Applications*, 19:127–153, 2001.

[ACL12]   Eric Angel, Romain Campigotto, and Christian Laforest. Implementation and comparison of heuristics for the vertex cover problem on huge graphs. In *International Symposium on Experimental Algorithms*, pages 39–50. Springer, 2012.

[ACLZ11]   Deepak Ajwani, Adan Cosgaya-Lozano, and Norbert Zeh. Engineering a topological sorting algorithm for massive graphs. In *Proceedings of the Meeting on Algorithm Engineering & Experimiments*, pages 139–150, 2011.

[ACP$^+$07]   H. Ahn, O. Cheong, C. Park, C. Shin, and A. Vigneron. Maximizing the overlap of two planar convex sets under rigid motions. *Computational Geometry: Theory and Applications*, 37:3–15, 2007.

[ADGM07]   Lyudmil Aleksandrov, Hristo Djidjev, Hua Guo, and Anil Maheshwari. Partitioning planar graphs with costs and weights. *Journal of Experimental Algorithmics (JEA)*, 11:1–5, 2007.

[ADKF70]   V. Arlazarov, E. Dinic, M. Kronrod, and I. Faradzev. On economical construction of the transitive closure of a directed graph. *Soviet Mathematics, Doklady*, 11:1209–1210, 1970.

[Adl94]   L. M. Adleman. Molecular computations of solutions to combinatorial problems. *Science*, 266:1021–1024, November 11, 1994.

[AE04]   G. Andrews and K. Eriksson. *Integer Partitions*. Cambridge University Press, 2004.

[AF96]   D. Avis and K. Fukuda. Reverse search for enumeration. *Disc. Applied Math.*, 65:21–46, 1996.

[AFGW10]   Ittai Abraham, Amos Fiat, Andrew V Goldberg, and Renato F Werneck. Highway dimension, shortest paths, and provably efficient algorithms. In *Proceedings of the Twenty-First Annual ACM-SIAM Symposium on Discrete Algorithms*, pages 782–793, 2010.

[AFH02]   P. Agarwal, E. Flato, and D. Halperin. Polygon decomposition for efficient construction of Minkowski sums. *Computational Geometry: Theory and Applications*, 21:39–61, 2002.

[AG00]   H. Alt and L. Guibas. Discrete geometric shapes: Matching, interpolation, and approximation. In J. Sack and J. Urrutia, editors, *Handbook of Computational Geometry*, pages 121–153. Elsevier, 2000.

[Aga18]   P. Agarwal. Range searching. In J. Goodman, J. O'Rourke, and C. Toth, editors, *Handbook of Discrete and Computational Geometry*, pages 1057–1093. CRC Press, 2018.

[AGR01]   N. Amato, M. Goodrich, and E. Ramos. A randomized algorithm for triangulating a simple polygon in linear time. *Discrete & Computational Geometry*, 26(2):245–265, 2001.

[AGSS89]   A. Aggarwal, L. Guibas, J. Saxe, and P. Shor. A linear-time algorithm for computing the Voronoi diagram of a convex polygon. *Discrete & Computational Geometry*, 4:591–604, 1989.

[AGU72]   A. Aho, M. Garey, and J. Ullman. The transitive reduction of a directed graph. *SIAM J. Computing*, 1:131–137, 1972.

[AHK$^+$08] E. Arkin, G. Hart, J. Kim, I. Kostitsyna, J. Mitchell, G. Sabhnani, and S. Skiena. The embroidery problem. In *20th Canadian Conf. Computational Geometry (CCCG)*, 2008.

[Aho90]   A. Aho. Algorithms for finding patterns in strings. In J. van Leeuwen, editor, *Handbook of Theoretical Computer Science: Algorithms and Complexity*, volume A, pages 255–300. MIT Press, 1990.

[AHU74]   A. Aho, J. Hopcroft, and J. Ullman. *The Design and Analysis of Computer Algorithms*. Addison-Wesley, 1974.

[AHU83]   A. Aho, J. Hopcroft, and J. Ullman. *Data Structures and Algorithms*. Addison-Wesley, 1983.

[AI06]   Alexandr Andoni and Piotr Indyk. Near-optimal hashing algorithms for approximate nearest neighbor in high dimensions. In *Foundations of Computer Science, 2006. FOCS'06. 47th Annual IEEE Symposium on*, pages 459–468. IEEE, 2006.

[AI08]   Alexandr Andoni and Piotr Indyk. Near-optimal hashing algorithms for approximate nearest neighbor in high dimensions. *Communications of the ACM*, 51(1):117, 2008.

[Aig88]   M. Aigner. *Combinatorial Search*. Wiley-Teubner, 1988.

[AIR18]   Alexandr Andoni, Piotr Indyk, and Ilya Razenshteyn. Approximate nearest-neighbor search in high dimensions. *arXiv preprint arXiv:1806.09823*, 2018.

[AITT00]   Y. Asahiro, K. Iwama, H. Tamaki, and T. Tokuyama. Greedily finding a dense subgraph. *J. Algorithms*, 34:203–221, 2000.

[AJNP18]   Nicolas Auger, Vincent Jugé, Cyril Nicaud, and Carine Pivoteau. On the worst-case complexity of timsort. *arXiv preprint arXiv:1805.08612*, 2018.

[AK89]   E. Aarts and J. Korst. *Simulated Annealing and Boltzman Machines: A Stochastic Approach to Combinatorial Optimization and Neural Computing*. John Wiley & Sons, 1989.

[AKL13]   Franz Aurenhammer, Rolf Klein, and Der-Tsai Lee. *Voronoi Diagrams and Delaunay Triangulations*. World Scientific Publishing Company, 2013.

[AKS04]   M. Agrawal, N. Kayal, and N. Saxena. PRIMES is in P. *Annals of Mathematics*, 160:781–793, 2004.

[AL97]   E. Aarts and J.K. Lenstra. *Local Search in Combinatorial Optimization*. John Wiley & Sons, 1997.

[AM93]   S. Arya and D. Mount. Approximate nearest neighbor queries in fixed dimensions. In *Proc. Fourth ACM-SIAM Symp. Discrete Algorithms (SODA)*, pages 271–280, 1993.

[AMN+98]   S. Arya, D. Mount, N. Netanyahu, R. Silverman, and A. Wu. An optimal algorithm for approximate nearest neighbor searching in fixed dimensions. *J. ACM*, 45:891–923, 1998.

[AMO93]   R. Ahuja, T. Magnanti, and J. Orlin. *Network Flows*. Prentice Hall, 1993.

[And98]   G. Andrews. *The Theory of Partitions*. Cambridge Univ. Press, 1998.

[And05]   A. Andersson. Searching and priority queues in $o(\log n)$ time. In D. Mehta and S. Sahni, editors, *Handbook of Data Structures and Applications*, pages 39:1–39:14. Chapman & Hall/CRC Press, 2005.

[ANN+18]   Alexandr Andoni, Assaf Naor, Aleksandar Nikolov, Ilya Razenshteyn, and Erik Waingarten. Hölder homeomorphisms and approximate nearest neighbors. In *2018 IEEE 59th Annual Symposium on Foundations of Computer Science (FOCS)*, pages 159–169. IEEE, 2018.

[ANOY14]   Alexandr Andoni, Aleksandar Nikolov, Krzysztof Onak, and Grigory Yaroslavtsev. Parallel algorithms for geometric graph problems. In *Proceedings of the forty-sixth annual ACM symposium on Theory of computing*, pages 574–583. ACM, 2014.

[AP72]   A. Aho and T. Peterson. A minimum distance error-correcting parser for context-free languages. *SIAM J. Computing*, 1:305–312, 1972.

[AP14]   Pankaj K Agarwal and Jiangwei Pan. Near-linear algorithms for geometric hitting sets and set covers. In *Proceedings of the Thirtieth Annual Symposium on Computational Geometry*, page 271. ACM, 2014.

[APT79]   B. Aspvall, M. Plass, and R. Tarjan. A linear-time algorithm for testing the truth of certain quantified boolean formulas. *Info. Proc. Letters*, 8:121–123, 1979.

[Aro98]   S. Arora. Polynomial time approximations schemes for Euclidean TSP and other geometric problems. *J. ACM*, 45:753–782, 1998.

[AS00]   P. Agarwal and M. Sharir. Arrangements. In J. Sack and J. Urrutia, editors, *Handbook of Computational Geometry*, pages 49–119. Elsevier, 2000.

[Ata83]   M. Atallah. A linear time algorithm for the Hausdorff distance between convex polygons. *Info. Proc. Letters*, 8:207–209, 1983.

[B+16]   Albert-László Barabási et al. *Network Science*. Cambridge University Press, 2016.

[Bab16]   László Babai. Graph isomorphism in quasipolynomial time. In *Proceedings of the Forty-Eighth Annual ACM Symposium on Theory of Computing*, pages 684–697. ACM, 2016.

[Bap10]   Ravindra B Bapat. *Graphs and Matrices*, volume 27. Springer, 2010.

[Bar03]   A. Barabasi. *Linked: How Everything Is Connected to Everything Else and What It Means*. Plume, 2003.

[BB14]   Charles H Bennett and Gilles Brassard. Quantum cryptography: public key distribution and coin tossing. *Theor. Comput. Sci.*, 560(12):7–11, 2014.

[BBF99]   V. Bafna, P. Berman, and T. Fujito. A 2-approximation algorithm for the undirected feedback vertex set problem. *SIAM J. Discrete Math.*, 12:289–297, 1999.

[BBPP99]   I. Bomze, M. Budinich, P. Pardalos, and M. Pelillo. The maximum clique problem. In D.-Z. Du and P.M. Pardalos, editors, *Handbook of Combinatorial Optimization*, volume A sup., pages 1–74. Kluwer, 1999.

[BC19]   Stephan Beyer and Markus Chimani. Strong steiner tree approximations in practice. *Journal of Experimental Algorithmics (JEA)*, 24(1):1–7, 2019.

[BCGM13]   Gunnar Brinkmann, Kris Coolsaet, Jan Goedgebeur, and Hadrien Mélot. House of graphs: a database of interesting graphs. *Discrete Applied Mathematics*, 161(1-2):311–314, 2013.

[BCGR92]   D. Berque, R. Cecchini, M. Goldberg, and R. Rivenburgh. The SetPlayer system for symbolic computation on power sets. *J. Symbolic Computation*, 14:645–662, 1992.

[BCW90]   T. Bell, J. Cleary, and I. Witten. *Text Compression*. Prentice Hall, 1990.

[BD99]   R. Bubley and M. Dyer. Faster random generation of linear extensions. *Disc. Math.*, 201:81–88, 1999.

[BD11]   Joseph Blitzstein and Persi Diaconis. A sequential importance sampling algorithm for generating random graphs with prescribed degrees. *Internet Mathematics*, 6(4):489–522, 2011.

[BDH97]   C. Barber, D. Dobkin, and H. Huhdanpaa. The Quickhull algorithm for convex hulls. *ACM Trans./ on Mathematical Software*, 22:469–483, 1997.

[BDN01]   G. Bilardi, P. D'Alberto, and A. Nicolau. Fractal matrix multiplication: a case study on portability of cache performance. In *Workshop on Algorithm Engineering (WAE)*, 2001.

[BDY06]   K. Been, E. Daiches, and C. Yap. Dynamic map labeling. *IEEE Trans./ Visualization and Computer Graphics*, 12:773–780, 2006.

[BEB12]   Tyson Brochu, Essex Edwards, and Robert Bridson. Efficient geometrically exact continuous collision detection. *ACM Transactions on Graphics (TOG)*, 31(4):96, 2012.

[Bel58]   R. Bellman. On a routing problem. *Quarterly of Applied Mathematics*, 16:87–90, 1958.

[Ben75]   J. Bentley. Multidimensional binary search trees used for associative searching. *Communications of the ACM*, 18:509–517, 1975.

[Ben90]   J. Bentley. *More Programming Pearls*. Addison-Wesley, 1990.

[Ben92a]   J. Bentley. Fast algorithms for geometric traveling salesman problems. *ORSA J. Computing*, 4:387–411, 1992.

[Ben92b]   J. Bentley. Software exploratorium: The trouble with qsort. *UNIX Review*, 10(2):85–93, February 1992.

[Ben99]   J. Bentley. *Programming Pearls*. Addison-Wesley, second edition, 1999.

[Ber82]   D. Bertsekas. *Constrained Optimization and Lagrange Multiplier Methods*. Academic Press, 1982.

[Ber89]   C. Berge. *Hypergraphs*. North-Holland, 1989.

[Ber02]   M. Bern. Adaptive mesh generation. In T. Barth and H. Deconinck, editors, *Error Estimation and Adaptive Discretization Methods in Computational Fluid Dynamics*, pages 1–56. Springer-Verlag, 2002.

[Ber04]   D. Bernstein.   Fast   multiplication   and   its   applications. http://cr.yp.to/arith.html, 2004.

[Ber15]   D. Bertsekas. *Convex Optimization Algorithms*. Athena Scientific Belmont, 2015.

[Ber19]   Chris Bernhardt. *Quantum Computing for Everyone*. MIT Press, 2019.

[BETT99]   G. Di Battista, P. Eades, R. Tamassia, and I. Tollis. *Graph Drawing: Algorithms for the Visualization of Graphs*. Prentice-Hall, 1999.

[BF00]   M. Bender and M. Farach. The LCA problem revisited. In *Proc. 4th Latin American Symp. on Theoretical Informatics*, pages 88–94. Springer-Verlag LNCS vol. 1776, 2000.

[BFH$^+$18]   Alon Baram, Efi Fogel, Dan Halperin, Michael Hemmer, and Sebastian Morr. Exact minkowski sums of polygons with holes. *Computational Geometry*, 73:46–56, 2018.

[BFP$^+$72]   M. Blum, R. Floyd, V. Pratt, R. Rivest, and R. Tarjan. Time bounds for selection. *J. Computer and System Sciences*, 7:448–461, 1972.

[BFS12]   G. Blelloch, J. Fineman, and J. Shun. Greedy sequential maximal independent set and matching are parallel on average. In *Proceedings of the Twenty-Fourth Annual ACM Symposium on Parallelism in Algorithms and Architectures*, pages 308–317. ACM, 2012.

[BFV07]   G. Brodal, R. Fagerberg, and K. Vinther. Engineering a cache-oblivious sorting algorithm. *ACM J. of Experimental Algorithmics*, 12, 2007.

[BG95]   J. Berry and M. Goldberg. Path optimization and near-greedy analysis for graph partitioning: An empirical study. In *Proc. 6th ACM-SIAM Symposium on Discrete Algorithms*, pages 223–232, 1995.

[BGJ$^+$12]   Arnab Bhattacharyya, Elena Grigorescu, Kyomin Jung, Sofya Raskhodnikova, and David P Woodruff. Transitive-closure spanners. *SIAM Journal on Computing*, 41(6):1380–1425, 2012.

[BGS95]   M Bellare, O. Goldreich, and M. Sudan. Free bits, PCPs, and non-approximability—towards tight results. In *Proc. IEEE 36th Symp. Foundations of Computer Science*, pages 422–431, 1995.

[BH90]   F. Buckley and F. Harary. *Distances in Graphs*. Addison-Wesley, 1990.

[BH01]   G. Barequet and S. Har-Peled. Efficiently approximating the minimum-volume bounding box of a point set in three dimensions. *J. Algorithms*, 38:91–109, 2001.

[BHK04]   Andreas Björklund, Thore Husfeldt, and Sanjeev Khanna. Approximating longest directed paths and cycles. In *International Colloquium on Automata, Languages, and Programming*, pages 222–233. Springer, 2004.

[BHR00]   L. Bergroth, H. Hakonen, and T. Raita. A survey of longest common subsequence algorithms. In *Proc. String Processing and Information Retreival (SPIRE)*, pages 39–48, 2000.

[BHvM09]   Armin Biere, Marijn Heule, and Hans van Maaren. *Handbook of Satisfiability*, volume 185. IOS press, 2009.

[BI15]   Arturs Backurs and Piotr Indyk. Edit distance cannot be computed in strongly subquadratic time (unless SETH is false). In *Proc. Forty-Seventh ACM Symposium on Theory of Computing*, pages 51–58. ACM, 2015.

[BIK$^+$04] H. Bronnimann, J. Iacono, J. Katajainen, P. Morin, J. Morrison, and G. Toussaint. Space-efficient planar convex hull algorithms. *Theoretical Computer Science*, 321:25–40, 2004.

[Bis06]   Christopher M Bishop. *Pattern Recognition and Machine Learning*. Springer, 2006.

[BJL$^+$94] A. Blum, T. Jiang, M. Li, J. Tromp, and M. Yanakakis. Linear approximation of shortest superstrings. *J. ACM*, 41:630–647, 1994.

[BJL06]   C. Buchheim, M. Jünger, and S. Leipert. Drawing rooted trees in linear time. *Software: Practice and Experience*, 36:651–665, 2006.

[BJLM83]   J. Bentley, D. Johnson, F. Leighton, and C. McGeoch. An experimental study of bin packing. In *Proc. 21st Allerton Conf. on Communication, Control, and Computing*, pages 51–60, 1983.

[BK04]   Y. Boykov and V. Kolmogorov. An experimental comparison of min-cut/max-flow algorithms for energy minimization in vision. *IEEE Trans./ Pattern Analysis and Machine Intelligence (PAMI)*, 26:1124–1137, 2004.

[BK18]   Karl Bringmann and Marvin Künnemann. Multivariate fine-grained complexity of longest common subsequence. In *Proceedings of the Twenty-Ninth Annual ACM-SIAM Symposium on Discrete Algorithms*, pages 1216–1235, 2018.

[BKRV00]   A. Blum, G. Konjevod, R. Ravi, and S. Vempala. Semi-definite relaxations for minimum bandwidth and other vertex-ordering problems. *Theoretical Computer Science*, 235:25–42, 2000.

[BL30]   Edward Bulwer Lytton. *Paul Clifford*. Henry Colburn and Richard Bentley, London, 1830.

[BL76]   K. Booth and G. Lueker. Testing for the consecutive ones property, interval graphs, and planarity using PQ-tree algorithms. *J. Computer System Sciences*, 13:335–379, 1976.

[BLS91]   D. Bailey, K. Lee, and H. Simon. Using Strassen's algorithm to accelerate the solution of linear systems. *J. Supercomputing*, 4:357–371, 1991.

[BLT12]   Gerth Stølting Brodal, George Lagogiannis, and Robert E. Tarjan. Strict fibonacci heaps. In *Proceedings of the Forty-Fourth Annual ACM Symposium on Theory of Computing*, pages 1177–1184. ACM, 2012.

[Blu67]   H. Blum. A transformation for extracting new descriptions of shape. In W. Wathen-Dunn, editor, *Models for the Perception of Speech and Visual Form*, pages 362–380. MIT Press, 1967.

[BLW76]   N. L. Biggs, E. K. Lloyd, and R. J. Wilson. *Graph Theory 1736-1936.*
           sndon Press, 1976.

[BM53]   G. Birkhoff and S. MacLane. *A Survey of Modern Algebra.* Macmillian,
           1953.

[BM77]   R. Boyer and J. Moore. A fast string-searching algorithm. *Communica-
           tions of the ACM*, 20:762–772, 1977.

[BM01]   E. Bingham and H. Mannila. Random projection in dimensionality re-
           duction: applications to image and text data. In *Proc. ACM Conf.
           Knowledge Discovery and Data Mining (KDD)*, pages 245–250, 2001.

[BM05]   A. Broder and M. Mitzenmacher. Network applications of bloom filters:
           A survey. *Internet Mathematics*, 1:485–509, 2005.

[BMRS16]   Kevin Buchin, Wouter Meulemans, André Van Renssen, and Bettina
           Speckmann. Area-preserving simplification and schematization of polyg-
           onal subdivisions. *ACM Transactions on Spatial Algorithms and Sys-
           tems*, 2(1):2, 2016.

[BMS+16]   Aydın Buluç, Henning Meyerhenke, Ilya Safro, Peter Sanders, and Chris-
           tian Schulz. Recent advances in graph partitioning. In *Algorithm Engi-
           neering*, pages 117–158. Springer, 2016.

[BMSW13]   David A Bader, Henning Meyerhenke, Peter Sanders, and Dorothea
           Wagner. *Graph partitioning and graph clustering*, volume 588. American
           Mathematical Society, 2013.

[BN09]   Albert Boggess and Francis J Narcowich. *A First Course in Wavelets
           with Fourier Analysis.* John Wiley & Sons, second edition, 2009.

[BO79]   J. Bentley and T. Ottmann. Algorithms for reporting and counting
           geometric intersections. *IEEE Transactions on Computers*, C-28:643–
           647, 1979.

[BO83]   M. Ben-Or. Lower bounds for algebraic computation trees. In *Proc.
           Fifteenth ACM Symp. on Theory of Computing*, pages 80–86, 1983.

[Bol01]   B. Bollobas. *Random Graphs.* Cambridge Univ. Press, second edition,
           2001.

[Bot12]   Léon Bottou. Stochastic gradient descent tricks. In *Neural Networks:
           Tricks of the Trade*, pages 421–436. Springer, 2012.

[BP76]   E. Balas and M. Padberg. Set partitioning—a survey. *SIAM Review*,
           18:710–760, 1976.

[BR95]   A. Binstock and J. Rex. *Practical Algorithms for Programmers.* Addison-
           Wesley, 1995.

[Bra99]   R. Bracewell. *The Fourier Transform and its Applications.* McGraw-Hill,
           third edition, 1999.

[Bra08]   Peter Brass. *Advanced Data Structures.* Cambridge University Press,
           2008.

[Brè79]   D. Brèlaz. New methods to color the vertices of a graph. *Comm. ACM*,
           22:251–256, 1979.

[Bri88]   E. Brigham. *The Fast Fourier Transform.* Prentice, facimile edition,
           1988.

[Bro95]   F. Brooks. *The Mythical Man-Month.* Addison-Wesley, twentieth an-
            niversary edition, 1995.

[Bro97]   Andrei Z. Broder. On the resemblance and containment of documents. In
            Proceedings. Compression and Complexity of SEQUENCES 1997, pages
            21–29. IEEE, 1997.

[BRS+10]   Lawrence E. Bassham, Andrew L. Rukhin, Juan Soto, James R. Nech-
            vatal, Miles E. Smid, Stefan D. Leigh, M. Levenson, M. Vangel,
            Nathanael A. Heckert, and D.L. Banks. A statistical test suite for ran-
            dom and pseudorandom number generators for cryptographic applica-
            tions. Technical report, NIST, 2010.

[Bru07]   P. Brucker. *Scheduling Algorithms.* Springer-Verlag, fifth edition, 2007.

[Brz64]   J. Brzozowski. Derivatives of regular expressions. *J. ACM*, 11:481–494,
            1964.

[BS76]   J. Bentley and M. Shamos. Divide-and-conquer in higher-dimensional
            space. In *Proc. Eighth ACM Symp. Theory of Computing*, pages 220–230,
            1976.

[BS86]   G. Berry and R. Sethi. From regular expressions to deterministic au-
            tomata. *Theoretical Computer Science*, 48:117–126, 1986.

[BS96]   E. Bach and J. Shallit. *Algorithmic Number Theory: Efficient Algo-
            rithms*, volume 1. MIT Press, 1996.

[BS97]   R. Bradley and S. Skiena. Fabricating arrays of strings. In *Proc. First
            Int. Conf. Computational Molecular Biology (RECOMB '97)*, pages 57–
            66, 1997.

[BS07]   A. Barvinok and A. Samorodnitsky. Random weighting, asymptotic
            counting and inverse isoperimetry. *Israel Journal of Mathematics*,
            158:159–191, 2007.

[BS13]   Charles-Edmond Bichot and Patrick Siarry. *Graph partitioning.* John
            Wiley & Sons, 2013.

[BSA18]   M. Bern, J. Shewchuk, and N. Amenta. Triangulations and mesh gen-
            eration. In J. Goodman, J. O'Rourke, and C. Toth, editors, *Handbook
            of Discrete and Computational Geometry*, pages 763–786. CRC Press,
            2018.

[BT92]   J. Buchanan and P. Turner. *Numerical Methods and Analysis.* McGraw-
            Hill, 1992.

[But11]   Giorgio C Buttazzo. *Hard Real-Time Computing Systems: Predictable
            Scheduling Algorithms and Applications*, volume 24. Springer Science &
            Business Media, 2011.

[BV04]   Stephen Boyd and Lieven Vandenberghe. *Convex Optimization.* Cam-
            bridge University Press, 2004.

[BvG99]   S. Baase and A. van Gelder. *Computer Algorithms.* Addison-Wesley,
            third edition, 1999.

[BW91]   G. Brightwell and P. Winkler. Counting linear extensions. *Order*, 3:225–
            242, 1991.

[BW94]   M. Burrows and D. Wheeler. A block sorting lossless data compression
            algorithm. Technical Report 124, Digital Equipment Corporation, 1994.

[BW00]   R. Borndorfer and R. Weismantel. Set packing relaxations of some integer programs. *Math. Programming A*, 88:425–450, 2000.

[BZ10]   Richard P Brent and Paul Zimmermann. *Modern computer Arithmetic*, volume 18. Cambridge University Press, 2010.

[Can87]   J. Canny. *The complexity of robot motion planning*. MIT Press, 1987.

[Cas95]   G. Cash. A fast computer algorithm for finding the permanent of adjacency matrices. *J. Mathematical Chemistry*, 18:115–119, 1995.

[CB04]   C. Cong and D. Bader. The Euler tour technique and parallel rooted spanning tree. In *Int. Conf. Parallel Processing (ICPP)*, pages 448–457, 2004.

[CC92]   S. Carlsson and J. Chen. The complexity of heaps. In *Proc. Third ACM-SIAM Symp. on Discrete Algorithms*, pages 393–402, 1992.

[CC97]   W. Cook and W. Cunningham. *Combinatorial Optimization*. John Wiley & Sons, 1997.

[CC16]   S. Chapra and R. Canale. *Numerical Methods for Engineers*. McGraw-Hill, seventh edition, 2016.

[CCDG82]   P. Chinn, J. Chvátolvá, A. K. Dewdney, and N. E. Gibbs. The bandwidth problem for graphs and matrices—a survey. *J. Graph Theory*, 6:223–254, 1982.

[CCL15]   Yixin Cao, Jianer Chen, and Yang Liu. On feedback vertex set: new measure and new structures. *Algorithmica*, 73(1):63–86, 2015.

[CD85]   B. Chazelle and D. Dobkin. Optimal convex decompositions. In G. Toussaint, editor, *Computational Geometry*, pages 63–133. North-Holland, 1985.

[CDG+18]   Diptarka Chakraborty, Debarati Das, Elazar Goldenberg, Michal Koucky, and Michael Saks. Approximating edit distance within constant factor in truly sub-quadratic time. In *2018 IEEE 59th Annual Symposium on Foundations of Computer Science (FOCS)*, pages 979–990. IEEE, 2018.

[CDL86]   B. Chazelle, R. Drysdale, and D. Lee. Computing the largest empty rectangle. *SIAM J. Computing*, 15:300–315, 1986.

[CE92]   B. Chazelle and H. Edelsbrunner. An optimal algorithm for intersecting line segments. *J. ACM*, 39:1–54, 1992.

[CFT99]   A. Caprara, M. Fischetti, and P. Toth. A heuristic method for the set covering problem. *Operations Research*, 47:730–743, 1999.

[CFT00]   A. Caprara, M. Fischetti, and P. Toth. Algorithms for the set covering problem. *Annals of Operations Research*, 98:353–371, 2000.

[CG94]   B. Cherkassky and A. Goldberg. On implementing push-relabel method for the maximum flow problem. Technical Report 94-1523, Department of Computer Science, Stanford University, 1994.

[CGJ98]   C.R. Coullard, A.B. Gamble, and P.C. Jones. Matching problems in selective assembly operations. *Annals of Operations Research*, 76:95–107, 1998.

[CGJ⁺13]   Markus Chimani, Carsten Gutwenger, Michael Jünger, Gunnar W. Klau, Karsten Klein, and Petra Mutzel. The Open Graph Drawing Framework (OGDF). *Handbook of Graph Drawing and Visualization*, pages 543–569, 2013.

[CGK⁺97]   C. Chekuri, A. Goldberg, D. Karger, M. Levine, and C. Stein. Experimental study of minimum cut algorithms. In *Proc. Symp. on Discrete Algorithms (SODA)*, pages 324–333, 1997.

[CGL85]   B. Chazelle, L. Guibas, and D. T. Lee. The power of geometric duality. *BIT*, 25:76–90, 1985.

[CGM⁺98]   B. Cherkassky, A. Goldberg, P. Martin, J. Setubal, and J. Stolfi. Augment or push: a computational study of bipartite matching and unit-capacity flow algorithms. *J. Experimental Algorithmics*, 3, 1998.

[CGR99]   B. Cherkassky, A. Goldberg, and T. Radzik. Shortest paths algorithms: theory and experimental evaluation. *Math. Prog.*, 10:129–174, 1999.

[CGS99]   B. Cherkassky, A. Goldberg, and C. Silverstein. Buckets, heaps, lists, and monotone priority queues. *SIAM J. Computing*, 28:1326–1346, 1999.

[CH06]   D. Cook and L. Holder. *Mining Graph Data*. John Wiley & Sons, 2006.

[CH09]   Graham Cormode and Marios Hadjieleftheriou. Finding the frequent items in streams of data. *Communications of the ACM*, 52(10):97–105, 2009.

[Cha91]   B. Chazelle. Triangulating a simple polygon in linear time. *Discrete & Computational Geometry*, 6:485–524, 1991.

[Cha00]   B. Chazelle. A minimum spanning tree algorithm with inverse-Ackerman type complexity. *J. ACM*, 47:1028–1047, 2000.

[Che85]   L. P. Chew. Planing the shortest path for a disc in $O(n^2 \lg n)$ time. In *Proc. First ACM Symp. Computational Geometry*, pages 214–220, 1985.

[Che15]   Chi-hau Chen. *Handbook of Pattern Recognition and Computer Vision*. World Scientific, 2015.

[CHL07]   M. Crochemore, C. Hancart, and T. Lecroq. *Algorithms on Strings*. Cambridge University Press, 2007.

[Chr76]   N. Christofides. Worst-case analysis of a new heuristic for the traveling salesman problem. Technical report, Graduate School of Industrial Administration, Carnegie-Mellon University, Pittsburgh PA, 1976.

[Chu97]   F. Chung. *Spectral Graph Theory*. AMS, 1997.

[Chv83]   V. Chvatal. *Linear Programming*. Freeman, 1983.

[CIPR01]   M. Crochemore, C. Iliopolous, Y. Pinzon, and J. Reid. A fast and practical bit-vector algorithm for the longest common subsequence problem. *Info. Processing Letters*, 80:279–285, 2001.

[CJCG⁺13]   Edward G. Coffman Jr, János Csirik, Gábor Galambos, Silvano Martello, and Daniele Vigo. Bin packing approximation algorithms: survey and classification. *Handbook of Combinatorial Optimization*, pages 455–531, 2013.

[CJL+16]   Lily Chen, Stephen Jordan, Yi-Kai Liu, Dustin Moody, Rene Peralta, Ray Perlner, and Daniel Smith-Tone. *Report on Post-Quantum Cryptography*. US Department of Commerce, National Institute of Standards and Technology, 2016.

[CK94]   A. Chetverin and F. Kramer. Oligonucleotide arrays: New concepts and possibilities. *Bio/Technology*, 12:1093–1099, 1994.

[CK97]   Pierluigi Crescenzi and Viggo Kann. Approximation on the web: A compendium of np optimization problems. In *International Workshop on Randomization and Approximation Techniques in Computer Science*, pages 111–118. Springer, 1997.

[CK05]   Chandra Chekuri and Sanjeev Khanna. A polynomial time approximation scheme for the multiple knapsack problem. *SIAM Journal on Computing*, 35(3):713–728, 2005.

[CK12]   W. Cheney and D. Kincaid. *Numerical Mathematics and Computing*. Centage Learning, seventh edition, 2012.

[CKPT17]   Henrik I. Christensen, Arindam Khan, Sebastian Pokutta, and Prasad Tetali. Approximation and online algorithms for multidimensional bin packing: A survey. *Computer Science Review*, 24:63–79, 2017.

[CL98]   M. Crochemore and T. Lecroq. Text data compression algorithms. In M. J. Atallah, editor, *Algorithms and Theory of Computation Handbook*, pages 12.1–12.23. CRC Press, 1998.

[CL13]   Ángel Corberán and Gilbert Laporte. *Arc Routing: Problems, Methods, and Applications*. SIAM, 2013.

[Cla92]   K. L. Clarkson. Safe and effective determinant evaluation. In *Proc. 31st IEEE Symposium on Foundations of Computer Science*, pages 387–395, Pittsburgh, PA, 1992.

[CLP11]   Timothy M Chan, Kasper Green Larsen, and Mihai Pătraşcu. Orthogonal range searching on the ram, revisited. In *Proceedings of the Twenty-Seventh Annual Symposium on Computational Geometry*, pages 1–10. ACM, 2011.

[CLRS09]   T. Cormen, C. Leiserson, R. Rivest, and C. Stein. *Introduction to Algorithms*. MIT Press, third edition, 2009.

[CM69]   E. Cuthill and J. McKee. Reducing the bandwidth of sparse symmetric matrices. In *Proc. 24th Nat. Conf. ACM*, pages 157–172, 1969.

[CM96]   J. Cheriyan and K. Mehlhorn. Algorithms for dense graphs and networks on the random access computer. *Algorithmica*, 15:521–549, 1996.

[CM99]   G. Del Corso and G. Manzini. Finding exact solutions to the bandwidth minimization problem. *Computing*, 62:189–203, 1999.

[CM13]   Sergio Cabello and Bojan Mohar. Adding one edge to planar graphs makes crossing number and 1-planarity hard. *SIAM Journal on Computing*, 42(5):1803–1829, 2013.

[CN18]   Timothy M. Chan and Yakov Nekrich. Towards an optimal method for dynamic planar point location. *SIAM Journal on Computing*, 47(6):2337–2361, 2018.

[CNAO85]   Norishige Chiba, Takao Nishizeki, Shigenobu Abe, and Takao Ozawa. A linear algorithm for embedding planar graphs using pq-trees. *Journal of Computer and System Sciences*, 30(1):54–76, 1985.

[Coh94]   E. Cohen. Estimating the size of the transitive closure in linear time. In *35th Annual Symposium on Foundations of Computer Science*, pages 190–200. IEEE, 1994.

[Con71]   J.H̃. Conway. *Regular Algebra and Finite Machines*. Chapman $ Hall, 1971.

[Coo71]   S. Cook. The complexity of theorem proving procedures. In *Third ACM Symp. Theory of Computing*, pages 151–158, 1971.

[Coo11]   William J Cook. *In Pursuit of the Traveling Salesman: Mathematics at the Limits of Computation*. Princeton University Press, 2011.

[CP90]   R. Carraghan and P. Pardalos. An exact algorithm for the maximum clique problem. In *Operations Research Letters*, volume 9, pages 375–382, 1990.

[CP05]   R. Crandall and C. Pomerance. *Prime Numbers: A Computational Perspective*. Springer, second edition, 2005.

[CP18]   Phillip Compeau and P.A. Pevzner. *Bioinformatics Algorithms: An Active Learning Approach*. Active Learning, third edition, 2018.

[CPARS18] Haochen Chen, Bryan Perozzi, Rami Al-Rfou, and Steven Skiena. A tutorial on network embeddings. *arXiv preprint arXiv:1808.02590*, 2018.

[CPW98]   B. Chen, C. Potts, and G. Woeginger. A review of machine scheduling: Complexity, algorithms and approximability. In D.-Z. Du and P. Pardalos, editors, *Handbook of Combinatorial Optimization*, volume 3, pages 21–169. Kluwer, 1998.

[CR76]   J. Cohen and M. Roth. On the implementation of Strassen's fast multiplication algorithm. *Acta Informatica*, 6:341–355, 1976.

[CR99]   W. Cook and A. Rohe. Computing minimum-weight perfect matchings. *INFORMS Journal on Computing*, 11:138–148, 1999.

[CR03]   M. Crochemore and W. Rytter. *Jewels of Stringology*. World Scientific, 2003.

[CS93]   J. Conway and N. Sloane. *Sphere Packings, Lattices, and Groups*. Springer-Verlag, 1993.

[CSG05]   A. Caprara and J. Salazar-González. Laying out sparse graphs with provably minimum bandwidth. *INFORMS J. Computing*, 17:356–373, 2005.

[CT65]   J. Cooley and J. Tukey. An algorithm for the machine calculation of complex Fourier series. *Mathematics of Computation*, 19:297–301, 1965.

[CT92]   Y. Chiang and R. Tamassia. Dynamic algorithms in computational geometry. *Proc. IEEE*, 80:1412–1434, 1992.

[CW90]   D. Coppersmith and S. Winograd. Matrix multiplication via arithmetic progressions. *J. Symbolic Computation*, pages 251–280, 1990.

[CW16]   Timothy M Chan and Bryan T Wilkinson. Adaptive and approximate orthogonal range counting. *ACM Transactions on Algorithms (TALG)*, 12(4):45, 2016.

[Dan63]   G. Dantzig. *Linear Programming and Extensions*. Princeton University Press, 1963.

[Dan94]   V. Dancik. Expected length of longest common subsequences. PhD. thesis, Univ. of Warwick, 1994.

[DB74]   G. Dahlquist and A. Bjorck. *Numerical Methods*. Prentice-Hall, 1974.

[DB86]   G. Davies and S. Bowsher. Algorithms for pattern matching. *Software—Practice and Experience*, 16:575–601, 1986.

[dBDK$^+$98] M. de Berg, O. Devillers, M. Kreveld, O. Schwarzkopf, and M. Teillaud. Computing the maximum overlap of two convex polygons under translations. *Theoretical Computer Science*, 31:613–628, 1998.

[dBvKOS08]   M. de Berg, M. van Kreveld, M. Overmars, and O. Schwarzkopf. *Computational Geometry: Algorithms and Applications*. Springer-Verlag, third edition, 2008.

[DCSL18]   James C Davis, Christy A Coghlan, Francisco Servant, and Dongyoon Lee. The impact of regular expression denial of service (redos) in practice: an empirical study at the ecosystem scale. In *Proceedings of the 2018 26th ACM Joint Meeting on European Software Engineering Conference and Symposium on the Foundations of Software Engineering*, pages 246–256. ACM, 2018.

[DDS09]   Christopher Dyken, Morten Dæhlen, and Thomas Sevaldrud. Simultaneous curve simplification. *Journal of geographical systems*, 11(3):273–289, 2009.

[Dem02]   Erik D Demaine. Cache-oblivious algorithms and data structures. *Lecture Notes from the EEF Summer School on Massive Data Sets*, 8(4):1–249, 2002.

[Den05]   L. Y. Deng. Efficient and portable multiple recursive generators of large order. *ACM Trans./ on Modeling and Computer Simulation*, 15:1–13, 2005.

[Dey06]   T. Dey. *Curve and Surface Reconstruction: Algorithms with Mathematical Analysis*. Cambridge Univ. Press, 2006.

[DF79]   E. Denardo and B. Fox. Shortest-route methods: 1. reaching, pruning, and buckets. *Operations Research*, 27:161–186, 1979.

[DF08]   Sanjoy Dasgupta and Yoav Freund. Random projection trees and low dimensional manifolds. In *Proceedings of the Fortieth Annual ACM Symposium on Theory of Computing*, pages 537–546. ACM, 2008.

[DF12]   Rodney G. Downey and Michael Ralph Fellows. *Parameterized Complexity*. Springer Science & Business Media, 2012.

[DFJ54]   G. Dantzig, D. Fulkerson, and S. Johnson. Solution of a large-scale traveling-salesman problem. *Operations Research*, 2:393–410, 1954.

[dFPP90]   H. de Fraysseix, J. Pach, and R. Pollack. How to draw a planar graph on a grid. *Combinatorica*, 10:41–51, 1990.

[DFU11]   Chandan Dubey, Uriel Feige, and Walter Unger. Hardness results for approximating the bandwidth. *Journal of Computer and System Sciences*, 77(1):62–90, 2011.

[DGH+02]   E. Dantsin, A. Goerdt, E. Hirsch, R. Kannan, J. Kleinberg, C. Papadim-
           itriou, P. Raghavan, and U. Schöning. A deterministic $(2 - 2/(k+1))n$
           algorithm for k-SAT based on local search. *Theoretical Computer Sci-
           ence*, 289:69–83, 2002.

[DGKK79]   R. Dial, F. Glover, D. Karney, and D. Klingman. A computational anal-
           ysis of alternative algorithms and labeling techniques for finding shortest
           path trees. *Networks*, 9:215–248, 1979.

[DH92]   D. Du and F. Hwang. A proof of Gilbert and Pollak's conjecture on the
           Steiner ratio. *Algorithmica*, 7:121–135, 1992.

[DH00]   Dingzhu Du and Frank Hwang. *Combinatorial Group Testing and its
           Applications*, volume 12. World Scientific, 2000.

[DHS00]   R. Duda, P. Hart, and D. Stork. *Pattern Classification*. Wiley-
           Interscience, second edition, 2000.

[Die04]   M. Dietzfelbinger. *Primality Testing in Polynomial Time: From Ran-
           domized Algorithms to "PRIMES Is in P"*. Springer, 2004.

[Dij59]   E. W. Dijkstra. A note on two problems in connection with graphs.
           Numerische Mathematik, 1:269–271, 1959.

[DIM16]   Maxence Delorme, Manuel Iori, and Silvano Martello. Bin packing and
           cutting stock problems: Mathematical models and exact algorithms. *Eu-
           ropean Journal of Operational Research*, 255(1):1–20, 2016.

[DIM18]   Maxence Delorme, Manuel Iori, and Silvano Martello. Bpplib: a li-
           brary for bin packing and cutting stock problems. *Optimization Letters*,
           12(2):235–250, 2018.

[DJ92]   G. Das and D. Joseph. Minimum vertex hulls for polyhedral domains.
           Theoret. Comput. Sci., 103:107–135, 1992.

[Dji00]   H. Djidjev. Computing the girth of a planar graph. In *Proc. 27th Int.
           Colloquium on Automata, Languages and Programming (ICALP)*, pages
           821–831, 2000.

[DJP04]   E. Demaine, T. Jones, and M. Patrascu. Interpolation search for non-
           independent data. In *Proc. 15th ACM-SIAM Symp. Discrete Algorithms
           (SODA)*, pages 522–523, 2004.

[DKR10]   Prasun Dutta, S. Pratik Khastgir, and Anushree Roy. Steiner trees
           and spanning trees in six-pin soap films. *American Journal of Physics*,
           78(2):215–221, 2010.

[DL76]   D. Dobkin and R. Lipton. Multidimensional searching problems. *SIAM
           J. Computing*, 5:181–186, 1976.

[DLR79]   D. Dobkin, R. Lipton, and S. Reiss. Linear programming is log-space
           hard for P. *Info. Processing Letters*, 8:96–97, 1979.

[DM80]   D. Dobkin and J.Ĩ. Munro. Determining the mode. *Theoretical Computer
           Science*, 12:255–263, 1980.

[DM97]   K. Daniels and V. Milenkovic. Multiple translational containment. part
           I: an approximation algorithm. *Algorithmica*, 19:148–182, 1997.

[DMBS70]   J. Dongarra, C. Moler, J. Bunch, and G. Stewart. *LINPACK User's
           Guide*. SIAM Publications, 1979.

[dMPF09]   Francisco de Moura Pinto and Carla Maria Dal Sasso Freitas. Fast medial axis transform for planar domains with general boundaries. In *2009
XXII Brazilian Symposium on Computer Graphics and Image Processing*, pages 96–103. IEEE, 2009.

[DMR97]   K. Daniels, V. Milenkovic, and D. Roth. Finding the largest area axis-
parallel rectangle in a polygon. *Computational Geometry: Theory and
Applications*, 7:125–148, 1997.

[DN07]   P. D'Alberto and A. Nicolau. Adaptive Strassen's matrix multiplication.
In *Proc. 21st Int. Conf. on Supercomputing*, pages 284–292, 2007.

[DN09]   Paolo D'Alberto and Alexandru Nicolau. Adaptive winograd's matrix
multiplications. *ACM Transactions on Mathematical Software (TOMS)*,
36(1):3, 2009.

[DNMB18]   Wouter De Nooy, Andrej Mrvar, and Vladimir Batagelj. *Exploratory
Social Network Analysis with Pajek: Revised and Expanded Edition for
Updated Software*, volume 46. Cambridge University Press, 2018.

[DP73]   D. H. Douglas and T. K. Peucker. Algorithms for the reduction of the
number of points required to represent a digitized line or its caricature.
Canadian Cartographer, 10(2):112–122, December 1973.

[DP18]   Samuel Dittmer and Igor Pak. Counting linear extensions of restricted
posets. *arXiv preprint arXiv:1802.06312*, 2018.

[DPS02]   J. Diaz, J. Petit, and M. Serna. A survey of graph layout problems.
ACM Computing Surveys, 34:313–356, 2002.

[DPV08]   Sanjoy Dasgupta, Christos H Papadimitriou, and Umesh Virkumar Vazirani. *Algorithms*. McGraw-Hill Higher Education, 2008.

[DR90]   N. Dershowitz and E. Reingold. Calendrical calculations. *Software—
Practice and Experience*, 20:899–928, 1990.

[DR02]   N. Dershowitz and E. Reingold. *Calendrical Tabulations: 1900–2200*.
Cambridge University Press, 2002.

[DRR+95]   S. Dawson, C. R. Ramakrishnan, I. V. Ramakrishnan, K. Sagonas,
S. Skiena, T. Swift, and D. S. Warren. Unification factoring for efficient
execution of logic programs. In *22nd ACM Symposium on Principles of
Programming Languages (POPL '95)*, pages 247–258, 1995.

[DS05]   Irit Dinur and Samuel Safra. On the hardness of approximating minimum
vertex cover. *Annals of Mathematics*, pages 439–485, 2005.

[DSR00]   D. Du, J. Smith, and J. Rubinstein. *Advances in Steiner Trees*. Kluwer,
2000.

[DT04]   M. Dorigo and T.Stutzle. *Ant Colony Optimization*. MIT Press, 2004.

[dVS82]   G. de V. Smit. A comparison of three string matching algorithms.
Software—Practice and Experience, 12:57–66, 1982.

[dVV03]   S. de Vries and R. Vohra. Combinatorial auctions: A survey. *Informs J.
Computing*, 15:284–309, 2003.

[DY94]   Y. Deng and C. Yang. Waring's problem for pyramidal numbers. *Science
in China (Series A)*, 37:377–383, 1994.

[DZ99]   D. Dor and U. Zwick. Selecting the median. *SIAM J. Computing*, pages 1722–1758, 1999.

[DZ01]   D. Dor and U. Zwick. Median selection requires $(2+\epsilon)n$ comparisons. *SIAM J. Discrete Math.*, 14:312–325, 2001.

[Ebe88]   J. Ebert. Computing Eulerian trails. *Info. Proc. Letters*, 28:93–97, 1988.

[ECW92]   V. Estivill-Castro and D. Wood. A survey of adaptive sorting algorithms. *ACM Computing Surveys*, 24:441–476, 1992.

[Ede87]   H. Edelsbrunner. *Algorithms for Combinatorial Geometry*. Springer-Verlag, 1987.

[Ede06]   H. Edelsbrunner. *Geometry and Topology for Mesh Generation*. Cambridge Univ. Press, 2006.

[Edm65]   J. Edmonds. Paths, trees, and flowers. *Canadian J. Math.*, 17:449–467, 1965.

[Edm71]   J. Edmonds. Matroids and the greedy algorithm. *Mathematical Programming*, 1:126–136, 1971.

[EE99]   D. Eppstein and J. Erickson. Raising roofs, crashing cycles, and playing pool: applications of a data structure for finding pairwise interactions. *Disc. Comp. Geometry*, 22:569–592, 1999.

[EG60]   P. Erdős and T. Gallai. Graphs with prescribed degrees of vertices. *Mat. Lapok (Hungarian)*, 11:264–274, 1960.

[EG89]   H. Edelsbrunner and L. Guibas. Topologically sweeping an arrangement. *J. Computer and System Sciences*, 38:165–194, 1989.

[EG91]   H. Edelsbrunner and L. Guibas. Corrigendum: Topologically sweeping an arrangement. *J. Computer and System Sciences*, 42:249–251, 1991.

[EGI98]   David Eppstein, Zvi Galil, and Giuseppe F Italiano. Dynamic graph algorithms. In *Algorithms and Theory of Computation Handbook*, pages 181–205. CRC Press, 1998.

[EGIN97]   David Eppstein, Zvi Galil, Giuseppe F Italiano, and Amnon Nissenzweig. Sparsification: a technique for speeding up dynamic graph algorithms. *Journal of the ACM (JACM)*, 44(5):669–696, 1997.

[EGS86]   H. Edelsbrunner, L. Guibas, and J. Stolfi. Optimal point location in a monotone subdivision. *SIAM J. Computing*, 15:317–340, 1986.

[EH10]   Herbert Edelsbrunner and John Harer. *Computational Topology: An Introduction*. American Mathematical Society, 2010.

[EHM+12]   Yuval Emek, Magnús M Halldórsson, Yishay Mansour, Boaz Patt-Shamir, Jaikumar Radhakrishnan, and Dror Rawitz. Online set packing. *SIAM Journal on Computing*, 41(4):728–746, 2012.

[EJ73]   J. Edmonds and E. Johnson. Matching, Euler tours, and the Chinese postman. *Math. Programming*, 5:88–124, 1973.

[EK72]   J. Edmonds and R. Karp. Theoretical improvements in the algorithmic efficiency for network flow problems. *J. ACM*, 19:248–264, 1972.

[EK10]   David Easley and Jon Kleinberg. *Networks, Crowds, and Markets*. Cambridge University Press, 2010.

[EKA84]   M. I. Edahiro, I. Kokubo, and T. Asano. A new point location algorithm
            and its practical efficiency – comparison with existing algorithms. *ACM
            Trans./ Graphics*, 3:86–109, 1984.

[EKS83]   H. Edelsbrunner, D. Kirkpatrick, and R. Seidel. On the shape of a set of
            points in the plane. *IEEE Trans./ on Information Theory*, IT-29:551–
            559, 1983.

[EM94]   H. Edelsbrunner and E. Mücke. Three-dimensional alpha shapes. *ACM
            Transactions on Graphics*, 13:43–72, 1994.

[ENSS98]   G. Even, J. Naor, B. Schieber, and M. Sudan. Approximating minimum
            feedback sets and multi-cuts in directed graphs. *Algorithmica*, 20:151–
            174, 1998.

[Epp98]   D. Eppstein. Finding the k shortest paths. *SIAM J. Computing*, 28:652–
            673, 1998.

[ES86]   H. Edelsbrunner and R. Seidel. Voronoi diagrams and arrangements.
            Discrete & Computational Geometry, 1:25–44, 1986.

[ESS93]   H. Edelsbrunner, R. Seidel, and M. Sharir. On the zone theorem for
            hyperplane arrangements. *SIAM J. Computing*, 22:418–429, 1993.

[ESV96]   F. Evans, S. Skiena, and A. Varshney. Optimizing triangle strips for fast
            rendering. In *Proc. IEEE Visualization '96*, pages 319–326, 1996.

[Eve79]   G. Everstine. A comparison of three resequencing algorithms for the
            reduction of matrix profile and wave-front. *Int. J. Numerical Methods
            in Engr.*, 14:837–863, 1979.

[Eve11]   Shimon Even. *Graph Algorithms*. Cambridge University Press, second
            edition, 2011.

[F48]   I. Fáry. On straight line representation of planar graphs. *Acta. Sci.
            Math. Szeged*, 11:229–233, 1948.

[FAKM14]   Bin Fan, Dave G. Andersen, Michael Kaminsky, and Michael D. Mitzen-
            macher. Cuckoo filter: practically better than Bloom. In *Proceedings
            of the 10th ACM International on Conference on Emerging Networking
            Experiments and Technologies*, pages 75–88. ACM, 2014.

[Fei98]   U. Feige. A threshold of $\ln n$ for approximating set cover. *J. ACM*,
            45:634–652, 1998.

[FF62]   L. Ford and D. R. Fulkerson. *Flows in Networks*. Princeton University
            Press, 1962.

[FG95]   U. Feige and M. Goemans. Approximating the value of two prover proof
            systems, with applications to max 2sat and max dicut. In *Proc. 3rd Israel
            Symp. on Theory of Computing and Systems*, pages 182–189, 1995.

[FH06]   E. Fogel and D. Halperin. Exact and efficient construction of Minkowski
            sums for convex polyhedra with applications. In *Proc. 6th Workshop on
            Algorithm Engineering and Experiments (ALENEX)*, 2006.

[FHT01]   Jerome Friedman, Trevor Hastie, and Robert Tibshirani. *The Elements
            of Statistical Learning*. Springer, 2001.

[FHW07]   E. Fogel, D. Halperin, and C. Weibel. On the exact maximum com-
            plexity of minkowski sums of convex polyhedra. In *Proc. 23rd Symp.
            Computational Geometry*, pages 319–326, 2007.

[FHW12]   Efi Fogel, Dan Halperin, and Ron Wein. *CGAL Arrangements and Their Applications: A Step-by-Step Guide*, volume 7. Springer Science & Business Media, 2012.

[FJ05]   M. Frigo and S. Johnson. The design and implementation of FFTW3. *Proc. IEEE*, 93:216–231, 2005.

[FJMO93] M. Fredman, D. Johnson, L. McGeoch, and G. Ostheimer. Data structures for traveling salesmen. In *Proc. 4th 7th Symp. Discrete Algorithms (SODA)*, pages 145–154, 1993.

[FK10]   Alireza Fathi and John Krumm. Detecting road intersections from gps traces. In *International Conference on Geographic Information Science*, pages 56–69. Springer, 2010.

[FK15]   Alan Frieze and Michał Karoński. *Introduction to Random Graphs*. Cambridge University Press, 2015.

[FKN+08] Michael R. Fellows, Christian Knauer, Naomi Nishimura, Prabhakar Ragde, F. Rosamond, Ulrike Stege, Dimitrios M. Thilikos, and Sue Whitesides. Faster fixed-parameter tractable algorithms for matching and packing problems. *Algorithmica*, 52(2):167–176, 2008.

[FL07]   Uriel Feige and James R. Lee. An improved approximation ratio for the minimum linear arrangement problem. *Information Processing Letters*, 101(1):26–29, 2007.

[Fle74]   H. Fleischner. The square of every two-connected graph is Hamiltonian. *J. Combinatorial Theory, B*, 16:29–34, 1974.

[Flo62]   R. Floyd. Algorithm 97 (shortest path). *Communications of the ACM*, 7:345, 1962.

[Flo64]   R. Floyd. Algorithm 245 (treesort). *Communications of the ACM*, 18:701, 1964.

[FLPR99] M. Frigo, C. Leiserson, H. Prokop, and S. Ramachandran. Cache-oblivious algorithms. In *Proc. 40th Symp. Foundations of Computer Science*, 1999.

[FM71]   M. Fischer and A. Meyer. Boolean matrix multiplication and transitive closure. In *IEEE 12th Symp. on Switching and Automata Theory*, pages 129–131, 1971.

[FM82]   C. Fiduccia and R. Mattheyses. A linear time heuristic for improving network partitions. In *Proc. 19th IEEE Design Automation Conf.*, pages 175–181, 1982.

[FN04]   K. Fredriksson and G. Navarro. Average-optimal single and multiple approximate string matching. *ACM J. of Experimental Algorithmics*, 9, 2004.

[For87]   S. Fortune. A sweepline algorithm for Voronoi diagrams. *Algorithmica*, 2:153–174, 1987.

[For13]   Lance Fortnow. *The Golden Ticket: P, NP, and the Search for the Impossible*. Princeton University Press, 2013.

[For18]   S. Fortune. Voronoi diagrams and Delauney triangulations. In J. Goodman, J. O'Rourke, and C. Toth, editors, *Handbook of Discrete and Computational Geometry*, pages 705–722. CRC Press, 2018.

[FPR99]   P. Festa, P. Pardalos, and M. Resende. Feedback set problems. In D.-Z. Du and P.M. Pardalos, editors, *Handbook of Combinatorial Optimization*, volume A. Kluwer, 1999.

[FPR01]   P. Festa, P. Pardalos, and M. Resende. Algorithm 815: Fortran subroutines for computing approximate solution to feedback set problems using GRASP. *ACM Transactions on Mathematical Software*, 27:456–464, 2001.

[FR75]   R. Floyd and R. Rivest. Expected time bounds for selection. *Communications of the ACM*, 18:165–172, 1975.

[FR94]   M. Fürer and B. Raghavachari. Approximating the minimum-degree Steiner tree to within one of optimal. *J. Algorithms*, 17:409–423, 1994.

[Fre62]   E. Fredkin. Trie memory. *Communications of the ACM*, 3:490–499, 1962.

[Fre76]   M. Fredman. How good is the information theory bound in sorting? *Theoretical Computer Science*, 1:355–361, 1976.

[FS03]   N. Ferguson and B. Schneier. *Practical Cryptography*. John Wiley, 2003.

[FSV01]   P. Foggia, C. Sansone, and M. Vento. A performance comparison of five algorithms for graph isomorphism. In *3rd IAPR TC-15 Workshop on Graph-based Representations in Pattern Recognition*, 2001.

[FT87]   M. Fredman and R. Tarjan. Fibonacci heaps and their uses in improved network optimization algorithms. *J. ACM*, 34:596–615, 1987.

[FvW93]   S. Fortune and C. van Wyk. Efficient exact arithmetic for computational geometry. In *Proc. 9th ACM Symp. Computational Geometry*, pages 163–172, 1993.

[FW77]   S. Fiorini and R. Wilson. *Edge-Colourings of Graphs*. Research Notes in Mathematics 16, Pitman, 1977.

[FW93]   M. Fredman and D. Willard. Surpassing the information theoretic bound with fusion trees. *J. Computer and System Sci.*, 47:424–436, 1993.

[FWH04]   E. Fogel, R. Wein, and D. Halperin. Code flexibility and program efficiency by genericity: Improving CGAL's arrangements. In *Proc. 12th European Symposium on Algorithms (ESA'04)*, pages 664–676, 2004.

[Gab77]   H. Gabow. Two algorithms for generating weighted spanning trees in order. *SIAM J. Computing*, 6:139–150, 1977.

[GAD+13]   Jared L. Gearhart, Kristin L. Adair, Richard J. Detry, Justin D. Durfee, Katherine A. Jones, and Nathaniel Martin. Comparison of open-source linear programming solvers. *Sandia National Laboratories, SAND2013-8847*, 2013.

[Gal86]   Z. Galil. Efficient algorithms for finding maximum matchings in graphs. *ACM Computing Surveys*, 18:23–38, 1986.

[Gal90]   K. Gallivan. *Parallel Algorithms for Matrix Computations*. SIAM, 1990.

[GALL13]   Mukulika Ghosh, Nancy M. Amato, Yanyan Lu, and Jyh-Ming Lien. Fast approximate convex decomposition using relative concavity. *Computer-Aided Design*, 45(2):494–504, 2013.

[Gas03]   S. Gass. *Linear Programming: Methods and Applications*. Dover, fifth edition, 2003.

[GBC16]   Ian Goodfellow, Yoshua Bengio, and Aaron Courville. *Deep Learning*. MIT Press, 2016.

[GBY91]   G. Gonnet and R. Baeza-Yates. *Handbook of Algorithms and Data Structures*. Addison-Wesley, second edition, 1991.

[GE19]   Craig Gidney and Martin Ekerå. How to factor 2048 bit RSA integers in 8 hours using 20 million noisy qubits. *arXiv preprint arXiv:1905.09749*, 2019.

[Gen06]   James E Gentle. *Random Number Generation and Monte Carlo Methods*. Springer Science & Business Media, 2006.

[GGJ77]   M. Garey, R. Graham, and D. Johnson. The complexity of computing Steiner minimal trees. *SIAM J. Appl. Math.*, 32:835–859, 1977.

[GGJK78]   M. Garey, R. Graham, D. Johnson, and D. Knuth. Complexity results for bandwidth minimization. *SIAM J. Appl. Math.*, 34:477–495, 1978.

[GH85]   R. Graham and P. Hell. On the history of the minimum spanning tree problem. *Annals of the History of Computing*, 7:43–57, 1985.

[GH06]   P. Galinier and A. Hertz. A survey of local search methods for graph coloring. *Computers and Operations Research*, 33:2547–2562, 2006.

[Gha16]   Mohsen Ghaffari. An improved distributed algorithm for maximal independent set. In *Proceedings of the Twenty-Seventh Annual ACM-SIAM Symposium on Discrete Algorithms*, pages 270–277, 2016.

[GHHP13]   Philippe Galinier, Jean-Philippe Hamiez, Jin-Kao Hao, and Daniel Porumbel. Recent advances in graph vertex coloring. In *Handbook of optimization*, pages 505–528. Springer, 2013.

[GHMS93]   L. J. Guibas, J. E. Hershberger, J. S. B. Mitchell, and J. S. Snoeyink. Approximating polygons and subdivisions with minimum link paths. *Internat. J. Comput. Geom. Appl.*, 3(4):383–415, December 1993.

[GHR95]   R. Greenlaw, J. Hoover, and W. Ruzzo. *Limits to Parallel Computation: P-completeness Theory*. Oxford University Press, 1995.

[GI89]   D. Gusfield and R. Irving. *The Stable Marriage Problem: Structure and Algorithms*. MIT Press, 1989.

[GI91]   Z. Galil and G. Italiano. Data structures and algorithms for disjoint set union problems. *ACM Computing Surveys*, 23:319–344, 1991.

[Gin18]   M. Ginsberg. *Factor Man*. Zowie Press, 2018.

[GJ77]   M. Garey and D. Johnson. The rectilinear Steiner tree problem is NP-complete. *SIAM J. Appl. Math.*, 32:826–834, 1977.

[GJ79]   M. R. Garey and D. S. Johnson. *Computers and Intractability: A Guide to the theory of NP-completeness*. W. H. Freeman, 1979.

[GJM02]   M. Goldwasser, D. Johnson, and C. McGeoch, editors. *Data Structures, Near Neighbor Searches, and Methodology: Fifth and Sixth DIMACS Implementation Challenges*, volume 59. AMS, 2002.

[GJPT78]   M. Garey, D. Johnson, F. Preparata, and R. Tarjan. Triangulating a simple polygon. *Info. Proc. Letters*, 7:175–180, 1978.

[GK95]   A. Goldberg and R. Kennedy. An efficient cost scaling algorithm for the assignment problem. *Math. Programming*, 71:153–177, 1995.

[GK98]   S. Guha and S. Khuller. Approximation algorithms for connected dominating sets. *Algorithmica*, 20:374–387, 1998.

[GKK74]   F. Glover, D. Karney, and D. Klingman. Implementation and computational comparisons of primal-dual computer codes for minimum-cost network flow problems. *Networks*, 4:191–212, 1974.

[GKP89]   R. Graham, D. Knuth, and O. Patashnik. *Concrete Mathematics*. Addison-Wesley, 1989.

[GKS95]   S. Gupta, J. Kececioglu, and A. Schäffer. Improving the practical space and time efficiency of the shortest-paths approach to sum-of-pairs multiple sequence alignment. *J. Computational Biology*, 2:459–472, 1995.

[GKSS08]   Carla P. Gomes, Henry Kautz, Ashish Sabharwal, and Bart Selman. Satisfiability solvers. *Foundations of Artificial Intelligence*, 3:89–134, 2008.

[GKT05]   D. Gibson, R. Kumar, and A. Tomkins. Discovering large dense subgraphs in massive graphs. In *Proc. 31st Int. Conf on Very Large Data Bases*, pages 721–732, 2005.

[GKW06]   A. Goldberg, H. Kaplan, and R. Werneck. Reach for A*: Efficient point-to-point shortest path algorithms. In *Proc. 8th Workshop on Algorithm Engineering and Experimentation (ALENEX)*, 2006.

[GL96]   G. Golub and C. Van Loan. *Matrix Computations*. Johns Hopkins University Press, third edition, 1996.

[GM86]   G. Gonnet and J.Ĩ. Munro. Heaps on heaps. *SIAM J. Computing*, 15:964–971, 1986.

[GM91]   S. Ghosh and D. Mount. An output-sensitive algorithm for computing visibility graphs. *SIAM J. Computing*, 20:888–910, 1991.

[GM12]   Gary Gordon and Jennifer McNulty. *Matroids: a geometric introduction*. Cambridge University Press, 2012.

[GMPV06]   F. Gomes, C. Meneses, P. Pardalos, and G. Viana. Experimental analysis of approximation algorithms for the vertex cover and set covering problems. *Computers and Operations Research*, 33:3520–3534, 2006.

[GO95]   Anka Gajentaan and Mark H Overmars. On a class of O(n^2) problems in computational geometry. *Computational Geometry*, 5(3):165–185, 1995.

[GO14]   Gene H. Golub and James M. Ortega. *Scientific Computing: An Introduction with Parallel Computing*. Elsevier, 2014.

[Goe97]   M. Goemans. Semidefinite programming in combinatorial optimization. *Mathematical Programming*, 79:143–161, 1997.

[Gol93]   L. Goldberg. *Efficient Algorithms for Listing Combinatorial Structures*. Cambridge University Press, 1993.

[Gol97]   A. Goldberg. An efficient implementation of a scaling minimum-cost flow algorithm. *J. Algorithms*, 22:1–29, 1997.

[Gol01]   A. Goldberg. Shortest path algorithms: Engineering aspects. In *12th International Symposium on Algorithms and Computation*, number 2223 in LNCS, pages 502–513. Springer, 2001.

[Gol04]   M. Golumbic. *Algorithmic Graph Theory and Perfect Graphs*, volume 57 of *Annals of Discrete Mathematics*. North Holland, second edition, 2004.

[Gon08]   Georges Gonthier. Formal proof–the four-color theorem. *Notices of the AMS*, 55(11):1382–1393, 2008.

[Gon18]   T. Gonzalez. *Handbook of Approximation Algorithms and Metaheuristics*. Chapman-Hall/CRCPress, second edition, 2018.

[GP68]   E. Gilbert and H. Pollak. Steiner minimal trees. *SIAM J. Applied Math.*, 16:1–29, 1968.

[GP79]   B. Gates and C. Papadimitriou. Bounds for sorting by prefix reversals. *Discrete Mathematics*, 27:47–57, 1979.

[GP07]   G. Gutin and A. Punnen. *The Traveling Salesman Problem and Its Variations*. Springer, 2007.

[GPS76]   N. Gibbs, W. Poole, and P. Stockmeyer. A comparison of several bandwidth and profile reduction algorithms. *ACM Trans./ Math. Software*, 2:322–330, 1976.

[Gra53]   F. Gray. Pulse code communication. US Patent 2632058, March 17, 1953.

[Gra72]   R. Graham. An efficient algorithm for determining the convex hull of a finite planar point set. *Info. Proc. Letters*, 1:132–133, 1972.

[Gri89]   D. Gries. *The Science of Programming*. Springer-Verlag, 1989.

[GS62]   D. Gale and L. Shapely. College admissions and the stability of marriages. *American Math. Monthly*, 69:9–14, 1962.

[GT94]   T. Gensen and B. Toft. *Graph Coloring Problems*. John Wiley & Sons, 1994.

[GT01]   Jonathan L. Gross and Thomas W. Tucker. *Topological Graph Theory*. Courier Corporation, 2001.

[GT14]   Andrew V. Goldberg and Robert E. Tarjan. Efficient maximum flow algorithms. *Communications of the ACM*, 57(8):82–89, 2014.

[GTG14]   Michael T. Goodrich, Roberto Tamassia, and Michael H. Goldwasser. *Data Structures and Algorithms in Java*. John Wiley & Sons, sixth edition, 2014.

[Gup66]   R. P. Gupta. The chromatic index and the degree of a graph. *Notices of the Amer. Math. Soc.*, 13:719, 1966.

[Gus94]   D. Gusfield. Faster implementation of a shortest superstring approximation. *Info. Processing Letters*, 51:271–274, 1994.

[Gus97]   D. Gusfield. *Algorithms on Strings, Trees, and Sequences: Computer Science and Computational Biology*. Cambridge University Press, 1997.

[GW95]   M. Goemans and D. Williamson. .878-approximation algorithms for MAX CUT and MAX 2SAT. *J. ACM*, 42:1115–1145, 1995.

[GW96]   I. Goldberg and D. Wagner. Randomness and the Netscape browser. *Dr. Dobb's Journal*, pages 66–70, 1996.

[GW97]   T. Grossman and A. Wool. Computational experience with approximation algorithms for the set covering problem. *European J. Operational Research*, 101, 1997.

[Ham87]   R. Hamming. *Numerical Methods for Scientists and Engineers*. Dover, second edition, 1987.

[Has82]   H. Hastad. Clique is hard to approximate within $n^{1-\epsilon}$. *Acta Mathematica*, 182:105–142, 182.

[HD80]   P. Hall and G. Dowling. Approximate string matching. *ACM Computing Surveys*, 12:381–402, 1980.

[HDD03]   M. Hilgemeier, N. Drechsler, and R. Drechsler. Fast heuristics for the edge coloring of large graphs. In *Proc. Euromicro Symp. on Digital Systems Design*, pages 230–239, 2003.

[HdlT01]   J. Holm, K. de lichtenberg, and M. Thorup. Poly-logarithmic deterministic fully-dynamic algorithms for connectivity, minimum spanning tree, 2-edge, and biconnectivity. *J. ACM*, 48:723–760, 2001.

[Hel01]   M. Held. VRONI: An engineering approach to the reliable and efficient computation of Voronoi diagrams of points and line segments. *Computational Geometry: Theory and Applications*, 18:95–123, 2001.

[HFN05]   H. Hyyro, K. Fredriksson, and G. Navarro. Increased bit-parallelism for approximate and multiple string matching. *ACM J. of Experimental Algorithmics*, 10, 2005.

[HG97]   P. Heckbert and M. Garland. Survey of polygonal surface simplification algorithms. SIGGRAPH 97 Course Notes, 1997.

[HH00]   I. Hanniel and D. Halperin. Two-dimensional arrangements in CGAL and adaptive point location for parametric curves. In *Proc. 4th International Workshop on Algorithm Engineering (WAE), LNCS v. 1982*, pages 171–182, 2000.

[HH09]   Idit Haran and Dan Halperin. An experimental study of point location in planar arrangements in CGAL. *Journal of Experimental Algorithmics (JEA)*, 13:3, 2009.

[HHL09]   Aram W. Harrow, Avinatan Hassidim, and Seth Lloyd. Quantum algorithm for linear systems of equations. *Physical Review lLetters*, 103(15):150502, 2009.

[HHS98]   T. Haynes, S. Hedetniemi, and P. Slater. *Fundamentals of Domination in Graphs*. CRC Press, Boca Raton, 1998.

[HIKP12]   Haitham Hassanieh, Piotr Indyk, Dina Katabi, and Eric Price. Simple and practical algorithm for sparse fourier transform. In *Proceedings of the Twenty-Third Annual ACM-SIAM Symposium on Discrete Algorithms*, pages 1183–1194, 2012.

[Hir75]   D. Hirschberg. A linear-space algorithm for computing maximum common subsequences. *Communications of the ACM*, 18:341–343, 1975.

[HK73]   J. Hopcroft and R. Karp. An $n^{5/3}$ algorithm for maximum matchings in bipartite graphs. *SIAM J. Computing*, 2:225–231, 1973.

[HK90]   D. P. Huttenlocher and K. Kedem. Computing the minimum Hausdorff distance for point sets under translation. In *Proc. 6th Annu. ACM Sympos. Comput. Geom.*, pages 340–349, 1990.

[HK11]   Markus Holzer and Martin Kutrib. Descriptional and computational complexity of finite automata—a survey. *Information and Computation*, 209(3):456–470, 2011.

[HKH12]   Michael Hemmer, Michal Kleinbort, and Dan Halperin. Improved implementation of point location in general two-dimensional subdivisions. In *European Symposium on Algorithms*, pages 611–623. Springer, 2012.

[HLD04]   W. Hörmann, J. Leydold, and G. Derflinger. *Automatic Nonuniform Random Variate Generation*. Springer, 2004.

[HM83]   S. Hertel and K. Mehlhorn. Fast triangulation of simple polygons. In *Proc. 4th Internat. Conf. Found. Comput. Theory*, pages 207–218. Lecture Notes in Computer Science, Vol. 158, 1983.

[HM99]   X. Huang and A. Madan. Cap3: A DNA sequence assembly program. *Genome Research*, 9:868–877, 1999.

[HMS03]   J. Hershberger, M. Maxel, and S. Suri. Finding the k shortest simple paths: A new algorithm and its implementation. In *Proc. 5th Workshop on Algorithm Engineering and Experimentation (ALENEX)*, 2003.

[HMU06]   J. Hopcroft, R. Motwani, and J. Ullman. *Introduction to Automata Theory, Languages, and Computation*. Addison-Wesley, third edition, 2006.

[HNP91]   Michael T. Heath, Esmond Ng, and Barry W Peyton. Parallel algorithms for sparse linear systems. *SIAM Review*, 33(3):420–460, 1991.

[HNSS18] Monika Henzinger, Alexander Noe, Christian Schulz, and Darren Strash. Practical minimum cut algorithms. *Journal of Experimental Algorithmics (JEA)*, 23(1):1–8, 2018.

[Hoa61]   C. A. R. Hoare. Algorithm 63 (partition) and algorithm 65 (find). *Communications of the ACM*, 4:321–322, 1961.

[Hoa62]   C. A. R. Hoare. Quicksort. *Computer Journal*, 5:10–15, 1962.

[Hoc96]   D. Hochbaum, editor. *Approximation Algorithms for NP-hard Problems*. PWS Publishing, 1996.

[Hof82]   C. M. Hoffmann. *Group-theoretic algorithms and graph isomorphism*. Lecture Notes in Computer Science. Springer-Verlag Inc., 1982.

[Hol75]   J.H. Holland. *Adaptation in Natural and Artificial Systems*. University of Michigan Press, 1975.

[Hol81]   I. Holyer. The NP-completeness of edge colorings. *SIAM J. Computing*, 10:718–720, 1981.

[Hol92]   J.H. Holland. Genetic algorithms. *Scientific American*, 267(1):66–72, 1992.

[Hop71]   J. Hopcroft. An $n \log n$ algorithm for minimizing the states in a finite automaton. In Z. Kohavi, editor, *The Theory of Machines and Computations*, pages 189–196. Academic Press, New York, 1971.

[Hor80]   R. N. Horspool. Practical fast searching in strings. *Software—Practice and Experience*, 10:501–506, 1980.

[HP73]   F. Harary and E. Palmer. *Graphical Enumeration*. Academic Press, New York, 1973.

[HP09]   Karla Hoffman and Manfred Padberg. Set covering, packing and partitioning problems. *Encyclopedia of Optimization*, pages 3482–3486, 2009.

[HPS+05]　M. Holzer, G. Prasinos, F. Schulz, D. Wagner, and C. Zaroliagis. Engineering planar separator algorithms. In *Proc. 13th European Symp. on Algorithms (ESA)*, pages 628–637, 2005.

[HRW92]　R. Hwang, D. Richards, and P. Winter. *The Steiner Tree Problem*, volume 53 of *Annals of Discrete Mathematics*. North Holland, 1992.

[HRW17]　Monika Henzinger, Satish Rao, and Di Wang. Local flow partitioning for faster edge connectivity. In *Proceedings of the Twenty-Eighth Annual ACM-SIAM Symposium on Discrete Algorithms*, pages 1919–1938, 2017.

[HS77]　J. Hunt and T. Szymanski. A fast algorithm for computing longest common subsequences. *Communications of the ACM*, 20:350–353, 1977.

[HS94]　J. Hershberger and J. Snoeyink. An $O(n \log n)$ implementation of the Douglas-Peucker algorithm for line simplification. In *Proc. 10th Annu. ACM Sympos. Comput. Geom.*, pages 383–384, 1994.

[HS98]　J. Hershberger and J. Snoeyink. Cartographic line simplification and polygon CSG formulae in $O(n \log^* n)$ time. *Computational Geometry: Theory and Applications*, 11:175–185, 1998.

[HS99]　J. Hershberger and S. Suri. An optimal algorithm for Euclidean shortest paths in the plane. *SIAM J. Computing*, 28:2215–2256, 1999.

[HS18]　D. Halperin and M. Sharir. Arrangements. In J. Goodman, J. O'Rourke, and C. Toth, editors, *Handbook of Discrete and Computational Geometry*, pages 723–762. CRC Press, 2018.

[HSS87]　J. Hopcroft, J. Schwartz, and M. Sharir. *Planning, Geometry, and Complexity of Robot Motion*. Ablex Publishing, Norwood NJ, 1987.

[HSS07]　R. Hardin, N. Sloane, and W. Smith. Maximum volume spherical codes. http://www.research.att.com/~njas/maxvolumes/, 2007.

[HSS18]　D. Halperin, O. Salzman, and M. Sharir. Algorithmic motion planning. In J. Goodman, J. O'Rourke, and C. Toth, editors, *Handbook of Discrete and Computational Geometry*, pages 1311–1343. CRC Press, 2018.

[HSWW05]　M. Holzer, F. Schultz, D. Wagner, and T. Willhalm. Combining speed-up techniques for shortest-path computations. *ACM J. of Experimental Algorithmics*, 10, 2005.

[HT73a]　J. Hopcroft and R. Tarjan. Dividing a graph into triconnected components. *SIAM J. Computing*, 2:135–158, 1973.

[HT73b]　J. Hopcroft and R. Tarjan. Efficient algorithms for graph manipulation. *Communications of the ACM*, 16:372–378, 1973.

[HT74]　J. Hopcroft and R. Tarjan. Efficient planarity testing. *J. ACM*, 21:549–568, 1974.

[HT84]　D. Harel and R. E. Tarjan. Fast algorithms for finding nearest common ancestors. *SIAM J. Comput.*, 13:338–355, 1984.

[Hub06]　M. Huber. Fast perfect sampling from linear extensions. *Disc. Math.*, 306:420–428, 2006.

[Huf52]　D. Huffman. A method for the construction of minimum-redundancy codes. *Proc. of the IRE*, 40:1098–1101, 1952.

[HUW02]   E. Haunschmid, C. Ueberhuber, and P. Wurzinger. Cache oblivious high performance algorithms for matrix multiplication. Vienna University of Technology, Tech. Report AURORA TR2002-08, 2002.

[HVDH19] David Harvey and Joris Van Der Hoeven. Integer multiplication in time $O(n \log n)$, 2019.

[HVDHL16] David Harvey, Joris Van Der Hoeven, and Grégoire Lecerf. Even faster integer multiplication. *Journal of Complexity*, 36:1–30, 2016.

[HW74]   J.Ẽ. Hopcroft and J. K. Wong. Linear time algorithm for isomorphism of planar graphs. In *Proc. Sixth Annual ACM Symposium on Theory of Computing*, pages 172–184, 1974.

[HWA+03] X. Huang, J. Wang, S. Aluru, S. Yang, and L. Hillier. PCAP: A whole-genome assembly program. *Genome Research*, 13:2164–2170, 2003.

[IK75]   O. Ibarra and C. Kim. Fast approximation algorithms for knapsack and sum of subset problems. *J. ACM*, 22:463–468, 1975.

[IMS18]   P. Indyk, J. Matousek, and A. Sidiropoulous. Low-distortion embeddings of finite metric spaces. In J. Goodman, J. O'Rourke, and C. Toth, editors, *Handbook of Discrete and Computational Geometry*, pages 211–232. CRC Press, 2018.

[Ind98]   P. Indyk. Faster algorithms for string matching problems: matching the convolution bound. In *Proc. 39th Symp. Foundations of Computer Science*, 1998.

[IR78]   A. Itai and M. Rodeh. Finding a minimum circuit in a graph. *SIAM J. Computing*, 7:413–423, 1978.

[IR09]   Costas S. Iliopoulos and M. Sohel Rahman. A new efficient algorithm for computing the longest common subsequence. *Theory of Computing Systems*, 45(2):355–371, 2009.

[Ita78]   A. Itai. Two commodity flow. *J. ACM*, 25:596–611, 1978.

[Iwa16]   Yoichi Iwata. Linear-time kernelization for feedback vertex set. *arXiv preprint arXiv:1608.01463*, 2016.

[J92]   J. JáJá. *An Introduction to Parallel Algorithms*. Addison-Wesley, 1992.

[Jac89]   G. Jacobson. Space-efficient static trees and graphs. In *Proc. Symp. Foundations of Computer Science (FOCS)*, pages 549–554, 1989.

[JAMS91] D. Johnson, C. Aragon, C. McGeoch, and D. Schevon. Optimization by simulated annealing: an experimental evaluation; part II, graph coloring and number partitioning. In *Operations Research*, volume 39, pages 378–406, 1991.

[Jar73]   R. A. Jarvis. On the identification of the convex hull of a finite set of points in the plane. *Info. Proc. Letters*, 2:18–21, 1973.

[JB19]   Riko Jacob and Gerth Stølting Brodal. Dynamic planar convex hull. *arXiv preprint arXiv:1902.11169*, 2019.

[JD88]   A. Jain and R. Dubes. *Algorithms for Clustering Data*. Prentice-Hall, 1988.

[JLR00]   S. Janson, T. Luczak, and A Rucinski. *Random Graphs*. John Wiley & Sons, 2000.

[JM93]   D. Johnson and C. McGeoch, editors. *Network Flows and Matching: First DIMACS Implementation Challenge*, volume 12. American Mathematics Society, 1993.

[JM12]   Michael Jünger and Petra Mutzel. *Graph drawing software.* Springer Science & Business Media, 2012.

[Joh63]   S. M. Johnson. Generation of permutations by adjacent transpositions. *Math. Computation*, 17:282–285, 1963.

[Joh74]   D. Johnson. Approximation algorithms for combinatorial problems. *J. Computer and System Sciences*, 9:256–278, 1974.

[Jon86]   D. W. Jones. An empirical comparison of priority-queue and event-set implementations. *Communications of the ACM*, 29:300–311, 1986.

[Jos12]   N. Josuttis. *The C++ Standard Library: A Tutorial and Reference.* Addison-Wesley, second edition, 2012.

[JR93]   T. Jiang and B. Ravikumar. Minimal NFA problems are hard. *SIAM J. Computing*, 22:1117–1141, 1993.

[JS91]   Brian Johnson and Ben Shneiderman. *Tree-Maps: A Space-Filling Approach to the Visualization of Hierarchical Information Structures.* IEEE, 1991.

[JS01]   A. Jagota and L. Sanchis. Adaptive, restart, randomized greedy heuristics for maximum clique. *J. Heuristics*, 7:1381–1231, 2001.

[JSV04]   Mark Jerrum, Alistair Sinclair, and Eric Vigoda. A polynomial-time approximation algorithm for the permanent of a matrix with nonnegative entries. *Journal of the ACM (JACM)*, 51(4):671–697, 2004.

[JT96]   D. Johnson and M. Trick. *Cliques, Coloring, and Satisfiability: Second DIMACS Implementation Challenge*, volume 26. AMS, 1996.

[JWWZ18]   Daniel Juhl, David M Warme, Pawel Winter, and Martin Zachariasen. The geosteiner software package for computing steiner trees in the plane: an updated computational study. *Mathematical Programming Computation*, 10(4):487–532, 2018.

[KA03]   P. Ko and S. Aluru. Space-efficient linear time construction of suffix arrays,. In *Proc. 14th Symp. on Combinatorial Pattern Matching (CPM)*, pages 200–210. Springer-Verlag LNCS, 2003.

[KA10]   S.R. Kodituwakku and U.S. Amarasinghe. Comparison of lossless data compression algorithms for text data. *Indian Journal of Computer Science and Engineering*, 1(4):416–425, 2010.

[Kah67]   D. Kahn. *The Code Breakers: The Story of Secret Writing.* Macmillan, New York, 1967.

[Kar72]   R. M. Karp. Reducibility among combinatorial problems. In R. Miller and J. Thatcher, editors, *Complexity of Computer Computations*, pages 85–103. Plenum Press, 1972.

[Kar84]   N. Karmarkar. A new polynomial-time algorithm for linear programming. *Combinatorica*, 4:373–395, 1984.

[Kar96]   H. Karloff. How good is the Goemans-Williamson MAX CUT algorithm? In *Proc. Twenty-Eighth Annual ACM Symposium on Theory of Computing*, pages 427–434, 1996.

[Kar00]   D. Karger. Minimum cuts in near-linear time. *J. ACM*, 47:46–76, 200.

[KBFS12]   Alexander Kröller, Tobias Baumgartner, Sándor P Fekete, and Christiane Schmidt. Exact solutions and bounds for general art gallery problems. *Journal of Experimental Algorithmics (JEA)*, 17:2–3, 2012.

[KBV09]   Yehuda Koren, Robert Bell, and Chris Volinsky. Matrix factorization techniques for recommender systems. *Computer*, 42(8):30–37, 2009.

[Kei00]   M. Keil. Polygon decomposition. In J.R. Sack and J. Urrutia, editors, *Handbook of Computational Geometry*, pages 491–518. Elsevier, 2000.

[KF11]   Sertac Karaman and Emilio Frazzoli. Sampling-based algorithms for optimal motion planning. *The International Journal of Robotics Research*, 30(7):846–894, 2011.

[KGV83]   S. Kirkpatrick, C. D. Gelatt, Jr., and M. P. Vecchi. Optimization by simulated annealing. *Science*, 220:671–680, 1983.

[Kha79]   L. Khachian. A polynomial algorithm in linear programming. *Soviet Math. Dokl.*, 20:191–194, 1979.

[Kir79]   D. Kirkpatrick. Efficient computation of continuous skeletons. In *Proc. 20th IEEE Symp. Foundations of Computing*, pages 28–35, 1979.

[Kir83]   D. Kirkpatrick. Optimal search in planar subdivisions. *SIAM J. Computing*, 12:28–35, 1983.

[KKT95]   D. Karger, P. Klein, and R. Tarjan. A randomized linear-time algorithm to find minimum spanning trees. *J. ACM*, 42:321–328, 1995.

[KL70]   B. W. Kernighan and S. Lin. An efficient heuristic procedure for partitioning graphs. *The Bell System Technical Journal*, pages 291–307, 1970.

[KLM+14]   Raimondas Kiveris, Silvio Lattanzi, Vahab Mirrokni, Vibhor Rastogi, and Sergei Vassilvitskii. Connected components in mapreduce and beyond. In *Proceedings of the ACM Symposium on Cloud Computing*, pages 1–13. ACM, 2014.

[KM72]   V. Klee and G. Minty. How good is the simplex algorithm. In *Inequalities III*, pages 159–172. Academic Press, 1972.

[KM95]   J.Ď. Kececioglu and E. W. Myers. Combinatorial algorithms for DNA sequence assembly. *Algorithmica*, 13(1/2):7–51, 1995.

[KMP77]   D. Knuth, J. Morris, and V. Pratt. Fast pattern matching in strings. *SIAM J. Computing*, 6:323–350, 1977.

[KMP+04]   L. Kettner, K. Mehlhorn, S. Pion, S. Schirra, and C. Yap. Classroom examples of robustness problems in geometric computations. In *Proc. 12th European Symp. on Algorithms (ESA'04)*, pages 702–713. www.mpi-inf.mpg.de/~mehlhorn/ftp/ClassRoomExamples.ps, 2004.

[KMR97]   David Karger, Rajeev Motwani, and Gurumurthy D.S. Ramkumar. On approximating the longest path in a graph. *Algorithmica*, 18(1):82–98, 1997.

[KMS97]   S. Khanna, M. Muthukrishnan, and S. Skiena. Efficiently partitioning arrays. In *Proc. ICALP '97*, volume 1256, pages 616–626. Springer-Verlag LNCS, 1997.

[KMS98]   János Komlós, Yuan Ma, and Endre Szemerédi. Matching nuts and bolts
           in $O(n \log n)$ time. *SIAM Journal on Discrete Mathematics*, 11(3):347–
           372, 1998.

[Knu94]   D. Knuth. *The Stanford GraphBase: A Platform for Combinatorial
           Computing*. ACM Press, New York, 1994.

[Knu97a]   D. Knuth. *The Art of Computer Programming, Volume 1: Fundamental
           Algorithms*. Addison-Wesley, third edition, 1997.

[Knu97b]   D. Knuth. *The Art of Computer Programming, Volume 2: Seminumer-
           ical Algorithms*. Addison-Wesley, third edition, 1997.

[Knu98]   D. Knuth. *The Art of Computer Programming, Volume 3: Sorting and
           Searching*. Addison-Wesley, second edition, 1998.

[Knu11]   D. Knuth. *The Art of Computer Programming, Volume 4A: Combina-
           torial Algorithms, Part 1*. Addison-Wesley Professional, 2011.

[Knu15]   D. Knuth. *The Art of Computer Programming, Volume 4, Fascicle 6:
           Satisfiability*. Addison-Wesley Professional, 2015.

[KO63]   A. Karatsuba and Yu. Ofman. Multiplication of multi-digit numbers on
           automata. *Sov. Phys. Dokl.*, 7:595–596, 1963.

[Koe05]   H. Koehler. A contraction algorithm for finding minimal feedback sets.
           In *Proc. 28th Australasian Computer Science Conference (ACSC)*, pages
           165–174, 2005.

[KOS91]   A. Kaul, M. A. O'Connor, and V. Srinivasan. Computing Minkowski
           sums of regular polygons. In *Proc. 3rd Canad. Conf. Comput. Geom.*,
           pages 74–77, 1991.

[KP98]   J. Kececioglu and J. Pecqueur. Computing maximum-cardinality match-
           ings in sparse general graphs. In *Proc. 2nd Workshop on Algorithm
           Engineering*, pages 121–132, 1998.

[KPP04]   H. Kellerer, U. Pferschy, and P. Pisinger. *Knapsack Problems*. Springer,
           2004.

[KR87]   R. Karp and M. Rabin. Efficient randomized pattern-matching algo-
           rithms. *IBM J. Research and Development*, 31:249–260, 1987.

[KR91]   A. Kanevsky and V. Ramachandran. Improved algorithms for graph
           four-connectivity. *J. Comp. Sys. Sci.*, 42:288–306, 1991.

[KR08]   Subhash Khot and Oded Regev. Vertex cover might be hard to ap-
           proximate to within 2- ε. *Journal of Computer and System Sciences*,
           74(3):335–349, 2008.

[Kru56]   J.B̃. Kruskal. On the shortest spanning subtree of a graph and the
           traveling salesman problem. *Proc. of the American Mathematical Society*,
           7:48–50, 1956.

[KS74]   D.E. Knuth and J.L. Szwarcfiter. A structured program to generate all
           topological sorting arrangements. *Information Processing Letters*, 2:153–
           157, 1974.

[KS85]   M. Keil and J.R̃. Sack. Minimum decomposition of geometric objects.
           Machine Intelligence and Pattern Recognition, 2:197–216, 1985.

[KS86]   D. Kirkpatrick and R. Siedel. The ultimate planar convex hull algorithm? *SIAM J. Computing*, 15:287–299, 1986.

[KS99]   D. Kreher and D. Stinson. *Combinatorial Algorithms: Generation, Enumeration, and Search*. CRC Press, 1999.

[KS02]   M. Keil and J. Snoeyink. On the time bound for convex decomposition of simple polygons. *Int. J. Comput. Geometry Appl.*, 12:181–192, 2002.

[KS05a]   H. Kaplan and N. Shafrir. The greedy algorithm for shortest superstrings. *Info. Proc. Letters*, 93:13–17, 2005.

[KS05b]   J. Kelner and D. Spielman. A randomized polynomial-time simplex algorithm for linear programming. *Electronic Colloquim on Computational Complexity*, 156:17, 2005.

[KS07]   H. Kautz and B. Selman. The state of SAT. *Disc. Applied Math.*, 155:1514–1524, 2007.

[KSB06]   Juha Kärkkäinen, Peter Sanders, and Stefan Burkhardt. Linear work suffix array construction. *Journal of the ACM (JACM)*, 53(6):918–936, 2006.

[KSBD07]   H. Kautz, B. Selman, R. Brachman, and T. Dietterich. *Satisfiability Testing*. Morgan and Claypool, 2007.

[KSPP03]   D. Kim, J. Sim, H. Park, and K. Park. Linear-time construction of suffix arrays. In *Proc. 14th Symp. Combinatorial Pattern Matching (CPM)*, pages 186–199, 2003.

[KST93]   J. Köbler, U. Schöning, and J. Túran. *The Graph Isomorphism Problem: Its Structural Complexity*. Birhauser, 1993.

[KSV97]   D. Keyes, A. Sameh, and V. Venkatarishnan. *Parallel Numerical Algorithms*. Springer, 1997.

[KT06]   J. Kleinberg and E. Tardos. *Algorithm Design*. Addison Wesley, 2006.

[Kur30]   K. Kuratowski. Sur le problème des courbes gauches en topologie. *Fund. Math.*, 15:217–283, 1930.

[KW01]   M. Kaufmann and D. Wagner. *Drawing Graphs: Methods and Models*. Springer-Verlag, 2001.

[Kwa62]   M. Kwan. Graphic programming using odd and even points. *Chinese Math.*, 1:273–277, 1962.

[LA04]   J. Leung and J. Anderson, editors. *Handbook of Scheduling: Algorithms, Models, and Performance Analysis*. CRC Press/Chapman & Hall, 2004.

[LA06]   J. Lien and N. Amato. Approximate convex decomposition of polygons. *Computational Geometry: Theory and Applications*, 35:100–123, 2006.

[Lam92]   J.-L. Lambert. Sorting the sums $(x_i + y_j)$ in $O(n^2)$ comparisons. *Theoretical Computer Science*, 103:137–141, 1992.

[Lat91]   J.-C. Latombe. *Robot Motion Planning*. Kluwer, 1991.

[Lau98]   J. Laumond. *Robot Motion Planning and Control*. Springer-Verlag, Lectures Notes in Control and Information Sciences, Volume 229, 1998.

[LaV06]   S. LaValle. *Planning Algorithms*. Cambridge University Press, 2006.

[Law11]   E. Lawler. *Combinatorial Optimization: Networks and Matroids*. Dover Publications, 2011.

[LD03]   R. Laycock and A. Day. Automatically generating roof models from building footprints. In *Proc. 11th Int. Conf. Computer Graphics, Visualization and Computer Vision (WSCG)*, 2003.

[L'E12]   Pierre L'Ecuyer. Random number generation. In *Handbook of Computational Statistics*, pages 35–71. Springer, 2012.

[Lec95]   T. Lecroq. Experimental results on string matching algorithms. *Software—Practice and Experience*, 25:727–765, 1995.

[Lee82]   D. T. Lee. Medial axis transformation of a planar shape. *IEEE Trans./ Pattern Analysis and Machine Intelligence*, PAMI-4:363–369, 1982.

[Len87a]   T. Lengauer. Efficient algorithms for finding minimum spanning forests of hierarchically defined graphs. *J. Algorithms*, 8, 1987.

[Len87b]   H. W. Lenstra. Factoring integers with elliptic curves. *Annals of Mathematics*, 126:649–673, 1987.

[Len89]   T. Lengauer. Hierarchical planarity testing algorithms. *J. ACM*, 36(3):474–509, July 1989.

[Len90]   T. Lengauer. *Combinatorial Algorithms for Integrated Circuit Layout*. John Wiley & Sons, 1990.

[LL96]   A. LaMarca and R. Ladner. The influence of caches on the performance of heaps. *ACM J. Experimental Algorithmics*, 1, 1996.

[LL99]   A. LaMarca and R. Ladner. The influence of caches on the performance of sorting. *J. Algorithms*, 31:66–104, 1999.

[LLCC12]   Cheng-Hung Lin, Chen-Hsiung Liu, Lung-Sheng Chien, and Shih-Chieh Chang. Accelerating pattern matching using a novel parallel algorithm on gpus. *IEEE Transactions on Computers*, 62(10):1906–1916, 2012.

[LLK83]   J.K̃. Lenstra, E. L. Lawler, and A. Rinnooy Kan. *Theory of Sequencing and Scheduling*. John Wiley & Sons, 1983.

[LLKS85]   E. Lawler, J. Lenstra, A. Rinnooy Kan, and D. Shmoys. *The Traveling Salesman Problem*. John Wiley & Sons, 1985.

[LLS92]   L. Lam, S.-W. Lee, and C. Suen. Thinning methodologies–a comprehensive survey. *IEEE Trans./ Pattern Analysis and Machine Intelligence*, 14:869–885, 1992.

[LMK18]   M. Lin, D. Manocha, and Y. Kim. Collision and proximity queries. In J. Goodman, J. O'Rourke, and C. Toth, editors, *Handbook of Discrete and Computational Geometry*, pages 1029–1056. CRC Press, 2018.

[LMM⁺18]   Daniel Lokshtanov, Pranabendu Misra, Joydeep Mukherjee, Geevarghese Philip, Fahad Panolan, and Saket Saurabh. A 2-approximation algorithm for feedback vertex set in tournaments. *arXiv preprint arXiv:1809.08437*, 2018.

[LMS06]   L. Lloyd, A. Mehler, and S. Skiena. Identifying co-referential names across large corpora. In *Combinatorial Pattern Matching (CPM 2006)*, pages 12–23. Lecture Notes in Computer Science, v.4009, 2006.

[LP02]   W. Langdon and R. Poli. *Foundations of Genetic Programming*. Springer, 2002.

[LP07]   A. Lodi and A. Punnen. TSP software. In G. Gutin and A. Punnen, editors, *The Traveling Salesman Problem and Its Variations*, pages 737–749. Springer, 2007.

[LP09]   László Lovász and Michael D. Plummer. *Matching Theory*, volume 367. American Mathematical Soc., 2009.

[LP16]   Adi Livnat and Christos H. Papadimitriou. Sex as an algorithm: the theory of evolution under the lens of computation. *Commun. ACM*, 59(11):84–93, 2016.

[LPW79]   T. Lozano-Perez and M. Wesley. An algorithm for planning collision-free paths among polygonal obstacles. *Comm. ACM*, 22:560–570, 1979.

[LR93]   K. Lang and S. Rao. Finding near-optimal cuts: An empirical evaluation. In *Proc. 4th Annual ACM-SIAM Symposium on Discrete Algorithms (SODA '93)*, pages 212–221, 1993.

[LRSV18]   Philippe Laborie, Jérôme Rogerie, Paul Shaw, and Petr Vilím. Ibm ilog cp optimizer for scheduling. *Constraints*, 23(2):210–250, 2018.

[LS87]   V. Lumelski and A. Stepanov. Path planning strategies for a point mobile automaton moving amidst unknown obstacles of arbitrary shape. *Algorithmica*, 3:403–430, 1987.

[LS95]   Y.-L. Lin and S. Skiena. Algorithms for square roots of graphs. *SIAM J. Discrete Mathematics*, 8:99–118, 1995.

[LSCK02]   P. L'Ecuyer, R. Simard, E. Chen, and W. D. Kelton. An object-oriented random-number package with many long streams and substreams. *Operations Research*, 50:1073–1075, 2002.

[LST14]   Daniel H. Larkin, Siddhartha Sen, and Robert E. Tarjan. A back-to-basics empirical study of priority queues. In *2014 Proceedings of the Sixteenth Workshop on Algorithm Engineering and Experiments (ALENEX)*, pages 61–72. SIAM, 2014.

[LSZ+15]   Liang Liu, Yuning Song, Haiyang Zhang, Huadong Ma, and Athanasios V Vasilakos. Physarum optimization: A biology-inspired algorithm for the steiner tree problem in networks. *IEEE Transactions on Computers*, 64(3):818–831, 2015.

[LT79]   R. Lipton and R. Tarjan. A separator theorem for planar graphs. *SIAM Journal on Applied Mathematics*, 36:346–358, 1979.

[LT80]   R. Lipton and R. Tarjan. Applications of a planar separator theorem. *SIAM J. Computing*, 9:615–626, 1980.

[Luc91]   E. Lucas. *Récréations Mathématiques*. Gauthier-Villares, Paris, 1891.

[Luk80]   E. M. Luks. Isomorphism of bounded valence can be tested in polynomial time. In *Proc. of the 21st Annual Symposium on Foundations of Computing*, pages 42–49. IEEE, 1980.

[LV88]   G. Landau and U. Vishkin. Fast string matching with k differences. *J. Comput. System Sci.*, 37:63–78, 1988.

[LV97]   M. Li and P. Vitányi. *An Introduction to Kolmogorov Complexity and its Applications*. Springer-Verlag, New York, second edition, 1997.

[LW77]   D. T. Lee and C. K. Wong. Worst-case analysis for region and partial region searches in multidimensional binary search trees and balanced quad trees. *Acta Informatica*, 9:23–29, 1977.

[LW88]   T. Lengauer and E. Wanke. Efficient solution of connectivity problems on hierarchically defined graphs. *SIAM J. Computing*, 17:1063–1080, 1988.

[M+13]   Aranyak Mehta et al. Online matching and ad allocation. *Foundations and Trends in Theoretical Computer Science*, 8(4):265–368, 2013.

[Mah76]   S. Maheshwari. Traversal marker placement problems are NP-complete. Technical Report CU-CS-09276, Department of Computer Science, University of Colorado, Boulder, 1976.

[Mai78]   D. Maier. The complexity of some problems on subsequences and supersequences. *J. ACM*, 25:322–336, 1978.

[Mak02]   R. Mak. *Java Number Cruncher: The Java Programmer's Guide to Numerical Computing.* Prentice Hall, 2002.

[Man89]   U. Manber. *Introduction to Algorithms.* Addison-Wesley, 1989.

[Man12]   Toufik Mansour. *Combinatorics of Set Partitions.* Chapman & Hall/CRC Press, 2012.

[MAR+17] Amgad Madkour, Walid G. Aref, Faizan Ur Rehman, Mohamed Abdur Rahman, and Saleh Basalamah. A survey of shortest-path algorithms. *arXiv preprint arXiv:1705.02044*, 2017.

[Mat87]   D. W. Matula. Determining edge connectivity in $O(nm)$. In *28th Ann. Symp. Foundations of Computer Science*, pages 249–251. IEEE, 1987.

[McC76]   E. McCreight. A space-economical suffix tree construction algorithm. *J. ACM*, 23:262–272, 1976.

[McK81]   B. McKay. Practical graph isomorphism. *Congressus Numerantium*, 30:45–87, 1981.

[MDS01]   D. Musser, G. Derge, and A. Saini. *STL Tutorial and Reference Guide: C++ Programming with the Standard Template Library.* Addison-Wesley Professional, second edition, 2001.

[Men27]   K. Menger. Zur allgemeinen Kurventheorie. *Fund. Math.*, 10:96–115, 1927.

[Mey01]   S. Meyers. *Effective STL: 50 Specific Ways to Improve Your Use of the Standard Template Library.* Addison-Wesley Professional, 2001.

[MF00]   Z. Michalewicz and D. Fogel. *How To Solve It: Modern Heuristics.* Springer, 2000.

[MG92]   J. Misra and D. Gries. A constructive proof of Vizing's theorem. *Info. Processing Letters*, 41:131–133, 1992.

[MG07]   Jiri Matousek and Bernd Gärtner. *Understanding and Using Linear Programming.* Springer Science & Business Media, 2007.

[MH78]   R. Merkle and M. Hellman. Hiding and signatures in trapdoor knapsacks. *IEEE Trans./ Information Theory*, 24:525–530, 1978.

[MH08]   Laurens van der Maaten and Geoffrey Hinton. Visualizing data using t-SNE. *Journal of Machine Learning Research*, 9(Nov):2579–2605, 2008.

[Mil76]   G. Miller. Riemann's hypothesis and tests for primality. *J. Computer and System Sciences*, 13:300–317, 1976.

[Mil97]   V. Milenkovic. Multiple translational containment. part II: exact algorithms. *Algorithmica*, 19:183–218, 1997.

[Min78]   H. Minc. *Permanents*, volume 6 of *Encyclopedia of Mathematics and its Applications*. Addison-Wesley, 1978.

[Mit99]   J. Mitchell. Guillotine subdivisions approximate polygonal subdivisions: A simple polynomial-time approximation scheme for geometric TSP, k-mst, and related problems. *SIAM J. Computing*, 28:1298–1309, 1999.

[MKT07]   E. Mardis, S. Kim, and H. Tang, editors. *Advances in Genome Sequencing Technology and Algorithms*. Artech House Publishers, 2007.

[ML14]   Marius Muja and David G Lowe. Scalable nearest neighbor algorithms for high dimensional data. *IEEE Transactions on Pattern Analysis & Machine Intelligence*, 36(11):2227–2240, 2014.

[MLL16]   Javier Minguez, Florant Lamiraux, and Jean-Paul Laumond. Motion planning and obstacle avoidance. In *Springer Handbook of Robotics*, pages 1177–1202. Springer, 2016.

[MM93]   U. Manber and G. Myers. Suffix arrays: A new method for on–line string searches. *SIAM J. Computing*, pages 935–948, 1993.

[MM96]   K. Mehlhorn and P. Mutzel. On the embedding phase of the Hopcroft and Tarjan planarity testing algorithm. *Algorithmica*, 16:233–242, 1996.

[MMI72]   D. Matula, G. Marble, and J. Isaacson. Graph coloring algorithms. In R. C. Read, editor, *Graph Theory and Computing*, pages 109–122. Academic Press, 1972.

[MN98]   M. Matsumoto and T. Nishimura. Mersenne twister: A 623-dimensionally equidistributed uniform pseudorandom number generator. *ACM Trans./ on Modeling and Computer Simulation*, 8:3–30, 1998.

[MN99]   K. Mehlhorn and S. Naher. *LEDA: A Platform for Combinatorial and Geometric Computing*. Cambridge University Press, 1999.

[MN07]   V. Makinen and G. Navarro. Compressed full text indexes. *ACM Computing Surveys*, 39, 2007.

[MNM+16]   Thomas Monz, Daniel Nigg, Esteban A. Martinez, Matthias F. Brandl, Philipp Schindler, Richard Rines, Shannon X. Wang, Isaac L. Chuang, and Rainer Blatt. Realization of a scalable Shor algorithm. *Science*, 351(6277):1068–1070, 2016.

[MO63]   L. E. Moses and R. V. Oakford. *Tables of Random Permutations*. Stanford University Press, 1963.

[Moo59]   E. F. Moore. The shortest path in a maze. In *Proc. International Symp. Switching Theory*, pages 285–292. Harvard University Press, 1959.

[MOS11]   Kurt Mehlhorn, Ralf Osbild, and Michael Sagraloff. A general approach to the analysis of controlled perturbation algorithms. *Computational Geometry*, 44(9):507–528, 2011.

[Mou18]   D. Mount. Geometric intersection. In J. Goodman, J. O'Rourke, and C. Toth, editors, *Handbook of Discrete and Computational Geometry*, pages 1113–1134. CRC Press, 2018.

[MOV96]   A. Menezes, P. Oorschot, and S. Vanstone. *Handbook of Applied Cryptography.* CRC Press, 1996.

[MP80]   W. Masek and M. Paterson. A faster algorithm for computing string edit distances. *J. Computer and System Sciences*, 20:18–31, 1980.

[MP14]   Brendan D McKay and Adolfo Piperno. Practical graph isomorphism, ii. *Journal of Symbolic Computation*, 60:94–112, 2014.

[MPC+06]   S. Mueller, D. Papamichial, J.R. Coleman, S. Skiena, and E. Wimmer. Reduction of the rate of poliovirus protein synthesis through large scale codon deoptimization causes virus attenuation of viral virulence by lowering specific infectivity. *J. of Virology*, 80:9687–96, 2006.

[MPT99]   S. Martello, D. Pisinger, and P. Toth. Dynamic programming and strong bounds for the 0–1 knapsack problem. *Management Science*, 45:414–424, 1999.

[MPT00]   S. Martello, D. Pisinger, and P. Toth. New trends in exact algorithms for the 0–1 knapsack problem. *European Journal of Operational Research*, 123:325–332, 2000.

[MR95]   R. Motwani and P. Raghavan. *Randomized Algorithms.* Cambridge University Press, 1995.

[MR01]   W. Myrvold and F. Ruskey. Ranking and unranking permutations in linear time. *Info. Processing Letters*, 79:281–284, 2001.

[MR06]   W. Mulzer and G. Rote. Minimum weight triangulation is NP-hard. In *Proc. 22nd ACM Symp. on Computational Geometry*, pages 1–10, 2006.

[MR10]   Nabil H Mustafa and Saurabh Ray. Improved results on geometric hitting set problems. *Discrete & Computational Geometry*, 44(4):883–895, 2010.

[MRRT53]   N. Metropolis, A. W. Rosenbluth, M. N. Rosenbluth, and A. H. Teller. Equation of state calculations by fast computing machines. *Journal of Chemical Physics*, 21(6):1087–1092, June 1953.

[MS91]   B. Moret and H. Shapiro. *Algorithm from P to NP: Design and Efficiency.* Benjamin/Cummings, 1991.

[MS95a]   D. Margaritis and S. Skiena. Reconstructing strings from substrings in rounds. Proc. 36th IEEE Symp. Foundations of Computer Science (FOCS), 1995.

[MS95b]   J. S. B. Mitchell and S. Suri. Separation and approximation of polyhedral objects. *Comput. Geom. Theory Appl.*, 5:95–114, 1995.

[MS00]   M. Mascagni and A. Srinivasan. Algorithm 806: Sprng: A scalable library for pseudorandom number generation. *ACM Trans./ Mathematical Software*, 26:436–461, 2000.

[MS18]   D. Mehta and S. Sahni. *Handbook of Data Structures and Applications.* Chapman & Hall/CRC Press, second edition, 2018.

[MT85]   S. Martello and P. Toth. A program for the *0-1* multiple knapsack problem. *ACM Trans. Math. Softw.*, 11(2):135–140, June 1985.

[MT87]   S. Martello and P. Toth. Algorithms for knapsack problems. In S. Martello, editor, *Surveys in Combinatorial Optimization*, volume 31 of *Annals of Discrete Mathematics*, pages 213–258. North-Holland, 1987.

[MT90a]   S. Martello and P. Toth. *Knapsack Problems: Algorithms and Computer Implementations.* Wiley & Sons, 1990.

[MT90b]   K. Mehlhorn and A. Tsakalidis. Data structures. In J. van Leeuwen, editor, *Handbook of Theoretical Computer Science: Algorithms and Complexity*, volume A, pages 301–341. MIT Press, 1990.

[MT10]   Enrico Malaguti and Paolo Toth. A survey on vertex coloring problems. *International Transactions in Operational Research*, 17(1):1–34, 2010.

[MT12]   Bernhard Meindl and Matthias Templ. Analysis of commercial and free and open source solvers for linear optimization problems. *Eurostat and Statistics Netherlands within the project ESSnet on common tools and harmonised methodology for SDC in the ESS*, 20, 2012.

[MU17]   M. Mitzenmacher and E. Upfal. *Probability and Computing: Randomized Algorithms and Probabilistic Analysis.* Cambridge University Press, second edition, 2017.

[Muc13]   Marcin Mucha. Lyndon words and short superstrings. In *Proceedings of the Twenty-Fourth Annual ACM-SIAM Symposium on Discrete Algorithms*, pages 958–972. SIAM, 2013.

[Mul94]   K. Mulmuley. *Computational Geometry: An Introduction Through Randomized Algorithms.* Prentice-Hall, New York, 1994.

[Mut05]   S. Muthukrishnan. *Data Streams: Algorithms and Applications.* Now Publishers, 2005.

[MV80]   S. Micali and V. Vazirani. An $O(\sqrt{|V|}|E|)$ algorithm for finding maximum matchings in general graphs. In *Proc. 21st. Symp. Foundations of Computing*, pages 17–27, 1980.

[MV99]   B. McCullough and H. Vinod. The numerical reliability of econometric software. *J. Economic Literature*, 37:633–665, 1999.

[MY07]   K. Mehlhorn and C. Yap. *Robust Geometric Computation.* manuscript, http://cs.nyu.edu/yap/book/egc/, 2007.

[Mye86]   E. Myers. An $O(nd)$ difference algorithm and its variations. *Algorithmica*, 1:514–534, 1986.

[Mye99a]   E. Myers. Whole-genome DNA sequencing. *IEEE Computational Engineering and Science*, 3:33–43, 1999.

[Mye99b]   G. Myers. A fast bit-vector algorithm for approximate string matching based on dynamic progamming. *J. ACM*, 46:395–415, 1999.

[Nav01a]   G. Navarro. A guided tour to approximate string matching. *ACM Computing Surveys*, 33:31–88, 2001.

[Nav01b]   G. Navarro. Nr-grep: a fast and flexible pattern matching tool. *Software Practice and Experience*, 31:1265–1312, 2001.

[NC02]   Michael A Nielsen and Isaac Chuang. *Quantum Computation and Quantum Information.* AAPT, 2002.

[Nes13]   Yurii Nesterov. *Introductory Lectures on Convex Optimization: A Basic Course*, volume 87. Springer Science & Business Media, 2013.

[Neu63]   J. Von Neumann. Various techniques used in connection with random digits. In A. H. Traub, editor, *John von Neumann, Collected Works*, volume 5. Macmillan, 1963.

[New18]   Mark Newman. *Networks*. Oxford university press, 2018.

[NH19]   M. Needham and A. Hodler. *Graph Algorithms: Practical Examples in Apache Spark and Neo4J*. O'Reilly, 2019.

[NI08]   Hiroshi Nagamochi and Toshihide Ibaraki. *Algorithmic Aspects of Graph Connectivity*. Cambridge University Press, 2008.

[NLKB11]   Sadegh Nobari, Xuesong Lu, Panagiotis Karras, and Stéphane Bressan. Fast random graph generation. In *Proceedings of the 14th International Conference on Extending Database Technology*, pages 331–342. ACM, 2011.

[Not02]   C. Notredame. Recent progress in multiple sequence alignment: a survey. *Pharmacogenomics*, 3:131–144, 2002.

[NR00]   G. Navarro and M. Raffinot. Fast and flexible string matching by combining bit-parallelism and suffix automata. *ACM J. of Experimental Algorithmics*, 5, 2000.

[NR04]   T. Nishizeki and S. Rahman. *Planar Graph Drawing*. World Scientific, 2004.

[NR07]   G. Navarro and M. Raffinot. *Flexible Pattern Matching in Strings: Practical On-Line Search Algorithms for Texts and Biological Sequences*. Cambridge University Press, 2007.

[NS07]   G. Narasimhan and M. Smid. *Geometric Spanner Networks*. Cambridge Univ. Press, 2007.

[Nuu95]   E. Nuutila. Efficient transitive closure computation in large digraphs. http://www.cs.hut.fi/~enu/thesis.html, 1995.

[NW78]   A. Nijenhuis and H. Wilf. *Combinatorial Algorithms for Computers and Calculators*. Academic Press, second edition, 1978.

[NZ02]   S. Näher and O. Zlotowski. Design and implementation of efficient data types for static graphs. In *European Symposium on Algorithms (ESA)*, pages 748–759, 2002.

[NZM91]   I. Niven, H. Zuckerman, and H. Montgomery. *An Introduction to the Theory of Numbers*. John Wiley & Sons, New York, fifth edition, 1991.

[OBSC00]   A. Okabe, B. Boots, K. Sugihara, and S. Chiu. *Spatial Tessellations: Concepts and Applications of Voronoi Diagrams*. John Wiley & Sons, 2000.

[Ogn93]   R. Ogniewicz. *Discrete Voronoi Skeletons*. Hartung-Gorre Verlag, 1993.

[Omo89]   Stephen M. Omohundro. *Five Balltree Construction Algorithms*. International Computer Science Institute Berkeley, 1989.

[O'R85]   J. O'Rourke. Finding minimal enclosing boxes. *Int. J. Computer and Information Sciences*, 14:183–199, 1985.

[O'R87]   J. O'Rourke. *Art Gallery Theorems and Algorithms*. Oxford University Press, 1987.

[O'R01]   J. O'Rourke. *Computational Geometry in C*. Cambridge University Press, second edition, 2001.

[Orl13]   James B. Orlin. Max flows in $O(nm)$ time, or better. In *Proceedings of the Forty-Fifth annual ACM Symposium on Theory of Computing*, pages 765–774. ACM, 2013.

[OST18]   J. O'Rourke, S. Suri, and C. Toth. Polygons. In J. Goodman, J. O'Rourke, and C. Toth, editors, *Handbook of Discrete and Computational Geometry*, pages 787–810. CRC Press, 2018.

[O08]   Owen OMalley. Terabyte sort on apache hadoop. `http://sortbenchmark.org/Yahoo-Hadoop.pdf`, 2008.

[Pal14]   Katarzyna Paluch. Better approximation algorithms for maximum asymmetric traveling salesman and shortest superstring. arXiv 1401.3670, 2014.

[Pan06]   R. Panigrahy. *Hashing, Searching, Sketching*. PhD thesis, Stanford University, 2006.

[Pap76a]   C. Papadimitriou. The complexity of edge traversing. *J. ACM*, 23:544–554, 1976.

[Pap76b]   C. Papadimitriou. The NP-completeness of the bandwidth minimization problem. *Computing*, 16:263–270, 1976.

[Par90]   G. Parker. A better phonetic search. *C Gazette*, 5–4, June/July 1990.

[PARS14]   Bryan Perozzi, Rami Al-Rfou, and Steven Skiena. Deepwalk: Online learning of social representations. In *Proc. 20th ACM SIGKDD Conf. on Knowledge Discovery and Data Mining*, pages 701–710. ACM, 2014.

[Pas97]   V. Paschos. A survey of approximately optimal solutions to some covering and packing problems. *Computing Surveys*, s171–209:171–209, 1997.

[Pas03]   V. Paschos. Polynomial approximation and graph-coloring. *Computing*, 70:41–86, 2003.

[Pat13]   Maurizio Patrignani. Planarity testing and embedding, 2013.

[Pav82]   T. Pavlidis. *Algorithms for Graphics and Image Processing*. Computer Science Press, Rockville MD, 1982.

[PČY+16]   Brian Paden, Michal Čáp, Sze Zheng Yong, Dmitry Yershov, and Emilio Frazzoli. A survey of motion planning and control techniques for self-driving urban vehicles. *IEEE Transactions on intelligent vehicles*, 1(1):33–55, 2016.

[Pec04]   M. Peczarski. New results in minimum-comparison sorting. *Algorithmica*, 40:133–145, 2004.

[Pec07]   M. Peczarski. The Ford-Johnson algorithm still unbeaten for less than 47 elements. *Info. Processing Letters*, 101:126–128, 2007.

[Pet03]   J. Petit. Experiments on the minimum linear arrangement problem. *ACM J. of Experimental Algorithmics*, 8, 2003.

[PFTV07]   W. Press, B. Flannery, S. Teukolsky, and W. T. Vetterling. *Numerical Recipes: The Art of Scientific Computing*. Cambridge University Press, third edition, 2007.

[PH80]   M. Padberg and S. Hong. On the symmetric traveling salesman problem: a computational study. *Math. Programming Studies*, 12:78–107, 1980.

[PIA78] Y. Perl, A. Itai, and H. Avni. Interpolation search–a log log n search. *Comm. ACM*, 21:550–554, 1978.

[Pin16] M. Pinedo. *Scheduling: Theory, Algorithms, and Systems*. Prentice Hall, fifth edition, 2016.

[PL94] P. A. Pevzner and R. J. Lipshutz. Towards DNA sequencing chips. In *19th Int. Conf. Mathematical Foundations of Computer Science*, volume 841, pages 143–158, Lecture Notes in Computer Science, 1994.

[PLM06] F. Panneton, P. L'Ecuyer, and M. Matsumoto. Improved long-period generators based on linear recurrences modulo 2. *ACM Trans./ Mathematical Software*, 32:1–16, 2006.

[PM88] S. Park and K. Miller. Random number generators: Good ones are hard to find. *Communications of the ACM*, 31:1192–1201, 1988.

[PN18] Shortest Paths and Networks. J. Mitchell. In J. Goodman, J. O'Rourke, and C. Toth, editors, *Handbook of Discrete and Computational Geometry*, pages 811–848. CRC Press, 2018.

[Pol57] G. Polya. *How to Solve It*. Princeton University Press, second edition, 1957.

[Pom84] C. Pomerance. The quadratic sieve factoring algorithm. In T. Beth, N. Cot, and I. Ingemarrson, editors, *Advances in Cryptology*, volume 209, pages 169–182. Lecture Notes in Computer Science, Springer-Verlag, 1984.

[PP06] M. Penner and V. Prasanna. Cache-friendly implementations of transitive closure. *ACM J. of Experimental Algorithmics*, 11, 2006.

[PR86] G. Pruesse and F. Ruskey. Generating linear extensions fast. *SIAM J. Computing*, 23:1994, 373-386.

[PR02] S. Pettie and V. Ramachandran. An optimal minimum spanning tree algorithm. *J. ACM*, 49:16–34, 2002.

[Pra75] V. Pratt. Every prime has a succinct certificate. *SIAM J. Computing*, 4:214–220, 1975.

[Pri57] R. C. Prim. Shortest connection networks and some generalizations. *Bell System Technical Journal*, 36:1389–1401, 1957.

[Prü18] H. Prüfer. Neuer Beweis eines Satzes über Permutationen. *Arch. Math. Phys.*, 27:742–744, 1918.

[PS85] F. Preparata and M. Shamos. *Computational Geometry*. Springer-Verlag, New York, 1985.

[PS98] C. Papadimitriou and K. Steiglitz. *Combinatorial Optimization: Algorithms and Complexity*. Dover Publications, 1998.

[PS02] H. Prömel and A. Steger. *The Steiner Tree Problem: A Tour Through Graphs, Algorithms, and Complexity*. Friedrick Vieweg and Son, 2002.

[PS03] S. Pemmaraju and S. Skiena. *Computational Discrete Mathematics: Combinatorics and Graph Theory with Mathematica*. Cambridge University Press, 2003.

[PSL90] A. Pothen, H. Simon, and K. Liou. Partitioning sparse matrices with eigenvectors of graphs. *SIAM J. Matrix Analysis*, 11:430–452, 1990.

[PSS07]   F. Putze, P. Sanders, and J. Singler. Cache-, hash-, and space-efficient bloom filters. In *Proc. 6th Workshop on Experimental Algorithms (WEA), LNCS 4525*, pages 108–121, 2007.

[PST07]   S. Puglisi, W. Smyth, and A. Turpin. A taxonomy of suffix array construction algorithms. *ACM Computing Surveys*, 39, 2007.

[PSW92]   T. Pavlides, J. Swartz, and Y. Wang. Information encoding with two-dimensional bar-codes. *IEEE Computer*, 25:18–28, 1992.

[PT05]   A. Pothen and S. Toledo. Cache-oblivious data structures. In D. Mehta and S. Sahni, editors, *Handbook of Data Structures and Applications*, pages 59:1–59:29. CChapman & Hall/CRC Press, 2005.

[PTUW11]   Rina Panigrahy, Kunal Talwar, Lincoln Uyeda, and Udi Wieder. Heuristics for vector bin packing. *research.microsoft.com*, 2011.

[Pug86]   G. Allen Pugh. Partitioning for selective assembly. *Computers and Industrial Engineering*, 11:175–179, 1986.

[PV96]   M. Pocchiola and G. Vegter. Topologically sweeping visibility complexes via pseudo-triangulations. *Discrete & Computational Geometry*, 16:419–543, 1996.

[Rab80]   M. Rabin. Probabilistic algorithm for testing primality. *J. Number Theory*, 12:128–138, 1980.

[Ram05]   R. Raman. Data structures for sets. In D. Mehta and S. Sahni, editors, *Handbook of Data Structures and Applications*, pages 33:1–33:22. Chapman & Hall/CRC Press, 2005.

[Raw92]   G. Rawlins. *Compared to What?* Computer Science Press, New York, 1992.

[RBT04]   H. Romero, C. Brizuela, and A. Tchernykh. An experimental comparison of approximation algorithms for the shortest common superstring problem. In *Proc. Fifth Mexican Int. Conf. in Computer Science (ENC'04)*, pages 27–34, 2004.

[RC55]   Rand-Corporation. *A Million Random Digits with 100,000 Normal Deviates*. The Free Press, 1955.

[RD18]   Edward M. Reingold and Nachum Dershowitz. *Calendrical Calculations: The Ultimate Edition*. Cambridge University Press, 2018.

[RDC93]   E. Reingold, N. Dershowitz, and S. Clamen. Calendrical calculations II: Three historical calendars. *Software—Practice and Experience*, 22:383–404, 1993.

[Rei72]   E. Reingold. On the optimality of some set algorithms. *J. ACM*, 19:649–659, 1972.

[Rei91]   G. Reinelt. TSPLIB–a traveling salesman problem library. *ORSA J. Computing*, 3:376–384, 1991.

[Rei94]   G. Reinelt. The traveling salesman problem: Computational solutions for TSP applications. In *Lecture Notes in Computer Science 840*, pages 172–186. Springer-Verlag, 1994.

[RF06]   S. Roger and T. Finley. *JFLAP: An Interactive Formal Languages and Automata Package*. Jones and Bartlett, 2006.

[RFS98]   M. Resende, T. Feo, and S. Smith. Algorithm 787: Fortran subroutines for approximate solution of maximum independent set problems using GRASP. *ACM Transactions on Mathematical Software*, 24:386–394, 1998.

[RHG07]   S. Richter, M. Helert, and C. Gretton. A stochastic local search approach to vertex cover. In *Proc. 30th German Conf. on Artificial Intelligence (KI-2007)*, 2007.

[RHS89]   A. Robison, B. Hafner, and S. Skiena. Eight pieces cannot cover a chessboard. *Computer Journal*, 32:567–570, 1989.

[Riv92]   R. Rivest. The MD5 message digest algorithm. RFC 1321, 1992.

[Rou17]   T. Roughgarden. *Algorithms Illuminated*, volume 1. Soundlikeyourself Publishing, 2017.

[RR99]   C.C. Ribeiro and M.G.C. Resende. Algorithm 797: Fortran subroutines for approximate solution of graph planarization problems using GRASP. *ACM Transactions on Mathematical Software*, 25:341–352, 1999.

[RS96]   H. Rau and S. Skiena. Dialing for documents: an experiment in information theory. *Journal of Visual Languages and Computing*, pages 79–95, 1996.

[RSA78]   R. Rivest, A. Shamir, and L. Adleman. A method for obtaining digital signatures and public-key cryptosystems. *Communications of the ACM*, 21:120–126, 1978.

[RSL77]   D. Rosenkrantz, R. Stearns, and P. M. Lewis. An analysis of several heuristics for the traveling salesman problem. *SIAM J. Computing*, 6:563–581, 1977.

[RSS02]   E. Rafalin, D. Souvaine, and I. Streinu. Topological sweep in degenerate cases. In *Proc. 4th Workshop on Algorithm Engineering and Experiments (ALENEX)*, pages 273–295, 2002.

[RSST96]   N. Robertson, D. Sanders, P. Seymour, and R. Thomas. Efficiently four-coloring planar graphs. In *Proc. 28th ACM Symp. Theory of Computing*, pages 571–575, 1996.

[Rus03]   F. Ruskey. Combinatorial Generation. Preliminary working draft. University of Victoria, Victoria, BC, Canada. Draft available at `http://www-csc.uvic.ca/home/fruskey/cgi-bin/html/main.html`, 2003.

[RW09]   Frank Ruskey and Aaron Williams. The coolest way to generate combinations. *Discrete Mathematics*, 309(17):5305–5320, 2009.

[RW18]   Sharath Raghvendra and Mariëtte C Wessels. A grid-based approximation algorithm for the minimum weight triangulation problem. In *Proceedings of the Twenty-Ninth Annual ACM-SIAM Symposium on Discrete Algorithms*, pages 101–120. SIAM, 2018.

[Ryt85]   W. Rytter. Fast recognition of pushdown automata and context-free languages. *Information and Control*, 67:12–22, 1985.

[RZ05]   G. Robins and A. Zelikovsky. Improved Steiner tree approximation in graphs. *Tighter Bounds for Graph Steiner Tree Approximation*, pages 122–134, 2005.

[SA95]   M. Sharir and P. Agarwal. *Davenport-Schinzel Sequences and Their Geometric Applications*. Cambridge University Press, 1995.

[Sah05]   S. Sahni. Double-ended priority queues. In D. Mehta and S. Sahni, editors, *Handbook of Data Structures and Applications*, pages 8:1–8:23. Chapman & Hall/CRC Press, 2005.

[Sal06]   D. Salomon. *Data Compression: The Complete Reference*. Springer-Verlag, fourth edition, 2006.

[Sam05]   H. Samet. Multidimensional spatial data structures. In D. Mehta and S. Sahni, editors, *Handbook of Data Structures and Applications*, pages 16:1–16:29. Chapman & Hall/CRC Press, 2005.

[Sam06]   H. Samet. *Foundations of Multidimensional and Metric Data Structures*. Morgan Kaufmann, 2006.

[San00]   P. Sanders. Fast priority queues for cached memory. *ACM Journal of Experimental Algorithmics*, 5, 2000.

[Sav97]   C. Savage. A survey of combinatorial gray codes. *SIAM Review*, 39:605–629, 1997.

[Sax80]   J.B. Saxe. Dynamic programming algorithms for recognizing small-bandwidth graphs in polynomial time. *SIAM J. Algebraic and Discrete Methods*, 1:363–369, 1980.

[Say17]   K. Sayood. *Introduction to Data Compression*. Morgan Kaufmann, fifth edition, 2017.

[SB01]   A. Samorodnitsky and A. Barvinok. The distance approach to approximate combinatorial counting. *Geometric and Functional Analysis*, 11:871–899, 2001.

[SB19]   Yihan Sun and Guy E. Blelloch. Parallel range, segment and rectangle queries with augmented maps. In *2019 Proceedings of the Twenty-First Workshop on Algorithm Engineering and Experiments (ALENEX)*, pages 159–173. SIAM, 2019.

[SBdB16] Punam K. Saha, Gunilla Borgefors, and Gabriella Sanniti di Baja. A survey on skeletonization algorithms and their applications. *Pattern Recognition Letters*, 76:3–12, 2016.

[SBW17]   Sima Soltanpour, Boubakeur Boufama, and Q.M. Jonathan Wu. A survey of local feature methods for 3d face recognition. *Pattern Recognition*, 72:391–406, 2017.

[Sch98]   A. Schrijver. Bipartite edge-coloring in $O(\delta\, m)$ time. *SIAM J. Computing*, 28:841–846, 1998.

[Sch11]   Hans-Jorg Schulz. Treevis.net: A tree visualization reference. *IEEE Computer Graphics and Applications*, 31(6):11–15, 2011.

[Sch15]   B. Schneier. *Applied Cryptography: Protocols, Algorithms, and Source Code in C*. John Wiley & Sons, twentieth anniversary edition, 2015.

[SD75]   M. Syslo and J. Dzikiewicz. Computational experiences with some transitive closure algorithms. *Computing*, 15:33–39, 1975.

[SD76]   D. C. Schmidt and L. E. Druffel. A fast backtracking algorithm to test directed graphs for isomorphism using distance matrices. *J. ACM*, 23:433–445, 1976.

[SDC16]   Jonathan Shewchuk, Tamal K Dey, and Siu-Wing Cheng. *Delaunay mesh generation*. Chapman & Hall/CRC Press, 2016.

[SDK83]   M. Syslo, N. Deo, and J. Kowalik. *Discrete Optimization Algorithms with Pascal Programs*. Prentice Hall, 1983.

[Sed77]   R. Sedgewick. Permutation generation methods. *Computing Surveys*, 9:137–164, 1977.

[Sed78]   R. Sedgewick. Implementing quicksort programs. *Communications of the ACM*, 21:847–857, 1978.

[Sed98]   R. Sedgewick. *Algorithms in C++, Parts 1–4: Fundamentals, Data Structures, Sorting, Searching, and Graph Algorithms*. Addison-Wesley, third edition, 1998.

[Sei18]   R. Seidel. Convex hull computations. In J. Goodman, J. O'Rourke, and C. Toth, editors, *Handbook of Discrete and Computational Geometry*, pages 687–704. CRC Press, 2018.

[SFG82]   M. Shore, L. Foulds, and P. Gibbons. An algorithm for the Steiner problem in graphs. *Networks*, 12:323–333, 1982.

[SH75]   M. Shamos and D. Hoey. Closest point problems. In *Proc. Sixteenth IEEE Symp. Foundations of Computer Science*, pages 151–162, 1975.

[SH99]   W. Shih and W. Hsu. A new planarity test. *Theoretical Computer Science*, 223(1–2):179–191, 1999.

[Sha87]   M. Sharir. Efficient algorithms for planning purely translational collision-free motion in two and three dimensions. In *Proc. IEEE Internat. Conf. Robot. Autom.*, pages 1326–1331, 1987.

[She97]   J.R. Shewchuk. Robust adaptive floating-point geometric predicates. *Disc. Computational Geometry*, 18:305–363, 1997.

[SHG09]   Nadathur Satish, Mark Harris, and Michael Garland. Designing efficient sorting algorithms for manycore gpus. In *2009 IEEE International Symposium on Parallel & Distributed Processing*, pages 1–10. IEEE, 2009.

[Sho99]   Peter W. Shor. Polynomial-time algorithms for prime factorization and discrete logarithms on a quantum computer. *SIAM Review*, 41(2):303–332, 1999.

[Sho09]   V. Shoup. *A Computational Introduction to Number Theory and Algebra*. Cambridge University Press, second edition, 2009.

[Si15]   Hang Si. Tetgen, a delaunay-based quality tetrahedral mesh generator. *ACM Transactions on Mathematical Software (TOMS)*, 41(2):11, 2015.

[Sin12]   Alistair Sinclair. *Algorithms for Random Generation and Counting: A Markov Chain Approach*. Springer Science & Business Media, 2012.

[Sip05]   M. Sipser. *Introduction to the Theory of Computation*. Course Technology, second edition, 2005.

[SK86]   T. Saaty and P. Kainen. *The Four-Color Problem*. Dover, New York, 1986.

[SK99]   D. Sankoff and J. Kruskal. *Time Warps, String Edits, and Macromolecules: The Theory and Practice of Sequence Comparison*. CSLI Publications, Stanford University, 1999.

[SK00]   R. Skeel and J. Keiper. *Elementary Numerical Computing with Mathe-matica.* Stipes Pub Llc., 2000.

[Ski88]   S. Skiena. Encroaching lists as a measure of presortedness. *BIT*, 28:775–784, 1988.

[Ski90]   S. Skiena. *Implementing Discrete Mathematics.* Addison-Wesley, 1990.

[Ski99]   S. Skiena. Who is interested in algorithms and why?: lessons from the stony brook algorithms repository. *ACM SIGACT News*, pages 65–74, September 1999.

[Ski12]   S. Skiena. Redesigning viral genomes. *IEEE Computer*, 45:47–53, March 2012.

[SL07]   M. Singh and L. Lau. Approximating minimum bounded degree spanning tree to within one of optimal. In *Proc. 39th Symp. Theory Computing (STOC)*, pages 661–670, 2007.

[SLL02]   J. Siek, L. Lee, and A. Lumsdaine. *The Boost Graph Library: User Guide and Reference Manual.* Addison Wesley, 2002.

[SLW⁺12]   Y. Song, Y. Liu, C. Ward, S. Mueller, B. Futcher, S. Skiena, A. Paul, and E. Wimmer. Identification of two functionally redundant RNA elements in the coding sequence of poliovirus using computer-generated design. *Proc. National Academy of Sciences*, 109(36):14301–14307, 2012.

[SM73]   L. Stockmeyer and A. Meyer. Word problems requiring exponential time. In *Proc. Fifth ACM Symp. Theory of Computing*, pages 1–9, 1973.

[SM10]   David Salomon and Giovanni Motta. *Handbook of Data Compression.* Springer Science & Business Media, 2010.

[ŞMK12]   Ioan A. Şucan, Mark Moll, and Lydia E. Kavraki. The Open Motion Planning Library. *IEEE Robotics & Automation Magazine*, 19(4):72–82, 2012. http://ompl.kavrakilab.org.

[Sno18]   J. Snoeyink. Point location. In J. Goodman, J. O'Rourke, and C. Toth, editors, *Handbook of Discrete and Computational Geometry*, pages 1005–1028. CRC Press, 2018.

[SP08]   Kaleem Siddiqi and Stephen Pizer. *Medial Representations: Mathematics, Algorithms and Applications*, volume 37. Springer Science & Business Media, 2008.

[SP15]   Jared T. Simpson and Mihai Pop. The theory and practice of genome sequence assembly. *Annual Review of Genomics and Human Genetics*, 16:153–172, 2015.

[SR83]   K. Supowit and E. Reingold. The complexity of drawing trees nicely. *Acta Informatica*, 18:377–392, 1983.

[SR03]   S. Skiena and M. Revilla. *Programming Challenges: The Programming Contest Training Manual.* Springer-Verlag, 2003.

[SRM14]   George M Slota, Sivasankaran Rajamanickam, and Kamesh Madduri. Bfs and coloring-based parallel algorithms for strongly connected components and related problems. In *2014 IEEE 28th International Parallel and Distributed Processing Symposium*, pages 550–559. IEEE, 2014.

[SS71]   A. Schönhage and V. Strassen. Schnelle Multiplikation grosser Zahlen. *Computing*, 7:281–292, 1971.

[SS07] K. Schurmann and J. Stoye. An incomplex algorithm for fast suffix array construction. *Software: Practice and Experience*, 37:309–329, 2007.

[SSTF12] Michael Stiebitz, Diego Scheide, Bjarne Toft, and Lene M Favrholdt. *Graph Edge Coloring: Vizing's Theorem and Goldberg's Conjecture*, volume 75. John Wiley & Sons, 2012.

[ST04] D. Spielman and S. Teng. Smoothed analysis: Why the simplex algorithm usually takes polynomial time. *J. ACM*, 51:385–463, 2004.

[Sta06] W. Stallings. *Cryptography and Network Security: Principles and Practice*. Prentice Hall, fourth edition, 2006.

[Str69] V. Strassen. Gaussian elimination is not optimal. *Numerische Mathematik*, 14:354–356, 1969.

[STV17] Ola Svensson, Jakub Tarnawski, and László A Végh. A constant-factor approximation algorithm for the asymmetric traveling salesman problem. *arXiv preprint arXiv:1708.04215*, 2017.

[SV87] J. Stasko and J. Vitter. Pairing heaps: Experiments and analysis. *Communications of the ACM*, 30(3):234–249, 1987.

[SV88] B. Schieber and U. Vishkin. On finding lowest common ancestors: simplification and parallelization. *SIAM J. Comput.*, 17(6):1253–1262, December 1988.

[SW86] Q. Stout and B. Warren. Tree rebalancing in optimal time and space. *Comm. ACM*, 29:902–908, 1986.

[SW11] R. Sedgewick and K. Wayne. *Algorithms in Java, Parts 1-4: Fundamentals, Data Structures, Sorting, Searching, and Graph Algorithms*. Addison-Wesley Professional, fourth edition, 2011.

[SW13] Maxim Sviridenko and Justin Ward. Large neighborhood local search for the maximum set packing problem. In *International Colloquium on Automata, Languages, and Programming*, pages 792–803. Springer, 2013.

[SWA03] S. Schlieimer, D. Wilkerson, and A. Aiken. Winnowing: Local algorithms for document fingerprinting. In *Proc. ACM SIGMOD Int. Conf. on Management of data*, pages 76–85, 2003.

[SWM95] J. Shallit, H. Williams, and F. Moraine. Discovery of a lost factoring machine. *The Mathematical Intelligencer*, 17-3:41–47, Summer 1995.

[SY18] V. Sharma and C. Yap. Robust geometric computation. In J. Goodman, J. O'Rourke, and C. Toth, editors, *Handbook of Discrete and Computational Geometry*, pages 1189–1224. CRC Press, 2018.

[Szp03] G. Szpiro. *Kepler's Conjecture: How Some of the Greatest Minds in History Helped Solve One of the Oldest Math Problems in the World*. John Wiley & Sons, 2003.

[TABN16] Krisnaiyan Thulasiraman, Subramanian Arumugam, Andreas Brandstädt, and Takao Nishizeki. *Handbook of Graph Theory, Combinatorial Optimization, and Algorithms*. Chapman & Hall/CRC Press, 2016.

[Tam13] Roberto Tamassia. *Handbook of Graph Drawing and Visualization*. Chapman & Hall/CRC Press, 2013.

[Tar95]   G. Tarry. Le problème de labyrinthes. *Nouvelles Ann. de Math.*, 14:187, 1895.

[Tar72]   R. Tarjan. Depth-first search and linear graph algorithms. *SIAM J. Computing*, 1:146–160, 1972.

[Tar75]   R. Tarjan. Efficiency of a good but not linear set union algorithm. *J. ACM*, 22:215–225, 1975.

[Tar79]   R. Tarjan. A class of algorithms which require non-linear time to maintain disjoint sets. *J. Computer and System Sciences*, 18:110–127, 1979.

[Tar83]   R. Tarjan. *Data Structures and Network Algorithms*. Society for Industrial and Applied Mathematics, 1983.

[TDS$^+$16]   Andrea Tagliasacchi, Thomas Delame, Michela Spagnuolo, Nina Amenta, and Alexandru Telea. 3d skeletons: A state-of-the-art report. In *Computer Graphics Forum*, volume 35, pages 573–597. Wiley Online Library, 2016.

[TH03]   R. Tam and W. Heidrich. Shape simplification based on the medial axis transform. In *Proc. 14th IEEE Visualization (VIS-03)*, pages 481–488, 2003.

[THG94]   J. Thompson, D. Higgins, and T. Gibson. CLUSTAL W: improving the sensitivity of progressive multiple sequence alignment through sequence weighting, position-specific gap penalties and weight matrix choice. *Nucleic Acids Research*, 22:4673–80, 1994.

[Thi03]   H. Thimbleby. The directed Chinese postman problem. *Software Practice and Experience,*, 33:1081–1096, 2003.

[Tho68]   K. Thompson. Regular expression search algorithm. *Communications of the ACM*, 11:419–422, 1968.

[Tin90]   G. Tinhofer. Generating graphs uniformly at random. *Computing*, 7:235–255, 1990.

[TOG18]   C. Toth, J. O'Rourke, and J. Goodman, editors. *Handbook of Discrete and Computational Geometry*. CRC Press, third edition, 2018.

[Tro62]   H. F. Trotter. Perm (algorithm 115). *Comm. ACM*, 5:434–435, 1962.

[Tur88]   J. Turner. Almost all k-colorable graphs are easy to color. *J. Algorithms*, 9:63–82, 1988.

[TV01]   R. Tamassia and L. Vismara. A case study in algorithm engineering for geometric computing. *Int. J. Computational Geometry and Applications*, 11(1):15–70, 2001.

[TV14]   Paolo Toth and Daniele Vigo. *Vehicle Routing: Problems, Methods, and Applications*. SIAM, 2014.

[TW88]   R. Tarjan and C. Van Wyk. An O($n \lg \lg n$) algorithm for triangulating a simple polygon. *SIAM J. Computing*, 17:143–178, 1988.

[Ukk92]   E. Ukkonen. Constructing suffix trees on-line in linear time. In *Intern. Federation of Information Processing (IFIP '92)*, pages 484–492, 1992.

[Val79]   L. Valiant. The complexity of computing the permanent. *Theoretical Computer Science*, 8:189–201, 1979.

[Val02]   G. Valiente. *Algorithms on Trees and Graphs*. Springer, 2002.

[Van98]   B. Vandegriend. Finding hamiltonian cycles: Algorithms, graphs and performance. M.S. Thesis, Dept. of Computer Science, Univ. of Alberta, 1998.

[Vaz04]   V. Vazirani. *Approximation Algorithms*. Springer, 2004.

[VB96]   L. Vandenberghe and S. Boyd. Semidefinite programming. *SIAM Review*, 38:49–95, 1996.

[vEBKZ77]   P. van Emde Boas, R. Kaas, and E. Zulstra. Design and implementation of an efficient priority queue. *Math. Systems Theory*, 10:99–127, 1977.

[Vel01]   Remco C Veltkamp. Shape matching: Similarity measures and algorithms. In *Proceedings International Conference on Shape Modeling and Applications*, pages 188–197. IEEE, 2001.

[Vit01]   J. Vitter. External memory algorithms and data structures: Dealing with massive data. *ACM Computing Surveys*, 33:209–271, 2001.

[Viz64]   V. Vizing. On an estimate of the chromatic class of a p-graph (in Russian). *Diskret. Analiz*, 3:23–30, 1964.

[vL90a]   J. van Leeuwen. Graph algorithms. In J. van Leeuwen, editor, *Handbook of Theoretical Computer Science: Algorithms and Complexity*, volume A, pages 525–631. MIT Press, 1990.

[vL90b]   J. van Leeuwen, editor. *Handbook of Theoretical Computer Science: Algorithms and Complexity*, volume A. MIT Press, 1990.

[VL05]   Fabien Viger and Matthieu Latapy. Efficient and simple generation of random simple connected graphs with prescribed degree sequence. In *International Computing and Combinatorics Conference*, pages 440–449. Springer, 2005.

[Vos92]   S. Voss. Steiner's problem in graphs: heuristic methods. *Discrete Applied Mathematics*, 40:45 – 72, 1992.

[War62]   S. Warshall. A theorem on boolean matrices. *J. ACM*, 9:11–12, 1962.

[Wat03]   B. Watson. A new algorithm for the construction of minimal acyclic DFAs. *Science of Computer Programming*, 48:81–97, 2003.

[Wat04]   D. Watts. *Six Degrees: The Science of a Connected Age*. W.W. Norton, 2004.

[WC04]   B. Wu and K. Chao. *Spanning Trees and Optimization Problems*. Chapman-Hall/CRC Press, 2004.

[WCL$^+$14]   Thomas Weise, Raymond Chiong, Jorg Lassig, Ke Tang, Shigeyoshi Tsutsui, Wenxiang Chen, Zbigniew Michalewicz, and Xin Yao. Benchmarking optimization algorithms: An open source framework for the traveling salesman problem. *IEEE Computational Intelligence Magazine*, 9(3):40–52, 2014.

[Wei73]   P. Weiner. Linear pattern-matching algorithms. In *Proc. 14th IEEE Symp. on Switching and Automata Theory*, pages 1–11, 1973.

[Wei11]   M. Weiss. *Data Structures and Algorithm Analysis in Java*. Pearson, third edition, 2011.

[Wel84]   T. Welch. A technique for high-performance data compression. *IEEE Computer*, 17-6:8–19, 1984.

[Wel13]   René Weller. *New Geometric Data Structures for Collision Detection and Haptics*. Springer Science & Business Media, 2013.

[Wes89]   Jeffery R Westbrook. Algorithms and data structures for dynamic graph problems. Princeton University, 1989.

[Wes00]   D. West. *Introduction to Graph Theory*. Prentice-Hall, second edition, 2000.

[WF74]   R. A. Wagner and M. J. Fischer. The string-to-string correction problem. *J. ACM*, 21:168–173, 1974.

[WH15]   Qinghua Wu and Jin-Kao Hao. A review on algorithms for maximum clique problems. *European Journal of Operational Research*, 242(3):693–709, 2015.

[Whi32]   H. Whitney. Congruent graphs and the connectivity of graphs. *American J. Mathematics*, 54:150–168, 1932.

[Whi12]   Tom White. *Hadoop: The Definitive Guide*. O'Reilly Media, 2012.

[WHS07]   Gerhard Wäscher, Heike Haußner, and Holger Schumann. An improved typology of cutting and packing problems. *European Journal of Operational Research*, 183(3):1109–1130, 2007.

[Wig83]   A. Wigerson. Improving the performance guarantee for approximate graph coloring. *J. ACM*, 30:729–735, 1983.

[Wil64]   J. W. J. Williams. Algorithm 232 (heapsort). *Communications of the ACM*, 7:347–348, 1964.

[Wil84]   H. Wilf. Backtrack: An O(1) expected time algorithm for graph coloring. *Info. Proc. Letters*, 18:119–121, 1984.

[Wil85]   D. E. Willard. New data structures for orthogonal range queries. *SIAM J. Computing*, 14:232–253, 1985.

[Wil89]   H. Wilf. *Combinatorial Algorithms: An Update*. SIAM, 1989.

[Wil12]   Virginia Vassilevska Williams. Multiplying matrices faster than Coppersmith-Winograd. In *Proceedings of the Forty-Fourth Annual ACM Symposium on Theory of Computing*, pages 887–898. ACM, 2012.

[Wil19]   D. Williamson. *Network Flow Algorithms*. Cambridge University Press, 2019.

[Win68]   S. Winograd. A new algorithm for inner product. *IEEE Trans./ Computers*, C-17:693–694, 1968.

[WM92a]   S. Wu and U. Manber. Agrep–a fast approximate pattern-matching tool. In *Usenix Winter 1992 Technical Conference*, pages 153–162, 1992.

[WM92b]   S. Wu and U. Manber. Fast text searching allowing errors. *Comm. ACM*, 35:83–91, 1992.

[Woe03]   G. Woeginger. Exact algorithms for NP-hard problems: A survey. In *Combinatorial Optimization–Eureka! You Shrink!*, volume 2570 Springer-Verlag LNCS, pages 185–207, 2003.

[Wol79]   T. Wolfe. *The Right Stuff*. Bantam Books, Toronto, 1979.

[WS11]   David P. Williamson and David B Shmoys. *The Design of Approximation Algorithms*. Cambridge University Press, 2011.

[WSR13]   Kai Wang, Steven Skiena, and Thomas G Robertazzi. Phase balancing algorithms. *Electric Power Systems Research*, 96:218–224, 2013.

[WSSJ14]   Jingdong Wang, Heng Tao Shen, Jingkuan Song, and Jianqiu Ji. Hashing for similarity search: A survey. *arXiv preprint arXiv:1408.2927*, 2014.

[WW95]   F. Wagner and A. Wolff. Map labeling heuristics: provably good and practically useful. In *Proc. 11th ACM Symp. Computational Geometry*, pages 109–118, 1995.

[WY05]   X. Wang and H. Yu. How to break MD5 and other hash functions. In *EUROCRYPT, LNCS v. 3494*, pages 19–35, 2005.

[XWW11]   Zhong Xie, Huimin Wang, and Liang Wu. The improved Douglas-Peucker algorithm based on the contour character. In *2011 19th International Conference on Geoinformatics*, pages 1–5. IEEE, 2011.

[Yan03]   S. Yan. *Primality Testing and Integer Factorization in Public-Key Cryptography*. Springer, 2003.

[Yao81]   A. C. Yao. A lower bound to finding convex hulls. *J. ACM*, 28:780–787, 1981.

[YLCZ05]   R. Yeung, S-Y. Li, N. Cai, and Z. Zhang. *Network Coding Theory*. http://www.nowpublishers.com/, Now Publishers, 2005.

[YLDF16]   Minghe Yu, Guoliang Li, Dong Deng, and Jianhua Feng. String similarity search and join: a survey. *Frontiers of Computer Science*, 10(3):399–417, 2016.

[YM08]   Noson S. Yanofsky and Mirco A. Mannucci. *Quantum Computing for Computer Scientists*. Cambridge University Press, 2008.

[You67]   D. Younger. Recognition and parsing of context-free languages in time $O(n^3)$. *Information and Control*, 10:189–208, 1967.

[YS96]   F. Younas and S. Skiena. Randomized algorithms for identifying minimal lottery ticket sets. *Journal of Undergraduate Research*, 2-2:88–97, 1996.

[YZ99]   E. Yang and Z. Zhang. The shortest common superstring problem: Average case analysis for both exact and approximate matching. *IEEE Trans./ Information Theory*, 45:1867–1886, 1999.

[ZL78]   J. Ziv and A. Lempel. A universal algorithm for sequential data compression. *IEEE Trans./ Information Theory*, IT-23:337–343, 1978.

[ZS04]   Z. Zaritsky and M. Sipper. The preservation of favored building blocks in the struggle for fitness: The puzzle algorithm. *IEEE Trans./ Evolutionary Computation*, 8:443–455, 2004.

[Zwi01]   U. Zwick. Exact and approximate distances in graphs – a survey. In *Proc. 9th Euro. Symp. Algorithms (ESA)*, pages 33–48, 2001.

Index

© The Editor(s) (if applicable) and The Author(s), under exclusive license to
Springer Nature Switzerland AG 2020
S. S. Skiena, *The Algorithm Design Manual*, Texts in Computer Science,
https://doi.org/10.1007/978-3-030-54256-6

Printed in the United States
by Baker & Taylor Publisher Services